Student Solutions Manual

Physics THIRD EDITION

Richard Wolfson

Middlebury College

Jay M. Pasachoff

Williams College

 **ADDISON-WESLEY**

An Imprint of Addison Wesley Longman, Inc.

Menlo Park, California • Reading, Massachusetts
New York • Harlow, England • Don Mills, Ontario • Sydney
Mexico City • Madrid • Amsterdam

Publisher: Robin Heyden
Sponsoring Editor: Sami Iwata
Developmental Editor: Catherine Flack
Publishing Associate: Bridget Biscotti Bradley
Senior Production Editor: Larry Olsen
Production Service: Curt Cowan
Composition: ICC Oregon
Cover Designer: Yvo Riezebos
Cover Photo: © Image Bank

ISBN 0-321-03575-5

4 5 6 7 8 9 10—CRS—03 02 01 00

ADDISON WESLEY

2725 Sand Hill Road
Menlo Park, California 94025

CONTENTS

PREFACE

The decision to compile a manual of worked-out solutions to the end-of-chapter problems in Wolfson and Pasachoff's textbook arose from a consideration of the crucial importance of homework problems in the learning of physics. The process of applying physical concepts to construct explanations and answer questions is an excellent means of increasing one's understanding and confirming one's knowledge of those very concepts. (Learning physics without doing problems is like learning to swim without going in the water!) These solutions are designed to aid this process by supplementing, reinforcing, and sometimes replacing classroom discussions.

This *Student Solutions Manual* contains solutions to every odd-numbered problem in the text. In order for the reader to achieve the most benefit from using this manual, most of the solutions have been written in a form that requires his or her active participation. Such an actively engaged reader simultaneously reads and works through each solution with pencil and paper, calculator, and open textbook. Intermediate steps, such as algebraic manipulations, substitution of variables or data, diagrams, conversion of units, etc., are frequently and intentionally omitted. They must be supplied by an engaged reader, using the manual's solutions as guideposts and textual materials as references. Wolfson and Pasachoff intended their problems to help readers understand and use concepts presented in their textbook, and develop problem-solving skills. I have integrated references to the text into my solutions in fulfillment of this goal.

My solutions are not necessarily the only ones possible (or even the best, simplest, etc.). In some cases, alternate methods are indicated; in others, a different point of view from the text's is deliberately presented. Students should remember that the logical equivalence of different approaches to a problem has to be demonstrated by means other than just getting the same answer.

Many authors have enumerated helpful suggestions for students to follow when solving problems. My own version of such a list is the following:

- Carefully READ and understand the question. It may be necessary to visualize a situation or device, as described in the text, or as known from personal experience. The context of the question, i.e., the chapter or section in which it occurs, is often of help.

- THINK about how the quantities that determine the sought-for answer are physically related to the quantities that are either given in the question, or obtainable from them and other sources of data. Construction of a simple, physical, conceptual model, to represent the situation in the problem, may often be necessary.

- Write down physical equations involving the relevant quantities and relations, perhaps using approximations where appropriate, and SOLVE for the desired variables. For some problems, this is the most difficult step.

- CONSIDER the reasonableness of your answer. Does it make sense? Is it consistent with your initial expectations? Does it change suitably when the conditions in the problem are altered? If not, the previous steps may require repetition or verification.

In the real world of experimental science, numerical results reflect the precision of actual measurements and theoretical uncertainties in their interpretation. Such subjects are probably more appropriate to laboratory or advanced courses. By default, I have adopted the once-standard convention of regarding most numbers in problems as accurate to three significant figures, unless otherwise indicated by the context. I think common sense is a better guide than consistency at this level. On a different aspect of accuracy, I must admit to the responsibility for any errors which inevitably are present in a work of this size, and will make any corrections brought to my attention.

Although I personally have solved every problem in this manual, I acknowledge a great debt to other authors, my colleagues, former teachers, and students. Wolfson and Pasachoff have produced a scholarly, interesting, and well-written textbook with problems to match. I have tried to emulate their high standards in this accompanying manual of solutions, which I hope will be a practical aid to both students and instructors.

Edw. S. Ginsberg
University of Massachusetts Boston
Boston, MA 02125-3393
edw.ginsberg@umb.edu

Additional Supplements to Wolfson and Pasachoff,
Physics for Scientists and Engineers, Third Edition

Student Study Guide Volume 1, with *ActivPhysics1* (ISBN: 0-321-05148-3)
Student Study Guide Volume 2, with *ActivPhysics2* (ISBN: 0-321-05147-5)

CHAPTER 1   DOING PHYSICS

Section 1-3:   Measurement Systems

Problem

1. What is your mass in (a) kg; (b) g; (c) Gg; (d) fg?

Solution

(a) Most people in the United States know their weight in pounds. One lb has a mass of about 0.454 kg. My weight, 167 lb, is equivalent to a mass of $(167 \text{ lb})(0.454 \text{ kg/lb}) = 75.8$ kg. Using prefixes from Table 1-1 and scientific notation, we can express this mass as: (b) $(75.8 \text{ kg})(10^3 \text{ g/kg}) = 7.58 \times 10^4$ g, (c) $(7.58 \times 10^4 \text{ g})(1 \text{ Gg}/10^9 \text{ g}) = 7.58 \times 10^{-5}$ Gg, (d) 7.58×10^{19} fg.

Problem

3. The diameter of a hydrogen atom is about 0.1 nm, and the diameter of a proton is about 1 fm. How much bigger is a hydrogen atom than a proton?

Solution

A nanometer is 10^{-9} m, and a femtometer is 10^{-15} m (see Table 1-1), so the ratio of the diameters of a hydrogen atom and a proton (its nucleus) is $0.1 \times 10^{-9} \text{ m}/10^{-15} \text{ m} = 10^{-1-9-(-15)} = 10^5$. The atom is about 100,000 times larger than its nucleus.

Problem

5. How long, in nanoseconds, is the period of the cesium radiation used to define the second?

Solution

By definition, 1 s = 9,192,631,770 periods of a cesium atomic clock, so 1 period = 1 s/9,192,631,770 = $1.087827757 \times 10^{-10}$ s = 0.1087827757 ns. (This is an alternative definition of the second; the other definition is really in terms of the frequency of the cesium-133 hyperfine transition, which is the reciprocal of the period.)

Problem

7. A hydrogen atom is about 0.1 nm in diameter. How many hydrogen atoms lined up side by side would make a line 1 cm long?

Solution

The desired number of atoms is the length of the line divided by the diameter of one atom, or $(1 \text{ cm}) \div (0.1 \text{ nm}) = (10^{-2} \text{ m})/(10^{-10} \text{ m}) = 10^8$. (See Table 1-1 for SI prefixes.)

Problem

9. Making a turn, a jetliner flies 2.1 km on a circular path of radius 3.4 km. Through what angle does it turn?

Solution

The angle in radians is the circular arc length divided by the radius, or $\theta = s/r = 2.1 \text{ km}/3.4 \text{ km} = 0.62$ radians. This corresponds to $(0.618)(180°/\pi) = 35.4°$, or about 35°. (See the tip on intermediate results in Section 1-7.)

Section 1-4:   Changing Units

Problem

11. I have enough postage for a 1-oz letter but only a metric scale. What's the maximum mass for my letter, in grams?

Solution

The conversion from ounces to grams is given in Appendix C; however, the calculation might be based on another easily remembered conversion factor between the metric and English systems, namely that 1 lb has a mass of approximately 0.454 kg. Then 1 oz = 1 lb/16 corresponds to a mass of 454 g/16 = 28.4 g.

Problem

13. How many cubic centimeters (cm^3) are there in a cubic meter (m^3)?

Solution

$1 \text{ m}^3 = (10^2 \text{ cm})^3 = 10^6 \text{ cm}^3$.

Problem

15. By what percentage do the 1500-m and 1-mile foot races differ?

Solution

The percent difference between 1 mi = 1609 m and 1500 m is $100(1609 - 1500)/1609 = 6.8\% \simeq 7\%$. (There is no general convention regarding which denominator to use in calculating a percent difference. Using 1500, one obtains $7.3\% \simeq 7\%$. For small differences, the distinction is not important.)

Problem

17. Superhighways in Canada have speed limits of 100 km/h. How does this compare with the 65-mi/h speed limit common in the U.S.?

Solution

$(100 \text{ km/h})(1 \text{ mi}/1.609 \text{ km}) = 62.2 \text{ mi/h}$. The speed limit in Canada is about 2.8 mi/h less than in the United States.

Problem

19. Express Neptune's distance from the Sun in astronomical units (see Fig. 1-16). *Hint:* Consult Appendices C and E.

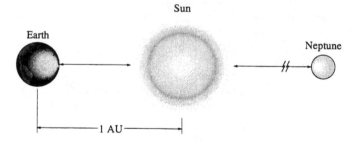

FIGURE 1-16 Problem 19 (sketch is not to scale).

Solution

Appendix E gives the mean distances from the Sun for Earth and Neptune as 150 and 4.50×10^3 respectively, in units of 10^6 km. The former is defined as 1 AU. Neptune's mean distance from the Sun is thus $(4.50 \times 10^3)(1 \text{ AU}/150) = 30 \text{ AU}$. (Note: In this case, in order to obtain a value accurate to three significant figures, we need to use data accurate to four significant figures in the calculation of the ratio, $4.497 \times 10^3 / 149.6 = 30.1$.)

Problem

21. A 3.0-lb box of grass seed will seed 2100 ft^2 of lawn. Express this coverage in m^2/kg.

Solution

The area covered per mass of seed is:

$$\frac{(2100 \text{ ft}^2)(9.290 \times 10^{-2} \text{ m}^2/\text{ft}^2)}{(3 \text{ lb})(0.454 \text{ kg/lb})} = 143 \text{ m}^2/\text{kg}$$

(We used conversion factors from Appendix C.)

Problem

23. A degree is how many rad?

Solution

$(1°)(\pi \text{ rad}/180°) = 1.75 \times 10^{-2} \text{ rad}$.

Section 1-5: Dimensional Analysis

Problem

25. An equation you'll encounter in the next chapter is $x = \frac{1}{2}at^2$, where x is distance, and t is time. Use dimensional analysis to find the dimensions of a.

Solution

We can first solve for $a = 2x/t^2$ to find its dimensions (which are usually denoted by square brackets). Thus $[a] = [x]/[t^2] = LT^{-2}$.

Problem

27. The speed of waves in deep water depends only on the gravitational acceleration g, with dimensions L/T^2 and on the wavelength λ, with dimension L. Which of these formulas *could* be the right one for the wave speed? (a) $v = \sqrt{g/\lambda}$; (b) $v = \lambda g^2$; (c) $v = \sqrt{\lambda g}$.

Solution

The dimensions of the righthand sides of the three formulas are: $[\sqrt{g/\lambda}] = ((L/T^2)(1/L))^{1/2} = 1/T$, $[\lambda g^2] = L(L/T^2)^2 = L^3/T^4$, $[\sqrt{\lambda g}] = (L(L/T^2))^{1/2} = L/T$. Since $[v] = L/T$, only the third formula is dimensionally correct. (Note: $[\ldots]$ means the dimensions of "...".)

Section 1-6: Scientific Notation

Problem

29. Add 3.6×10^5 m and 2.1×10^3 km.

Solution

$3.6 \times 10^5 \text{ m} + 2.1 \times 10^3 \text{ km} = (0.36 + 2.10) \times 10^3 \text{ km} = 2.46 \times 10^3$ km. (Note: We displayed the manipulation of numbers in scientific notation, assuming that all quantities are to be expressed to three significant figures; to two significant figures, our result would have been 2.5×10^3 km. See Section 1-7.)

Problem

31. Divide 4.2×10^3 m/s by 0.57 ms, and express your answer in m/s².

Solution

$(4.2 \times 10^3 \text{ m/s})/(0.57 \times 10^{-3} \text{ s}) = 7.37 \times 10^6 \text{ m/s}^2$

Problem

33. If there are 1 million electronic components on a semiconductor chip that measures 5.0 mm by 5.0 mm, (a) how much area does each component occupy? (b) If the individual components are square, how long is each on a side?

Solution

(a) The area of each (identical) component is the total area of the chip divided by the number of components, or $(5 \text{ mm} \times 5 \text{ mm})/10^6 = 2.5 \times 10^{-5} \text{ mm}^2$. (b) The side of a square is the square root of its area, so the side of one component is $\sqrt{2.5 \times 10^{-5} \text{ mm}^2} = 5.0 \times 10^{-3} \text{ mm} = 5.0 \ \mu\text{m}$.

Problem

35. In Chapter 19, we will show that the average temperature of the Earth should be given roughly by $T = \sqrt[4]{S/4\sigma}$, where T is the temperature in kelvins, $S = 1.4 \times 10^3 \text{ kg} \cdot \text{s}^{-3}$ is the intensity of sunlight, and $\sigma = 5.7 \times 10^{-8} \text{ kg} \cdot \text{s}^{-3} \cdot \text{K}^{-4}$ is a constant. Find the value of T.

Solution

Substitute the given values into the equation for the temperature

$$T = \left(\frac{S}{4\sigma}\right)^{1/4} = \left(\frac{1.4 \times 10^3 \text{ kg} \cdot \text{s}^{-3}}{4 \times 5.7 \times 10^{-8} \text{ kg} \cdot \text{s}^{-3} \cdot \text{K}^{-4}}\right)^{1/4}$$

$$= 2.8 \times 10^2 \text{ K}$$

The result is expressed in scientific notation with two significant figures. (Note: In a hand calculation or estimate, before taking the root of a number, one expresses it to a power of ten which is divisible by that root, $(6.14 \times 10^9)^{1/4} = (61.4 \times 10^8)^{1/4}$ in this case; an electronic calculator does this automatically.)

Section 1-7:   Accuracy and Significant Figures

Problem

37. A 3.6-cm-long radio antenna is added to the front of an airplane 41 m long (Fig. 1-17). What is the overall length?

FIGURE 1-17 Problem 37.

Solution

The overall length of the airplane could be increased by as much as 3.6 cm, depending on how the antenna is attached. However, to two significant figures, the airplane is still 41 m long. That is because in this context, 41 m means a length greater than or equal to 40.5 m, but less than 41.5 m, and 41 m + 3.6 cm = 41.036 m satisfies this condition.

Problem

39. "Machine epsilon" is a computer term describing the minimum quantity that can be added to 1 to give a different number. If machine epsilon for a certain computer is 0.000001, roughly how many significant figures does the computer carry in its calculations?

Solution

If 1 and $1 + \epsilon$ are to be different machine numbers, the 1 must be expressed to the same number of decimal places as ϵ, $1.000000 + 0.000001 = 1.000001$, or seven significant decimal figures in this case.

Section 1-8:   Estimation

Problem

41. Paper is made from wood pulp. Estimate the number of trees that must be cut to make one day's run of a big city's daily newspaper. Assume no recycling.

Solution

A typical nightly run of a big city daily might have a circulation of about 500,000 and consist of newspapers weighing about 1 lb each, thus consuming 5×10^5 lb of paper. Newsprint is mostly made from wood pulp, a suspension of ground-up trees, so roughly 5×10^5 lb of trees are needed. (Actually, newsprint consists of about 80% wood pulp, 15% cellulose, and 5% glue and fillers.) The size of trees used in the paper industry varies widely, but is smallish compared to trees used for lumber. A tree 1 ft in diameter at the base and 40 ft tall (a conical volume of $\frac{1}{3}\pi r^2 h = \frac{1}{3}\pi \left(\frac{1}{2} \text{ ft}\right)^2 \times (40 \text{ ft}) \approx 10 \text{ ft}^3$) with density slightly less than water (approximately 60 lb/ft³—recall that logs float) would weigh about $(60 \text{ lb/ft}^3)(10 \text{ ft}^3) = 600$ lb. Therefore, about 5×10^5 lb/600 lb $\approx 10^3$ or a thousand trees go into a weekday run of a large newspaper. (A Sunday edition might use four times this number.)

Problem

43. How many Earths would fit inside the Sun?

Solution

We can estimate the number of Earth-sized planets that would fit inside the Sun by dividing the volume of the Sun by the volume of the Earth (since both are spheres, the factor $\frac{4}{3}\pi$ cancels out): $(R_\odot/R_E)^3 = (696/6.37)^3 \simeq 1.3\times10^6$ (values from Appendix E).

Problem

45. The average American uses electrical energy at the rate of about 3 kilowatts (kW). Solar energy reaches Earth's surface at an average rate of about 300 watts on every square meter. What fraction of the United States' land area would have to be covered with solar cells to provide all our electrical energy? Assume the cells are 20% efficient at converting sunlight to electricity.

Solution

The electrical power consumed by the entire population of the United States, divided by the power converted by one square meter of solar cells, is the area required by this question: $(250\times10^6\times3$ kW$)/(20\%\times 0.3$ kW/m$^2) = 1.25\times10^{10}$ m$^2 \approx 10^4$ km^2. (We assume that 3 kW is a per capita average over 24 h periods of all types of weather.) The land area of the United States is approximately the area of a rectangle the size of the distance from New York to Los Angeles by the distance from New York to Miami, or 5000 km$\times$ 2000 km $= 10^7$ km^2. (See the figure for the preceding problem.) Then the fraction of area to be covered by solar cells would be only 10^4 km$^2/10^7$ km$^2 = 0.1\%$, comparable to the fraction of land now covered by airports.

Problem

47. (a) Estimate the volume of water in Earth's oceans. (b) If scientists succeed in harnessing nuclear fusion as an energy source, each gallon of sea water will be equivalent to 340 gallons of gasoline. At our present rate of gasoline consumption (see Example 1-6), how long would the oceans supply our fuel needs?

Solution

(a) Oceans cover about 70% of the Earth's surface area $(4\pi R_E^2)$ to an average depth of about 4 km.

Therefore, the volume of ocean water is about $4\pi(6.37\times10^6$ m$)^2(0.7)(4\times10^3$ m$) \simeq 1.4\times10^{18}$ m^3.
(b) This amount of water is equivalent to the energy in $340\times(1.4\times10^{18}$ m$^3)(264.2$ gal/m$^3) \simeq 1.3\times10^{23}$ gal of gasoline. At the consumption rate of Example 1-6, this would last $(1.3\times10^{23}$ gal$)/(10^{11}$ gal/y$) = 1.3\times10^{12}$ y. This is about 50 to 100 times the present age of the universe.

Problem

49. The density of interstellar space is about 1 atom per cubic cm. Stars in our galaxy are typically a few light-years apart (one light-year is the distance light travels in one year) and have typical masses of 10^{30} kg. Estimate whether there is more matter in the stars or in the interstellar gas.

Solution

From Table 1-2, the distance to the nearest star is 4×10^{16} m. Suppose there is one star of mass 10^{30} kg per cubical volume of this dimension. Then the density of star mass is $(10^{30}$ kg$)/(4\times10^{16}$ m$)^3 \sim 1.6\times10^{-20}$ kg/m^3. Interstellar gas is mostly hydrogen, so its density is (1 atom/ cm$^3)(1$ g/6.02×10^{23} atoms$)$ $(10^3$ kg/m$^3)/(1$ g/cm$^3) \simeq \frac{1}{6}\times10^{-20}$ kg/m^3, or about ten times smaller than the star mass density.

Supplementary Problems
Problem

51. A human hair is about 100 μm across. Estimate the number of hairs in a typical braid.

Solution

A typical braid might have a cross-sectional area of a few cm^2, whereas the cross-sectional area of a single hair is about $(100 \mu$m$)^2 = (10^{-2}$ cm$)^2 = 10^{-4}$ cm^2. The number of hairs in a braid is approximately the ratio of the cross-sectional areas (compare this reasoning with that used in Problem 43), or a few times 10^4.

Problem

53. The moon barely covers the Sun at a solar eclipse. Given that the moon is 4×10^5 km from Earth and that the Sun is 1.5×10^8 km from Earth, determine how much bigger the Sun's diameter is than the moon's. If the moon's radius is 1800 km, how big is the Sun?

Solution

The Sun and the moon subtend the same angle (about $\frac{1}{2}$° when viewed from Earth; therefore $\theta = s/r \approx$

diameter/distance $= d_{moon}/4{\times}10^5$ km $= d_{Sun} \div 1.5{\times}10^8$ km. (The small angle approximation is justified since the diameter is much smaller than the distance of each body.) Thus, the ratio of the diameters is approximately $d_{Sun}/d_{moon} = (1.5{\times}10^8 \div 4{\times}10^5) \approx 3.8{\times}10^2$. Of course, the radii are in the same ratio as the diameters, so if $R_{moon} \approx 1800$ km, then $R_{Sun} \approx (3.8{\times}10^2){\times}(1800$ km$) \approx 680,000$ km.

Problem

55. Estimate the number of (a) atoms and (b) cells in your body.

Solution

(a) Human tissue is mostly water, so for a rough estimate we could consider the human body to contain about as many atoms as an equivalent amount of water. One mole of water (H_2O) is 18 g $= 0.018$ kg and contains Avogadro's number of molecules, or about $3{\times}6{\times}10^{23}$ atoms. An average-sized human of 65 kg (Table 1-3) would contain about $(65$ kg $\div 0.018$ kg$){\times}(3{\times}6{\times}10^{23}) \simeq 6.5{\times}10^{27}$ atoms. (b) With an average density of 1 kg/L (same as water), the volume of an average-sized human is 65 L (volume $=$ mass/density). The volume of an average-sized cell (a red blood cell, Table 1-1) is about $(8~\mu m)^3 \simeq 5{\times}10^{-13}$ L, so an average human body might contain approximately $65/5{\times}10^{-13} = 1.3{\times}10^{14}$ cells.

Problem

57. A good-size nuclear weapon has an explosive yield equivalent to one million tons (1 megaton) of the chemical explosive TNT. Estimate the length of a train of boxcars needed to carry 1 megaton of TNT. (A 1-megaton nuclear weapon, in contrast, is on the order of 1 m long and may have a mass of a few hundred kg.)

Solution

A typical railroad boxcar is about 60 ft long and can carry around 100 tons. Therefore, $10^6/10^2 = 10^4$ or ten thousand boxcars would be needed to transport one megaton of TNT. A train with this many boxcars would be $60{\times}10^4$ ft ≈ 114 mi ≈ 180 km long, not including the engine!

Problem

59. As you will see in Chapter 16, the speed of waves on a stretched string depends only on the tension, F_0, in the string and on the mass per unit length of string μ. Tension has the dimensions $M{\cdot}L/T^2$ and mass per unit length obviously has dimensions M/L. Find a combination of F_0 and μ that has the units of speed; this combination must play a prominent role in the expression for the speed of waves on the string.

Solution

Suppose that the speed of sound is the product of powers of the tension and mass per unit length; dimensionally $[v] = [F_0]^x[\mu]^y$. Then $LT^{-1} = (M{\cdot}L/T^2)^x(M/L)^y = M^{x+y}L^{x-y}T^{-2x}$. Comparison of the exponents of the basic dimensions M, L, and T leads to $x + y = 0$, $x - y = 1$, and $-2x = -1$, or $x = \frac{1}{2} = -y$. Thus, a dimensionally correct combination is $[v] = [F_0^{1/2}\mu^{-1/2}] = [\sqrt{F_0/\mu}]$.

PART 1   MECHANICS

CHAPTER 2   MOTION IN A STRAIGHT LINE

ActivPhysics can help with these problems:
Activities 1.2–1.9

Section 2-1:   Distance, Time, Speed, and Velocity

Problem

1. In 1996 Donovan Bailey of Canada set a world record in the 100-m dash, with a time of 9.84 s. What was his average speed?

Solution

Bailey's average speed was (Equation 2-1)
$\bar{v} = \Delta x/\Delta t = 100$ m/9.84 s $= 10.16$ m/s. (One can assume that the race distance was known to more than four significant figures.)

Problem

3. In 1996, Josia Thugwame of South Africa won the Olympic Marathon, completing the 26-mi, 385-yd course in 2 h 12 min 36 s. What was Thugwame's average speed, in meters per second?

Solution

$$\bar{v} = \frac{\Delta r}{\Delta t} = \frac{(26 + 385/1760) \text{ mi}}{(2 + 756/3600) \text{ h}} = 11.9 \frac{\text{mi}}{\text{h}} \approx 5.30 \frac{\text{m}}{\text{s}},$$

or a little over half the speed of Bailey's 100 m dash in Problem 1. (Runners usually compute their average pace, $1/\bar{v}$, which in this case was 5 min 3.4 s per mile. See Appendix C for the appropriate conversion factors.)

Problem

5. Starting from home, you bicycle 24 km north in 2.5 h, then turn around and pedal straight home in 1.5 h. What are your (a) displacement at the end of the first 2.5 h, (b) average velocity over the first 2.5 h, (c) average velocity for the homeward leg of the trip, (d) displacement for the entire trip, and (e) average velocity for the entire trip?

Solution

(a) $\Delta r_{\text{out}} = 24$ km (north). (b) $v_{\text{out}} = 24$ km$\times$(north)/2.5 h $= 9.6$ km/h(north). (c) $v_{\text{back}} =$ 24 km(south)/1.5 h $= 16$ km/h (south). (d) $\Delta r_{\text{out and back}} = 0$. (e) $v_{\text{round trip}} = 0$.

Problem

7. Australian Chris McCormack won the 1997 world triathlon championship, completing the 1500-m swim, 40-km bicycle ride, and 10-km run in 1 h, 48 min, 29 s. What was McCormack's average speed?

Solution

$\bar{v} = \Delta r/\Delta t = (1.5 + 40 + 10)$ km/$(1 + 48/60 +$ $29/3600)$ h $= (28.5$ km/h$)(1$ m/s$)(3.6$ km/h$) =$ 7.91 m/s.

Problem

9. You allow yourself 40 min to drive 25 mi to the airport, but are caught in heavy traffic and average only 20 mi/h for the first 15 min. What must your average speed be on the rest of the trip if you are to get there on time?

Solution

At an average speed of 20 mi/h for the first 15 min $= \frac{1}{4}$ h, you travel only $(20$ mi/h$)(\frac{1}{4}$ h$) = 5$ mi. Therefore, you must cover the remaining $(25 - 5)$ mi $= 20$ mi in $(40 - 15)$ min $= 25$ min $= \frac{5}{12}$ h. This implies an average speed of 20 mi/$(\frac{5}{12}$ h$) =$ 48 mi/h. (Note that your overall average speed was pre-determined to be 25 mi/$(40$ h$/60) = 37.5$ mi/h, and that this equals the time-weighted average of the average speeds for the two parts of the trip: $(15$ min/40 min$)(20$ mi/h$) + (25$ min/40 min$)\times$ $(48$ mi/h$)$.)

Problem

11. What is the conversion factor from meters per second to miles per hour?

Solution

$1 \text{ mi/h} = 1609 \text{ m}/3600 \text{ s} = 0.447 \text{ m/s} = (2.24)^{-1} \text{ m/s}.$

Problem

13. A fast base-runner can get from first to second base in 3.4 s. If he leaves first base as the pitcher throws a 90 mi/h fastball the 61-ft distance to the catcher, and if the catcher takes 0.45 s to catch and rethrow the ball, how fast does the catcher have to throw the ball to second base to make an out? Home plate to second base is the diagonal of a square 90 ft on a side.

Solution

At 90 mi/h $= 132$ ft/s, the ball takes 61 ft$/(132$ ft/s$) = 0.462$ s to travel from the pitcher to the catcher. (We are keeping extra significant figures in the intermediate calculations as suggested in Section 1-7.) After the catcher throws the ball, it has $3.4 \text{ s} - 0.462 \text{ s} - 0.45 \text{ s} = 2.49$ s to reach second base at the same time as the runner. The distance is $\sqrt{2}(90$ ft), so the minimum speed is $\bar{v} = \sqrt{2}(90 \text{ ft})/2.49 \text{ s} = 51.2$ ft/s $= 35$ mi/h. A prudent catcher would allow extra time for the player covering second base to make the tag.

Problem

15. If you drove the 4600 km from coast to coast of the United States at 65 mi/h (105 km/h), stopping an average of 30 min for rest and refueling after every 2 h of driving, (a) What would be your average velocity for the entire trip? (b) How long would it take?

Solution

If you stopped 30 min for every 2 h of driving at 105 km/h, your average speed would be $\bar{v} = (2 \text{ h}/2.5 \text{ h})(105 \text{ km/h}) + (0.5 \text{ h}/2.5 \text{ h})(0) = 84.0$ km/h, and a coast-to-coast trip would take $(4600 \text{ km}) \div (84.0 \text{ km/h}) = 54.8$ h. However, this is only approximate, because the exact travel time does not include a 30-min stop after the final segment. (a) To find the total time, note that every 2 h 30 min you would cover a distance $x = \bar{v}t = (105 \text{ km/h})(2 \text{ h}) = 210$ km, so it would take you $21 \times 2.5 \text{ h} = 52.5$ h to travel $21 \times 210 \text{ km} = 4410$ km. You could drive the final 190 km in $(190 \text{ km})/(105 \text{ km/h}) = 1.81$ h, so the complete trip would take 54.3 h. (b) Overall, $\bar{v} = 4600 \text{ km}/54.3 \text{ h} = 84.7$ km/h.

Problem

17. A jetliner leaves San Francisco for New York, 4600 km away. With a strong tailwind, its speed is 1100 km/h. At the same time, a second jet leaves New York for San Francisco. Flying into the wind, it makes only 700 km/h. When and where do the two planes pass each other?

Solution

When the planes pass, the total distance traveled by both is 4600 km. Therefore, 4600 km $= (1100 \text{ km/h})\Delta t + (700 \text{ km/h})\Delta t$, or $\Delta t = 4600 \text{ km} \div (1800 \text{ km/h}) = 2.56$ h. (The planes meet 2.56 h after taking off.) The encounter occurs at a point about $(700 \text{ km/h})(2.56 \text{ h}) \approx 1790$ km from New York City or $(1100 \text{ km/h})(2.56 \text{ h}) \approx 2810$ km from San Francisco.

Section 2-2: Instantaneous Velocity

Problem

19. On a single graph, plot distance versus time for the two trips from Houston to Des Moines described on page 24. For each trip, identify graphically the average velocity and, for each segment of the trip, the instantaneous velocity.

Solution

Both trips start at the same place (Houston, point A) $x_A = -1000$ km at time $t_A = 0$, and end at the same place (Des Moines, point B) $x_B = 300$ km at $t_B = 2.6$ h. (We are using the coordinate system in Fig. 2-2.) They have the same overall displacement, $\Delta x = x_B - x_A = 1300$ km, in the same time period, $\Delta t = t_B - t_A = 2.6$ h, and thus the same average velocity $\bar{v}_{AB} = 500$ km/h, as explained in the text. $\bar{v}_{AB}$ is the slope of the straight line AB. AB is also the graph of the first trip, a direct flight at constant velocity, $x_1(t) = x_A + \bar{v}_{AB}t$ for $0 \le t \le 2.6$ h. (Short intervals of acceleration at takeoff and landing are ignored.) The second trip, using a faster plane (steeper slopes when flying), stops for a while in Minneapolis at $x_C = 650$ km (this segment is flat) and then proceeds south to Des Moines (negative velocity and slope). This trip is shown by three straight segments

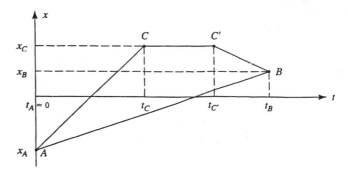

Problem 19 Solution.

$ACC'B$, and is given analytically by the equations

$$x_2(t) = \begin{cases} x_A + \left(\dfrac{x_C - x_A}{t_C - 0}\right)t = x_A + \bar{v}_{AC}t, \\ \qquad\qquad\qquad \text{for } t_A = 0 \le t \le t_C \\ x_C, \quad \text{for} \quad t_C \le t \le t_{C'} \\ x_C + \left(\dfrac{x_B - x_C}{t_B - t_{C'}}\right)(t - t_{C'}) = x_C + \bar{v}_{C'B}(t - t_{C'}), \\ \qquad\qquad\qquad \text{for } t_{C'} \le t \le t_B. \end{cases}$$

(In the graph, we assumed each segment of the second trip was executed with constant velocity and ignored takeoffs and landings as before. The times t_C and $t_{C'}$ and velocities $\bar{v}_{AC}$ and $\bar{v}_{C'B}$ were chosen arbitrarily.)

Problem

21. Figure 2-21 shows the position of an object as a function of time. From the graph, determine the instantaneous velocity at (a) 1.0 s; (b) 2.0 s; (c) 3.0 s; (d) 4.5 s. (e) What is the average velocity over the interval shown?

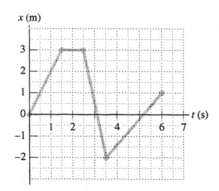

FIGURE 2-21 Problem 21.

Solution

The instantaneous velocity at a particular time is the slope of the graph of x versus t at that point, $v(t) = dx/dt$. For a straight line segment of graph, v equals the average velocity over that segment, $\bar{v} = \Delta x/\Delta t$. Each of the times specified in this problem falls on a different straight segment of the graph in Fig. 2-21, whose slopes we determine from the coordinates of the endpoints of that segment.

(a) $v(1\text{ s}) = (3 - 0)\text{ m}/(1.5 - 0)\text{ s} = 2\text{ m/s}$;

(b) $v(2\text{ s}) = (3 - 3)\text{ m}/(2.5 - 1.5)\text{ s} = 0$;

(c) $v(3\text{ s}) = (-2 - 3)\text{ m}/(3.5 - 2.5)\text{ s} = -5\text{ m/s}$;

(d) $v(4.5\text{ s}) = [1 - (-2)]\text{ m}/(6 - 3.5)\text{ s} = 1.2\text{ m/s}$.

(e) The overall average velocity is $\bar{v} = [x(6\text{ s}) - x(0)] \div (6\text{ s} - 0) = 1\text{ m}/6\text{ s} = 0.167\text{ m/s}$.

Problem

23. A model rocket is launched straight upward; its altitude y as a function of time is given by $y = bt - ct^2$, where $b = 82$ m/s, $c = 4.9$ m/s^2, t is the time in seconds, and y is in meters. (a) Use differentiation to find a general expression for the rocket's velocity as a function of time. (b) When is the velocity zero?

Solution

(a) Equation 2-3 can be used to find the derivative of each term in the altitude: $v(t) = dy/dt = b - 2ct$.
(b) The velocity is zero when $b = 2ct$, or $t = b/2c = (82\text{ m/s})/2(4.9\text{ m/s}^2) = 8.37$ s.

Problem

25. The position of an object is given by $x = bt^3 - ct^2 + dt$, with x in meters and t in seconds. The constants b, c, and d are $b = 3.0$ m/s^3, $c = 8.0$ m/s^2, and $d = 1.0$ m/s. (a) Find all times when the object is at position $x = 0$. (b) Determine a general expression for the instantaneous velocity as a function of time, and from it find (c) the initial velocity and (d) all times when the object is instantaneously at rest. (e) Graph the object's position as a function of time, and identify on the graph the quantities you found in (a) to (d).

Solution

(a) With the aid of the quadratic formula and factorization, $x = t(bt^2 - ct + d) = 0$ implies $t = 0$, or $t = (c \pm \sqrt{c^2 - 4bd})/2b$. Substituting the given constants, $t = 0$, $t = (4 \pm \sqrt{13})\text{ s}/3 = 0.131$ s and 2.54 s. (b) $v(t) = dx/dt = 3bt^2 - 2ct + d$. (c) When $t = 0, v(0) = d = 1$ m/s. (d) $v = 3bt^2 - 2ct + d = 0$ implies $t = (c \pm \sqrt{c^2 - 3bd})/3b = (8 \pm \sqrt{55})\text{ s}/9 = 64.9$ ms and 1.71 s. (e) The graph of this cubic has roots from part (a), slope at the origin from part (c), and relative maximum and minimum from part (d), as shown.

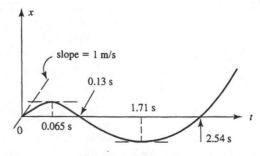

Problem 25 Solution.

Section 2-3: Acceleration

Problem

27. A giant eruption on the Sun propels solar material from rest to a final speed of 450 km/s over a period of 1 h. What is the average acceleration of this material, in m/s²?

Solution

Equation 2-4, the definition of average linear acceleration, for this one-hour time interval, gives $\bar{a} = \Delta v/\Delta t = (v_2 - v_1)/(t_2 - t_1) = (450 \text{ km/s} - 0)/3600 \text{ s} = 125 \text{ m/s}^2$.

Problem

29. A space shuttle's main engines cut off 8.5 min after launch, at which time the shuttle's speed is 7.6 km/s. What is the shuttle's average acceleration during this interval?

Solution

The average acceleration of the shuttle along its trajectory, from liftoff until its main engines stop, was (Equation 2-4) $\bar{a} = \Delta v/\Delta t = (7.6{\times}10^3 \text{ m/s} - 0)/(8.5{\times}60 \text{ s}) = 14.9 \text{ m/s}^2 \approx 1.5 \text{ g}$, where g = 9.8 m/s², the acceleration due to gravity at the surface of the Earth, is a frequently used anthropomorphic unit.

Problem

31. An airplane's takeoff speed is 320 km/h. If its average acceleration is 2.9 m/s², how long is it on the runway after starting its takeoff roll?

Solution

If we assume that the airplane starts from rest, $\Delta v = 320$ km/h = 88.9 m/s at the end of a time interval Δt, during which the average acceleration was $\bar{a} = 2.9 \text{ m/s}^2 = \Delta v/\Delta t$. Solving for Δt, we find $\Delta t = (88.9 \text{ m/s})/(2.9 \text{ m/s}^2) = 30.7 \text{ s}$.

Problem

33. Your plane reaches its takeoff runway and then holds for 4.0 min because of air-traffic congestion. The plane then heads down the runway with an average acceleration of 3.6 m/s². It is airborne 35 s later. What are (a) its takeoff speed and (b) its average acceleration from the time it reaches the takeoff runway until it's airborne?

Solution

(a) During the 35 s the plane is actually taking off, Equation 2-4 gives $\Delta v = v - 0 = \bar{a}\Delta t =$

$(3.6 \text{ m/s}^2)(35 \text{ s}) = 126 \text{ m/s} = 454$ km/h. (b) If we include the four-minute wait before taking off, the average acceleration for the entire interval on the runway is only $\bar{a} = \Delta v/\Delta t = (126 \text{ m/s} - 0) \div (4 \text{ min} + 35 \text{ s}) = (126 \text{ m/s})/(275 \text{ s}) = 0.458 \text{ m/s}^2$.

Problem

35. Determine the instantaneous acceleration as a function of time for the motion in Problem 25.

Solution

From the answer to Problem 25(b), we find: $a(t) = dv/dt = (d/dt)(3bt^2 - 2ct + d) = 6bt - 2c$.

Section 2-4: Constant Acceleration

Problem

37. A car accelerates from rest to 25 m/s in 8.0 s. Determine the distance it travels in two ways: (a) by multiplying the average velocity given in Equation 2-8 by the time and (b) by calculating the acceleration from Equation 2-7 and using the result in Equation 2-10.

Solution

(a) For constant acceleration, Equation 2-8 can be combined with Equation 2-1 to yield $\Delta x = \bar{v}\,\Delta t = \frac{1}{2}(v_0 + v)\Delta t = \frac{1}{2}(0 + 25 \text{ m/s})(8.0 \text{ s}) = 100$ m.
(b) Alternatively, for constant acceleration, Equation 2-7 gives $a = \bar{a} = \Delta v/\Delta t = (25 \text{ m/s} - 0) \div 8.0 \text{ s} = 3.13 \text{ m/s}^2$, so Equation 2-10 yields $\Delta x = x - x_0 = v_0 t + \frac{1}{2}at^2 = 0 + \frac{1}{2}(3.13 \text{ m/s}^2)(8.0 \text{ s})^2 = 100$ m.

Problem

39. If you square Equation 2-7, you'll have an expression for v^2. Equation 2-11 also gives an expression for v^2. Equate the two expressions for v^2, and show that the resulting equation reduces to Equation 2-10.

Solution

Squaring Equation 2-7, $v^2 = (v_0 + at)^2 = v_0^2 + 2v_0at + a^2t^2$, and equating to Equation 2-11, $v^2 = v_0^2 + 2a \times (x - x_0)$, one finds $2v_0at + a^2t^2 = 2a(x - x_0)$, or $x - x_0 = v_0t + \frac{1}{2}at^2$, which is Equation 2-10.

Section 2-5: Using the Equations of Motion

Problem

41. A rocket rises with constant acceleration to an altitude of 85 km, at which point its speed is 2.8 km/s. (a) What is its acceleration? (b) How long does the ascent take?

Solution

(a) In Equation 2-11 (with x positive upward) we are given that $x - x_0 = 85$ km, $v_0 = 0$ (the rocket starts from rest), and $v = 2.8$ km/s. Therefore, we can solve for the acceleration, $a = (v^2 - v_0^2)/2(x - x_0) = (2.8 \text{ km/s})^2/2(85 \text{ km}) = 46.1$ m/s^2 (note the change of units). (b) From Equation 2-9, we can solve for the time of flight, $t = 2(x - x_0)/(v_0 + v) = 2(85 \text{ km}) \div (2.8 \text{ km/s}) = 60.7$ s. (We chose to relate t directly to the given data, but once the acceleration is known, Equation 2-7 or 2-10 could have been used to find $t = v/a$ or $t = \sqrt{2(x - x_0)/a}$, respectively.)

Problem

43. Starting from rest, a car accelerates at a constant rate, reaching 88 km/h in 12 s. (a) What is its acceleration? (b) How far does it go in this time?

Solution

(a) From Equation 2-7, $a = (v - v_0)/t = (88 \text{ km/h} - 0)/12 \text{ s} = 7.33$ km/h/s $= 2.04$ m/s^2. (b) From Equation 2-9, $x - x_0 = \frac{1}{2}(v_0 + v)t = \frac{1}{2}(0 + 88 \text{ km/h})(12 \text{ s}) = 147$ m. (Note the change in units. Again, we chose equations that relate the answers directly to the given data; see solution to Problem 41.)

Problem

45. In an X-ray tube, electrons are accelerated to a velocity of 10^8 m/s, then slammed into a tungsten target. The electrons undergo rapid deceleration, producing X rays. If the stopping time for an electron is on the order of 10^{-19} s, approximately how far does an electron move while decelerating? Assume constant deceleration.

Solution

Assuming the electrons travel in a straight line while coming to rest ($v = 0$), Equation 2-9 gives $x - x_0 = \frac{1}{2}(v_0 + v)t = \frac{1}{2}(10^8 \text{ m/s})(10^{-19} \text{ s}) = 5 \times 10^{-12}$ m for the stopping distance. (The X rays emitted are called *bremsstrahlung*.)

Problem

47. The Barringer meteor crater in northern Arizona is 180 m deep and 1.2 km in diameter. The fragments of the meteor lie just below the bottom of the crater. If these fragments decelerated at a constant rate of 4×10^5 m/s^2 as they ploughed through the Earth in forming the crater, what was the speed of the meteor's impact at Earth's surface?

Solution

For a particular fragment (which followed a straight-line path to the bottom, perpendicular to the desert surface), we can use Equation 2-11 to find the initial speed: $-v_0^2 = 2(-4 \times 10^5 \text{ m/s}^2)(180 \text{ m})$ or $v_0 = \sqrt{(144 \times 10^6 \text{ m}^2/\text{s}^2)} = 12$ km/s.

Problem

49. A hockey puck moving at 32 m/s slams through a wall of snow 35 cm thick. It emerges moving at 18 m/s. (a) How much time does it spend in the snow? (b) How thick a wall of snow would be needed to stop the puck entirely?

Solution

(a) If we assume a constant linear deceleration for the puck, Equation 2-9 can be used to find the time it spends traversing 35 cm of snow: $t = 2(x - x_0)/(v_0 + v) = 2(0.35 \text{ m})/(32 + 18)(\text{m/s}) = 14$ ms. (b) If we assume the same deceleration for penetrating any wall of snow, Equation 2-11, with $v = 0$, gives the thickness necessary to stop a puck moving with the same initial speed: $x - x_0 = -v_0^2/2a$. The acceleration (which is negative when the puck is decelerating) can be found from Equation 2-7 with the time from part (a) (or from a second application of Equation 2-11 with data from part (a), etc.): $a = (18 - 32)(\text{m/s}) \div (0.014 \text{ s}) = -10^3$ m/s^2. Then any wall of snow thicker than $-(32 \text{ m/s})^2/2(-10^3 \text{ m/s}^2) = 51.2$ cm would stop this puck.

Problem

51. A jetliner touches down at 220 km/h, reverses its engines to provide braking, and comes to a halt 29 s later. What is the shortest runway on which this aircraft can land, assuming constant deceleration starting at touchdown?

Solution

From Equation 2-9 with $v = 0$, we find $x - x_0 = \frac{1}{2}v_0 t = \frac{1}{2}(220 \text{ km/h})(29 \text{ h}/3600) = 886$ m (over half a mile).

Problem

53. The maximum acceleration that a human being can survive even for a short time is about $200g$. In a highway accident, a car moving at 88 km/h slams into a stalled truck. The front end of the car is squashed by 80 cm on impact. If the deceleration during the collision is constant, will a passenger wearing a seatbelt survive?

Solution

The passenger, originally moving with velocity $v_0 = 88$ km/h $= 24.4$ m/s, comes to rest, $v = 0$, in a distance $x - x_0 = 0.8$ m, so the acceleration (from Equation 2-11) was $a = (v^2 - v_0^2)/2(x - x_0) = -(24.4 \text{ m/s})^2/1.6 \text{ m} = -373 \text{ m/s}^2 = -38.1g$. Such a person could survive. Without a seatbelt, however, the stopping distance would not have been 0.8 m (think about it!) and the passenger would surely not survive the secondary collision with the interior of the car (see Problem 86).

Problem

55. The maximum deceleration of a car on a dry road is about 8 m/s^2. If two cars are moving head-on toward each other at 88 km/h (55 mi/h), and their drivers apply their brakes when they are 85 m apart, will they collide? If so, at what relative speed? If not, how far apart will they be when they stop? On the same graph, plot distance versus time for both cars.

Solution

The minimum distance a car needs to stop $(v = 0)$ from an initial speed $v_0 = 88$ km/h $= 24.4$ m/s, with a constant acceleration $a = -8$ m/s^2, is (Equation 2-11) $x - x_0 = -v_0^2/2a = -(24.4 \text{ m/s})^2/2(-8 \text{ m/s}^2) = 37.3$ m (positive in the direction of v_0). Since 85 m is greater than twice this distance, the cars can avoid a collision, and they will be $85 \text{ m} - 2(37.3 \text{ m}) = 10.3$ m apart when stopped.

To plot x versus t, using Equation 2-10 for each car, we need to choose an origin, say $x = 0$ at the midpoint of the separation between the cars, with positive x in the direction of the initial velocity of the first car, and $t = 0$ when the brakes are applied. Then

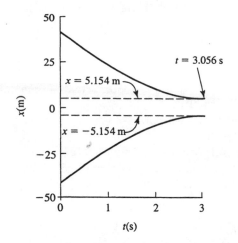

Problem 55 Solution.

$x_{10} = -42.5$ m $= -x_{20}, v_{10} = 24.4$ m/s $= -v_{20}$, and $a_1 = -8$ m/s $= -a_2$. A graph of $x_1(t)$ and $x_2(t)$ is as shown.

Problem

57. After 35 minutes of running, at the 9-km point in a 10-km race, you find yourself 100 m behind the leader and moving at the same speed. What should your acceleration be if you are to catch up by the finish line? Assume that the leader maintains constant speed throughout the entire race.

Solution

Taking $x_0 = 0$ and $t = 0$ at the 9-km point (and assuming a straight path to the finish), we can express your position (runner A) and that of the leader (runner B) as $x_A = v_0 t + \frac{1}{2}at^2$, and $x_B = 100 \text{ m} + v_0 t$. Since B's speed was constant, $v_0 = \Delta x/\Delta t = (9 \text{ km} + 100 \text{ m})/35 \text{ min} = 0.26$ km/min. If both runners finish simultaneously, $x_A = x_B = 1$ km, so $a = 2(1 \text{ km} - v_0 t)v_0^2/(v_0 t)^2$. We multiplied and divided by v_0 because the quantity $v_0 t$ (at the finish) equals $x_B - 100 \text{ m} = 1 \text{ km} - 100 \text{ m} = 0.9$ km. Therefore

$$a = 2(1 \text{ km} - 0.9 \text{ km})(0.26 \text{ km/min})^2/(0.9 \text{ km})^2$$
$$= 1.67 \times 10^{-2} \text{ km/min}^2 = 4.64 \times 10^{-3} \text{ m/s}^2.$$

Problem

59. Repeat the preceding problem, now assuming your initial speed is 95 km/h.

Solution

The position as a function of time for either car, moving with constant acceleration, is given by Equation 2-10. Let us choose our origin $t = 0$ and $x = 0$ at the time and place the speeding driver in car number one notices car number two in front and applies the brakes, with the direction of initial motion positive. Then $x_{10} = 0$, $x_{20} = 10$ m, $v_{10} > v_{20} = 60$ km/h $= 16.7$ m/s, $a_1 = -4.2$ m/s^2, and $a_2 = 0$. The position of the cars is $x_1(t) = v_{10}t + \frac{1}{2}a_1t^2$ and $x_2(t) = x_{20} + v_{20}t$, valid for $0 \leq t \leq t^*$, where t^* is the time for which the accelerations remain constant. (Thus, t^* is either the time the cars collide, if this happens, or the time when car number one stops decelerating.) The distance between the cars is $x_{21}(t) = x_2(t) - x_1(t)$. The condition for a collision is that the quadratic equation $x_{21}(t) = 0$ have a real root (in which case the smaller root is t^*), and the condition for no collision is that this equation have no real roots. The solution of the equation $x_{21}(t) = 0 = -\frac{1}{2}a_1t^2 - (v_{10} - v_{20})t + x_{20}$ follows from the quadratic formula,

$t = [(v_{10} - v_{20}) \pm \sqrt{(v_{10} - v_{20})^2 - 2 |a_1| x_{20}}]/|a_1|$.
(Since a_1 is negative, we wrote it explicitly as $a_1 = -|a_1|$.) Thus, if $(v_{10} - v_{20})^2 \geq 2 |a_1| x_{20}$, there is a collision at time $t^* = [(v_{10} - v_{20}) - \sqrt{(v_{10} - v_{20})^2 - 2 |a_1| x_{20}}]/|a_1|$, from which the relative velocity at collision, $v_1(t^*) - v_{20}$, can be calculated. On the other hand, if $(v_{10} - v_{20})^2 < 2 |a_1| x_{20}$, there is no collision, and the minimum distance x_{21} can be found by setting the derivative of $x_{21}(t)$ equal to zero, or by physical reasoning.

When $v_{10} = 95$ km/h $= 26.4$ m/s, $(v_{10} - v_{20})^2 = (26.4 \text{ m/s} - 16.7 \text{ m/s})^2 = 94.5 \text{ m}^2/\text{s}^2 > 2(4.2 \text{ m/s}^2)(10 \text{ m}) = 84 \text{ m}^2/\text{s}^2$, so there is a collision at $t^* = (9.72 \text{ m/s} - \sqrt{10.5 \text{ m}^2/\text{s}^2})/(4.2 \text{ m/s}^2) = 1.54$ s. The relative speed at collision is $v_1(t^*) - v_{20} = v_{10} - v_{20} - |a_1| t^* = 9.72$ m/s $- (4.2 \text{ m/s}^2)(1.54 \text{ s}) = 3.24$ m/s $= 11.7$ km/h, where we used Equation 2-7 for the velocities.

When $v_{10} = 85$ km/h, $(v_{10} - v_{20})^2 = 25$ km/h$)^2 = 48.2 \text{ m}^2/\text{s}^2 < 2 |a_1| x_{20} = 84 \text{ m}^2/\text{s}^2$, and there is no collision. The relative distance is the quadratic $x_{21}(t) = \frac{1}{2}|a_1|t^2 - (v_{10} - v_{20})t + x_{20}$. One way to obtain the distance of closest approach is to minimize this function of time. Setting the derivative equal to zero gives us $dx_{21}/dt = |a_1|t - (v_{10} - v_{20}) = 0$, or $t_{min} = (v_{10} - v_{20})/|a_1|$. Then $x_{21}(t_{min}) = \frac{1}{2}|a_1|t_{min}^2 - (v_{10} - v_{20})t_{min} + x_{20} = x_{20} - (v_{10} - v_{20})^2/2|a_1| = 10$ m $- (48.2 \text{ m}^2/\text{s}^2)/2(4.2 \text{ m/s}^2) = 4.26$ m. This is in fact a minimum because $d^2x_{21}/dt^2 = |a_1| > 0$.

Another way to obtain the minimum x_{21}, without using calculus, relies on purely physical reasoning. As long as the velocity of car number one, $v_1(t)$, is greater than 60 km/h (the velocity of car number two), it is gaining ground on car number two, so the relative distance x_{21} is decreasing. When $v_1(t)$ falls below 60 km/h, car number one loses ground to car number two and x_{21} starts increasing. Therefore, the closest approach occurs when $v_1(t) = v_{10} - |a_1|t = v_{20} = 60$ km/h, which gives the same t_{min} as above.

Section 2-6: The Constant Acceleration of Gravity

Problem

61. Your friend is sitting 6.5 m above you in a tree branch. How fast should you throw an apple so that it just reaches her?

Solution

Equation 2-11 describes the vertical motion of the apple, whose acceleration is $-g$ (positive upward), if one ignores air resistance, intervening leaves, etc. The difference in height between your friend and you is $y - y_0 = 6.5$ m, v_0 is the initial velocity we desire, and

v is the velocity of the apple when it reaches your friend. If the apple *just* reaches her, $v = 0$. Then $v_0^2 = 0 + 2g(y - y_0)$, or $v_0 = \sqrt{2(9.8 \text{ m/s}^2)(6.5 \text{ m})} = 11.3$ m/s. (We chose the positive square root because v_0 is upward.)

Problem

63. A foul ball leaves the bat going straight upward at 23 m/s. (a) How high does it rise? (b) How long is it in the air? Neglect the distance between the bat and the ground.

Solution

(a) At the maximum height, $v^2 = 0 = v_0^2 - 2g \times (y_{max} - y_0)$, so $y_{max} - y_0 = v_0^2/2g = (23 \text{ m/s})^2 \div 2(9.8 \text{ m/s}^2) = 27.0$ m. (b) If we neglect the distance between the bat and the ground (and assume that the foul ball is not caught), the flight of the ball lasts until it falls back to its initial height. Then $y - y_0 = 0 = v_0t - \frac{1}{2}gt^2$, or $t = 2(23 \text{ m/s})/(9.8 \text{ m/s}^2) = 4.69$ s.

Problem

65. Space pirates kidnap an earthling and hold him imprisoned on one of the planets of the solar system. With nothing else to do, the prisoner amuses himself by dropping his watch from eye level (170 cm) to the floor. He observes that the watch takes 0.95 s to fall. On what planet is he being held? *Hint:* Consult Appendix E.

Solution

The planet's surface gravity can be found, since 1.7 m $= \frac{1}{2}g(0.95 \text{ s})^2$, or $g = 3.77 \text{ m/s}^2$. This is closest to the value listed for Mars, in Appendix E.

Problem

67. A falling object travels one-fourth of its total distance in the last second of its fall. From what height was it dropped?

Solution

The total distance traveled by a falling object in a time t is given by Equation 2-10, with $a = -g$ and $v_0 = 0$ (the meaning of dropped). Thus $y_0 - y(t) = \frac{1}{2}gt^2$. The distance fallen during the last second (an interval from $t - 1$ s to t) is $y(t - 1 \text{ s}) - y(t) = \frac{1}{2}gt^2 - \frac{1}{2}g(t - 1 \text{ s})^2$. The latter is one-fourth of the former when (cancel off the common factors of $\frac{1}{2}g$) $\frac{1}{4}t^2 = t^2 - (t - 1 \text{ s})^2$. Then $t - 1 \text{ s} = \pm\sqrt{\frac{3}{4}}t$, or $t = 1 \text{ s}/(1 - \sqrt{\frac{3}{4}}) = 7.46$ s. (We discarded the negative square root because t is obviously greater than 1 s.) Substituting this value of t into the equation for the total distance fallen, we find

$y_0 - y(t) = \frac{1}{2}(9.8 \text{ m/s}^2)(7.46 \text{ s})^2 = 273$ m. (In a real fall from this height, air resistance should be considered.)

Problem

69. A kingfisher is 30 m above a lake when it accidentally drops the fish it is carrying. A second kingfisher 5 m above the first dives toward the falling fish. What initial speed should it have if it is to reach the fish before the fish hits the water?

Solution

We are concerned with just the vertical motion of the fish and bird, which we describe with the y-axis positive upward and origin at the water's surface. If the fish is dropped at $t = 0$, its position is $y_{\text{Fish}} = 30 \text{ m} - (4.9 \text{ m/s}^2)t^2$, where "drop" means $v_{0,\text{Fish}} = 0$, and we substituted the standard value for $\frac{1}{2}g$. Suppose the second bird starts its dive at t_1, with an initial velocity v_0. Its position is $y_{\text{Bird}} = 35 \text{ m} + v_0(t - t_1) - (4.9 \text{ m/s}^2)(t \cdot t_1)^2$, where we assume the dive is a free-fall. The bird catches the fish before either hits the water if $y_{\text{Fish}} \geq 0$ when $y_{\text{Bird}} = y_{\text{Fish}}$. If the dive started without delay, $t_1 = 0$, and the equation $y_{\text{Bird}} = y_{\text{Fish}} = 30 \text{ m} - (4.9 \text{ m/s}^2)t^2 = 35 \text{ m} + v_0 t - (4.9 \text{ m/s}^2)t^2$ gives $v_0 t = -5$ m. When this is substituted into the inequality $y_{\text{Fish}} \geq 0$, one obtains $30 \text{ m} - (4.9 \text{ m/s}^2) \times (-5 \text{ m/}v_0)^2 \geq 0$, or $|v_0| \geq 2.02$ m/s. This is the minimum downward speed necessary for the bird to catch the fish.

Problem

71. A balloon is rising at 10 m/s when its passenger throws a ball straight up at 12 m/s. How much later does the passenger catch the ball?

Solution

The initial (positive upward) velocity of the ball is 12 m/s relative to the passenger who throws it. Because the passenger is moving upward with constant velocity of 10 m/s, the initial velocity of the ball relative to the ground is 22 m/s. Assuming the ball is acted upon only by gravity (after being thrown at $t = 0$), we can write its vertical position as $y_B(t) = y_0 + (22 \text{ m/s})t - \frac{1}{2}gt^2$. The balloon carrying the passenger is acted upon by the buoyant force of the air, in addition to gravity, so that it ascends with constant velocity (see Section 18-3). Thus, the vertical position of the passenger (in the same coordinate system used for the ball) is $y_P(t) = y_0 + (10 \text{ m/s})t$. The passenger catches the ball when $y_B(t) = y_P(t)$ for $t > 0$. This implies $y_0 + (22 \text{ m/s})t - \frac{1}{2}gt^2 = y_0 + (10 \text{ m/s})t$, or $t = 2(12 \text{ m/s})/(9.8 \text{ m/s}^2) = 2.45$ s.

(Because the balloon is moving with constant velocity, a coordinate system attached to the passenger, $y_P' = 0$, is an inertial frame (see Section 3-5) in which the ball's position is $y_B' = (12 \text{ m/s})t - \frac{1}{2}gt^2$. Setting $y_B' = y_P'$ gives one the same time of flight.)

Paired Problems

Problem

73. You drive 14 km to the next town, maintaining a speed of 50 km/h except for a stop lasting 4.1 min at a red light. You shop for 20 min, then head back toward your starting point at a steady 70 km/h. You stop at a gas station 4.4 km from the town. What are (a) your average speed and (b) the magnitude of your average velocity between your starting point and the gas station?

Solution

(a) The average speed is the total distance traveled, 14 km + 4.4 km, divided by the total time spent, 14 km/(50 km/h) + 4.1 min + 20 min + 4.4 km ÷ (70 km/h). Thus, the average speed is 18.4 km ÷ 0.745 h = 24.7 km/h. (b) The average velocity is the total displacement divided by the total time. If we take the origin at the starting point and the positive direction toward the next town, $\Delta x = 14 \text{ km} - 4.4 \text{ km}$, while Δt is the same as in part (a). Thus, the average velocity is 9.6 km/0.745 h = 12.9 km/h.

Problem

75. A skier starts from rest, and heads downslope with a constant acceleration of 1.9 m/s^2. How long does it take her to go 20 m, and what is her speed at that point?

Solution

The equations for linear motion with constant acceleration are summarized in Table 2-1. Since the initial velocity is zero, $x(t) - x_0 = \frac{1}{2}at^2$, and the time to travel 20 m is $t = \sqrt{2(20 \text{ m})/(1.9 \text{ m/s}^2)} = 4.59$ s. The velocity at this time is $v = at = (1.9 \text{ m/s}^2) \times (4.59 \text{ s}) = 8.72$ m/s.

Problem

77. A frustrated student drops a book out of his dormitory window, releasing it from rest. After falling 2.3 m, it passes the top of a 1.5-m high window on a lower floor. How long does it take to cross the window?

Solution

Equation 2-10 gives the distance fallen by the book when dropped ($v_0 = 0$) at time $t = 0$: $y_0 - y(t) = \frac{1}{2}gt^2$. The book passes the top of the lower-floor window at time t_1, given by $y_0 - y(t_1) = 2.3$ m $= \frac{1}{2}gt_1^2$, or $t_1 = \sqrt{2(2.3\ \text{m})/(9.8\ \text{m/s}^2)} = 0.685$ s. It passes the bottom of the window at time t_2, given by $y_0 - y(t_2) = 2.3$ m $+ 1.5$ m $= \frac{1}{2}gt_2^2$, or $t_2 = 0.881$ s. The time to cross the window is $t_2 - t_1 = 0.196$ s.

Problem

79. A subway train is traveling at 80 km/h when it approaches a slower train 50 m ahead traveling in the same direction at 25 km/h. If the faster train begins decelerating at 2.1 m/s^2, while the slower train continues at constant speed, how soon and at what relative speed will they collide?

Solution

Take the origin $x = 0$ and $t = 0$ at the point where the first train begins decelerating, with positive x in the direction of motion. Equation 2-10 gives the instantaneous position of each train, with $x_{10} = 0$, $v_{10} = 80$ km/h, $a_1 = -2.1$ m/s^2, $x_{20} = 50$ m, $v_{20} = 25$ km/h, and $a_2 = 0$ given. Thus $x_1(t) = v_{10}t + \frac{1}{2}a_1t^2$, and $x_2(t) = x_{20} + v_{20}t$. The trains collide at the first time that $x_1 = x_2$, or when $x_{20} - (v_{10} - v_{20})t - \frac{1}{2}a_1t^2 = 0$. Using the quadratic formula to solve for the smaller root, we find $t = [(v_{10} - v_{20}) - \sqrt{(v_{10} - v_{20})^2 + 2a_1x_{20}}]/(-a_1) = [(55\ \text{m}/3.6\ \text{s}) - \sqrt{(55\ \text{m}/3.6\ \text{s})^2 + 2(-2.1\ \text{m/s}^2)(50\ \text{m})}]/(2.1\ \text{m/s}^2) = 4.97$ s. The velocity of the first train at the time of the collision is $v_1 = v_{10} + a_1t = (80\ \text{km/h}) - (2.1\ \text{m/s}^2) \times (4.97\ \text{s})(3.6\ \text{km/h/m/s}) = 42.4$ km/h. Therefore, the relative speed at impact is $v_1 - v_2 = 42.4$ km/h $- 25$ km/h $= 17.4$ km/h.

Problem

81. You toss a hammer over the 3.7-m-high wall of a construction site, starting your throw at a height of 1.2 m above the sidewalk. On the other side of the wall, the hammer falls to the bottom of an excavation 7.9 m below the sidewalk (see Fig. 2-23). (a) What is the minimum speed at which you must throw the hammer for it to clear the wall? (b) Assuming it's thrown with the speed given in part (a), when will it hit the bottom of the excavation?

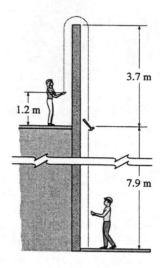

FIGURE 2-23 Problem 81.

Solution

If we consider just the vertical motion of the hammer, and ignore air resistance, etc., the equations in Table 2-1 (with y replacing x and $a = -g$) apply. (a) Equation 2-11 evaluated at the highest point of the hammer's trajectory gives $v_0^2 = 2g(y_{\text{top}} - y_0)$, since the instantaneous vertical velocity at the highest point is zero. In order to clear the top of the wall, $y_{\text{top}} - y_0 \geq (3.7 - 1.2)$ m $= 2.5$ m (from Fig. 2-23), so $v_0 \geq \sqrt{2(9.8\ \text{m/s}^2)(2.5\ \text{m})} = 7.00$ m/s. (b) From Equation 2-10, the hammer hits bottom when $y_0 - y_{\text{bot}} = 1.2$ m $+ 7.9$ m $= \frac{1}{2}gt^2 - (7\ \text{m/s})t$, where the displacement $y_0 - y_{\text{bot}}$ is shown in Fig. 2-23, and we used the minimum initial velocity from part (a). The time in this equation is measured from $t = 0$ when the hammer is thrown; therefore $t > 0$ at the bottom. The positive root of this quadratic equation is $t = [(7\ \text{m/s}) + \sqrt{(7\ \text{m/s})^2 + 2(9.8\ \text{m/s}^2)(9.1\ \text{m})}] \div (9.8\ \text{m/s}) = 2.25$ s, which is the time of flight to the bottom.

Supplementary Problems

Problem

83. A car accelerates away from a red light at 2.5 m/s^2 until its speed reaches 10 m/s. It travels at that speed for 8.0 s, then brakes to a stop at the next red light with deceleration 4.0 m/s^2. What is the distance between lights?

Solution

The distance covered by the car, accelerating from rest ($v_0 = 0$) to a speed $v = 10$ m/s, away from the first stoplight, is $(x - x_0)_{\text{accel.}} = (v^2 - v_0^2)/2a = (10\ \text{m/s})^2/2(2.5\ \text{m/s}^2) = 20$ m (see Equation 2-11).

Traveling at a constant speed of $v = 10$ m/s for the next $t = 8.0$ s, the car covers a distance of $(x - x_0)_{\text{no accel.}} = vt = (10 \text{ m/s})(8.0 \text{ s}) = 80$ m (see Equation 2-10 with $a = 0$). Finally, the distance covered decelerating ($a = -4.0$ m/s^2) from speed $v_0 = 10$ m/s to rest ($v = 0$), at the second stoplight, is $(x - x_0)_{\text{decel.}} = (0 - (10 \text{ m/s})^2)/2(-4.0 \text{ m/s}^2) = 12.5$ m (see Equation 2-11 again). The total distance covered between the stoplights is the sum of these three distances, or approximately 113 m. (Note: we redefined $t = 0$ and x_0 for each of the segments of the car's motion.)

Problem

85. You see the traffic light ahead of you is about to turn from red to green, so you slow to a steady speed of 10 km/h and cruise to the light, reaching it just as it turns green. You accelerate to 60 km/h in the next 12 s, then maintain constant speed. At the light, you pass a Porsche that has stopped. Just as you pass (and the light turns green) the Porsche begins accelerating, reaching 65 km/h in 6.9 s, then maintaining constant speed. (a) Plot the motions of both cars on a graph showing the 10-s period after the light turns green. (b) How long after the light turns green does the Porsche pass you? (c) How far are you from the light when the Porsche passes you?

Solution

(a) Let the stoplight be at $x = 0$ and turn green at $t = 0$. Then

$$x_{\text{You}}(t) = \begin{cases} (10 \text{ km/h})t + \dfrac{1}{2}\left(\dfrac{60 \text{ km/h} - 10 \text{ km/h}}{12 \text{ s}}\right)t^2, \\ \qquad\qquad 0 \le t \le 12 \\ 116.7 \text{ m} + (60 \text{ km/h})(t - 12 \text{ s}), \quad t \ge 12 \text{ s}. \end{cases}$$

$$x_{\text{Porsche}}(t) = \begin{cases} \dfrac{1}{2}\left(\dfrac{65 \text{ km/h}}{6.9 \text{ s}}\right)t^2, \quad 0 \le t \le 6.9 \text{ s} \\ 62.29 \text{ m} + (65 \text{ km/h})(t - 6.9 \text{ s}), \\ \qquad\qquad t \ge 6.9 \text{ s}. \end{cases}$$

Before plotting x versus t, we first calculate that

$$x_{\text{You}}(6.9 \text{ s}) = 46.72 \text{ m}, \ x_{\text{You}}(12 \text{ s}) = 116.7 \text{ m},$$
$$x_{\text{Porsche}}(6.9 \text{ s}) = 62.29 \text{ m}, \ x_{\text{Porsche}}(12 \text{ s}) = 154.4 \text{ m}.$$

(b) Evidentally, the Porsche passes you before 6.9 s, while both cars are accelerating, so $x_{\text{You}} = x_{\text{Porsche}}$ implies:

$$(10 \text{ km/h})t + \frac{1}{2}\left(\frac{50 \text{ km/h}}{12 \text{ s}}\right)t^2 = \frac{1}{2}\left(\frac{65 \text{ km/h}}{6.9 \text{ s}}\right)t^2,$$

or $t = 3.81$ s.

(c) When the cars pass, both are at the same position:

$$x_{\text{You}}(3.81 \text{ s}) = x_{\text{Porsche}}(3.81 \text{ s})$$
$$= \frac{1}{2}\left(\frac{65 \text{ km/h}}{6.9 \text{ s}}\right)(3.81 \text{ s})^2 = 19.0 \text{ m}$$

from the green light.

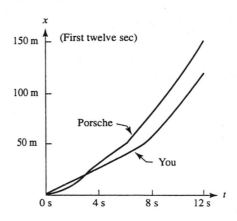

Problem 85 Solution.

Problem

87. The position of a particle as a function of time is given by $x = x_0 \sin \omega t$, where x_0 and ω are constants. (a) Take derivatives to find expressions for the velocity and acceleration. (b) What are the maximum values of velocity and acceleration? *Hint:* Consult the table of derivatives in Appendix A.

Solution

(a) For $x(t) = x_0 \sin \omega t$, $dx/dt = v(t) = \omega x_0 \cos \omega t$ and $dv/dt = d^2x/dt^2 = a(t) = -\omega^2 x_0 \sin \omega t = -\omega^2 x(t)$. (b) Since the maximum value of the sine or cosine functions is 1, $v_{\max} = \omega x_0$ and $a_{\max} = \omega^2 x_0$. (The motion described by $x(t)$ is called simple harmonic motion; see Chapter 15.)

Problem

89. A faucet leaks water at the rate of 15 drops per second. At the instant one drop leaves the faucet, another strikes the sink below, and two additional drops are in between on the way down. How far is it from the faucet to the sink bottom?

Solution

Drops appear at the faucet every 1/15 of a second. Under the conditions stated (one drop at the faucet, two in the air, and one striking the sink), the time of fall for one drop is $3(1/15)$ s $= (1/5)$ s, so the distance fallen (starting from rest) is $y_0 - y = \frac{1}{2}gt^2 = \frac{1}{2}(9.8 \text{ m/s}^2)(0.2 \text{ s})^2 = 19.6$ cm.

Problem

91. A student is staring idly out her dormitory window when she sees a water balloon fall past. If the balloon takes 0.22 s to cross the 130-cm-high window, from what height above the top of the window was it dropped?

Solution

If the balloon was dropped from height y_0 at time $t = 0$, then its height at any later time is $y = y_0 - \frac{1}{2}gt^2$. When it passes the top of the window, $y_1 = y_0 - \frac{1}{2}gt_1^2$, and when passing the bottom, $y_2 = y_0 - \frac{1}{2}gt_2^2$. The length of the window is $1.3 \text{ m} = y_1 - y_2 = \frac{1}{2}g(t_2^2 - t_1^2) = \frac{1}{2}g(t_2 - t_1)(t_2 + t_1)$. But $t_2 - t_1 = 0.22$ s (the time required to cross the window), so $t_2 + t_1 = 2(1.3 \text{ m})/(9.8 \text{ m/s}^2)(0.22 \text{ s}) = 1.21$ s. Combined with the value of the difference in times, we find that $t_1 = \frac{1}{2}(1.21 \text{ s} - 0.22 \text{ s}) = 0.493$ s. Finally, the height above the top of the window is $y_0 - y_1 = \frac{1}{2}gt_1^2 = \frac{1}{2}(9.8 \text{ m/s}^2)(0.493 \text{ s})^2 = 1.19$ m.

CHAPTER 3   THE VECTOR DESCRIPTION OF MOTION

ActivPhysics can help with these problems:
Activity 4.1

Section 3-2:   Vector Arithmetic

Problem

1. You walk west 220 m, then north 150 m. What are the magnitude and direction of your displacement vector?

Solution

The triangle formed by the two displacement vectors and their sum is a right triangle, so the Pythagorean Theorem gives the magnitude $C = \sqrt{A^2 + B^2} = \sqrt{(220 \text{ m})^2 + (150 \text{ m})^2} = 266$ m, and the basic definition of the tangent gives $\beta = \tan^{-1}(150 \text{ m}/220 \text{ m}) = 34.3°$. The direction of $\mathbf{C}$ can be specified as $34.3°$ N of W, or $55.7°$ W of N, or by the azimuth $304.3°$ (CW from N), etc.

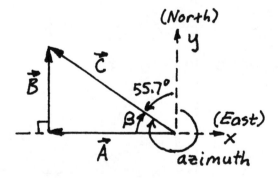

Problem 1 Solution.

Problem

3. A migrating whale follows the west coast of Mexico and North America toward its summer home in Alaska. It first travels 360 km due northwest to just off the coast of Northern California and then turns due north and travels 400 km toward its destination. Determine graphically the magnitude and direction of its displacement vector.

Solution

We can find the magnitude and direction of the vector sum of the two displacements either using geometry and a diagram, or by adding vector components. From the law of cosines:

$$C = \sqrt{A^2 + B^2 - 2AB\cos\gamma}$$
$$= \sqrt{(360 \text{ km})^2 + (400 \text{ km})^2 - 2(360 \text{ km})(400 \text{ km})\cos 135°}$$
$$= 702 \text{ km}.$$

From the law of sines: $C/\sin\gamma = B/\sin\beta$, or

$$\beta = \sin^{-1}\left(\frac{B\sin\gamma}{C}\right) = \sin^{-1}\left[\left(\frac{400 \text{ m}}{702 \text{ m}}\right)\sin 135°\right] = 23.7°.$$

The direction of $\mathbf{C}$ can be specified as $45° + 23.7° = 68.7°$ N of W, or $180° - 68.7° = 111°$ CCW from the x-axis (east) in the illustration.

In a coordinate system with x-axis east and y-axis north, the first displacement is 360 km ($\hat{\imath}\cos 135° + \hat{\jmath}\sin 135°$) and the second simply 400 km $\hat{\jmath}$. Their sum is $(-255\hat{\imath} + 255\hat{\jmath} + 400\hat{\jmath})$ km $= (-255\hat{\imath} + 655\hat{\jmath})$ km, which is the total displacement. Its magnitude is $\sqrt{(-255)^2 + (655)^2}$ km/702 km and its direction (measured CCW from the x-axis) is $\theta_x = \cos^{-1}(-255 \text{ km}/702 \text{ km}) = 111°$, as above. (Note that since $C_y > 0$ and $C_x < 0$, θ_x is in the second quadrant.)

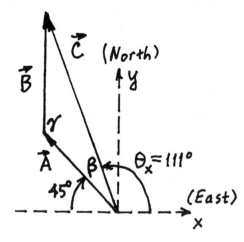

Problem 3 Solution.

Problem

5. Vector $\mathbf{A}$ has magnitude 3.0 m and points to the right; vector $\mathbf{B}$ has magnitude 4.0 m and points vertically upward. Find the magnitude and direction of a vector $\mathbf{C}$ such that $\mathbf{A} + \mathbf{B} + \mathbf{C} = 0$.

Solution

The vectors $\mathbf{A}, \mathbf{B}$, and $\mathbf{C}$ form a 3-4-5 right triangle, as shown in the sketch. Therefore, $C = 5$ m, and the direction of $\mathbf{C}$, measured CCW from the direction of $\mathbf{A}$, is $180° + \tan^{-1}(B/A) = 180° + 53.1° = 233°$ (other angles could have been chosen). $\mathbf{C}$ could also be determined algebraically from components, with x-axis parallel to $\mathbf{A}$ and y-axis parallel to $\mathbf{B}$. Then $\mathbf{A} = (3.0 \text{ m})\hat{\imath}$, $\mathbf{B} = (4.0 \text{ m})\hat{\jmath}$, and $\mathbf{C} = -(\mathbf{A} + \mathbf{B}) = (-3.0 \text{ m})\hat{\imath} + (-4.0 \text{ m})\hat{\jmath} = C_x\hat{\imath} + C_y\hat{\jmath}$. The magnitude of $\mathbf{C}$ is $\sqrt{C_x^2 + C_y^2} = \sqrt{(-3.0 \text{ m})^2 + (-4.0 \text{ m})^2}$, and the angle that $\mathbf{C}$ makes with the x-axis is $\cos^{-1}(C_x/C) = \cos^{-1}(-3.0 \text{ m}/5.0 \text{ m})$, which is in the third quadrant, as calculated above. (Note: the angle of $\mathbf{C}$ could also be specified as $-127°$, or CW from the x-axis.)

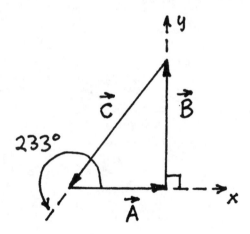

Problem 5 Solution.

Problem

7. Vectors $\mathbf{A}$ and $\mathbf{B}$ in Fig. 3-22 have the same magnitude, A. Find the magnitude and direction of (a) $\mathbf{A} - \mathbf{B}$ and (b) $\mathbf{A} + \mathbf{B}$.

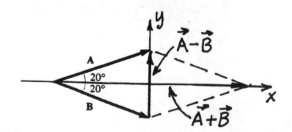

FIGURE 3-22 Problem 7 Solution.

Solution

The vectors $\mathbf{A}$ and $\mathbf{B}$ in Fig. 3-22 form two sides of a parallelogram, in which $\mathbf{A} - \mathbf{B}$ and $\mathbf{A} + \mathbf{B}$ are the diagonals, as shown. Since the magnitudes of $\mathbf{A}$ and $\mathbf{B}$ are equal, the parallelogram is a rhombus, and the diagonals are perpendicular (the converse of this is also true; see Problem 60). Then $\mathbf{A} + \mathbf{B}$ is along the perpendicular bisector of the base $\mathbf{A} - \mathbf{B}$ of an isosceles triangle, and vice versa. Using the given angles, we find the magnitudes (a) $|\mathbf{A} - \mathbf{B}| = 2A \sin 20° = 0.684A$, and (b) $|\mathbf{A} + \mathbf{B}| = 2A \cos 20° = 1.88A$. In Fig. 3-22, (a) $\mathbf{A} - \mathbf{B}$ is up, and (b) $\mathbf{A} + \mathbf{B}$ is to the right, but the directions could be specified relative to $\mathbf{A}, \mathbf{B}$, or some other coordinate system. (This problem can also be readily solved with components and unit vectors. Figure 3-22 suggests a coordinate system with x-axis to the right and y-axis up, as shown. Then $\mathbf{A} = A(\hat{\imath} \cos 20° + \hat{\jmath} \sin 20°)$ and $\mathbf{B} = A(\hat{\imath} \cos 20° - \hat{\jmath} \sin 20°)$, from which $\mathbf{A} \pm \mathbf{B}$ are easily obtained.)

Problem

9. Three vectors $\mathbf{A}$, $\mathbf{B}$, and $\mathbf{C}$ have the same magnitude L and form an equilateral triangle, as shown in Fig. 3-23. Find the magnitude and direction of the vectors (a) $\mathbf{A} + \mathbf{B}$, (b) $\mathbf{A} - \mathbf{B}$, (c) $\mathbf{A} + \mathbf{B} + \mathbf{C}$, (d) $\mathbf{A} + \mathbf{B} - \mathbf{C}$.

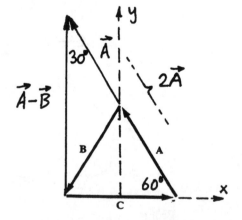

FIGURE 3-23 Problems 9, 16, and 22.

Solution

(c) Since the vectors form a closed figure (a triangle), their sum is zero, i.e., $\mathbf{A} + \mathbf{B} + \mathbf{C} = 0$. (a) The vector equation in part (c) has solution $\mathbf{A} + \mathbf{B} = -\mathbf{C}$. Thus $|\mathbf{A} + \mathbf{B}| = |-\mathbf{C}| = |\mathbf{C}| = L$ and the direction of $\mathbf{A} + \mathbf{B}$ is opposite to the direction of $\mathbf{C}$, or $180°$ from $\mathbf{C}$. (d) Similarly $(\mathbf{A} + \mathbf{B} - \mathbf{C}) = (-\mathbf{C}) - \mathbf{C} = -2\mathbf{C}$, and so has magnitude $2L$ and direction opposite to $\mathbf{C}$. (b) Finally, $(\mathbf{A} - \mathbf{B}) = \mathbf{A} - (-\mathbf{A} - \mathbf{C}) = 2\mathbf{A} + \mathbf{C}$, so these vectors form a $30° - 60° - 90°$ right triangle, as shown. Then $|\mathbf{A} - \mathbf{B}| = \sqrt{3}|\mathbf{C}| = \sqrt{3}L$, and its direction is $90°$ CCW from $\mathbf{C}$.

Of course, this problem can be solved readily using components and a coordinate system with x-axis parallel to $\mathbf{C}$ and y-axis perpendicular, as shown superposed on Fig. 3-23. Then $\mathbf{A}=|\mathbf{A}|(\hat{\imath}\cos 120°+\hat{\jmath}\sin 120°)=L(-\hat{\imath}+\sqrt{3}\hat{\jmath})/2$, $\mathbf{B}=L(-\hat{\imath}-\sqrt{3}\hat{\jmath})/2$, and $\mathbf{C}=L\hat{\imath}$. It is a simple matter to find
(a) $\mathbf{A}+\mathbf{B}=-L\hat{\imath}$, (b) $\mathbf{A}-\mathbf{B}=\sqrt{3}L\hat{\jmath}$,
(c) $\mathbf{A}+\mathbf{B}+\mathbf{C}=0$, and (d) $\mathbf{A}+\mathbf{B}-\mathbf{C}=-2L\hat{\imath}$.
The magnitudes and directions are as above.

Section 3-3:   Coordinate Systems, Vector Components, and Unit Vectors

Problem

11. Vector $\mathbf{V}$ represents a displacement of 120 km at 29° counterclockwise from the x-axis. Write $\mathbf{V}$ in unit vector notation.

Solution

Take the y-axis 90? CCW from the x-axis, as in Figs. 3-10 and 11. Then $\mathbf{V}=V_x\hat{\imath}+V_y\hat{\jmath}=V(\hat{\imath}\cos\theta_x+\hat{\jmath}\cos\theta_y)=V(\hat{\imath}\cos\theta_x+\hat{\jmath}\sin\theta_x)=(120\text{ km})\times(\hat{\imath}\cos 29°+\hat{\jmath}\sin 29°)=(105\hat{\imath}+58.2\hat{\jmath})$ km. (Note: The component of a vector along an axis is defined in terms of the cosine of the angle it makes with that axis. In two dimensions, $\theta_y=|\theta_x-90°|$, and $\cos\theta_y=\sin\theta_x$.)

Problem

13. Express each of the vectors of Fig. 3-24 in unit vector notation, with the x-axis horizontally to the right and the y-axis vertically upward.

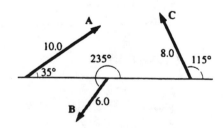

FIGURE 3-24 Problems 13, 19, and 20.

Solution

Take the x-axis to the right and the y-axis 90° counterclockwise from it. Then

$$\mathbf{A} = 10(\hat{\imath}\cos 35°+\hat{\jmath}\sin 35°)=8.19\hat{\imath}+5.74\hat{\jmath}$$
$$\mathbf{B} = 6(\hat{\imath}\cos 235°+\hat{\jmath}\sin 235°)=-3.44\hat{\imath}-4.91\hat{\jmath}$$
$$\mathbf{C} = 8(\hat{\imath}\cos 115°+\hat{\jmath}\sin 115°)=-338\hat{\imath}+7.25\hat{\jmath}$$

Problem

15. Repeat Problem 3, using unit vector notation.

Solution

See solution to Problem 3.

Problem

17. Let $\mathbf{A}=15\hat{\imath}-40\hat{\jmath}$ and $\mathbf{B}=31\hat{\jmath}+18\hat{\mathbf{k}}$. Find a vector $\mathbf{C}$ such that $\mathbf{A}+\mathbf{B}+\mathbf{C}=0$.

Solution

$\mathbf{C}=-\mathbf{A}-\mathbf{B}=-(15\hat{\imath}-40\hat{\jmath})-(31\hat{\jmath}+18\hat{\mathbf{k}})=-15\hat{\imath}+9\hat{\jmath}-18\hat{\mathbf{k}}$. (Since $\mathbf{A}$ and $\mathbf{B}$ are specified in terms of unit vectors, this form is also appropriate for $\mathbf{C}$.)

Problem

19. Use the result of Problem 13 to find the vectors
(a) $\mathbf{A}+\mathbf{B}+\mathbf{C}$, (b) $\mathbf{A}-\mathbf{B}+\mathbf{C}$, and
(c) $\mathbf{A}+1.5\mathbf{B}-2.2\mathbf{C}$.

Solution

The components of $\mathbf{A}, \mathbf{B}$, and $\mathbf{C}$ are given in the solution to Problem 13. A component of the sum (or difference) is the sum (or difference) of the components; the component of a scalar multiple is the scalar multiple of the component. For example, the x-component of $\mathbf{A}+1.5\mathbf{B}-2.2\mathbf{C}$ is $A_x+1.5B_x-2.2C_x$. Therefore (a) $\mathbf{A}+\mathbf{B}+\mathbf{C}=1.37\hat{\imath}+8.07\hat{\jmath}=8.19(\hat{\imath}\cos 80.4°+\hat{\jmath}\sin 80.4°)$ (b) $\mathbf{A}-\mathbf{B}+\mathbf{C}=8.25\hat{\imath}+17.9\hat{\jmath}=19.7(\hat{\imath}\cos 65.3°+\hat{\jmath}\sin 65.3°)$ (c) $\mathbf{A}+1.5\mathbf{B}-2.2\mathbf{C}=10.5\hat{\imath}-17.6\hat{\jmath}=20.5(\hat{\imath}\cos 301°+\hat{\jmath}\sin 301°)$ The first form is the vector in components, the second gives the magnitude $(\sqrt{x^2+y^2})$ and direction $\theta_x=\tan^{-1}(y/x)$.

Problem

21. You're trying to reach a pond that lies 3.5 km to the northeast of your starting point. You first follow a logging road that runs east for 0.80 km. Then you follow a deer trail heading northeast for 2.1 km. From there you bushwack straight to the pond. Describe your final displacement vector, (a) in unit vector notation; and (b) as a magnitude and compass direction.

Solution

The desired total displacement, $\mathbf{R}=3.5$ km NE, is the sum of three displacements, $\mathbf{R}_1=0.80$ kmE, $\mathbf{R}_2=2.1$ km NE, and $\mathbf{R}_3=\mathbf{R}-\mathbf{R}_1-\mathbf{R}_2$ to be found. (a) With x-axis E and y-axis N, $\mathbf{R}_1=0.80$km, $\mathbf{R}_2=(2.1\text{ km})(\hat{\imath}\cos 45°+\hat{\jmath}\sin 45°)=1.48\hat{\imath}+1.48\hat{\jmath}$ km, and $\mathbf{R}=2.47\hat{\imath}+2.47\hat{\jmath}$ km. Therefore $\mathbf{R}_3=(2.47-0.80-1.48)\hat{\imath}+(2.47-1.48)\hat{\jmath}$ km $=0.190\hat{\imath}+0.990\hat{\jmath}$ km $=(1.01\text{ km})(\hat{\imath}\cos 79.1°+$

$\hat{\jmath}\sin 79.1°$). (See Equations 3-1 and 3-2 or the solution to Problem 19 for the last step.)

Problem

23. In Fig. 3-14 the angle between x- and x'-axes is 21°, the angle between the vector **A** and the x-axis is 54°, and **A**'s magnitude is 10 units. (a) Find the components of **A** in both coordinate systems shown. (b) Verify that the magnitude of **A**, computed using Equation 3-1, is the same in both coordinate systems.

Solution

(a) In the x-y system, $A_x = A\cos\theta = 10\cos 54° = 5.88$ and $A_y = 10\sin 54° = 8.09$. In the x'-y' system, $\theta' = 54° - 21° = 33°$ so $A'_x = 10\cos 33° = 8.39$ and $A'_y = 10\sin 33° = 5.45$ (b) Direct calculation shows that $\sqrt{(5.88)^2 + (8.09)^2} = \sqrt{(8.39)^2 + (5.45)^2} = 10$, which reflects the fact that $\sin^2 + \cos^2 = 1$. (The mathematical definition of a two-dimensional vector is a pair of numbers (V_x, V_y) which transform like the position vector (x, y) when the coordinate axes are rotated.)

Problem

25. Express the sum of the unit vectors $\hat{\imath}$, $\hat{\jmath}$, and $\hat{k}$ in unit vector notation, and determine its magnitude.

Solution

$$\mathbf{r} = \hat{\imath} + \hat{\jmath} + \hat{k}; \quad |\mathbf{r}| = \sqrt{1^2 + 1^2 + 1^2} = \sqrt{3}.$$

Problem

27. In Fig. 3-15, suppose that vectors **A** and **C** both make 30° angles with the horizontal while **B** makes a 60° angle, and that $A = 2.3$ km, $B = 1.0$ km, and $C = 2.9$ km. (a) Express the displacement vector $\Delta\mathbf{r}$ from start to summit in each of the coordinate systems shown, and (b) determine its length.

Solution

Using the x-y system in Fig. 3-15, we have θ_A $\theta_C = 30°$, and $\theta_B = 60°$. Then $\mathbf{A} = A(\hat{\imath}\cos\theta_A + \hat{\jmath}\sin\theta_A) = (2.3\text{ km})(\hat{\imath}\cos 30° + \hat{\jmath}\sin 30°) = (1.99\hat{\imath} + 1.15\hat{\jmath})$km, $\mathbf{B} = (1\text{ km})(\hat{\imath}\cos 60° + \hat{\jmath}\sin 60°) = (0.50\hat{\imath} + 0.87\hat{\jmath})$ km, and $\mathbf{C} = (2.9\text{ km})(\hat{\imath}\cos 30° + \hat{\jmath}\sin 30°) = (2.51\hat{\imath} + 1.45\hat{\jmath})$km. Thus, $\Delta\mathbf{r} = \mathbf{A} + \mathbf{B} + \mathbf{C} = (5.00\hat{\imath} + 3.47\hat{\jmath})$ km, and $\Delta r = \sqrt{(5.00)^2 + (3.47)^2}$ km $= 6.09$ km. A similar calculation in the x'-y' system, with $\theta'_A = \theta'_C = 0$ and $\theta'_B = 30°$, yields $\mathbf{A} = 2.3\hat{\imath}'$ km, $\mathbf{B} = (1\text{ km}) \times (\hat{\imath}'\cos 30° + \hat{\jmath}'\sin 30°) = (0.87\hat{\imath}' + 0.50\hat{\jmath}')$ km, $\mathbf{C} =$

2.9$\hat{\imath}'$ km, $\Delta\mathbf{r} = (6.07\hat{\imath}' + 0.50\hat{\jmath}')$ km, and $\sqrt{(6.07)^2 + (0.50)^2} = 6.09$, of course.

Section 3-4: Velocity and Acceleration Vectors

Problem

29. A car drives north at 40 mi/h for 10 min, then turns east and goes 5.0 mi at 60 mi/h. Finally, it goes southwest at 30 mi/h for 6.0 min. Draw a vector diagram and determine (a) the car's displacement and (b) its average velocity for this trip.

Solution

Take a coordinate system with x-axis east, y-axis north, and origin at the starting point. The first segment of the trip can be represented by a displacement vector in the y direction of length (40 mi/h)(10 min), or $\mathbf{r}_1 = (20/3)\hat{\jmath}$ mi. For the second segment, $\mathbf{r}_2 = 5\hat{\imath}$ mi. The time spent on this segment is $t_2 = 5$ mi/(60 mi/h) $= 5$ min. The final segment has length (30 mi/h)(6 min). A unit vector in the southwest direction is displacements and their sum are shown in the sketch. (a) The total displacement is $\mathbf{r}_{\text{tot}} = \mathbf{r}_1 + \mathbf{r}_2 + \mathbf{r}_3 = [(20/3)\hat{\jmath} + 5\hat{\imath} - (3/\sqrt{2})(\hat{\imath}+\hat{\jmath})]$ mi $= (2.88\hat{\imath} + 4.55\hat{\jmath})$ mi. (b) The total time is 10 min $+$ 5 min $+$6 min $= 21$ min, so the average velocity for the trip is $\bar{v} = \mathbf{r}_{\text{tot}}/t_{\text{tot}} = (2.88\hat{\imath} + 4.55\hat{\jmath})$ mi/(21/60) h $= (8.22\hat{\imath} + 13.0\hat{\jmath})$ mi/h. (Note: Instead of unit vector notation, $\mathbf{r}_{\text{tot}}$ and $\bar{v}$ could be specified by their magnitudes $\sqrt{(2.88)^2 + (4.55)^2}$ mi $= 5.38$ mi and 15.4 mi/h, respectively, and common direction, $\theta = \tan^{-1}(4.55/2.88) = 57.7°$ N of E.)

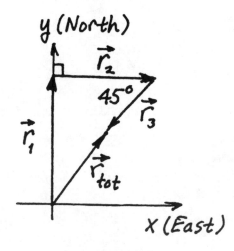

Problem 29 Solution.

Problem

31. The Orlando-to-Atlanta flight described in Problem 10 takes 2.5 h. What is the average velocity? Express (a) as a magnitude and direction, and (b) in unit vector notation with the x-axis east and the y-axis north.

Solution

(b) The displacement from Orlando to Atlanta calculated in Problem 10 was $\mathbf{A} = (-320\hat{\imath} + 577\hat{\jmath})$ km in a coordinate system with x-axis east and y-axis north. If this trip took 2.5 h, the average velocity was $\hat{\mathbf{v}} = \mathbf{A}/2.5$ h $= (-128\hat{\imath} + 231\hat{\jmath})$ km/h. (a) This has magnitude $\sqrt{(-128)^2 + (231)^2}$ km/h $= 264$ km/h and direction $\theta = \tan^{-1}(231/-128) = 119°$ (which was given).

Problem

33. A hot-air balloon rises vertically 800 m over a period of 10 min, then drifts eastward 14 km in 27 min. Then the wind shifts, and the balloon moves northeastward for 15 min, at a speed of 24 km/h. Finally, it drops vertically in 5 min until it is 250 m above the ground. Express the balloon's average velocity in unit vector notation, using a coordinate system with the x-axis eastward, the y-axis northward, and the z-axis upward.

Solution

The displacement for the first segment of the balloon's excursion is $\mathbf{r}_1 = 0.8\hat{\mathbf{k}}$ km, in the coordinate system specified, and for the second segment $\mathbf{r}_2 = 14\hat{\imath}$ km. The third segment has length $(24 \text{ km/h})(15 \text{ min}) = 6$ km in the northeast direction $\hat{\imath}\cos 45° + \hat{\jmath}\sin 45°$, so $\mathbf{r}_3 = (6 \text{ km})(\hat{\imath} + \hat{\jmath})/\sqrt{2} = (4.24\hat{\imath} + 4.24\hat{\jmath})$ km. Finally, the last segment's displacement is $\mathbf{r}_4 = (250 - 800 \text{ m})\hat{\mathbf{k}} = -0.55\hat{\mathbf{k}}$ km (this is a drop of 550 m from the preceding altitude). The total displacement $\Delta\mathbf{r} = \mathbf{r}_1 + \mathbf{r}_2 + \mathbf{r}_3 + \mathbf{r}_4 = [(14 + 4.24)\hat{\imath} + 4.24\hat{\jmath} + (0.8 - 0.55)\hat{\mathbf{k}}]$ km $= (18.2\hat{\imath} + 4.24\hat{\jmath} + 0.25\hat{\mathbf{k}})$ km is accomplished in total time $\Delta t = (10 + 27 + 15 + 5)$ min $= 0.950$ h, so the average velocity is $\Delta\mathbf{r}/\Delta t = (19.2\hat{\imath} + 4.47\hat{\jmath} + 0.263\hat{\mathbf{k}})$ km/h.

Problem

35. An object's position as a function of time is given by $\mathbf{r} = 12t\hat{\imath} + (15t - 5.0t^2)\hat{\jmath}$ m, where t is time in s. (a) What is the object's position at $t = 2.0$ s? (b) What is its average velocity in the interval from $t = 0$ to $t = 2.0$ s? (c) What is its instantaneous velocity at $t = 2.0$ s?

Solution

(a) The object's position is given as a function of time, so when $t = 2$ s, this is $\mathbf{r}(2 \text{ s}) = (12 \text{ m/s})(2 \text{ s})\hat{\imath} + [(15 \text{ m/s})(2 \text{ s}) - (5.0 \text{ m/s}^2)(2 \text{ s})^2]\hat{\jmath} = 24\hat{\imath} + 10\hat{\jmath}$ m, where we explicitly displayed the units of the coefficients in the intermediate step. (b) Since $\mathbf{r}(0) = 0$, the average velocity for this interval is $\bar{\mathbf{v}} = [\mathbf{r}(2 \text{ s}) - \mathbf{r}(0)]/(2 \text{ s} - 0) = 12\hat{\imath} + 5\hat{\jmath}$ m/s. (c) The instantaneous velocity at any time is $d\mathbf{r}/dt = (12 \text{ m/s})\mathbf{l} + [(15 \text{ m/s}) - (5.0 \text{ m/s}^2)2t]\hat{\jmath} = \mathbf{v}(t)$ (see Appendix A-2 for the derivative of t^n), so when $t = 2$ s, $\mathbf{v}(2s) = 12\hat{\imath} - 5\hat{\jmath}$ m/s.

Problem

37. A car, initially going eastward, rounds a 90° bend and ends up heading southward. If the speedometer reading remains constant, what is the direction of the car's average acceleration vector?

Solution

Since the speed is constant, the change in velocity for the 90° turn is $\Delta\mathbf{v} = -v\hat{\jmath} - (v\hat{\imath}) = -v(\hat{\imath} + \hat{\jmath})$, where $\hat{\imath}$ is east and $\hat{\jmath}$ is north. The direction of the average acceleration is the same as that of $\Delta\mathbf{v}$, which is parallel to $-(\hat{\imath} + \hat{\jmath})$ or southwest. ($\theta = \tan^{-1}(-1/-1) = 225°$.)

Problem

39. What are (a) the average velocity and (b) the average acceleration of the tip of the 2.4-cm-long hour hand of a clock in the interval from 12 P.M. to 6 P.M.? Express in unit vector notation, with the x-axis pointing toward 3 P.M. and the y-axis toward 12 P.M.

Solution

The position and velocity of the tip of the hour hand are shown in the diagram for 12 P.M. and 6 P.M., in

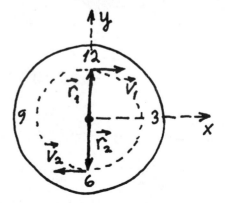

Problem 39 Solution.

the coordinate system specified. The magnitude of the position is a constant, namely, the 2.4-cm radius. The magnitude of the velocity is also constant, namely, the circumference divided by 12 h, or $2\pi(2.4 \text{ cm})/12 \text{ h} = 1.26 \text{ cm}/h$. (a) $\bar{\mathbf{v}} = (\mathbf{r}_2 - \mathbf{r}_1)/6 \text{ h} = (-2.4\hat{\jmath} - 2.4\hat{\jmath}) \text{ cm}/6 \text{ h} = -0.8\hat{\jmath} \text{ cm}/h$. (b) $\bar{\mathbf{a}} = (\mathbf{v}_2 - \mathbf{v}_1)/6 \text{ h} = (-1.26\hat{\imath} - 1.26\hat{\imath})(\text{cm}/h)/6 \text{ h}. = -0.419\hat{\imath} \text{ cm}/h^2$.

Problem

41. An object undergoes acceleration of $2.3\hat{\imath} + 3.6\hat{\jmath} \text{ m}/s^2$ over a 10-s interval. At the end of this time, its velocity is $33\hat{\imath} + 15\hat{\jmath} \text{ m}/s$. (a) What was its velocity at the beginning of the 10-s interval? (b) By how much did its speed change? (c) By how much did its direction change? (d) Show that the speed change is *not* given by the magnitude of the acceleration times the time. Why not?

Solution

(a) $\mathbf{v} = \mathbf{v}_0 + \mathbf{a}t$, so $\mathbf{v}_0 = \mathbf{v} - \mathbf{a}t$, or $\mathbf{v}_0 = (33\hat{\imath} + 15\hat{\jmath}) \text{ m}/s - (2.3\hat{\imath} + 3.6\hat{\jmath}) \text{ m}/s^2(10 \text{ s}) = (10\hat{\imath} - 21\hat{\jmath}) \text{ m}/s$. (b) $v_0 = \sqrt{(10)^2 + (-21)^2} = 23.3 \text{ m}/s$, and $v = \sqrt{(33)^2 + (15)^2} = 36.2 \text{ m}/s$, so the change in speed is $\Delta v = v - v_0 = 13.0 \text{ m}/s$ (we did not round off before subtracting). (c) $\theta = \tan^{-1}(15/33) = 24.4°$ and $\theta_0 = \tan^{-1}(-21/10) = 295° = -64.5°$ (positive angles CCW, negative angles CW, from x-axis) so the direction changed by $\Delta\theta = \theta - \theta_0 = 89.0°$. (d) $at = \sqrt{(2.3)^2 + (3.6)^2} \text{ m}/s^2(10 \text{ s}) = 42.7 \text{ m}/s \neq \Delta v$. The difference between $at = |\mathbf{v} - \mathbf{v}_0|$ and $\Delta v = v - v_0$ can be seen from the triangle inequality: $|v - v_0| \leq |\mathbf{v} - \mathbf{v}_0| \leq v + v_0$.

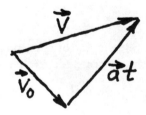

Problem 41 Solution.

Problem

43. The position of an object is given by $\mathbf{r} = (ct - bt^3)\hat{\imath} + dt^2\hat{\jmath}$, with constants $c = 6.7 \text{ m}/s$, $b = 0.81 \text{ m}/s^3$, and $d = 4.5 \text{ m}/s^2$. (a) Determine the object's velocity at time $t = 0$. (b) How long does it take for the direction of motion to change by 90°? (c) By how much does the speed change during this time?

Solution

(a) $\mathbf{v}(t) = d\mathbf{r}/dt = (c - 3bt^2)\hat{\imath} + 2dt\hat{\jmath}$, so $\mathbf{v}(0) = (6.7 \text{ m}/s)\hat{\imath}$. (b) The direction of $\mathbf{v}(t)$ is $\theta(t) = \tan^{-1}(2dt/(c - 3bt^2))$, measured CCW from the x-axis, so $\theta(0) = 0°$. $\theta(t) = 90°$ when the argument of the arctan is ∞, or $t = \sqrt{c/3b}$. Thus, $t = \sqrt{(6.7 \text{ m}/s)/3(0.81 \text{ m}/s^3)} = 1.66 \text{ s}$. (The direction of motion was $-90°$ at $t = -1.66 \text{ s}$.) (c) Since $v_x = 0$ when $t = \sqrt{c/3b}$ the speed at $t = 1.66 \text{ s}$ is $2d\sqrt{c/3b} = 14.9 \text{ m}/s$. The speed at $t = 0$ is $c = 6.7 \text{ m}/s$, so the change was 8.24 m/s.

Section 3-5: Relative Motion

Problem

45. A dog paces around the perimeter of a rectangular barge that is headed up a river at 14 km/h relative to the riverbank. The current in the water is at 3.0 km/h. If the dog walks at 4.0 km/h, what are its speeds relative to (a) the shore and (b) the water as it walks around the barge?

Solution

(a) Let S be a frame of reference fixed on the shore, with x-axis upstream, and let S' be a frame attached to the barge. The velocity of S' relative to S is $\mathbf{V} = (14 \text{ km}/h)\hat{\imath}$. The velocity of the dog relative to the shore is (from Equation 3-25) $\mathbf{v} = \mathbf{v}' + \mathbf{V}$, and its speed is $v = |\mathbf{v}' + \mathbf{V}|$, where $v' = 4 \text{ km}/h$. When the dog is walking upstream, $\mathbf{v}' \parallel \mathbf{V}$, $v = (4 + 14) \text{ km}/h = 18 \text{ km}/h$, and when walking downstream, $-\mathbf{v}' \parallel \mathbf{V}$, and $v = (14 - 4) \text{ km}/h = 10 \text{ km}/h$. When $\mathbf{v}' \perp \mathbf{V}$, $v = \sqrt{14^2 + 4^2} = 14.7 \text{ km}/h$. (In general, $v^2 = v'^2 + V^2 + 2v'V \cos\theta'$, where θ' is the angle between $\mathbf{v}'$ and $\mathbf{V}$ in S'.) (b) Since the current flows downstream, according to Equation 3-25:

$$\begin{pmatrix} \text{vel. of barge} \\ \text{rel. to water} \end{pmatrix} = \begin{pmatrix} \text{vel. of barge} \\ \text{rel. to shore} \end{pmatrix} - \begin{pmatrix} \text{vel. of water} \\ \text{rel. to shore} \end{pmatrix}$$
$$= 14\hat{\imath} - (-3\hat{\imath}) \text{ km}/h.$$

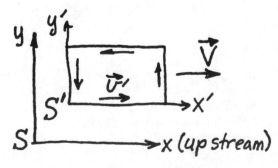

Problem 45 Solution.

Going through the same steps as in part (a), for a new frame S moving with the water, with a new relative velocity $\mathbf{V} = (17 \text{ km/h})\hat{\imath}$, we find the speed of the dog relative to the water to be $(4+17) = 21$ km/h, $(17-4) = 13$ km/h, and $\sqrt{4^2 + 17^2} = 17.4$ km/h, for the corresponding segments of the barge's perimeter.

Problem

47. A spacecraft is launched toward Mars at the instant Earth is moving in the $+x$ direction at its orbital speed of 30 km/s, in the Sun's frame of reference. Initially the spacecraft is moving at 40 km/s relative to Earth, in the $+y$ direction. At the launch time, Mars is moving in the $-y$ direction at its orbital speed of 24 km/s. Find the spacecraft's velocity relative to Mars.

Solution

Equation 3-10 says that the velocity of the spacecraft relative to Mars, $\mathbf{v}_{CM}$, equals the difference of the velocities of each relative to the Sun, $\mathbf{v}_{CS}, -\mathbf{v}_{MS}$. (Our notation $\mathbf{v}_{CM}$ means the velocity of C relative to M.) $\mathbf{v}_{MS}$ is given as $-(24 \text{ km/s})\hat{\jmath}$ in the Sun's reference frame, the x-y coordinates in the problem. The Earth has velocity $\mathbf{v}_{ES} = (30 \text{ km/s})\hat{\imath}$ in the Sun's frame, and $\mathbf{v}_{CE} = (40 \text{ km/s})\hat{\jmath}$, where the y-axes in the Earth's and Sun's frames are assumed to be parallel. A second application of Equation 3-10 gives $\mathbf{v}_{CS} = \mathbf{v}_{CE} + \mathbf{v}_{ES} = (40\hat{\jmath} + 30\hat{\imath})$ km/s; therefore $\mathbf{v}_{CM} = \mathbf{v}_{CS} - \mathbf{v}_{MS} = (30\hat{\imath} = 40\hat{\jmath})$ km/s $- (-24\hat{\jmath})$ km/s $= (30\hat{\imath} + 64\hat{\jmath})$ km/s.

Problem

49. You're on an airport "people mover," a conveyor belt going at 2.2 m/s through a level section of the terminal. A button falls off your coat and drops freely 1.6 m, hitting the belt 0.57 s later. What are the magnitude and direction of the button's displacement and average velocity during its fall in (a) the frame of reference of the "people mover" and (b) the frame of reference of the airport terminal? (c) As it falls, what is its acceleration in each frame of reference?

Solution

(c) Let S be the frame of reference of the airport and S' the frame of reference of the conveyor belt. If the velocity of S' relative to S is a constant, 2.2 m/s in the x-x' direction, then the acceleration of gravity is the same in both systems. (a) In S', the initial velocity of the button is zero, and it falls vertically downward. Its displacement is simply $\Delta y' = -1.6$ m, and its average velocity is $\Delta y'/\Delta t = -1.6$ m/0.57 s $= 2.81$ m/s. (b) In S, the initial velocity of the button is not zero

(it is 2.2 m/s in the x direction), and so the button follows a projectile trajectory to be described in Chapter 4. Here, we observe that while the button falls vertically through a displacement $\Delta y = \Delta y' = -1.6$ m, it also moves horizontally (in the direction of the conveyor belt) through a displacement $\Delta x = (2.2 \text{ m/s})(0.57 \text{ s}) = 1.25$ m. Its net displacement in S is $\Delta\mathbf{r} = \Delta x\hat{\imath} + \Delta y\hat{\jmath} = (1.25\hat{\imath} - 1.60\hat{\jmath})$ m, so its average velocity is $\bar{\mathbf{v}} = \Delta\mathbf{r}/\Delta t = (1.25\hat{\imath} - 1.60\hat{\jmath})$ m/0.57 s $= (2.20\hat{\imath} - 2.8\hat{\jmath})$ m/s. These have magnitudes $|\Delta\mathbf{r}| = \sqrt{(1.25)^2 + (-1.60)^2}$ m $= 2.03$ m and $|\mathbf{v}| = \sqrt{(2.20)^2 + (-2.81)^2}$ m/s $= 3.57$ m/s, and the same direction $\theta = \tan^{-1}(-1.60/1.25) = -51.9°$ from the horizontal, or 38.1° from the downward vertical.

Paired Problems

Problem

51. A rabbit scurries across a field, going eastward 21.0 m. It then turns and darts southwestward for 8.50 m. Then it pops down a rabbit hole, 1.10 m vertically downward. What is the magnitude of the displacement from its starting point?

Solution

In a coordinate system with x-axis east, y-axis north, and z-axis up, the three displacements can be written in terms of their components and the unit vectors: $\mathbf{r}_1 = 21.0\hat{\imath}$ m, $\mathbf{r}_2 = (8.50 \text{ m})(\hat{\imath}\cos 225° + \hat{\jmath}\sin 225°) = -(6.01)(\hat{\imath} + \hat{\jmath})$ m, and $\mathbf{r}_3 = -1.10\hat{\mathbf{k}}$ m. (Note that southwest is $180° + 45°$ CCW from east.) The total displacement is the vector sum $\mathbf{r}_1 + \mathbf{r}_2 + \mathbf{r}_3 = [(21.0 - 6.01)\hat{\imath} - 6.01\hat{\jmath} - 1.10\hat{\mathbf{k}}]$ m, and its magnitude is the square root of the sum of its components, $\sqrt{(15.0)^2 + (-6.01)^2 + (-1.10)^2}$ m $= 16.2$ m.

Problem

53. A car is heading into a turn at 85 km/h. It enters the turn, slows to 55 km/h, and emerges 28 s later at 35° to its original direction, still moving at 55 km/h. What are (a) the magnitude and (b) the direction of its average acceleration, the latter measured with respect to the car's original direction?

Solution

The initial and final velocities have magnitudes of 85 km/h and 55 km/h, respectively, and make an angle of 35? as shown, where we chose the x-axis parallel to $\mathbf{v}_i$ and the y-axis in the direction of the turn. The change in velocity is $\Delta\mathbf{v} = \mathbf{v}_f - \mathbf{v}_i = (55 \text{ km/h})(\hat{\imath}\cos 35° + \hat{\jmath}\sin 35°) - (85 \text{ km/h})\hat{\imath} = (-39.9\hat{\imath} + 31.5\hat{\jmath})$ km/h: (a) The

magnitude of the average acceleration is
$\bar{a} = |\Delta\mathbf{v}|/\Delta t = \sqrt{(-39.9)^2 + (31.5)^2}$ (km/h)/28 s =
1.82 km/h/s = 0.505 m/s². (b) The direction of
$\bar{\mathbf{a}}$ is the same as the direction of $\Delta\mathbf{v}$, or $\theta =$
$\tan^{-1}(31.5/-39.9) = 142°$. (This problem could also
have been solved using the laws of cosines and sines;
see the solution to the next problem.)

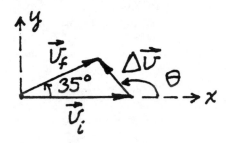

Problem 53 Solution.

Problem

55. The sweep-second hand of a clock is 3.1 cm long.
What are the magnitude of (a) the average
velocity and (b) the average acceleration of the
hand's tip over a 5.0-s interval? (c) What is the
angle between the average velocity and
acceleration vectors?

Solution

There will be numerous occasions to use vector
components to analyze circular motion in later
chapters (or see the solutions to Problems 18, 32, 38,
and 39), so let's use geometry to solve this problem.
(a) The angular displacement of the hand during a 5 s
interval is $\theta = (5/60)(360°) = 30°$. The position
vectors (from the center hub) of the tip at the
beginning and end of the interval, $\mathbf{r}_1$ and $\mathbf{r}_2$, form the
sides of an isosceles triangle whose base is the
magnitude of the displacement, $|\Delta\mathbf{r}| = 2|\mathbf{r}|\sin\frac{1}{2}\theta =$
$2(3.1$ cm$)\sin(30°/2) = 1.60$ cm, and whose base angle
is $\frac{1}{2}(180° - 30°) = 75°$. Thus, the average velocity has
magnitude $|\Delta\mathbf{r}|/\Delta t = 1.60$ cm/5 s = 0.321 cm/s and
direction $180° - 75° = 105°$CW from $\mathbf{r}_1$. (b) The
instantaneous speed of the tip of the second-hand is a
constant and equal to the circumference divided by 60
s, or $v = 2\pi(3.1$ cm$)/60$ s = 0.325 cm/s. The direction
of the velocity of the tip is tangent to the
circumference, or perpendicular to the radius, in the
direction of motion (CW). The angle between two
tangents is the same as the angle between the two
corresponding radii, so $\mathbf{v}_1$, $\mathbf{v}_2$ and $\Delta\mathbf{v}$ form an
isosceles triangle similar to the one in part (a). Thus
$|\Delta\mathbf{v}| = 2|\mathbf{v}|\sin\frac{1}{2}\theta = 2(0.325$ cm/s$)\sin(\frac{1}{2} \times 30°) =$
0.168 cm/s. The magnitude of the average acceleration

is $|\Delta\mathbf{v}|/\Delta t = (0.168$ cm/s$)/5$ s = 3.36×10^{-2} cm/s²,
and its direction is 105° CW from the direction of $\mathbf{v}_1$,
or 195° CW from the direction of $\mathbf{r}_1$. (c) The angle
between $\bar{\mathbf{a}}$ and $\bar{\mathbf{v}}$, from parts (a) and (b), is
$195° - 105° = 90°$. (Note: This is the geometry used
in Section 4.4 to discuss centripetal acceleration.)

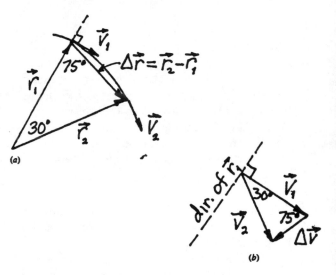

Problem 55 Solution.

Problem

57. A ferryboat sails between two towns directly
opposite one another on a river. If the boat sails
at 15 km/h relative to the water, and if the
current flows at 6.3 km/h, at what angle should
the boat head?

Solution

The velocity of the boat relative to the ground, $\mathbf{v}$, is
perpendicular to the velocity of the water relative to
the ground, the current velocity $\mathbf{V}$, which form a right
triangle with hypotenuse $\mathbf{v}'$ equal to the velocity of the
boat relative to the water, as shown in the diagram
and as required by Equation 3-10. The heading
upstream is $\theta = \sin^{-1}(|\mathbf{V}|/|\mathbf{v}'|) = \sin^{-1}(6.3/15) =$
24.8°.

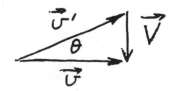

Problem 57 Solution.

Supplementary Problems

Problem

59. A satellite is in a circular orbit 240 km above Earth's surface, moving at a constant speed of 7.80 km/s. A tracking station picks up the satellite when it is 5.0° above the horizon, as shown in Fig. 3-26. The satellite is tracked until it is directly overhead. What are the magnitudes of (a) its displacement, (b) its average velocity, and (c) its average acceleration during the tracking interval? Is the value of the average acceleration approximately familiar?

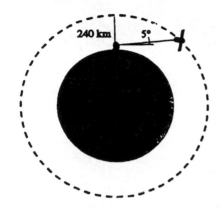

FIGURE 3-26 Problem 59 (figure is not to scale).

Solution

Once we know the angular displacement, $\Delta\theta$, of the satellite in its orbit, we can calculate the magnitudes of the displacement from its initial position P_1 to its final position P_2, and its average velocity and acceleration, during the same tracking interval, by using the geometrical analysis in the solution to Problem 55, to which the reader is referred. In the diagram based on Fig. 3-26, O is the center of the Earth, $R_E = 6370$ km is the average radius of the Earth, A is the tracking station, and $r = R_E + h = 6370$ km + 240 km = 6610 km is the radius of the orbit. We apply the law of cosines to triangle OAP_1 to find AP_1, and then the law of sines to find $\Delta\theta$. Thus,

$$(6610 \text{ km})^2 = (6370 \text{ km})^2 + (AP_1)^2$$
$$- 2(6370 \text{ km})(AP_1)\cos 95°.$$

This is a quadratic equation with (positive) solution

$$AP_1 = (6370 \text{ km})\cos 95°$$
$$+ \sqrt{(6370 \text{ km})^2 \cos^2 95° + (6610 \text{ km})^2 - (6370 \text{ km})^2}$$
$$= 1295 \text{ km}.$$

(We are keeping four significant figures in the intermediate results, but will round off to three at the end.) Then $\sin\Delta\theta/AP_1 = \sin 95°/r$ gives $\Delta\theta = \sin^{-1}(1295 \sin 95°/6610) = 11.26°$.

(a) Now that we know $\Delta\theta$, the magnitude of the displacement can be found from the isosceles triangle OP_1P_2 as in Problem 55. $P_1P_2 = 2r\sin\frac{1}{2}\Delta\theta = 2(6610 \text{ km})\sin\frac{1}{2}(11.26°) = 1296 \text{ km} \approx 1300 \text{ km}$.

(b) To find the magnitude of the average velocity, we first need to find the tracking interval Δt. The time for a complete orbit (called the period) is the orbital circumference divided by the speed, or $T = 2\pi r/v = 2\pi(6610 \text{ km})/(7.8 \text{ km/s}) = 5.325\times10^3$ s. During the tracking interval, the satellite completes only a fraction $\Delta\theta/360°$ of a complete orbit, so $\Delta t = (11.26°/360°)(5.325\times10^3 \text{ s}) = 166.5$ s. Thus, $|\bar{\mathbf{v}}| = P_1P_2/\Delta t = 1296 \text{ km}/166.5 \text{ s} = 7.78$ km/s.

(c) As shown in the solution to Problem 55, the velocity vectors at P_1 and P_2 form an isosceles triangle similar to OP_1P_2. Thus $|\Delta\mathbf{v}| = 2|\mathbf{v}|\sin\frac{1}{2}\Delta\theta = 2(7.8 \text{ km/s})\sin\frac{1}{2}(11.26°) = 1.530$ km/s, and the magnitude of the average acceleration is $|\bar{\mathbf{a}}| = |\Delta\mathbf{v}|/\Delta t = (1.530 \text{ km/s})/166.5 \text{ s} = 9.19 \text{ m/s}^2$. When we recall that the Earth's gravity holds the satellite in its orbit, it is not surprising that this magnitude is close to g. Of course, at an altitude of 240 km, the Earth's gravitational field is only 9.11 m/s² (compared to 9.81 m/s² at the surface). We must also remember that there is a discrepancy between the average and instantaneous accelerations due to the finite size of Δt.

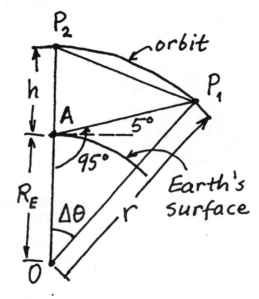

Problem 59 Solution

Problem

61. Find two vectors in the x-y plane that are perpendicular to the vector $a\hat{\imath} + b\hat{\jmath}$.

Solution

One vector perpendicular to the given vector, $\mathbf{A} = a\hat{\imath} + b\hat{\jmath}$, is a vector $\mathbf{B}$ of the same magnitude, $|\mathbf{B}| = |\mathbf{A}|$, but pointing at an angle 90° more than the angle of $\mathbf{A}$ with the x-axis, as shown in the sketch, i.e., $\theta_B = 90° + \theta_A$. The components of $\mathbf{A}$ are $A_x = |\mathbf{A}| \cos\theta_A = a$ and $A_y = |\mathbf{A}| \sin\theta_A = b$, while the components of $\mathbf{B}$ are $B_x = |\mathbf{B}| \cos\theta_B = |\mathbf{A}| \cos(90° + \theta_A) = -|\mathbf{A}| \sin\theta_A = -b$ and $B_y = |\mathbf{B}| \sin(90° + \theta_A) = |\mathbf{A}| \cos\theta_A = a$. Thus, $\mathbf{B} = -b\hat{\imath} + a\hat{\jmath}$ is perpendicular to $A = a\hat{\imath} + b\hat{\jmath}$. Clearly, $-\mathbf{B} = b\hat{\imath} - a\hat{\jmath}$ is also perpendicular to $\mathbf{A}$, as is any scalar multiple of $\mathbf{B}$. (Another way to do this problem is to use the scalar product defined in Section 7.2.)

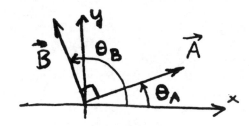

Problem 61 Solution.

Problem

63. Write an expression for a unit vector that lies at 45° between the positive x- and y-axes.

Solution

A vector of unit magnitude, making a 45° angle CCW with the x-axis, can be expressed as $1 \cdot \cos 45°\hat{\imath} + 1 \cdot \sin 45°\hat{\jmath} = (\hat{\imath} + \hat{\jmath})/\sqrt{2}$. (A unit vector in any direction in the x-y plane is therefore $\hat{n} = \hat{\imath}\cos\theta + \hat{\jmath}\sin\theta$.)

Problem

65. Figure 3-27 shows two arbitrary vectors $\mathbf{A}$ and $\mathbf{B}$ that sum to a third vector $\mathbf{C}$. By working with components, prove the law of cosines: $C^2 = A^2 + B^2 - 2AB \cos\gamma$.

Solution

Adding some lines parallel to the axes and labeling the angles of the vectors in Fig. 3-27, one sees that $C_x = A_x + B_x = A\cos\theta_A + B\cos\theta_B$ and $C_y = A\sin\theta_A + B\sin\theta_B$. Squaring, adding, and using two trigonometric identities from Appendix A, one

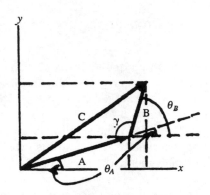

FIGURE 3-27 Problem 65.

gets $C^2 = C_x^2 + C_y^2 = A^2(\cos^2\theta_A + \sin^2\theta_A) + B^2(\cos^2\theta_B + \sin^2\theta_B) + 2AB(\cos\theta_A\cos\theta_B + \sin\theta_A\sin\theta_B) = A^2 + B^2 + 2AB\cos(\theta_B - \theta_A)$. This is the law of cosines when we replace $\theta_B - \theta_A$ by $180° - \gamma$, where γ is the angle between sides A and B in the vector triangle.

Problem

67. Town B is located across the river from town A and at a 40.0° angle upstream from A, as shown in Fig. 3-28. A ferryboat travels from A to B; it sails at 18.0 km/h relative to the water. If the current in the river flows at 5.60 km/h, at what angle should the boat head? What will be its speed relative to the ground? *Hint:* Set up Equation 3-10 for this situation. Each component of Equation 3-10 yields two equations in the unknowns v, the magnitude of the boat's velocity relative to the ground, and ϕ, the unknown angle. Solve the x equation for $\cos\phi$ and substitute into the second equation, using the relation $\sin\phi = \sqrt{1 - \cos^2\phi}$. You can then solve for v, then go back and get ϕ from your first equation.

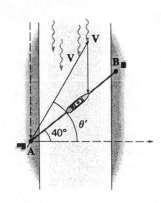

FIGURE 3-28 Problem 67.

Solution

The diagram shows the velocities and coordinate axes added to Fig. 3-28. $\mathbf{v'}$ (the velocity of the boat relative to the water), $\mathbf{v}$ (the velocity of the boat relative to the ground), and $\mathbf{V}$ (the velocity of the water relative to the ground) are related by Equation 3-10, $\mathbf{v'} = \mathbf{v} - \mathbf{V}$. This vector equation is equivalent to two scalar equations, one for each component, $v'_x = v_x - V_x$ and $v'_y = v_y - V_y$. The components of each vector, in terms of its magnitude and angle, are $v'_x = v' \cos \theta' \times$ $v'_y = v' \sin \theta'$, $v_x = v \cos 40°$, $v_y = v \sin 40°$, and $V_x = 0$, $V_y = -V$. Therefore, the x and y component equations are $v' \cos \theta' = v \cos 40°$ and $v' \sin \theta' = v \sin 40° + V$. We can eliminate θ' by

squaring and adding (since $\sin^2 + \cos^2 = 1$): $v'^2(\cos^2 \theta' + \sin^2 \theta') = v^2(\cos^2 40° + \sin^2 40°) + 2vV \sin 40° + V^2$, or $v^2 + 2vV \sin 40° - v'^2 + V^2 = 0$. The positive root of this quadratic for v (appropriate for a magnitude) is $v = -V \sin 40° + \sqrt{V^2 \sin^2 40° + v'^2 - V^2}$. If the given values $v' = 18.0$ km/h and $V = 5.60$ km/h are substituted, we find the speed relative to the ground is $v = 13.9$ km/h. Going back to the x component equation, we find the heading $\theta' = \cos^{-1}(v \cos 40°/v') = \cos^{-1}(13.9 \cos 40°/18.0) = 53.8°$. (Equations similar to these will be solved when collisions in two-dimensions and the conservation of momentum are discussed in Chapter 11.)

CHAPTER 4 MOTION IN MORE THAN ONE DIMENSION

ActivPhysics can help with these problems: All Activities in Section 3, Projectile Motion

Section 4-1: Velocity and Acceleration

Problem

1. A skater is gliding along the ice at 2.4 m/s, when she undergoes an acceleration of magnitude 1.1 m/s² for 3.0 s. At the end of that time she is moving at 5.7 m/s. What must be the angle between the acceleration vector and the initial velocity vector?

Solution

For constant acceleration, Equation 4-3 shows that the vectors $\mathbf{v}_0$, $\mathbf{a}\Delta t$ and $\mathbf{v}$ form a triangle as shown. The law of cosines gives $v^2 = v_0^2 + (a\Delta t)^2 - 2v_0 a\Delta t \times \cos(180° - \theta_0)$. When the given magnitudes are substituted, one can solve for θ_0: $(5.7 \text{ m/s})^2 = (2.4 \text{ m/s})^2 + (1.1 \text{ m/s}^2)^2(3.0 \text{ s})^2 + 2(2.4 \text{ m/s}) \times (1.1 \text{ m/s}^2)(3.0 \text{ s})\cos\theta_0$, or $\cos\theta_0 = 1.00$ (exactly), and $\theta_0 = 0°$. Since $\mathbf{v}_0$ and $\mathbf{a}$ are colinear, the change in speed is maximal.

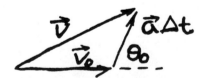

Problem 1 Solution.

Problem

3. An object is moving in the x direction at 1.3 m/s when it is subjected to an acceleration given by $\mathbf{a} = 0.52\hat{\mathbf{j}}$ m/s². What is its velocity vector after 4.4 s of acceleration?

Solution

From Equation 4-3, $\mathbf{v} = \mathbf{v}_0 + \mathbf{a}\Delta t = (1.30 \text{ m/s})\hat{\mathbf{i}} + (0.52\hat{\mathbf{j}} \text{ m/s}^2)(4.4\text{s}) = (1.30\hat{\mathbf{i}} + 2.29\hat{\mathbf{j}})$ m/s.

Section 4-2: Constant Acceleration

Problem

5. The position of an object as a function of time is given by $\mathbf{r} = (3.2t + 1.8t^2)\hat{\mathbf{i}} + (1.7t - 2.4t^2)\hat{\mathbf{j}}$ m, where t is the time in seconds. What are the magnitude and direction of the acceleration?

Solution

One can always find the acceleration by taking the second derivative of the position, $\mathbf{a}(t) = d^2\mathbf{r}(t)/dt^2$. However, collecting terms with the same power of t, one can write the position in meters as $\mathbf{r}(t) = (3.2\hat{\mathbf{i}} + 1.7\hat{\mathbf{j}})t + (1.8\hat{\mathbf{i}} - 2.4\hat{\mathbf{j}})t^2$. Comparison with Equation 4-4 shows that this represents motion with constant acceleration equal to twice the coefficient of the t^2 term, or $\mathbf{a} = (3.6\hat{\mathbf{i}} - 4.8\hat{\mathbf{j}})$ m/s². The magnitude and direction of $\mathbf{a}$ are $\sqrt{(3.6)^2 + (-4.8)^2}$ m/s² = 4.49 m/s², and $\tan^{-1}(-4.8/3.6) = 307°$ (CCW from the x-axis, in the fourth quadrant) or $-53.1°$ (CW from the x-axis).

Problem

7. An asteroid is heading toward Earth at a steady 21 km/s. To save their planet, astronauts strap a giant rocket to the asteroid, giving it an acceleration of 0.035 km/s² at right angles to its original motion. If the rocket firing lasts 250 s, (a) by what angle does the direction of the asteroid's motion change? (b) How far does it move during the firing?

Solution

(a) The change in velocity, $\Delta\mathbf{v} = \mathbf{a}\Delta t$, is at right angles to the initial velocity, $\mathbf{v}_0$, so they form the legs of a right triangle with hypotenuse $\mathbf{v}$ (see Fig. 4-2a). The angle between $\mathbf{v}$ and $\mathbf{v}_0$ is $\theta = \tan^{-1}(|\Delta\mathbf{v}| \div |\mathbf{v}_0|) = \tan^{-1}(0.035 \text{ km/s})(250 \text{ s})/(21 \text{ km/s}) = 22.6°$. (b) For constant acceleration, the displacement is (see Equation 4-4) $\Delta\mathbf{r} = \mathbf{r} - \mathbf{r}_0 = \mathbf{v}_0 t + \frac{1}{2}\mathbf{a}t^2$. Again, the two vectors $\mathbf{v}_0 t$ and $\frac{1}{2}\mathbf{a}t^2$ form the legs of a right triangle, whose hypotenuse, $\Delta\mathbf{r}$, has magnitude

$$\sqrt{(21 \text{ km/s})^2(250 \text{ s})^2 + (\tfrac{1}{2})^2(0.035 \text{ km/s}^2)^2(250 \text{ s})^4} =$$

5.36×10^3 km. (Note: this is the asteroid's

displacement during the rocket firing, not the pathlength of its trajectory.)

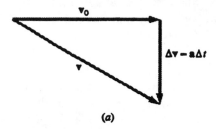

(a)

FIGURE 4-2*a* Problem 7 Solution.

Problem

9. A hockey puck is moving at 14.5 m/s when a stick imparts a constant acceleration of 78.2 m/s^2 at a 90.0° angle to the original direction of motion. If the acceleration lasts 0.120 s, what is the magnitude of the puck's displacement during this time?

Solution

Take the x-axis in the direction of the initial velocity, $\mathbf{v}_0 = 14.5$ m/s, and the y-axis in the direction of the acceleration, $\mathbf{a} = 78.2\hat{\jmath}$ m/s^2. The displacement during the 0.120 s interval of constant acceleration is (Equation 4-4) $\Delta\mathbf{r} = \mathbf{r} - \mathbf{r}_0 = \mathbf{v}_0 t + \frac{1}{2}\mathbf{a}t^2 =$ $(14.5 \text{ m/s})(0.120 \text{ s})\hat{\imath} + \frac{1}{2}(78.2 \text{ m/s}^2)(0.120 \text{ s})^2\hat{\jmath} =$ $(1.74\hat{\imath} + 0.563\hat{\jmath})$ m. This has magnitude $\sqrt{(1.74)^2 + (0.563)^2}$ m $= 1.83$ m (and makes an angle of $\tan^{-1}(0.563/1.74) = 17.9°$ with the direction of the initial velocity).

Problem

11. A particle leaves the origin with initial velocity $\mathbf{v}_0 = 11\hat{\imath} + 14\hat{\jmath}$ m/s. It undergoes a constant acceleration given by $\mathbf{a} = -1.2\hat{\imath} + 0.26\hat{\jmath}$ m/s^2.
    (a) When does the particle cross the y-axis?
    (b) What is its y-coordinate at the time? (c) How fast is it moving, and in what direction, at that time?

Solution

(a) Since the particle leaves from the origin ($\mathbf{r}_0 = 0$), its position is $\mathbf{r}(t) = \mathbf{v}_0 t + \frac{1}{2}\mathbf{a}t^2$. It crosses the y-axis when $x(t) = v_{0x}t + \frac{1}{2}a_x t^2 = 0$, or $t = -2v_{0x}/a_x =$ $-2(11 \text{ m/s})/(-1.2 \text{ m/s}^2) = 18.3$ s. (b) $y(t) = v_{0y}t + \frac{1}{2}a_y t^2 = [14 \text{ m/s} + \frac{1}{2}(0.26 \text{ m/s}^2)(18.3 \text{ s})](18.3 \text{ s}) =$ 300 m. (c) $\mathbf{v}(t) = \mathbf{v}_0 + \mathbf{a}t = (11\hat{\imath} + 14\hat{\jmath})$ m/s $+$ $(-1.2\hat{\imath} + 0.26\hat{\jmath})(18.3)$m/s $= (-11\hat{\imath} + 18.8\hat{\jmath})$ m/s. Then $|\mathbf{v}(t)| = \sqrt{(-11)^2 + (18.8)^2}$ m/s $= 21.8$ m/s and $\theta_x = \tan^{-1}(18.8/(-11)) = 120°$.

Problem

13. Figure 4-27 shows a cathode-ray tube, used to display electrical signals in oscilloscopes and other scientific instruments. Electrons are accelerated by the electron gun, then move down the center of the tube at 2.0×10^9 cm/s. In the 4.2-cm-long deflecting region they undergo an acceleration directed perpendicular to the long axis of the tube. The acceleration "steers" them to a particular spot on the screen, where they produce a visible glow. (a) What acceleration is needed to deflect the electrons through 15°, as shown in the figure? (b) What is the shape of an electron's path in the deflecting region?

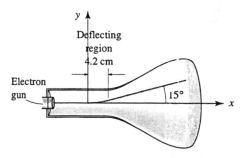

FIGURE 4-27 Problem 13 Solution.

Solution

(a) With x-y axes as drawn on Fig. 4-27, the electrons emerge from the deflecting region with velocity $\mathbf{v} = v_0\hat{\imath} + at\hat{\jmath}$, after a time $t = x/v_0$, where $x = 4.2$ cm and $v_0 = 2 \times 10^9$ cm/s. The angle of deflection (direction of $\mathbf{v}$) is 15°, so $\tan 15° = v_y/v_x = at/v_0 =$ ax/v_0^2. Thus, $a = v_0^2 \tan 15°/x = 2.55 \times 10^{17}$ cm/s^2 (when values are substituted). (b) Since the acceleration is assumed constant, the electron trajectory is parabolic in the deflecting region.

Section 4-3:   Projectile Motion

Problem

15. A carpenter tosses a shingle off a 8.8-m-high roof, giving it an initially horizontal velocity of 11 m/s. (a) How long does it take to reach the ground? (b) How far does it move horizontally in this time?

Solution

(a) The shingle reaches the ground when $y(t) = 0 = y_0 - \frac{1}{2}gt^2$, or

$$t = \sqrt{\frac{2y_0}{g}} = \sqrt{\frac{2(8.8 \text{ m})}{(9.8 \text{ m/s}^2)}} = 1.34 \text{ s}.$$

(b) The horizontal displacement is $x = v_0t = (11 \text{ m/s})(1.34 \text{ s}) = 14.7 \text{ m}$.

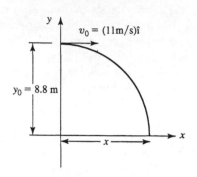

Problem 15 Solution.

Problem

17. A kid fires water horizontally from a squirt gun held 1.6 m above the ground. It hits another kid 2.1 m away square in the back, at a point 0.93 m above the ground (see Fig. 4-28). What was the initial speed of the water?

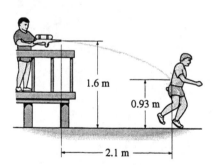

FIGURE 4-28 Problem 17.

Solution

Since the water was fired horizontally ($v_{0y} = 0$), the time it takes to fall from $y_0 = 1.6 \text{ m}$ to $y = 0.93 \text{ m}$ is given by Equation 4-8, $t = \sqrt{2(y_0 - y)/g} = \sqrt{2(1.6 - 0.93) \text{ m}/(9.8 \text{ m/s}^2)} = 0.370 \text{ s}$. Its initial speed, $v_0 = v_{0x}$, can be found from Equation 4-7, $v_0 = (x - x_0)/t = 2.1 \text{ m}/0.370 \text{ s} = 5.68 \text{ m/s}$.

Problem

19. Ink droplets in an ink-jet printer are ejected horizontally at 12 m/s, and travel a horizontal distance of 1.0 mm to the paper. How far do they fall in this interval?

Solution

From $x - x_0 = v_{0x}t$, the time of flight can be found. Substitution into $y_0 - y = \frac{1}{2}gt^2$ (recall that

$v_{0y} = 0$) yields $y_0 - y = \frac{1}{2}g(x - x_0)^2/v_0^2 = \frac{1}{2}(9.8 \text{ m/s}^2)(10^{-3} \text{ m})^2/(12 \text{ m/s})^2 = 3.40 \times 10^{-8} \text{ m} = 34 \text{ nm}$ for the distance fallen, practically negligible. Note that this analysis is equivalent to using Equation 4-9 with $\theta_0 = 0$.

Problem

21. You're standing on the ground 3.0 m from the wall of a building, and you want to throw a package from your 1.5-4.2m shoulder level to someone in a second-floor window above the ground. At what speed and angle should you throw it so it just barely reaches the window?

Solution

We suppose that "just barely" means that the maximum height of the package equals the height of the window sill. When the package reaches the sill (in the coordinate system shown), $v_y^2 = 0 = v_{0y}^2 - 2gy$, so $v_{0y} = \sqrt{2(9.8 \text{ m/s}^2)(2.7 \text{ m})} = 7.27 \text{ m/s}$. Since $v_y = 0 = v_{0y} - gt$, the time of flight is $t = v_{0y}/g$. Therefore $v_{0x} = x/t = (3.0 \text{ m})(9.8 \text{ m/s}^2)/(7.27 \text{ m/s}) = 4.04 \text{ m/s}$. From these components, we find: $v_0 = \sqrt{v_{0x}^2 + v_{0y}^2} = 8.32 \text{ m/s}$, and $\theta_0 = \tan^{-1}(v_{0y}/v_{0x}) = 60.9°$.

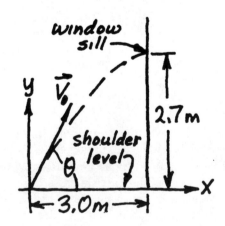

Problem 21 Solution.

Problem

23. A car moving at 40 km/h strikes a pedestrian a glancing blow, breaking both the car's front signal light lens and the pedestrian's hip. Pieces of the lens are found 4.0 m down the road from the center of a 1.2-m-wide crosswalk, and a lawsuit hinges on whether or not the pedestrian was in the crosswalk at the time of the accident. Assuming that the lens was initially 63 cm off the ground, and that the lens pieces continued moving

horizontally with the car's speed at the time of the impact, was the pedestrian in the crosswalk?

Solution

What is an issue here is the horizontal range of a piece of signal light lens in projectile motion, starting from a height of $y_0 - y = 0.63$ m off the ground, with an initial horizontal velocity of $v_0 = v_{0x} = (40 \text{ m}/3.6 \text{ s})$, and $v_{0y} = 0$. Eliminating the time of flight from Equations 4-7 and 4-8 (see the solution to Problem 19), one obtains $x - x_0 = v_0\sqrt{2(y_0 - y)/g} =$ $(11.1 \text{ m/s})\sqrt{2(0.63 \text{ m})/(9.8 \text{ m/s}^2)} = 3.98$ m. If the pieces of lens did not bounce very far from the point where they hit the ground, this places the point of impact of the accident just $4.00 \text{ m} - 3.98 \text{ m} = 2$ cm from the center of the crosswalk. (Forensic physics is crucial to the prosecution's case.)

Problem

25. In part (b) of the exercise following Example 4-5, what is the vertical component of the velocity with which the cyclist strikes the ground?

Solution

One first needs to work out part (a) of the exercise, which is similar to the first part of Example 4-5. At the minimum speed at takeoff, the cyclist covers a horizontal range $x = 48$ m, and a vertical drop $y = -5.9$ m (recall that $x_0 = y_0 = 0$ in Equation 4-9). Then

$$v_{0,\text{min}} = \left[\frac{(9.8 \text{ m/s}^2)(48 \text{ m})^2}{2(\cos^2 15°)[(48 \text{ m})\tan 15° - (-5.9 \text{ m})]}\right]^{1/2}$$
$$= 25.4 \text{ m/s}.$$

The cyclist's actual initial speed is $v_0 = (1 + 50\%)(25.4 \text{ m/s}) = 38.1$ m/s. Equation 2-11 gives the vertical component of the velocity (negative downward) for the given drop, $v_y = -\sqrt{v_{0y}^2 - 2gy} =$ $-\sqrt{(38.1 \text{ m/s})^2 \sin^2 15° - 2(9.8 \text{ m/s}^2)(-5.9 \text{ m})} = -14.6 \text{ m/s}$

Problem

27. A submarine-launched missile has a range of 4500 km. (a) What launch speed is needed for this range when the launch angle is 45°? (b) What is the total flight time? (c) What would be the minimum launch speed at a 20° launch angle, used to "depress" the trajectory so as to foil a space-based antimissile defense?

Solution

(a) Assuming Equation 4-10 applies (i.e., the trajectory begins and ends at the same height, or $y(t) = y_0$), one finds $v_0 = \sqrt{xg/\sin 2\theta} =$ $\sqrt{(4500 \text{ km})(0.0098 \text{ km/s}^2)/\sin 90°} = 6.64$ km/s. (b) The time of flight is the positive solution of Equation 4-8 when $y(t) - y_0 = 0 = v_{0y}t - \frac{1}{2}gt^2$. Thus $t = 2v_{0y}/g = 2(6.64 \text{ km/s})\sin 45°/(9.8 \text{ m/s}^2) = 958$ s $= 16.0$ min. (c) At a 20° launch angle, $$v_0 = \sqrt{(4500 \text{ km})(0.0098 \text{ km/s}^2)/\sin 40°} = 8.28 \text{ km/s}.$$

Problem

29. At a circus, a human cannonball is shot from a cannon at 35 km/h at an angle of 40°. If he leaves the cannon 1.0 m off the ground, and lands in a net 2.0 m off the ground, how long is he in the air?

Solution

The time of flight can be calculated from Equation 4-8, where the trajectory begins at $y_0 = 1.0$ m and $t = 0$, ends at $y(t) = 2.0$ m, and $v_{0y} = v_0 \sin\theta_0 = (35 \text{ m}/3.6 \text{ s})\sin 40° = 6.25$ m/s. The equation is a quadratic, $\frac{1}{2}gt^2 - v_{0y}t + (y - y_0) = 0$, with solutions $t = [v_{0y} \pm \sqrt{v_{0y}^2 - 2g(y - y_0)}]/g =$ $[6.25 \text{ m/s} \pm \sqrt{(6.25 \text{ m/s})^2 - 2(9.8 \text{ m/s}^2)(2 \text{ m} - 1 \text{ m})}] \div (9.8 \text{ m/s}^2) = 0.188$ s or 1.09 s. The trajectory crosses the height 2.0 m twice, once going up, at the smaller time of flight, and once going down into the net, at the larger time of flight. The latter is the answer to the question asked here.

Problem

31. If you can hit a golf ball 180 m on Earth, how far can you hit it on the moon? (Your answer is an underestimate, because the distance on Earth is restricted by air resistance as well as by a larger g.)

Solution

For given $\mathbf{v}_0$, the horizontal range is inversely proportional to g. With surface gravities from Appendix E, we find $x_{\text{moon}} = (g_{\text{Earth}}/g_{\text{moon}})x_{\text{Earth}} = (9.81/1.62)(180 \text{ m}) = 1090$ m.

Problem

33. A projectile launched at an angle θ_0 to the horizontal reaches a maximum height h. Show that its horizontal range is $4h/\tan\theta_0$.

Solution

The intermediate expression for the horizontal range (when the initial and final heights are equal) is

$x = 2v_0^2 \sin\theta_0 \cos\theta_0/g = 2v_{0x}v_{0y}/g$ (see the equation before Equation 4-10). The components of the initial velocity are related by $v_{0y}/v_{0x} = \tan\theta_0$. The maximum height, $h = y_{\max} - y_0$, can be found from Equation 2-11 (when $v_y = 0$) or $v_{0y}^2 = 2gh$. Then $x = 2v_{0y}v_{0x}/g = 2v_{0y}(v_{0y}/\tan\theta_0)/g = 2(2gh)/g\tan\theta_0 = 4h/\tan\theta_0$. (This result reflects a classical geometrical property of the parabola, namely, that the latus rectum is four times the distance from vertex to focus.)

Problem

35. A circular fountain has jets of water directed from the circumference inward at an angle of 45°. Each jet reaches a maximum height of 2.2 m. (a) If all the jets converge in the center of the circle and at their initial height, what is the radius of the fountain? (b) If one of the jets is aimed at 10° too low, how far short of the center does it fall?

Solution

(a) The radius is the horizontal range, $r = v_0^2/g$ (Equation 4-10 with $\theta = 45°$). The maximum height is $h = v_{0y}^2/2g = v_0^2/4g$ (Equation 2-11 with $v_y = 0$ and $v_{0y} = v_0 \cos 45° = v_0/\sqrt{2}$). Therefore, $r = (4gh)/g = 4h = 4(2.2 \text{ m}) = 8.8 \text{ m}$. (b) If one jet is directed at 35° with the same initial speed ($v_0^2 = rg$), it would fall short by $r - x$, where x is given by Equation 4-10. Therefore, $r - x = r - (v_0^2/g)\sin(2 \times 35°) = (8.8 \text{ m})(1 - \sin 70°) = 0.531 \text{ m}$.

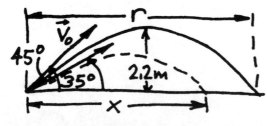

Problem 35 Solution.

Problem

37. In 1991 Mike Powell shattered Bob Beamon's 1968 world long jump record with a leap of 8.95 m (see Fig. 4-31). Studies show that Powell jumps at 22° to the vertical. Treating him as a projectile, at what speed did Powell begin his jump?

Solution

The horizontal range formula (Equation 4-10) gives $v_0 = \sqrt{xg/\sin 2\theta_0} =$
$\sqrt{(8.95 \text{ m})(9.8 \text{ m/s}^2)/\sin 2(90° - 22°)} = 11.2 \text{ m/s}$
(Air resistance and body control are important factors in the long jump also.)

Problem

39. Show that, for a given initial speed, the horizontal range of a projectile is the same for launch angles $45° + \alpha$ and $45° - \alpha$, where α is between 0° and 45°.

Solution

The trigonometric identity in Appendix A for the sine of the sum of two angles shows that $\sin 2(45° \pm \alpha) = \sin(90° \pm 2\alpha) = \sin 90° \cos 2\alpha \pm \cos 90° \sin 2\alpha = \cos 2\alpha$, so the horizontal range formula (Equation 4-10) gives the same range for either launch angle, at the same initial speed.

Problem

41. A basketball player is 15 ft horizontally from the center of the basket, which is 10 ft off the ground. At what angle should the player aim the ball if it is thrown from a height of 8.2 ft with a speed of 26 ft/s?

Solution

With origin at the point from which the ball is thrown, the equation of the trajectory (Equation 4-9), evaluated at the basket, becomes

$$y = (10 - 8.2)\text{ft} = (15 \text{ ft})\tan\theta_0 - \frac{(32 \text{ ft/s}^2)(15 \text{ ft})^2}{2(26 \text{ ft/s})^2 \cos^2\theta_0},$$

or $1.8 = 15\tan\theta_0 - 5.33/\cos^2\theta_0$. Using the trigonometric identity $1 + \tan^2\theta_0 = 1/\cos^2\theta_0$, we can convert this equation into a quadratic in $\tan\theta_0$: $7.13 - 15\tan\theta_0 + 5.33\tan^2\theta_0 = 0$, so

$$\theta_0 = \tan^{-1}\left[\frac{15 \pm \sqrt{15^2 - 4(5.33)(7.13)}}{2(5.33)}\right]$$

$$= 31.2° \text{ or } 65.7°$$

Like the horizontal range formula (see Fig. 4-13), for given v_0 there are two launch angles whose trajectories pass through the basket, although in this case they are not symmetrically placed about 45°.

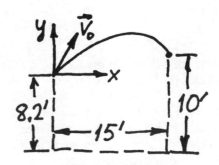

Problem 41 Solution.

Section 4-4: Circular Motion

Problem

43. Estimate the acceleration of the moon, which completes a nearly circular orbit of 385,000 km radius in 27 days.

Solution

The centripetal acceleration is given in terms of the period for uniform circular motion by Equation 4-12 in Example 4-8. In the case of the moon, $a = 4\pi^2 r/T^2 = 4\pi^2(3.85\times10^8 \text{ m})/(27.3\times86,400 \text{ s})^2 = 2.73\times10^{-3} \text{ m/s}^2$, where we used more accurate data from Appendix E. (Note: "centripetal" is a purely kinematic adjective descriptive of circular motion. In this case, the origin of the moon's centripetal acceleration is the gravitational attraction of the Earth.)

Problem

45. When Apollo astronauts landed on the moon, they left one astronaut behind in a circular orbit around the moon. For the half of the orbit spent over the far side of the moon, that individual was completely cut off from communication with the rest of humanity. How long did this lonely state last? Assume a sufficiently low orbit that you can use the moon's surface gravitational acceleration (see Appendix E) for the spacecraft.

Solution

Consider a circular orbit around the moon with radius slightly larger than the lunar radius, $r = 1.74\times10^6$ m, and centripetal acceleration approximately equal to the lunar surface gravity, $a_c = 1.62 \text{ m/s}^2$ (see Appendix E). The orbital period is related to r and a_c by Equation 4-12 as in Example 4-8, $T = 2\pi\sqrt{r/a_c}$, and radio communications with Earth were blocked for half of this period, or $\frac{1}{2}T = \pi\sqrt{(1.74\times10^6 \text{ m})/(1.62 \text{ m/s}^2)} = 3.26\times10^3$ s or about 54.3 min.

Problem

47. A jet is diving vertically downward at 1200 km/h (see Fig. 4-32). If the pilot can withstand a maximum acceleration of $5g$ (i.e., 5 times Earth's gravitational acceleration) before losing consciousness, at what height must the plane start a quarter turn to pull out of the dive? Assume the speed remains constant.

Solution

The height at the start of the 90°-turn must be greater than the radius of the turn, in order to avoid

hitting the ground. The radius of the turn must be great enough that the centripetal acceleration not exceed $5g$, i.e., $a_c = v^2/r \leq 5g$ or $r \geq v^2/5\,g = (1200 \text{ m}/3.6 \text{ s})^2 = 5(9.8 \text{ m/s}^2) = 2.27$ km.

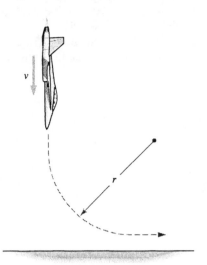

FIGURE 4-32 Problem 47 Solution.

Problem

49. How long would a day last if Earth were rotating so fast that the acceleration of an object on the equator were equal to g?

Solution

The Earth's equitorial radius is about 6378 km. If the centripetal acceleration at the equator were $a_c = 9.8 \text{ m/s}^2$, the Earth's period of rotation would have to be $T = 2\pi\sqrt{R_E/a_c} = 2\pi\sqrt{(6378 \text{ km})/(0.0098 \text{ km/s}^2)} = 5.07\times10^3$ s = 1 h 24.5 min. (See Equation 4-12.)

Section 4-5: Nonuniform Circular Motion

Problem

51. A space station 120 m in diameter is set rotating in order to give its occupants "artificial gravity." Over a period of 5.0 min, small rockets bring the station steadily to its final rotation rate of 1 revolution every 20 s. What are the radial and tangential accelerations of a point on the rim of the station 2.0 min after the rockets start firing?

Solution

If the rotation rate increases steadily from 0 to 1 revolution in 20 s, over a 5-minute interval, the rotation rate after 2 minutes is $(2/5)(1 \text{ rev}/20 \text{ s}) = 1 \text{ rev}/50$ s, or the (instantaneous) period after

2 minutes is 50 s. Thus, by Equation 4-12, the centripetal (radial) acceleration is $a_c = 4\pi^2 r/T^2 = 4\pi^2(60 \text{ m})/(50 \text{ s})^2 = 0.947 \text{ m/s}^2$. The tangential speed increases steadily from 0 to $2\pi(60 \text{ m})/20 \text{ s} = 6\pi$ m/s, in 5 minutes, so the tangential acceleration is $a_t = (6\pi \text{ m/s})/5 \min = 6.28 \times 10^{-2} \text{ m/s}^2$. (The solution to this problem may appear more straightforward after the discussion of angular motion in Chapter 12.)

Problem

53. An object is set into motion on a circular path of radius r by giving it a constant tangential acceleration a_t. Derive an expression for the time t when the acceleration vector points at 45° to the direction of motion.

Solution

The tangential acceleration (in the direction of motion) is perpendicular to the radial acceleration, so the resultant total acceleration (their vector sum) is at 45° between them at the instant when $a_t = a_r = v^2/r$. The linear speed along the circle depends on a_t only (since a_c is perpendicular to the velocity), so $v = a_t t$ for constant a_t (provided the object is "set into motion" with $v_0 = 0$ at $t = 0$). Thus, $a_t = (a_t t)^2/r$ or $t = \sqrt{r/a_t}$.

Paired Problems

Problem

55. An alpine rescue team is using a slingshot to send an emergency medical packet to climbers stranded on a ledge, as shown in Fig. 4-34. What should be the launch speed from the slingshot?

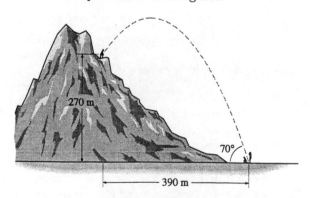

270 m

70°

390 m

FIGURE 4-34 Problem 55.

Solution

If we take the origin of coordinates at the slingshot and the stranded climbers at $x = 390$ m and $y = 270$ m, we can use Equation 4-9 for the trajectory

to solve for v_0:

$$v_0 = \frac{x}{\cos\theta_0}\sqrt{\frac{g}{2(x\tan\theta_0 - y)}}$$

$$= \frac{390 \text{ m}}{\cos 70°}\sqrt{\frac{9.8 \text{ m/s}^2}{2(390 \text{ m} \cdot \tan 70° - 270 \text{ m})}} = 89.2 \text{ m/s}.$$

Problem

57. If you can throw a stone straight up to a height of 16 m, how far could you throw it horizontally over level ground? Assume the same throwing speed and optimum launch angle.

Solution

To throw an object vertically to a maximum height of $h = 16 \text{ m} = y_{\max} - y_0$ requires an initial speed of $v_0 = \sqrt{2g(y_{\max} - y_0)} = \sqrt{2gh}$. With this value of v_0 and the optimum launch angle $\theta_0 = 45°$, Equation 4-10 gives a maximum horizontal range on level ground of $x = v_0^2/g = 2h = 32$ m. (The maximum horizontal range on level ground is twice the maximum height for vertical motion with the same initial speed. This result holds in the approximation of constant g and no air resistance.)

Problem

59. I can kick a soccer ball 28 m on level ground, giving it an initial velocity at 40° to the horizontal. At the same initial speed and angle to the horizontal, what horizontal distance can I kick the ball on a 15° upward slope?

Solution

We need to find the intersection of the trajectory of the ball (Equation 4-9) with a 15° slope through the same origin, $y = x \tan 15°$. The appearance of the trajectory equation can be simplified by use of the fact that $y = 0$ when $x = 28$ m and $\theta_0 = 40°$. Thus, $y = 0 = x\tan\theta_0 - (g/2v_{0x}^2)x^2 = (28 \text{ m})[\tan 40° - (g/2v_{0x}^2)(28 \text{ m})]$, or the coefficient $(g/2v_{0x}^2)$ equals $\tan 40°/28$ m. The trajectory equation simplifies to $y = x\tan 40° - x^2(\tan 40°/28 \text{ m}) =$

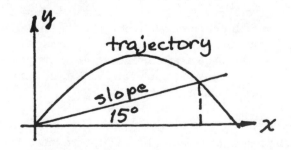

Problem 59 Solution.

$x(1 - x/28 \text{ m}) \tan 40°$. The intersection of this with the slope occurs when y also equals $x \tan 15°$, or $x \tan 15° = x(1 - x/28 \text{ m}) \tan 40°$. The x-coordinates of the two points of intersection are $x = 0$ (the origin) and $x = (28 \text{ m})(1 - \tan 15°/\tan 40°) = 19.1 \text{ m}$ (the horizontal distance queried in this problem).

Problem

61. A fireworks rocket is 73 m above the ground when it explodes. Immediately after the explosion, one piece is moving at 51 m/s at 23° to the upward vertical direction. A second piece is moving at 38 m/s at 11° below the horizontal direction. At what horizontal distance from the explosion site does each piece land?

Solution

In the trajectory equation (Equation 4-9) with origin at the position of the explosion, the coefficients are known for each piece. One can solve this quadratic equation for x, when $y = -73$ m (ground level), and select the positive root (since the trajectories start at $x = 0$ and end on the ground in the direction of v_{0x}, which is chosen positive).

For the first piece, $\tan 23° = 0.424$ and $g/2v_{0x}^2 = (9.8 \text{ m/s}^2)/2(51 \cos 23° \text{ m/s})^2 = 2.22 \times 10^{-3} \text{ m}^{-1}$, so the quadratic is $-73 \text{ m} = 0.424x - (2.22 \times 10^{-3} \text{ m}^{-1})x^2$. This has positive root $x = [0.424 + \sqrt{(0.424)^2 + 4(73 \text{ m})(2.22 \times 10^{-3} \text{ m}^{-1})}] \div (4.44 \times 10^{-3} \text{ m}^{-1}) = 300 \text{ m}$.

Similarly, for the second piece, $\tan(-11°) = -0.194$ and $g/2v_{0x}^2 = (9.8 \text{ m/s}^2) \div 2(38 \cos(-11°) \text{ m/s})^2 = 3.52 \times 10^{-3} \text{ m}^{-1}$, so $x = [-0.194 + \sqrt{(-0.194)^2 + 4(73 \text{ m})(3.52 \times 10^{-3} \text{ m}^{-1})}] \div (7.04 \times 10^{-3} \text{ m}^{-1}) = 119 \text{ m}$.

Problem

63. You toss a chocolate bar to your hiking companion located 8.6 m up a 39° slope, as shown in Fig. 4-36. Determine the initial velocity vector so

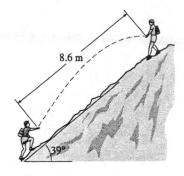

8.6 m

39°

FIGURE 4-36 Problem 63.

that the chocolate bar will reach your friend moving horizontally.

Solution

The candy bar moves horizontally only at the apex of its trajectory (where $v_x = v_{0x}$ and $v_y = 0$). Thus, $y_{\max} - y_0 = (8.6 \text{ m}) \sin 39° = 5.41 \text{ m}$, and $v_{0y} = \sqrt{2g(y_{\max} - y_0)} = \sqrt{2(9.8 \text{ m/s}^2)(5.41 \text{ m})} = 10.3 \text{ m/s}$ (see Equation 2-11). The time to reach the apex is $t = v_{0y}/g$, so $v_{0x} = (x - x_0)/t = (x - x_0)g/v_{0y}$ (see Equations 4-6 and 4-7). The horizontal distance from apex to origin is $x - x_0 = (8.6 \text{ m}) \cos 39° = 6.68 \text{ m}$, so $v_{0x} = (6.68 \text{ m})(9.8 \text{ m/s}^2)/(10.3 \text{ m/s}) = 6.36 \text{ m/s}$. $\mathbf{v_0}$ can be expressed in unit vector notation as $(6.36\hat{\imath} + 10.3\hat{\jmath})$ m/s, or by its magnitude $\sqrt{v_{0x}^2 + v_{0y}^2} = 12.1$ m/s and direction $\theta = \tan^{-1}(v_{0y}/v_{0x}) = 58.3°$ (CCW from the x-axis).

Problem

65. After takeoff, a plane makes a three-quarter circle turn of radius 7.1 km, maintaining constant altitude but steadily increasing its speed from 390 to 740 km/h. Midway through the turn, what is the angle between the plane's velocity and acceleration vectors?

Solution

The plane's velocity is tangent to its circular path in the direction of motion and so is the tangential acceleration a_t. The radial (centripetal) acceleration a_c is perpendicular to this, so the angle between the total acceleration and the velocity is $\theta = \tan^{-1}(a_c/a_t)$, inclined toward the center of the turn. For the linear motion over a circular distance of $s = \frac{3}{4}(2\pi r) = 1.5\pi(7.1 \text{ km}) = 33.5 \text{ km}$, Equation 2-11 can be used to find the constant a_t, $a_t = (v_f^2 - v_i^2)/2s = [(740/3.6)^2 - (390/3.6)^2](\text{m}^2/\text{s}^2)/2(33.5 \text{ km}) = 0.456 \text{ m/s}^2$. We can find $a_c = v^2/r$ midway through the turn by using Equation 2-11 again, since $v^2 = v_i^2 + 2a_t(s/2) = v_i^2 + \frac{1}{2}(v_f^2 - v_i^2) = \frac{1}{2}(v_i^2 + v_f^2)$. Then $a_c = \frac{1}{2}[(390/3.6)^2 + (740/3.6)^2](\text{m}^2/\text{s}^2)/(7.1 \text{ km}) = 3.80 \text{ m/s}^2$. Finally, $\theta = \tan^{-1}(3.80/0.456) = 83.2°$. (Instead of working out each component of acceleration numerically, we could have written the final result symbolically as follows:

$$\frac{a_c}{a_t} = \left(\frac{v_i^2 + v_f^2}{2r}\right) \cdot \frac{2(3/4)(2\pi r)}{(v_f^2 - v_i^2)} = \frac{3\pi}{2}\left(\frac{v_f^2 + v_i^2}{v_f^2 - v_i^2}\right),$$

and $\theta = \tan^{-1}[3\pi(740^2 + 390^2)/2(740^2 - 390^2)]$ as before.)

Supplementary Problems

Problem

67. Verify the maximum altitude and flight time for the 15° launch angle trajectory of the missile described at the end of the Application: Ballistic Missile Defense (page 79).

Solution

The missile has a range (on level ground) of $R = x_{max} - x_0 = 1000$ km at a launch angle of $\theta_0 = 15°$, so Equation 4-10 gives the launch speed as $v_0^2 = gR/\sin 2\theta_0$. (Numerically, this is
$$\sqrt{(0.0098\ \text{km/s}^2)(1000\ \text{km})/\sin 2(15°)} = 4.43\ \text{km/s},$$
as stated in the text.) The maximum altitude can be found from Equation 2-11, $h = y_{max} - y_0 = v_{0y}^2/2g$, since $v_y = 0$ at this point in the trajectory. With the value of v_0 above, we find $h = v_0^2 \sin^2 \theta_0/2g = (gR/\sin 2\theta_0)\sin^2\theta_0/2g = R\sin^2\theta_0/2\sin 2\theta_0 = \frac{1}{4}R\tan\theta_0 = \frac{1}{4}(1000\ \text{km})\tan 15° = 67.0$ km, in agreement with the text. (To answer just this question, one might have substituted numbers into the first expression for h, i.e., $(4.43\ \text{km/s})^2 \sin^2 15°/2g$, but working algebraically, the simpler expression from Problem 33 is obtained. Moreover, in this particular case, $2\sin 2\theta_0 = 2\sin 2(15°) = 1$, so the numerical calculation is faster too.) The time of flight can be found from Equation 4-7, $t = (x_{max} - x_0)/v_{0x} = R/v_0 \cos\theta_0 = (1000\ \text{km})/(4.43\ \text{km/s})\cos 15° = 234$ s $= 3.90$ min, completing the verification of the text. (Again, alternate expressions could have been used, e.g., $t = 2v_{0y}/g$, from Equation 4-6, or $t = \sqrt{R/2g}\ (\cos 15°)^{-1}$.)

Problem

69. A monkey is hanging from a branch a height h above the ground. A naturalist stands a horizontal distance d from a point directly below the monkey. The naturalist aims a tranquilizer dart directly at the monkey, but just as she fires the monkey lets go. Show that the dart will nevertheless hit the monkey, provided its initial speed exceeds $\sqrt{(d^2 + h^2)g/2h}$.

Solution

Gravity accelerates the dart and the monkey equally, so both fall the same vertical distance from the point of aim (the monkey's original position) resulting in a hit, provided the initial speed of the dart is sufficient to reach the monkey before the monkey reaches the ground. To prove this assertion, let the dart be fired from ground level ($y = 0$) with speed v_0 and direction $\theta_0 = \tan^{-1}(h/d)$ (line of sight from naturalist N to

monkey M), while the monkey drops from height h at $t = 0$. The vertical height of each is $y_{monkey} = h - \frac{1}{2}gt^2$ and $y_{dart} = v_{0y}t - \frac{1}{2}gt^2$, where $v_{0y} = v_0 \sin\theta_0 = v_0 h/\sqrt{d^2 + h^2}$. (The term $-\frac{1}{2}gt^2$ represents the effect of gravity, which appears the same way in both y-coordinate equations.) The dart strikes the monkey when $y_{monkey} = y_{dart}$, which implies $h = v_{0y}t$, or $t = \sqrt{d^2 + h^2}/v_0$. This must be less than the time required for the monkey to fall to the ground, which is $\sqrt{2h/g}$ (from $y_{monkey} = 0$). Thus $\sqrt{d^2 + h^2}/v_0 < \sqrt{2h/g}$ or $v_0 > \sqrt{g(d^2 + h^2)/2h}$. (This condition can also be understood from the horizontal range formula, Equation 4-10. The range of the dart has to be greater than the horizontal distance to the monkey, $d < v_0^2 \sin 2\theta_0/g = v_0^2 2hd/(d^2 + h^2)g$.)

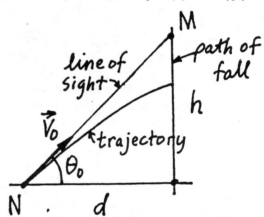

Problem 69 Solution.

Problem

71. A diver leaves a 3-m board on a trajectory that takes her 2.5 m above the board, and then into the water a horizontal distance of 2.8 m from the end of the board. At what speed and angle did she leave the board?

Solution

Since we are given the maximum height (at which point $v_y = 0$), Equation 2-11 can be used to find the y component of the diver's initial velocity, $0 = v_{0y}^2 - 2g(y_{max} - y_0)$ or $v_{0y} = \sqrt{2(9.8\ \text{m/s}^2)(2.5\ \text{m})} = 7.00$ m/s. (We take the positive square root because the diver springs upward off the board.) The x-component of $\mathbf{v}_0$ can be found from Equation 4-7, once the time of flight is known. The latter is the positive root (the dive begins at $t = 0$) of the quadratic Equation 4-8, when $y_0 - y = 3$ m (a 3-m board is 3 m above the water level). Thus, $t = [v_{0y} + \sqrt{v_{0y}^2 + 2g(y_0 - y)}]/g = [7\ \text{m/s} +$

$\sqrt{49 \text{ m}^2/\text{s}^2 + 2(9.8 \text{ m/s}^2)(3 \text{ m})]}/(9.8 \text{ m/s}^2) = 1.77$ s,
and $v_{0x} = (x - x_0)/t = 2.8 \text{ m}/1.77 \text{ s} = 1.58$ m/s.
From v_{0x} and v_{0y} we find the magnitude
$v_0 = \sqrt{v_{0x}^2 + v_{0y}^2} = 7.18$ m/s and direction
$\theta_0 = \tan^{-1}(v_{0y}/v_{0x}) = 77.3°$.

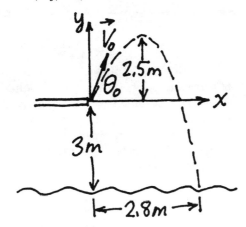

Problem 71 Solution.

Problem

73. A projectile is launched with speed v_0 from the edge of a cliff of height h; the ground below the cliff is flat. Using the technique of the preceding problem, show that the maximum range occurs when the launch angle is

$$\theta_0 = \tan^{-1}\left(\frac{v_0}{\sqrt{2gh + v_0^2}}\right).$$

Solution

The quadratic formula can be used to solve Equation 4-9 for the horizontal range of a projectile with positive v_{0x}, whose trajectory begins at the origin and ends at the point $(x, y) : x = (v_0^2 \cos^2\theta_0/g)[\tan\theta_0 \pm \sqrt{\tan^2\theta_0 - 2gy/v_0^2\cos^2\theta_0}]$. When the origin is on a cliff of height h above where the projectile lands, $y = -h$, and only the solution with the positive sign before the square root corresponds to a positive range. (For $y > 0$, both solutions might be possible.) Thus, the horizontal range appropriate to the situation in this problem is $x = (v_0^2 \cos\theta_0/g)[\sin\theta_0 + \sqrt{\sin^2\theta_0 + 2gh/v_0^2}]$. We chose to multiply through one factor of $\cos\theta_0$ in order to simplify the θ_0-dependence of each term before differentiating; other choices also work.

Inspection of the trajectory shows that it is reasonable to expect one maximum value of x, with v_0 constant, for $0 < \theta_0 < 90°$. It can be found by the method suggested, or by other methods, such as

Lagrange multipliers. In taking the derivative, we use the Product rule, the Chain rule, and the derivatives of the sine, cosine, and square root ($\sqrt{z} = z^{1/2}$) given in Appendix A. Then,

$$\frac{g}{v_0^2}\frac{dx}{d\theta_0} = [\sin\theta_0 + \sqrt{\cdots}](-\sin\theta_0)$$
$$+ \cos\theta_0\left[\cos\theta_0 + \frac{\sin\theta_0\cos\theta_0}{\sqrt{\cdots}}\right]$$
$$= [\sin\theta_0 + \sqrt{\cdots}]\left[-\sin\theta_0 + \frac{\cos^2\theta_0}{\sqrt{\cdots}}\right] = 0,$$

where $\sqrt{\cdots} = \sqrt{\sin^2\theta_0 + 2gh/v_0^2}$. The first factor is never zero, so $\sqrt{\cdots} = \cos^2\theta_0/\sin\theta_0 = (1/\sin\theta_0) - \sin\theta_0$. Squaring and simplifying, we find that

$$(\sqrt{\cdots})^2 = \sin^2\theta_0 + \frac{2gh}{v_0^2} = \frac{1}{\sin^2\theta_0} - 2\left(\frac{1}{\sin\theta_0}\right)(\sin\theta_0)$$
$$+ \sin^2\theta_0, \quad \text{or} \quad 2 + \frac{2gh}{v_0^2}$$
$$= \frac{1}{\sin^2\theta_0} = 1 + \frac{1}{\tan^2\theta_0},$$

or $\tan\theta_0 = \dfrac{v_0}{\sqrt{v_0^2 + 2gh}}$, as stated. (With this value for θ_0, the maximum value of x turns out to be $(v_0^2/g)\sqrt{1 + 2gh/v_0^2}$. For $h = 0$, these values of θ_0 and x reduce to the case discussed in the text.)

Problem

75. A well-engineered ski jump is less dangerous than it looks because skiers hit the ground with very small velocity components perpendicular to the ground. Skiers leave the Olympic ski jump in Lake Placid, New York, at an angle of 9.5° below the horizontal. Their landing zone is a horizontal distance of 55 m from the end of the jump. The ground at the point is contoured so skiers' trajectories make an angle of only 3.0° with the ground on landing, as suggested in Fig. 4-38. What is the slope of the ground in the landing zone?

Solution

The direction of the skier's velocity is $\theta = \tan^{-1}(v_y/v_x)$, where angles are measured CCW from the x-axis, chosen horizontal to the right in Fig. 4-38 with the y-axis upward. In the landing zone, θ is in the fourth quadrant, which can be represented by a negative angle below the x-axis. The slope of the ground at this point can be represented by a similar angle θ_g, and for the safety of ski jumpers, $\theta_g - \theta = 3.0°$.

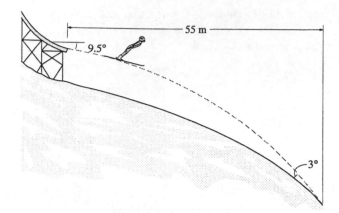

FIGURE 4-38 Problem 75.

The slope $v_y/v_x = (dy/dt)/(dx/dt) = dy/dx$ can be calculated by differentiating the trajectory equation, but it is just as easy in this problem to use Equations 4-5, 4-6 and 4-7. Thus, $v_x = v_{0x} = v_0 \cos \theta_0$, and $v_y = v_{0y} - gt = v_0 \sin \theta_0 - gt$. The time of flight can be eliminated, since $x - x_0 = v_{0x}t = 55$ m is given, so $v_y/v_x = (v_{0y}/v_{0x}) - g(x - x_0)/v_{0x}^2 = \tan \theta_0 - g(x - x_0)/v_0^2 \cos^2 \theta_0 = \tan(-9.5°) - (9.8 \text{ m/s}^2)(55 \text{ m})/(28 \cos(-9.5°) \text{ m/s})^2 = -0.874$. Finally, $\theta = \tan^{-1}(-0.874) = -41.2°$, and $\theta_g = \theta + 3.0° = -38.2°$.

Problem

77. Derive a general expression for the flight time of a projectile launched on level ground with speed v_0 and launch angle θ_0.

Solution

Equation 4-8 gives the height of a projectile above its launch site, $y - y_0 = (v_{0y} - \frac{1}{2}gt)t$. Here, we factored the time to show that the two solutions for zero height are the launch time $t = 0$, and the time of flight, given by $v_{0y} - \frac{1}{2}gt = 0$, or $t = 2v_{0y}/g = 2v_0 \sin \theta_0/g$.

Problem

79. In the Olympic hammer throw, contestants whirl a 7.3-kg ball on the end of a 1.2-m-long steel wire before releasing it. In a particular throw, the hammer is released from a height of 1.3 m while moving in a direction 24° above the horizontal. If it travels 84 m horizontally, what is its radial acceleration just before release (see Fig. 4-39)?

Solution

Just before release, the hammer ball is traveling in a circle (approximately of radius $r = 1.2$ m), so its radial

(or centripetal) acceleration is $a_c = v_0^2/r$, where v_0 is the launch speed. We can determine v_0 from Equation 4-9 and the data given for the throw, $\theta_0 = 24°$, $x = 84$ m, $y = -1.3$ m, as in Example 4-5, thus:

$$a_c = \frac{v_0^2}{r} = \frac{1}{r} \left[\frac{gx^2}{2 \cos^2 \theta_0 (x \tan \theta_0 - y)} \right] = 892 \text{ m/s}^2.$$

Problem

81. Two golfers stand equal distances on opposite sides of a hole, as shown in Fig. 4-40. Golfer A hits his ball at a 50° angle to the horizontal. At the instant she hears A's club hit the ball, Golfer B hits her ball at the same speed as A, but at a 40° angle. If the two balls reach the hole simultaneously, how far apart are the golfers? The speed of sound is 340 m/s.

Solution

Since both golfers and the hole are on level ground, we can use the result of Problem 77 to determine their separation, R. If golfer A's ball is hit at $t = 0$, it will reach the hole at $t = 2v_0 \sin 50°/g$. Golfer B's ball has a time of flight of $2v_0 \sin 40°/g$, but it starts after a delay of $R/(340 \text{ m/s})$, due to the sound travel-time. Since both balls arrive at the hole simultaneously (a double hole-in-one!), $2v_0 \sin 40°/g + R/(340 \text{ m/s}) = 2v_0 \sin 50°/g$, or $R = (340 \text{ m/s})(2v_0/g) \times (\sin 50° - \sin 40°)$. We can eliminate v_0 by using Equation 4-10, with range $R/2$, since the launch angles are complementary. Then, $R/2 = (v_0^2/g) \sin 2\theta_0 = (2v_0^2/g) \sin 40° \sin 50°$, or $2v_0 = \sqrt{gR/\sin 40° \sin 50°}$. Substituting above, we find $R = (340 \text{ m/s}) \times (\sqrt{gR/\sin 40° \sin 50°}/g)(\sin 50° - \sin 40°)$, or

$$R = \frac{(340 \text{ m/s})^2 (\sin 50° - \sin 40°)^2}{(9.8 \text{ m/s}^2) \sin 40° \sin 50°} = 364 \text{ m}.$$

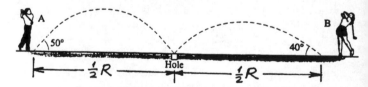

FIGURE 4.40 Problem 81 Solution

Problem

83. A projectile is launched with initial speed v_0 at an angle θ_0 to the horizontal. Find expressions for the angle the trajectory makes with the horizontal (a) as a function of time and (b) as a function of position.

Solution

(a) The slope of the trajectory is v_y/v_x, and the angle it makes with the x-axis is $\theta = \tan^{-1}(v_y/v_x)$. For projectile motion, $v_x = v_{0x} = v_0 \cos\theta_0$ is a constant, and $v_y = v_{0y} - gt = v_0 \sin\theta_0 - gt$; therefore, $v_y/v_x = (v_{0y}/v_{0x}) - (gt/v_{0x}) = \tan\theta_0 - gt/v_0\cos\theta_0$. (b) We can eliminate t from the expression for the slope by using $x - x_0 = v_{0x}t$, or $t = (x - x_0)/v_{0x}$. Thus, $v_y/v_x = \tan\theta_0 - g(x - x_0)/v_0^2\cos^2\theta_0$. (The expression for the angles is the inverse tangent of the slopes above.)

CHAPTER 5   FORCE AND MOTION

ActivPhysics can help with these problems:
All activities in Section 2, Forces and Motion

Section 5-4:   Newton's Second Law

Problem

1. A subway train has a mass of 1.5×10^6 kg. What force is required to accelerate the train at 2.5 m/s²?

Solution

$F = ma = (1.5\times10^6$ kg$)(2.5$ m/s²$) = 3.75$ MN. (This is the magnitude of the net force acting; see Table 1-1 for SI prefixes.)

Problem

3. A small plane starts down the runway with acceleration 7.2 m/s². If the force provided by its engine is 1.1×10^4 N, what is the plane's mass?

Solution

If we assume that the runway is horizontal (so that the vertical force of gravity and the normal force of the surface cancel) and neglect aerodynamic forces (which are small just after the plane begins to move) then the net force equals the engine's thrust and is parallel to the acceleration. The horizontal component of Equation 5-3 gives the airplane's mass, $m = F/a = (1.1\times10^4$ N$)/(7.2$ m/s²$) = 1.53\times10^3$ kg.

Problem

5. In an x-ray tube, electrons are accelerated to speeds on the order of 10^8 m/s, then slammed into a target where they come to a stop in about 10^{-18} s. Estimate the average stopping force on each electron.

Solution

The magnitude of the average force is $\bar{F} = m\bar{a} = m|\Delta v/\Delta t| = (9.11\times10^{-31}$ kg$)(10^8$ m/s$)/(10^{-18}$ s$) \simeq 9\times10^{-5}$ N. Compared to the TV tube in Example 5-2, the electron in an x-ray tube experiences a force billions of times greater. It is a result of the violence of this interaction that x rays, called *bremsstrahlung*, are emitted (see Problem 2-45).

Problem

7. Object A accelerates at 8.1 m/s² when a 3.3-N force is applied. Object B accelerates at 2.7 m/s² when the same force is applied. (a) How do the masses of the two objects compare? (b) If A and B were stuck together and accelerated by the 3.3-N force, what would be the acceleration of the composite object?

Solution

In this idealized one-dimensional situation, the applied force of $F = 3.3$ N is the only force acting. (a) When applied to either object, Newton's second law gives $F = m_A a_A$ and $F = m_B a_B$, so $m_B/m_A = a_A/a_B = (8.1$ m/s²$)/(2.7$ m/s²$) = 3$. (For constant net force, mass is inversely proportional to acceleration.) (b) When F is applied to the combined object, $F = (m_A + m_B)a$. Since $F = m_A a_A$, and $m_B = 3m_A$, one finds $a = F/(m_A + m_B) = m_A a_A/4m_A = \frac{1}{4}(8.1$ m/s²$) = 2.03$ m/s². (Note: It was not necessary to calculate the masses, which are $m_A = (3.3$ N$) \div (8.1$ m/s²$) = 0.407$ kg, and $m_B = (3.3$ N$) \div (2.7$ m/s²$) = 1.22$ kg.)

Problem

9. By how much does the force required to stop a car increase if the initial speed is doubled and the stopping distance remains the same?

Solution

The average net force on a car of given mass is proportional to the average acceleration, $F_{av} \sim a_{av}$. To stop a car in a given distance, $(x - x_0), a_{av} = (0 - v_0^2)/2(x - x_0)$, so $F_{av} \sim v_0^2$. Doubling v_0 quadruples the magnitude of F_{av}, a fact that is important to remember when driving at high speeds.

Problem

11. The maximum braking force of a 1400-kg car is about 8.0 kN. Estimate the stopping distance when the car is traveling (a) 40 km/h; (b) 60 km/h; (c) 80 km/h; (d) 55 mi/h.

Solution

The maximum braking acceleration is $a = F/m = -8.0\times10^3$ N$/1400$ kg $= -5.71$ m/s². (We expressed the braking force as negative because it is opposite to the direction of motion.) (a) On a straight horizontal road, a car traveling at a velocity of $v_{0x} = 40$ km/h can stop in a minimum distance found from Equation 2-11 and the maximum deceleration just calculated: $x - x_0 = -v_{0x}^2/2a = -(40$ m/3.6 s$)^2 \div 2(-5.71$ m/s$^2) = 10.8$ m. For the other initial velocities, the stopping distance is (b) 24.3 m, (c) 43.2 m, and (d) 52.9 m.

Problem

13. A car moving at 50 km/h collides with a truck, and the front of the car is crushed 1.1 m as it comes to a complete stop. The driver is wearing a seatbelt, but the passenger is not. The passenger, obeying Newton's first law, keeps moving and slams into the dashboard after the car has stopped. If the dashboard compressed 5.0 cm on impact, find and compare the forces exerted on the driver by the seatbelt and on the passenger by the dashboard. Assume the two have the same 65-kg mass.

Solution

The seatbelt constrains the driver to have the same average acceleration as the car, while stopping. Assuming the passenger compartment stays intact and is stopped after moving a horizontal distance of $\Delta x_d = 1.1$ m, we can estimate the driver's average acceleration from Equation 2-11 as $a_d = -v_0^2/2\Delta x_d$. Then the magnitude of the average force stopping the driver is $F_d = m|a_d| = \frac{1}{2}(65$ kg$)(50$ m/3.6 s$)^2 \div (1.1$ m$) = 5.70$ kN (over half a ton!), which is exerted mostly by the seatbelt. A passenger striking a stationary dashboard with the same speed v_0, which yields a horizontal distance $\Delta x_p = 0.05$ m before bringing her/him to a stop (probably permanently), experiences an average acceleration of $a_p = -v_0^2 \div 2\Delta x_p = (\Delta x_d/\Delta x_p)a_d = (1.1$ m/0.05 m$)a_d = 22.0a_d$. Therefore, the average force stopping the passenger is 22 times greater than that for the driver, or 22×5.70 kN $= 125$ kN. Buckle-up for safety—it's a law of physics!

Problem

15. A 1.25-kg object is moving in the x direction at 17.4 m/s. 3.41 s later, it is moving at 26.8 m/s at 34.0° to the x-axis. What are the magnitude and direction of the force applied during this time?

Solution

Newton's second law says that the average force acting is equal to the rate of change of momentum, $\mathbf{F}_{av} =$ $m\Delta\mathbf{v}/\Delta t$ (as explained following Equation 5-2). The initial velocity is $17.4\hat{\imath}$ m/s, and the final velocity is $(26.8$ m/s$)(\hat{\imath} \cos 34° + \hat{\jmath} \sin 34°) = (22.2\hat{\imath} + 15.0\hat{\jmath})$ m/s, so $\mathbf{F}_{av} = (1.25$ kg$)[(22.2 - 17.4)\hat{\imath} + 15.0\hat{\jmath}](m/s)/(3.41$ s$) = (1.77\hat{\imath} + 5.49\hat{\jmath})$ N. This has magnitude 5.77 N and direction 72.2° CCW to the x-axis.

Section 5-5:   Mass and Weight: The Force of Gravity

Problem

17. My spaceship crashes on one of the Sun's nine planets. Fortunately, the ship's scales are intact, and show that my weight is 532 N. If I know my mass to be 60 kg, where am I? *Hint:* Consult Appendix E.

Solution

The surface gravity of the planet is $g - W/m = 532$ N/60 kg $= 8.87$ m/s^2, precisely the value for Venus in Appendix E.

Problem

19. A cereal box says "net weight 340 grams." What is the actual weight (a) in SI units? (b) in ounces?

Solution

(a) The actual weight (equal to the force of gravity at rest on the surface of the Earth) is $mg = (0.340$ kg$) \times (9.81$ m/s$^2) = 3.33$ N. (b) With reference to Appendix C, $(3.33$ N$)(0.2248 \times 16$ oz/N$) = 12.0$ oz. (The word "net" in net weight means just the weight of the contents; gross weight includes the weight of the container, etc. This may be compared with the use of the word in net force, which means the sum of all the forces or the resultant force. A net weight, profit, or amount is the resultant after all corrections have been taken into account.)

Problem

21. A bridge specifies a maximum load of 10 tons. What's the maximum mass, in kilograms, that the bridge can carry?

Solution

The conversion between mass and (ordinary) weight is $m = W/g$. Because the English unit of mass (the slug) is rarely used, the direct equivalence between mass in SI units and weight (force) in English units is usually given, as in Appendix C. Thus 10 tons $= 2 \times 10^4$ lb is equivalent to the weight of $(2 \times 10^4$ lb$)(0.4536$ kg/lb$) = 9.07 \times 10^3$ kg.

Problem

23. A neutron star is a fantastically dense object with the mass of a star crushed into a region about 10 km in diameter. If my mass is 75 kg, and if I would weigh 5.8×10^{14} N on a certain neutron star, what is the acceleration of gravity on the neutron star?

Solution

If we define weight on a neutron star analogously to its definition on Earth, the surface gravity of the neutron star is an enormous $g = W/m = (5.8 \times 10^{14} \text{ N}) \div (75 \text{ kg}) = 7.73 \times 10^{12} \text{ m/s}^2$, nearly 10^{12} times g.

Section 5-6:   Adding Forces

Problem

25. A rope can withstand a maximum tension force of 450 N before breaking. The rope is used to pull a 32-kg bucket of water upward. What is the maximum upward acceleration if the rope is not to break?

Solution

Assume that the only vertical forces acting on the bucket are the upward (positive) force exerted by the rope, $F_{rope} \leq 450$ N, and the downward force of gravity, $F_{grav} = -mg$. (Air resistance is ignored.) The vertical component of Equation 5-3 applied to the bucket is $F_{net} = F_{rope} + F_{grav} = ma$, so $a = (F_{rope} - mg)/m \leq (450 \text{ N}/32 \text{ kg}) - 9.8 \text{ m/s}^2 = 4.26 \text{ m/s}^2$, which is the maximum upward acceleration achievable with this rope. (Note that if the tensile strength of the rope were less than the weight of the bucket, 314 N, the maximum acceleration would have been downward; see Problem 60.)

Problem

27. An elevator accelerates downward at 2.4 m/s^2. What force does the floor of the elevator exert on a 52-kg passenger?

Solution

The passenger also accelerates downward with $a_y = -2.4 \text{ m/s}^2$ (y-axis positive upward), so the vertical component of the net force on the passenger is $F_{net,y} = ma_y$. The only vertical forces acting on the passenger are the force of gravity, $F_{g,y} = -W = -mg$, and the normal force of the floor, $F_{norm,y} = N$. Therefore, $F_{net,y} = -mg + N = ma_y$, or $N = m(g + a_y) = (52 \text{ kg})(9.8 - 2.4) \text{ m/s}^2 = 385$ N.

Problem

29. An airplane encounters sudden turbulence, and you feel momentarily lighter. If your apparent weight seems to be about 70% of your normal weight, what are the magnitude and direction of the plane's acceleration?

Solution

The vertical forces acting on you are gravity downward ($-mg$) and the normal force of your seat ($N \geq 0$). The latter is what you experience as your apparent weight. During the turbulence, the vertical component of Newton's second law gives $F_{net,y} = -mg + N = -mg + 70\%mg = ma_y$; therefore $a_y = -30\%g = -2.94 \text{ m/s}^2$. If the horizontal components of the acceleration are zero, then the plane's acceleration is 2.94 m/s^2 downward.

Problem

31. At liftoff, a space shuttle with 2.0×10^6 kg total mass undergoes an upward acceleration of $0.60g$. (a) What is the total thrust force developed by its engines? (b) What force does the seat exert on a 60-kg astronaut during liftoff?

Solution

(a) At liftoff, the only significant vertical forces on the space shuttle are gravity ($F_{g,y} = -mg$ downward) and the thrust ($F_y > 0$ upward). (Air resistance can be neglected because the initial velocity is zero.) Therefore, $F_{net,y} = F_y - mg = ma_y$, or $F_y = m(g + a_y) = mg(1 + 0.6) = (2 \times 10^6 \text{ kg})(9.8 \text{ m/s}^2) \times (1.6) = 31.4$ MN. (b) The vertical forces on an astronaut are her/his weight and the normal force of the seat back ($N > 0$ upward; the astronauts are seated facing upward at liftoff). The acceleration of the astronaut is the same as that of the shuttle, so with reasoning analogous to part (a) above, $N = m(g + a_y) = (60 \text{ kg})(9.8 \text{ m/s}^2)(1.6) = 941$ N.

Problem

33. An elevator moves upward at 5.2 m/s. What is the minimum stopping time it can have if the passengers are to remain on the floor?

Solution

There are two vertical forces on a passenger, gravity downward (equal to his or her weight) $-mg$, and the upward normal force of the floor $N \geq 0$ (we take the y-axis positive upward). The latter is a contact force and always acts in a direction away from the surface of contact. (Otherwise, we would be dealing with an adhesive force.) The condition $N = 0$ is the limit that the surfaces remain in contact. Of course, as long as the passenger is in contact with the floor, his or her vertical acceleration is the same as the floor's and the elevator's. Using the y component of Newton's second law $N - mg = ma_y$, we may express the condition for contact as $N = m(a_y + g) \geq 0$, or since m is positive, $a_y \geq -g$. In words, the passenger remains in contact with the floor as long as the vertical acceleration of the elevator is either upward, or downward with magnitude less than g. To come to a stop ($v_y = 0$) from an initial upward velocity ($v_{0y} = 5.2$ m/s) requires a time $t = (0 - v_{0y})/a_y = v_{0y}/(-a_y)$, and therefore $t \geq v_{0y}/g = (5.2 \text{ m/s})/(9.8 \text{ m/s}^2) = 0.531$ s is the condition for the passengers to stay on the floor.

Section 5-7:   Newton's Third Law

Problem

35. What upward force does a 5600-kg elephant exert on Earth?

Solution

The upward force that the elephant exerts on the Earth is equal in magnitude to the downward force of gravity that the Earth exerts on the elephant (Newton's third law). The latter is $mg = (5600 \text{ kg}) \times (9.8 \text{ m/s}^2) = 54.9$ kN.

Problem

37. Repeat the preceding problem, now with the left-right order of the blocks reversed. (That is, find the force on the rightmost mass, now 1.0 kg.)

Solution

Assume that the table surface is horizontal and frictionless so that the only horizontal forces are the applied force and the contact forces between the blocks (positive to the right). The latter we denote by F_{12}, etc., which means the force block 1 exerts on block 2, etc. Since the blocks are in contact, they all have the same acceleration a, to the right. Newton's second law applied to each block separately is $F_{\text{app}} + F_{23} = m_3 a$, $F_{32} + F_{12} = m_2 a$, and $F_{21} = m_1 a$ (where we just consider horizontal components). Similarly, Newton's third law applied to each pair of contact forces is $F_{32} + F_{23} = 0$, and $F_{21} + F_{12} = 0$. Adding the second-law equations and using the third-law equations, we find $F_{\text{app}} + (F_{23} + F_{32}) + (F_{12} + F_{21}) = F_{\text{app}} = (m_1 + m_2 + m_3)a$, or $a = F_{\text{app}}/(m_1 + m_2 + m_3)$. Substituting this into the second-law equation for m_1, we find $F_{21} = m_1 a = m_1 F_{\text{app}}/(m_1 + m_2 + m_3) = (1 \text{ kg})(12 \text{ N}) \div (1 + 2 + 3) \text{ kg} = 2$ N. This is the force that the middle

block exerts on the block to its right; the other contact force, $F_{23} = -6$ N, can be found from a similar procedure.

Problem 37 Solution.

Problem

39. I have a mass of 65 kg. If I jump off a 120-cm-high table, how far toward me does Earth move during the time I fall?

Solution

As mentioned at the end of Section 5-7, during free fall, the accelerations of you and the Earth have magnitudes $mg/m = g$ and mg/M_E, respectively, in opposite directions. If you and the Earth both start from rest, the distance fallen by each is $d = \frac{1}{2}gt^2$ and $d_E = \frac{1}{2}(mg/M_E)t^2 = (m/M_E)d$, where $d + d_E = 1.2$ m. Solving for d_E, we find $d_E = (m/M_E)(1.2 \text{ m} - d_E)$, or $d_E = (1.2 \text{ m})(m/M_E)(1 + m/M_E)^{-1}$. Since $m/M_E = 65/5.97 \times 10^{24} = 1.09 \times 10^{-23}$ is so small, $d_E \approx (1.09 \times 10^{-23})(1.2 \text{ m}) \approx 1.3 \times 10^{-23}$ m. (This is about 10^5 times smaller than the smallest physically meaningful distances studied to date.)

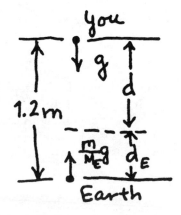

Problem 39 Solution.

Problem

41. A 2200-kg airplane is pulling two gliders, the first of mass 310 kg and the second of mass 260 kg, down the runway with an acceleration of 1.9 m/s^2 (Fig. 5-37). Neglecting the mass of the two ropes and any frictional forces, determine (a) the

horizontal thrust of the plane's propeller; (b) the tension force in the first rope; (c) the tension force in the second rope; and (d) the net force on the first glider.

Solution

Assuming a level runway (as shown in Fig. 5-37), we may write the horizontal component (positive in direction of **a**) of the equations of motion (Newton's second law) for the three planes (all assumed to have the same **a**) as follows: $F_{th} - T_1 = m_1 a$ (airplane), $T_1 - T_2 = m_2 a$ (first glider), $T_2 = m_3 a$ (second glider). (Note: the tension has the same magnitude at every point in a rope of negligible mass.) (a) Add all the equations of motion (the tensions cancel in pairs due to Newton's third law): $F_{th} = (m_1 + m_2 + m_3)a = (2200 + 310 + 260)$ kg $(1.9$ m/s$^2) = 5.26$ kN. (b) $T_1 = F_{th} - m_1 a = (m_2 + m_3)a = (570$ kg$)\times (1.9$ m/s$^2) = 1.08$ kN. (c) $T_2 = m_3 a = (260$ kg$)\times (1.9$ m/s$^2) = 494$ N. (d) $m_2 a = (310$ kg$)(1.9$ m/s$^2) = 589$ N.

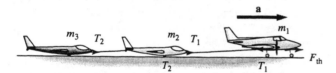

FIGURE 5-37 Problem 41 Solution.

Section 5-8: Measuring Force

Problem

43. What force is necessary to stretch a spring 48 cm, if the spring constant is 270 N/m?

Solution

If the spring obeys Hooke's law, the necessary applied force is $kx = (270$ N/m$)(0.48$ m$) = 130$ N. (We did not use the minus sign in Equation 5-9 because the force applied to the spring is the negative of the force exerted by the spring, by Newton's third law, i.e., $F_{app} = -F_{spr} = kx$.)

Problem

45. A spring sketches 22 cm when a 40-N force is applied. If a 6.1-kg mass is suspended from the spring, how much will it stretch?

Solution

Hooke's law relates the magnitude of the applied force (the "reaction" to the spring force in Equation 5-9) to the magnitude of the stretch; hence $k = 40$ N $\div$ 0.22 m $= 182$ N/m is the spring constant for this

spring. If a 6.1 kg mass is suspended from the spring, it will exert a force equal to its weight, so the stretch produced in the spring will be $x = (6.1$ kg$)\times (9.8$ m/s$^2)/(182$ N/m$) = 32.9$ cm.

Problem

47. A father pulls his 27-kg daughter across frictionless ice, using a horizontal spring with spring constant 160 N/m. If the spring is stretched 32 cm from its equilibrium position, what is the child's acceleration?

Solution

The tension in the stretched spring is $|kx| = (160$ N/m$)(0.32$ m$) = 51.2$ N, and if we assume this is the only horizontal force acting on the child, it will impart an acceleration of $a = F/m = 51.2$ N/27 kg $= 1.90$ m/s^2 to her, in the direction her father pulls.

Problem

49. A biologist is studying the growth of rats in an orbiting space station. To determine a rat's mass, she puts it in a 320-g cage, attaches a spring scale, and pulls so the scale reads 0.46 N. If the resulting acceleration of the rat and cage is 0.40 m/s^2, what is the rat's mass?

Solution

According to the scale, a force of 0.46 N applied to the cage and rat produces an acceleration of 0.40 m/s^2, so their combined mass is $F/a = (0.46$ N$)/(0.40$ m/s$^2) = 1.15$ kg. The rat's mass is this minus the cage's, or $1150 - 320 = 830$ g.

Problem

51. A 7.2-kg mass is hanging from the ceiling of an elevator by a spring of spring constant 150 N/m whose unstretched length is 80 cm. What is the overall length of the spring when the elevator (a) starts moving upward with acceleration 0.95 m/s^2; (b) moves upward at a steady 14 m/s; (c) comes to a stop while moving upward at 14 m/s, taking 9.0 s to do so? (d) If the elevator measures 3.2 m from floor to ceiling, what is the maximum acceleration it could undergo without the 7.2-kg mass hitting the floor?

Solution

There are two vertical forces acting on the mass, the spring force F_s (positive upward) and gravity $-mg$. Newton's second law for the mass yields $F_s - mg = ma$, or $F_s = m(g + a)$. (a is the vertical acceleration of the elevator; we assume that the spring responds so quickly that the vertical acceleration of the mass is the same.) The stretch in the spring (assumed to obey Hooke's law) is $y = -F_s/k$ (y is negative measured downward from the lower end of the unstretched spring), so the total length of the spring is $L = 80 \text{ cm} + |y| = 80 \text{ cm} + m(g+a)/k$. (a) If we substitute numerical values for m, k, and a, we find $L = 0.80 \text{ m} + (7.2 \text{ kg})(9.8 + 0.95)(\text{m/s}^2)/(150 \text{ N/m}) = 1.32 \text{ m}$. (b) When the velocity is constant, $a = 0$, and $L = 1.27 \text{ m}$. (c) The acceleration (assumed constant) is $a = \Delta v/\Delta t = (0 - 14 \text{ m/s})/9 \text{ s} = -1.56 \text{ m/s}^2$, which yields $L = 1.20 \text{ m}$. (d) The total length of the spring has to be less than or equal to the height of the elevator compartment, or $3.2 \text{ m} \geq L = 0.80 \text{ m} + m(g+a)/k$. This inequality can be solved for a: $a \leq [(2.40 \text{ m}) k/m] - g = [(2.40 \text{ m})(150 \text{ N/m})/7.2 \text{ kg}] - 9.8 \text{ m/s}^2 = 40.2 \text{ m/s}^2$. This is the maximum upward acceleration, provided the elastic behavior of the spring can be extrapolated to four times its unstretched length.

Paired Problems

Problem

53. Starting from rest, a 940-kg racing car covers 400 m in 4.95 s. What is the average force acting on the car?

Solution

The car's average acceleration (for straight-line motion) is $a_{av} = 2(x - x_0)/t^2$ (see Equation 2-10 with $v_0 = 0$). Newton's second law gives the average net force on the car as $F_{av} = ma_{av} = 2(940 \text{ kg})(400 \text{ m}) \div (4.95 \text{ s})^2 = 30.7 \text{ kN}$, in the direction of the motion.

Problem

55. In an egg-dropping contest, a student encases an 85-g egg in a styrofoam block. If the force on the egg is not to exceed 1.5 N, and if the block hits the ground at 1.2 m/s, by how much must the styrofoam crush?

Solution

In order that the average net force on the egg not exceed the stated limit, the magnitude of the deceleration should satisfy $a_{av} \leq F_{max}/m = 1.5 \text{ N}/0.085 \text{ kg} = 17.6 \text{ m/s}^2$. From an initial speed of $v_0 = 1.2 \text{ m/s}$, this implies a stopping distance $x - x_0$ satisfying $a_{av} = |0 - v_0^2|/2(x - x_0) \leq 17.6 \text{ m/s}^2$, or $x - x_0 \geq (1.2 \text{ m/s})^2/(35.3 \text{ m/s}^2) = 4.08 \text{ cm}$. (Note: in an actual case, the stopping force is the magnitude of the vector sum of the force of the styrofoam block and gravity acting on the egg.)

Problem

57. You step into an elevator, and it accelerates to a downward speed of 9.2 m/s in 2.1 s. How does your apparent weight during this acceleration time compare with your actual weight?

Solution

Your apparent weight is the force you exert on the floor of the elevator, equal and opposite to the upward force the floor exerts on you. The latter and gravity, which equals your actual weight, are the two forces which determine your vertical acceleration, a_y. If we take the positive direction vertically upward, then $W_{app} - W = ma_y = (W/g)a_y$, or $W_{app}/W = 1 + a_y/g$. (We expressed your mass in terms of your actual weight in order to facilitate comparison with your apparent weight.) As expected, $W_{app}/W > 1$ for upward acceleration $a_y > 0$, and vice versa for downward acceleration as in this problem. If the elevator starts from rest, its acceleration is $(v - 0)/t = (-9.2 \text{ m/s})/(2.1 \text{ s}) = -4.38 \text{ m/s}^2$ and $W_{app}/W = 1 - 4.38/9.8 = 55.3\%$.

Problem

59. A 20-kg fish at the end of a fishing line is being yanked vertically into a boat. When its acceleration reaches 2.2 m/s^2, the line breaks, and the fish goes free. What is the maximum tension the line can tolerate without breaking?

Solution

Consider the vertical motion of the fish, acted on by two forces, the tension in the line, T (positive upward), and gravity, $-mg$. Newton's second law gives $T - mg = ma_y$, or $T = m(g + a_y)$. The maximum upward acceleration the line can tolerate when pulling on a 20-kg fish is 2.2 m/s^2; hence $T < 20 \text{ kg} (9.8 + 2.2) \text{ m/s}^2 = 240 \text{ N}$.

Problem

61. A 2.0-kg mass and a 3.0-kg mass are on a horizontal frictionless surface, connected by a massless spring with spring constant $k = 140 \text{ N/m}$. A 15-N force is applied to the larger mass, as shown in Fig. 5-39. How much does the spring stretch from its equilibrium length?

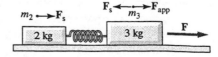

FIGURE 5-39 Problem 61 Solution.

Solution

The spring stretches until the acceleration of both masses is the same (positive in the direction of the applied force in Fig. 5-39). Since the spring is assumed massless, the tension in it is the same at both ends. (If this were not so, there would be a non-zero net horizontal force on the spring; hence its acceleration would be infinite, $a = (F \neq 0)/(m = 0)$.) The magnitude of the spring tension is given by Hooke's law, $F_s = k|x|$, where $|x|$ is the stretch. The horizontal component of Newton's second law applied to each mass is $F_{app} - F_s = m_3 a$ and $F_s = m_2 a$, as indicated in the sketch. Eliminating the acceleration between these equations gives us $F_{app} - F_s = m_3(F_s/m_2)$ or $F_s = k|x| = F_{app}/(1 + m_3/m_2)$, from which the stretch is easily found: $|x| = F_{app}/k(1 + m_3/m_2) = (15 \text{ N}/140 \text{ N/m})/(1 + 3/2) = 4.29$ cm.

Supplementary Problems

Problem

63. In throwing a 200-g ball, your hand exerts a constant upward force of 9.4 N for 0.32 s. How high does the ball rise after leaving your hand?

Solution

The maximum height to which the ball rises after leaving your hand can be calculated from its initial velocity v_{0y} (since $v_y = 0$ when $y = y_{max}$ in Equation 2-11, and we neglect air resistance), $y_{max} - y_0 = v_{0y}^2/2g$. To find v_{0y}, the vertical forces acting on the ball during the 0.32 s it is thrown must be considered. These are the applied force of your hand, $F_{app} = 9.4$ N (positive upward), and the weight of the ball, $-mg$. They impart an upward acceleration of $(F_{app} - mg)/m$ to the ball (Newton's second law), which results in an initial upward speed for the throw of $v_{0y} = (F_{app} - mg)t/m$ (definition of acceleration). Putting this together and substituting the numerical values given, we get

$$y_{max} - y_0 = \left(\frac{F_{app}}{m} - g\right)^2 \frac{t^2}{2g}$$

$$= \left(\frac{9.4 \text{ N}}{0.2 \text{ kg}} - 9.8\frac{\text{m}}{\text{s}^2}\right)^2 \frac{(0.32 \text{ s})^2}{2(9.8 \text{ m/s}^2)} = 7.23 \text{ m}.$$

Problem

65. What engine thrust (force) is needed to accelerate a rocket of mass m (a) downward at $1.40g$ near Earth's surface; (b) upward at $1.40g$ near Earth's surface; (c) at $1.40g$ in interstellar space far from any star or planet?

Solution

(a) Near the Earth's surface (constant g), but with the neglect of atmospheric friction, a vertical thrust produces an acceleration given by Newton's second law: $F_{th} - mg = ma$, or $F_{th} = mg(1 + a/g)$, (positive upward), where mg is the weight of the rocket. If $a = -1.40\, g$, $F_{th} = -0.40mg$ (the minus sign means F_{th} is downward). (b) If $a = 1.40g$, then $F_{th} = 2.40mg$. (c) Away from any significant gravitational attraction, $F_{th} = ma = 1.40mg$ in the direction of a.

Problem

67. You have a mass of 60 kg, and you jump from a 78-cm-high table onto a hard floor. (a) If you keep your legs rigid, you come to a stop in a distance of 2.9 cm, as your body tissues compress slightly. What force does the floor exert on you? (b) If you bend your knees when you land, the bulk of your body comes to a stop over a distance of 0.54 m. Now estimate the force exerted on you by the floor. Neglect the fact that your legs stop in a shorter distance than the rest of you.

Solution

Suppose the motion is purely vertical. You would hit the floor with velocity $v = -\sqrt{2gh}$ (positive upward), where $h = 78$ cm. Your average acceleration while stopping is $a = v^2/2d$, where d is the distance over which your body comes to rest. The net force exerted on you is $ma = m(v^2/2d) = m(2gh/2d) = mg(h/d)$, while the force exerted on you by the floor is $F - mg = ma$, or $F = (1 + h/d)mg$, expressed as a factor times your weight, $mg = (60 \text{ kg})(9.8 \text{ m/s}^2) = 588$ N. (a) With your legs kept rigid, $F = (588 \text{ N}) \times (1 + 78/2.9) = 16.4$ kN (within one order of magnitude of producing certain injury). (b) Under the conditions stated, a much safer value of $F = (588 \text{ N})(1 + 78/54) = 1.44$ kN is sustained.

Problem

69. A spider of mass m_s drapes a silk thread of negligible mass over a stick with its far end a distance h off the ground, as shown in Fig. 5-40. The stick is lubricated by a drop of dew, so that there is essentially no friction between silk and stick. The spider waits on the ground until a fly of mass $m_f (m_f > m_s)$ lands on the other end of the silk and sticks to it. The spider immediately begins to climb her end of the silk. (a) With what acceleration must she climb to keep the fly from falling? (b) If she climbs with acceleration a_s, at what height y above the ground will she encounter the fly?

FIGURE 5-40 Problem 69.

Solution

Assuming vertical forces only, we can write the equations of motion of the spider and fly as $T - m_s g = m_s a_s$, and $T - m_f g = m_f a_f$, where T is the tension in the thread; a_s and a_f the accelerations of the spider and fly, and positive is up. (a) If "keep the fly from falling" means that the fly is stationary, then $a_f = 0$. Thus $T = m_f g$, and $a_s = (T - m_s g) \div m_s = g(m_f - m_s)/m_s$. (Note: If the spider starts on the floor, $a_s > 0$, which is possible only if $m_f > m_s$. If both were initially suspended, the spider could accelerate downwards and keep the fly stationary even if $m_f < m_s$.) (b) It is important to realize that a_s can be different from the value found in part (a); here $a_f \neq 0$. Starting at $t = 0$, the spider's height is $y_s = \frac{1}{2}a_s t^2$, and the fly's height is $y_f = h + \frac{1}{2}a_f t^2 = h + (a_f/a_s)y_s$ (we used y_s to eliminate t). The fly is encountered when $y_f = y_s = h + (a_f/a_s)y_s$, or $y_s = h/(1 - a_f/a_s)$. (Note that if the fly is stationary, $a_f = 0$ and $y_s = h$ as expected.) We can find the ratio of the accelerations by eliminating T from the equations of motion: $(m_f - m_s)g = m_s a_s - m_f a_f$, or

$$\frac{a_f}{a_s} = \frac{m_s}{m_f} - \left(\frac{m_f - m_s}{m_f}\right)\frac{g}{a_s}. \text{ Then } 1 - \frac{a_f}{a_s}$$

$$= \left(1 - \frac{m_s}{m_f}\right)\left(1 + \frac{g}{a_s}\right), \text{ and}$$

$$y_s = \frac{m_f a_s h}{(m_f - m_s)(g + a_s)}.$$

Problem

71. Three identical massless springs of unstretched length ℓ and spring constant k are connected to three equal masses m as shown in Fig. 5-41. A force is applied at the top of the upper spring to give the whole system the same acceleration a. Determine the length of each spring.

Solution

Neglect the mass of the springs and let T_n, $n = 1, 2, 3$, be the tensions, starting from the bottom spring. With positive components upward, the equations of motion are $T_1 - mg = ma$, $T_2 - T_1 - mg = ma$, and $T_3 - T_2 - mg = ma$. Therefore $T_1 = m(g + a)$, $T_2 = T_1 + m(g + a) = 2m(g + a)$, and $T_3 = T_2 + m(g + a) = 3m(g + a)$. The stretch in each spring is $x_n = T_n/k$, and its length is $\ell + x_n = \ell + nm(g + a)/k$, where $n = 1, 2, 3$ is the spring's number.

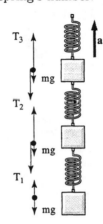

FIGURE 5-41 Problem 71 Solution.

Problem

73. Two springs have the same unstretched length but different spring constants k_1 and k_2. (a) If they are connected side-by-side and stretched a distance x, as shown in Fig. 5-42a, show that the force exerted by the combination is $(k_1 + k_2)x$. (b) If they are now connected end-to-end and the combination is stretched a distance x (Fig. 5-42b), show that they exert a force $k_1 k_2 x/(k_1 + k_2)$.

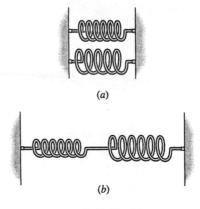

FIGURE 5-42 Problem 73.

Solution

(a) When connected in "parallel," $F = F_1 + F_2$, and $x = x_1 = x_2$, where F and x are the (magnitude of the) force and stretch of the combination, and subscripts 1 and 2 refer to the individual springs. From Hooke's law, $F_1 = k_1 x_1$, and $F_2 = k_2 x_2$; therefore $F = k_1 x_1 + k_2 x_2 = (k_1 + k_2)x$. (If we write $F = k_{||}x$, then $k_{||} = k_1 + k_2$.) (b) When connected in "series," the tension is the same in both springs, $F = F_1 = F_2$ (true for "massless" springs), while the total stretch is the sum of the individual stretches, $x = x_1 + x_2$. Again using Hooke's Law, we find $x = F_1/k_1 + F_2/k_2 = F(k_1^{-1} + k_2^{-1})$, or $F = k_1 k_2 x/(k_1 + k_2)$. (If $F = k_s x$, then $k_s^{-1} = k_1^{-1} + k_2^{-1}$.)

Problem

75. Although we usually write Newton's law for one-dimensional motion in the form $F = ma$, the most basic version of the law reads $F = d(mv)/dt$. The simpler form holds only when the mass is constant. (a) Consider an object whose mass may be changing, and show that the rule for the derivative of a product (see Appendix A) can be used to write Newton's law in the form $F = ma + v(dm/dt)$. (b) A railroad car is being pulled beneath a grain elevator that dumps in grain at the rate of 450 kg/s. What force must be applied to keep the car moving at a constant 2.0 m/s?

Solution

(a) $F_{net} = d(mv)/dt = m(dv/dt) + (dm/dt)v = ma + v(dm/dt)$. (b) We can apply Newton's second law in the above form to the horizontal motion of the rail car. At a steady speed, $a = dv/dt = 0$, so the net horizontal force that must act on the car, whose mass is increasing at a rate of $dm/dt = 450$ kg/s, is $F_{net} = v(dm/dt) = (2$ m/s$)(450$ kg/s$) = 900$ N. If the grain does not exert any horizontal force on the car as it falls in, then all of the net force above must be applied to the car by a locomotive.

CHAPTER 6   USING NEWTON'S LAWS

ActivPhysics can help with these problems: All Activities in Section 2 "Forces and Motion" and Section 4 "Circular Motion."

Section 6-1:   Using Newton's Second Law

Problem

1. Two forces, both in the x-y plane, act on a 1.5-kg mass, which accelerates at 7.3 m/s^2 in a direction 30° counter-clockwise from the x-axis. One force has magnitude 6.8 N and points in the $+x$ direction. Find the other force.

Solution

Newton's second law for this mass says $\mathbf{F}_{net} = \mathbf{F}_1 + \mathbf{F}_2 = m\mathbf{a}$, where we assume no other significant forces are acting. Since the acceleration and the first force are given, one can solve for the second, $\mathbf{F}_2 = m\mathbf{a} - \mathbf{F}_1 = (1.5 \text{ kg})(7.3 \text{ m/s}^2)(\hat{\imath}\cos 30° + \hat{\jmath}\sin 30°) - (6.8 \text{ N})\hat{\imath} = (2.68\hat{\imath} + 5.48\hat{\jmath})$ N. This has magnitude 6.10 N and direction 63.9° CCW from the x-axis.

Problem

3. A 3700-kg barge is being pulled along a canal by two mules, as shown in Fig. 6-59. The tension in each tow rope is 1100 N, and the ropes make 25° angles with the forward direction. What force does the water exert on the barge (a) if it moves with constant velocity and (b) if it accelerates forward at 0.16 m/s^2?

Solution

The horizontal forces on the barge are the two tensions and the resistance of the water, as shown on Figure 6-59. The net force is in the x direction, so

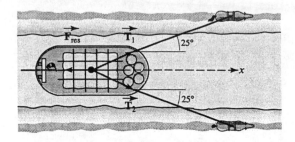

FIGURE 6-59 Problem 3 Solution.

$2T\cos 25° - F_{res} = ma_x$, since $T_1 = T_2 = T$.
(a) If $a_x = 0$, $F_{res} = 2(1100 \text{ N})\cos 25° = 1.99$ kN.
(b) If $a_x = 0.16 \text{ m/s}^2$, $F_{res} = 1.99 \text{ kN} - (3700 \text{ kg}) \times (0.16 \text{ m/s}^2) = 1.40$ kN.

Problem

5. A block of mass m slides with acceleration a down a frictionless slope that makes an angle θ to the horizontal; the only forces acting on it are the force of gravity $\mathbf{F}_g$ and the normal force $\mathbf{N}$ of the slope. Show that the magnitude of the normal force is given by $N = m\sqrt{g^2 - a^2}$.

Solution

Choose the x-axis down the slope (parallel to the acceleration) and the y-axis parallel to the normal. Then $a_x = a, a_y = 0, N_x = 0, N_y = N, F_{gx} = F_g\cos(90° - \theta) = mg\sin\theta$, and $F_{gy} = -mg\cos\theta$. Newton's second law, $\mathbf{N} + \mathbf{F}_g = m\mathbf{a}$, in components gives $mg\sin\theta = ma$, and $N - mg\cos\theta = 0$. Eliminate θ (using $\sin^2\theta + \cos^2\theta = 1$) to find $(a/g)^2 + (N/mg)^2 = 1$, or $N = m\sqrt{g^2 - a^2}$.

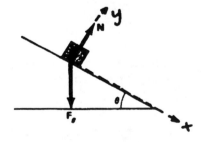

Problem 5 Solution.

Problem

7. A block is launched up a frictionless ramp that makes an angle of 35° to the horizontal. If the block's initial speed is 2.2 m/s, how far up the ramp does it slide?

Solution

The acceleration up the ramp is $-g\sin 35°$, so the block goes a distance in this direction calculated from the equation $v_{0x}^2 - 2g\sin 35°(x - x_0) = 0$. Thus, $x - x_0 = (2.2 \text{ m/s})^2/2(9.8 \text{ m/s}^2)\sin 35° = 43.1$ cm.

Problem

9. A 15-kg monkey hangs from the middle of a massless rope as shown in Fig. 6-60. What is the tension in the rope? Compare with the monkey's weight.

Solution

The sum of the forces at the center of the rope (shown on Fig. 6-60) is zero (if the monkey is at rest), $\mathbf{T}_1 + \mathbf{T}_2 + \mathbf{W} = 0$. The x component of this equation requires that the tension is the same on both sides: $T_1 \cos 8° + T_2 \cos 172° = 0$, or $T_1 = T_2$. The y component gives $2T \sin 8° = W$, or $T = W/2 \sin 8° = 3.59W = 3.59(15 \text{ kg})(9.8 \text{ m/s}^2) = 528 \text{ N}$.

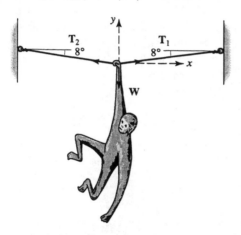

FIGURE 6-60 Problem 9 Solution.

Problem

11. A 10-kg mass is suspended at rest by two strings attached to walls, as shown in Fig. 6-62. Find the tension forces in the two strings.

Solution

The force diagram is superimposed on Fig. 6-62. Since the mass is at rest, the sum of the forces is zero, $\mathbf{T}_1 + \mathbf{T}_2 + \mathbf{W} = 0$, which is true for the x and

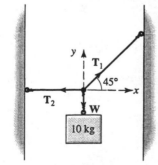

FIGURE 6-62 Problem 11 Solution.

y components separately, $T_1 \cos 45° - T_2 = 0$, and $T_1 \sin 45° - W = 0$. Solving for the magnitudes of the tensions, and substituting 98 N for the weight, we find $T_1 = \sqrt{2} \ 98 \text{ N} = 139 \text{ N}$, and $T_2 = T_1/\sqrt{2} = 98 \text{ N}$.

Problem

13. A camper hangs a 26-kg pack between two trees, using two separate pieces of rope of different lengths, as shown in Fig. 6-64. What is the tension in each rope?

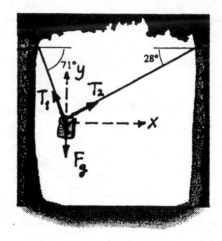

FIGURE 6-64 Problem 13.

Solution

The sum of the forces acting on the pack (gravity and the tension along each rope) is zero, since it is at rest, $\mathbf{F}_g + \mathbf{T}_1 + \mathbf{T}_2 = 0$. In a coordinate system with x-axis horizontal to the right and y-axis vertical upward, the x and y components of the net force are $-T_1 \cos 71° + T_2 \cos 28° = 0$, and $-26 \times 9.8 \text{ N} + T_1 \sin 71° + T_2 \sin 28° = 0$ (see Example 6-3). Solving for T_1 and T_2, one finds $T_1 = (26 \times 9.8 \text{ N})(\sin 71° + \tan 28° \cos 71°) = 228 \text{ N}$ and $T_2 = (26 \times 9.8 \text{ N}) \div (\sin 28° + \tan 71° \cos 28°) = 84.0 \text{ N}$.

Section 6-2: Multiple Objects

Problem

15. Your 12-kg baby sister is hanging on the bottom of the tablecloth with all her weight. In the middle of the table, 60 cm from each edge, is a 6.8-kg roast turkey. (a) What is the acceleration of the turkey? (b) From the time she starts pulling, how long do you have to intervene before the turkey goes over the edge of the table?

Solution

The vertical motion of your baby sister and the horizontal motion of the turkey are analogous to the climber and rock in Example 6-5. If we assume that both have accelerations of the same magnitude as the tablecloth, which has negligible mass, no friction with the table, etc., then (a) $a = a_{rx} = m_c g/(m_c + m_r) =$ $(12 \text{ kg})(9.8 \text{ m/s}^2)/(12 \text{ kg} + 6.8 \text{ kg}) = 6.26 \text{ m/s}^2$ is the turkey's horizontal acceleration, and (b) $t = \sqrt{2x/a} =$ $\sqrt{2(60 \text{ cm})/(6.26 \text{ m/s}^2)} = 0.438$ s is the time you have to save the turkey from going over the edge. (The assumptions relevant to Example 6-5 might be somewhat over-restrictive in this situation.)

Problem

17. If the left-hand slope in Fig. 6-56 makes a 60° angle with the horizontal, and the right-hand slope makes a 20° angle, how should the masses compare if the objects are not to slide along the frictionless slopes?

Solution

The free-body force diagrams for the left- and right-hand masses are shown in the sketch, where there is only a normal contact force since each slope is frictionless, and we indicate separate parallel and perpendicular x-y-axes. If the masses don't slide, the net force on each must be zero, or $T_\ell - m_\ell g \sin 60° = 0$, and $m_r g \sin 20° - T_r = 0$ (we only need the parallel components in this problem). If the masses of the string and pulley are negligible and there is no friction, then $T_\ell = T_r$. Adding the force equations, we find $m_r g \sin 20° - m_\ell g \sin 60° = 0$, or the mass ratio must be $m_r/m_\ell = \sin 60°/\sin 20° = 2.53$ for no motion.

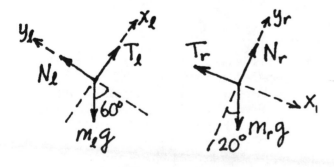

Problem 17 Solution.

Problem

19. Suppose the angles shown in Fig. 6-56 are 60° and 20°. If the left-hand mass is 2.1 kg, what should be the right-hand mass in order that
    (a) it accelerates downslope at 0.64 m/s^2;
    (b) it accelerates upslope at 0.76 m/s^2?

Solution

With reference to the solution to Problem 17, the parallel component of the equations of motion for the masses are $T_\ell - m_\ell g \sin 60° = m_\ell a_\ell$ and $m_r g \sin 20° - T_r = m_r a_r$. The accelerations and tensions are equal, respectively, provided the string doesn't stretch and the other assumptions in Problem 17 hold. Then $m_r a_r + m_\ell a_\ell = (m_r + m_\ell)a =$ $m_r g \sin 20° - T_r + T_\ell - m_\ell g \sin 60° = m_r g \sin 20° -$ $m_\ell g \sin 60°$, or $m_r = m_\ell (g \sin 60° + a)/(g \sin 20° - a)$. (a) A downslope right-hand acceleration is positive for the coordinate systems we have chosen, so substituting $a = 0.64 \text{ m/s}^2$ and $m_\ell = 2.1$ kg, we find $m_r = 7.07$ kg. (b) If $a = -0.76 \text{ m/s}^2$, then $m_r = 3.95$ kg.

Problem

21. In a florist's display, hanging plants of mass 3.85 kg and 9.28 kg are suspended from an essentially massless wire, as shown in Fig. 6-67. Find the tension in each section of the wire.

Solution

Let the tensions in each section of wire be denoted by T_1, T_2, and T_3 as shown in the figure. The horizontal and vertical components of the net force on the junction of the wire with each plant are equal to zero, since the system is stationary. Thus:

$$T_1 \sin 54.0° - T_2 \sin 13.9° - (3.85 \times 9.8) \text{ N} = 0$$
$$-T_1 \cos 54.0° + T_2 \cos 13.9° = 0$$
$$T_2 \sin 13.9° - T_3 \sin 68.0° - (9.28 \times 9.8) \text{ N} = 0$$
$$-T_2 \cos 13.9° + T_3 \cos 68.0° = 0$$

One can solve any three of these equations for the unknown tensions, perhaps using the fourth equation as a check (if you do, remember not to round off). For example, $T_1 = (3.85 \times 9.8) \text{ N}/(\sin 54.0° - \cos 54.0° \times \tan 13.9°) = 56.9 \text{ N}, T_2 = T_1 \cos 54.0°/\cos 13.9° = 34.4 \text{ N}$, and $T_3 = T_2 \cos 13.9°/\cos 68.0° = 89.2 \text{ N}$.

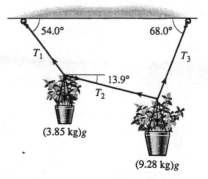

FIGURE 6-67 Problem 21 Solution.

(Note that the given angles and weights are not independent of one another.)

Section 6-3:   Circular Motion

Problem

23. A simplistic model for the hydrogen atom pictures its single electron in a circular orbit of radius 0.0529 nm about the fixed proton. If the electron's orbital speed is 2.18×10^6 m/s, what is the magnitude of the force between the electron and the proton?

Solution

For a particle in uniform circular motion, the net force equals the mass times the centripetal acceleration, $F = mv^2/r = (9.11\times10^{-31}$ kg$)(2.18\times10^6$ m/s$)^2 \div (5.29\times10^{-11}$ m$) = 8.18\times10^{-8}$ N.

Problem

25. Show that the force needed to keep a mass m in a circular path of radius r with period T is $4\pi^2mr/T^2$.

Solution

For an object of mass m in uniform circular motion, the net force has magnitude mv^2/r (Equation 6-1). The period of the motion (time for one revolution) is $T = 2\pi r/v$, so the centripetal force can also be written as $m(2\pi r/T)^2/r = mr(2\pi/T)^2 = 4\pi^2mr/T^2$ (see Equation 4-18).

Problem

27. A 940-g rock is whirled in a horizontal circle at the end of a 1.3-m-long string. (a) If the breaking strength of the string is 120 N, what is the maximum allowable speed of the rock? (b) At this maximum speed, what angle does the string make with the horizontal?

Solution

The situation is the same as described in Example 6-6. The horizontal component of the tension is the centripetal force, $T\cos\theta = mv^2/r = mv^2/\ell\cos\theta$, and the vertical component balances the weight, $T\sin\theta = mg$. (b) At the maximum speed, the tension in the string is at its breaking strength, $T_{\max} = 120$ N; therefore the minimum angle the string makes with the horizontal is given by $\sin\theta_{\min} = mg/T_{\max}$, or $\theta_{\min} = \sin^{-1}(0.940\times9.8$ N$/120$ N$) = 4.40°$. (a) At these values of T and θ, the speed is $v_{\max} = \sqrt{T_{\max}\ell\cos^2\theta_{\min}/m} = \sqrt{(120$ N$)(1.3$ m$)(\cos 4.40°)^2/(0.940$ kg$)} = 12.8$ m/s.

Problem

29. A subway train rounds an unbanked curve at 67 km/h. A passenger hanging onto a strap notices that an adjacent unused strap makes an angle of 15° to the vertical. What is the radius of the turn?

Solution

The net force on the unused strap is the vector sum of the tension in the strap (acting along its length at 15° to the vertical) and its weight. This must equal the mass times the horizontal centripetal acceleration. The free-body diagram for the strap is the same as Fig. 6-18, except that the angle from the vertical is now given, as shown. Thus, $T\cos\theta = mg$, and $T\sin\theta = mv^2/r$. Dividing these equations to eliminate T, and solving for the radius of the turn, one finds $r = v^2/g\tan\theta = (67$ m$/3.6$ s$)^2/(9.8\tan 15°$m/s$^2) = 132$ m.

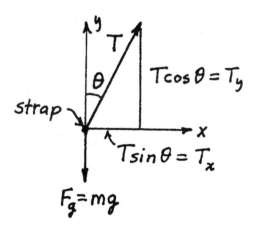

FIGURE 6-18 Problem 29 Solution.

Problem

31. Riders on the "Great American Revolution" loop-the-loop roller coaster of Example 6-8 wear seatbelts as the roller coaster negotiates its 6.3-m-radius loop with a speed of 9.7 m/s. At the top of the loop, what are the magnitude and direction of the force exerted on a 60-kg rider (a) by the roller-coaster seat and (b) by the seatbelt? (c) What would happen if the rider unbuckled at this point?

Solution

(a) As shown in Example 6-8, at the top of the loop, $N + mg = mv^2/r$, so $N = (60$ kg$)[-9.8$ m/s$^2 + (9.7$ m/s$)^2/6.3$ m$] = 308$ N. (b) Actually, 308 N is the difference between the normal force of the seat and the force exerted by the seatbelt, i.e., $N = 308$ N $+ F_{\text{belt}}$. The seatbelt, firmly adjusted, perhaps adds a few pounds (1 lb = 4.45 N), providing a feeling of security. (c) The seatbelt is required in case of accidents or rapid tangential decelerations; it is not needed to contribute to the centripetal force.

Problem

33. An indoor running track is square-shaped with rounded corners; each corner has a radius of 6.5 m on its inside edge. The track includes six 1.0-m-wide lanes. What should be the banking angles on (a) the innermost and (b) the outermost lanes if the design speed of the track is 24 km/h?

Solution

The banking angle is $\theta = \tan^{-1}(v^2/gr)$ (see Example 6-7). A competitive runner rounds a turn on the inside edge of his or her lane. (a) $\theta = \tan^{-1}[(24 \text{ m}/3.6 \text{ s})^2 \div (9.8 \text{ m/s}^2)(6.5 \text{ m})] = 34.9°$. (b) The radius of the inside edge of the outermost lane is 6.5 m + 5(1 m) = 11.5 m, so $\theta = 21.5°$. (This type of banking is shown in Fig. 6-58.)

Problem

35. You're a passenger in a car rounding a turn with radius 180 m. You take your keys from your pocket and dangle them from the end of your keychain. They make an 18° angle with the vertical, as shown in Fig. 6-70. What is the car's speed?

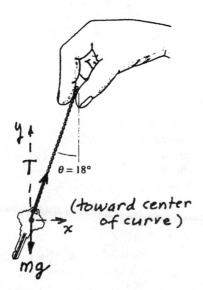

FIGURE 6-70 Problem 35.

Solution

You and your keys are rounding the curve along with the car, so all have the same centripetal acceleration,

$a_c = v^2/r$, which we assume is horizontal and directed toward the center of the curve. The forces on the keys are the tension along the direction of the keychain and gravity downward, as shown on Fig. 6-70, so the situation is just like Example 6-6, where the horizontal component of the tension supplies the centripetal acceleration, and the vertical component balances the weight of the keys. Then the horizontal and vertical components of Newton's second law for the keys are: $T\cos(90° - 18°) = T\sin 18° = ma_c = mv^2/r$, and $T\cos 18° - mg = 0$. Eliminating the tension, as in the sixth step of Example 6-6, we find $mg\tan 18° = mv^2/r$, or $v = \sqrt{gr\tan 18°} = \sqrt{(9.8 \text{ m/s}^2)(180 \text{ m})\tan 18°} = 23.9 \text{ m/s} = 86.2 \text{ km/h}$. (This problem could also be approached using an accelerated coordinate system at rest relative to the car, and introducing a fictitious force, $-ma_c$, called the centrifugal force, to account for the accelerated motion. The beginning student is advised to stick with inertial coordinate systems, however, in which there is less chance for confusion.)

Problem

37. A 1200-kg car drives on the country road shown in Fig. 6-71. The radius of curvature of the crests and dips is 31 m. What is the maximum speed at which the car can maintain road contact at the crests?

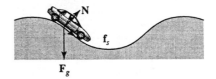

FIGURE 6-71 Problem 37.

Solution

If air resistance is ignored, the forces on the car are gravity and the contact force of the road, which is represented by the sum of the normal force (perpendicularly away from the road) and friction between the tires and the road (parallel to the road in the direction of motion). Newton's second law for the car is $\mathbf{F}_g + \mathbf{N} + \mathbf{f}_s = m\mathbf{a}$. At a crest, $\mathbf{N}$ is vertically upward, $\mathbf{f}_s$ is horizontal, and the vertical component of $\mathbf{a}$ is the radial acceleration $-v^2/r$ (downward in this case). The vertical component of Newton's second law is then $-mg + N = -mv^2/r$. As long as the car is in contact with the road, $N \geq 0$; thus, $v \leq \sqrt{gr} = \sqrt{(9.8 \text{ m/s}^2)(31 \text{ m})} = 17.4 \text{ m/s} = 62.7 \text{ km/h}$.

Section 6-4:   Friction

Problem

39. Movers slide a file cabinet along a floor. The mass of the cabinet is 73 kg, and the coefficient of kinetic friction between cabinet and floor is 0.81. What is the frictional force on the cabinet?

Solution

If the floor is level, the normal force on the cabinet is equal in magnitude to its weight, so the frictional force has magnitude $f_k = \mu_k N = \mu_k mg = (0.81)(73 \text{ kg}) \times (9.8 \text{ m/s}^2) = 579 \text{ N}$. The direction of sliding friction opposes the motion.

Problem

41. Eight 80-kg rugby players climb on a 70-kg "scrum machine," and their teammates proceed to push them with constant velocity across a field. If the coefficient of kinetic friction between scrum machine and field is 0.78, with what force must they push?

Solution

As in Problem 40(a), a horizontal applied force must have magnitude equal to the frictional force in order to push an object at constant velocity along a level surface. The total weight on the scrum machine is $(8 \times 80 \text{ kg} + 70 \text{ kg})(9.8 \text{ m/s}^2) = 6.96 \text{ kN}$; therefore $F_{\text{app}} = f_k = \mu_k N = (0.78)(6.96 \text{ kN}) = 5.43 \text{ kN}$.

Problem

43. A child sleds down a 12° slope at constant speed. What is the coefficient of friction between slope and sled?

Solution

The frictional force must balance the downslope component of gravity on the sled to produce a constant speed. The normal force on the sled must balance the perpendicular component of gravity if just gravity and the contact force are acting. Thus, $mg \sin \theta = f_k = \mu_k N$, and $mg \cos \theta = N$, or $\mu_k = \tan 12° = 0.21$ (as in Example 6-10).

Problem

45. Repeat Example 6-5, now assuming that the coefficient of kinetic friction between rock and ice is 0.057.

Solution

If there is friction between the rock and the ice, we must modify the rock's equation of motion, $\mathbf{T}_r +$

$\mathbf{F}_{gr} + \mathbf{N} + \mathbf{f}_k = m\mathbf{a}_r$. Since the ice surface is horizontal, only the rock's x-equation changes, $T_{rx} - \mu_k N = m_r a_{rx}$. Now we need to use the rock's y-equation to eliminate N, obtaining $T_{rx} - \mu m_r g = m_r a_{rx}$. Solving for a_{rx} as before, we find $\mu_k m_r g - m_r a_{cy} - m_c g = m_c a_{cy}$, or $a_{rx} = -a_{cy} = (m_c - \mu_k m_r)g/(m_c + m_r) = (70 \text{ kg} - 0.057 \times 940 \text{ kg}) \times (9.8 \text{ m/s}^2)/(1010 \text{ kg}) = 0.159 \text{ m/s}^2$. Now the climber has more time, $t = \sqrt{2(51 \text{ m})/(0.159 \text{ m/s}^2)} = 25.3 \text{ s}$, to pray for rescue.

Problem

47. A bat crashes into the vertical front of an accelerating subway train. If the coefficient of friction between bat and train is 0.86, what is the minimum acceleration of the train that will allow the bat to remain in place?

Solution

Since $\mathbf{N}$ is parallel to the acceleration, but perpendicular to gravity and friction, $N = ma$, and $f_s = mg \leq \mu_s N = \mu_s ma$. Therefore, in order to remain in place, $a \geq g/\mu_s = (9.8 \text{ m/s}^2)/0.86 = 11.4 \text{ m/s}^2$.

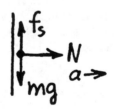

Problem 47 Solution.

Problem

49. The coefficient of static friction between steel train wheels and steel rails is 0.58. The engineer of a train moving at 140 km/h spots a stalled car on the tracks 150 m ahead. If he applies the brakes so that the wheels do not slip, will the train stop in time?

Solution

When stopping on a level track, the maximum acceleration due to friction is $a = -\mu_s g$, as explained in Example 6-12. The minimum stopping distance from an initial speed of $(140/3.6)$ m/s is $\Delta x = v_0^2/(-2a) = (38.9 \text{ m/s})^2/(2 \times 0.58 \times 9.8 \text{ m/s}^2) = 133 \text{ m}$. With split-second timing, an accident could be averted.

Problem

51. A bug crawls outward from the center of a compact disc spinning at 200 revolutions per minute. The coefficient of static friction between the bug's sticky feet and the disc surface is 1.2. How far does the bug get from the center before slipping?

Solution

Assume that the disc is level. Then the frictional force produces the (centripetal) acceleration of the bug, and the normal force equals its weight. Thus, $f_s = m(v^2/r) = mr(2\pi/T)^2 \le \mu_s N = \mu_s mg$, or $r \le \mu_s g(T/2\pi)^2 = (1.2)(9.8 \text{ m/s}^2)(60 \text{ s}/2\pi(200))^2 = 2.68$ cm. Note that the period of revolution is 60 s divided by the number of revolutions per minute.

Problem

53. A 2.5-kg block and a 3.1-kg block slide down a 30° incline as shown in Fig. 6-73. The coefficient of kinetic friction between the 2.5-kg block and the slope is 0.23; between the 3.1-kg block and the slope it is 0.51. Determine the (a) acceleration of the pair and (b) the force the lighter block exerts on the heavier one.

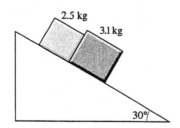

FIGURE 6-73 Problem 53.

Solution

The forces on the blocks are as shown. (Since $\mu_{k2} > \mu_{k1}$, there will be a contact force of magnitude F_c, such that the acceleration of both blocks down the incline is a.) The x and y components of Newton's second law for each block are

$$m_1 g \sin 30° - \mu_{k1} N_1 - F_c$$
$$= m_1 a, \ N_{k1} - m_1 g \cos 30° = 0$$
$$m_2 g \sin 30° - \mu_{k2} N_2 + F_c$$
$$= m_2 a, \ N_2 - m_2 g \cos 30° = 0.$$

(a) To solve for a, add the x equations and use values of N from the y equations: $a = [(m_1 + m_2)g \sin 30° - (m_1\mu_{k1} + m_2\mu_{k2})g \cos 30°]/(m_1 + m_2)$, or $a = 1.63 \text{ m/s}^2$, when the given m's and μ_k's are substituted. (b) To solve for F_c, divide each x

equation by the corresponding m, and subtract: $F_c = (\mu_{k2} - \mu_{k1})m_1 m_2 g \cos 30°/(m_1 + m_2) = 3.29$ N.

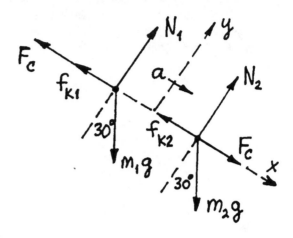

Problem 53 Solution.

Problem

55. In a typical front-wheel-drive car, 70% of the car's weight rides on the front wheels. If the coefficient of friction between tires and road is 0.61, what is the maximum acceleration of the car?

Solution

On a level road, the maximum acceleration from static friction between the tires and the road is $a_{\max} = \mu_s N/m$ (see Example 6-12). In this case, the normal force on the front tires (the ones producing the frictional force which accelerates the car) is 70% of mg, whereas the whole mass must be accelerated. Thus, $a_{\max} = (0.61)(0.70)(9.8 \text{ m/s}^2) = 4.18 \text{ m/s}^2$.

Problem

57. A police officer investigating an accident estimates from the damage done that a moving car hit a stationary car at 25 km/h. If the moving car left skid marks 47 m long, and if the coefficient of kinetic friction is 0.71, what was the initial speed of the moving car?

Solution

On a level road, the acceleration of a skidding car is $-\mu_k g$ (see Example 6-12). From the kinematics of the reconstructed accident, $v^2 = v_0^2 - 2\mu_k g(x - x_0)$, from which we calculate that
$$v_0 = \sqrt{(25 \text{ m/3.6 s})^2 + 2(0.71)(9.8 \text{ m/s}^2)(47 \text{ m})} = 26.5 \text{ m/s} = 95.4 \text{ km/h} = 59.3 \text{ mi/h}. \text{ (Add speeding to the traffic citation!)}$$

Problem

59. You try to push a heavy trunk, exerting a force at an angle of 50° below the horizontal (Fig. 6-74). Show that, no matter how hard you try to push, it is impossible to budge the trunk if the coefficient of static friction exceeds 0.84.

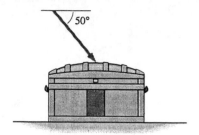

FIGURE 6-74 Problem 59.

Solution

The trunk remains at rest if the sum of the forces on it (x and y components) is 0: $F_a \cos 50° - f_s = 0$, $N - mg - F_a \sin 50° = 0$. Since $f_s = F_a \cos 50° \leq \mu_s N = \mu_s(mg + F_a \sin 50°)$, the condition for equilibrium can be written $F_a(\cos 50° - \mu_s \sin 50°) \leq \mu_s mg$, or $(\cos 50°/\mu_s) - \sin 50° \leq (mg/F_a)$. The right-hand side is always positive (F_a and mg are magnitudes), but the left-hand side can be positive or negative. If it is negative, the trunk does not move, independent of F_a. Thus, the equilibrium condition will always be satisfied if $\sin 50° > \cos 50°/\mu_s$, or $\mu_s > \cot 50° = 0.84$.

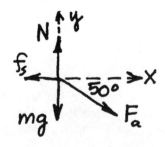

Problem 59 Solution.

Problem

61. A block is shoved down a 22° slope with an initial speed of 1.4 m/s. If it slides 34 cm before stopping, what is the coefficient of friction?

Solution

The acceleration down the slope can be found from kinematics (Equation 2-11) and from Newton's second law (as in Example 6-10): $a = g \sin \theta - \mu_k g \cos \theta =$

$-v_0^2/2\ell$. In this equation, the values of θ, v_0, and ℓ are given; therefore we can solve for μ_k:

$$\mu_k = \tan 22° + \frac{(1.4 \text{ m/s})^2}{2(0.34 \text{ m})(9.8 \text{ m/s}^2) \cos 22°} = 0.72.$$

Section 6-5: Drag Forces

Problem

63. Find the drag force on a 7.4-cm-diameter baseball moving through air (density 1.2 kg/m³) at 45 m/s. The drag coefficient is 0.50.

Solution

The cross-sectional area of the baseball (a sphere) is $\frac{1}{4}\pi d^2$, so Equation 6-4 and the other given quantities result in a drag force of magnitude $F_D = \frac{1}{2}C\rho Av^2 = \frac{1}{2}(0.50)(1.2 \text{ kg/m}^3)\frac{1}{4}\pi(7.4 \text{ cm})^2(45 \text{ m/s})^2 = 2.61$ N.

Problem

65. Find the terminal speed of a 1.0-mm-diameter spherical raindrop in air. The densities of air and water are 1.2 kg/m³ and 1000 kg/m³, respectively, and the drag coefficient is 0.50.

Solution

The expression for the terminal speed found in Example 6-15 can be used, $v_t = \sqrt{2 \, mg/C\rho A}$, where the mass of the drop is its volume, V, times the density of water, ρ_w. The volume and cross-sectional area of a spherical drop, in terms of the diameter, are $\frac{1}{6}\pi d^3$ and $\frac{1}{4}\pi d^2$, respectively, so their ratio is $\frac{2}{3}d$. Then the terminal speed becomes $v_t = \sqrt{4 \, gd \, (\rho_w/\rho)/3C} = \sqrt{4(9.8 \text{ m/s}^2)(10^{-3}\text{m})(1000/1.2)/3(0.50)} = 4.67$ m/s, where we canceled identical units in the density ratio.

Paired Problems

Problem

67. In Fig. 6-76, suppose $m_1 = 5.0$ kg and $m_2 = 2.0$ kg, and that the surface and pulley are frictionless. Determine the magnitude and direction of m_2's acceleration.

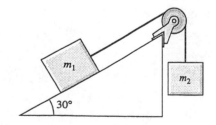

FIGURE 6-76 Problems 67, 68.

Solution

Since we are not interested in the tension in the rope connecting the masses, this is a good opportunity to take advantage of the type of shortcut mentioned in the solution to Problem 18. (For a similar solution using equations of motion for each object, see Problem 17 or 19.) Being tied together by a rope (which is assumed to be unextensible), m_1 and m_2 move as a unit, with the same acceleration (in magnitude) which we'll choose to be positive for m_2 upward and m_1 downslope. Gravity acts positive downslope on $m_1(m_1 g \sin 30°)$ and negative downward on $m_2(-m_2 g)$, so the net force on the system of masses, in the positive direction of motion, equal to the total mass times the acceleration, is $m_1 g \sin 30° - m_2 g = (m_1 + m_2)a$. (Here, we neglect the mass of the rope and pulley, which are also accelerated, and any frictional forces.) Thus, $a = (5 \text{ kg} \sin 30° - 2 \text{ kg}) \times (9.8 \text{ m/s}^2)/(5 \text{ kg} + 2 \text{ kg}) = 0.700 \text{ m/s}^2$.

Problem

69. A tetherball on a 1.7-m rope is struck so it goes into circular motion in a horizontal plane, with the rope making a 15° angle to the horizontal. What is the ball's speed?

Solution

The tetherball whirling in a horizontal circle is analogous to the mass on a string in Example 6-6. From step 6, $v = \sqrt{g\ell \cos^2 \theta / \sin \theta} = \sqrt{(9.8 \text{ m/s}^2)(1.7 \text{ m}) \cos^2 15° / \sin 15°} = 7.75 \text{ m/s}$.

Problem

71. Starting from rest, a skier slides 100 m down a 28° slope. How much longer does the run take if the coefficient of kinetic friction is 0.17 instead of 0?

Solution

If air resistance is ignored, the forces acting on the skier are analogous to those on the sled in Example 6-10, so the downslope acceleration is $a_\| = g(\sin \theta - \mu_k \cos \theta)$. Starting from rest, the time needed to coast a distance Δx downslope is $t = \sqrt{2\Delta x / a_\|}$. With no friction, $t = \sqrt{2(100 \text{ m})/(9.8 \text{ m/s}^2) \sin 28°} = 6.59 \text{ s}$. If $\mu_k = 0.17$, $t' = \sqrt{2(100 \text{ m})/(9.8 \text{ m/s}^2)(\sin 28° - 0.17 \cos 28°)} = 7.99 \text{ s}$, or about 1.40 s longer.

Problem

73. A car moving at 40 km/h negotiates a 130-m-radius banked turn designed for 60 km/h. (a) What coefficient of friction is needed to keep the car on the road? (b) To which side of the curve would it move if it hit an essentially frictionless icy patch?

Solution

The forces on a car (in a plane perpendicular to the velocity) rounding a banked curve at arbitrary speed are analyzed in detail in the solution to Problem 81 below. (a) It is shown there that to prevent skidding, $\mu_s \geq \left| v^2 - v_d^2 \right| / gR(1 + v^2 v_d^2/g^2 R^2)$, where R is the radius of the curve, and v_d is the design speed for the proper banking angle, $\tan \theta_d = v_d^2/gR$. In this problem, $v_d = (60/3.6) \text{ m/s}, v = (40/3.6) \text{ m/s}$, and $R = 130$ m, so $\mu_s \geq 0.12$. (b) Since $v < v_d$, the car would slide down the bank of the curve in the absence of friction.

Supplementary Problems

Problem

75. A space station is in the shape of a hollow ring, 450 m in diameter (Fig. 6-77). At how many revolutions per minute should it rotate in order to simulate Earth's gravity—that is, so that the normal force on an astronaut at the outer edge would be the astronaut's weight on Earth?

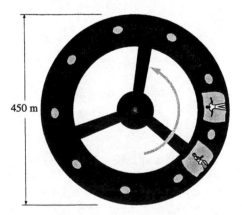

FIGURE 6-77 Problem 75.

Solution

Standing on the outer edge of the space station, rotating with it, the astronaut experiences a normal force equal to the centripetal force, $N = ma_c = m \, 4\pi^2 r/T^2$, where T is the period of rotation (see Example 4-8). Since T is the time per revolution, the number of revolutions per unit time is $1/T$ (called the frequency of revolution). If the normal force is to duplicate Earth's gravity, $a_c = g$, and $1/T = (1/2\pi) \times$

$\sqrt{g/r} = (1/2\pi)\sqrt{(9.8\,\text{m/s}^2)/(450\,\text{m}/2)} = (3.32 \times 10^{-2}\,\text{rev/s})(60\,\text{s/min}) = 1.99\,\text{rpm}.$

Problem

77. In the loop-the-loop track of Fig. 6-25, show that the car leaves the track at an angle ϕ given by $\cos\phi = v^2/rg$, where ϕ is the angle made by a vertical line through the center of the circular track and a line from the center to the point where the car leaves the track.

Solution

The angle ϕ and the forces acting on the car are shown in the sketch. The radial component of the net force (towards the center of the track) equals the mass times the centripetal acceleration, $N + mg\cos\phi = mv^2/r$. (The tangential component is not of interest in this problem.) The car leaves the track when $N = (mv^2/r) - mg\cos\phi = 0$ (no more contact) or $\cos\phi = v^2/gr$. This implies that the car leaves the track at real angles for $v^2 < gr$; otherwise, the car never leaves the track, as in Example 6-8.

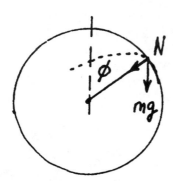

Problem 77 Solution.

Problem

79. You stand on a spring scale at the north pole and again at the equator. (a) Which scale reading will be lower, and why? (b) By what percentage will the lower reading differ from the higher one? (Here you're neglecting variations in g due to geological factors.)

Solution

When standing on the Earth's surface, you are rotating with the Earth about its axis through the poles, with a period of $1\,d$. The radius of your circle of rotation (your perpendicular distance to the axis) is $r = R_E\cos\theta$, where R_E is the radius of the Earth (constant if geographical variations are neglected) and θ is your lattitude. Your centripetal acceleration has

magnitude $a_c = (2\pi/T)^2 r$ and is directed toward the axis of rotation (see Example 4-8). We assume there are only two forces acting on you, gravity, $\mathbf{F}_g$ (magnitude mg approximately constant, directed towards the center of the Earth), and the force exerted by the scale, $\mathbf{F}_s$. Newton's second law requires that $\mathbf{F}_g + \mathbf{F}_s = m\mathbf{a}_c$. (a) At the north pole, $\mathbf{a}_c = 0$, so the magnitudes of $\mathbf{F}_g$ and $\mathbf{F}_s$ are equal, or $F_s = mg$; but at the equator, $\mathbf{a}_c$ has a maximum magnitude, equal to the difference in the magnitudes of $\mathbf{F}_g$ and $\mathbf{F}_s$, or $F_s = mg - m(2\pi/T)^2 R_E$. Therefore F_s (your "weight") is lower at the equator than at the pole. (b) The fractional difference of these two values is $(F_{s,\text{pole}} - F_{s,\text{eq.}})/F_{s,\text{pole}} = (2\pi/T)^2 R_E/g = (2\pi/86,400\,\text{s})^2(6.37\times10^6\,\text{m})/(9.81\,\text{m/s}^2) = 0.34\%.$

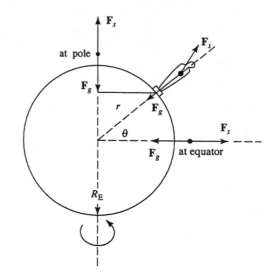

Problem 79 Solution.

Problem

81. A highway turn of radius R is banked for a design speed v_d. If a car enters the turn at speed $v = v_d + \Delta v$, where Δv can be positive or negative, show that the minimum coefficient of static friction needed to prevent slipping is

$$\mu_s = \frac{|\Delta v|}{gR} \frac{(2v_d + \Delta v)}{[1 + (v_d v/gR)^2]}.$$

Solution

The equation of motion for a car rounding a banked turn is $\mathbf{N} + m\mathbf{g} + \mathbf{f}_s = m\mathbf{a}_r$, where $a_r = v^2/R$ is the radical acceleration (assumed horizontal and constant in magnitude) and the forces are as shown. Note that the frictional force changes direction for v greater or less than the design speed. Taking components parallel and perpendicular to the road, we find $N - mg\cos\theta = m(v^2/R)\sin\theta$, $mg\sin\theta \pm f_s = (v^2/R)\cos\theta$, where the upper sign is for $v > v_d$, and the lower for $v < v_d$. (We chose these components because the solution for N and f_s is direct.) This argument applies if the car does not skid (otherwise $\mathbf{a} \neq \mathbf{a}_r$) so $f_s \leq \mu_s N$. Therefore:

$$\mu_s \geq \frac{f_s}{N} = \frac{\mp g\sin\theta \pm v^2\cos\theta/R}{g\cos\theta + v^2\sin\theta/R} = \frac{\mp g\tan\theta \pm v^2/R}{g + v^2\tan\theta/R}$$

$$= \frac{\mp v_d^2 \pm v^2}{gR(1 + v^2 v_d^2/g^2 R^2)}$$

since $v_d^2/gR = \tan\theta$. If we set $\Delta v = v - v_d$, the condition on μ_s becomes:

$$\mu_s \geq \pm\frac{\Delta v(2v_d + \Delta v)}{gR(1 + v_d^2 v^2/g^2 R^2)} = \frac{|\Delta v|(2v_d + \Delta v)}{gR(1 + v_d^2 v^2/g^2 R^2)}.$$

(Note that $|\Delta v|$ is $+\Delta v$ for $v > v_d$ and $-\Delta v$ for $v < v_d$.) The expression for the minimum coefficient of friction is not particularly simple, but for $v = 0$ (car at rest) it reduces to $|-v_d|(2v_d - v_d)/gR = v_d^2/gR = \tan\theta$, as in Example 6-14.

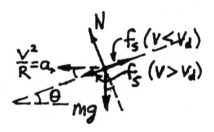

Problem 81 Solution.

Problem

83. A block is projected up an incline making an angle θ with the horizontal. It returns to its initial position with half its initial speed. Show that the coefficient of kinetic friction is $\mu_k = \frac{3}{5}\tan\theta$.

Solution

Going up the incline, the block's acceleration (positive down the incline) is $g(\sin\theta + \mu_k\cos\theta) = a_{\text{up}}$,

whereas going down, the acceleration is $g(\sin\theta - \mu_k\cos\theta) = a_{\text{down}}$. If the block slides up a distance ℓ, its initial speed upward was $\sqrt{2a_{\text{up}}\ell}$, whereas, sliding down the same distance, it returns to the bottom with speed $\sqrt{2a_{\text{down}}\ell}$. Given that the latter speed is half the former, $4a_{\text{down}} = a_{\text{up}} = 4g(\sin\theta - \mu_k\cos\theta) = g(\sin\theta + \mu_k\cos\theta)$. Therefore $3\sin\theta = 5\mu_k\cos\theta$, or $\mu_k = \frac{3}{5}\tan\theta$.

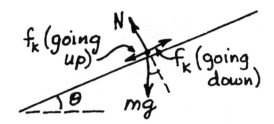

Problem 83 Solution.

Problem

85. A 2.1-kg mass is connected to a spring of spring constant $k = 150$ N/m and unstretched length 18 cm. The pair are mounted on a frictionless air table, with the free end of the spring attached to a frictionless pivot. The mass is set into circular motion at 1.4 m/s. Find the radius of its path.

Solution

Since the airtable is frictionless, the only horizontal force acting on the mass is the spring force, of magnitude $k(\ell - \ell_0)$ and in the direction of the centripetal acceleration v^2/ℓ. Here, the radius of the circle is ℓ, the length of the spring, while ℓ_0 is the unstretched length. Therefore, $k(\ell - \ell_0) = mv^2/\ell$. This is a quadratic equation for ℓ, $\ell^2 - \ell_0\ell - mv^2/k = 0$, with positive solution $\ell = \frac{1}{2}[\ell_0 + \sqrt{\ell_0^2 + 4mv^2/k}] = \frac{1}{2}[0.18\text{ m} + \sqrt{(0.18\text{ m})^2 + 4(2.1\text{ kg})(1.4\text{ m/s})^2/(150\text{ N/m})]} = 27.9$ cm.

CHAPTER 7 WORK, ENERGY, AND POWER

ActivPhysics can help with these problems:
Activity 5.1

Section 7-1: Work

Problem

1. How much work do you do as you exert a 75-N force to push a shopping cart through a 12-m-long supermarket aisle?

Solution

If the force is constant and parallel to the displacement, $W = \mathbf{F} \cdot \Delta \mathbf{r} = F \Delta r = (75 \text{ N})(12 \text{ m}) = 900$ J.

Problem

3. A crane lifts a 650-kg beam vertically upward 23 m, then swings it eastward 18 m. How much work does the crane do? Neglect friction, and assume the beam moves with constant speed?

Solution

Lifting the beam at constant speed, the crane exerts a constant force vertically upward and equal in magnitude to the weight of the beam. During the horizontal swing, the force is the same, but is perpendicular to the displacement. The work done is $\mathbf{F} \cdot \Delta \mathbf{r} = (mg\hat{\jmath}) \cdot (\Delta y \hat{\jmath} + \Delta x \hat{\imath}) = mg \, \Delta y = (650 \text{ kg}) \times (9.8 \text{ m/s}^2)(23 \text{ m}) = 147$ kJ.

Problem

5. The world's highest waterfall, the Cherun-Meru in Venezuela, has a total drop of 980 m. How much work does gravity do on a cubic meter of water dropping down the Cherun-Meru?

Solution

The force of gravity at the Earth's surface on a cubic meter of water is $F_g = mg = 9.8$ kN vertically downward (see the inside book cover of the text for the density of water). The displacement of the water is parallel to this, so the work done by gravity on the water is $W_g = F_g \, \Delta y = (9.8 \text{ kN})(980 \text{ m}) = 9.6$ MJ.

Problem

7. You slide a box of books at constant speed up a 30° ramp, applying a force of 200 N directed up the slope. The coefficient of sliding friction is 0.18. (a) How much work have you done when the box has risen 1 m vertically? (b) What is the mass of the box?

Solution

(a) The displacement up the ramp (parallel to the applied force) is $\Delta r = 1 \text{ m}/ \sin 30° = 2$ m, so $W_a = \mathbf{F}_a \cdot \Delta \mathbf{r} = (200 \text{ N})(2 \text{ m}) = 400$ J. (b) We could easily solve Newton's second law, with zero acceleration, to find the mass, $m = F_a/g(\sin\theta + \mu_k \cos\theta)$, but it is instructive to obtain the same result using the concept of work. The work done by gravity is $\mathbf{F}_g \cdot \Delta \mathbf{r} = \mathbf{F}_g \cdot (\Delta x \hat{\imath} + \Delta y \hat{\jmath}) = -mg \, \Delta y = -m(9.8 \text{ m/s}^2)(1 \text{ m}) = -m(9.8 \text{ J/kg})$ (see Example 7-9). The work done by friction is $\mathbf{f}_k \cdot \Delta \mathbf{r} = -f_k \Delta r = -\mu_k N \Delta r = -\mu_k \times (mg \cos 30°) \Delta r = -0.18m(9.8 \text{ m/s}^2)(2 \text{ m}) \cos 30° = -m(3.06 \text{ J/kg})$. The total work is zero ($\mathbf{v}$ is constant in the work-energy theorem), so $0 = 400 \text{ J} - m \times (9.8 + 3.06) \text{ J/kg}$, or $m = 400 \text{ J}/(12.86 \text{ J/kg}) = 31.1$ kg.

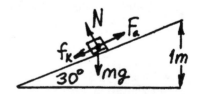

Problem 7 Solution.

Problem

9. A locomotive does 7.9×10^{11} J of work in pulling a 3.4×10^5-kg train 180 km. What is the average force in the coupling between the locomotive and the rest of the train?

Solution

If we define the average force by $W = F_{av} \Delta r$, then $F_{av} = 7.9\times10^{11}$ J$/180$ km $= 4.39$ MN. (The train's mass is not required to answer this question.)

Section 7-2: Work and the Scalar Product

Problem

11. Show that the scalar product obeys the distributive law: $\mathbf{A} \cdot (\mathbf{B}+\mathbf{C}) = \mathbf{A} \cdot \mathbf{B} + \mathbf{A} \cdot \mathbf{C}$.

Solution

This follows easily from the definition of the scalar product in terms of components: $\mathbf{A} \cdot (\mathbf{B} + \mathbf{C}) = A_x(B_x + C_x) + A_y(B_y + C_y) + A_z(B_z + C_z) = A_xB_x + A_yB_y + A_zB_z + A_xC_x + A_yC_y + A_zC_z = \mathbf{A} \cdot \mathbf{B} + \mathbf{A} \cdot \mathbf{C}$.

With more effort, it also follows from trigonometry. First:

$$D\sin(\theta_D - \theta_B) = C\sin(\theta_C - \theta_B) \quad \text{(law of sines)},$$

$$D = \sqrt{B^2 + C^2 + 2BC\cos(\theta_C - \theta_B)} \quad \text{(law of cosines)},$$

and

$$\cos\alpha = \sqrt{1 - \sin^2\alpha}$$

together give

$$D\cos(\theta_D - \theta_B) = B + C\cos(\theta_C - \theta_B).$$

Second:

$$\begin{aligned}
\mathbf{A} \cdot \mathbf{D} &= AD\cos\theta_D \\
&= AD[\cos(\theta_D - \theta_B)\cos\theta_B - \sin(\theta_D - \theta_B)\sin\theta_B] \\
&= A[B + C\cos(\theta_C - \theta_B)]\cos\theta_B \\
&\quad - A[C\sin(\theta_C - \theta_B)]\sin\theta_B \\
&= AB\cos\theta_B + AC[\cos(\theta_C - \theta_B)\cos\theta_B \\
&\quad - \sin(\theta_C - \theta_B)\sin\theta_B] \\
&= \mathbf{A} \cdot \mathbf{B} + \mathbf{A} \cdot \mathbf{C}.
\end{aligned}$$

We used the identity for $\cos(\alpha + \beta)$ from Appendix A.

Problem

13. One vector has magnitude 15 units, and another 6.5 units. Find their scalar product if the angle between them is (a) 27° and (b) 78°.

Solution

(a) $\mathbf{A} \cdot \mathbf{B} = (15\text{u})(6.5\text{u})\cos 27° = 86.9\text{u}^2$. (b) $\mathbf{A} \cdot \mathbf{B} = (15\text{u})(6.5\text{u})\cos 78° = 20.3\text{u}^2$. (Note: We used u as the symbol for the unspecified units.)

Problem

15. (a) Find the scalar product of the vectors $a\hat{\imath} + b\hat{\jmath}$ and $b\hat{\imath} - a\hat{\jmath}$, and (b) determine the angle between them. (Here a and b are arbitrary constants.)

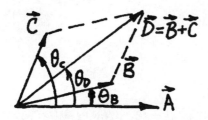

Problem 11 Solution.

Solution

(a) $(a\hat{\imath} + b\hat{\jmath}) \cdot (b\hat{\imath} - a\hat{\jmath}) = ab + b(-a) = 0$. (b) The vectors are perpendicular, so $\theta = 90°$.

Problem

17. Use Equations 7-3 and 7-4 to show that the angle between the vectors $\mathbf{A} = a_x\hat{\imath} + a_y\hat{\jmath}$, and $\mathbf{B} = b_x\hat{\imath} + b_y\hat{\jmath}$ is

$$\theta = \cos^{-1}\left\{\frac{a_xb_x + a_yb_y}{[(a_x^2 + a_y^2)(b_x^2 + b_y^2)]^{1/2}}\right\}.$$

Solution

$\mathbf{A} \cdot \mathbf{B} = a_xb_x + a_yb_y = \sqrt{a_x^2 + a_y^2}\sqrt{b_x^2 + b_y^2}\cos\theta$, from which the desired expression for θ follows directly.

Problem

19. Find the work done by a force $\mathbf{F} = 1.8\hat{\imath} + 2.2\hat{\jmath}$ N as it acts on an object moving from the origin to the point $\mathbf{r} = 56\hat{\imath} + 31\hat{\jmath}$ m.

Solution

The force is constant, and $\mathbf{r}$ is the displacement from the origin, so the work done by $\mathbf{F}$ is $W = \mathbf{F} \cdot \mathbf{r} = [(1.8)(56) + (2.2)(31)]$ N·m $= 169$ J.

Problem

21. A force $\mathbf{F} = 67\hat{\imath} + 23\hat{\jmath} + 55\hat{k}$ N is applied to a body as it moves in a straight line from $\mathbf{r}_1 = 16\hat{\imath} + 31\hat{\jmath}$ to $\mathbf{r}_2 = 21\hat{\imath} + 10\hat{\jmath} + 14\hat{k}$ m. How much work is done by the force?

Solution

$W = \mathbf{F} \cdot \Delta\mathbf{r} = \mathbf{F} \cdot (\mathbf{r}_2 - \mathbf{r}_1) = [67(21 - 16) + 23(10 - 31) + 55(14)]$ N·m $= 622$ J.

Section 7-3:   A Varying Force

Problem

23. Find the total work done by the force shown in Fig. 7-27 as the object on which it acts moves

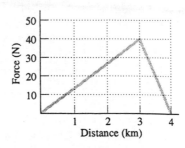

FIGURE 7-27 Problem 23.

(a) from $x = 0$ to $x = 3$ km; (b) from $x = 3$ km to $x = 4$ km.

Solution

The work done by a one-dimensional force is (Equation 7-8) $W = \int_{x_1}^{x_2} F(x)\, dx$. From Fig. 7-27, $F(x)$ is a linear function in the two intervals specified:

$$F(x) = \begin{cases} (40\text{ N}/3\text{ km})x, & \text{for } 0 \le x \le 3\text{ km.} \\ 40\text{ N} - (40\text{ N/km}) \times \\ (x - 3\text{ km}), & \text{for } 3\text{ km} \le x \le 4\text{ km.} \end{cases}$$

(Use the slope/intercept equation for a straight line, $y = mx + b$, to verify this.) Therefore:

(a) $W_{0\to3} = \int_0^{3\text{ km}} \left(\dfrac{40\text{ N}}{3\text{ km}}\right) x\, dx$

$= \left(\dfrac{40\text{ N}}{3\text{ km}}\right) \dfrac{(3\text{ km})^2}{2} = 60$ kJ,

(b) $W_{3\to4} = \int_{3\text{ km}}^{4\text{ km}} \left(\dfrac{40\text{ N}}{\text{km}}\right) (4\text{ km} - x)\, dx$

$= \left(\dfrac{40\text{ N}}{\text{km}}\right) \left| (4\text{ km})x - \dfrac{x^2}{2} \right|_{3\text{ km}}^{4\text{ km}}$

$= 20$ kJ.

(Of course, the triangular areas under the force vs distance curve could have been calculated in one's head; however, it is instructive to understand the general method for evaluating Equation 7-8.)

Problem

25. A force F acts in the x direction, its magnitude given by $F = ax^2$, where x is in meters, and a is exactly 5 N/m^2. (a) Find an exact value for the work done by this force as it acts on a particle moving from $x = 0$ to $x = 6$ m. Now find approximate values for the work by dividing the area under the force curve into rectangles of width (b) $\Delta x = 2$ m; (c) $\Delta x = 1$ m; (d) $\Delta x = \frac{1}{2}$ m with height equal to the magnitude of the force in the center of the interval. Calculate the percent error in each case.

Solution

(a) $W = \int_{x_1}^{x_2} F\, dx = \int_0^{6\text{ m}} ax^2\, dx = \frac{1}{3}a \left| x^3 \right|_0^{6\text{ m}} = \frac{1}{3}(5\text{ N/m}^2)(6\text{ m})^3 = 360$ J. (b) $W \approx \sum_{i=1}^3 F(x_i)\, \Delta x_i$, where $x_i = 1$ m, 3 m, 5 m are the midpoints, and $\Delta x_i = 2$ m. Then $W \approx (5\text{ N/m}^2)(1^2 + 3^2 + 5^2) \times$ m^2(2 m) $= 350$ J. The percent error is only $\delta = 100(360 - 350)/360 = 2.78\%$. (c) Now, $x_i = 0.5, 1.5, 2.5, 3.5, 4.5,$ and 5.5 (in meters), and $\Delta x_i = 1$ m. $W \approx \sum_{i=1}^6 F(x_i)\, \Delta x_i = (5\text{ N/m}^2)(0.5\text{ m})^2(1^2 + 3^2 + 5^2 + 7^2 + 9^2 + 11^2)(1\text{ m}) = 357.5$ J, and $\delta = 100(2.5) \div 360 = 0.694\%$. (d) $W \approx \sum_{i=1}^{12} F(x_i)\Delta x_i = (5\text{ N/m}^2) \times$

$(0.25\text{ m})^2 (1^2 + 3^2 + \cdots + 23^2)(0.5\text{ m}) = 359.375$ J with $\delta = 0.174\%$. (The direct calculation of the sum is tedious, but we can use the formula for the sum of the squares of the first n numbers, namely $\sum_1^n k^2 = \frac{1}{6}n(n+1)(n+2)$. The sum in question is $\sum_{k=1}^{12} \times (2k-1)^2 = \sum_{k=1}^{23} k^2 - \sum_{k=1}^{11}(2k)^2 = 4324 - 2024 = 2300$.)

Problem

27. A certain amount of work is required to stretch spring A a certain distance. Twice as much work is required to stretch spring B half that distance. Compare the spring constants of the two springs.

Solution

We are given $W_A = \frac{1}{2}k_A x^2$ and $W_B = \frac{1}{2}k_B(x/2)^2 = \frac{1}{8}k_B x^2 = 2W_A = k_A x^2$. Therefore $k_B = 8k_A$.

Problem

29. A force $\mathbf{F}$ acts in the x direction; its x component is given by $F = F_0 \cos(x/x_0)$, where $F_0 = 51$ N and $x_0 = 13$ m. Calculate the work done by this force acting on an object as it moves from $x = 0$ to $x = 37$ m. *Hint:* Consult Appendix A for the integral of the cosine function and treat the argument of the cosine as a quantity in radians.

Solution

From Equation 7-8:

$$W = \int_0^{37\text{ m}} F_0 \cos\frac{x}{x_0}\, dx = F_0 \left| x_0 \sin\frac{x}{x_0} \right|_0^{37\text{ m}}$$

$$= (51\text{ N})(13\text{ m})\sin(37/13) = 193 \text{ J}.$$

Problem

31. A force given by $F = a\sqrt{x}$ acts in the x direction, where $a = 9.5$ N/m$^{1/2}$. Calculate the work done by this force acting on an object as it moves (a) from $x = 0$ to $x = 3$ m; (b) from $x = 3$ m to $x = 6$ m; (c) from 6 m to 9 m.

Solution

From Equations 7-8 and 7-9:

$$W_{x_1 \to x_2} = \int_{x_1}^{x_2} ax^{1/2}dx = \frac{ax^{3/2}}{(3/2)}\bigg|_{x_1}^{x_2}$$

$$= \left(\frac{2a}{3}\right)(x_2^{3/2} - x_1^{3/2}).$$

Therefore, (a) $W_{0\to3} = (2/3)(9.5\text{ N/m}^{1/2})(3\text{ m})^{3/2} = 32.9$ J, (b) $W_{3\to6} = (6.33\text{ N·m}^{-1/2})[(6\text{ m})^{3/2} - (3\text{ m})^{3/2}] = 60.2$ J, and (c) $W_{6\to9} = 77.9$ J.

Problem

33. The force exerted by a rubber band is given approximately by

$$F = F_0 \left[\frac{\ell_0 + x}{\ell_0} - \frac{\ell_0^2}{(\ell_0 + x)^2} \right],$$

where ℓ_0 is the unstretched length, x the stretch, and F_0 is a constant (although F_0 varies with temperature). Find the work needed to stretch the rubber band a distance x.

Solution

$$W = \int_0^x F_0 \left[\frac{\ell_0 + x'}{\ell_0} - \frac{\ell_0^2}{(\ell_0 + x')^2} \right] dx'$$

$$= F_0 \left| \frac{1}{\ell_0} \left(\ell_0 x' + \frac{x'^2}{2} \right) + \frac{\ell_0^2}{\ell_0 + x'} \right|_0^x$$

$$= F_0 \left(x + \frac{x^2}{2\ell_0} + \frac{\ell_0^2}{\ell_0 + x} - \ell_0 \right).$$

(x' is a dummy variable)

Section 7-4: Force and Work in Three Dimensions

Problem

35. A particle moves from point A to point B along the semicircular path of radius R, as shown in Fig. 7-29. It is subject to a force of constant magnitude F. Find the work done by the force (a) if the force always points upward in Fig. 7-29, (b) if the force always points to the right in Fig. 7-29, and (c) if the force always points in the direction of the particle's motion.

Solution

(a) and (b) If the force is a constant vector (in magnitude and direction) it may be factored out from under the integral in the work (which is the limit of a sum) to yield $W = \int_A^B \mathbf{F} \cdot d\mathbf{r} = \mathbf{F} \cdot \int_A^B d\mathbf{r}$. The remaining integral is the sum of the displacements around a semicircle, which is just the total vector displacement along the diameter, from A to B in Fig. 7-29. If we introduce x-y coordinates to the right and upward, respectively, with origin at the center of the semicircle, then $\int_A^B d\mathbf{r} = \vec{AB} = 2R\hat{\imath}$. For $\mathbf{F} = F\hat{\jmath}$, $W = F\hat{\jmath} \cdot 2R\hat{\imath} = 0$, and for $\mathbf{F} = F\hat{\imath}$, $W = F\hat{\imath} \cdot 2R\hat{\imath} = 2RF$. (c) If the force is constant in magnitude and parallel to $d\mathbf{r}$ along the path, then $\mathbf{F} \cdot d\mathbf{r} = F|d\mathbf{r}|$, and $W = F \int_A^B |d\mathbf{r}|$, where the scalar integral is now just the length of the semicircular path. Thus $W = \pi RF$. (Note that the symbol $d\mathbf{r}$ is a displacement along the path, so $|d\mathbf{r}|$ is an element of path length, not the differential of the radius. Where confusion might arise,

one can use $d\ell$ for path element, as in Chapters 25, 30, and 31.)

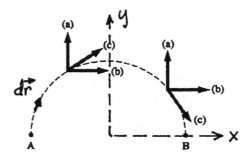

FIGURE 7-29 Problem 35 Solution.

Problem

37. A particle of mass m moves from the origin to the point $x = 3$ m, $y = 6$ m along the curve $y = ax^2 - bx$, where $a = 2$ m^{-1} and $b = 4$. It is subject to a force $\mathbf{F} = cxy\hat{\imath} + d\hat{\jmath}$ where $c = 10$ N/m^2, and $d = 15$ N. Calculate the work done by the force.

Solution

Since the equation for the curve gives y as a function of x, we can eliminate y and dy in the line integral for the work:

$$W = \int_{r_1}^{r_2} \mathbf{F} \cdot d\mathbf{r} = \int_{(0,0)}^{(3,6)} (F_x dx + F_y \, dy)$$

$$= \int_0^3 [cx(ax^2 - bx) + d(2ax - b)] \, dx$$

$$= \left| c \left(a\frac{x^4}{4} - b\frac{x^3}{3} \right) + d \left(2a\frac{x^2}{2} - bx \right) \right|_0^3$$

$$= (45 + 90) \text{ J} = 135 \text{ J},$$

where the given values for the constants were used, and all distances are in meters.

Problem

39. You put your little sister (mass m) on a swing whose chains have length ℓ, and pull slowly back until the swing makes an angle ϕ with the vertical. Show that the work you do is $mg\ell(1 - \cos\phi)$.

Solution

The path is a circular arc (of radius ℓ and differential arc length $|d\mathbf{r}| = \ell d\theta$) so that only the tangential (or parallel) components of any forces acting do work on the swing; the radial (or perpendicular) components do no work since the dot product with the path element is zero. Thus, the tension in the chains and the radial components of gravity or the applied force do no work.

If you pull slowly, so that the tangential acceleration is zero, then $F_{||} = mg \sin \theta$, and the work you do is

$$W = \int_0^\phi \mathbf{F} \cdot d\mathbf{r} = \int_0^\phi F_{||} \, |d\mathbf{r}| = \int_0^\phi mg \sin \theta \cdot \ell \, d\theta$$

$$= mg\ell \, |- \cos \theta|_0^\phi = mg\ell(1 - \cos \phi).$$

(Since forces perpendicular to the path of the swing do no work, all the work you do, for zero tangential acceleration, is done against gravity, $W = -W_g$. But $W_g = -mg \, \Delta y$ (as in Example 7-9), where Δy is the change in height of the swing. In this case, $\Delta y = \ell - \ell \cos \phi$, which gives the same W as above.)

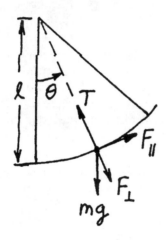

Problem 39 Solution.

Section 7-5:   Kinetic Energy

Problem

41. Electrons in a color TV tube are accelerated to 25% of the speed of light. How much work does the TV tube do on each electron? (At this speed, relativity introduces small but measurable corrections; here you neglect these effects.) See inside front cover for the electron mass.

Solution

From the work-energy theorem, $W_{\text{net}} = \Delta K = \frac{1}{2} m_e (0.25 \, c)^2 = 0.5(9.11 \times 10^{-31} \text{ kg})(0.25 \times 3 \times 10^8 \text{ m/s})^2 = 2.56 \times 10^{-15} \text{ J} = 16.0 \text{ keV}.$

Problem

43. At what speed must a 950-kg subcompact car be moving to have the same kinetic energy as a 3.2×10^4-kg truck going 20 km/h?

Solution

We want $\frac{1}{2} m_C v_C^2 = \frac{1}{2} m_T v_T^2$, so $v_C = v_T \sqrt{m_T/m_C} = (20 \text{ km/h}) \sqrt{(3.2 \times 10^4 \text{ kg})/(950 \text{ kg})} = 5.80 v_T = 116 \text{ km/h}.$

Problem

45. Two unknown elementary particles pass through a detection chamber. If they have the same kinetic energy and their mass ratio is 4 : 1, what is the ratio of their speeds?

Solution

If $m_1 = 4m_2$ and $K_1 = K_2$, then $v_1 = \sqrt{2K_1/m_1} = \sqrt{m_2 v_2^2/m_1} = v_2 \sqrt{m_2/m_1} = \frac{1}{2} v_2$; a mass ratio of 4 : 1 corresponds to a speed ratio of 1 : 2 if the kinetic energy is the same.

Problem

47. After a tornado, a 0.50-g drinking straw was found embedded 4.5 cm in a tree. Subsequent measurements showed that the tree would exert a stopping force of 70 N on the straw. What was the straw's speed when it hit the tree?

Solution

Since the stopping force (70 N) is so much larger than the weight of the straw (0.0049 N), we may assume that the net work done is essentially that done by just the stopping force, and use the work-energy theorem, $W_{\text{net}} = \Delta K$. The force is opposite to the displacement, so $-F \, \Delta r = 0 - \frac{1}{2} mv^2$, or $v = \sqrt{2F\Delta r/m} = \sqrt{2(70 \text{ N})(0.045 \text{ m})/0.5 \times 10^{-3} \text{ kg}} = 112 \text{ m/s}$ (~ 250 mi/h).

Problem

49. A hospital patient's leg slipped off the stretcher and his heel hit the concrete floor. As a physicist, you are called to testify about this accident. You estimate that the foot and leg had an effective mass of 8 kg, that they dropped freely a distance of 70 cm, and that the stopping distance was 2 cm. What force can you claim the floor exerted on the foot? Give your answer in pounds for the jury's sake.

Solution

This problem can be solved with the use of Newton's second law and kinematics (see Problem 5-67, for example), but alternatively, we may apply the work-energy theorem: $W_{\text{net}} = W_g + W_F = \Delta K$. The leg starts from rest at point A and stops at point B, so $\Delta K = 0$. The work done by gravity is $W_g = mg \times$

(72 cm), and that done by the stopping force is $W_F = -F(2 \text{ cm})$. Therefore, $mg(72 \text{ cm}) - F(2 \text{ cm}) = 0$, or $F = (8 \text{ kg})(9.8 \text{ m/s}^2)(72/2) = 2.82 \text{ kN} = 635 \text{ lbs}$.

Problem 49 Solution.

Problem

51. Catapults run by high-pressure steam from the ship's nuclear reactor are used on the aircraft carrier *Enterprise* to launch jet aircraft to takeoff speed in only 76 m of deck space. A catapult exerts a 1.1×10^6 N force on a 3.3×10^4 kg aircraft. What are (a) the kinetic energy and (b) the speed of the aircraft as it leaves the catapult? (c) How long does the catapulting operation take? (d) What is the acceleration of the aircraft?

Solution

(a) If we assume that the carrier deck and catapult force are horizontal, then the catapult force is the net force acting on the aircraft. The work-energy theorem (Equation 7-16) gives $\Delta K = K - 0 = W_{\text{net}} = F_{\text{net}} \Delta x = (1.1 \times 10^6 \text{ N})(76 \text{ m}) = 83.6 \text{ MJ}$, for an aircraft starting from rest. (b) From Equation 7-15, $v = \sqrt{2K/m} = \sqrt{2(83.6 \text{ MJ})/(3.3 \times 10^4 \text{ kg})} = 71.2 \text{ m/s} = 256 \text{ km/h}$. (c) From Equation 2-9, $t = 2(x - x_0)/(v_0 + v) = 2(76 \text{ m})/(0 + 71.2 \text{ m/s}) = 2.14 \text{ s}$. (d) The acceleration can be calculated from $F_{\text{net}}/m = 33.3 \text{ m/s}^2$, or from Equation 2-7, $a = v/t$. In fact, $v/t = v/(2 \Delta x/v) = v^2/2 \Delta x = (2K/m)/(2 \Delta x) = F_{\text{net}}/m$.

Section 7-6: Power

Problem

53. A typical car battery stores about 1 kWh of energy. What is its power output if it is drained completely in (a) 1 minute; (b) 1 hour; (c) 1 day?

Solution

The average power (Equation 7-17) is (a) $\bar{P} = (1 \text{ kWh})/(1 \text{ h}/60) = 60 \text{ kW}$; (b) 1 kWh/1 h = 1 kW; and (c) 1 kWh/24 h = 41.7 W.

Problem

55. How much work can a 3.5-hp lawnmower engine do in 1 h?

Solution

Working at constant power output, Equation 7-17 gives the total work (energy output) as $\Delta W = \bar{P} \Delta t = (3.5 \text{ hp}) (746 \text{ W/hp})(3600 \text{ s}) = 9.40 \text{ MJ}$. (Note the change to appropriate SI units.)

Problem

57. A "mass driver" is designed to launch raw material mined on the moon to a factory in lunar orbit. The driver can accelerate a 1000-kg package to 2.0 km/s (just under lunar escape speed) in 55 s. (a) What is its power output during a launch? (b) If the driver makes one launch every 30 min, what is its average power consumption?

Solution

(a) The work done by the driver during a launch is equal to the change in kinetic energy of the package, so the power output is $P = W/t = \Delta K/t = \frac{1}{2}(10^3 \text{ kg})(2 \text{ km/s})^2/55 \text{ s} = 36.4 \text{ MW}$. (b) The same work (one launch) spread over 30 min yields an average power of $P_{\text{av}} = \Delta W/\Delta t = 2 \times 10^9 \text{ J} \div 30 \times 60 \text{ s} = 1.11 \text{ MW}$. (Note: we are neglecting any work done by lunar gravity during the launch, which would be a small correction to W.)

Problem

59. Estimate your power output as you do deep knee bends at the rate of one per second.

Solution

The work done against gravity in raising or lowering a weight through a height, h, has magnitude mgh. The body begins and ends each deep knee bend at rest ($\Delta K = 0$), so the muscles do a total work (down and up) of $2mgh$ for each complete repetition. If we assume that the lower extremities comprise 35% of the body mass, and are not included in the moving mass, then mg, for a 75 kg person, is about $0.65(75 \text{ kg}) \times (9.8 \text{ m/s}^2) = 480 \text{ N}$. We guess that h is somewhat greater than 25% of the body height, or about 45 cm, so the muscle power output for one repetition per second is about $2(480 \text{ N})(0.45 \text{ m})/\text{s} = 430 \text{ W}$.

Problem

61. In midday sunshine, solar energy strikes Earth at the rate of about 1 kW/m^2. How long would it take a perfectly efficient solar collector of 15 m^2 area to collect 40 kWh of energy? (This is roughly the energy content of a gallon of gasoline.)

Solution

The average power received by the collector is $(1 \text{ kW/m}^2)(15 \text{ m}^2) = 15$ kW, so it would take $\Delta t = \Delta W / \bar{P} = 40 \text{ kWh}/15 \text{ kW} = 2.67$ h to collect the required energy. (The average intensity of sunlight is discussed in more detail in Chapter 34.)

Problem

63. The rate at which the United States imports oil, expressed in terms of the energy content of the imported oil, is nearly 700 GW. Using the "Energy Content of Fuels" table in Appendix C, convert this figure to gallons per day.

Solution

Appendix C lists the energy content of oil as 39 kWh/gal. Therefore, the import rate is $(700 \text{ GW})(1 \text{ gal}/39 \text{ kWh}) \times (24 \text{ h/d})$, or roughly 430 million gallons per day.

Solution

The energy consumption of each device is $\Delta W = \bar{P} \Delta t$; $(1.2 \text{ kW})(1 \text{ h}/6) = 0.2$ kWh for the hair dryer and slightly less, $(7 \text{ W})(24 \text{ h}) = 0.168$ kWh, for the night light.

Problem

65. By measuring oxygen uptake, sports physiologists have found that the power output of long-distance runners is given approximately by $P = m(bv - c)$, where m and v are the runner's mass and speed, respectively, and where b and c are constants given by $b = 4.27$ J/kg·m and $c = 1.83$ W/kg. Determine the work done by a 54-kg runner who runs a 10-km race at a speed of 5.2 m/s.

Solution

The runner's average power output is $\bar{P} = (54 \text{ kg}) \times$ [(4.27 J/kg·m)(5.2 m/s) − 1.83 W/kg] = 1.10 kW. Over the race time, $10 \text{ km}/(5.2 \text{ m/s}) = 1.92 \times 10^3$ s, the runner's work output is $\Delta W = \bar{P} \Delta t = 2.12$ MJ = 0.588 kWh.

Problem

67. A 1400-kg car ascends a mountain road at a steady 60 km/h. The force of air resistance on the car is 450 N. If the car's engine supplies energy to the drive wheels at the rate of 38 kW, what is the slope angle of the road?

Solution

At constant velocity, there is no change in kinetic energy, so the net work done on the car is zero. Therefore, the power supplied by the engine equals the power expended against gravity and air resistance. The latter can be found from Equation 7-21, since gravity makes an angle of $\theta + 90°$ with the velocity (where θ is the slope angle to the horizontal), while air resistance makes an angle of 180° to the velocity. Then 38 kW $= -\mathbf{F}_g \cdot \mathbf{v} - \mathbf{F}_{air} \cdot \mathbf{v} = -mgv \cos(\theta + 90°) - F_{air} v \cos(180°) = mgv \sin\theta + F_{air} v$, or $\theta = \sin^{-1} \times$ [((38 kW/60 km/h) − 450 N)/(1400 kg $\times$ 9.8 m/s^2)] = 7.67°. (See Example 7-14 and use care with SI units and prefixes.)

Paired Problems

Problem

69. You apply a 470-N force to push a stalled car at a 17° angle to its direction of motion, doing 860 J of work in the process. How far do you push the car?

Solution

Equation 7-2 or 7-5 gives the work done by the applied force; hence $\Delta r = W/F \cos\theta = 860 \text{ J}/(470 \text{ N} \times \cos 17°) = 1.91$ m.

Problem

71. A force pointing in the x direction is given by $F = F_0(x/x_0)$, where F_0 and x_0 are constants, and x is the position. Find an expression for the work done by this force as it acts on an object moving from $x = 0$ to $x = x_0$.

Solution

The work done in this one-dimensional situation is given by Equation 7-8:

$$W = \int_0^{x_0} \left(\frac{F_0}{x_0}\right) x\, dx = \left(\frac{F_0}{x_0}\right) \frac{x_0^2}{2} = \frac{1}{2} F_0 x_0.$$

Problem

73. Two vectors have equal magnitude, and their scalar product is one-third of the square of their magnitude. Find the angle between them.

Solution

We are given that $A^2 = B^2 = 3\mathbf{A} \cdot \mathbf{B} = 3AB \cos\theta$. Therefore $\theta = \cos^{-1}(1/3) = 70.5°$.

Problem

75. A 460-kg piano is pushed at constant speed up a ramp, raising it a vertical distance of 1.9 m (see Fig. 7-32). If the coefficient of friction between ramp and piano is 0.62, find the work done by the agent pushing the piano if the ramp angle is (a) 15° and (b) 30°. Assume the force is applied parallel to the ramp.

Solution

The usual relevant forces on an object pushed up an incline of length $\ell = h/\sin\theta$ by an applied force parallel to the slope are shown in the sketch. At constant velocity, the acceleration is zero, so the parallel and perpendicular components of Newton's second law, together with the empirical relation for kinetic friction, give $N = mg\cos\theta$, $f_k = \mu_k N$, and $F_{app} = mg\sin\theta + f_k = mg(\sin\theta + \mu_k\cos\theta)$. Thus, the work done by the applied force is $W_{app} = F_{app}\ell = F_{app}h/\sin\theta = mgh(1 + \mu_k\cot\theta)$, where h is the vertical rise. (a) Evaluating the above expression using the data supplied, we find $W_{app} = (460\text{ kg})\times(9.8\text{ m/s}^2)(1.9\text{ m})(1 + 0.62\cot 15°) = 28.4$ kJ. (b) When 15° is replaced by 30° in the above calculation, we find $W_{app} = 17.8$ kJ. (The work done against gravity is the same in parts (a) and (b) since h is the same, but the work done against friction is greater in (a) because the incline is longer and the normal force is greater; however F_{app} is less.)

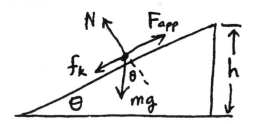

Problem 75 Solution.

Problem

77. (a) How much power is needed to push a 95-kg chest at 0.62 m/s along a horizontal floor where the coefficient of friction is 0.78? (b) How much work is done in pushing the chest 11 m?

Solution

If you push parallel to the floor at constant velocity, the normal force on the chest equals its weight, $N = mg$, and the applied force equals the frictional force, $F_{app} = f_k = \mu_k N = \mu_k mg$. (a) The power required is (Equation 7-21) $P_{app} = F_{app}v = (0.78)\times(95\text{ kg})(9.8\text{ m/s}^2)(0.62\text{ m/s}) = 450$ W, or about

0.6 hp. (b) The work done by the applied force acting over a displacement $\Delta x = 11$ m is $W_{app} = F_{app}\Delta x = P_{app}(\Delta x/v)$, where $t = \Delta x/v$ is the time over which the power is applied. Using either expression, we find $W_{app} = 7.99$ kJ.

Supplementary Problems

Problem

79. The power output of a machine of mass m increases linearly with time, according to the formula $P = bt$, where b is a constant. (a) Find an expression for the work done between $t = 0$ and some arbitrary time t. (b) Suppose the machine is initially at rest and all the work it supplies goes into increasing its own speed. Use the work-energy theorem to show that the speed increases linearly with time, and find an expression for the acceleration.

Solution

(a) From Equation 7-20, $W = \int_0^t P\,dt' = \int_0^t bt'\,dt' = \frac{1}{2}bt^2$. (We used t' for the dummy variable of integration.) (b) If we assume that $W = W_{net} = \Delta K$, then $\frac{1}{2}bt^2 = \frac{1}{2}mv^2$, since the machine starts from rest. Thus $v = \sqrt{b/m}\,t$ and $a = dv/dt = \sqrt{b/m}$. (v is the speed and a is the tangential acceleration along the path of the machine.)

Problem

81. The per-capita energy consumption rate plotted in Fig. 7-21 can be approximated by the expression $P = P_0 + at + bt^2 + ct^3$, where $P_0 = 4.4$ kW, $a = -5.57\times10^{-2}$ kW/y, $b = 3.84\times10^{-3}$ kW/y^2, $c = -2.79\times10^{-5}$ kW/y^3, and t is the time in years since 1900 (i.e., 1960 is $t = 60$). Integrate this expression to find approximate values for the energy used per capita during the decades (a) from 1940 to 1950 and (b) from 1960 to 1970. It's easiest to give your answer in kilowatt-years.

Solution

The energy used between times t_1 and t_2 is

$$W = \int_{t_1}^{t_2} P\,dt = \int_{t_1}^{t_2} (P_0 + at + bt^2 + ct^3)dt$$
$$= P_0(t_2 - t_1) + \tfrac{1}{2}a(t_2^2 - t_1^2) + \tfrac{1}{3}b(t_2^3 - t_1^3) + \tfrac{1}{4}c(t_2^4 - t_1^4).$$

(a) Using $t_1 = 40$ y, $t_2 = 50$ y, and the given coefficients, we find the energy used in the 1940s to be $W = 71.3$ kW·y. (b) A similar calculation for the 1960s gives $W = 93.3$ kW·y.

Problem

83. Figure 7-33 shows the power a baseball bat delivers to the ball, as a function of time. Use graphical integration to determine the total work the bat does on the ball.

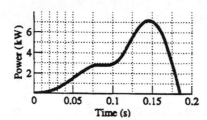

FIGURE 7-33 Problem 83.

Solution

The work done on the ball, $W = \int P\,dt$, is the area under the graph of power versus time, where each small rectangle in Figure 7-33 has an "area" of $(0.01\text{ s})(1\text{ kw}) = 10$ J. There are approximately 54 or 55 rectangles under the curve, so the work done was about 545 J.

Problem

85. An unusual spring has the force-distance curve shown in Fig. 7-34 and described by $F = 100x^2$ for $0 \le x \le 1$ and $F = 100(4x - x^2 - 2)$ for $1 \le x \le 2$, where x is the displacement in meters from the spring's unstretched length, and F is in newtons.

Find the work done in stretching this spring (a) from $x = 0$ to $x = 1$ m and (b) from $x = 1$ m to $x = 2$ m.

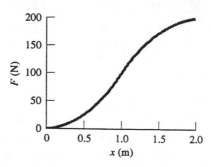

FIGURE 7-34 Problem 85.

Solution

The spring force is a restoring force (opposite to the displacement); the work we must do (against this force) to stretch the spring is $\int F\,dx$. (a) Between $x = 0$ and 1:

$$W_{(a)} = \int_0^1 100x^2\,dx = 100\left.\frac{1}{3}x^3\right|_0^1 = 33.3 \text{ J.}$$

(b) Between $x = 1$ and 2:

$$W_{(b)} = \int_1^2 100(4x - x^2 - 2)dx$$

$$= 100\left|2x^2 - \frac{1}{3}x^3 - 2x\right|_1^2 = 167 \text{ J.}$$

CHAPTER 8   CONSERVATION OF ENERGY

ActivPhysics can help with these Problems:
All Activities in Section 5, Work and Energy

Section 8-1:   Conservative and Nonconservative Forces

Problem

1. Determine the work done by the frictional force in moving a block of mass m from point 1 to point 2 over the two paths shown in Fig. 8-26. The coefficient of friction has the constant value μ over the surface. (The diagram lies in a horizontal plane.)

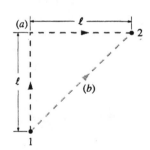

FIGURE 8-26 Problems 1, 2.

Solution

Figure 8-26 is a plane view of the horizontal surface over which the block is moved, showing the paths (a) and (b). The force of friction is μmg opposite to the displacement $(\mathbf{f} \cdot d\mathbf{r} = -f\,dr)$, so $W_{(a)} = -\mu mg(\ell + \ell) = -2\mu mg\ell$, and $W_{(b)} = -\mu mg\sqrt{\ell^2 + \ell^2} = -\sqrt{2}\mu mg\ell$. Since the work done depends on the path, friction is not a conservative force.

Problem

3. The force in Fig. 8-22a is given by $\mathbf{F} = F_0\hat{\mathbf{j}}$, where F_0 is a constant. The force in Fig. 8-22b is given by $\mathbf{F} = F_0(x/a)\hat{\mathbf{j}}$, where the origin is taken at the lower left corner of the box, a is the width of the square box, and the distance x increases horizontally to the right. Determine the work done by $\mathbf{F}$ on an object moved counterclockwise around each box, starting at the lower left corner.

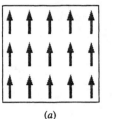 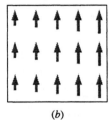

(a)   (b)

FIGURE 8-22 Problem 3.

Solution

The path around the square consists of four segments, each along a side, in the direction of a counterclockwise circulation. Thus

$$W = \oint \mathbf{F} \cdot d\mathbf{r} = \int_0^a \mathbf{F}(y=0) \cdot \hat{\mathbf{i}}\,dx + \int_0^a \mathbf{F}(x=a) \cdot \hat{\mathbf{j}}\,dy$$
$$+ \int_a^0 \mathbf{F}(y=a) \cdot \hat{\mathbf{i}}\,dx + \int_a^0 \mathbf{F}(x=0) \cdot \hat{\mathbf{j}}\,dy$$

(Note that a path parallel to the x-axis from right to left is represented by $d\mathbf{r} = \hat{\mathbf{i}}\,dx$ with x going from a to 0 in the limits of integration, etc.) (a) For $\mathbf{F} = F_0\hat{\mathbf{j}}$, the expression for the work becomes

$$W = 0 + F_0 \int_0^a dy + 0 + F_0 \int_a^0 dy = F_0a - F_0a = 0.$$

(b) For $\mathbf{F} = F_0(x/a)\hat{\mathbf{j}}$, the work around the square is

$$W = 0 + F_0\left(\frac{a}{a}\right)\int_0^a dy + 0 + 0 = F_0a.$$

Section 8-2:   Potential Energy

Problem

5. Find the potential energy of a 70-kg hiker (a) atop New Hampshire's Mount Washington, 1900 m above sea level, and (b) in Death Valley, California, 86 m below sea level. Take the zero of potential energy at sea level.

Solution

If we define the zero of potential energy to be at zero altitude ($y = 0$), then $U(0) = 0$, and Equation 8-3 (for the gravitational potential energy near the surface of the Earth, $|y| \ll 6370$ km) gives $U(y) - U(0) = U(y) = mg(y - 0) = mgy$. Therefore, (a) $U(1900$ m$) =$

$(70{\times}9.8$ N$)(1900$ m$) = 1.30$ MJ, and (b) $U(-86$ m$) = (70{\times}9.8$ N$)(-86$ m$) = -59.0$ kJ.

Problem

7. Show using Equation 8-2b that the potential energy difference between the ground and a distance h above the ground is mgh regardless of whether you choose the y-axis upward or downward.

Solution

Equation 8-2b gives the potential energy difference for a constant force in the y direction, $\Delta U = U(y_2) - U(y_1) = -F_y(y_2 - y_1)$. If you take the y-axis upward, with the ground at y_1, the gravitational force is $F_y = -mg$, while a point a distance h above ground is $y_2 = h + y_1$. Then $\Delta U = -(-mg)h = mgh$. On the other hand, if the y-axis is downward, then $F_y = mg$ but $y_2 = -h + y_1$, so $\Delta U = -(mg)(-h) = mgh$ is the same.

Problem

9. A 1.50-kg brick measures 20.0 cm$\times$8.00 cm$\times$ 5.50 cm. Taking the zero of potential energy when the brick lies on its broadest face, what is the potential energy (a) when the brick is standing on end and (b) when it is balanced on its 8-cm edge, with its center directly above that edge? *Note:* You can treat the brick as though all its mass is concentrated at its center.

Solution

The center of the brick is a distance $\Delta y = 10$ cm $- 2.75$ cm $= 7.25$ cm above the zero of potential energy in position (a), and $\Delta y = \frac{1}{2}\sqrt{(20\text{ cm})^2 + (5.5\text{ cm})^2} - 2.75$ cm $= 10.4$ cm $- 2.75$ cm $= 7.62$ cm in position (b) (see sketch; the center is midway along the diagonal of the face of the brick). From Equation 8-3, the gravitational potential energy is $U_a = mg\,\Delta y = (1.5{\times}9.8$ N$)(7.25$ cm$) = 1.07$ J and $U_b = (1.5{\times}9.8$ N$)(7.62$ cm$) = 1.12$ J above the zero energy.

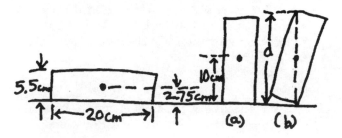

Problem 9 Solution.

Problem

11. How much energy can be stored in a spring with $k = 320$ N/m if the maximum allowed stretch is 18 cm?

Solution

From Equation 8-4, $U = \frac{1}{2}kx^2 = \frac{1}{2}(320$ N/m$){\times}$ $(0.18$ m$)^2 = 5.18$ J is the maximum potential energy of the spring whose stretch is not greater than 18 cm.

Problem

13. How far would you have to stretch a spring of spring constant $k = 1.4$ kN/m until it stored 210 J of energy?

Solution

Assuming one starts stretching from the unstretched position ($x = 0$), Equation 8-4 gives $x = \sqrt{2U/k} = \sqrt{2(210\text{ J})/(1.4\text{ kN/m})} = 54.8$ cm.

Problem

15. The force exerted by an unusual spring when it's compressed a distance x from equilibrium is given by $F = -kx - cx^3$, where $k = 220$ N/m, and $c = 3.6$ kN/m^3. Find the energy stored in this spring when it's been compressed 15 cm.

Solution

For a one-dimensional force, one can use Equation 8-2a to find $U(x) - U(0) = -\int_0^{15\,\text{cm}}(-kx - cx^3)dx = \left|\frac{1}{2}kx^2 + \frac{1}{4}cx^4\right|_0^{15\,\text{cm}} = \frac{1}{2}(220$ N/m$)(0.15$ m$)^2 + \frac{1}{4}(3.6$ kN/m$^3)(0.15$ m$)^4 = 2.93$ J. Since this energy could be recaptured from the spring, in the form of kinetic energy imparted to a mass, for example, it is referred to as stored energy.

Problem

17. A particle moves along the x-axis under the influence of a force $F = ax^2 + b$, where a and b are constants. Find its potential energy as a function of position, taking $U = 0$ at $x = 0$.

Solution

Equation 8-2a, with $U(0) = 0$, gives

$$U(x) = -\int_0^x F_x\,dx' = -\int_0^x (ax'^2 + b)dx'$$

$$= -\frac{1}{3}ax^3 - bx.$$

Problem

19. The force exerted by a rubber band is given approximately by

$$F = F_0 \left[\frac{\ell + x}{\ell} - \frac{\ell^2}{(\ell + x)^2} \right],$$

where ℓ is the unstretched length, and F_0 is a constant. Find the potential energy of the rubber band as a function of the distance x it is stretched. Take the zero of potential energy in the unstretched position.

Solution

Assuming that the direction of F is opposite to the displacement, we find:

$$U = -\int_0^x \mathbf{F} \cdot d\mathbf{x}' = F_0 \int_0^x \left[\frac{\ell + x'}{\ell} - \frac{\ell^2}{(\ell + x')^2} \right] dx'$$

$$= F_0 \left| x' + \frac{x'^2}{2\ell} + \frac{\ell^2}{(\ell + x')} \right|_0^x$$

$$= F_0 \left[x + \frac{x^2}{2\ell} + \frac{\ell^2}{(\ell + x)} - \ell \right].$$

(This problem is identical to Problem 7-33. x' is a dummy variable.)

Section 8-3:   Conservation of Mechanical Energy

Problem

21. A Navy jet of mass 10,000 kg lands on an aircraft carrier and snags a cable to slow it down. The cable is attached to a spring with spring constant 40,000 N/m. If the spring stretches 25 m to stop the plane, what was the landing speed of the plane?

Solution

If we assume no change in gravitational potential energy, $K_{initial} = \frac{1}{2}mv^2 = U_{final} = \frac{1}{2}kx^2$, or $v = \sqrt{k/m} \, x$. Thus, $v = \sqrt{(40,000 \text{ N/m})/(10,000 \text{ kg})} \times (25 \text{ m}) = 50$ m/s.

Problem

23. A 120-g arrow is shot vertically from a bow whose effective spring constant is 430 N/m. If the bow is drawn 71 cm before shooting the arrow, to what height does the arrow rise?

Solution

If we ignore losses in energy due to air resistance etc., the mechanical energy of the bow and arrow is the same just before shooting the arrow and when the arrow is at its maximum height, $U_0 + K_0 = U + K$. Before shooting, $K_0 = 0$, and $U_0 = \frac{1}{2}kx^2 + mgy_0$, the potential energy of the taut bow plus the gravitational energy of the arrow at the initial position. At the maximum height, $K = 0$ (instantaneously) and $U = mgy_{max}$. Therefore, $\frac{1}{2}kx^2 + mgy_0 = mgy_{max}$, or $y_{max} - y_0 = kx^2/2mg = (430 \text{ N/m})(0.71 \text{ m})^2 \div 2(0.12 \text{ kg})(9.8 \text{ m/s}^2) = 92.2$ m. (It is assumed that any change in the gravitational potential energy of the bow is negligible.)

Problem

25. With $x - x_0 = h$ and $a = g$, Equation 2-11 gives the speed of an object thrown downward with initial speed v_0 after it has dropped a distance h: $v = \sqrt{v_0^2 + 2gh}$. Use conservation of energy to derive the same result.

Solution

Gravity is a conservative force, so for free fall near the Earth's surface, $U_0 + K_0 = U + K$, or $\frac{1}{2}mv_0^2 + mgy_0 = \frac{1}{2}mv^2 + mgy$. (We neglect the effects of air resistance.) Here, the y-axis is positive upward, so that the gravitational potential energy, relative to zero potential energy at $y = 0$, has the usual form mgy. Then the distance dropped is $y_0 - y = h$, instead of $x - x_0 = h$, and $a_y = -g$. Canceling m, and rearranging terms, one recaptures the stated result.

Problem

27. Two clever kids use a huge spring with $k = 890$ N/m to launch their toboggan at the top of a 9.5-m-high hill (Fig. 8-28). The mass of kids plus toboggan is 80 kg. If the kids manage to compress the spring 2.6 m, (a) what will be their speed at the bottom of the hill? (b) What fraction of their final kinetic energy was initially stored in the spring? Neglect friction.

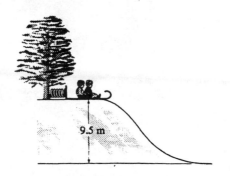

FIGURE 8-28 Problem 27.

Solution

If friction is neglected everywhere (in the spring, on the snow, through the air, etc.), the mechanical energy of the kids plus toboggan (including potential energy of gravitation and the spring, as well as kinetic energy) is conserved: $U_{\text{top}} + K_{\text{top}} = U_{\text{bot}} + K_{\text{bot}}$. At the top of the hill, $K_{\text{top}} = 0$ (the toboggan starts from rest), and $U_{\text{top}} = U_{\text{top},g} + U_{\text{top},s} = mgy_{\text{top}} + \frac{1}{2}kx^2$, while at the bottom, $K_{\text{bot}} = \frac{1}{2}mv^2$ and $U_{\text{bot}} = mgy_{\text{bot}}$ (since $U_{\text{bot},s} = 0$ when the spring is no longer compressed). (b) Using the given data, we find $K_{\text{bot}} = mg(y_{\text{top}} - y_{\text{bot}}) + \frac{1}{2}kx^2 = (80\text{ kg})(9.8\text{ m/s}^2)(9.5\text{ m}) + \frac{1}{2}(890\text{ N/m})(2.6\text{ m})^2 = 7.45\text{ kJ} + 3.01\text{ kJ} = 10.5\text{ kJ}$, so $U_{\text{top},s}/K_{\text{bot}} = 3.01\text{ kJ}/10.5\text{ kJ} = 28.8\%$. (a) The speed at the bottom is $v_{\text{bot}} = \sqrt{2K_{\text{bot}}/m} = \sqrt{2(10.5\text{ kJ})/(80\text{ kg})} = 16.2\text{ m/s}$.

Problem

29. An initial speed of 2.4 km/s (the "escape speed") is required for an object launched from the moon to get arbitrarily far from the moon. At a mining operation on the moon, 1000-kg packets of ore are to be launched to a smelting plant in orbit around the Earth. If they are launched with a large spring whose maximum compression is 15 m, what should be the spring constant of the spring?

Solution

For an ideal spring (without losses), $\frac{1}{2}ky^2 = \frac{1}{2}mv_B^2 + mgy$. (This is Equation 8-7 for points A and B in the sketch.) Since the gravitational potential energy change is negligible compared to the other terms, $k \simeq m(v_B/y)^2 = 10^3\text{ kg}(2.4\text{ km/s} \div 15\text{ m})^2 = 25.6\text{ MN/m}$. (Note: The surface gravity on the moon is 1.62 m/s^2, so the maximum change in potential energy of a packet is only $mgy = 24.3\text{ kJ}$, while its kinetic energy, $\frac{1}{2}mv_B^2 = 2.88\text{ GJ}$, is more than 10^5 times larger. Likewise, the gravitational potential energy of the spring is negligible.)

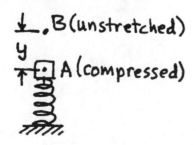

Problem 29 Solution.

Problem

31. A low-damage bumper on a 1400-kg car is mounted on springs whose total effective spring constant is 7.0×10^5 N/m. The springs can undergo a maximum compression of 16 cm without damage to the bumper, springs, or car. What is the maximum speed at which the car can collide with a stationary object without sustaining damage?

Solution

The maximum energy stored by the springs, $\frac{1}{2}kx^2$, without incurring damage, is equal to the maximum kinetic energy, $\frac{1}{2}mv^2$, of the car before a collision. (We assume a level road and horizontal collision so that there is no change in the car's gravitational potential energy.) Thus, $\frac{1}{2}kx^2 = \frac{1}{2}mv^2$, or $v = \sqrt{k/m}\, x = \sqrt{(7.0 \times 10^5\text{ N/m})/(1400\text{ kg})}\,(0.16\text{ m}) = 3.58\text{ m/s} = 12.9\text{ km/h}$.

Problem

33. Show that the rope in Example 8-6 can remain taut all the way to the top of its smaller loop only if $a \leq \frac{2}{5}\ell$. (Note that the maximum release angle is 90° for the rope to be taut on the way down.)

Solution

For the rope to be taut at the top of the small circle, the tension must be greater than (or equal to) zero; $T = mv_{\text{top}}^2/a - mg \geq 0$, or $v_{\text{top}}^2 \geq ga$. Therefore, the mechanical energy at the top is greater than (or equal to) a corresponding value: $E = K + U = \frac{1}{2}mv_{\text{top}}^2 + mg(2a) \geq \frac{1}{2}m(ga) + 2mga = \frac{5}{2}mga$, where the zero of potential energy is the lowest point (as in Example 8-6). Since energy is conserved, $E = U_0 = mg\ell(1 - \cos\theta_0)$, where θ_0 is the release angle (see Example 8-6 again), so $a \leq 2E/5mg = \frac{2}{5}\ell(1 - \cos\theta_0)$. The greatest upper limit for the radius a corresponds to the maximum release angle, as stated in the question above, since $\cos 90° = 0$.

Problem

35. A 2.0-kg mass rests on a frictionless table and is connected over a frictionless pulley to a 4.0-kg mass, as shown in Fig. 8-31. Use conservation of energy to calculate the speed of the masses after they have moved 50 cm.

Solution

Since friction is absent by assumption, mechanical energy is conserved, and $U_0 + K_0 = U_1 + K_1$, where the subscripts 0 and 1 refer to the initial position (both masses at rest) and the final position (after each

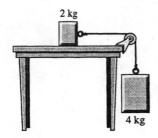

FIGURE 8-31 Problem 35.

has moved 50 cm), respectively. Then $U_0 - U_1 = K_1 =$ (4 kg)(9.8 m/s^2)(0.5 m) $= \frac{1}{2}$(2 kg + 4 kg)v^2, since the potential energy changes only for the falling mass, and both masses have the same speed. Solving for v, we find $v = 2.56$ m/s. (The kinetic energy of the pulley was assumed to be negligible.)

Problem

37. A mass m is dropped from a height h above the top of a spring of constant k that is mounted vertically on the floor (Fig. 8-33). Show that the maximum compression of the spring is given by $(mg/k)(1 + \sqrt{1 + 2kh/mg})$. What is the significance of the other root of the quadratic equation?

Solution

If the maximum compression is y, as shown, and we measure gravitational potential energy from the lowest point, B, then the conservation of energy between point A and B requires that $mg(h + y) = \frac{1}{2}ky^2$. The quadratic formula can be used to find $y = (mg/k)\times$ $(1 \pm \sqrt{1 + 2kh/mg})$. Only positive values of y are physically meaningful in this problem, because the spring is not compressed unless $y > 0$.

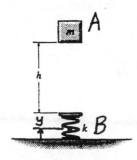

FIGURE 8-33 Problem 37 Solution.

Section 8-4:   Potential Energy Curves
Problem

39. A particle slides back and forth on a frictionless track whose height as a function of horizontal

position x is given by $y = ax^2$, where $a = 0.92$ m^{-1}. If the particle's maximum speed is 8.5 m/s, find the turning points of its motion.

Solution

With no friction, the only forces acting are gravity, a conservative force with potential energy $U = mgy$ above a reference level at $y = 0$, and the normal contact force of the track. Although the latter is nonconservative, it is always perpendicular to the displacement along the track (by definition); hence it does no work on the particle, and the mechanical energy is conserved: $\frac{1}{2}mv^2 + mgy =$ constant. Since the height $y = ax^2$ is given in terms of the horizontal displacement, we may write $\frac{1}{2}mv^2 + mgax^2 =$ constant, or $v^2 + 2gax^2 =$ (a different) constant. One can see that the maximum speed occurs when the displacement is a minimum (zero, in fact) and vice-versa. Thus, one can determine the constant two ways: $v^2 + 2gax^2 = v_{max}^2 = 2gax_{max}^2$. The turning points are the places where the velocity is instantaneously zero (a minimum) and hence are given by $x_{max} = \pm\sqrt{v_{max}^2/2ga} =$ $\pm\sqrt{(8.5 \text{ m/s})^2/2(9.8 \text{ m/s}^2)(0.92 \text{ m}^{-1})} = \pm 2.00$ m. (In this case, the particle's motion is oscillatory, back and forth between the turning points, with maximum speed, either forward or backward, at the middle, $x = 0$.)

Problem

41. The potential energy associated with a conservative force is shown in Fig. 8-35. Consider particles with total energies $E_1 = -1.5$ J, $E_2 = -0.5$ J, $E_3 = 0.5$ J, $E_4 = 1.5$ J, and $E_5 = 3.0$ J. Discuss the subsequent motion, including the approximate location of any turning points, if the particles are initially at point $x = 1$ m and moving in the $-x$ direction.

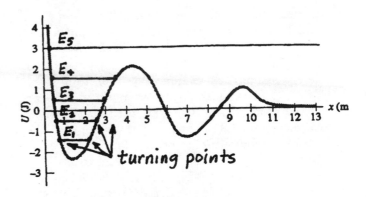

FIGURE 8-35 Problem 41 Solution.

Solution

All the particles start in the left-hand potential well and reverse direction when they hit the left-most (infinite) potential barrier. The first four particles have insufficient energy to escape from this well (the height of the next barrier is about 2.0 J $> E_4$) and experience a second turning point (between $x = 2.0$ m and 4.0 m, depending on energy). The fifth particle's motion is unbounded.

Problem

43. (a) Derive an expression for the potential energy of an object subject to a force $F_x = ax - bx^3$, where $a = 5$ N/m, and $b = 2$ N/m^3, taking $U = 0$ at $x = 0$. (b) Graph the potential energy curve for $x > 0$, and use it to find the turning points for an object whose total energy is -1 J.

Solution

(a) The force is conservative (any one-dimensional force given by an integrable function of position is), so the potential energy can be found from Equation 8-2a:

$$U(x_2) - U(x_1) = -\int_{x_1}^{x_2} (ax - bx^3)\, dx$$

$$= -\frac{a}{2}(x_2^2 - x_1^2) + \frac{b}{4}(x_2^4 - x_1^4).$$

If we define the zero of potential energy at $x = 0$, then $U(x) = -\frac{1}{2}ax^2 + \frac{1}{4}bx^4$. (b) A graph of $U(x)$ for $x \geq 0$, when $a = 5$ N/m, $b = 2$ N/m^3 and x is in meters, is shown. (Note that the potential energy is symmetric, $U(-x) = U(x)$, but that only positive displacements are considered in this problem.) The conservation of energy can be written in terms of the total energy, $E = \frac{1}{2}m(dx/dt)^2 + U(x)$, so that $dx/dt = \pm\sqrt{2[E - U(x)]/m}$. The maximum speed occurs when $U(x)$ is a minimum; i.e., $dU/dx = 0$, and $d^2U/dx^2 > 0$. Taking the derivative, one finds $0 = -ax + bx^3$, which has solutions $x = 0$ and $x = \pm\sqrt{a/b} = \pm\sqrt{5/2}$ m $= \pm 1.58$ m. The second derivative $d^2U/dx^2 = -a + 3bx^2$ is negative for $x = 0$, which is a local maximum, but is positive for $x = \pm\sqrt{a/b}$, which are minima with $U_{min} = U(\pm\sqrt{a/b}) = -a^2/4b = -(25/8)$ J $= -3.13$ J. There is real physical motion ($K \geq 0$) for total energy $E \geq U_{min}$. The turning points (where $dx/dt = 0$) can be found from the equation $U(x) = E$; there are four solutions (two positive) for energies with $U_{min} < E < 0$, and two solutions (one positive) for $E > 0$. The equation $U(x) - E = 0$ is equivalent to $x^4 - 2(a/b)x^2 - 4(E/b) = 0$. The quadratic formula gives $x = \pm\{(a/b) \pm [(a/b)^2 + 4(E/b)]^{1/2}\}^{1/2}$ for $U_{min} < E < 0$, and $x = \pm\{(a/b) + [(a/b)^2 + 4(E/b)]^{1/2}\}^{1/2}$ for $E > 0$. For the particular values given ($E = -1$ J), the

positive turning points are $x = \sqrt{(5 \pm \sqrt{17})/2}$ m $= 0.662$ m and 2.14 m, as can be seen in the graph.

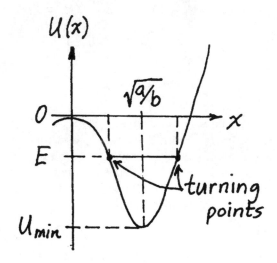

Problem 43 Solution.

Section 8-5: Force and Potential Energy

Note: In the following problems, motion is restricted to one dimension.

Problem

45. A particle is trapped in a potential well described by $U(x) = 2.6x^2 - 4$, where U is in joules, and x is in meters. Find the force on the particle when it's at (a) $x = 2.1$ m; (b) $x = 0$ m; and (c) $x = -1.4$ m.

Solution

For one-dimensional motion, Equation 8-8 gives the force $F_x = -dU/dx = -d/dx(1.6x^2 - 4) = -2(1.6)x = -3.2x$, where F_x is in newtons for x in meters. Therefore, (a) $F_x(2.1$ m$) = -(3.2)(2.1)$ N $= -6.72$ N (force in the negative x direction); (b) $F_x(0) = 0$; (c) $F_x(-1.4$ m$) = -(3.2)(-1.4)$ N $= 4.48$ N.

Problem

47. The potential energy associated with a certain conservative force is given by $U = bx^2$, where b is a constant. Show that the force always tends to accelerate a particle toward the origin if b is positive and away from the origin if b is negative.

Solution

The conservative force represented by the one-dimensional potential energy $U(x) = bx^2$ is given by Equation 8-8, $F_x = -dU/dx = -2bx$. x is the displacement from the origin, so this is towards the origin for $b > 0$ and away from the origin for $b < 0$. (The former is called a restoring force.)

Problem

49. The potential energy of a spring is given by $U = ax^2 - bx + c$, where $a = 5.20$ N/m, $b = 3.12$ N, and $c = 0.468$ J, and where x is the *overall* length of the spring (not the stretch). Find (a) the equilibrium length of the spring and (b) the spring constant.

Solution

(a) The natural, or unstretched, length of the spring is the value of x for which the spring force is zero. Thus, $F_x = -dU/dx = -2ax + b = 0$, when $x = b/2a = (3.12$ N$)/2(5.2$ N/m$) = 30$ cm. (b) Hooke's Law defines the spring constant in terms of the stretch (in this case $x - b/2a$). Since $F_x = -2a(x - b/2a)$, the spring constant is $k = 2a = 10.4$ N/m. (Alternatively, $k = -dF_x/dx = d^2U/dx^2$.)

Section 8-6: Nonconservative Forces

Problem

51. A basketball dropped from a height of 2.40 m rebounds to a maximum height of 1.55 m. What fraction of the ball's initial energy is lost to nonconservative forces? Take the zero of potential energy at the floor.

Solution

The energy principle in the form of Equation 8-5 is $W_{nc} = \Delta K + \Delta U$. Since $v = 0$ where the ball is dropped (point A) and at the maximum of its rebound (point B), one has $W_{nc} = \Delta U_{AB} = mg(y_B - y_A)$. This is a loss of energy since $y_B < y_A$. The initial energy was just mgy_A, relative to zero of potential energy on the floor at $y = 0$, so the fraction lost is $mg|y_B - y_A| \div mgy_A = 1 - y_B/y_A = 1 - 1.55/2.40 = 35.4\%$. (The words "loss" and "lost" replace the minus sign in the energy change; i.e., a "35% loss" equals a change of -35%.)

Problem

53. A pumped-storage reservoir sits 140 m above its generating station and holds 8.5×10^9 kg of water. The power plant generates 330 MW of electric power while draining the reservoir over an 8.0-hour period. What fraction of the initial potential energy is lost to nonconservative forces (i.e., does not emerge as electricity)?

Solution

If all the water fell through the same difference in height, the amount of gravitational potential energy released would be $\Delta U = mg\Delta y = (8.5 \times 10^9$ kg$) \times (9.8$ m/s$^2)(140$ m$) = 1.17 \times 10^{13}$ J. The energy generated by the power plant at an average power output of 330 MW over an 8 h period is $(330$ MW$) \times (8 \times 3600$ s$) = 9.50 \times 10^{12}$ J, so the fraction lost is $(11.7 - 9.50)/11.7 = 18.5\%$.

Problem

55. A 2.5-kg block strikes a horizontal spring at a speed of 1.8 m/s, as shown in Fig. 8-38. The spring constant is 100 N/m. If the maximum compression of the spring is 21 cm, what is the coefficient of friction between the block and the surface on which it is sliding?

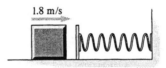

1.8 m/s

FIGURE 8-38 Problem 55.

Solution

The final kinetic energy, the initial potential energy of the spring, and the change in the gravitational potential energy are all zero. Therefore, $W_{nc} = -\mu_K mgx = \Delta K + \Delta U = -\frac{1}{2}mv_0^2 + \frac{1}{2}kx^2 = 0.5[(100$ N/m$)(0.21$ m$)^2 - (2.5$ kg$)(1.8$ m/s$)^2] = -1.85$ J, or $\mu_k = (1.85$ J$)/(2.5$ kg$)(9.8$ m/s$^2)(0.21$ m$) = 0.36$.

Problem

57. A surface is frictionless except for a region between $x = 1$ m, and $x = 2$ m, where the coefficient of friction is given by $\mu = ax^2 + bx + c$, with $a = -2$ m^{-2}, $b = 6$ m^{-1}, and $c = -4$. A block is sliding in the $-x$ direction when it encounters this region. What is the minimum speed it must have to get all the way across the region?

Solution

Assume that the surface is horizontal, so that there are no changes in the block's potential energy ($\Delta U = 0$) and the force of friction on it is $f_k = -\mu_k N = -\mu_k mg$ (opposite to the direction of motion along the x-axis). If the block crosses the entire region from $x_1 = 1$ m to $x_2 = 2$ m, the work-energy theorem demands that

$$W_{nc} = -\int_{x_1}^{x_2} \mu_k mg \, dx = \Delta K = \frac{1}{2}m(v_2^2 - v_1^2)$$

or

$$v_1^2 - v_2^2 = 2g \int_{x_1}^{x_2} (ax^2 + bx + c)\,dx$$

$$= 2g \left| \frac{a}{3}x^3 + \frac{b}{2}x^2 + cx \right|_{1\,\text{m}}^{2\,\text{m}} = 6.53 \text{ m}^2/\text{s}^2.$$

(The given values of a, b, and c were used.) The minimum speed at the start of the region is the value of v_1 when $v_2 = 0$, or $v_{1,\text{min}} = \sqrt{6.53 \text{ m}^2/\text{s}^2} = 2.56$ m/s.

Problem

59. A skier starts from rest at the top of the left-hand peak in Fig. 8-39. What is the maximum coefficient of kinetic friction on the slopes that would allow the skier to coast to the second peak? (Your answer, of course, neglects air resistance.)

Solution

Assume that the maximal frictional force of the slopes and gravity are the only significant forces acting on the skier, who starts from rest on the higher peak (point A) and finishes at rest just reaching the lower peak (point B). Then the work-energy theorem requires that $W_f = \Delta U = mg(y_B - y_A)$. The force of friction on each slope is $f_k = -\mu_k^{\text{max}} mg \cos\theta$, and the displacement along each slope is $\ell = y/\sin\theta$ (measured from the bottom of both slopes), so that work done by friction on each slope is $f_k \ell = (-\mu_k^{\text{max}} mg \cos\theta) \times (y/\sin\theta) = -\mu_k^{\text{max}} mgy \cot\theta$. Then the work-energy theorem becomes $-mg(y_A - y_B) = -\mu_k^{\text{max}} mg \times (y_A \cot\theta_A + y_B \cot\theta_B)$, from which μ_k^{max} can be determined: $\mu_k^{\text{max}} = (950 - 840)/(950 \cot 27° + 840 \cot 35°) = 0.036$.

FIGURE 8-39 Problem 59 Solution.

Problem

61. A 190-g block is launched by compressing a spring of constant $k = 200$ N/m a distance of 15 cm. The spring is mounted horizontally, and the surface directly under it is frictionless. But beyond the equilibrium position of the spring end, the surface

has coefficient of friction $\mu = 0.27$. This frictional surface extends 85 cm, followed by a frictionless curved rise, as shown in Fig. 8-41. After launch, where does the block finally come to rest? Measure from the left end of the frictional zone.

Solution

The energy of the block when it first encounters friction (at point O) is $K_0 = \frac{1}{2}(200 \text{ N/m})(0.15 \text{ m})^2 = 2.25$ J, if we take the zero of gravitational potential energy at that level. Crossing the frictional zone, the block loses energy $\Delta E = W_{nc} = -\mu_k mg\ell = -(0.27) \times (0.19 \text{ kg})(9.8 \text{ m/s}^2)(0.85 \text{ m}) = -0.427$ J. Since $K_0/|\Delta E| = 5.27$, five complete crossings are made, leaving the block with energy $K_0 - 5|\Delta E| = 0.113$ J on the curved rise side. This remaining energy is sufficient to move the block a distance $s = 0.113 \text{ J} \div \mu_k mg = 22.5$ cm towards point O, so the block comes to rest $85 - 22.5 = 62.5$ cm to the right of point O.

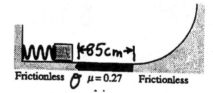

FIGURE 8-41 Problem 61 Solution.

Section 8-7: Conservation of Energy and Mass-Energy

Problem

63. A hypothetical power plant converts matter entirely into electrical energy. Each year, a worker at the plant buys a box of 1-g raisins, and each day drops one raisin into the plant's energy conversion unit. Estimate the average power output of the plant, and compare with a 500-MW coal-burning plant that consumes a 100-car trainload of coal every 3 days.

Solution

The energy equivalent of 1 g of mass is mc^2, so the power generated in one day is $(10^{-3}$ kg$) \times (3 \times 10^8 \text{ m/s})^2/(86,400 \text{ s}) = 1.04$ GW, or approximately twice the output of a 500-MW coal-burning plant. Thus, the mass-energy of one raisin is equivalent to the chemical energy (times the overall efficiency) released by burning 67 car-loads of coal.

Paired Problems

Problem

65. A block slides down a frictionless incline that terminates in a ramp pointing up at a 45° angle, as shown in Fig. 8-42. Find an expression for the horizontal range x shown in the figure, as a function of the heights h_1 and h_2 shown.

Solution

After leaving the ramp (at point 2), with speed v_2 at 45° to the horizontal, the block describes projectile motion with a horizontal range of $x = v_2^2/g$ (see Equation 4-10). Since the track is frictionless (and the normal force does no work), the mechanical energy of the block (kinetic plus gravitational potential) is conserved between point 2 and its start from rest at point 1. Then $\Delta K = \frac{1}{2}mv_2^2 - 0 = -\Delta U = mg(h_1 - h_2)$, and $x = 2(h_1 - h_2)$.

FIGURE 8-42 Problem 65 Solution.

Problem

67. A ball of mass m is being whirled around on a string of length R in a vertical circle; the string does no work on the ball. (a) Show from force considerations that the speed at the top of the circle must be at least $\sqrt{Rg}$ if the string is to remain taut. (b) Show that, as long as the string remains taut, the speed at the bottom of the circle can be no more than $\sqrt{5}$ times the speed at the top.

Solution

(a) At the top of the circle, the forces acting on the mass are gravity and the string tension, both downward and parallel to the centripetal acceleration. Thus $T + mg = mv_{\text{top}}^2/R$. Since $T_{\text{top}} \geq 0$ if the string is taut, $v_{\text{top}}^2 \geq gR$. (See Example 6-8.) (b) The mechanical energy of the mass is conserved, since the tension does no work (by assumption), gravity is conservative, and air resistance is ignored. Thus $U_{\text{top}} + K_{\text{top}} = U_{\text{bot}} + K_{\text{bot}} = \frac{1}{2}mv_{\text{top}}^2 + mgy_{\text{top}} = \frac{1}{2}mv_{\text{bot}}^2 + mgy_{\text{bot}}$, or $v_{\text{bot}}^2 = v_{\text{top}}^2 + 2g(y_{\text{top}} - y_{\text{bot}}) = v_{\text{top}}^2 + 4gR$ (since $y_{\text{top}} - y_{\text{bot}}$ is the diameter of the circle). The result of part (a) then leads to

$v_{\text{bot}}^2 \leq v_{\text{top}}^2 + 4v_{\text{top}}^2 = 5v_{\text{top}}^2$, equivalent to the assertion in the problem.

Problem

69. A pendulum consisting of a mass m on a string of length ℓ is pulled back so the string is horizontal, as shown in Fig. 8-45. The pendulum is then released. Find (a) the speed of the mass and (b) the magnitude of string tension when the string makes a 45° angle with the horizontal.

Solution

(a) We assume that the mechanical energy of the pendulum mass is conserved (neglect possible losses), and that this consists of kinetic and gravitational potential energy. At point 1 where the mass is released, $U_1 = mgy_1$ and $K_1 = 0$, while after a 45° swing to point 2, $U_2 = mgy_2$, and $K_2 = \frac{1}{2}mv_2^2$. Energy conservation implies $K_2 = U_1 - U_2$, or $v_2^2 = 2g(y_1 - y_2) = 2g\ell \sin 45° = \sqrt{2}g\ell$, where we used the trigonometry apparent in Fig. 8-45 to express the difference in height in terms of the length of the string and angle of swing. (b) The string tension is in the direction of the centripetal acceleration (also shown in Fig. 8-45 at point 2), so the radial component of Newton's second law gives $T_2 - mg \sin 45° = mv_2^2/\ell$. Using the result of part (a), we find $T_2 = (mv_2^2/\ell) + mg \sin 45° = (m\sqrt{2}g\ell/\ell) + (mg/\sqrt{2}) = 3mg/\sqrt{2}$.

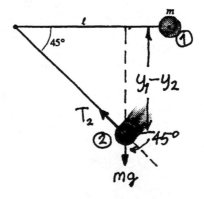

FIGURE 8-45 Problem 69 Solution.

Problem

71. A particle slides back and forth in a frictionless bowl whose height is given by $h(x) = 0.18x^2$, where x and h are both in meters. If the particle's maximum speed is 47 cm/s, find the x coordinates of its turning points.

Solution

This situation is exactly similar to that in Problem 39, where the energy principle led to $\frac{1}{2}mv_{max}^2 = mgh_{max} = mg(0.18/m)x_{max}^2$. Thus, the turning points occur at

$$x_{max} = \pm\sqrt{(0.47 \text{ m/s})^2/2(9.8 \text{ m/s}^2)(0.18/m)} = \pm 25.0 \text{ cm}.$$

Problem

73. A child sleds down a frictionless hill whose vertical drop is 7.2 m. At the bottom is a level but rough stretch where the coefficient of kinetic friction is 0.51. How far does she slide across the level stretch?

Solution

The child starts near the hilltop with $K_A = 0$ and stops on rough level ground, $K_B = 0$, after falling through a potential energy difference $\Delta U = U_B - U_A = -mg(y_A - y_B)$, where $y_A - y_B = 7.2$ m. The work done by friction (on level ground, $N = mg$) is $W_{nc} = -f_k x = -\mu_k mgx$, where x is the distance slid across the rough level stretch. The energy principle, Equation 8-5, relates these quantities: $W_{nc} = -\mu_k mgx = \Delta K + \Delta U = 0 - mg(y_A - y_B)$. Thus $x = (y_A - y_B)/\mu_k = 7.2 \text{ m}/0.51 = 14.1$ m.

Supplementary Problems

Problem

75. A uranium nucleus has a radius of 1.43×10^{-10} m. An alpha particle (mass 6.7×10^{-27} kg) leaves the surface of the nucleus with negligible speed, subject to a repulsive force whose magnitude is $F = A/x^2$, where $A = 4.1 \times 10^{-26}$ N·m^2, and where x is the distance from the alpha particle to the center of the nucleus. What is the speed of the alpha particle when it is (a) 4 nuclear radii from the nucleus; (b) 100 nuclear radii from the nucleus; (c) very far from the nucleus ($x \to \infty$)?

Solution

In Problem 16 above, we found the potential energy difference for a repulsive inverse square force,

$$U(x_2) - U(x_1) = -\int_{x_1}^{x_2} \frac{A}{x^2}dx = A\left.\frac{1}{x}\right|_{x_1}^{x_2}$$

$$= A\left(\frac{1}{x_2} - \frac{1}{x_1}\right),$$

for $x_2 < x_1$. If the kinetic energy of an alpha particle at the nuclear surface is zero, $K_2 = 0$, then its kinetic energy at some other point is $K_1 = U_2 - U_1$.

Therefore, its speed is

$$v_1 = \sqrt{\frac{2K_1}{m}} = \sqrt{\frac{2A}{mx_2}\left(1 - \frac{x_2}{x_1}\right)}$$

$$= \left[\frac{2(4.1 \times 10^{-26} \text{ N·m}^2)}{(6.7 \times 10^{-27} \text{ kg})(1.43 \times 10^{-10} \text{ m})}\left(1 - \frac{x_2}{x_1}\right)\right]^{1/2}$$

$$= (2.93 \times 10^5 \text{ m/s})\sqrt{1 - x_2/x_1},$$

where x_2 is the nuclear radius. (a) If $x_1 = 4x_2$, $v_1 = 2.53 \times 10^5$ m/s, and (b) if $x_1 = 100x_2$, $v_1 = 2.91 \times 10^5$ m/s. (c) Evidently, 2.93×10^5 m/s $= v_1$ for $x_1 \to \infty$.

Problem

77. With the brick of Problem 9 standing on end, what is the minimum energy that can be given the brick to make it fall over?

Solution

To fall over, the center of the brick must lie to the right of the vertical through its 8-cm edge, as shown. The potential energy of the brick would have to be increased by $U_b - U_a = mg(y_b - y_a)$. If we assume that energy is conserved between a and b, this is equal to the minimum kinetic energy sought ($K_a - K_b = U_b - U_a = K_a^{min}$, if $K_b - 0$). The numerical value is just the difference between the answers to Problems 9(b) and 9(a), but recalculating, we find $mg(y_b - y_a) = (1.5 \text{ kg})(9.8 \text{ m/s}^2)\frac{1}{2}[\sqrt{(20 \text{ cm})^2 + (5.5 \text{ cm})^2} - 20 \text{ cm}] = 54.6$ mJ.

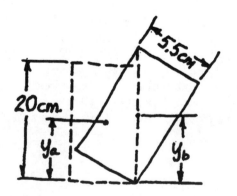

Problem 77 Solution.

Problem

79. Together, the springs in a 1200-kg car have an effective spring constant of 110,000 N/m and can compress a maximum distance of 40 cm. What is the maximum abrupt drop in road level (Fig. 8-48) that the car can tolerate without "bottoming out"—that is, without its springs reaching

maximum compression? Assume the car is driving fast enough that it becomes temporarily airborne.

FIGURE 8-48 Problem 79.

Solution

Let us neglect any possible losses in energy and assume that the kinetic energy associated with the car's horizontal motion is unchanged by the drop. Then the energy principle implies that $U_{grav}(A) = U_{grav}(B) + U_{spr}(B)$, where A is a point before the drop in the road, and B is the point after the drop where the springs are maximally compressed (the kinetic energy associated with the car's vertical motion on its springs is instantaneously zero at B). Thus, $U_{grav}(A) - U_{grav}(B) = mgh = U_{spr}(B) = \frac{1}{2}kx^2$, or $h = (110,000 \text{ N/m})(0.4 \text{ m})^2/2(1200 \times 9.8 \text{ N}) = 74.8$ cm.

Problem

81. An electron with kinetic energy 0.85 fJ enters a region where its potential energy as a function of position is $U = ax^2 - bx$, where $a = 2.7$ fJ/cm^2 and $b = 4.2$ fJ/cm. (a) How far into the region does the electron penetrate? (b) At what position does the electron have its maximum speed? (c) what is this maximum speed?

Solution

(a) The electron's total energy, $E = 0.85$ fJ, equals its initial kinetic energy upon entering the region at $x = 0$. At any other point in the region $x \geq 0$, $E = K + U$, so the point of maximum penetration (the turning point) can be found from the equation $K(x_m) = E -$

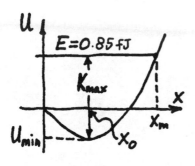

Problem 81 Solution.

$U(x_m) = 0$. Therefore, $ax_m^2 - bx_m - 0.85 \text{ fJ} = 0$, or $x_m = [4.2 + \sqrt{(4.2)^2 + 4(2.7)(0.85)}]/2(2.7) = 1.74$ cm. (The negative root can be discarded since $x \geq 0$.) (b) The maximum kinetic energy occurs at the point where U is a minimum; that is, $dU/dx = 2ax_0 - b = 0$, or $x_0 = b/2a = 0.778$ cm. (c) $v_{max} = \sqrt{2K_{max}/m_e} = \sqrt{2(E - U_{min})/m_e}$. Now, $U_{min} = x_0(ax_0 - b) = -b^2/4a = -(4.2 \text{ fJ/cm})^2/4(2.7 \text{ fJ/cm}^2) = -1.63$ fJ, so that $v_{max} = \sqrt{2(0.85 + 1.63) \text{ fJ}/(9.11 \times 10^{-31} \text{ kg})} = 7.38 \times 10^7$ m/s.

Problem

83. (a) Repeat the previous problem for the case of a force $\mathbf{F} = (ax - bx^3)\hat{\imath}$, where a and b are positive constants. (b) What is the significance of the negative square root that can occur for some values of x? (c) Find an expression for the particle's maximum speed.

Solution

(a) With the same reasoning in the previous problem's solution (i.e., conservation of energy $K_0 + U_0 = K + U$, with $K_0 = 0$ at $x = 0$), $U - U_0 = -\int_0^x (ax' - bx'^3)dx' = -(x^2/2)(a - bx^2/2) = -K = -\frac{1}{2}mv^2$, or $v = \sqrt{(x^2/m)(a - bx^2/2)}$. (b) The total energy of the particle is $E = U_0$, so the particle is confined to the region where $U(x) \leq E$. (This is the same as the condition $K \geq 0$.) $U(x) = E$ when $x = \pm\sqrt{2a/b}$, which are the turning points of the motion. Values of position with $|x| > \sqrt{2a/b}$ are forbidden, classically. (c) The maximum speed occurs at the minimum of potential energy. $dU/dx = bx^3 - ax = 0$ has roots at $x = 0$ and $x = \pm\sqrt{a/b}$. The former is a maximum and the latter are minima of U. Substitution into the expression for the speed yields $v_{max} = \sqrt{(a/bm)(a - ba/2b)} = a/\sqrt{2bm}$.

Problem

85. A force points in the $-x$ direction with magnitude given by $F = ax^b$, where a and b are constants. Evaluate the potential energy as a function of position, taking $U = 0$ at some point $x_0 > 0$. Use your result to show that an object of mass m released at $x = \infty$ will reach x_0 with finite velocity provided $b < -1$. Find the velocity for this case.

Solution

For one-dimensional motion, Equation 8-2a for the potential energy yields

$$U(x) = -\int_{x_0}^x F_x \, dx = -\int_{x_0}^x (-ax^b)dx$$
$$= a(b+1)^{-1}(x^{b+1} - x_0^{b+1}),$$

where $U(x_0) = 0$ and $x_0 > 0$. This result applies provided $b \neq -1$; otherwise $U(x)$ is logarithmic. If no other forces act, mechanical energy is conserved, so a particle released at $x = \infty$ with kinetic energy $K(\infty) = 0$, will have kinetic energy at x_0 of $K(x_0) = U(\infty)$. This is finite provided $b + 1 < 0$. (Otherwise, $x^{b+1} \to \infty$, as $x \to \infty$, or if $b = -1$, $\ln x \to \infty$ also.) In case $b + 1 < 0$, $|b + 1| = -(b + 1)$ and $\frac{1}{2}mv_0^2 = (a/|b+1|)x_0^{-|b+1|}$, or $v_0 = (2a/m|b+1|)^{1/2}x_0^{-|b+1|/2}$.

Problem

87. The climbing rope described in Example 8-3 is securely fastened at its lower end. At the upper end is a 65-kg climber. At a height 2.4 m directly below the climber, the rope passes through a carabiner (essentially a frictionless metal loop), as shown in Fig. 8-50. If the climber falls, through

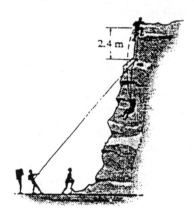

FIGURE 8-50 Problem 87.

what maximum total distance—including the stretching of the rope—does he drop? *Hint:* Your statement of energy conservation should result in a cubic equation, which you can solve graphically, numerically, or analytically.

Solution

If the climber falls vertically, the maximum change in height is 4.8 m plus the maximum stretch in the rope; i.e., $y_i - y_f = 4.8$ m $+ x_{\max}$. The subscripts i and f refer to the initial and final (i.e., at the maximum stretch of the rope where the velocity is instantaneously zero) positions of the climber, so $K_i = K_f = 0$. The carabiner is frictionless, so the conservation of energy implies that $\Delta K + \Delta U_{\text{grav}} + \Delta U_{\text{spr}} = 0$, or $mg \times (y_i - y_f) = mg(4.8$ m $+ x_{\max}) = \frac{1}{2}kx_{\max}^2 - \frac{1}{3}bx_{\max}^3$, where the potential energy stored in the rope was calculated in Example 8-3. Substitution of the values given for the constants leads to a cubic equation for the maximum stretch: $(2.145 \times 10^{-3})x_{\max}^3 - 0.175x_{\max}^2 + x_{\max} + 4.8 = 0$, where $x_{\max}$ is in meters. Without the cubic term, the quadratic formula gives $x_{\max} \approx 8.82$ m, so we expect $x_{\max}$ to be somewhat greater than this because the rope's potential energy is less than that of an ideal spring for the same stretch. We choose to use Newton's method, with a first guess of $x_{\max} = 9$, to solve the cubic (see Holder, De Franza & Pasachoff, *Single Variable Calculus*, 2nd Ed. Brooks/Cole, 1994, Section 4.5). After two iterations, we find $x_{\max} \approx 9.694$ m, or the maximum fall is about 4.8 m $+ x_{\max} = 14.5$ m. (This might be compared to the stretch of the rope when in equilibrium with the climber's weight, $x_{\text{eq}} \approx 3.02$ m, at a distance of 7.82 m below the start of the fall.)

CHAPTER 9   GRAVITATION

ActivPhysics can help with these problems:
Activity 4.8

Section 9-2:   The Law of Universal Gravitation

Problem

1. Space explorers land on a planet with the same mass as Earth, but they find they weigh twice as much as they would on Earth. What is the radius of the planet?

Solution

At rest on a uniform spherical planet, a body's weight is proportional to the surface gravity, $g = GM/R^2$. Therefore, $(g_p/g_E) = (M_p/M_E)(R_E/R_p)^2 = 2$. Since $M_p/M_E = 1$, $R_p = R_E/\sqrt{2}$.

Problem

3. To what fraction of its current radius would Earth have to be shrunk (with no change in mass) for the gravitational acceleration at its surface to triple?

Solution

If the surface gravity of the Earth were three times its present value, with no change in mass, then $GM_E/R^2 = 3GM_E/R_E^2$, where R_E is the present radius. Thus, $R/R_E = 1/\sqrt{3} = 57.7\%$ gives the new, shrunken radius.

Problem

5. Two identical lead spheres are 14 cm apart and attract each other with a force of 0.25 μN. What is their mass?

Solution

Newton's law of the universal gravitation (Equation 9-1), with $m_1 = m_2 = m$, gives

$$m = \sqrt{\frac{Fr^2}{G}} = \left(\frac{0.25 \times 10^{-6} \text{ N } (0.14 \text{ m})^2}{6.67 \times 10^{-11} \text{ N·m}^2/\text{kg}^2}\right)^{1/2}$$
$$= 8.57 \text{ kg}.$$

(Newton's law holds as stated, for two uniform spherical bodies, if the distance is taken between their centers.)

Problem

7. What is the approximate value of the gravitational force between a 67-kg astronaut and a 73,000-kg space shuttle when they're 84 m apart?

Solution

If we treat the astronaut and space shuttle as spherical bodies, Equation 9-1 can be used to calculate the approximate force:

$$F = (6.67 \times 10^{-11} \text{ N·m}^2/\text{kg}^2)(67 \text{ kg})(73,000 \text{ kg}) \div$$
$$(84 \text{ m})^2 = 46.2 \text{ nN}.$$

Problem

9. A sensitive gravimeter is carried to the top of Chicago's Sears Tower, where its reading for the acceleration of gravity is 0.00136 m/s^2 lower than at street level. Find the height of the building.

Solution

The difference in the acceleration of gravity between the bottom of the Sears Tower (distance R from the center of the Earth) and the top (distance $R + h$ from the Earth's center, where h is the height of the building) is

$$\Delta g = \frac{GM_E}{R^2} - \frac{GM_E}{(R+h)^2} = \frac{GM_E}{R^2}\frac{h(2R+h)}{(R+h)^2} \approx g_{\text{bot}}\frac{2h}{R}.$$

In the last step, g_{bot} is the value of the acceleration of gravity at the bottom of the building, and since $h \ll R$, we approximated the second fraction by neglecting h compared to R. If we use average values for g_{bot} (about 9.80 m/s^2 in Chicago) and R (approximately 6370 km) and the given Δg, then $h \approx (\Delta g/g_{\text{bot}})(R/2) = (0.00136/9.80)(6370 \text{ km}/2) = 442$ m. (The actual value of g_{bot}, for example, depends on the shape of the earth, the altitude, the distribution and type of underlying rocks, etc. Present gravimeters can measure differences in g as small as a few tenths of a milligal, where 1 milligal $= 10^{-5}$ m/s^2 is the unit used to measure gravity anomalies by geologists.)

Problem

11. If you're standing on the ground 15 m directly below the center of a spherical water tank containing 4×10^6 kg of water, by what fraction is

your weight reduced due to the gravitational attraction of the water?

Solution

The fraction by which your weight is reduced is equal to the gravitational attraction of the water tank on you divided by your weight, or $(Gmm'/r^2)/mg = Gm'/gr^2$. Numerically, this is $(6.67 \times 10^{-11} \text{ N·m}^2/\text{kg}^2)(4 \times 10^6 \text{ kg})/(9.8 \text{ m/s}^2) \times (15 \text{ m})^2 = 1.21 \times 10^{-7}$.

Section 9-3: Orbital Motion

Problem

13. Find the speed of a satellite in geosynchronous orbit.

Solution

Equation 9-3 and the radius of a geosynchronous orbit from Example 9-4 give $v = \sqrt{GM_E/r} = [(6.67 \times 10^{-11} \text{ N·m}^2/\text{kg}^2)(5.97 \times 10^{24} \text{ kg})/(42.2 \times 10^6 \text{ m})]^{1/2} = 3.07 \text{ km/s}$.

Problem

15. Calculate the orbital period for Jupiter's moon Io, which orbits 4.22×10^5 km from the center of the 1.9×10^{27}-kg planet.

Solution

From Equation 9-4 and the data given:

$$T = \left(\frac{4\pi^2 r^3}{GM}\right)^{1/2}$$

$$= \left[\frac{4\pi^2 (4.22 \times 10^8 \text{ m})^3}{(6.67 \times 10^{-11} \text{ N·m}^2/\text{kg}^2)(1.9 \times 10^{27} \text{ kg})}\right]^{1/2}$$

$$= 1.53 \times 10^5 \text{ s} = 1.77 \text{ d}.$$

Problem

17. During the Apollo moon landings, one astronaut remained with the command module in lunar orbit, about 130 km above the surface. For half of each orbit, this astronaut was completely cut off from the rest of humanity, as the spacecraft rounded the far side of the moon (see Fig. 9-34). How long did this period last?

Solution

The period of a circular orbit of altitude $h = 130$ km above the moon's surface is (see Equation 9-4) $T = 2\pi \sqrt{(R_m + h)^3/GM_m} = 2\pi[(1.74 + 0.13)^3 \times 10^{18} \text{ m}^3 \div$

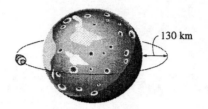

FIGURE 9-34 Problem 17.

$(6.67 \times 10^{-11} \text{ N·m}^2/\text{kg}^2)(7.35 \times 10^{22} \text{ kg})]^{1/2} = 7.26 \times 10^3 \text{ s} = 121 \text{ min}$. The command module was cut off from Earth communications for roughly half of this period, or about 1 h.

Problem

19. Given that our Sun orbits the galaxy with a period of 200 My at a distance of 2.6×10^{20} m from the galactic center, estimate the mass of the galaxy. Assume (incorrectly) that the galaxy is essentially spherical and that most of its mass lies interior to the Sun's orbit. To how many Sun-mass stars is your estimate equivalent?

Solution

For a circular orbit enclosing a spherically distributed mass M, Equation 9-4 gives $M = 4\pi^2 r^3/GT^2$. If we apply this to the Sun's orbit around the galactic center (and use the rough conversion $1 \text{ y} = \pi \times 10^7 \text{ s}$), we can estimate the mass of the Milky Way galaxy as

$$M \approx \left(\frac{2\pi}{2 \times 10^8 \times \pi \times 10^7 \text{ s}}\right)^2 \frac{(2.6 \times 10^{20} \text{ m})^3}{(6.67 \times 10^{-11} \text{ N·m}^2/\text{kg}^2)}$$

$$\approx 2.6 \times 10^{41} \text{ kg}$$

This is about 1.3×10^{11} solar masses ($m_\odot \approx 2 \times 10^{30}$ kg), which is not an unreasonable estimate of the number of stars in our galaxy. Astronomers plot the orbital velocity of objects (such as stars, clusters of stars, or clouds of hydrogen atoms) versus their distance from the galactic center to obtain "the rotation curve" for the galaxy. If most of the matter in a galaxy were concentrated within a certain distance of the center, then the rotation curve should fall off, as indicated in Equation 9-3, for greater distances. It doesn't. The identification of this so called "dark matter," which is not seen in telescopes by light it gives off, but which is present due to its gravitational effects, is currently a topic of great interest in astronomy.

Problem

21. Where should a satellite be placed to orbit the Sun in a circular orbit with a period of 100 days?

Solution

As long as the masses of the orbiting bodies (satellite or planet) are negligible compared to the mass of the Sun, Equation 9-4, as a ratio, can be used to compare the mean orbital radii and periods, $(T_1/T_2)^2 = (r_1/r_2)^3$ (this is, in fact, an approximate statement of Kepler's third law). The mean orbital radius of the Earth is 1 AU (an astronomical unit) $\approx 1.50 \times 10^8$ km, and its period is 1 y (a sidereal year) ≈ 365 d, so $(T/1\text{ y})^2 = (r/1\text{ AU})^3$ for any satellite of the Sun. If $T = 100$ d, then $(r/1\text{ AU}) = (100/365)^{2/3} = 0.422$, or $R = 0.422$ AU $= 6.3 \times 10^7$ km. Of course, Equation 9-4 could be used directly:

$$
\begin{aligned}
r &= [GM(T/2\pi)^2]^{1/3} \\
&= \left[\left(6.67 \times 10^{-11} \, \frac{\text{N·m}^2}{\text{kg}^2}\right)(1.99 \times 10^{30} \text{ kg}) \right.\\
&\quad \left. \times \left(\frac{8.64 \times 10^6 \text{ s}}{2\pi}\right)^2\right]^{1/3} = 6.31 \times 10^7 \text{ km.}
\end{aligned}
$$

Problem

23. How far from the Sun's center would a satellite in circular orbit be heliosynchronous? *Hint:* Consult Appendix E, and remember that a synchronous orbit must parallel the equator. So which rotation period is appropriate?

Solution

The equitorial rotational period of the Sun is 27 d (from Appendix E). The radius of a circular orbit (for a body with mass much smaller than the Sun) with the same orbital period can be found from Kepler's third law (Equation 9-4 in ratio form; see solution to Problem 21): $r = (T/1\text{ y})^{2/3}$ AU $= (27/365)^{2/3}$ AU $= 0.176$ AU $= 2.64 \times 10^7$ km.

Problem

25. The asteriod Pasachoff orbits the Sun with a period of 1417 days. What is the semimajor axis of its orbit? Determine using Kepler's third law in comparison with Earth's orbital radius and period.

Solution

As in the solution to Problem 21, $r = (T/1\text{ y})^{2/3} \times (1 \text{ AU}) = (1417/365)^{2/3}$ AU $= 2.47$ AU $= 3.71 \times 10^8$ km.

Problem

27. Our galaxy belongs to a large group of galaxies known as the Virgo Cluster. The cluster contains about 1000 times the mass of our galaxy, which in turn contains about 10^{11} times the Sun's mass. We're roughly 50 million light-years from the center of the approximately spherical cluster. Could our galaxy have completed a full orbit of the cluster since the universe began, some 15 billion years ago?

Solution

Our galaxy's distance from the center of the Virgo Cluster is about $r = (50 \times 10^6 \text{ ly})(9.46 \times 10^{15} \text{ m/ly}) = 4.73 \times 10^{23}$ m. If most of the cluster's mass, $M \approx 10^3 \times 10^{11}(2 \times 10^{30} \text{ kg}) = 2 \times 10^{44}$ kg, were spherically distributed inside this distance, and our galaxy were in a circular orbit around it, our period would be

$$
\begin{aligned}
T &= 2\pi(r^3/GM)^{1/2} = 2\pi[(4.73 \times 10^{23} \text{ m})^3 \\
&\quad \div (6.67 \times 10^{-11} \text{ N·m}^2/\text{kg}^2)(2 \times 10^{44} \text{ kg})]^{1/2} \\
&= 1.77 \times 10^{19} \text{ s} \approx 5.6 \times 10^{11} \text{ y.}
\end{aligned}
$$

This is a few tens times the age of the universe, so obviously a full orbit could not have been completed (at least under the present conditions). Even if the Virgo Cluster contained 100 times as much mass, perhaps as dark matter, our period would still be a few times greater than the age of the universe.

Section 9-4: Gravitational Energy

Problem

29. How much energy does it take to launch a 230-kg instrument package on a vertical trajectory that peaks at an altitude of 1800 km?

Solution

If we neglect any kinetic energy differences associated with the orbital or rotational motion of the Earth or package, the required energy is just the difference in potential energy of the Earth's gravity given by Equation 9-5, $\Delta U = GM_E m \, [R_E^{-1} - (R_E + h)^{-1}]$. In terms of the more convenient combination of constants $GM_E = gR_E^2$, $\Delta U = mgR_E h/(R_E + h) = (230 \times 9.81 \text{ N})(6370 \times 1800 \text{ km})/(8170) = 3.17$ GJ.

Problem

31. A rocket is launched vertically upward from Earth's surface at a speed of 3.1 km/s. What is its maximum altitude?

Solution

If we consider the Earth at rest as approximately an inertial system, then a vertically launched rocket would have zero kinetic energy (instantaneously) at its maximum altitude, and the situation is the same as Example 9-6. Conservation of energy gives $\frac{1}{2}mv_0^2 - GM_Em/R_E = -GM_Em/(R_E + h)$, or

$$h = \left[\frac{1}{R_E} - \frac{v_0^2}{2GM_E}\right]^{-1} - R_E = 530 \text{ km},$$

when the proper values are substituted.

Problem

33. Find the energy necessary to put 1 kg, initially at rest on Earth's surface, into geosynchronous orbit.

Solution

The energy of an object at rest on the Earth's surface is $U_0 = -GM_Em/R_E$ (neglect diurnal rotational energy, etc.), while its total mechanical energy in a circular orbit is $E = \frac{1}{2}U = -GM_Em/2r$ (Equation 9-9). The energy necessary to put a mass of $m = 1$ kg into a circular geosynchronous orbit with $r = 4.22\times10^7$ m (see Example 9-4) is the difference of these energies, $\Delta E = \frac{1}{2}U - U_0$, or

$$\Delta E = GM_E(1 \text{ kg})\left(\frac{1}{6.37\times10^6 \text{ m}} - \frac{1}{2(4.22\times10^7 \text{ m})}\right)$$

$$= 57.8 \text{ MJ}.$$

Problem

35. Neglecting air resistance, to what height would you have to fire a rocket for the constant acceleration equations of Chapter 2 to give a height that is in error by 1%? Would those methods over- or underestimate the height?

Solution

If the rocket has an initial vertical speed v_0, we can find the height, h, to which it can rise (where its kinetic energy is instantaneously zero) from the conservation of energy:

$$K_0 + U_0 = \frac{1}{2}mv_0^2 - \frac{GM_Em}{R_E} = K + U = -\frac{GM_Em}{R_E + h}$$

Before solving for h, we replace GM_E by gR_E^2, and let $h' = v_0^2/2g$. (h' is the maximum height for constant gravity, from Equation 2-11.) After some algebra, we find $h = h'R_E/(R_E - h')$. Since the factor multiplying h' is > 1, $h > h'$, and the equations of constant gravity underestimate the height. (This could have been anticipated because the force of gravity decreases with increasing altitude.) h' differs from h by 1% $((h - h')\div h = 0.01)$, if $h' = 0.99h$. This occurs for $h = 0.99hR_E\div (R_E - 0.99h)$, or $h = R_E/99 = 64.3$ km.

Problem

37. Drag due to small amounts of residual air causes satellites in low Earth orbit to lose energy and eventually spiral to Earth. What fraction of its orbital energy is lost as a satellite drops from 300 to 100 km, assuming its orbit remains essentially circular?

Solution

The fractional difference in orbital energy is $\Delta E/E = (E - E')/E$, where $E = \frac{1}{2}U = -GM_Em \div 2r \sim 1/(R_E + h)$ is the orbital energy, and h is the altitude of the circular orbit (see Equation 9-9). Thus $\Delta E/E = (r^{-1} - r'^{-1})/r^{-1} = (h' - h)\div (R_E + h') = (100 - 300) \text{ km}/(6370 + 100) \text{ km} = -2/64.7 = -3.09\times10^{-2} \approx -3\%$. (The energy difference is negative because energy is lost going from a higher to a lower orbit.)

Problem

39. By what factor must the speed of an object in circular orbit be increased to reach escape speed from its orbital altitude?

Solution

The escape speed at a distance r from the Earth's center is just $\sqrt{2}$ times the speed in a circular orbit of the same radius. Compare Equations 9-3 and 9-7.

Problem

41. The escape speed from a planet of mass 2.9×10^{24} kg is 7.1 km/s. What is the planet's radius?

Solution

Equation 9-7 implies $R = 2GM/v_{esc}^2 = 2(6.67\times10^{-11} \text{ N·m}^2/\text{kg}^2)(2.9\times10^{24} \text{ kg})\div (7.1\times10^3 \text{ m/s})^2 = 7.67\times10^6$ m.

Problem

43. Two meteoroids are 250,000 km from Earth and moving at 2.1 km/s. One is headed straight for Earth, while the other is on a path that will come within 8500 km of Earth's center (Fig. 9-36). (a) What is the speed of the first meteoroid when it strikes Earth? (b) What is the speed of the second meteoroid at its closest approach to Earth? (c) Will the second meteoroid ever return to Earth's vicinity?

Solution

(a) Conservation of energy applied to the first meteoroid gives: $\frac{1}{2}mv_0^2 - G(M_Em/r_0) = \frac{1}{2}mv^2 -$

$G(M_E m/R_E)$, or
$v = \sqrt{v_0^2 + 2GM_E(1/R_E - 1/r_0)}$. In the numerical
evaluation, replace GM_E by gR_E^2, to obtain:

$$v = \sqrt{\left(2.1\frac{km}{s}\right)^2 + 2\left(0.0098\frac{km}{s^2}\right)(6370\ km)\left(1 - \frac{6370}{250,000}\right)}$$

$$= 11.2\ km/s.$$

(b) For the second meteoroid,

$$v = \sqrt{v_0^2 + 2gR_E^2\left(\frac{1}{8500\ km} - \frac{1}{250,000\ km}\right)}$$

$$= 9.74\ km/s.$$

(c) The escape velocity at a distance of
$r = 8500$ km from the center of the earth is
$v_{esc} = \sqrt{2GM_E/r} =$
$\sqrt{2gR_E^2/r} = 9.67$ km/s, so the second meteoroid
will probably not return. Alternatively, v_{esc} at a
distance of 250,000 km is 1.78 km/s.

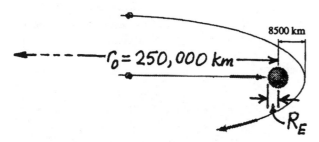

8500 km

$r_0 = 250,000$ km

R_E

FIGURE 9-36 Problem 43 Solution.

Problem

45. Neglecting Earth's rotation, show that the energy
needed to launch a satellite of mass m into circular
orbit at altitude h is

$$\left(\frac{GM_E m}{R_E}\right)\left(\frac{R_E + 2h}{2(R_E + h)}\right).$$

Solution

The energy of a satellite in a circular orbit is
$E = \frac{1}{2}U = -GM_E m/2r$, where $r = R_E + h$ (see
Equation 9-9). This is the energy in an
Earth-centered, non-rotating reference frame,
neglecting the gravitational influence of any other
body, e.g., the Sun. The energy of a satellite on the
Earth's surface depends on its location, because of the
Earth's diurnal rotation, $E_0 = U_0 + \frac{1}{2}mv_0^2$, where
$U_0 = -GM_E m/R_E$, and v_0 is the speed of the Earth's
surface at that location. If we neglect the extra kinetic
energy associated with v_0 (since $v_0 \leq 2\pi R_E/1d$,
$\frac{1}{2}mv_0^2 \leq 0.34\%\ |U_0|$), the energy required to launch the
satellite into circular orbit at altitude h is

$E - E_0 = \frac{1}{2}U - U_0 = GM_E m[R_E^{-1} - \frac{1}{2}(R_E + h)^{-1}] =$
$GM_E m(R_E + 2h)/2R_E(R_E + h)$ as claimed.

Problem

47. A projectile is launched vertically upward from a
planet of mass M and radius R; its initial speed is
twice the escape speed. Derive an expression for
its speed as a function of the distance r from the
center of the planet.

Solution

If we consider just the gravitational field of the planet
and neglect possible losses of energy (from atmospheric
drag, etc.), then the conservation of energy, $K + U =$
$K_0 + U_0$, gives the speed of an object at a radial
distance r (from the center of a spherical massive
body) as a function of its speed v_0 and distance r_0 at
a particular point: $v = \sqrt{v_0^2 - 2GM(r_0^{-1} - r^{-1})}$ (see
Example 9-7). If we set $v_0^2 = (2v_{esc})^2 = 8GM/R$ at
$r_0 = R$ (from Equation 9-7) for the projectile in this
problem, then $v = \sqrt{2GM(4R^{-1} - R^{-1} + r^{-1})} =$
$\sqrt{2GM(3R^{-1} + r^{-1})}$.

Problem

49. The *Pioneer* spacecraft left Earth's vicinity
moving at about 38 km/s relative to the Sun (this
figure combines the effect of rocket boost and
Earth's orbital motion). How far out in the solar
system could Pioneer get without additional
rocket power or use of the "gravitational
slingshot" effect?

Solution

If we consider just the gravitational field of the Sun,
which is the dominant source of potential energy for
the *Pioneer* spacecraft in this problem, then the
conservation of energy in the form $K + U = K_0 + U_0$
(as in Example 9-6) gives $\frac{1}{2}mv^2 - GM_\odot m/r =$
$\frac{1}{2}mv_0^2 - GM_\odot m/r_0$, or $r = [r_0^{-1} - (v_0^2 - v^2)/2GM_\odot]^{-1}$.
Here, v_0 is given as 38 km/s, and $r_0 = 1$ AU $=$
1.50×10^8 km. We do not know the actual shape of the
satellite's orbit, but $v^2 \ll v_0^2$ at the maximum distance
from the sun, so $r < r_{max} = [r_0^{-1} - v_0^2/2GM_\odot]^{-1} =$
$[(1.50\times10^{11}\ m)^{-1} - (38\times10^3\ m/s)^2/2(6.67\times$
$10^{-11}\ N\cdot m^2/kg^2)(1.99\times10^{30}\ kg)]^{-1} = 8.15\times10^{11}\ m =$
5.43 AU. This is a little farther than the orbit of
Jupiter.

Section 9-6:   Tidal Forces

Problem

51. Show that the force of the Sun's gravity on Earth
is nearly 200 times that of the moon's gravity, but

that the tidal force of the moon on Earth (see preceding problem) is about twice that of the Sun.

Solution

The ratio of the magnitude of the direct gravitational forces is

$$\frac{F_{\text{Sun}}}{F_{\text{moon}}} = \frac{GM_SM_E/r_{SE}^2}{GM_MM_E/r_{EM}^2} = \frac{M_S}{M_M}\left(\frac{r_{EM}}{r_{SE}}\right)^2$$

$$= \left(\frac{1.99\times10^{30}}{7.35\times10^{22}}\right)\left(\frac{0.385}{150}\right)^2 = 178$$

(see Appendix E), while the ratio of the differential tidal forces, as given in the preceding problem, is (the factors relating to the Earth, like m and a, cancel out even in a more exact calculation) $\Delta F_{\text{Sun}}/\Delta F_{\text{moon}} = (M_S/M_M)(r_{EM}/r_{SE})^3 = 0.458$.

Paired Problems

Problem

53. An astronaut hits a golf ball horizontally from the top of a lunar mountain so fast that it goes into circular orbit. What is its orbital period?

Solution

The period of a grazing orbit, $r \approx R$, around a spherical object of mass M and radius R (see Appendix E for lunar values) can be found from Equation 9-4, $T = 2\pi\sqrt{R^3/GM} = 2\pi[(1.74\times10^6 \text{ m})^3 \div (6.67\times10^{-11} \text{ N·m}^2/\text{kg}^2)(7.35\times10^{22} \text{ kg})]^{1/2} = 6.51\times10^3$ s $= 109$ min.

Problem

55. Two meteoroids are 160,000 km from Earth's center and heading straight toward Earth. One is moving at 10 km/s, the other at 20 km/s. At what speed will they strike Earth?

Solution

If we consider only the difference of potential energy of the Earth's gravitational field (since $r_0 = 160,000$ km is small compared to the distance from the Sun, approximately 150×10^6 km, the difference of potential energy of the meteoroids in the Sun's gravitational field is less than 2% of the Earth's), the conservation of energy can be applied as in Example 9-7 to determine the impact speed. (This estimate also neglects atmospheric effects or, alternatively, gives the speed of impact on the atmosphere, $r \approx R_E$, instead of the Earth's surface.) Then $v_{\text{impact}}^2 = v_0^2 + 2GM_E(r^{-1} - r_0^{-1}) = v_0^2 + 1.20\times10^8$ (m/s)^2, where we used data from Appendix E. (a) For $v_0 = 10$ km/s, $v_{\text{impact}} = 14.8$ km/s, and (b) for $v_0 = 20$ km/s, $v_{\text{impact}} = 22.8$ km/s.

Problem

57. A satellite is in an elliptical orbit at altitudes ranging from 230 to 890 km. At the high point it's moving at 7.23 km/s. How fast is it moving at the low point?

Solution

The conservation of energy is applied to a satellite in an elliptical Earth orbit in Example 9-7 (where it is a good approximation to neglect the gravitational influence of other bodies, atmospheric drag, etc.) to relate the speed and distance at perigee (the lowest point) to the same quantities at apogee (the highest point): $v_p^2 = v_a^2 + 2GM_E(r_p^{-1} - r_a^{-1})$. The calculation can be simplified by expressing the distances in terms of altitude above the Earth's surface and using a little algebra. Then $r = R_E(1 + h/R_E)$ and $(r_p^{-1} - r_a^{-1}) = (r_a - r_p)/r_ar_p = (h_a - h_p)/R_E^2(1 + h_a/R_E)(1 + h_p/R_E)$. Since $GM_E/R_E^2 = g, v_p^2 = (7.23 \text{ km/s})^2 + 2(9.81 \text{ m/s}^2)(890 \text{ km} - 230 \text{ km})/[(1 + 890/6370)(1 + 230/6370)]$, and $v_p = 7.95$ km/s. (This result also follows from the conservation of angular momentum or Kepler's second law, which implies that $v_ar_a = v_pr_p$.)

Problem

59. To what radius would Earth have to be shrunk, with no loss of mass, for escape speed at its surface to be 30 km/s?

Solution

The escape speed from the surface of a spherical body of mass M (equal to M_E in this case) is given by Equation 9-7, from which the radius can be found if v_{esc} is given: $R = 2GM_E/v_{\text{esc}}^2$. In the numerical calculation, we could use the known escape speed of the earth, $v_{\text{esc}} = 11.2$ km/s for $R = R_E$, to eliminate the constants $2GM_E = (11.2 \text{ km/s})^2R_E$, obtaining $R = (11.2/30)^2R_E = 0.139R_E \approx 888$ km. (Other combinations of constants, e.g., $GM_E = gR_E^2$, give the same result to within the accuracy of the data.)

Problem

61. A 720-kg spacecraft has total energy -5.3×10^{11} J and is in circular orbit about the Sun. Find (a) its orbital radius, (b) its kinetic energy, and (c) its speed.

Solution

For a small object (the spacecraft) in a circular orbit about a central massive body (the Sun), the total, kinetic, and potential energies are related by Equations 9-8 and 9, $E = -K = \frac{1}{2}U$. (b) Therefore $K = -E = 5.3\times10^{11}$ J. (c) Since $K = \frac{1}{2}mv^2, v =$

$\sqrt{2K/m} = [2(5.3\times10^{11} \text{ J})/720 \text{ kg}]^{1/2} = 38.4$ km/s.
(a) Since $U = 2E = -GM_{\odot}m/r, r = GM_{\odot}m/(-2E) =$
$(6.67\times10^{-11} \text{ N}\cdot\text{m}^2/\text{kg}^2)(1.99\times10^{30} \text{ kg})(720 \text{ kg})\div$
$(2 \times 5.3\times10^{11} \text{ J}) = 9.02\times10^{10}$ m $= 0.601$ AU.

Supplementary Problems

Problem

63. Mercury's orbital speed varies from 38.8 km/s at aphelion to 59.0 km/s at perihelion. If the planet is 6.99×10^{10} m from the Sun's center at aphelion, how far is it at perihelion?

Solution

Kepler's second law implies $r_p = r_a(v_a/v_p) =$
$(6.99\times10^{10} \text{ m})(38.8/59.0) = 4.60\times10^{10}$ m (see the solution to Problem 57). Otherwise, an equation similar to that resulting from the conservation of energy in Example 9-7, $r_p^{-1} - r_a^{-1} = (v_p^2 - v_a^2)\div$ $2GM_{\odot} = 7.44\times10^{-12}$ m^{-1}, gives the same value of r_p.

Problem

65. A black hole is an object so dense that its escape speed exceeds the speed of light. Although a full description of black holes requires general relativity, the radius of a black hole can be calculated using Newtonian theory. (a) Show that the radius of a black hole of mass M is $2GM/c^2$, where c is the speed of light. What are the radii of black holes with (b) the mass of the Earth and (c) the mass of the Sun?

Solution

(a) The so-called Schwarzchild radius of a black hole of mass M turns out to be the same as the radius given by Equation 9-7 with $v_{\text{esc}} = c = \sqrt{2GM/r_s}$. Thus $r_s = 2GM/c^2$. (b) $2GM_E/c^2 = 8.85$ mm, and (c) $2GM_{\odot}/c^2 = 2.95$ km.

Problem

67. Two satellites are in geosynchronous orbit, but in diametrically opposite positions (Fig. 9-39). Into how much lower a circular orbit should one spacecraft descend if it is to catch up with the other after 10 complete orbits? Neglect rocket firing times and time spent moving between the two circular orbits.

FIGURE 9-39 Problem 67.

Solution

In a lower circular orbit (smaller r) the orbital speed is faster (see Equation 9-3). The time for 10 complete orbits of the faster satellite must equal the time for $9\frac{1}{2}$ geosynchronous orbits, if the faster satellite is to catch up as described (it starts out one-half an orbit behind). Thus, $10T = 9.5(1 \text{ d})$, or $T = 0.95$ d, where T is the period of the lower, faster orbit. Kepler's third law (Equation 9-4) then gives $r = (GM_ET^2/4\pi^2)^{1/3} =$ $(0.95)^{2/3}r_{GS}$, since for the geosynchronous orbit, $(GM_E/4\pi^2)^{1/3} = r_{GS}/(1 \text{ d})^{2/3}$, where $r_{GS} =$ 42,200 km (see Example 9-4). The difference in the orbital radii is $r_{GS}, -r = [1 - (0.95)^{2/3}](42,200 \text{ km}) =$ 1420 km. (We neglected the time spent in changing orbits as suggested.)

CHAPTER 10   SYSTEMS OF PARTICLES

ActivPhysics can help with these problems: Activities 6.6, 6.7

Section 10-1:   Center of Mass

Problem

1. A 28-kg child sits at one end of a 3.5-m-long seesaw. Where should her 65-kg father sit so the center of mass will be at the center of the seesaw?

Solution

Take the x-axis along the seesaw in the direction of the father, with origin at the center. The center of mass of the child and her father is at the origin, so $x_{cm} = 0 = m_c x_c + m_f x_f$, where the masses are given, and $x_c = -(3.5 \text{ m})/2$ (half the length of the seesaw in the negative x direction). Thus, $x_f = -m_c x_c/m_f = (28/65)(1.75 \text{ m}) = 75.4$ cm from the center.

Problem

3. Four trucks, with masses indicated in Fig. 10-23, are on a rectangular barge of mass 35 Mg whose center of mass is at its center. The trucks' individual centers of mass are located 25 m apart on the barge's long dimension and 10 m apart on the short dimension, as shown. Where is the center of mass of the entire system? Express in relation to the truck at lower left.

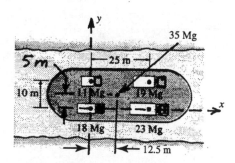

FIGURE 10-23 Top view of four trucks on a barge, with truck masses given. Dots mark individual trucks' centers of mass (Problems 3, 8).

Solution

As explained in the text (see Figs. 10-8 and 9), in order to find the center of mass of this system, the trucks and the barge can be treated as point masses located at their centers of mass. With x-y axes as shown superimposed on Fig. 10-23,

$$x_{cm} = \frac{(18 \text{ Mg})(0) + (23 \text{ Mg})(25 \text{ m}) + (19 \text{ Mg})(25 \text{ m}) + (11 \text{ Mg})(0) + (35 \text{ Mg})(}{(18 + 23 + 19 + 11 + 35) \text{ Mg}}$$

$$= 14.0 \text{ m},$$

$$y_{cm} = \frac{(19 \text{ Mg} + 11 \text{ Mg})(10 \text{ m}) + (35 \text{ Mg})(5 \text{ m})}{106 \text{ Mg}}$$

$$= 4.48 \text{ m}.$$

Problem

5. Three equal masses lie at the corners of an equilateral triangle of side ℓ. Where is the center of mass?

Solution

Take x-y coordinates with origin at the center of one side as shown. From the symmetry (for every mass at x, there is an equal mass at $-x$) $x_{cm} = 0$. Since $y = 0$ for the two masses on the x-axis, and $y = \ell \sin 60° = \ell\sqrt{3}/2$ for the other mass, Equation 10-2 gives $y_{cm} = m(\ell\sqrt{3}/2)/3m = \ell/2\sqrt{3} = 0.289\ell$.

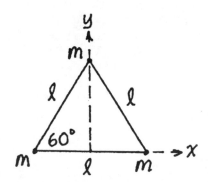

Problem 5 Solution.

Problem

7. Find the center of mass of the plane object shown in Fig. 10-9, assuming uniform density.

Solution

With origin at the lower left-hand corner, the coordinates of the centers of the three constituent rectangles are $CM_1 = (10, 42.5)$, $CM_2 = (57.5, 72.5)$, and $CM_3 = (82.5, 42.5)$, all dimensions in centimeters (see Fig. 10-9). The masses are proportional to the areas, $m_1 : m_2 : m_3 = (85 \times 20) : (75 \times 25) : (35 \times 25) = 1700 : 1875 : 875$. Thus,

$$x_{cm} = \frac{(1700)(10) + (1875)(57.5) + (875)(82.5)}{(1700 + 1875 + 875)}$$
$$= 44.3 \text{ cm},$$

and

$$y_{cm} = \frac{(1700)(42.5) + (1875)(72.5) + (875)(42.5)}{(1700 + 1875 + 875)}$$
$$= 55.1 \text{ cm}.$$

Since 20 cm $< x_{cm} <$ 75 cm, and $y_{cm} <$ 60 cm, the center of mass lies outside the object, as shown.

Problem

9. Find the center of mass of a pentagon of side a with one triangle missing, as shown in Fig. 10-24. *Hint:* See Example 10-3, and treat the pentagon as a group of triangles.

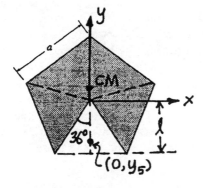

FIGURE 10-24 Problem 9 Solution.

Solution

Choose coordinates as shown. From symmetry, $x_{cm} = 0$. If the fifth isosceles triangle (with the same assumed uniform density) were present, the center of mass of the whole pentagon would be at the origin, so $0 = (my_5 + 4my_{cm})/5m$, where y_{cm} gives the position of the center of mass of the figure we want to find, and y_5 is the position of the center of mass of the fifth triangle. Of course, the mass of the figure is four times the mass of the triangle. In Example 10-3, the center of mass of an isosceles triangle is calculated, so $y_5 = -\frac{2}{3}\ell$, and from the geometry of a pentagon,

$\tan 36° = \frac{1}{2}a/\ell$. Therefore, $y_{cm} = -\frac{1}{4}y_5 = \frac{1}{6}\ell = \frac{1}{12}a \cot 36° = 0.115a$.

Problem

11. A water molecule (H_2O) consists of two hydrogen atoms, each of mass 1.0 u, and one oxygen atom of mass 16 u. The hydrogen atoms are 96 pm from the oxygen and are separated by an angle of 105° (Fig. 10-26). Where is the center of mass of the molecule?

Solution

Take the origin at the center of the oxygen atom with the y-axis along the axis of symmetry (so that $x_{cm} = 0$). Each hydrogen atom has a y coordinate of $(96 \text{ pm}) \cos(\frac{1}{2} 105°) = 58.4$ pm, so $y_{cm} = [(16 \text{ u})(0) + 2(1 \text{ u})(58.4 \text{ pm})]/(18 \text{ u}) = 6.49$ pm.

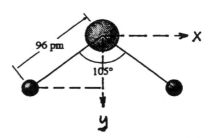

FIGURE 10-26 Problem 11 Solution.

Problem

13. Consider a system of three equal-mass particles moving in a plane; their positions are given by $a_i\hat{\mathbf{i}} + b_i\hat{\mathbf{j}}$, where a_i and b_i are functions of time with the units of position. Particle 1 has $a_1 = 3t^2 + 5$ and $b_1 = 0$; particle 2 has $a_2 = 7t + 2$ and $b_2 = 2$; particle 3 has $a_3 = 3t$ and $b_3 = 2t + 6$. Find the position, velocity, and acceleration of the center of mass as functions of time.

Solution

Since the particles have equal masses, Equation 10-2 gives $x_{cm} = \sum m_i x_i / \sum m_i = \frac{1}{3} \sum a_i$, and $y_{cm} = \frac{1}{3} \sum b_i$. Using the given values of the coefficients (position in meters and time in seconds), we find $x_{cm} = \frac{1}{3}(3t^2 + 5 + 7t + 2 + 3t) = t^2 + (10/3)t + 7/3$ and $y_{cm} = \frac{1}{3}(0 + 2 + 2t + 6) = (2/3)t + 8/3$. Differentiation yields $v_{cm,x} = dx_{cm}/dt = 2t + 10/3$, and $v_{cm,y} = 2/3$ (both in m/s), and $a_{cm,x} = dv_{cm,x}/dt = d^2x_{cm}/dt^2 = 2$, and $a_{cm,y} = 0$ (both in m/s^2).

Problem

15. Estimate how much Earth would move if its entire human population jumped simultaneously into the 1-mile-deep Grand Canyon.

Solution

If the entire human population (mass $m_p \approx 5 \times 10^9 \times$ 70 kg) were assembled at the same place, a distance r from the center of the Earth, the center of mass of the system would be at a distance $r_{cm} = m_p r / (M_E + m_p)$ as shown. Therefore, a change in r produces a change in r_{cm} of $dr_{cm} = m_p dr / (M_E + m_p)$, and the center of the Earth would shift by this amount, relative to the CM. Numerically, if $dr = 1$ mi, $dr_{cm} \approx (5 \times 10^9 \times$ 70 kg$)(1.6$ km$)/(5.97 \times 10^{24}$ kg$) \approx 9.4 \times 10^{-11}$ m, about the size of an atom. Actually, just assembling the world's population at the Grand Canyon would cause a much greater shift in the CM (assumed normally at the center of the Earth) of

$$r_{cm} = \frac{m_p R_E}{M_E + m_p} \approx \frac{(5 \times 10^9 \times 70 \text{ kg})(6370 \text{ km})}{(5.97 \times 10^{24} \text{ kg})}$$
$$\approx 373 \text{ nm}$$

about the wavelength of near ultraviolet light.

Problem 15 Solution.

Problem

17. A hemispherical bowl is at rest on a frictionless kitchen counter. A mouse drops onto the rim of the bowl from a cabinet directly overhead. The mouse climbs down the inside of the bowl to eat the crumbs at the bottom. If the bowl moves along the counter a distance equal to one-tenth of its diameter, how does the mouse's mass compare with the bowl's mass?

Solution

When the mouse starts at the rim, the center of mass of the mouse-bowl system has x component:

$$x_{cm} = (m_b x_b + m_m x_m)/(m_b + m_m)$$
$$= m_m R/(m_b + m_m),$$

since initially, $x_b = 0$, and $x_m = R$. Because there is no external horizontal force (no friction), x_{cm} remains

constant as the mouse climbs. When it reaches the center of the bowl, both have x coordinates equal to x_{cm}, which is, therefore, the distance moved by the bowl across the counter, given as $\frac{1}{5}R$. Thus, $x_{cm} = \frac{1}{5}R = m_m R/(m_b + m_m)$, or $m_m = \frac{1}{4}m_b$.

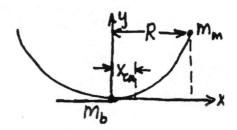

Problem 17 Solution.

Section 10-2: Momentum

Problem

19. A popcorn kernel in a hot pan bursts into two pieces, with masses of 91 mg and 64 mg. The more massive piece moves horizontally at 47 cm/s. Describe the motion of the second piece.

Solution

Suppose the popcorn kernel was initially at rest, and not subject to a net external force while bursting apart. The momentum of the two pieces will still be zero immediately afterwards (conservation of momentum during the break-up). Then $0 = m_1 \mathbf{v}_1 + m_2 \mathbf{v}_2$, where $\mathbf{v}_1$ and $\mathbf{v}_2$ are the final velocities of the two pieces, so $\mathbf{v}_2 = -(m_1/m_2)\mathbf{v}_1$. If $\mathbf{v}_1 = (47$ cm/s$)\hat{\imath}$ is the final velocity of the more massive piece, then $\mathbf{v}_2 = -(91$ mg$/64$ mg$)(47$ cm/s$)\hat{\imath} = -(66.8$ cm/s$)\hat{\imath}$; i.e., the less massive piece moves with speed 66.8 cm/s in the opposite direction to the more massive piece.

Problem

21. A firecracker, initially at rest, explodes into two fragments. The first, of mass 14 g, moves in the positive x direction at 48 m/s. The second moves at 32 m/s. Find its mass and the direction of its motion.

Solution

The momentum of the firecracker/fragments is approximately conserved during a short time interval around the explosion (as for the popcorn kernel in Problem 19), so $0 = m_1 \mathbf{v}_1 + m_2 \mathbf{v}_2$, or $m_2 \mathbf{v}_2 = -(14$ g$)(48\hat{\imath}$ m/s$)$. The direction of the second piece is opposite to that of the first, or $\mathbf{v}_2 = -32\hat{\imath}$ m/s. Then $m_2 = (14$ g$)(48/32) = 21$ g.

Problem

23. A 680-g wood block is at rest on a frictionless table, when a 27-g bullet is fired into it. If the block with the embedded bullet moves off at 19 m/s, what was the original speed of the bullet?

Solution

Since the table's surface is frictionless, there are no horizontal external forces on the block and bullet, and the component of their combined momentum, in the direction of the bullet's initial horizontal velocity v_0, is conserved. As in Example 10-6, $mv_0 = (m + M)V$, so $v_0 = (680 + 27)(19 \text{ m/s})/27 = 498 \text{ m/s}$.

Problem

25. A runaway toboggan of mass 8.6 kg is moving horizontally at 23 km/h. As it passes under a tree, 15 kg of snow drop onto it. What is its subsequent speed?

Solution

The horizontal component of the momentum of the toboggan and snow is conserved (no net external horizontal force), and the final velocity of each is the same: $m_t v_{ti} + 0 = (m_t + m_s)v_f$, or $v_f = m_t v_{ti} \div (m_t + m_s) = (8.6)(23 \text{ km/h})/(8.6 + 15) = 8.38 \text{ km/h}$.

Problem

27. During a heavy storm, rain falls at the rate of 2.0 cm/hour; the speed of the individual raindrops is 25 m/s. (a) If the rain strikes a flat roof and then flows off the roof with negligible speed, what is the force exerted per square meter of roof area? (b) How much water would have to stand on the roof to exert the same force? The density of water is 1.0 g/cm³.

Solution

(a) Suppose the number of drops (each with mass δm and speed v) falling per second is dN/dt. The momentum lost per second is $dp/dt = (dN/dt)\delta m \cdot v$. This equals the magnitude of the force exerted on the roof, so $F/A = (dN/dt)\delta m \cdot v/A$, where A is the roof area. The mass of water falling on the roof per unit time equals the volume rate of fall times the density, or $(dN/dt)\delta m = A(2 \text{ cm/h})(1 \text{ g/cm}^3)$. Therefore, $F/A = (2 \text{ cm}/3600 \text{ s})(1 \text{ g/cm}^3)(25 \text{ m/s}) = 0.139 \text{ N/m}^2$. (b) Water standing on area A to a depth h has weight $Mg = Ah\rho g$. Equating this to the force found in part (a), we find: $Ah\rho g/A = 0.139 \text{ N/m}^3$, or $h = (0.139 \text{ N/m}^3)/(9.8 \text{ m/s}^2)(10^3 \text{ kg/m}^3) = 0.0142 \text{ mm}$. (Even during a heavy storm, the force of the raindrops is quite small.)

Problem

29. An 11,000-kg freight car rests against a spring bumper at the end of a railroad track. The spring has constant $k = 3.2\times10^5$ N/m. The car is hit by a second car of 9400-kg mass moving at 8.5 m/s, and the two cars couple together. (a) What is the maximum compression of the spring? (b) What is the speed of the two cars together when they rebound from the spring?

Solution

(a) The motion of the center of mass of the two freight cars (on an assumed horizontal, frictionless track) is determined by the only horizontal external force, that of the spring. If this is conservative, the potential energy of the spring at maximum compression equals the kinetic energy of the center of mass prior to contact with the spring; i.e., $\frac{1}{2}kx_{max}^2 = \frac{1}{2}MV_{cm}^2$. Now $x_{cm} = (m_1x_1 + m_2x_2)/M$, so $V_{cm} = m_2v_2/M$, since the first car is initially at rest. Thus, $V_{cm} = (9,400 \text{ kg})\times(8.5 \text{ m/s}) = (11,000 + 9,400) \text{ kg} = 3.92 \text{ m/s}$, and $x_{max} = V_{cm}\sqrt{M/k} = (3.92 \text{ m/s})\times \sqrt{(20,400 \text{ kg})/(3.2\times10^5 \text{ N/m})} = 98.9 \text{ cm}$. (b) When the cars rebound, they are coupled together and both have the same velocity as their center of mass. Since the spring is ideal (by assumption), its maximum potential energy, $\frac{1}{2}kx_{max}^2$, is transformed back into kinetic energy of the cars, $\frac{1}{2}MV_{cm}^2$, so the rebound speed equals the initial V_{cm}, or 3.92 m/s. (Reconsider this problem after reading Chapter 11, especially Example 11-2.)

Problem

31. A 1600-kg automobile is resting at one end of a 4500-kg railroad flatcar that is also at rest. The automobile then drives along the flatcar at 15 km/h relative to the flatcar. Unfortunately, the flatcar brakes are not set. How fast does it move?

Solution

Suppose that the track is horizontal and that the horizontal component of the net external force on the flatcar/automobile system is negligible (e.g. rolling friction and air resistance). Then the total momentum along the track, initially zero, is conserved, or $P_{tot} = 0$. Let v_C be the velocity of the flatcar relative to the track, and $v_A = 15 \text{ km/h} + v_C$ be the velocity (positive in the direction of the auto's motion) of the auto relative to the track (see Equation 3-10). Then $0 = m_C v_C + m_A v_A = m_C v_C + m_A(15 \text{ km/h}) + m_A v_C$, or $v_C = -m_A(15 \text{ km/h})/(m_A + m_C) = -(16/61)\times(15 \text{ km/h}) = -3.93 \text{ km/h}$.

Problem

33. A 950-kg compact car is moving with velocity $\mathbf{v}_1 = 32\hat{\imath} + 17\hat{\jmath}$ m/s. It skids on a frictionless icy patch, and collides with a 450-kg hay wagon moving with velocity $\mathbf{v}_2 = 12\hat{\imath} + 14\hat{\jmath}$ m/s. If the two stay together, what is their velocity?

Solution

If there are no external horizontal forces acting on the car-wagon system, momentum (in the x-y plane) is conserved: $m_1\mathbf{v}_1 + m_2\mathbf{v}_2 = (m_1 + m_2)\mathbf{v}$. Therefore,

$$\mathbf{v} = [(950 \text{ kg})(32\hat{\imath} + 17\hat{\jmath})\text{m/s}$$
$$+ (450 \text{ kg})(12\hat{\imath} + 14\hat{\jmath}) \text{ m/s}]/(950 + 450) \text{ kg}$$
$$= (25.6\hat{\imath} + 16.0\hat{\jmath}) \text{ m/s, or } v = \sqrt{v_x^2 + v_y^2}$$
$$= 30.2 \text{ m/s,}$$

and $\theta = \tan^{-1}(v_y/v_x) = 32.1°$ (with the x-axis).

Problem

35. A biologist fires 20-g rubber bullets at a rhinoceros that is charging at 1.9 m/s. If the gun fires 5 bullets per second, with a speed of 1600 m/s, and if the biologist fires for 13 s to stop the rhino in its tracks, what is the mass of the rhino? Assume the bullets drop vertically after striking the rhino, and neglect forces exerted by the rhino's feet.

Solution

The problem describes a one-dimensional collision between a rhino and $5 \times 13 = 65$ rubber bullets, fired in the opposite direction to the rhino's charge (assumed to be horizontal). If the external horizontal forces are negligible, the horizontal momentum of the rhino and bullets is conserved. Since the bullets are supposed to drop vertically and the rhino is stopped, the final total horizontal momentum is zero, so initially $65(0.02 \text{ kg})(1600 \text{ m/s}) - m(1.9 \text{ m/s}) = 0$, or the rhino's mass is $m = 1.09$ metric tons. (Note that 1.9 m/s corresponds to a 14:07 min/mi pace, so a healthy biologist could outrun this rhino's charge.)

Problem

37. An Ariane rocket ejects 1.0×10^5 kg of fuel in the 90 s after launch. (a) How much thrust is developed if the fuel is ejected at 3.0 km/s with respect to the rocket? (b) What is the maximum total mass of the rocket if it is to get off the ground?

Solution

(a) The thrust, given by Equation 10-10, is

$$F_{\text{Th}} = -v_{\text{ex}}\frac{dM}{dt} = -(3 \times 10^3 \text{ m/s})\left(-\frac{10^5 \text{ kg}}{90 \text{ s}}\right)$$
$$= 3.33 \times 10^6 \text{ N.}$$

(b) The thrust must exceed the launch weight of the rocket, $F_{\text{Th}} \geq Mg$, or $M \leq F_{\text{Th}}/g = 3.40 \times 10^5$ kg.

Problem

39. If a rocket's exhaust speed is 200 m/s relative to the rocket, what fraction of its initial mass must be ejected to increase the rocket's speed by 50 m/s?

Solution

From Equation 10-11, $M_i/M_f = e^{(v_f - v_i)/v_{\text{ex}}} = e^{(50 \text{ m/s})/(200 \text{ m/s})} = e^{1/4}$. The mass fraction ejected is $(M_i - M_f)/M_i = 1 - e^{-1/4} = 0.221 = 22.1\%$.

Section 10-3: Kinetic Energy in Many-Particle Systems

Problem

41. Determine the center of mass and internal kinetic energies before and after decay of the lithium nucleus of Example 10-7. Treat the individual nuclei as point particles.

Solution

Before the decay, the system consists of one particle (the ^{5}Li-nucleus), so $K_{\text{int},i} = 0$, and $K_{cm} = K_i = \frac{1}{2}m_{\text{Li}}v_{\text{Li}}^2 = \frac{1}{2}(5 \times 1.67 \times 10^{-27} \text{ kg})(1.6 \times 10^6 \text{ m/s})^2 = 1.07 \times 10^{-14}$ J $= 66.8$ keV. Afterwards, K_{cm} is the same (since momentum is conserved), while $K_{\text{int},f} = K_f - K_{cm} = \frac{1}{2}m_H v_H^2 + \frac{1}{2}M_{\text{He}}v_{\text{He}}^2 - K_{cm} = \frac{1}{2}(1.67 \times 10^{-27} \text{ kg})[(4.5 \times 10^6 \text{ m/s})^2 + 4(1.4 \times 10^6 \text{ m/s})^2] - 66.8$ keV $= 79.8$ keV.

Problem

43. A 1200-kg car moving at 88 km/h collides with a 7600-kg truck moving in the same direction at 65 km/h. The two stick together, continuing in their original direction at 68 km/h. Determine the center of mass and internal energies of the (car + truck) system before and after the collision.

Solution

Since momentum is conserved (for a short enough interval of time around the collision) $K_{cm} = P_{\text{tot}}^2/2M$ is the same, before and after (denoted by subscripts i and f). $K_{cm,i} = [(1200 \text{ kg})(88 \text{ m/s}) + (7600 \text{ kg}) \times$

$(65 \text{ m}/3.6 \text{ s})]^2/2(1200 + 7600) \text{ kg} = 1.58 \text{ MJ} = K_{cm,f}$. (Note that all the momenta are colinear in this collision.) From Equation 10-13, $K_{int,i} = K_i - K_{cm} = \frac{1}{2}(1200 \text{ kg})(88 \text{ m}/3.6 \text{ s})^2 + \frac{1}{2}(7600 \text{ kg})(65 \text{ m}/3.6 \text{ s})^2 - 1.58 \text{ MJ} = 21.2 \text{ kJ}$, and since $K_f = K_{cm}$ (the vehicles stick together so both have zero velocity relative to their center of mass), $K_{int,f} = 0$. (Alternatively, one could determine $v_{cm} = 68.1 \text{ km/h}$ and the relative velocities of the vehicles to calculate $K_{int,i}$ directly.)

Paired Problems

Problem

45. A drinking glass is in the shape of a cylinder whose inside dimensions are 9.0 cm high and 8.0 cm in diameter as shown in Fig. 10-31. Its base has a mass of 140 g, while the mass of the curved, cylindrical sides is 85 g. (a) Where is its center of mass? (b) If the glass is three-quarters filled with juice (density 1.0 g/cm³), where is the center of mass of the glass-juice system? Assume the thickness of the glass is negligible.

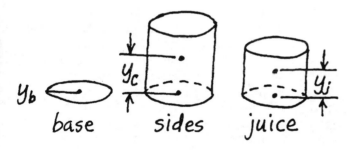

Problem 45 Solution.

Solution

(a) To find the CM of the empty glass, consider it to be composed of two subpieces (as suggested in the text following Example 10-3), namely, its base ($m_b = 140$ g, with CM at the center $y_b = 0$) and the curved cylindrical sides ($m_c = 85$ g, with CM at the midpoint of the axis $y_c = 4.5$ cm). Then

$$y_{cm} = \frac{m_b y_b + m_c y_c}{m_b + m_c} = \frac{85}{225}(4.5 \text{ cm}) = 1.70 \text{ cm}$$

(above the center of the base). (b) Now consider the glass of juice to be composed of two subpieces, namely the empty glass ($m_g = 225$ g, with CM at $y_g = 1.70$ cm) and a cylinder of juice (m_j = density × volume = $\rho \pi r^2 h = (1 \text{ g/cm}^3)\pi(4 \text{ cm})^2(\frac{3}{4} \times 9 \text{ cm}) = 339$ g, with CM at the midpoint on its axis $y_j = \frac{1}{2}(\frac{3}{4} \times 9 \text{ cm}) =$

3.38 cm). Then

$$y_{cm} = \frac{m_g y_g + m_j y_j}{m_g + m_j}$$
$$= \frac{(225)(1.70 \text{ cm}) + (339)(3.38 \text{ cm})}{225 + 339} = 2.71 \text{ cm}$$

is the height of the CM of the glass of juice above the center of the base.

Problem

47. Olympic-champion ice dancers Marina Klimova (mass 50 kg) and Sergei Ponomarenko (mass 70 kg) start from rest and push off each other on essentially frictionless ice. If Ponomarenko's speed is 2.8 m/s, how far apart are they after 3.0 s?

Solution

For frictionless ice, the horizontal momentum of the pair of skaters along the direction of motion is conserved, $P_{tot} = m_1 v_1 + m_2 v_2 = 0$ (since they start from rest). Then $v_2 = -(m_1/m_2)v_1$, and their relative speed of separation is $v_{rel} = v_1 - v_2 = (1 + m_1/m_2) \times v_1 = (1 + 70/50)(2.8 \text{ m/s}) = 6.72 \text{ m/s}$. After $t = 3$ s, they are $v_{rel}t = (6.72 \text{ m/s})(3 \text{ s}) = 20.2$ m apart.

Problem

49. A 42-g firecracker is at rest at the origin when it explodes into three pieces. The first, with mass 12 g, moves along the x-axis at 35 m/s. The second, with mass 21 g, moves along the y-axis at 29 m/s. Find the velocity of the third piece.

Solution

The instant after the explosion (before any external forces have had any time to act appreciably) the total momentum of the pieces of firecracker is still zero, $\mathbf{P}_{tot} = 0 = m_1\mathbf{v}_1 + m_2\mathbf{v}_2 + m_3\mathbf{v}_3$. Therefore

$$\mathbf{v}_3 = \frac{-(12 \text{ g})(35 \text{ m/s})\hat{\mathbf{i}} - (21 \text{ g})(20 \text{ m/s})\hat{\mathbf{j}}}{(42 - 12 - 21) \text{ g}}$$
$$= -(46.7\hat{\mathbf{i}} + 67.7\hat{\mathbf{j}}) \text{ m/s}.$$

Problem

51. A 4,100-kg bull elephant is at one end of a 16,000-kg circus train car initially at rest on a frictionless track. The elephant charges the 17-m length of the car at a speed of 6.5 m/s, then stops when it hits the far wall. (a) How far does the car move during the elephant's charge? (b) How fast is the car moving during the charge? (c) What is the car's speed after the elephant has stopped? Assume the center of mass of the car is at its center, and treat the elephant as a particle.

Solution

In the absence of external forces parallel to the track (assumed horizontal), the total momentum of the car and elephant along the track is zero (they are initially at rest), and their combined center of mass is stationary; i.e.,
$P_{tot} = MV_{cm} = 0$, or $m_E v_E + m_c v_c = 0$, and $m_E x_E + m_c x_c$ = constant. Here, we measure x and v on a fixed coordinate axis along the track, with positive direction parallel to the elephant's motion. (a) Let x denote a position at the start of the charge and x' one at the end. Then
$m_E x_E + m_c x_c = m_E x'_E + m_c x'_c$, or $m_c(x_c - x'_c) = m_E(x'_E - x_E)$. Now, $x_c - x'_c$ is the distance the car moves, $x'_E - x'_c$ is the distance from the far wall to the CM of the car, and $x_c - x_E + x'_E - x'_c = 17$ m, the length of the car (see sketch). Thus,
$x'_E - x_E = 17$ m $- (x_c - x'_c)$, and $m_c(x_c - x'_c) = m_E[17$ m $- (x_c - x'_c)]$, so $x_c - x'_c = m_E(17$ m$)/(m_E + m_c) = (41)(17$ m$)/(201) =$ 3.47 m. (Note: It was not necessary to assume that the CM of the car was at it's center.) (b) While the elephant is moving with speed 6.5 m/s relative to the car, $v_E = -m_c v_c/m_E$, and $v_{rel} = v_E - v_c$. Then $v_{rel} = -v_c - m_c v_c/m_E = -v_c(1 + m_c/m_E)$, or $v_c = -(6.5$ m/s$)/(1 + 160/41) =$ -1.33 m/s (the minus sign means that the car moves in the opposite direction to the elephant). (c) After the elephant has stopped, $v_E = 0 = v_c$.

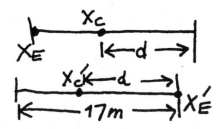

Problem 51 Solution.

Problem

53. Firefighters spray water horizontally at the rate of 41 kg/s from a nozzle mounted on a 12,000-kg fire truck. The water speed is 28 m/s relative to the truck. (a) Neglecting friction, what is the initial acceleration of the truck? (b) If the 12,000-kg truck mass includes 2,400 kg of water, how fast will the truck be moving when the water is exhausted? *Hint:* Think about rockets.

Solution

The fire truck behaves like a water rocket with thrust given by Equation 10-10b: $F_{thrust} = -v_{ex}(dM/dt) =$ $-(28$ m/s$)(-41$ kg/s$) = 1.15$ kN. (a) If the thrust is the only significant horizontal force, the truck's initial acceleration is $a_i = F_{thrust}/M_i = 1.15$ kN/12,000 kg $=$ 9.57 cm/s^2. (b) If the truck starts from rest, Equation 10-11, with $v_i = 0$, gives $v_f =$ $v_{ex} \ln(M_i/M_f) =$ $(28$ m/s$) \ln(12,000/(12,000 - 2,400)) = 6.25$ m/s.

Supplementary Problems

Problem

55. A 55-kg sprinter is standing at the left end of a 240-kg cart moving to the left at 7.6 m/s. She runs to the right end and continues horizontally off the cart. What should be her speed relative to the cart in order to leave the cart with no horizontal velocity component relative to the ground?

Solution

If there are no external forces on the cart/sprinter system, with components along the direction of motion (positive in the direction of the sprinter), then the total momentum in that direction is conserved, or $P_{tot} = (m_s + m_c)v_{cm} = m_s v_s + m_c v_c$ = constant. The velocity of the sprinter relative to the cart is $v_{rel} = v_s - v_c$, so $(m_s + m_c)v_{cm} = m_s v_s + m_c(v_s - v_{rel}) = (m_s + m_c)v_s - m_c v_{rel}$. v_{cm} is a constant -7.6 m/s, so $v_s = 0$ when $v_{rel} = -(m_s + m_c)v_{cm}/m_c =$ $-(55 + 240)(-7.6$ m/s$)/240 = 9.34$ m/s.

Problem

57. Figure 10-32 shows a paraboloidal solid of constant density ρ. It extends a height h along the z-axis, and is described by $z = ar^2$, where the units of a are m^{-1}, and r is the radius in a plane perpendicular to the z-axis. Find expressions for (a) the mass of the solid and (b) the z coordinate of its center of mass.

Solution

For mass elements, take disks parallel to the x-y plane, of radius $r = \sqrt{z/a}$ and thickness dz, as shown in Fig. 10-32 and also in Fig. 10-28 for Problem 18. Then $dm = \rho\pi r^2 \, dz = \rho\pi(z/a) \, dz$. (a) The total mass is $M = \int_0^h dm = \int_0^h (\rho\pi/a)z \, dz = \rho\pi h^2/2a$. (b) $z_{cm} = \int_0^h z \, dm/M = (2a/\rho\pi h^2) \int_0^h (\rho\pi/a)z^2 \, dz = (2/h^2) \times (h^3/3) = 2h/3$. (Since the paraboloid is symmetrical about the z-axis, $x_{cm} = y_{cm} = 0$.)

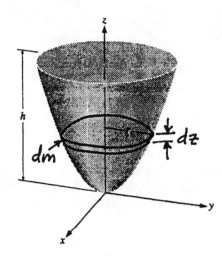

FIGURE 10-32 Problem 57.

Problem

59. Cookie sheets move along a conveyor belt toward an oven, with mounds of unbaked dough dropping vertically onto the sheets at the rate of one 12-g mound every two seconds. What average force must the conveyer belt exert on a cookie sheet to keep it moving at a constant 50 cm/s?

Solution

If we assume that the conveyor belt is horizontal and moving with speed $v = 50$ cm/s, and that the mounds of dough fall vertically, then the change in the horizontal momentum of each mound of mass Δm is $(\Delta m)v$. The average horizontal force needed is equal to the rate at which mounds are dropped (a number N in time Δt, or $N/\Delta t$) times the change in momentum of a mound, or $F_{av} = (N/\Delta t)(\Delta m)v$. (This is just the total change in momentum, $N(\Delta m)v$, divided by the time, Δt.) Therefore, $F_{av} = (1/2\text{ s})(12\text{ g})(50\text{ cm/s}) = 300$ dyne $= 3$ mN. (The dyne $= \text{g·cm/s}^2$ is the old cgs-unit for force; see Appendix C.)

Problem

61. While standing on frictionless ice, you (mass 65.0 kg) toss a 4.50-kg rock with initial speed of 12.0 m/s. If the rock is 15.2 m from you when it lands, (a) at what angle did you toss it? (b) How fast are you moving?

Solution

If we assume that the ice surface is horizontal, then the x component of the center of mass of you and the rock stays at rest (at the origin), and the horizontal momentum is conserved until the rock lands:

$$x_{cm} = \frac{m_1 x_1 + m_2 x_2}{m_1 + m_2} = 0, \qquad m_1 v_{1x} + m_2 v_{2x} = 0.$$

(a) The angle of elevation is $\theta_0 = \frac{1}{2}\sin^{-1}(gx_1/v_0^2)$, from Equation 4-10, if we neglect the initial height of the toss. Since $x_1 - x_2 = 15.2$ m, and $x_2 = -(4.5\text{ kg}/65\text{ kg})x_1$, we find $x_1 = (65/69.5)(15.2\text{ m}) = 14.2$ m, and $\theta_0 = \frac{1}{2}\sin^{-1}(9.8\text{ m/s}^2 \times 14.2\text{ m} \div (12.0\text{ m/s})^2) = 37.7°$. (b) $v_{2x} = -(m_1/m_2)v_{1x} = -(4.5/65)(12.0\text{ m/s})\cos 37.7° = -65.8$ cm/s.

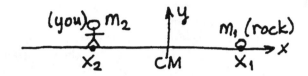

Problem 61 Solution.

Problem

63. A fireworks rocket is launched vertically upward at 40 m/s. At the peak of its trajectory, it explodes into two equal-mass fragments. One reaches the ground 2.87 s after the explosion. When does the second reach the ground?

Solution

The rocket reaches a maximum height (from Equation 2-11 with altered notation) of $y_0 = (40\text{ m/s})^2 \div 2(9.8\text{ m/s}^2) = 81.6$ m, at the time of the explosion ($t = 0$). The first piece falls to the ground ($y = 0$) in time $t_1 = 2.87$ s, so the y component of its initial velocity (just after the explosion) can be found from Equation 4-8: $0 = y_0 + (v_{0y})_1 t_1 - \frac{1}{2}gt_1^2$, or $(v_{0y})_1 = \frac{1}{2}gt_1 - y_0/t_1 = -14.4$ m/s. At $t = 0$, the y component of the velocity of the center of mass (which follows the original trajectory of the rocket) is zero; i.e., $0 = (m_1 + m_2)(v_{0y})_{cm} = m_1(v_{0y})_1 + m_2(v_{0y})_2$, or $(v_{0y})_2 = -(v_{0y})_1 = 14.4$ m/s (recall that the fragments have equal mass). The time for the second piece to hit the ground can also be found from Equation 4-8 (this time with the aid of the quadratic formula):

$$0 = y_0 + (v_{0y})_2 t_2 - \tfrac{1}{2}gt_2^2, \qquad \text{or}$$

$$t_2 = [(v_{0y})_2 + \sqrt{(v_{0y})_2^2 + 2gy_0}]/g = 5.80\text{ s}.$$

Problem

65. (a) Derive an expression for the thrust of a jet aircraft engine. Moving through the air with speed v, the engine takes in air at the rate dM_{in}/dt. It uses the air to burn fuel with a fuel/air ratio f (that is, f kg of fuel burned for each kg of air), and ejects the exhaust gases at speed v_{ex} with respect to the engine. (b) Use your result to find the thrust of a JT-8D engine on a Boeing 727

jetliner, for which $v_{\text{ex}} = 1034$ ft/s and $dM_{\text{in}}/dt = 323$ lb/s, and which consumes 3760 lb of fuel per hour while cruising at 605 mi/h.

Solution

(a) The operation of a ramjet engine is similar to the rockets described in Section 10-2, except that air, originally at rest, is taken in and exhausted along with the gas from fuel combustion. If we neglect gravity (e.g., in horizontal flight) and air resistance, the momentum of the engine-fuel-air system is conserved. At the beginning of a time interval dt, the engine and fuel have mass $m + dm_f$ and horizontal speed v, and the air is at rest, so $P_i = (m + dm_f)v$. After a time dt, the engine has mass m and speed $v + dv$, and combustion gas of mass dm_f and air of mass dM_{in} are ejected with speed $v - v_{\text{ex}}$ relative to the ground, so $P_f = m(v + dv) + (dm_f + dM_{\text{in}})(v - v_{\text{ex}})$. Equate P_i and P_f, expand the products, rearrange terms, and divide by dt: $0 = P_f - P_i = m(v + dv) + (dm_f + dM_{\text{in}})(v - v_{\text{ex}}) - (m + dm_f)v$, or $m(dv/dt) = v_{\text{ex}}(dm_f/dt) + (v_{\text{ex}} - v)(dM_{\text{in}}/dt) = $ thrust. (Note that if $dM_{\text{in}}/dt = 0$, we recover the rocket equation, since $dm_f = -dm$.) In terms of the fuel/air ratio, defined by $dm_f = f \, dM_{\text{in}}$, the thrust is $m(dv/dt) = [(1 + f)v_{\text{ex}} - v](dM_{\text{in}}/dt)$. (b) With the data given for the JT-8D engine, $v_{\text{ex}} = 1034$ ft/s, $v = 605$ mi/h $= 887.3$ ft/s, $(dM_{\text{in}}/dt)g = 323$ lb/s, $(g = 32.2$ ft/s$^2)$, $f = (3760 \text{ lb}/3600 \text{ s})/(323 \text{ lb/s}) = 3.234 \times 10^{-3}$, the thrust is $[(1 + 3.234 \times 10^{-3})(1034 \text{ ft/s}) - 887.3 \text{ ft/s}] \times (323 \text{ lb/s})/(32.2 \text{ ft/s}^2) = 1504$ lb. (Note that dM_{in}/dt is the rate of air-mass intake, while 323 lb/s is the rate of air-weight intake.)

Problem

67. An ideal spring of spring constant k rests on a frictionless surface. Blocks of mass m_1 and m_2 are pushed against the two ends of the spring until it is compressed a distance x from its equilibrium length. The blocks are then released. What are their speeds when they leave the spring?

Solution

If the surface is also level as well as frictionless, there are no external horizontal forces on the blocks or spring, so the horizontal momentum is conserved. Initially, everything is at rest so $P_{\text{tot}} = 0$, and when the blocks leave the spring, $m_1 v_1 + m_2 v_2 = 0$, or $v_2 = -m_1 v_1/m_2$, where we take the positive direction parallel to the motion of m_1. (Since there is no net horizontal force on the spring, its center of mass momentum is zero.) For an ideal spring, there is no loss of energy, so $\frac{1}{2}kx^2 = \frac{1}{2}m_1 v_1^2 + \frac{1}{2}m_2 v_2^2$. If we eliminate v_2 using conservation of momentum, then $kx^2 = m_1 v_1^2 + m_2(-m_1 v_1/m_2)^2 = m_1(1 + m_1/m_2)v_1^2$, so $v_1 = [m_2 kx^2/m_1(m_1 + m_2)]^{1/2}$. Finally, $v_2 = -m_1 v_1/m_2 = -[m_1 kx^2/m_2(m_1 + m_2)]^{1/2}$. (The speeds are the magnitudes of the velocities found above.)

CHAPTER 11   COLLISIONS

ActivPhysics can help with these problems:
Activities 6.2–6.6

Section 11-1:   Impulse and Collisions

Problem

1. What is the impulse associated with a 650-N force acting for 80 ms?

Solution

The impulse for a constant force (from Equation 11-1) is $I = \int F \, dt = (650 \text{ N})(0.08 \text{ s}) = 52$ N·s, in the direction of the force.

Problem

3. A 62-kg parachutist hits the ground moving at 35 km/h and comes to a stop in 140 ms. Find the average impulsive force on the chutist, and compare with the chutist's weight.

Solution

The average impulsive force (from Equations 11-1 and 2) is $\mathbf{F}_{av} = \mathbf{\Delta p}/\Delta t = (0 - m\mathbf{v}_i)/\Delta t = -(62 \text{ kg}) \times$ (35 m/3.6 s)/(0.14 s) = −4.31 kN (opposite to the direction of $\mathbf{v}_i$, or upward for most jumps under calm wind conditions). The magnitude is slightly over seven times the parachutist's weight, or 7.09 mg.

Problem

5. A 240-g ball is moving with velocity $\mathbf{v}_i = 6.7\hat{\imath}$ m/s when it undergoes a collision lasting 52 ms. After the collision its velocity is $\mathbf{v}_f = -4.3\hat{\imath} + 3.1\hat{\jmath}$ m/s. Find (a) the impulse and (b) the average impulsive force associated with this collision.

Solution

(a) From Equation 11-1, $\mathbf{I} = \mathbf{\Delta p} = m(\mathbf{v}_f - \mathbf{v}_i) =$ $(0.24 \text{ kg})(-4.3\hat{\imath} + 3.1\hat{\jmath} - 6.7\hat{\imath})(\text{m/s}) = (-2.64\hat{\imath} +$ $0.744\hat{\jmath})$ N·s. (b) From Equation 11-2, $\mathbf{F}_{av} =$ $\mathbf{I}/(0.052 \text{ s}) = (-50.8\hat{\imath} + 14.3\hat{\jmath})$ N.

Problem

7. Safety standards call for a 1900-kg car colliding at 12 m/s with a concrete wall to experience an average impulsive force not greater than 50 kN. What is the minimum permissible time for the car to come to a stop during such a collision?

Solution

From Equation 11-2, if $F_{av} = \Delta p/\Delta t \le 50$ kN, then $\Delta t \ge \Delta p/50$ kN. When the car is brought to rest in a collision with the wall, the magnitude of its momentum change is $\Delta p = |0 - m\mathbf{v}_i| = (1900 \text{ kg}) \times$ (12 m/s), so $\Delta t \ge 0.456$ s. (Note: if the car had bounced off the wall during the collision, Δp would have been larger, as well as Δt.)

Problem

9. A 727 jetliner in level flight with a total mass of 8.6×10^4 kg encounters a downdraft lasting 1.3 s. During this time, the plane acquires a downward velocity component of 85 m/s. Find (a) the impulse and (b) the average impulsive force on the plane.

Solution

(a) The impulse equals the downward component of the change in the plane's momentum (Equation 11-1): $I = \Delta p_y = mv_y = (8.6 \times 10^4 \text{ kg})(-85 \text{ m/s}) =$ -7.31×10^6 N·s. (b) The average impulsive force (Equation 11-2) is $F_{av} = I/\Delta t = I/1.3$ s $= -5.62$ MN. (Here, minus means downward.)

Problem

11. (a) Estimate the impulse imparted by the force shown in Fig. 11-18. (b) What is the average impulsive force?

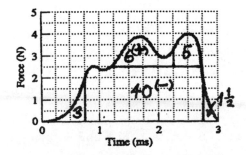

FIGURE 11-18 Problem 11 Solution.

Solution

(a) The impulse is the area under the curve in Fig. 11-18. By counting boxes, each of which has "area" $(0.25 \text{ ms})(0.5 \text{ N}) = 1.25 \times 10^{-4}$ N·s, we find $I \simeq$ $(55.5 \text{ boxes})(1.25 \times 10^{-4} \text{ N·s/box}) = 6.94 \times 10^{-3}$ N·s. (The direction of I is in the direction of F, assumed to be constant.) (b) $F_{av} = I/\Delta t = 6.94 \times 10^{-3}$ N·s $\div$ 3×10^{-3} s $= 2.31$ N.

Section 11-2:   Collisions and the Conservation Laws

Problem

13. At the peak of its trajectory, a 1.0-kg projectile moving horizontally at 15 m/s collides with a 2.0-kg projectile at the peak of a vertical trajectory. If the collision takes 0.10 s, how good is the assumption that momentum is conserved during the collision? To find out, compare the change in momentum of the colliding system with the system's total momentum.

Solution

The total momentum before the collision is due to just the first projectile (the second is instantaneously at rest), so $\mathbf{P}_i = m_1 \mathbf{v}_{1i} = (1 \text{ kg})(15 \text{ m/s})\hat{\mathbf{i}} = (15 \text{ kg·m/s})\hat{\mathbf{i}}$. The change in total momentum is due to the external force (gravity), so $\Delta \mathbf{P} = \mathbf{F}_g \Delta t = (1 \text{ kg} + 2 \text{ kg}) \times$ $(9.8 \text{ m/s}^2)(-\hat{\mathbf{j}})(0.1 \text{ s}) = -(2.94 \text{ kg·m/s})\hat{\mathbf{j}}$. This is almost 20% of $\mathbf{P}_i$ in magnitude. However, since $\Delta \mathbf{P}$ and $\mathbf{P}_i$ are perpendicular, $P_f = \sqrt{P_i^2 + \Delta P^2} \approx$ $P_i(1 + \frac{1}{2}(\Delta P/P_i)^2 + \cdots)$, which differs from P_i by slightly less than 2% in magnitude, and about 11° in direction.

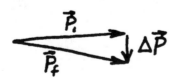

Problem 13 Solution.

Problem

15. An 1800-kg car moving at 25 m/s collides with an identical car moving in the same direction at 15 m/s. If an external frictional force of 6.1 kN acts on both cars, what is the minimum collision time that will ensure the system momentum changes by less than 0.1% during the collision?

Solution

The change in the total momentum of the cars equals the (external) impulse of friction, $\Delta P_x = F_{av}\Delta t$ (we suppose that the motion of the cars and the frictional force are along the x-axis). The total momentum of the (identical) cars is $P_x = mv_{1x} + mv_{2x}$. Then

$|\Delta P_x/P_x| < 0.1\%$ implies $|F_{av}\Delta t/P_x| < 0.1\%$, or $\Delta t < 10^{-3}|P_x/F_{av}| = (1800 \text{ kg})(25 + 15)(\text{m/s})10^{-3} \div$ $(6.1 \times 10^3 \text{ N}) = 11.8$ ms.

Section 11-3:   Inelastic Collisions

Problem

17. In a railroad switchyard, a 56-ton freight car is sent at 7.0 mi/h toward a 31-ton car that is moving in the same direction at 2.6 mi/h. (a) What is the speed of the pair after they couple together? (b) What fraction of the initial kinetic energy was lost in the collision?

Solution

(a) If we assume the switchyard track is straight and level, the collision is one-dimensional, totally inelastic, and Equation 11-4 applies: $v_f = [(56 \text{ T})(7.0 \text{ mi/h}) +$ $(31 \text{ T})(2.6 \text{ mi/h})]/(56 + 31)\text{T} = 5.43$ mi/h. (b) The initial and final kinetic energies are $K_i = \frac{1}{2}[(56 \text{ T}) \times$ $(7.0 \text{ mi/h})^2 + (31 \text{ T})(2.6 \text{ mi/h})^2] = 1477 \text{ T}(\text{mi/h})^2$; $K_f = \frac{1}{2}(56 + 31)\text{T} (5.43 \text{ mi/h})^2 = 1284 \text{ T}(\text{mi/h})^2$. The fraction lost is $(K_i - K_f)/K_i = 13.1\%$. (Note: It was not necessary to change to standard units to answer this question.)

Problem

19. A sled and child with a total mass of 33 kg are moving horizontally at 10 m/s when a second child leaps on with negligible speed. If the sled's speed drops to 6.4 m/s, what is the mass of the second child?

Solution

We may assume that the horizontal momentum is conserved and use the component of Equation 11-3 in the direction of motion. With $m_1 = 33$ kg and $v_{1i} = 10$ m/s (representing the original child and sled), and m_2 and $v_{2i} = 0$ (representing the second child), $m_1 v_{1i} = (m_1 + m_2)v_f$, or $m_2 = m_1(v_{1i} - v_f)/v_f =$ $33 \text{ kg } (10 - 6.4) \text{ m/s}/(6.4 \text{ m/s}) = 18.6$ kg.

Problem

21. A mass m collides totally inelastically with a mass M initially at rest. Show that a fraction $M/(m + M)$ of the initial kinetic energy is lost in the collision.

Solution

In a totally inelastic collision, when one of the two bodies is initially at rest, $v_f = mv_i/(m + M)$ (see Equation 11-4). Then $K_f = \frac{1}{2}(m + M)v_f^2 =$ $\frac{1}{2}m^2 v_i^2/(m + M) = K_i m/(m + M)$, so the fractional loss of kinetic energy is $(K_i - K_f)/K_i = 1 - K_f/K_i = 1 - m/(m + M) = M/(m + M)$.

Problem

23. Astronomers warn that there is a nonzero chance of Earth colliding inelastically with a substantial asteroid. Impact speed of the asteroid might be 10 km/s. Estimate the mass of an asteroid needed to alter Earth's orbital speed by 0.01%.

Solution

For a totally inelastic collision between an asteroid and the Earth, Equation 11-3 can be rearranged to give $M_E(\mathbf{v}_f - \mathbf{v}_{Ei}) = m_a(\mathbf{v}_{ai} - \mathbf{v}_f)$. Now $\mathbf{v}_f - \mathbf{v}_{Ei} = \Delta\mathbf{v}_E$ is the change in the velocity of the Earth, and $\mathbf{v}_{ai} - \mathbf{v}_f = \mathbf{v}_r$ is the velocity of the asteroid relative to the center of mass of the Earth-asteroid system. Since $M_E \gg m_a$, $\mathbf{v}_r$ is approximately the velocity of impact relative to the Earth. Then, in magnitude, $M_E |\Delta\mathbf{v}_E| = m_a |\mathbf{v}_r|$. For $|\Delta\mathbf{v}_E| \sim 10^{-4}(30 \text{ km/s})$, i.e., 0.01% of the Earth's orbital speed, and $|\mathbf{v}_r| \sim 10$ km/s, the asteroid's mass would have to be $m_a = M_E |\Delta\mathbf{v}_E| / |\mathbf{v}_r| \sim (6\times10^{24} \text{ kg})(30\times10^{-4})/10 \sim 10^{21}$ kg. (The mass of Ceres, the largest asteroid, is about 1.4×10^{21} kg, which is nearly half the mass of all asteroids.)

Problem

25. A neutron (mass 1 u) strikes a deuteron (mass 2 u), and the two combine to form a tritium nucleus. If the neutron's initial velocity was $28\hat{\imath} + 17\hat{\jmath}$ Mm/s and if the tritium nucleus leaves the reaction with velocity $12\hat{\imath} + 20\hat{\jmath}$ Mm/s, what was the velocity of the deuteron?

Solution

Since the masses and two of the three velocities in Equation 11-3, which describes the conservation of momentum for the totally inelastic collision producing the tritium nucleus, are known, the initial velocity of the deuteron can easily be found:

$$\mathbf{v}_d = \frac{m_t\mathbf{v}_t - m_n\mathbf{v}_n}{m_d}$$

$$= \frac{3 \text{ u}(12\hat{\imath} + 20\hat{\jmath}) - 1 \text{ u}(28\hat{\imath} + 17\hat{\jmath})}{2 \text{ u}} \frac{\text{Mm}}{\text{s}}$$

$$= (4\hat{\imath} + 21.5\hat{\jmath}) \text{ Mm/s}$$

Problem

27. Two identical pendulum bobs are suspended from strings of equal length, and one is released from a height h as shown in Fig. 11-19. When the first bob hits the second, the two stick together. Show that the maximum height to which the combination rises is $\frac{1}{4}h$.

Solution

In a totally inelastic collision between equal-mass objects, half the initial kinetic energy is lost. (See Problem 18; the result follows from Equation 11-4 with $m_1 = m_2$ and $v_{2i} = 0$, since then $v_f = \frac{1}{2}v_{1i}$ and $K_f = \frac{1}{2}(2 \text{ m})(\frac{1}{2}v_{1i})^2 = \frac{1}{2}K_i$.) The conservation of energy applied to a simple pendulum (see Example 8-6) requires that $\Delta U_{\max} = mgh = K_i$ before the collision, and $\Delta U'_{\max} = (2m)gh' = K_f$ after. Therefore $K_f = \frac{1}{2}K_i$ implies that $2mgh' = \frac{1}{2}mgh$, or $h' = \frac{1}{4}h$.

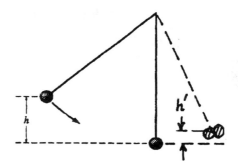

FIGURE 11-19 Problem 27 Solution.

Problem

29. A 400-mg popcorn kernel is skittering across a nonstick frying pan at 8.2 cm/s when it pops and breaks into two equal-mass pieces. If one piece ends up at rest, how much energy was released in the popping?

Solution

In order to conserve momentum, the piece which is not at rest must acquire a velocity equal to twice that of the original kernel, $m\mathbf{v}_i = (\frac{1}{2}m)\mathbf{v}_{1f}$. Therefore, the final energy is twice the initial energy, $K_f = \frac{1}{2}(\frac{1}{2}m)\times v_{1f}^2 = \frac{1}{2}(\frac{1}{2}m)(2v_i)^2 = mv_i^2 = 2K_i$. The energy released is $K_f - K_i = K_i = \frac{1}{2}(4\times10^{-4} \text{ kg})(0.082 \text{ m/s})^2 = 1.34\times10^{-6}$ J.

Problem

31. Two identical objects with the same initial speed collide and stick together. If the composite object moves with half the initial speed of either object, what was the angle between the initial velocities?

Solution

In Equation 11-4, set $m_1 = m_2$, $|\mathbf{v}_{1i}| = |\mathbf{v}_{2i}| = 2|\mathbf{v}_f|$, and square. After canceling the masses, one obtains $v_f^2 = \frac{1}{4}v_{1i}^2 = \frac{1}{4}(v_{1i}^2 + 2\mathbf{v}_{1i}\cdot\mathbf{v}_{2i} + v_{2i}^2) = \frac{1}{4}2v_{1i}^2(1 + \cos\theta)$. Solving for the angle, we find: $\theta = \cos^{-1}(-\frac{1}{2}) = 120°$.

Problem

33. While playing ball in the street, a child accidentally tosses a ball at 18 m/s toward the front of a car moving toward him at 14 m/s. What is the speed of the ball after it rebounds elastically from the car?

Solution

In a head-on elastic collision, the relative velocity of separation is equal to the negative of the relative velocity of approach (see Equation 11-8). If $m_2 \gg m_1$ (a car versus a ball), then $v_{2f} \approx v_{2i} = 14$ m/s, and $v_{1f} = v_{2f} + v_{2i} - v_{1i} = 14$ m/s $+ 14$ m/s $- (-18$ m/s$) = 46$ m/s. (We chose positive velocities in the direction of the car.)

Problem

35. A proton moving at 6.9 Mm/s collides elastically and head-on with a second proton moving in the opposite direction at 11 Mm/s. Find their velocities after the collision.

Solution

An elastic, head-on collision between two protons is described by Equations 11-9a and b with $m_1 = m_2$. Therefore, $v_{1f} = v_{2i} = -11$ Mm/s, and $v_{2f} = v_{1i} = 6.9$ Mm/s; i.e., the protons simply exchange places.

Problem

37. Two objects, one initially at rest, undergo a one-dimensional elastic collision. If half the kinetic energy of the initially moving object is transferred to the other object, what is the ratio of their masses?

Solution

If one sets $v_{2i} = 0$ in Equations 11-9a and b, one obtains $v_{1f} = (m_1 - m_2)v_{1i}/(m_1 + m_2)$ and $v_{2f} = 2m_1v_{1i}/(m_1 + m_2)$. (This describes a one-dimensional elastic collision between two objects, one initially at rest.) If half the kinetic energy of the first object is transferred to the second, $\frac{1}{2}K_{1i} = \frac{1}{4}m_1v_{1i}^2 = K_{2f} = \frac{1}{2}m_2[2m_1v_{1i}/(m_1 + m_2)]^2$, or $8m_1m_2 = (m_1 + m_2)^2$. The resulting quadratic equation, $m_1^2 - 6m_1m_2 + m_2^2 = 0$, has two solutions, $m_1 = (3 \pm \sqrt{8})m_2 = 5.83m_2$ or $(5.83)^{-1}m_2$. Since the quadratic is symmetric in m_1 and m_2, one solution equals the other with m_1 and m_2 interchanged. Thus, one object is 5.83 times more massive than the other.

Problem

39. Blocks B and C have masses $2m$ and m, respectively, and are at rest on a frictionless surface. Block A, also of mass m, is heading at speed v toward block B as shown in Fig. 11-20. If all subsequent collisions are elastic, determine the final velocity of each block.

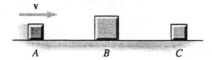

FIGURE 11-20 Problem 39.

Solution

Applying Equations 11-9a and b to the collision between A and B (with $m_B = 2m_A, v_{Ai} = v, v_{Bi} = 0$), we find

$$v_{Af} = [(m_A - 2m_A/3m_A)]v = -v/3,$$
$$v_{Bf} = (2m_A/3m_A)v = 2v/3 \quad \text{(intermediate value)}.$$

Repeating for the collision between B and C (with $m_B = 2m_C, v_{Ci} = 0, v_{Bi} = \frac{2}{3}v$), we find

$$v_{Bf} = (2v/3)(2m_C - m_C)/3m_C = 2v/9$$
$$v_{Cf} = (2v/3)2(2m_C)/3m_C = 8v/9$$

Problem

41. Rework Example 11-5 using a frame of reference in which the carbon nucleus is initially at rest.

Solution

We use Equation 3-10 to express velocities in the rest system of the carbon nucleus. Thus, $v'_{2i} = 0$, $v'_{1i} = (460 - 220)$ km/s $= 240$ km/s, and $v'_{2f} = v_{2f} - v_{2i} = 120$ km/s. From Equation 11-9b, $m_1 = m_2v'_{2f} \div (2v'_{1i} - v'_{2f}) = (12$ u$)(120)/(2\times240 - 120) = 4$ u (as in Example 11-5). From Equation 11-9a, $v'_{1f} = [(4 - 12)/(4 + 12)]v'_{1i} = -\frac{1}{2}v'_{1i} = -120$ km/s (also consistent with Example 11-5, since in the lab system, $v_{1f} = v'_{1f} + 220$ km/s $= 100$ km/s).

Problem

43. An object collides elastically with an equal-mass object initially at rest. If the collision is not head-on, show that the final velocity vectors are perpendicular.

Solution

In a two-body elastic collision when one body is initially at rest, the conservation of momentum and kinetic energy take the form $\mathbf{p}_{1i} = \mathbf{p}_{1f} + \mathbf{p}_{2f}$ and $(p_{1i}^2/2m_1) = (p_{1f}^2/2m_1) + (p_{2f}^2/2m_2)$. If $m_1 = m_2$, then $p_{1i}^2 = p_{1f}^2 + p_{2f}^2 = |\mathbf{p}_{1f} + \mathbf{p}_{2f}|^2 = p_{1f}^2 + 2\mathbf{p}_{1f} \cdot \mathbf{p}_{2f} + p_{2f}^2$, or $\mathbf{p}_{1f} \cdot \mathbf{p}_{2f} = 0$. Therefore, the final velocities are perpendicular unless $\mathbf{p}_{1f} = 0$, as for a head-on collision. (Recall that for a particle, $\mathbf{p} = m\mathbf{v}$, and $K = p^2/2m$.)

Problem

45. Two pendulums of equal length $\ell = 50$ cm are suspended from the same point. The pendulum bobs are steel spheres with masses of 140 and 390 g. The more massive bob is drawn back to make a 15° angle with the vertical (Fig. 11-21). When it is released, the bobs collide elastically. What is the maximum angle made by the less massive pendulum?

FIGURE 11-21 Problem 45.

Solution

The speed of the larger bob before the collision is (from conservation of energy) $v_{1i} = \sqrt{2g\ell(1 - \cos 15°)}$ (see Example 8-6), while for the smaller bob, $v_{2i} = 0$. After the collision, the smaller bob will reach a maximum angle given by (again from conservation of energy) $v_{2f}^2 = 2g\ell(1 - \cos\theta_2)$, where v_{2f} can be found from Equation 11-9b. Therefore,

$$v_{2f}^2 = \left(\frac{2m_1}{m_1 + m_2}\right)^2 v_{1i}^2 = \left(\frac{2(390)}{530}\right)^2 2g\ell(1 - \cos 15°)$$
$$= 2g\ell(1 - \cos\theta_2), \quad \text{or}$$
$$\cos\theta_2 = 0.926, \text{ and } \theta_2 = 22.2°.$$

Problem

47. A particle of mass m is moving along the x-axis when it collides elastically with a particle of mass $3m$. The more massive particle moves off at 1.2 m/s at 17° to the x-axis, while the less massive particle moves off at 0.92 m/s at 48° to the x-axis, as shown in Fig. 11-22. Find two sets of possible values for the initial speeds of both particles and the initial direction of the more massive particle.

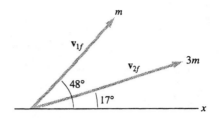

FIGURE 11-22 Problem 47.

Solution

The x and y components of momentum conservation, with particle 1 of mass m originally moving in the x direction, are

$$mv_{1i} + 3mv_{2i}\cos\theta = m(0.92 \text{ m/s})\cos 48°$$
$$+ 3m(1.2 \text{ m/s})\cos 17°,$$
$$3mv_{2i}\sin\theta = m(0.92 \text{ m/s})\sin 48°$$
$$+ 3m(1.2 \text{ m/s})\sin 17°.$$

(We assume all angles are measured counter-clockwise from the x direction, as shown.) Conservation of kinetic energy, for an elastic collision, leads to

$$\tfrac{1}{2}mv_{1i}^2 + \tfrac{1}{2}(3m)v_{2i}^2 = \tfrac{1}{2}m(0.92 \text{ m/s})^2 + \tfrac{1}{2}(3m)(1.2 \text{ m/s})^2.$$

With m divided out, the numbers evaluated, and speeds in m/s, these equations can be arranged to yield

$$3v_{2i}\cos\theta = 4.06 - v_{1i},$$
$$3v_{2i}\sin\theta = 1.74,$$
$$9v_{2i}^2 = 3(5.17 - v_{1i}^2).$$

We rewrote the equations so that v_{2i} and θ can be eliminated by squaring and adding the first two equations and setting the result equal to the third

Problem 47 Solution.

equation:

$$(3v_{2i})^2(\cos^2\theta + \sin^2\theta) = (4.06 - v_{1i})^2 + (1.74)^2$$
$$= 3(5.17 - v_{1i}^2) = 9v_{2i}^2.$$

This is a quadratic equation in v_{1i}: $4v_{1i}^2 - 2(4.06)v_{1i} + 3.99 = 0$, with solutions: $v_{1i} = \frac{1}{4}[4.06 \pm \sqrt{(4.06)^2 - 4(3.99)}] = 0.833$ m/s or 1.20 m/s. The corresponding solutions for v_{2i} are $\sqrt{\frac{1}{3}(5.17 - v_{1i}^2)} = 1.22$ m/s or 1.12 m/s, respectively. The angle θ can be found from the second equation above, $\theta = \sin^{-1}(1.74/3v_{2i}) = 28.3°$ or $31.2°$, respectively. (Note: Although we display numerical constants to three significant figures, we did not round off until after the entire computation.)

Problem

49. A tennis ball moving at 18 m/s strikes the 45° hatchback of a car moving away at 12 m/s, as shown in Fig. 11-23. Both speeds are given with respect to the ground. What is the velocity of the ball with respect to the ground after it rebounds elastically from the car? *Hint:* Work in the frame of reference of the car; then transform to the ground frame.

Solution

Let the x-y frame be fixed to the ground, and the x'-y' frame fixed to the car. Then $\mathbf{v}' = \mathbf{v} - \mathbf{v}_{\text{car}}$ is the transformation between the two frames, where $\mathbf{v}_{\text{car}} = (12\,\text{m/s})\hat{\mathbf{i}}$. In the primed frame, the initial velocity of the tennis ball is $\mathbf{v}'_{1i} = (6\,\text{m/s})\hat{\mathbf{i}}$, and, of course, $\mathbf{v}'_{2i} = 0$ for the car. Since the mass of the car is so much larger than the mass of the ball, in an elastic collision, the speed of the ball is unchanged, and it rebounds at a 45° angle to the hatchback, or $\mathbf{v}'_{1f} = (6\,\text{m/s})\hat{\mathbf{j}}$. Thus, the ball's final velocity relative to the ground is $\mathbf{v}_{1f} = \mathbf{v}'_{1f} + \mathbf{v}_{\text{car}} = (6\,\text{m/s})\hat{\mathbf{j}} + (12\,\text{m/s})\hat{\mathbf{i}}$, or $v_{1f} = \sqrt{6^2 + 12^2}$ m/s $= 13.4$ m/s, at an angle (with the x-axis) of $\theta = \tan^{-1}(6/12) = 26.6°$.

FIGURE 11-23 Problem 49 Solution.

Problem

51. Two identical billiard balls are initially at rest when they are struck symmetrically by a third identical ball moving with velocity $\mathbf{v}_0 = v_0\hat{\mathbf{i}}$, as shown in Fig. 11-24. Find the velocities of all three balls after they undergo an elastic collision.

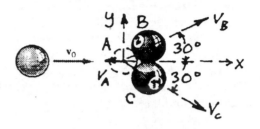

FIGURE 11-24 Problem 51 Solution.

Solution

In a symmetrical collision, the lines of contact between balls B and C (originally at rest), and ball A, make angles of 30° on either side of the direction of $\mathbf{v}_0$ (the x-axis). The final velocities, $\mathbf{v}_B$ and $\mathbf{v}_C$, are along the 30° directions and have equal magnitudes, while $\mathbf{v}_A$ has only an x component. The x component of the conservation of momentum (with the identical masses canceled out) is

$$v_0 = v_{A,x} + v_B\cos 30° + v_C\cos 30° = v_{A,x} + \sqrt{3}\,v_B.$$

(The y component of momentum conservation just confirms our assumptions of symmetry; i.e., $v_B = v_C$, and both angles are 30°.) The conservation of kinetic energy for this elastic collision, with the factors of $\frac{1}{2}m$ canceled, is $v_0^2 = v_A^2 + v_B^2 + v_C^2 = v_{A,x}^2 + 2v_B^2$.

Eliminating v_B, $v_0^2 = v_{A,x}^2 + 2\left(\frac{v_0 - v_{A,x}}{\sqrt{3}}\right)^2$, and we are led to the equation $(5v_{A,x} + v_0)(v_{A,x} - v_0) = 0$. This has solution $v_{A,x} = -\frac{1}{5}v_0$, and thus $v_B = v_C = \sqrt{\frac{1}{2}(v_0^2 - v_{A,x}^2)} = \frac{2}{5}\sqrt{3}v_0$. (The solution $v_{A,x} = v_0$, $v_B = v_C = 0$ corresponds to no collision.) (Since we know the angles, $v_{B,x} = v_{C,x} = \frac{2}{5}\sqrt{3}v_0\cos 30° = \frac{3}{5}v_0$, $v_{B,y} = -v_{C,y} = \frac{2}{5}\sqrt{3}v_0\sin 30° = \frac{1}{5}\sqrt{3}v_0$, so the solution in terms of unit vectors is $\mathbf{v}_A = -\frac{1}{5}v_0\hat{\mathbf{i}}$, $\mathbf{v}_B = \frac{3}{5}v_0\hat{\mathbf{i}} + \frac{1}{5}\sqrt{3}v_0\hat{\mathbf{j}}$, and $\mathbf{v}_C = \frac{3}{5}v_0\hat{\mathbf{i}} - \frac{1}{5}\sqrt{3}v_0\hat{\mathbf{j}}$.)

Paired Problems

Problem

53. A 590-g basketball is moving at 9.2 m/s when it hits a backboard at 45°. It bounces off at a 45° angle, still moving at 9.2 m/s. If the ball is in contact with the backboard for 22 ms, find the average impulsive force on the ball.

Solution

The normal force of the backboard acts to reverse the perpendicular component of the basketball's momentum. From Equation 11-2, $F_{av}\,\Delta t = |\mathbf{p}_f - \mathbf{p}_i| = \sqrt{2}\,|\mathbf{p}_i|$, since $|\mathbf{p}_f| = |\mathbf{p}_i|$ and their directions are perpendicular. Therefore $F_{av} = \sqrt{2}(0.590 \text{ kg})(9.2 \text{ m/s})/(0.022 \text{ s}) = 349$ N (away from the backboard).

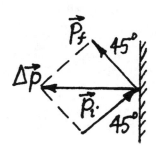

Problem 53 Solution.

Problem

55. A 2100-kg car is moving east at 75 km/h when it undergoes a totally inelastic collision with a 7900-kg truck moving northeast at 50 km/h. Find the final velocity of the combined wreckage.

Solution

The final velocity in a two-body totally inelastic collision is given by Equation 11-4 (which applies to a sufficiently brief interval of time around the collision such that any external forces have a negligible effect). If we take the x-axis East and the y-axis North, the car's initial velocity is $\mathbf{v}_{1i} = 75\hat{\imath}$ km/h and the truck's is $\mathbf{v}_{2i} = (\cos 45°\hat{\imath} + \sin 45°\hat{\jmath})(50 \text{ km/h})$. The x and y components of Equation 11-4 are $v_x = [(2100)(75) + (7900)(50)\cos 45°](\text{km/h})/(2100 + 7900) = 43.7$ km/h, and $v_y = (7900)(50)\sin 45°(\text{km/h})/(2100 + 7900) = 27.9$ km/h. The magnitude and direction are $v = 51.8$ km/h and $\theta = 32.6°$ (N of E).

Problem

57. A truck slams into a parked car, and the resulting one-dimensional collision is elastic and transfers 41% of the truck's kinetic energy to the car. Compare the masses of the two vehicles.

Solution

In a two-body, one-dimensional elastic collision with one body initially at rest (say $v_{2i} = 0$ for the car), Equations 11-9a and b give $v_{1f}/v_{2f} = (m_1 - m_2)/2m_1$. But 41% of the initial kinetic energy is transferred to m_2 and 59% to m_1, so $m_1 v_{1f}^2 / m_2 v_{2f}^2 = 59/41$.

Combining these equations, one obtains $(m_1 - m_2)^2 \div 4m_1 m_2 = 59/41$ or $(m_2/m_1)^2 - 2(100 + 59)(41)^{-1} \times (m_2/m_1) + 1 = 0$. (We wrote the quadratic in this way so that $(100 + 59)/41 = (K_i + K_{1f})/K_{2f}$ can be used in the next problem.) The quadratic formula yields $(m_2/m_1) = (159/41) \pm \sqrt{(159/41)^2 - 1} = 7.62$ or $0.131 = (7.62)^{-1}$. (The two solutions are reciprocals, since the original quadratic was symmetric in m_1 and m_2; see solution to Problem 37.) It is more likely that the truck has the greater mass, so $m_1 = 7.62m_2$ is the best choice here, but the sign of v_{1f}, if known, would determine whether m_1 was greater or less than m_2. Finally, we note that the above solution can be written in various equivalent forms, one of the simplest being $m_2/m_1 = [(\sqrt{K_i} \pm \sqrt{K_{1f}})/\sqrt{K_{2f}}]^2$, where $K_i = K_{1f} + K_{2f}$ represents the sharing of kinetic energy. This follows from the conservation of momentum, $p_{1i} = p_{1f} + p_{2f}$, written in terms of kinetic energies, $\sqrt{2m_1 K_i} = \pm\sqrt{2m_1 K_{1f}} + \sqrt{2m_2 K_{2f}}$, after solving for the mass ratio $\sqrt{m_2/m_1}$.

Problem

59. A head-on, elastic collision between two particles with equal initial speed v leaves the more massive particle (mass m_1) at rest. Find (a) the ratio of the particle masses and (b) the final speed of the less massive particle.

Solution

Given that $v_{1i} = -v_{2i} = v$, $v_{1f} = 0$, and $m_1 > m_2$, Equations 11-9a and b imply

$$0 = \frac{m_1 - m_2}{m_1 + m_2}v - \frac{2m_2}{m_1 + m_2}v, \quad \text{or} \quad m_1 - m_2 = 2m_2,$$

or $m_1 = 3m_2$, and $v_{2f} = \dfrac{2m_1}{m_1 + m_2}v + \dfrac{m_1 - m_2}{m_1 + m_2}v$

$$= \left(\frac{6}{4} + \frac{2}{4}\right)v = 2v.$$

Problem

61. A billiard ball moving at 1.8 m/s strikes an identical ball initially at rest as illustrated in Fig. 11-25. They undergo an elastic collision, and the first ball moves off at 23° counter-clockwise from its original direction. Find the final speeds of both balls and the direction of the second ball's motion.

Solution

The kinematics of this collision is the same as that in Example 11-6. Therefore $v_{1f} = v_{1i}\cos\theta_1 = (1.8 \text{ m/s})\cos 23° = 1.66$ m/s, $v_{2f} = \sqrt{v_{1i}^2 - v_{1f}^2} =$

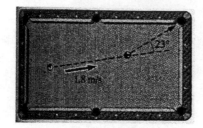

FIGURE 11-25 Problem 61.

0.703 m/s, and $\theta_2 = \sin^{-1}(-v_{1f} \sin \theta_1/v_{2f}) = -67.0°$. (This last result could have been anticipated from the statement of Problem 43 or the conclusion of Example 11-6.)

Supplementary Problems

Problem

63. A 114-g Frisbee is lodged on a tree branch 7.65 m above the ground. To free it, you lob a 240-g wad of mud vertically upward. The mud leaves your hand at a point 1.23 m above the ground, moving at 17.7 m/s. It sticks to the Frisbee. Find (a) the maximum height reached by the Frisbee-mud combination and (b) the speed with which the combination hits the ground.

Solution

Assume that momentum is conserved during this one-dimensional, vertical, totally inelastic collision between the mud and the Frisbee (i.e., suppose the Frisbee is merely resting on a branch, that there is no interference from leaves, etc.). Just before the collision, the initial velocity of the mud is $v_y = \sqrt{v_{0y}^2 - 2g(y - y_0)} =$ $\sqrt{(17.7 \text{ m/s})^2 - 2(9.8 \text{ m/s}^2)(7.65 \text{ m} - 1.23 \text{ m})} =$ 13.7 m/s (positive upward, see Equation 2-11). Just after the collision, the final velocity of the combination of Frisbee and mud is $V_y = mv_y/(m + M) =$ (240/354)(13.7 m/s) = 9.28 m/s (see Equation 11-4 or Example 11-2). (a) Therefore, the combination travels $h = V_y^2/2g = 4.40$ m higher, to a total height of 4.40 m + 7.65 m = 12.0 m above the ground. (b) An object falling from this height, unimpeded by air resistance or other obstacles, would attain a speed of $\sqrt{2(9.8 \text{ m/s}^2)(12.0 \text{ m})} = 15.4$ m/s when it reached the ground.

Problem

65. A 1400-kg car moving at 75 km/h runs into a 1200-kg car moving in the same direction at 50 km/h (Fig. 11-26). The two cars lock together and both drivers immediately slam on their

brakes. If the cars come to rest in a distance of 18 m, what is the coefficient of friction?

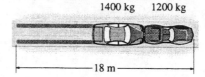

FIGURE 11-26 Problem 65.

Solution

This is a one-dimensional, totally inelastic collision, so the final velocity (positive in the initial direction of motion) is

$$v_f = \frac{(1400 \text{ kg})(75 \text{ km/h}) + (1200 \text{ kg})(50 \text{ km/h})}{1400 \text{ kg} + 1200 \text{ kg}}$$

$$= 63.5 \text{ km/h} = 17.6 \text{ m/s}.$$

On a horizontal road (normal force equals weight, and no change in gravitational potential energy), the work-energy theorem implies $W_{nc} = -\mu_k mgx = \Delta K = -\frac{1}{2}mv_f^2$; therefore

$$\mu_k = \frac{v_f^2}{2gx} = \frac{(17.6 \text{ m/s})^2}{2(9.8 \text{ m/s}^2)(18 \text{ m})} = 0.88.$$

Problem

67. Consider a one-dimensional elastic collision with $m_1 \ll m_2$ and m_2 initially at rest. In discussing this extreme case, we neglected m_1 altogether and showed that m_1 then rebounds with its initial speed. Now use the binomial theorem (see Appendix A) to show that a better approximation gives a rebound speed that is less than the incident speed by an amount $2m_1v_1/m_2$ for the case $m_1 \ll m_2$. In applying the binomial theorem, keep terms of order m_1/m_2, but neglect terms of order m_1^2/m_2^2.

Solution

The binomial expansion applied to Equation 11-9a, with $v_{2i} = 0$, $m_1 \ll m_2$, and the omission of terms of order m_1^2/m_2^2, yields

$$v_{1f} = -\left(\frac{1 - m_1/m_2}{1 + m_1/m_2}\right)v_{1i}$$

$$\simeq -v_{1i}\left(1 - \frac{m_1}{m_2}\right)\left(1 - \frac{m_1}{m_2} + \cdots\right)$$

$$= -v_{1i}(1 - 2m_1/m_2 + \cdots).$$

Problem

69. A 1.0-kg particle is moving in the $+x$ direction at 4.0 m/s when it collides elastically with a 4.0-kg particle moving in the $-x$ direction at 1.0 m/s. After colliding, the 1-kg particle moves off at $130°$ counterclockwise from the positive x axis. Find the final speeds of both particles and the direction of the more massive one.

Solution

Although this two-dimensional elastic collision can be analyzed as in Example 11-6, it is much simpler to realize that since the total momentum is zero $((1 \text{ kg})(4\hat{\imath} \text{ m/s}) + (4 \text{ kg})(-\hat{\imath} \text{ m/s}) = 0)$, the collision is specified in the center-of-mass system and thus characterized by the given scattering angle. The final velocities of the particles are oppositely directed, $\theta_{1f} = 130°$, so $\theta_{2f} = -50°$, and have the same magnitudes as the initial velocities, $v_{1f} = 4$ m/s and $v_{2f} = 1$ m/s.

Problem

71. A 1200-kg car moving at 25 km/h undergoes a one-dimensional collision with an 1800-kg car initially at rest. The collision is neither elastic nor totally inelastic; the kinetic energy lost is 5800 J. Find the speeds of both cars after the collision.

Solution

Take positive velocities in the initial direction of motion of the 1200-kg car, so that $m_1/m_2 = 2/3$, $v_{1i} = 25$ km/h, and $v_{2i} = 0$. The conservation of momentum, $m_1 v_{1i} = m_1 v_{1f} + m_2 v_{2f}$, and the given change in kinetic energy, $\Delta K = -5800 \text{ J} = \frac{1}{2}m_1 v_{1f}^2 + \frac{1}{2}m_2 v_{2f}^2 - \frac{1}{2}m_1 v_{1i}^2$, provide two equations from which the unknown final velocities can be determined. Solve for v_{2f} in the first equation, $v_{2f} = (m_1/m_2)(v_{1i} - v_{1f})$, and substitute into the second equation, $m_1 v_{1f}^2 + m_2(m_1/m_2)^2(v_{1i} - v_{1f})^2 - m_1 v_{1i}^2 - 2\Delta K = 0$. This quadratic in v_{1f} can be written as $v_{1f}^2 - 2bv_{1f} - c = 0$, where $b = m_1 v_{1i}/(m_1 + m_2) = (2/5)(25 \text{ km/h}) = 10$ km/h, and $c = [(m_2 - m_1)v_{1i}^2 + 2m_2\Delta K/m_1] \times (m_1 + m_2)^{-1} = (1/5)[(25 \text{ km/h})^2 - (5800 \text{ J}/200 \text{ kg}) \times (3.6 \text{ km/h})^2(\text{m/s})^{-2}] = 49.83 \text{ (km/h)}^2$. The roots are $v_{1f} = b \pm \sqrt{b^2 + c} = (10 \pm 12.24)$ km/h, so that $v_{2f} = (2/3)(15 \mp 12.24)$ km/h. For a collision between cars, the first car cannot penetrate through the car it strikes; therefore $v_{1f} < v_{2f}$, and one of the roots is unphysical. The other solution, $v_{1f} = -2.24$ km/h and $v_{2f} = 18.2$ km/h, is the appropriate one for this problem.

Problem

73. A block of mass M is moving at speed v_0 on a frictionless surface that ends in a rigid wall. Farther from the wall is a more massive block of mass αM, initially at rest (Fig. 11-28). The less massive block undergoes elastic collisions with the other block and with the wall, and the motion of both blocks is confined to one dimension. (a) Show that the two blocks will undergo only one collision if $\alpha \leq 3$. (b) Show that the two blocks will undergo two collisions if $\alpha = 4$, and determine their final speeds. (c) Find out how many collisions the two blocks will undergo if $\alpha = 10$, and determine their final speeds.

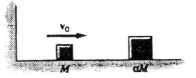

FIGURE 11-28 Problem 73.

Solution

(a) In the first head-on elastic collision between the blocks, the initial velocities (positive to the right) are $v_{1i}^{(1)} = v_0, v_{2i}^{(1)} = 0$, and the final velocities (from Equations 11-9a and b) are $v_{1f}^{(1)} = -v_0(\alpha - 1) \div (\alpha + 1)$, $v_{2f}^{(1)} = 2v_0/(\alpha + 1)$. We measured masses in units of M, so $m_1 = 1$ and $m_2 = \alpha > 1$. The parenthetical superscript is the number of the collision between the blocks. In an elastic collision with the wall, the velocity of the first block reverses direction with no loss in speed, so there will be no second collision between the blocks if $v_{2f}^{(1)} \geq \left|v_{1f}^{(1)}\right|$ (if the second block is going faster, the first will never catch it). This condition is satisfied if $2 \geq \alpha - 1$, or $\alpha \leq 3$.

(b) If $\alpha > 3$, there will be a second collision between the blocks, with $v_{1i}^{(2)} = \left|v_{1f}^{(1)}\right| = v_0(\alpha - 1) \div (\alpha + 1)$, and $v_{2i}^{(2)} = v_{2f}^{(1)} = 2v_0/(\alpha + 1)$. The final velocities (from Equations 11-9a and b) are (after substitution for $v_{1i}^{(2)}$ and $v_{2i}^{(2)}$)

$$v_{1f}^{(2)} = -\left(\frac{\alpha - 1}{\alpha + 1}\right)^2 v_0 + \frac{4\alpha}{(\alpha + 1)^2}v_0, \quad v_{2f}^{(2)} = \frac{4(\alpha - 1)}{(\alpha + 1)}v_0.$$

There will be no third collision if $v_{2f}^{(2)} \geq \left|v_{1f}^{(2)}\right|$, or $4(\alpha - 1) \geq \left|4\alpha - (\alpha - 1)^2\right|$. For $\alpha = 4$, this condition is satisfied ($4 \times 3 \geq |16 - 9|$), so the blocks undergo just two collisions with $v_{1f}^{(2)} = (-\frac{9}{25} + \frac{16}{25})v_0 = \frac{7}{25}v_0$, and $v_{2f}^{(2)} = \frac{12}{25}v_0$.

(c) The general condition for no third collision was

$4(\alpha - 1) \geq \left|6\alpha - \alpha^2 - 1\right|$. This is not satisfied for $\alpha = 10$, so there is a third collision for this value. (One can solve the inequality in general to show that if $\alpha > 5 + 2\sqrt{5} = 9.472 \ldots$ there are three or more collisions.) When $\alpha = 10$, the initial velocities for the third collision are

$$v_{1i}^{(3)} = \left|v_{1f}^{(2)}\right| = \left|-\left(\frac{9}{11}\right)^2 + \frac{40}{11^2}\right|v_0 = \frac{41}{121}v_0,$$

$$v_{2i}^{(3)} = v_{2f}^{(2)} = \frac{4(9)}{11^2}v_0 = \frac{36}{121}v_0.$$

The final velocities are

$$v_{1f}^{(3)} = -\left(\frac{9}{11}\right)\left(\frac{41}{121}\right)v_0 + \left(\frac{20}{11}\right)\left(\frac{36}{121}\right)v_0,$$

$$v_{2f}^{(3)} = \left(\frac{2}{11}\right)\left(\frac{41}{121}\right)v_0 + \left(\frac{9}{11}\right)\left(\frac{36}{121}\right)v_0,$$

or $v_{1f}^{(3)} = 0.264v_0$, and $v_{2f}^{(3)} = 0.305v_0$. Since $v_{2f}^{(3)} > \left|v_{1f}^{(3)}\right|$ (in fact both are in the same direction), there will be no fourth collision for $\alpha = 10$.

CHAPTER 12   ROTATIONAL MOTION

ActivPhysics can help with these problems:
Activities 7.7, 7.8, 7.10, 7.12–7.15

Section 12-1:   Angular Velocity and Acceleration

Problem

1. Determine the angular speed, in rad/s, of (a) Earth about its axis; (b) the minute hand of a clock; (c) the hour hand of a clock; (d) an egg beater turning at 300 rpm.

Solution

The angular speed is $\omega = \Delta\theta/\Delta t$. (a) $\omega_E = 1$ rev$\div$ 1 d $= 2\pi/86,400$ s $= 7.27\times10^{-5}$ s^{-1}. (b) $\omega_{min} =$ 1 rev/1 h $= 2\pi/3600$ s $= 1.75\times10^{-3}$ s^{-1}. (c) $\omega_{hr} =$ 1 rev/12 h $= \frac{1}{12}\omega_{min} = 1.45\times10^{-4}$ s^{-1}. (d) $\omega =$ 300 rev/min $= 300\times2\pi/60$ s $= 31.4$ s^{-1}. (Note: Radians are a dimensionless angular measure, i.e., pure numbers; therefore angular speed can be expressed in units of inverse seconds.)

Problem

3. Express each of the following in radians per second: (a) 720 rpm; (b) 50°/h; (c) 1000 rev/s; (d) 1 rev/year (which is the angular speed of Earth in its orbit).

Solution

(a) (720 rev/min)(2π/rev)(min /60 s) $= 24\pi$ s^{-1} = 75.4 s^{-1}. (b) (50°/h)(π/180°)(h/3600 s) $= 2.42\times$ 10^{-4} s^{-1}. (c) (1000 rev/s)(2π/rev) $= 2000\pi$ s$^{-1} \approx$ 6.28×10^3 s^{-1}. (d) (1 rev/y) $\approx 2\pi/(\pi\times10^7$ s) $=$ 2×10^{-7} s^{-1}. (See note in solution to Problem 1. The approximate value for 1 y used in part (d) is often handy for estimates, and is fairly accurate; see Chapter 1, Problem 12.)

Problem

5. A wheel turns through 2.0 revolutions while being accelerated from rest at 18 rpm/s. (a) What is the final angular speed? (b) How long does it take to turn the 2.0 revolutions?

Solution

For constant angular acceleration, (a) Equation 12-11 gives $\omega_f = \sqrt{\omega_0^2 + 2\alpha(\theta_f - \theta_0)} =$

$\sqrt{0 + 2(18 \text{ rev}\times60/\text{min}^2)(2 \text{ rev})} = 65.7$ rpm, and (b) Equation 12-10 gives $\theta_f - \theta_0 = 0 + \frac{1}{2}\alpha t^2$, or $t = \sqrt{2(2 \text{ rev})/(18 \text{ rev}/60 \text{ s}^2)} = 3.65$ s.

Problem

7. A compact disc (CD) player varies the rotation rate of the disc in order to keep the part of the disc from which information is being read moving at a constant linear speed of 1.30 m/s. Compare the rotation rates of a 12.0-cm-diameter CD when information is being read from (a) its outer edge and (b) a point 3.75 cm from the center. Give your answers in rad/s and rpm.

Solution

Equation 12-4 gives the relation between linear speed and angular speed, $\omega = v/r$, where r is the distance from the center of rotation. With v a constant 130 cm/s, (a) $\omega = (130 \text{ cm/s})/(6 \text{ cm}) = 21.7$ s$^{-1} =$ 207 rpm for a point on the CD's outer edge, and (b) $\omega = (130 \text{ cm/s})/(3.75 \text{ cm}) = 34.7$ s$^{-1} = 331$ rpm at the second point specified. (See the note to the solution of Problem 1 and solution of Problem 3 for the conversion of angular units.)

Problem

9. The rotation rate of a compact disc varies from about 200 rpm to 500 rpm (see Problem 7). If the disc plays for 74 min, what is its average angular acceleration in (a) rpm/s and (b) rad/s^2?

Solution

From Equation 12-5 (before the limit is taken), $\alpha_{av} = \Delta\omega/\Delta t = (500 - 200)$ rpm/(74$\times$60 s) $=$ 6.76×10^{-2} rpm/s $= 7.08\times10^{-3}$ s^{-2}. (Recall that 1 rpm $= 2\pi/60$ s.)

Problem

11. During startup of a power plant, a turbine accelerates from rest at 0.52 rad/s^2. (a) How long does it take to reach its 3600-rpm operating speed? (b) How many revolutions does it make during this time?

Solution

(a) From Equation 12-9 with $\omega_0 = 0$, $t = \omega/\alpha = (3600 \times \pi/30)$ s$^{-1}/(0.52$ s$^{-2}) = 725$ s $= 12.1$ min.
(b) From Equation 12-11, $\Delta\theta = \omega^2/2\alpha = (3600$ rev$/60$ s$)^2/2(0.52$ rev$/2\pi$ s$^2) = 2.17 \times 10^4$ rev.

Problem

13. A piece of machinery is spinning at 680 rpm. When a brake is applied, its rotation rate drops to 440 rpm while it turns through 180 revolutions. What is the magnitude of the angular deceleration?

Solution

If the angular acceleration is constant, Equation 12-11 gives $\alpha = [(440\pi/30$ s$)^2 - (680\pi/30$ s$)^2]/2(180 \times 2\pi) = -1.30$ s^{-2}. (A negative acceleration is a positive deceleration.)

Problem

15. The angular acceleration of a wheel in rad/s^2 is given by $24t^2 - 16t^3$, where t is the time in seconds. The wheel starts from rest at $t = 0$. (a) When is it again at rest? (b) How many revolutions has it turned between $t = 0$ and when it is again at rest?

Solution

(a) Integrating Equation 11-5 (with $\omega_0 = 0$), we find $\omega(t) = \int_0^t \alpha(t')\, dt' = \int_0^t (24t'^2 - 16t'^3)\, dt' = 8t^3 - 4t^4$. The wheel is at rest when $\omega(t) = (8 - 4t)t^3 = 0$, or at $t = 0$ and $t = 2$ s. (b) Integrating Equation 11-2 for the angular displacement (in radians), we find $\theta(t) - \theta_0 = \int_0^t \omega(t')\, dt' = \int_0^t (8t'^3 - 4t'^4)\, dt' = 2t^4 - 4t^5/5$. For $t = 2$ s, $\Delta\theta = 2^5(1 - 4/5) = 6.4$ radians $= 1.02$ rev.

Section 12-2: Torque

Problem

17. A torque of 110 N·m is required to start a revolving door rotating. If a child can push with a maximum force of 90 N, how far from the door's rotation axis must she apply this force?

Solution

If the force is applied perpendicular to the door, the radial distance should be $r = \tau/F = 110$ N·m$\div$ 90 N $= 1.22$ m from the axis. (See Equation 12-12 with $\theta = 90°$.)

Problem

19. A 55-g mouse runs out to the end of the 17-cm-long minute hand of a grandfather clock when the clock reads 10 minutes past the hour. What torque does the mouse's weight exert about the rotation axis of the clock hand?

Solution

The angle between the minute hand (for the clock in upright position) and the weight of the mouse (vertically downward) is 120° at ten past the hour, so the magnitude of the torque exerted is $\tau = rmg \sin\theta = (17$ cm$)(55$ g$)(9.8$ m/s$^2) \sin 120° = 7.94 \times 10^{-2}$ N·m. The direction of the torque is clockwise.

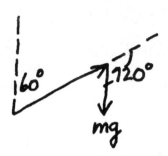

Problem 19 Solution.

Problem

21. A pulley 12 cm in diameter is free to rotate about a horizontal axle. A 220-g mass and a 470-g mass are tied to either end of a massless string, and the string is hung over the pulley. If the string does not slip, what torque must be applied to keep the pulley from rotating?

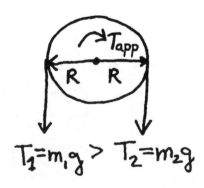

Problem 21 Solution.

Solution

If the pulley and string are not moving (no rotation and no slipping) the net torque on the pulley is zero, and the tensions in the string on either side are equal to the weights tied on either end. The tensions are perpendicular to the radii to each point of application, so the torques due to the tensions have magnitude $TR = mgR$, but are in opposite directions. Thus, $0 = \tau_{app} - m_1gR + m_2gR$, where we chose positive torques in the direction of the applied torque, which is opposite to the torque produced by the greater mass. Numerically, $\tau_{app} = (m_1 - m_2)gR = (0.47 - 0.22)$ kg× $(9.8$ m/s$^2)(0.06$ m$) = 0.147$ N·m. (Note: τ_{app} equals the torque which would be produced by balancing the pulley with 250 g added to the side with lesser mass.)

Problem

23. A 1.5-m-diameter wheel is mounted on an axle through its center. (a) Find the net torque about the axle due to the forces shown in Fig. 12-42. (b) Are there any forces that don't contribute to the torque?

Solution

The displacements from the center of the wheel to the points of application of the forces, and the angles between these displacements and the forces, are shown on Fig. 12-42. (a) If we take positive torques counterclockwise, the net torque (sum of Equation 12-12) is $\tau_{net} = \sum r_i F_i \sin\theta_i = r_1 F_1 \sin 0° +$ $(0.25$ m$)(3.0$ N$)\sin 45° - (0.50$ m$)(2.0$ N$)\sin 90° +$ $(0.75$ m$)(1.8$ N$)\sin 60° + 0 \times F_5 = 0.699$ N·m. (b) Evidently, $\mathbf{F}_1$ and $\mathbf{F}_5$, whose lines of action pass through the center of the wheel, produce no torque about the center.

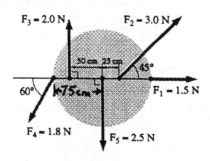

FIGURE 12-42 Problem 23 Solution.

Section 12-3: Rotational Inertia and the Analog of Newton's Law

Problem

25. The shaft connecting a turbine and electric generator in a power plant is a solid cylinder of mass 6.8×10^3 kg and diameter 85 cm. Find its rotational inertia.

Solution

The rotational inertia of a solid cylinder or disk about its axis is $I = \frac{1}{2}MR^2$ (see Example 12-8 or

Table 12-2), so $I = (0.5)(6.8 \times 10^3$ kg$)(0.85$ m/2$)^2 = 6.14 \times 10^2$ kg·m^2.

Problem

27. A square frame is made from four thin rods, each of length ℓ and mass m. Calculate its rotational inertia about the three axes shown in Fig. 12-43.

Solution

(a) The rotational inertia of each of the rods perpendicular to the axis is $\frac{1}{12}m\ell^2$, and that of each rod parallel to the axis is $m(\frac{1}{2}\ell)^2 = \frac{1}{4}m\ell^2$. The total is $I_a = [2(\frac{1}{12}) + 2(\frac{1}{4})]m\ell^2 = \frac{2}{3}m\ell^2$. (b) The rotational inertia of a uniform thin rod, about an axis through one end, making an angle θ with the length of the rod, is

$$I = \int y^2\, dm = \int_0^{\ell\cos\theta} (x\tan\theta)^2 \frac{M\, dx}{\ell\cos\theta}$$
$$= \frac{M\tan^2\theta}{\ell\cos\theta} \left.\frac{x^3}{3}\right|_0^{\ell\cos\theta} = \frac{1}{3}M\ell^2\sin^2\theta.$$

Since there are four rods, all making the same angle, $\theta = 45°$, with the axis, the total I_b is $4(\frac{1}{3}m\ell^2 \times \sin^2 45°) = \frac{2}{3}m\ell^2$. (c) The rotational inertia of one rod (from the parallel axis theorem) is $I_1 = \frac{1}{12}m\ell^2 + m(\frac{1}{2}\ell)^2 = \frac{1}{3}m\ell^2$, and there are four rods, so $I_c = \frac{4}{3}m\ell^2$.

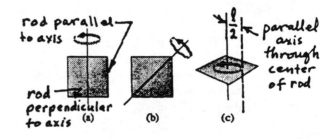

FIGURE 12-43 Problem 27 Solution.

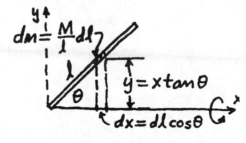

Problem 27 Solution.

Problem

29. The wheel shown in Fig. 12-40 consists of a 120-kg outer rim 55 cm in radius, connected to the center by five 18-kg spokes. Treating the rim as a thin ring and the spokes as thin rods, determine the rotational inertia of the wheel.

Solution

The sum of the rotational inertias of the parts is $I = M_{\text{rim}}R_{\text{rim}}^2 + 5 \times \frac{1}{3}M_{\text{rod}}\ell_{\text{rod}}^2$, where $R_{\text{rim}} = \ell_{\text{rod}}$. Thus, $I = (120 \text{ kg} + 5 \times 18 \text{ kg}/3)(0.55 \text{ m})^2 = 45.4 \text{ kg·m}^2$ (about an axis perpendicular to the wheel through its center).

Problem

31. A uniform, rectangular, flat plate has mass M and dimensions a by b. Use the parallel-axis theorem in conjunction with Table 12-2 to show that its rotational inertia about the side of length b is $\frac{1}{3}Ma^2$.

Solution

Table 12-2 lists the rotational inertia of a flat plate about a central axis (in the plane of the plate and through its center of mass) as $I_{cm} = \frac{1}{12}Ma^2$, when the axis is parallel to a side of length b. The CM is a perpendicular distance $\frac{a}{2}$ from this side, so the parallel-axis theorem gives $I_{\text{side } b} = \frac{1}{12}Ma^2 + M(\frac{1}{2}a)^2 = \frac{1}{3}Ma^2$.

Problem

33. (a) Estimate the rotational inertia of Earth, assuming it to be a uniform solid sphere.
(b) What torque would have to be applied to Earth to cause the length of the day to change by one second every century?

Solution

(a) For a uniform solid sphere, and an axis through the center, $I_E = \frac{2}{5}M_E R_E^2 = \frac{2}{5}(5.97 \times 10^{24} \text{ kg}) \times (6.37 \times 10^6 \text{ m})^2 = 9.69 \times 10^{37} \text{ kg·m}^2$. (The Earth has a core of denser material, so its actual rotational inertia is less than this.) (b) The angular speed of rotation of Earth is $\omega = 2\pi/T$, where the period $T = 1 \text{ d} = 86,400 \text{ s}$. If the period were to change by 1 s per century, $|dT/dt| = 1 \text{ s}/(100 \times 3.16 \times 10^7 \text{ s})$, this would correspond to an angular acceleration of $\alpha = d\omega/dt = d(2\pi/T)/dt = -(2\pi/T^2)(dT/dt)$. Therefore, a torque of magnitude $I|\alpha| = I(2\pi/T^2)|dT/dt| = (9.69 \times 10^{37} \text{ kg·m}^2) \cdot (2\pi/(86,400 \text{ s})^2)(1/3.16 \times 10^9) = 2.58 \times 10^{19} \text{ N·m}$ would be required. (Such a torque is actually generated by tidal friction between the moon and the Earth.)

Problem

35. A neutron star is an extremely dense, rapidly spinning object that results from the collapse of a star at the end of its life. A neutron star of 1.8 times the Sun's mass has an approximately uniform density of $1 \times 10^{18} \text{ kg/m}^3$. (a) What is its rotational inertia? (b) The neutron star's spin rate slowly decreases as a result of torque associated with magnetic forces. If the spin-down rate is $5 \times 10^{-5} \text{ rad/s}^2$, what is the magnetic torque?

Solution

(a) For a uniform sphere about an axis through its center, $I = \frac{2}{5}MR^2$. Since the density is $\rho = M \div (\frac{4}{3}\pi R^3)$, the radius can be eliminated to yield $I = \frac{2}{5}M(3M/4\pi\rho)^{2/3} = 0.4M^{5/3}(0.75/\pi\rho)^{2/3} = 0.4[(1.8)(1.99 \times 10^{30} \text{ kg})]^{5/3}[0.75/\pi(10^{18} \text{ kg/m}^3)]^{2/3} = 1.29 \times 10^{38} \text{ kg·m}^2$. (b) The torque responsible for the spin-down rate (i.e., the angular deceleration) is $\tau = I\alpha = (1.29 \times 10^{38} \text{ kg·m}^2)(-5 \times 10^{-5} \text{ s}^{-2}) = -6.45 \times 10^{33} \text{ N·m}$, the negative value indicating a direction opposite to the angular velocity.

Problem

37. Proof of the parallel-axis theorem: Fig. 12-44 shows an object of mass M with axes through the center of mass and through an arbitrary point A. Both axes are perpendicular to the page. Let $\mathbf{h}$ be a vector from the axis through the center of mass to the axis through point A, $\mathbf{r}_{cm}$ a vector from the axis through the CM to an arbitrary mass element dm, and $\mathbf{r}$ a vector from the axis through point A to the mass element dm, as shown. (a) Use the law of cosines to show that

$$r^2 = r_{cm}^2 + h^2 - 2\mathbf{h} \cdot \mathbf{r}_{cm}.$$

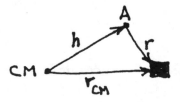

FIGURE 12-44 Problem 37.

(b) Use this result in the expression $I = \int r^2 \, dm$ to calculate the rotational inertia of the object about the axis through A. Each of the three terms in your expression for r^2 leads to a separate integral. Identify one as the rotational inertia about the

CM, another as the quantity Mh^2, and show that the third is zero because it involves the position of the center of mass relative to itself. Your result is then a statement of the parallel-axis theorem (Equation 12-20).

Solution

(a) Since $\mathbf{r} = \mathbf{r}_{cm} - \mathbf{h}$, $r^2 = (\mathbf{r}_{cm} - \mathbf{h}) \cdot (\mathbf{r}_{cm} - \mathbf{h}) = r_{cm}^2 + h^2 - 2\mathbf{r}_{cm} \cdot \mathbf{h}$ (which is equivalent to the law of cosines; $r^2 = r_{cm}^2 + h^2 - 2r_{cm}h\cos\theta$, where θ is the angle between the sides r_{cm} and h in the triangle in Fig. 12-44). (b) $I_A = \int r^2\, dm = \int r_{cm}^2\, dm + h^2 \int dm - 2\mathbf{h} \cdot \int \mathbf{r}_{cm}\, dm$. The first integral is I_{cm}, the second is Mh^2, and the third is zero (from the definition of the center of mass—in the CM frame, the origin *is* the CM, so $0 = \int \mathbf{r}_{cm}\, dm / \int dm$).

Problem

39. A space station is constructed in the shape of a wheel 22 m in diameter, with essentially all of its 5.0×10^5-kg mass at the rim (Fig. 12-45). Once the station is completed, it is set rotating at a rate that requires an object at the rim to have radial acceleration g, thereby simulating Earth's surface gravity. This is accomplished using two small rockets, each with 100 N thrust, that are mounted on the rim of the station as shown. (a) How long will it take to reach the desired spin rate? (b) How many revolutions will the station make in this time?

Solution

(a) Suppose that any difference in the distance to the axis from the center to the rockets, or any part of the ring, is negligible compared to the radius of the ring, $R = 11$ m. The angular acceleration (about the axis perpendicular to the ring and through its center) of the ring is $\alpha = \tau/I = 2FR/MR^2$. Starting from rest $(\omega_0 = 0)$, it would take time $t = \omega/\alpha$ for the ring to reach a final angular speed ω. Since $g = \omega^2 R$ specifies the desired rate of rotation, we find

$$t = \frac{\omega}{\alpha} = \sqrt{\frac{g}{R}}\left(\frac{MR}{2F}\right)$$

$$= \frac{(5 \times 10^5\ \text{kg})\sqrt{(9.8\ \text{m/s}^2)(11\ \text{m})}}{2(100\ \text{N})}$$

$$= 2.60 \times 10^4\ \text{s} = 7.21\ \text{h}.$$

(b) From either Equation 12-8 or 12-10, $\theta - \theta_0 = \frac{1}{2}\alpha t^2 = \frac{1}{2}\omega t = \frac{1}{2}\sqrt{g/R}\,t = \frac{1}{2}\sqrt{(9.8\ \text{m/s}^2)/(11\ \text{m})} \times (2.60 \times 10^4\ \text{s})(1\ \text{rev}/2\pi) \cong 1950\ \text{rev}.$

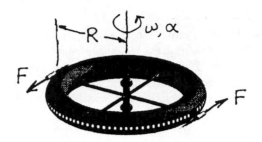

FIGURE 12-45 Problem 39 Solution.

Problem

41. The crane shown in Fig. 12-46 contains a hollow drum of mass 150 kg and radius 0.80 m that is driven by an engine to wind up a cable. The cable passes over a solid, cylindrical, 30-kg pulley 0.30 m in radius to lift a 2000-N weight. How much torque must the engine apply to the drum to lift the weight with an acceleration of $1.0\ \text{m/s}^2$? Neglect the rotational inertia of the engine and the mass of the cable.

Solution

The equations of motion of the drum, pulley, and weight are $\tau - T_1 R_1 = I_1\alpha_1$, $(T_1 - T_2)R_2 = I_2\alpha_2$, and $T_2 - mg = ma$, where subscripts 1 and 2 refer to the drum and pulley, respectively, and τ is the engine's torque. We assume $I_1 = M_1 R_1^2$ and $I_2 = \frac{1}{2}M_2 R_2^2$. If the cable does not slip or stretch, $a = \alpha_1 R_1 = \alpha_2 R_2$. When all these results are combined, we may solve for the torque: $\tau = (M_1 a + T_1)R_1$; $T_1 = \frac{1}{2}M_2 a + T_2$; $T_2 = m(g + a)$; therefore $\tau = [M_1 + \frac{1}{2}M_2 + m(1 + g/a)]R_1 a$. Numerically,

$$\tau = \left[150\ \text{kg} + \frac{30\ \text{kg}}{2} + \frac{2000\ \text{kg}}{9.8}(1 + 9.8)\right](0.8\ \text{m})\left(1\frac{\text{m}}{\text{s}^2}\right)$$

$$= 1.90\ \text{kN·m}.$$

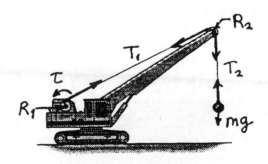

FIGURE 12-46 Problem 41 Solution.

Problem

43. A bicycle is under repair and is upside-down with its 66-cm-diameter wheel spinning freely at 230 rpm. The mass of the wheel is 1.9 kg and is concentrated mostly at the rim. The cyclist holds a wrench against the tire for 3.1 s, with a normal force of 2.7 N. If the coefficient of friction between the wrench and the tire is 0.46, what is the final angular speed of the wheel?

Solution

Since the wheel was spinning freely, the frictional torque of the wrench is the only one acting about its axis of rotation, and $\tau_f = I\alpha$, where $I = MR^2$ is the rotational inertia of the wheel (whose mass is concentrated at its rim). The frictional force is, of course, tangential, so $\tau_f = -f_k R = -\mu_k NR$, where N is the normal force, and we chose the positive sense of rotation in the direction of the wheel's motion. Thus, $\alpha = -\mu_k NR/MR^2 = -\mu_k N/MR$, and Equation 12-9 gives the final angular speed, $\omega_f = \omega_0 - (\mu_k N/MR)t =$ 230 rpm $- [(0.46)(2.7\ \text{N})/(1.9\ \text{kg})(0.33\ \text{m})](3.1\ \text{s})\times$ $(60\ \text{rpm}/2\pi\ \text{s}^{-1}) = 230\ \text{rpm} - 58.6\ \text{rpm} = 171\ \text{rpm}$.

Problem

45. Two blocks of mass m_1 and m_2 are connected by a massless string that passes over a solid cylindrical pulley, as shown in Fig. 12-48. The surface under the block m_2 is frictionless, and the pulley rides on frictionless bearings. The string passes over the pulley without slipping. When released, the masses accelerate at $\frac{1}{3}g$. The tension in the lower half of the string is 2.7 N and that in the upper half, 1.9 N. What are the masses of the pulley and of the two blocks?

Solution

The equations of motion for the blocks and pulley are $m_1 g - T_1 = m_1 a_1$, $T_2 = m_2 a_2$, and $(T_1 - T_2)R = I\alpha$.

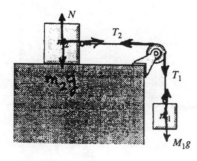

FIGURE 12-48 Problem 45 Solution.

(The equation for the vertical forces on block two is not needed, since the horizontal surface on which it slides is frictionless.) Here, R is the radius of the pulley, and $I = \frac{1}{2}MR^2$. The given constraints require that $a_1 = a_2$ (inextensible string) $= \alpha R$ (no slipping) $= \frac{1}{3}g$. Substituting these into the equations of motion, we find $m_1 = 3T_1/2g = 413$ g, $m_2 = 3T_2/g = 582$ g, and $M = 6(T_1 - T_2)/g = 490$ g. ("g" is also the abbreviation for gram.)

Problem

47. A 25-cm-diameter circular saw blade has a mass of 0.85 kg, distributed uniformly as in a disk. (a) What is its rotational kinetic energy at 3500 rpm? (b) What average power must be applied to bring the blade from rest to 3500 rpm in 3.2 s?

Solution

(a) $K = \frac{1}{2}I\omega^2 = \frac{1}{2}(\frac{1}{2}MR^2)\omega^2 = \frac{1}{4}(0.85\ \text{kg})\times$ $(0.125\ \text{m})^2(3500\pi/30\ \text{s})^2 = 446$ J. (b) $P_{\text{av}} =$ $W/t = \Delta K/t = 446\ \text{J}/3.2\ \text{s} = 139$ W.

Problem

49. A 150-g baseball is pitched at 33 m/s, spinning at 42 rad/s. What fraction of its kinetic energy is rotational? Treat the baseball as a uniform solid sphere of radius 3.7 cm.

Solution

The total kinetic energy is part center-of-mass energy and part internal rotational energy associated with spin about the center of mass. From Equation 12-24, $K_{\text{rot}}/K_{\text{tot}} = I_{cm}\omega^2/(Mv^2 + I_{cm}\omega^2)$. For a solid sphere, $I_{cm} = 0.4MR^2$, so the ratio becomes $0.4(R\omega)^2/[v^2 + 0.4(R\omega)^2] = 0.4(0.037\times42)^2/[(33)^2 + 0.4(0.037\times42)^2] = 8.86\times10^{-4} \approx 0.089\%$. (Note: We canceled the mass and common units in the ratio.)

Problem

51. A ship's anchor weighs 5000 N. Its cable passes over a roller of negligible mass and is wound around a hollow cylindrical drum of mass 380 kg and radius 1.1 m, as shown in Fig. 12-49. The drum is mounted on a frictionless axle. The anchor is released and drops 16 m to the water. Use energy considerations to determine the drum's rotation rate when the anchor hits the water. Neglect the mass of the cable.

Solution

The only significant energy changes in this problem are the change in the anchor's gravitational potential

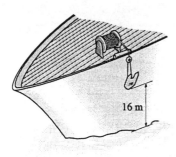

FIGURE 12-49 Problem 51.

energy, $mg\,\Delta y = (5000\text{ N})(16\text{ m}) = 80$ kJ, and the changes in the drum's and anchor's kinetic energies, $\frac{1}{2}I\omega^2$ and $\frac{1}{2}mv^2$ respectively. (This is because we are neglecting the mass of the roller and cable, and friction.) If the cable doesn't slip over the drum, the drum's tangential speed equals the anchor's speed, so $\omega R = v$. For a hollow cylinder, $I = MR^2$, so the total kinetic energy change is $\frac{1}{2}(MR^2)\omega^2 + \frac{1}{2}m(\omega R)^2 = \frac{1}{2}R^2\omega^2(M + m)$, which equals the 80 kJ change in potential energy. Therefore $\omega = [2(80\text{ kJ})/(380\text{ kg} + 5000\text{ N}/(9.8\text{ m/s}^2))]^{1/2}/(1.1\text{ m}) = 12.2$ s^{-1}.

Section 12-5:   Rolling Motion

Problem

53. A basketball rolls down a 30° incline. If it starts from rest, what is its speed after it's gone 8.4 m along the incline? (The basketball is hollow.)

Solution

The ball rolls through a vertical height $h = \ell\sin\theta$, where ℓ is the distance along the incline, which makes an angle θ with the horizontal. For a hollow ball, the rotational inertia factor $\alpha = I_{cm}/MR^2$ is $\frac{2}{3}$ (see Table 12-2), so we can use the result of Problem 55 to find $v_{cm} = \sqrt{2gh/(1+\alpha)} = \sqrt{2(9.8\text{ m/s})(8.4\text{ m})\sin 30°/(1+\frac{2}{3})} = 7.03$ m/s.

Problem

55. As long as its mass is distributed symmetrically about a central axis, a round object like a sphere or wheel may be characterized by a rotational inertia of the form $I = \alpha MR^2$, where α is a constant. Derive a formula for the final speed of such an object if it starts from rest and rolls without slipping down an incline of vertical height h.

Solution

Example 12-14 shows that the conservation of mechanical energy, applied to an object starting from rest and rolling without slipping down an incline, yields $Mgh = \frac{1}{2}I_{cm}\omega^2 + \frac{1}{2}Mv_{cm}^2$, where $\omega = v_{cm}/R$. If we also substitute $I_{cm} = \alpha MR^2$, we obtain $2Mgh = \alpha MR^2(v_{cm}/R)^2 + Mv_{cm}^2$, or $v_{cm} = \sqrt{2gh/(1+\alpha)}$.

Problem

57. The rotational kinetic energy of a rolling automobile wheel is 40% of its translational kinetic energy. The wheel is then redesigned to have 10% lower rotational inertia and 20% less mass, while keeping its radius the same. By what percentage does its total kinetic energy at a given speed decrease?

Solution

As in Example 12-14, the kinetic energy of the wheel, rolling without slipping, can be written as $K = \frac{1}{2}Mv_{cm}^2 + \frac{1}{2}I_{cm}(v_{cm}/R)^2 = (1 + I_{cm}/MR^2)(\frac{1}{2}Mv_{cm}^2)$. Initially, $I_{cm}/MR^2 = 40\%$, and after redesign, $I'_{cm}/M'R^2 = 0.9I_{cm}/0.8MR^2 = (9/8)40\% = 45\%$. The fractional decrease in kinetic energy is

$$\frac{K - K'}{K} = 1 - \frac{(1 + I'_{cm}/M'R^2)M'}{(1 + I_{cm}/MR^2)M}$$
$$= 1 - \frac{(1.45)0.8M}{(1.40)M} = 17.1\%$$

Problem

59. A ball rolls without slipping down a slope of vertical height 34 cm, and reaches the bottom moving at 2.0 m/s. Is the ball hollow or solid?

Solution

Reference to the solution of Problem 55 shows that the speed of the ball at the bottom of the slope is $v_{cm} = \sqrt{2gh/(1+\alpha)}$, or $\alpha = (2gh/v_{cm}^2) - 1 = [2(9.8\text{ m/s}^2)(0.34\text{ m})/(2\text{ m/s})^2] - 1 = 0.666$. For a solid sphere, $\alpha = \frac{2}{5}$, while for a hollow one, $\alpha = \frac{2}{3}$, which is almost exactly the value calculated for this ball.

Paired Problems

Problem

61. A merry-go-round starts from rest and accelerates with angular acceleration 0.010 rad/s^2 for 14 s. (a) How many revolutions does it make during this time? (b) What is its average angular speed during the spin-up time?

Solution

The kinematics of motion with constant angular acceleration are summarized in Table 12-1, where the initial angular velocity at $t = 0$ is zero if the merry-go-round starts at rest. (a) Equation 12-10 gives the angular displacement: $\Delta\theta = \theta - \theta_0 = \frac{1}{2}\alpha t^2 = \frac{1}{2}(0.01\ \text{s}^2)(14\ \text{s})^2 = (0.98\ \text{radians}) \times (1\ \text{rev}/2\pi) = 0.156\ \text{rev} = 56.1°$. (b) Equations 12-8 and 9 give the average angular speed: $\omega_{av} = \frac{1}{2}\omega_f = \frac{1}{2}\alpha t = \frac{1}{2}(0.01\ \text{s}^{-2})(14\ \text{s}) = 0.07\ \text{s}^{-1} = 0.668\ \text{rpm}$. (Note that the average angular speed is always equal to $\Delta\theta/\Delta t$, in this case $\frac{1}{2}\alpha t^2/t$, but that Equations 12-8 to 11 apply only to constant α.)

Problem

63. A disk of radius R has an initial mass M. Then a hole of radius $\frac{1}{4}R$ is drilled, with its edge at the disk center (Fig. 12-51). Find the new rotational inertia about the central axis. *Hint:* Find the rotational inertia of the missing piece, and subtract from that of the whole disk. You'll need to determine what fraction the missing mass is of the total M, and you'll need to use the parallel-axis theorem.

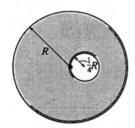

FIGURE 12-51 Problems 63, 76.

Solution

Equation 12-15 shows that the rotational inertia of an object is the sum of the rotational inertias of the pieces, so $I_{disk} = I_{hole} + I_{remainder}$ (the hint expresses this fact as $I_{remainder} = I_{disk} - I_{hole}$). I_{disk} is the rotational inertia of the whole disk about an axis perpendicular to the disk and through its center, which is $\frac{1}{2}MR^2$ (see Example 12-8). I_{hole} is the rotational inertia of the material removed to form the hole, which the parallel axis theorem gives as $I_{hole} = M_{hole}h^2 + I_{cm}$. Here, $h = R/4$ is the distance of the hole's CM from the axis of the disk, and $I_{cm} = \frac{1}{2}M_{hole}(R/4)^2$ is the rotational inertia of the hole material about a parallel axis through its CM. Since the planar mass density of the disk (assumed to be uniform) is $M/\pi R^2$, the mass of the hole material is $M_{hole} = (M/\pi R^2)\pi(R/4)^2 = M/16$. Therefore, $I_{hole} = (M/16)(R/4)^2 + \frac{1}{2}(M/16)(R/4)^2 = 3MR^2/512$,

and $I_{remainder} = \frac{1}{2}MR^2 - 3MR^2/512 = (253/512) \times MR^2 = 0.494MR^2$.

Problem

65. A 50-kg mass is tied to a massless rope wrapped around a solid cylindrical drum. The drum is mounted on a frictionless horizontal axle. When the mass is released, it falls with acceleration $a = 3.7\ \text{m/s}^2$. Find (a) the tension in the rope and (b) the mass of the drum.

Solution

The situation is similar to Example 12-11. (a) The equation of motion (Newton's second law) of the falling mass is $mg - T = ma$ (positive "a" downward), so $T = m(g - a) = (50\ \text{kg})(9.8 - 3.7)\ \text{m/s}^2 = 305\ \text{N}$. (b) The equation of motion of the rotating drum (rotational analog of Newton's second law) is $\tau = RT = I\alpha = (\frac{1}{2}MR^2)(a/R)$, so $M = 2T/a = 2(350\ \text{N})/(3.7\ \text{m/s}^2) = 165\ \text{kg}$ (since $\tau = RT$ and $a = \alpha R$ as explained in Example 12-11).

Problem

67. Your little sister is building a toy car. She attaches four spools to a 48-g milk carton. The spools are essentially solid cylinders, each with mass 25 g and radius 1.8 cm and are mounted on frictionless axles of negligible mass. Starting from rest, the contraption rolls without slipping down a slope. (a) How fast is it going after it's dropped a vertical distance of 85 cm? (b) What per cent of its total kinetic energy is in the translation motion of the milk carton alone?

Solution

(a) If we neglect air resistance of the milk carton, as well as friction of the axles etc., the conservation of mechanical energy for the toy car requires that $\Delta U + \Delta K = 0 = -m_{total}gh + K_{carton} + K_{wheels}$, where K is the kinetic energy of a component of the car after falling through a vertical distance h. For the carton, $K_{carton} = \frac{1}{2}M_{carton}v^2$. For the spool wheels, $K_{wheels} = 4 \cdot \frac{1}{2}I\omega^2$, where $\omega = v/R$ (for a spool of radius R rolling without slipping), and $I = \frac{3}{2}m_{wheel}R^2$ (from the parallel-axis theorem for the rotational inertia about the point of contact of each wheel with the ground, which is instantaneously at rest). Therefore, $M_{tot}gh = \frac{1}{2}m_{carton}v^2 + 3m_{wheel}v^2$, or $v = \sqrt{2ghm_{tot}/(m_{carton} + 6m_{wheel})} = \sqrt{2(9.8)(0.85)\text{m}^2/\text{s}^2}\sqrt{148/198} = 3.53\ \text{m/s}$. (b) The total kinetic energy is $K_{tot} = m_{tot}gh$, so $K_{carton} \div K_{tot} = m_{carton}v^2/2m_{tot}gh = (0.48\ \text{kg})(3.53\ \text{m/s})^2 \div 2(0.148\ \text{kg})(9.8\ \text{m/s}^2)(0.85\ \text{m}) = 24.2\%$.

Supplementary Problems

Problem

69. A solid marble starts from rest and rolls without slipping on the loop-the-loop track shown in Fig. 12-53. Find the minimum starting height of the marble from which it will remain on the track through the loop. Assume the marble radius is small compared with R.

Solution

The CM of the marble travels in a circle of radius $R - r$ inside the loop-the-loop, so at the top, $mg + N = mv^2/(R - r)$. To remain in contact with the track, $N \geq 0$, or $v^2 \geq g(R - r)$. If we assume that energy is conserved between points A (the start) and B (the top of the loop), and use $K_A = 0$ and $K_B = (1 + \frac{2}{5})\frac{1}{2}mv_B^2$, we find $U_A + K_A = mg(h + r) = U_B + K_B = mg(2R - r) + \frac{7}{10}mv_B^2$, or $h = 2(R - r) + \frac{7}{10}(v_B^2/g) \geq 2(R - r) + \frac{7}{10}(R - r) = 2.7(R - r)$.

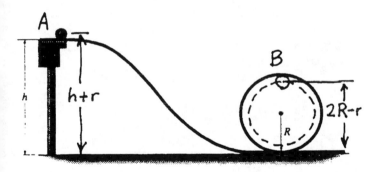

FIGURE 12-53 Problem 69 Solution.

Problem

71. An object of rotational inertia I is initially at rest. A torque is then applied to the object, causing it to begin rotating. The torque is applied for only one-quarter of a revolution, during which time its magnitude is given by $\tau = A \cos \theta$, where A is a constant and θ is the angle through which the object has rotated. What is the final angular speed of the object?

Solution

If we assume that the object can only rotate about a fixed axis (for which the rotational inertia I is given), then the work-energy theorem (Equation 12-22b) gives

$$W = \int_0^{\theta_f} \tau \, d\theta = \int_0^{\pi/2} A \cos \theta \, d\theta = |A \sin \theta|_0^{\pi/2}$$
$$= A = \Delta K = \frac{1}{2}I\omega_f^2.$$

Thus, $\omega_f = \sqrt{2A/I}$.

Problem

73. A thin, uniform vane of mass M is in the shape of a right triangle, as shown in Fig. 12-54. Find the rotational inertia about a vertical axis through its apex, as shown in the figure. Express your answer in terms of the triangle's base width b and its mass M.

Solution

The integral in Equation 12-16 for the rotational inertia of a 2-dimensional object is really a double integral over the area of the object. If we choose the x- and y-axis to coincide with the base of the triangular vane and the axis of rotation, respectively, then the rotational inertia of an infinitesimal mass element $\rho \, dx \, dy$, located at position (x, y), is $x^2\rho \, dx \, dy$, since x is the perpendicular distance to the y-axis. ρ is the planar density of the triangle, in this case assumed to be uniform and equal to $M/(\frac{1}{2}bh)$. Therefore Equation 12-16 becomes $I_y = (2M/bh) \times \int\int x^2 \, dx \, dy$, where the integration is over the area of the triangle. If we choose to integrate over y first, the area can be specified by letting y go from 0 to xh/b and x from 0 to b. Then

$$I_y = \left(\frac{2M}{bh}\right)\int_0^b x^2 \, dx \int_0^{xh/b} dy$$
$$= \left(\frac{2M}{bh}\right)\int_0^b \left(\frac{h}{b}\right)x^3 \, dx = \left(\frac{2M}{b^2}\right)\frac{b^4}{4} = \frac{1}{2}Mb^2.$$

Alternatively, if we integrate over x first, the limits of integration are yb/h to b for x, and 0 to h for y, so

$$I_y = \left(\frac{2M}{bh}\right)\int_0^h dy \int_{yb/h}^b x^2 \, dx$$
$$= \left(\frac{2M}{bh}\right)\int_0^h dy \frac{1}{3}\left[b^3 - \left(\frac{by}{h}\right)^3\right]$$
$$= \left(\frac{2M}{bh}\right)\frac{1}{3}\left[b^3 h - \left(\frac{b}{h}\right)^3 \frac{h^4}{4}\right] = \frac{1}{2}Mb^2.$$

(Note: The first approach amounts to choosing 1-dimensional mass elements as vertical strips of mass $dm = \rho y \, dx = (2M/bh)(xh/b) \, dx = (2M/b^2)x \, dx$ and rotational inertia $dI_y = x^2 \, dm = (2M/b^2)x^3 \, dx$. Then a 1-dimensional integration from $x = 0$ to $x = b$ gives $I_y = \frac{1}{2}Mb^2$. The second approach corresponds to choosing horizontal strips of length $\ell = b - by/h$ and mass $dm = \rho\ell \, dy$. The rotational inertia of these strips must be calculated from the parallel-axis theorem, $dI_y = (x_{cm}^2 + \frac{1}{12}\ell^2) \, dm$, where $x_{cm} = (by/h) + \ell/2$. A little algebra transforms this to the second integrand above, or it may be integrated directly over y, from 0 to h, to get the same $I_y = \frac{1}{2}Mb^2$.)

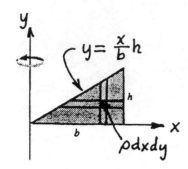

FIGURE 12-54 Problem 73 Solution.

Problem

75. A uniform disk of radius R is free to rotate about a horizontal axis at its edge. If it's at its lowest position, as shown in Fig. 12-56, what is the minimum angular speed ω necessary to ensure that its motion describes a complete circle?

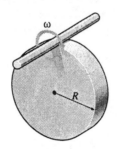

FIGURE 12-56 Problem 75.

Solution

In order to make a complete circle, the kinetic energy at the lowest point must be greater than the gain in potential energy at the highest point, or $K_{bot} > U_{top} - U_{bot}$. (This follows from the conservation of mechanical energy, if we neglect possible losses due to friction on the axis or air resistance, since $K_{top} > 0$ is the condition to make a complete circle.) The CM of the disk rises a distance of $2R$, so $U_{top} - U_{bot} = 2MgR$. Equation 12-21 gives $K_{bot} = \frac{1}{2}I_{axis}\omega_{bot}^2$, where $I_{axis} = MR^2 + \frac{1}{2}MR^2 = \frac{3}{2}MR^2$ from Equation 12-20. Therefore, $\frac{1}{2}(\frac{3}{2}MR^2)\omega_{bot}^2 > 2MgR$, or $\omega_{bot} > \sqrt{8g/3R}$.

Problem

77. A thin solid rod of length ℓ and mass M is free to pivot about one end. If it makes an angle ϕ with the horizontal, find the torque due to gravity about the pivot point. You'll need to integrate the torques on the individual mass elements comprising the rod.

Solution

Let ℓ' be the distance along the rod from the pivot at O. The force of gravity on each mass element of rod dm makes an angle of $90° - \phi$ with the displacement along ℓ', so the torque of gravity on this element has magnitude $d\tau_g = \ell'g\,dm\sin(90° - \phi) = \ell'g\cos\phi\,dm$ in a clockwise direction. The total torque of gravity is the sum of these over the whole rod, or $\tau_g = \int_0^\ell \ell'g\cos\phi\,dm$, clockwise. The mass element is the linear density (assumed uniform) times the length element, $dm = (M/\ell)\,d\ell'$, so $\tau_g = (Mg\cos\phi/\ell)\times \int_0^\ell \ell'\,d\ell' = \frac{1}{2}Mg\ell\cos\phi$, clockwise. (Note: The center of gravity of an object is defined as the point at which the total weight of the object would produce the same torque, i.e., $\tau_g = Mg\ell_{CG}\sin(90° - \phi_{cm})$. For the rod, all parts are at the same angle ϕ, so $\ell_{CG} = \int_0^\ell \ell'd(mg)/Mg$, and if g is constant, $\ell_{CG} = \int_0^\ell \ell'dm/M = \ell_{cm} = \frac{1}{2}\ell$, the same as the center of mass. See Section 14-2.)

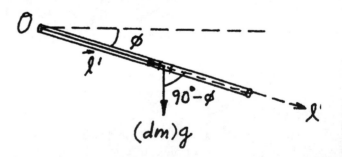

Problem 77 Solution.

CHAPTER 13 ROTATIONAL VECTORS AND ANGULAR MOMENTUM

ActivPhysics can help with these problems:
Activities 7.1, 7.16, 7.17

Section 13-1: Angular Velocity and Acceleration Vectors

Problem

1. A car is headed north at 70 km/h. Give the magnitude and direction of the angular velocity of its 62-cm-diameter wheels.

Solution

If we assume that the wheels are rolling without slipping (see Section 12-5), the magnitude of the angular velocity is $\omega = v_{cm}/r = (70 \text{ m}/3.6 \text{ s}) \div (0.31 \text{ m}) = 62.7 \text{ s}^{-1}$. With the car going north, the axis of rotation of the wheels is east-west. Since the top of a wheel is going in the same direction as the car, the right-hand rule gives the direction of ω as west.

Problem

3. A wheel is spinning at 45 rpm with its spin axis vertical. After 15 s, it's spinning at 60 rpm with its axis horizontal. Find (a) the magnitude of its average angular acceleration and (b) the angle the average angular acceleration vector makes with the horizontal.

Solution

Suppose that the x axis is horizontal in the direction of the final angular velocity ($\omega_f = (60 \text{ rpm})\hat{\imath}$) and the y axis is vertical in the direction of the initial angular velocity ($\omega_i = (45 \text{ rpm})\hat{\jmath}$). Equation 13-1 implies that $\alpha_{av} = (\omega_f - \omega_i)/\Delta t = (60\hat{\imath} - 45\hat{\jmath}) \text{ rpm}/15 \text{ s} = (4\hat{\imath} - 3\hat{\jmath}) \text{ rpm/s}$. Its magnitude is $\alpha_{av} = \sqrt{(4)^2 + (-3)^2} \text{ rpm/s} = 5 \text{ rpm/s} = 5(\pi/30) \text{ s}^{-2} = 0.524 \text{ s}^{-2}$, at an angle $\theta = \tan^{-1}(-\frac{3}{4}) = -36.9°$ to the x axis (i.e., below the horizontal).

Problem

5. A wheel is spinning with angular speed $\omega = 5.0$ rad/s, when a constant angular acceleration $\alpha = 0.85 \text{ rad/s}^2$ is applied at right angles to the initial angular velocity. How long does it take for the angular speed to increase by 10 rad/s?

Solution

For constant angular acceleration, $\omega_f = \omega_i + \alpha t$. If α is perpendicular to ω_i, $\omega_f^2 = \omega_i^2 + (\alpha t)^2$, or $t = \sqrt{\omega_f^2 - \omega_i^2}/\alpha$. Therefore, for $\omega_i = 5 \text{ s}^{-1}$ to increase to $\omega_f = 15 \text{ s}^{-1}$ under the given conditions requires time $t = \sqrt{(15)^2 - (5)^2} \text{ s}^{-1}/(0.85 \text{ s}^{-2}) = 16.6$ s. (Note: ω_f is the total angular velocity of the wheel, not just its spin.)

Section 13-2: Torque and the Vector Cross Product

Problem

7. A coordinate system lies with its x-y plane in the plane of this page, and its z axis coming out of the page. A force $\mathbf{F}$ is applied at the point $x = 1$, $y = 1$. What is the direction of the torque about the origin if $\mathbf{F}$ points (a) in the x direction, (b) in the y direction, and (c) in the z direction?

Solution

The displacement from the origin to the point of application of the force is $\mathbf{r} = \hat{\imath} + \hat{\jmath}$ (in distance units), so the direction of the torque is the direction of the cross product $(\hat{\imath} + \hat{\jmath}) \times \hat{\mathbf{F}}$, where $\hat{\mathbf{F}}$ is a unit vector in the direction of the force. (a) For $\hat{\mathbf{F}} = \hat{\imath}$, $(\hat{\imath} + \hat{\jmath}) \times \hat{\imath} = \hat{\imath} \times \hat{\imath} + \hat{\jmath} \times \hat{\imath} = 0 - \hat{\mathbf{k}} = -\hat{\mathbf{k}}$. (b) For $\hat{\mathbf{F}} = \hat{\jmath}$, $(\hat{\imath} + \hat{\jmath}) \times \hat{\jmath} = \hat{\imath} \times \hat{\jmath} + \hat{\jmath} \times \hat{\jmath} = \hat{\mathbf{k}} + 0 = \hat{\mathbf{k}}$. (c) For $\hat{\mathbf{F}} = \hat{\mathbf{k}}$, $(\hat{\imath} + \hat{\jmath}) \times \hat{\mathbf{k}} = \hat{\imath} \times \hat{\mathbf{k}} + \hat{\jmath} \times \hat{\mathbf{k}} = -\hat{\jmath} + \hat{\imath}$, which makes an angle of $-45°$ (or $315°$) CCW from the x axis. (Inspection of Fig. 13-9 and the right-hand rule provides the cross products of the rectangular unit vectors.)

Problem

9. A 12-N force is applied at the point $x = 3$ m, $y = 1$ m. Find the torque about the origin if the force points in (a) the x direction, (b) the y direction, and (c) the z direction.

Solution

Equation 13-2 and the definition of the cross product give: (a) $\tau = \mathbf{r} \times \mathbf{F} = (3\hat{\imath} + \hat{\jmath}) \times 12\hat{\imath} \text{ N·m} = -12\hat{\mathbf{k}} \text{ N·m}$, (b) $(3\hat{\imath} + \hat{\jmath}) \times 12\hat{\jmath} \text{ N·m} = 36\hat{\mathbf{k}} \text{ N·m}$, (c) $(3\hat{\imath} + \hat{\jmath}) \times$

$12\hat{\mathbf{k}}$ N·m $=(12\hat{\mathbf{i}} - 36\hat{\mathbf{j}})$ N·m, (magnitude 37.9 N·m in the x-y plane, 71.6° clockwise from the x axis).

Problem

11. You slip a wrench over a bolt. Taking the origin at the bolt, the other end of the wrench is at $x = 18$ cm, $y = 5.5$ cm. You apply a force $\mathbf{F} = 88\hat{\mathbf{i}} - 23\hat{\mathbf{j}}$ N to the end of the wrench. What is the torque on the bolt?

Solution

The displacement from the bolt (the origin) to the point of application of the force is $\mathbf{r} = (0.18\hat{\mathbf{i}} + 0.055\hat{\mathbf{j}})$m, so the torque is $\tau = \mathbf{r} \times \mathbf{F} = (0.18\hat{\mathbf{i}} + 0.055\hat{\mathbf{j}})(88\hat{\mathbf{i}} - 23\hat{\mathbf{j}})$(N·m) $= [(0.18)(-23) - (0.055) \times (88)]\hat{\mathbf{k}}$ (N·m) $= -8.98\hat{\mathbf{k}}$ N·m. The cross products of unit vectors can be calculated as in Problem 7, or the general result in Problem 14 can be used. (The orientation of the axes are not specified in this problem, but with the x-axis to the right and the y-axis up in the plane of the page, the z-axis is out of, and the above torque is into, the page, respectively.)

Problem

13. Vector $\mathbf{A}$ points 30° counterclockwise from the x axis. Vector $\mathbf{B}$ is twice as long as $\mathbf{A}$. Their product $\mathbf{A} \times \mathbf{B}$ has length A^2, and points in the negative z direction. What is the direction of vector $\mathbf{B}$?

Solution

The cross product, $\mathbf{A} \times \mathbf{B} = -A^2\hat{\mathbf{k}}$, is perpendicular to the plane of $\mathbf{A}$ and $\mathbf{B}$, so these vectors lie in the x-y plane. The right-hand rule implies that the angle between $\mathbf{A}$ and $\mathbf{B}$, measured clockwise from $\mathbf{A}$, is less than 180°, i.e., $\theta_A - \theta_B < 180°$, or $-150° < \theta_B < 30° = \theta_A$, where θ_A and θ_B are measured counterclockwise from the x axis. The magnitude of $\mathbf{A} \times \mathbf{B}$ is $AB\sin(\theta_A - \theta_B) = 2A^2\sin(\theta_A - \theta_B) = A^2$ (as given, with $B = 2A$), so $\sin(\theta_A - \theta_B) = \frac{1}{2}$ or $\theta_A - \theta_B = 30°$ or 150°. When this is combined with the given value of θ_A and the range of θ_B, one finds that $\theta_B = 0°$ or $-120°$ (i.e., along the x axis or 120° clockwise from the x axis).

Problem

15. Show that $\mathbf{A} \cdot (\mathbf{A} \times \mathbf{B}) = 0$ for any vectors $\mathbf{A}$ and $\mathbf{B}$.

Solution

$\mathbf{A} \times \mathbf{B}$ is perpendicular to $\mathbf{A}$ (or $\mathbf{B}$) so $\mathbf{A} \cdot (\mathbf{A} \times \mathbf{B}) = 0$ (or $\mathbf{B} \cdot (\mathbf{A} \times \mathbf{B}) = 0$). (Alternatively, in components (see Problem 14), $\mathbf{A} \cdot (\mathbf{A} \times \mathbf{B}) = A_x(A_yB_z - A_zB_y) +$

$A_y(A_zB_x - A_xB_z) + A_z(A_xB_y - A_yB_x) = (A_yA_z - A_zA_y)B_x + (A_zA_x - A_xA_z)B_y + (A_xA_y - A_yA_x)B_z = (\mathbf{A} \times \mathbf{A}) \cdot \mathbf{B} = 0$. In general, $\mathbf{A} \cdot (\mathbf{B} \times \mathbf{C})$ is called the triple scalar product and $\mathbf{A} \cdot (\mathbf{B} \times \mathbf{C}) = (\mathbf{A} \times \mathbf{B}) \cdot \mathbf{C}$, i.e., the "dot" and the "cross" in the triple scalar product can be interchanged. This is equivalent to a cyclic permutation of the three vectors, $\mathbf{A} \cdot (\mathbf{B} \times \mathbf{C}) = \mathbf{C} \cdot (\mathbf{A} \times \mathbf{B}) = \mathbf{B} \cdot (\mathbf{C} \times \mathbf{A})$. On the other hand, interchanging any two vectors introduces a minus sign, $\mathbf{A} \cdot (\mathbf{B} \times \mathbf{C}) = -\mathbf{C} \cdot (\mathbf{B} \times \mathbf{A}) = -\mathbf{B} \cdot (\mathbf{A} \times \mathbf{C}) = -\mathbf{A} \cdot (\mathbf{C} \times \mathbf{B})$.)

Problem

17. Find a force vector that, when applied at the point $x = 2.0$ m, $y = 1.5$ m, will produce a torque $\tau = 4.7\hat{\mathbf{k}}$ N·m about the origin. (There are many answers.)

Solution

$\mathbf{F}$ is perpendicular to the torque and thus lies in the x-y plane (i.e., $\mathbf{F} = \hat{\mathbf{i}}F_x + \hat{\mathbf{j}}F_y$ and $F_z = 0$). Specifically, $\tau = 4.7\hat{\mathbf{k}}$ N·m $= \mathbf{r} \times \mathbf{F} = (2\hat{\mathbf{i}} + 1.5\hat{\mathbf{j}})(F_x\hat{\mathbf{i}} + F_y\hat{\mathbf{j}})$ m $= (2F_y - 1.5F_x)\hat{\mathbf{k}}$ m. Therefore, any force with components $2F_y - 1.5F_x = 4.7$ N and $F_z = 0$ will produce the given torque about the origin, when applied at the given point.

Section 13-3: Angular Momentum

Problem

19. In the Olympic hammer throw (Fig. 13-29), a contestant whirls a 7.3-kg steel ball on the end of a 1.2-m cable. If the contestant's arms reach an additional 90 cm from his axis of rotation and if the speed of the ball just prior to release is 27 m/s, what is the magnitude of its angular momentum?

Solution

We assume the ball is traveling in a circle of radius $r = (1.2 + 0.9)$ m with speed $v = 27$ m/s. Since $\mathbf{v}$ is perpendicular to $\mathbf{r}$, the magnitude of the angular momentum about the center is $L = |\mathbf{r} \times \mathbf{p}| = mvr = (7.3 \text{ kg})(27 \text{ m/s})(2.1 \text{ m}) = 4.14$ J·s. (The direction is along the axis of rotation.)

Problem

21. A gymnast of rotational inertia 62 kg·m² is tumbling head over heels. If her angular momentum is 470 kg·m²/s, what is her angular speed?

Solution

If we regard the gymnast as a rigid body rotating about a fixed (instantaneous) axis, Equation 13-4 gives

$\omega = L/I = (470 \text{ kg}\cdot\text{m}^2/\text{s})/(62 \text{ kg}\cdot\text{m}^2) = 7.58 \text{ s}^{-1} =$ 1.21 rev/s as her angular speed about that axis.

Problem

23. A 7.4-cm-diameter baseball has a mass of 150 g and is spinning at 210 rad/s. What is the magnitude of its angular momentum? Treat the baseball as a solid sphere.

Solution

Equation 13-4 and Table 12-2 give the magnitude of the angular momentum about the spin axis of the baseball as $L = I\omega = \frac{2}{5}mr^2\omega = 0.4(0.15 \text{ kg}) \times$ $(0.037 \text{ m})^2(210 \text{ s}^{-1}) = 1.72 \times 10^{-2} \text{ J}\cdot\text{s}$. (Recall that units J·s = kg·m²/s.)

Problem

25. A weightlifter's barbell consists of two 25-kg masses on the ends of a 15-kg rod 1.6 m long. The weightlifter holds the rod at its center and spins it at 10 rpm about an axis perpendicular to the rod. What is the magnitude of the barbell's angular momentum?

Solution

The rotational inertia of the weights and bar about the specified axis is $I = 2m_{\text{wt}}(\ell/2)^2 + m_{\text{bar}}\ell^2/12$ (see Table 12-2) so the angular momentum about this axis is $L = I\omega = [2(25 \text{ kg})(0.8 \text{ m})^2 + (15 \text{ kg})(1.6 \text{ m})^2/12] \times$ $(10\pi/30 \text{ s}) = 36.9 \text{ J}\cdot\text{s}$.

Problem

27. Two identical 1800-kg cars are traveling in opposite directions at 90 km/h. Each car's center of mass is 3.0 m from the center of the highway (Fig. 13.31). What are the magnitude and direction of the angular momentum of the system consisting of the two cars, about a point on the center-line of the highway?

Solution

The position of a particle with respect to some point can always be expressed in terms of components perpendicular and parallel to its direction of motion (i.e., momentum $\mathbf{p}$). Thus, Equation 13-3 can be written as $\mathbf{L} = \mathbf{r} \times \mathbf{p} = (\mathbf{r}_\perp + \mathbf{r}_\parallel) \times \mathbf{p} = \mathbf{r}_\perp \times \mathbf{p}$, since $\mathbf{r}_\parallel \times \mathbf{p} = 0$. For straight line motion, $\mathbf{r}_\perp$ is a constant (this is an alternate solution to Problem 26). For the two cars (regarded as particles located at their respective centers of mass) $\mathbf{p}_1 = -\mathbf{p}_2$ and $\mathbf{r}_{1\perp} = -\mathbf{r}_{2\perp}$ (for any point on the center line), so the total angular momentum of their centers of mass is $\mathbf{L} = \mathbf{L}_1 + \mathbf{L}_2 = \mathbf{r}_{1\perp} \times \mathbf{p}_1 + (-\mathbf{r}_{1\perp}) \times (-\mathbf{p}_1) =$

$2\mathbf{r}_{1\perp} \times \mathbf{p}_1$. This has magnitude $2r_{1\perp}mv_1 = 2(3 \text{ m}) \times$ $(1800 \text{ kg})(90 \text{ m}/3.6 \text{ s}) = 2.70 \times 10^5 \text{ J}\cdot\text{s}$ and direction out of the plane of Fig. 13-30.

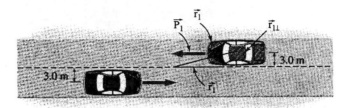

FIGURE 13-30 Problem 27 Solution.

Problem

29. Engineers redesign a car's wheels, with the goal of dropping each wheel's angular momentum by 30% for a given linear speed for the car. Other design considerations require that the wheel diameter go from 38 cm to 35 cm. If the old wheel had a rotational inertia of 0.32 kg·m², what must be the new wheel's rotational inertia if the engineers are to achieve their goal?

Solution

The specifications require that $L_2 = I_2\omega_2 = 0.7L_1 = 0.7I_1\omega_1$, for $v_1 = \omega_1 R_1 = v_2 = \omega_2 R_2$ (see Equations 13-4 and 12-25). Then $I_2 = 0.7 I_1(\omega_1/\omega_2) = 0.7I_1(2R_2/2R_1) = (0.7)(0.32 \text{ kg}\cdot\text{m}^2)(35 \text{ cm}/38 \text{ cm}) = 0.206 \text{ kg}\cdot\text{m}^2$.

Section 13-4:   Conservation of Angular Momentum

Problem

31. A 3.0-m-diameter merry-go-round with rotational inertia 120 kg·m² is spinning freely at 0.50 rev/s. Four 25-kg children sit suddenly on the edge of the merry-go-round. (a) Find the new angular speed, and (b) determine the total energy lost to friction between the children and the merry-go-round.

Solution

(a) If we assume conservation of angular momentum (about the vertical axis of the merry-go-round), $I_0\omega_0 = I\omega = (I_0 + 4mr^2)\omega$, where $I_0 = 120 \text{ kg}\cdot\text{m}^2$ is the rotational inertia of the merry-go-round, and $4mr^2 = 4(25 \text{ kg})(1.5 \text{ m})^2 = 225 \text{ kg}\cdot\text{m}^2$ is that of the four children. Then $\omega = (120/345)(0.5 \text{ rev/s}) = 0.174 \text{ rev/s}$. (b) The change in kinetic energy is $\Delta K = \frac{1}{2}I\omega^2 - \frac{1}{2}I_0\omega_0^2 = \frac{1}{2}(I^{-1} - I_0^{-1})L_0^2$, where we used the conservation of angular momentum above, and set

$L_0 = I_0\omega_0$. Changing 0.5 rev/s to π rad/s, we find

$$\Delta K = \frac{1}{2}\left(\frac{1}{345 \text{ kg·m}^2} - \frac{1}{120 \text{ kg·m}^2}\right)$$
$$\times (120 \text{ kg·m}^2 \times \pi \text{ s}^{-1})^2$$
$$= 386 \text{ J}.$$

Problem

33. In Fig. 13-33 the lower disk, of mass 440 g and radius 3.5 cm, is rotating at 180 rpm on a frictionless shaft of negligible radius. The upper disk, of mass 270 g and radius 2.3 cm, is initially not rotating. It drops freely down the shaft onto the lower disk, and frictional forces act to bring the two disks to a common rotational speed. (a) What is that speed? (b) What fraction of the initial kinetic energy is lost to friction?

Solution

(a) There are no external torques acting about the frictionless shaft, so the total angular momentum of the two-disk system in this direction is conserved, or $\mathbf{L}_1 + \mathbf{L}_2 = \text{const.}$ (The frictional torques between the disks are internal to the system.) Initially, $L_1 = I_1\omega_{1i}$ and $L_2 = 0$, and finally $L_1 + L_2 = (I_1 + I_2)\omega_f$, therefore $\omega_f = \omega_{1i}I_1/(I_1 + I_2) = (180 \text{ rpm})\times [440\times3.5^2/(440\times3.5^2 + 270\times2.3^2)] = 142 \text{ rpm}$. (We canceled common factors of $\frac{1}{2}$ g·cm^2 from the rotational inertias.) (b) The fraction of initial kinetic energy lost is

$$\frac{K_i - K_f}{K_i} = 1 - \frac{K_f}{K_i} = 1 - \frac{I_1}{I_1 + I_2} = 1 - \frac{\omega_f}{\omega_{1i}}$$
$$= 1 - 0.791 = 20.9\%.$$

(We used $K = \frac{1}{2}I\omega^2 = L^2/2I$ for the kinetic energy of a rigid body rotating about a fixed axis, and the result of part (a).)

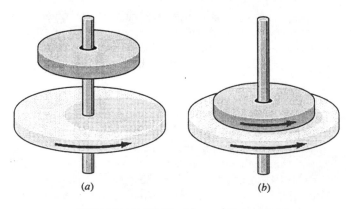

(a)   (b)

FIGURE 13-33 Problems 33 and 34.

Problem

35. A uniform, spherical cloud of interstellar gas has mass 2.0×10^{30} kg and radius 1.0×10^{13} m, and it is rotating with period 1.4×10^6 years. If the cloud collapses to form a star 7.0×10^8 m in radius, what will be the star's rotation period?

Solution

If we assume there are no external torques and no mass loss during the collapse of the star forming cloud, its angular momentum is conserved. For a uniform sphere of constant mass, this implies that $I_i\omega_i = \frac{2}{5}MR_i^2\omega_i = I_f\omega_f = \frac{2}{5}MR_f^2\omega_f$, or $\omega_i/\omega_f = (R_f/R_i)^2$. Since the angular velocity is one rotation divided by one period, $\omega = 2\pi/T$, and $T_f = T_i(R_f/R_i)^2 = (7\times10^8/10^{13} \text{ m})^2(1.4\times10^6 \text{ y})(365 \text{ d/y}) = 2.5$ d. (In current models of star formation, the collapsing cloud does not maintain a spherical shape, forming a flattened disk instead, and the central star retains just a fraction of the original cloud's mass.)

Problem

37. A turntable of radius 25 cm and rotational inertia 0.0154 kg·m^2 is spinning freely at 22.0 rpm about its central axis, with a 19.5-g mouse on its outer edge. The mouse walks from the edge to the center. Find (a) the new rotation speed and (b) the work done by the mouse.

Solution

(a) We suppose that "spinning freely" means that there are no external torques acting about the axis of rotation, so that the total angular momentum of the turntable and mouse in this direction is constant, or $I_i\omega_i = I_f\omega_f$. Initially, the rotational inertia of the mouse, mr^2, must be added to that of the turntable, I_0, but we assume the mouse contributes nothing when at the center, so $\omega_f = \omega_i(I_i/I_f) = (22 \text{ rpm})\times [0.0154 \text{ kg·m}^2 + 0.0195 \text{ kg}(0.25 \text{ m})^2]\div (0.0154 \text{ kg·m}^2) = 23.7 \text{ rpm}$. (b) The work done by the mouse (when it exerts reaction forces to friction between its feet and the turntable) can be found from the work-energy theorem, $W_{nc} = K_f - K_i = \frac{1}{2}I_f\omega_f^2 - \frac{1}{2}I_i\omega_i^2 = \frac{1}{2}I_f\omega_f^2(1 - \omega_i/\omega_f)$, where we used the conservation of angular momentum from part (a). Numerically, $W_{nc} = \frac{1}{2}(0.0154)(23.7\pi/30)^2$ J$\times (1 - 22/23.7) = 3.49$ mJ.

Problem

39. Two small beads of mass m are free to slide on a frictionless rod of mass M and length ℓ, as shown in Fig. 13-34. Initially the beads are held together at the rod center, and the rod is set spinning freely

with initial angular speed ω_0 about a vertical axis coming out of the page in Fig. 13-34. The beads are released, and they slide to the ends of the rod and then off. Find the expressions for the angular speed of the rod (a) when the beads are halfway to the ends of the rod, (b) when they're at the ends, and (c) after the beads are gone. *Hint:* Two of the answers are the same. Why?

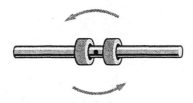

FIGURE 13-34 Problem 39.

Solution

If the rod and beads are spinning free of external torques about the axis of rotation, the vertical component of their total angular momentum is conserved. (Fig. 13-34 is a top view.) Initially, $L_z = \frac{1}{12}M\ell^2\omega_0$ (since the rotational inertia of the beads at the center is zero), but as the beads move away from the center (presumably symmetrically) they contribute to the rotational inertia, and $L_z = (\frac{1}{12}M\ell^2 + 2mr^2)\omega$, for $0 \le r \le \ell/2$. Therefore, $\omega = \omega_0(1 + 24mr^2/M\ell^2)^{-1}$ when (a) $r = \ell/4$, $\omega_a = \omega_0(1 + 3m/2M)^{-1}$, and (b) $r = \ell/2$, $\omega_b = \omega_0 \times (1 + 6m/M)^{-1}$. (c) When the beads leave the rod, they no longer exert any influence on it and its angular speed stays constant at ω_b. (The beads leave the rod moving on parallel tangents in opposite directions with vertical component of angular momentum about the midpoint of their perpendicular separation, $\frac{1}{2}\ell = b$, of $2m(\ell/2)^2\omega_b = 2m(\ell/2)(\omega_b\ell/2) = 2mvb$, as in Problems 26 and 27.)

Problem

41. Eight 60-kg skaters join hands and skate down an ice rink at 4.6 m/s. Side by side, they form a line 12 m long. The skater at one end stops abruptly, and the line proceeds to rotate rigidly about that skater. Find (a) the angular speed, (b) the linear speed of the outermost skater, and (c) the force that must be exerted on the outermost skater.

Solution

The force that abruptly stops the skater at one end exerts no torque about that skater (point P), so the total angular momentum about a vertical axis through P is conserved. Initially, the angular momentum of each of the other seven skaters about P is mv_0b, where

b is the perpendicular distance from the original straight-line motion to the point P (see Problems 26 or 27). For these seven skaters, $b_n = n\ell/7$, where $n = 1, 2, \ldots, 7$ and $\ell = 12$ m, so $L = \sum_1^7 mv_0(n\ell/7) = (mv_0\ell/7)\sum_1^7 n = (mv_0\ell/7)(7 \times 8/2) = 4mv_0\ell$. (a) When the skaters rotate rigidly about P with angular speed ω, their angular momentum is $L = I\omega = (\sum_1^7 mb_n^2)\omega = (m\ell^2\omega/49)\sum_1^7 n^2 = (m\ell^2\omega/49) \times (7 \times 8 \times 15/6) = 20m\ell^2\omega/7$. Since L is constant, $4mv_0\ell = 20m\ell^2\omega/7$ implies $\omega = (7v_0/5\ell) = (1.4) \times (4.6 \text{ m/s})/(12 \text{ m}) = 0.537 \text{ s}^{-1}$. (b) The linear speed of the outermost skater is $v_7 = \omega b_7 = \omega\ell = (0.537 \text{ s}^{-1}) \times (12 \text{ m}) = 6.44$ m/s. (c) The force is the centripetal force, $mv_7^2/\ell = m\ell\omega^2 = (60 \text{ kg})(12 \text{ m})(0.537 \text{ s}^{-1})^2 = 207$ N. (Note: the sums of integers and squares of integers can be found in mathematical tables.)

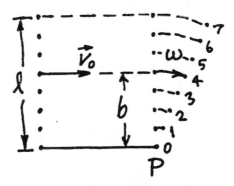

Problem 41 Solution.

Section 13-5:   Rotational Dynamics in Three Dimensions

Problem

43. A gyroscope consists of a solid disk of radius 7.5 cm, mounted on one end of a shaft of negligible mass. The far end of the shaft is placed on a frictionless pivot and the disk set spinning at 950 rpm. The gyroscope then precesses at 2.1 rad/s. How long is the shaft?

Solution

The situation described is similar to that in Example 13-5, except that the disk is mounted at one end of its shaft, so $d = \ell = L\Omega/mg = I\omega\Omega/mg = \frac{1}{2}mR^2\omega\Omega \div mg = \omega\Omega R^2/2 \text{ g} = (950\,\pi/30 \text{ s})(2.1 \text{ s}^{-1})(7.5 \text{ cm})^2 \div 2(980 \text{ cm/s}^2) = 6.00$ cm.

Problem

45. A gyroscope consists of a disk and shaft mounted on frictionless bearings in a frame of diameter d, as shown in Fig. 13-35. Initially the gyroscope is

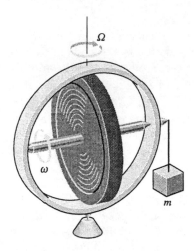

FIGURE 13-35 Problem 45.

spinning with angular speed ω and is perfectly balanced so it's not precessing. When a mass m is hung from the frame at one of the shaft bearings, the gyroscope precesses about a vertical axis with angular speed Ω. Find an expression for the rotational inertia of the gyroscope.

Solution

The spin axis is horizontal and the gravitational torque in Fig. 13-35 is $mg(d/2)$, so Equation 13-7, for a gyroscope under these circumstances, gives $\Omega = \tau/L = (mgd/2)/I\omega$ or $I = mgd/2\omega\Omega$.

Paired Problems

Problem

47. The dot product of a pair of vectors is twice the magnitude of their cross product. What is the angle between the vectors?

Solution

The given relationship means that $\mathbf{A}\cdot\mathbf{B} = AB\cos\theta = 2|\mathbf{A}\times\mathbf{B}| = 2AB\sin\theta$, so $\theta = \tan^{-1}(\frac{1}{2}) = 26.6°$.

Problem

49. A student's rotational inertia about a vertical axis through his center is 4.5 kg·m² with his arms held to his chest and 5.6 kg·m² with his arms outstretched. In a physics demonstration, the student stands on a turntable rotating at 1.0 rev/s, clutching two 7.5-kg weights to his chest. The turntable's rotational inertia is 4.0 kg·m². If the student extends his arms fully so the weights are each 95 cm from his rotation axis, what will be his new angular speed?

Solution

We suppose that the angular momentum of the student, weights, and turntable, about the vertical axis of rotation of the turntable, is conserved, or $I_i\omega_i = I_f\omega_f$. If the weights start out on the rotation axis, they do not contribute to the initial rotational inertia, whereas with the student's arms extended, they contribute $2mr^2$. When this fact is combined with the other given rotational inertias, we find $\omega_f = (1 \text{ rev/s})[4.5 + 4.0]/[5.6 + 4.0 + 2(7.5)(0.95)^2] = 0.367$ rev/s.

Problem

51. The turntable in Fig. 13-36 has rotational inertia 0.021 kg·m², and is rotating at 0.29 rad/s about a frictionless vertical axis. A wad of clay is tossed onto the turntable and sticks 15 cm from the rotation axis. The clay impacts with horizontal velocity component 1.3 m/s, at right angles to the turntable's radius, and in a direction that opposes the rotation, as suggested in Fig. 13-36. After the clay lands, the turntable has slowed to 0.085 rad/s. Find the mass of the clay.

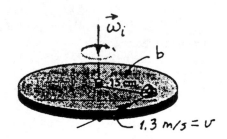

FIGURE 13-36 Problem 51 Solution.

Solution

There are no external torques in the direction of the turntable's axis, so the vertical angular momentum of the turntable/clay system is conserved. (The horizontal forces, which cause the clay to stick to the turntable, are internal forces.) If we take the sense of rotation of the turntable to define the positive direction of vertical angular momentum, then the system's initial angular momentum is $I\omega_i - mvb$, where I is the rotational inertia of the turntable, v the horizontal component of the velocity of the mass, m, of clay, and b the perpendicular distance to the axis of rotation. After the clay lands, this angular momentum equals $(I + mb^2)\omega_f$, so solving for the mass, we find $m = I(\omega_i - \omega_f)/(vb + b^2\omega_f) = (0.021 \text{ kg·m}^2)\times(0.29 \text{ s}^{-1} - 0.085 \text{ s}^{-1})/[(1.3 \text{ m/s})(0.15 \text{ m}) + (0.15 \text{ m})^2\times(0.085 \text{ s}^{-1})] = 21.9$ g. (We assumed ω_f and ω_i have the same sense. If the sense of rotation of the turntable were reversed after impact of the clay, i.e., $\omega_f = -0.085$ s^{-1}, then $m = 40.8$ g.)

Problem

53. A dog of mass m is standing on the edge of a stationary, frictionless turntable of rotational inertia I and radius R. The dog walks once around the turntable. What fraction of a full circle does the dog's motion make with respect to the ground?

Solution

Walking once around relative to the turntable, the dog describes an angular displacement of $\Delta\theta_D$ relative to the ground, and the turntable one of $\Delta\theta_T$ in the opposite direction, such that $\Delta\theta_D + |\Delta\theta_T| = 2\pi$. We suppose that the vertical component of the angular momentum of the dog and turntable is conserved (which was zero initially), so that $0 = I_D\omega_D + I_T\omega_T = I_D(\Delta\theta_D/\Delta t) - I_T|\Delta\theta_T/\Delta t|$, where $I_D = mR^2 = (17\text{ kg})(1.81\text{ m})^2$ and $I_T = 95$ kg·m^2 are the rotational inertias about the axis of rotation, and we wrote the angular velocities (which are in opposite directions) in terms of the angular displacements and the common time interval. The numerical values are from Problem 38. Eliminating $\Delta\theta_T$, we find $I_D\Delta\theta_D - I_T(2\pi - \Delta\theta_D) = 0$, or $(\Delta\theta_D/2\pi) = (1 + I_D/I_T)^{-1} = (1 + 55.7/95)^{-1} = 0.630$. ($\Delta\theta_D$ is 63% of a full circle relative to the ground, for the numerical data given in Problem 38.)

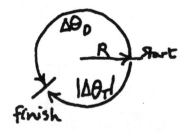

Problem 53 Solution.

Supplementary Problems
Problem

55. An advanced civilization lives on a solid spherical planet of uniform density. Running out of room for their expanding population, the civilization's government calls an engineering firm specializing in planetary reconfiguration. Without adding any material or angular momentum, the engineers reshape the planet into a hollow shell whose thickness is one-fifth of its outer radius. Find the ratio of the new to the old (a) surface area and (b) length of day.

Solution

Since the planetary mass (M) and density (ρ) are constant, the volume of the original sphere (radius R_0) is the same as the volume of the new spherical shell (outer radius R, inner radius $R - \frac{1}{5}R = \frac{4}{5}R$). Thus, $\frac{4}{3}\pi R_0^3 = \frac{4}{3}\pi R^3(1 - 0.8^3)$, or $R/R_0 = (1 - 0.512)^{-1/3} = 1.27$. (a) The ratio of the outer surface area of the new to old planet is $4\pi R^2/4\pi R_0^2 = (R/R_0)^2 = (1.27)^2 = 1.61$. (b) Since the angular momentum is also constant, $I_0\omega_0 = I_0(2\pi/T_0) = I\omega = I(2\pi/T)$, or $T/T_0 = I/I_0$, where T is the planetary rotation period (length of day) and I the rotational inertia about the axis of rotation. For a solid uniform sphere, $I_0 = \frac{2}{5}MR_0^2 = \frac{2}{5}(\frac{4}{3}\pi\rho)R_0^5$, while for the uniform shell, $I = \frac{2}{5}(\frac{4}{3}\pi\rho)\times[R^5 - (0.8R)^5]$, so the ratio is $T/T_0 = (R/R_0)^5 \times (1 - 0.8^5) = (1.27)^5(1 - 0.8^5) = 2.22$.

Problem

57. If you're familiar with determinants, show that the cross product $\mathbf{A} \times \mathbf{B}$ can be written as a determinant:

$$\mathbf{A} \times \mathbf{B} = \begin{vmatrix} \hat{\mathbf{i}} & \hat{\mathbf{j}} & \hat{\mathbf{k}} \\ A_x & A_y & A_z \\ B_x & B_y & B_z \end{vmatrix}.$$

Solution

Expand the determinant along the first row:

$$\mathbf{A} \times \mathbf{B} = \hat{\mathbf{i}}\begin{vmatrix} A_y & A_z \\ B_y & B_z \end{vmatrix} - \hat{\mathbf{j}}\begin{vmatrix} A_x & A_z \\ B_x & B_z \end{vmatrix} + \hat{\mathbf{k}}\begin{vmatrix} A_x & A_y \\ B_x & B_y \end{vmatrix}$$
$$= \hat{\mathbf{i}}(A_yB_z - A_zB_y) + \hat{\mathbf{j}}(A_zB_x - A_xB_z) + \hat{\mathbf{k}}(A_xB_y - A_yB_x).$$

This is the expression derived from the cross products of the unit vectors found in Problem 14.

Problem

59. About 99.9% of the solar system's total mass lies in the Sun. Using data from Appendix E, estimate what fraction of the solar system's angular momentum about its center is associated with the Sun. Where is most of the rest of the angular momentum?

Solution

The planets orbit the Sun in planes approximately perpendicular to the Sun's rotation axis, so most of the angular momentum in the solar system is in this direction. We can estimate the orbital angular momentum of a planet by mvr, where m is its mass, v its average orbital speed, and r its mean distance from the Sun (see the solution to Problem 24, where one also finds that $L_{orb} \gg L_{rot}$). Compared to the orbital angular momentum of the four giant planets, everything else is negligible, except for the rotational angular momentum of the Sun itself. The latter can be estimated by assuming the Sun to be a uniform sphere rotating with an average period of $\frac{1}{2}(27 + 36)$ d. (This is an overestimate since most of the Sun's mass is in its core, with radius about $\frac{1}{4}R_{\odot}$.) The numerical data in Appendix E results in the following estimates:

ORBITAL ANGULAR MOMENTUM (mvr)		%
Jupiter	19.2×10^{42} J·s	59.7
Saturn	7.85×10^{42} J·s	24.4
Uranus	1.69×10^{42} J·s	5.2
Neptune	2.52×10^{42} J·s	7.8

ROTATIONAL ANGULAR MOMENTUM ($\frac{2}{5}MR^2\omega$)		
Sun	0.89×10^{42} J·s	2.8
Total	32.2×10^{42} J·s	99.9

Problem

61. A rod of length ℓ and mass M is suspended from a pivot, as shown in Fig. 13-39. The rod is struck midway along its length by a wad of putty of mass m moving horizontally at speed v. The putty sticks to the rod. Find an expression for the minimum speed v that will result in the rod's making a complete circle rather than swinging like a pendulum.

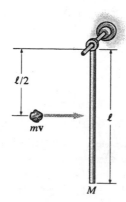

FIGURE 13-39 Problem 61.

Solution

Assume that during the collision of the putty with the rod, the angular momentum about the pivot is conserved. Before the collision (when just the putty is moving) the angular momentum is $mv(\ell/2)$ (see Problem 27), and just afterwards it is $I\omega_0$. Here, $I = \frac{1}{3}M\ell^2 + m(\frac{1}{2}\ell)^2$ is the rotational inertia of the

system, after the collision, about the pivot and ω_0 is the angular speed when the system is at the lowest point of its subsequent swing. If the pivot is frictionless, the mechanical energy will be conserved during this swing, or $K_0 + U_0 = K + U$. The center of mass is at the center of the rod (since the putty is stuck there) and so the maximum change in potential energy, when the rod is at the highest point of its swing, is $(m + M)g\ell$. If there is also kinetic energy at this point, the rod will make a complete circle, i.e., $K_{top} \geq 0$ implies $K_0 \geq U_{top} - U_0$ or $\frac{1}{2}I\omega_0^2 \geq (m + M)g\ell$. Since $\omega_0 = mv\ell/2I$, from above, this condition becomes $\omega_0^2 = (mv\ell/2I)^2 \geq 2(m + M)g\ell/I$, or $v^2 \geq (8I/\ell^2)(m + M)g\ell/m^2 = (\frac{1}{4}m + \frac{1}{3}M) \times 8(m + M)g\ell/m^2$.

Problem

63. A solid disk of mass M and radius R is initially stationary on a frictionless horizontal axle. A blob of putty with mass m falls and strikes a small protrusion on the side of the disk, as shown in Fig. 13-40. The protrusion is initially level with the axle, and the putty sticks to it. The protrusion has negligible mass. Find (a) the angular speed of the disk immediately after the putty hits and (b) the lowest angular speed the disk subsequently obtains. (c) Find an expression for the minimum h such that the disk will rotate rather than oscillate back and forth after the putty hits.

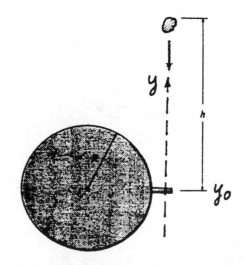

FIGURE 13-40 Problem 63.

Solution

(a) Assume that the totally inelastic collision between the putty and disk happens so quickly that the torque of gravity has no time to act appreciably. Then the angular momentum of the putty and disk about the horizontal axle is conserved. Initially, the putty has angular momentum $m\sqrt{2gh}R$, where $\sqrt{2gh}$ is its speed after falling through a height h, and the disk has angular momentum zero. Immediately after the collision, the putty and disk rotate with angular speed ω_0, and their rotational inertia about the axle is $I = (\frac{1}{2}M + m)R^2$, so $m\sqrt{2gh}R = I\omega_0$, or $\omega_0 = m\sqrt{2gh}/(\frac{1}{2}M + m)R$. (b) After the collision, the angular momentum changes because gravity exerts a net torque on the putty, but since the axle is frictionless, the mechanical energy of the putty and disk is conserved. Their kinetic energy is $\frac{1}{2}I\omega^2$ and their potential energy is mgy, where y is the vertical position of the putty and the potential energy of the disk is a constant which can be ignored. (The constant is Mgy_0, where y_0 is the vertical position of both the axle and the putty immediately after the collision, which is the beginning of the putty-disk system's subsequent motion.) Then $\frac{1}{2}I\omega^2 + mgy = \frac{1}{2}I\omega_0^2 + mgy_0$, or $\omega^2 = \omega_0^2 - 2mg(y - y_0)/I$. The largest value of $y - y_0$ is R, so ω^2 has a lower limit of $\omega_0^2 - 2mgR \div I = 2mg[mh - (\frac{1}{2}M + m)R] / (\frac{1}{2}M + m)^2 R^2$, where ω_0 and I were substituted from part (a). If this lower limit is positive, then the lowest angular speed is its square root, and the putty and disk will keep rotating. (c) For this to occur, $mh - (\frac{1}{2}M + m)R \geq 0$, or $h \geq (1 + M/2m)R$. (If h is less than this, the minimum ω is zero, and the motion is oscillatory.)

Problem

65. A solid ball of mass M and radius R is spinning with angular velocity ω_0 about a horizontal axis. It drops vertically onto a surface where the coefficient of kinetic friction with the ball is μ_k (Fig. 13-42). Find expressions for (a) the final angular velocity once it's achieved pure rolling

motion and (b) the time it takes until it's in pure rolling motion.

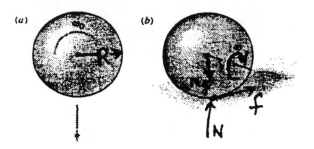

FIGURE 13-42 Problem 65.

Solution

When the ball comes in contact with the horizontal surface, the frictional force $f_k = \mu_k mg$ acts to accelerate the ball's center of mass velocity, $a_{cm} = f_k/m = \mu_k g$. It also produces a torque, about the horizontal axis through the ball's center of mass, which independently acts to decelerate the spin, $\alpha_{cm} = \tau_{cm}/I_{cm} = -\mu_k mgR/\frac{2}{5}MR^2 = -\frac{5}{2}(a_{cm}/R)$. The angular speed about the center of mass decreases from $\omega_0 (\omega = \omega_0 + \alpha_{cm}t)$ and v_{cm} increases from zero $(v_{cm} = a_{cm}t)$ until finally $v_{cm,f} = \omega_f R$, when the ball begins rolling without slipping, and the frictional force is also (practically) zero. Then $(\omega_0 - \omega_f)/\omega_f = -\alpha_{cm}t/(v_{cm}/R) = -\alpha_{cm}/(a_{cm}/R) = \frac{5}{2}$, or $\omega_f = \frac{2}{7}\omega_0$. This occurs after a time $t = v_{cm}/a_{cm} = \omega_f R/a_{cm} = \frac{2}{7}\omega_0 R/\mu_k g$. (Note: The fact that the horizontal axis of rotation does not change direction in this case, allowed us to use the same equations for the rotational motion about the center of mass, as derived for rotation about a fixed axis.)

CHAPTER 14 STATIC EQUILIBRIUM

ActivPhysics can help with these Problems:
Activities 7.1–7.6

Section 14-1:   Conditions for Equilibrium

Problem

1. Five forces act on a rod, as shown in Fig. 14-25. Write the torque equations that must be satisfied for the rod to be in static equilibrium taking the torques (a) about the top of the rod and (b) about the center of the rod.

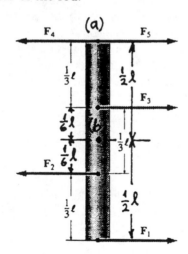

FIGURE 14-25 Problem 1.

Solution

All of the forces lie in the plane of Fig. 14-25, so all of the torques about any point on the rod are into or out of the page. Suppose the latter direction, out of the page or counterclockwise, is positive. Moreover, all of the forces are perpendicular to the rod, so their lever arms about any point on the rod (recall that the magnitude of the torque is force times lever arm) can easily be read-off from Fig. 14-25. (a) About the top of the rod, $\mathbf{F}_4$ and $\mathbf{F}_5$ contribute zero torque, and Equation 14-2 becomes $0 = \frac{1}{3}\ell F_3 - \frac{2}{3}\ell F_2 + \ell F_1$.
(b) About the center of the rod, the perpendicular distances to $\mathbf{F}_2$ and $\mathbf{F}_3$ are $\frac{1}{6}\ell$, and to $\mathbf{F}_1$, $\mathbf{F}_4$ and $\mathbf{F}_5$ are $\frac{1}{2}\ell$, so $0 = \frac{1}{2}\ell(F_1 + F_4 - F_5) - \frac{1}{6}\ell(F_2 + F_3)$.

Problem

3. Suppose the force $\mathbf{F}_3$ in the preceding problem is doubled so the forces no longer balance and the

body is therefore accelerating. Show that (a) the torque about the point $(-7\,\text{m}, 1\,\text{m})$ is still zero, but that (b) the torque about the origin is no longer zero. What is the torque about the origin?

Solution

(a) Since $\mathbf{r}_3 = (-7\hat{\mathbf{i}} + \hat{\mathbf{j}})$ m is the point of application of $\mathbf{F}_3$, the total torque about $\mathbf{r}_3$ is just due to $\mathbf{F}_1$ and $\mathbf{F}_2$: $(\sum \boldsymbol{\tau}_i)_3 = (\mathbf{r}_1 - \mathbf{r}_3) \times \mathbf{F}_1 + (\mathbf{r}_2 - \mathbf{r}_3) \times \mathbf{F}_2 = [(2\hat{\mathbf{i}} + 7\hat{\mathbf{i}} - \hat{\mathbf{j}}) \times (2\hat{\mathbf{i}} + 2\hat{\mathbf{j}}) + (-\hat{\mathbf{i}} + 7\hat{\mathbf{i}} - \hat{\mathbf{j}}) \times (-2\hat{\mathbf{i}} - 3\hat{\mathbf{j}})]\text{N·m} = [(9 \times 2)\hat{\mathbf{k}} - (1 \times 2)(-\hat{\mathbf{k}}) + 6 \times (-3)\hat{\mathbf{k}} + (1 \times 2)(-\hat{\mathbf{k}})]$ N·m $= 0$. (b) $(\sum \boldsymbol{\tau}_i)_0 = \sum (\mathbf{r}_i \times \mathbf{F}_i) = [2\hat{\mathbf{i}} \times (2\hat{\mathbf{i}} + 2\hat{\mathbf{j}}) + (-\hat{\mathbf{i}}) \times (2\hat{\mathbf{i}} - 3\hat{\mathbf{j}}) + (-7\hat{\mathbf{i}} + \hat{\mathbf{j}}) \times (2\hat{\mathbf{j}})]$ N·m $= [4\hat{\mathbf{k}} + 3\hat{\mathbf{k}} - 14\hat{\mathbf{k}}]$ N·m $= -7\hat{\mathbf{k}}$ N·m.

Problem

5. In Fig. 14-26 the forces shown all have the same magnitude F. For each of the cases shown, is it possible to place a third force so the three forces

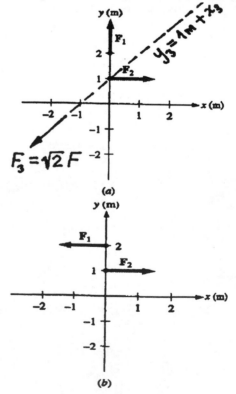

FIGURE 14-26 Problem 5.

meet both conditions for static equilibrium? If so, specify the force and a suitable application point; if not, why not?

Solution

The conditions for static equilibrium, under the action of three forces, can be written as: $\mathbf{F}_3 = -(\mathbf{F}_1 + \mathbf{F}_2)$ and $\mathbf{r}_3 \times \mathbf{F}_3 = -(\mathbf{r}_1 \times \mathbf{F}_1 + \mathbf{r}_2 \times \mathbf{F}_2)$. (a) In this case, $\mathbf{F}_1 = F\hat{\jmath}$, $\mathbf{r}_1 = (2 \text{ m})\hat{\jmath}$, $\mathbf{F}_2 = F\hat{\imath}$, and $\mathbf{r}_2 = (1 \text{ m})\hat{\jmath}$. Thus, $\mathbf{F}_3 = -F(\hat{\imath} + \hat{\jmath})$, which is a force of magnitude $\sqrt{2}F$, 45° down into the third quadrant ($\theta_x = 225°$ or $-135°$ CCW from the x axis). The point of application, $\mathbf{r}_3$, can be found from the second condition, $\mathbf{r}_3 \times \mathbf{F}_3 = (x_3\hat{\imath} + y_3\hat{\jmath}) \times (-F\hat{\imath} - F\hat{\jmath}) = (-x_3 + y_3)F\hat{\mathbf{k}} = -\mathbf{r}_1 \times \mathbf{F}_1 - \mathbf{r}_2 \times \mathbf{F}_2 = 0 - (1 \text{ m})\hat{\jmath} \times F\hat{\imath} = (1 \text{ m})F\hat{\mathbf{k}}$. Thus, $-x_3 + y_3 = 1$ m, or the line of action of $\mathbf{F}_3$ passes through the point of application of $\mathbf{F}_2$ (the point $(0, 1$ m$)$). Any point on this line is a suitable point of application for $\mathbf{F}_3$ (e.g. the point $(0, 1$ m$)$). (b) In this case, $\mathbf{F}_1 = -\mathbf{F}_2$ so $\mathbf{F}_3 = 0$, but $\mathbf{r}_1 \times \mathbf{F}_1 + \mathbf{r}_2 \times \mathbf{F}_2 = (\mathbf{r}_2 - \mathbf{r}_1) \times \mathbf{F}_2 \neq 0$ so $\mathbf{r}_3 \times \mathbf{F}_3 \neq 0$. Thus there is no single force that can be added to produce static equilibrium.

Problem

7. Four forces act on a body, as shown in Fig. 14-27. Write the set of scalar equations that must hold for the body to be in equilibrium, evaluating the torques (a) about point O and (b) about point P.

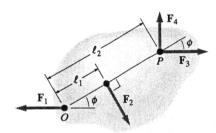

FIGURE 14-27 Problem 7.

Solution

All of the forces lie in the same plane, which includes the points O and P, so there are two independent components of the force condition (Equation 14-1) and one component of the torque condition (Equation 14-2). Taking the x axis to the right, the y axis up and the z axis out of the page in Figure 14-27, we have: $\sum F_x = 0 = -F_1 + F_2 \sin\phi + F_3$, $\sum F_y = 0 = -F_2 \cos\phi + F_4$, $(\sum \tau_z)_0 = 0 = -\ell_1 F_2 - \ell_2 F_3 \sin\phi + \ell_2 F_4 \cos\phi$, and $(\sum \tau_z)_P = -\ell_2 F_1 \sin\phi + (\ell_2 - \ell_1) \times F_2 = 0$. (The lever arms of all the forces about either O or P should be evident from Fig. 14-27).

Section 14-2:   Center of Gravity

Problem

9. Figure 14-29a shows a thin, uniform square plate of mass m and side ℓ. The plate is in a vertical plane. Find the magnitude of the gravitational torque on the plate about each of the three points shown.

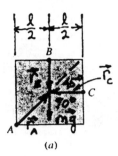

(a)

FIGURE 14-29(a) Problem 9 Solution.

Solution

The center of gravity is at the center of a uniform plate. In calculating the gravitational torque, one may consider the entire weight as acting at the center of gravity. (a) $\mathbf{r}_A = \sqrt{2}\ell/2$ at 135° from the weight of the plate, so $\tau_A = (\sqrt{2}\ell/2)mg \sin 135° = \frac{1}{2}mg\,\ell$. (b) $\mathbf{r}_B$ is colinear with the weight, so $\tau_B = 0$. (c) $\tau_C = \frac{1}{2}\ell\, mg \sin 90° = \frac{1}{2}mg\,\ell$ (but note that $\tau_C = -\tau_A$). (We also assumed that B and C are at the centers of their respective sides. Alternatively, the torques can be found from the lever arms shown.)

Problem

11. Three identical books of length L are stacked over the edge of a table as shown in Fig. 14-30. The top book overhangs the middle one by $\frac{1}{2}L$, so it just barely avoids falling. The middle book overhangs the bottom one by $\frac{1}{4}L$. How much of the bottom book can be allowed to overhang the edge of the table without the books falling?

Solution

In equilibrium, the farthest right the center of mass of the combination of three books can lie is directly above the edge of the table. (This is unstable equilibrium, since the slightest disturbance to the right would cause the books to fall.) The center of mass of each book is at its center, so if we take the origin at the edge with positive to the right, this condition becomes

$$0 = x_{cm}$$
$$= \frac{1}{3m}\left[mx_1 + m(x_1 + \frac{1}{4}L) + m(x_1 + \frac{1}{4}L + \frac{1}{2}L)\right],$$

where x_1 is the horizontal position of the center of the bottom book, and the centers of the other books are displaced as given. Therefore, $3x_1 + L = 0$, or $x_1 = -\frac{1}{3}L$. If the center of the bottom book is $\frac{1}{2}L$ to the left of the edge, then only $\frac{1}{2}L - \frac{1}{3}L = \frac{1}{6}L$ can overhang on the right. (An argument based on torques is equivalent, since at the farthest right position, the normal contact force on the books acts essentially just at the table's edge.)

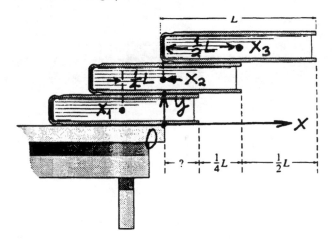

FIGURE 14-30 Problem 11 Solution.

Section 14-3:   Examples of Static Equilibrium
Problem

13. Where should the child in Fig. 14-31 sit if the scale is to read (a) 100 N and (b) 300 N?

Solution

If we consider torques about the pivot point (so that the force exerted by the pivot does not contribute) then Equation 14-2 is sufficient to determine the position of the child. As shown on Fig. 14-31, the weight of the tabletop (acting at its center of gravity), the weight of the child (acting a distance x from the left end), and the scale force, F_s, produce zero torque about the pivot:

$$(1/g)\left(\sum \tau\right)_P = 0 =$$
$$(F_s/g)(160 \text{ cm}) - (60 \text{ kg})(40 \text{ cm}) + (40 \text{ kg})(80 \text{ cm} - x).$$

Therefore, $x = 20$ cm $+ (F_s/9.8 \text{ N})4$ cm. If (a) $F_s =$ 100 N, then $x = 20$ cm $+ (400/9.8)$ cm $= 60.8$ cm, and if (b) $F_s = 300$ N, $x = 142$ cm. (Note that the child is on opposite sides of the pivot in parts (a) and (b), since without the child, $F_s = 147$ N.)

Problem

15. Two pulleys are mounted on a horizontal axle, as shown in Fig. 14-32. The inner pulley has a

diameter of 6.0 cm, the outer a diameter of 20 cm. Cords are wrapped around both pulleys so they don't slip. In the configuration shown, with what force must you pull on the outer rope in order to support the 40-kg mass?

Solution

Since each cord is tangent to its respective pulley, the lever arms are just the radii, as shown on the figure. The two torques are equal in magnitude, $R_1 F_1 = R_2 F_2$, so that

$$F_1 = (40 \text{ kg})(9.8 \text{ m/s}^2)(6 \text{ cm}) = (20 \text{ cm}) = 118 \text{ N}.$$

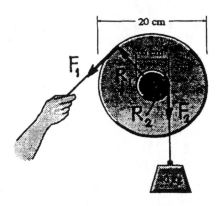

FIGURE 14-32 Problem 15 Solution.

Problem

17. Figure 14-34 shows a traffic signal, with masses and positions of its various members indicated. The structure is mounted with two bolts, located symmetrically about the vertical member's centerline, as indicated. What tension force must the left-hand bolt be capable of withstanding?

Solution

The forces on the traffic signal structure, and their lever arms about point 0 (on the vertical member's centerline between the bolts) are shown on Fig. 14-34. The normal forces exerted by the bolts and the ground on the vertical member are designated by N_ℓ and N_r, measured positive upward. (Of course, the ground can only make a positive contribution, and the bolts only a negative contribution, to these normal forces.) The two conditions of static equilibrium needed to determine N_ℓ and N_r are: $\sum F_y = 0 = N_\ell + N_r -$ (9.8)(320 + 170 + 65) N (the vertical component of Equation 14-1, positive up) and $\left(\sum \tau_z\right)_0 = 0 =$ $(N_r - N_\ell)(0.38 \text{ m}) - (170 \times 3.5 + 65 \times 8)(9.8 \text{ N·m})$ (the out-of-the-page-component of Equation 14-2, positive

CCW). These can be written as $N_r + N_\ell = 5.44$ kN, and $N_r - N_\ell = 28.8$ kN. Thus $N_\ell = -11.7$ kN, which is downward and must be exerted by the bolt. The reaction force on the bolt is upward and is a tensile force. (Really, N_ℓ is the difference between the downward force exerted by the bolt and the upward force exerted by the ground. Tightening the bolt increases the tensile force it must withstand beyond the minimum value calculated above, under the assumption that the ground exerts no force.)

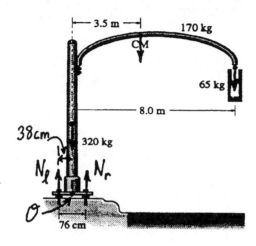

FIGURE 14-34 Problem 17 Solution.

Problem

19. Figure 14-36a shows an outstretched arm with a mass of 4.2 kg The arm is 56 cm long, and its center of gravity is 21 cm from the shoulder. The hand at the end of the arm holds a 6.0-kg mass. (a) What is the torque about the shoulder due to the weights of the arm and the 6.0-kg mass? (b) If the arm is held in equilibrium by the deltoid muscle, whose force on the arm acts 5.0° below the horizontal at a point 18 cm from the shoulder joint (Fig. 14-36b), what is the force exerted by the muscle?

Solution

(a) The magnitude of the (external) torque on the arm is $\tau_0 = [(4.2\,\text{kg})(0.21\,\text{m}) + (6\,\text{kg})(0.56\,\text{m})] \times (9.8\,\text{m/s}^2)\sin 105° = 40.2$ N·m. The direction is clockwise (into the page) about the shoulder joint. (b) The deltoid muscle exerts a counterclockwise torque of magnitude $Fr\sin\theta = F(0.18\,\text{m})\sin 170°$, which, under equilibrium conditions, equals the magnitude of the torque in part (a). Thus, $F = 40.2$ N·m$/(0.18\,\text{m})\sin 170° = 1.28$ kN, underscoring the comment at the end of Example 14-5. The

skeleto-muscular structure of the human extremities evolved for speed and range of motion, not mechanical advantage.

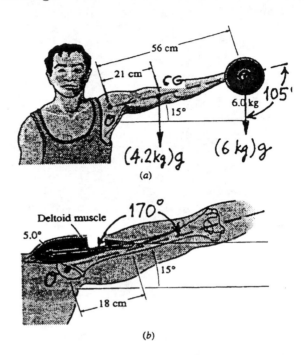

FIGURE 14-36 Problem 19 Solution.

Problem

21. A 15.0-kg door measures 2.00 m high by 75.0 cm wide. It hangs from hinges mounted 18.0 cm from top and bottom. Assuming that each hinge carries half the door's weight, determine the horizontal and vertical forces that the door exerts on each hinge.

Solution

If the door is properly hung, all the forces *on* the door are coplanar. We assume that the CM is at the geometrical center of the door. The conditions for equilibrium are:

$$\sum F_x = 0 = F_{Ax} + F_{Bx},$$
$$\sum F_y = 0 = F_{Ay} + F_{By} - Mg,$$
$$(\sum \tau)_B = 0 = \mathbf{r}_A \times \mathbf{F}_A + \mathbf{r}_{cm} \times M\mathbf{g}$$
$$= (164\,\text{cm}\hat{\mathbf{j}}) \times (F_{Ax}\hat{\mathbf{i}} + F_{Ay}\hat{\mathbf{j}})$$
$$+ (37.5\,\text{cm}\hat{\mathbf{i}} + 82\,\text{cm}\hat{\mathbf{j}}) \times (-Mg\hat{\mathbf{j}}),$$

where we chose to calculate torques about the lower hinge at B. Expanding the cross products, $0 = 164F_{Ax}(-\hat{\mathbf{k}}) - 37.5Mg(\hat{\mathbf{k}})$, we find $F_{Ax} = -(37.5)\times (15\,\text{kg})(9.8\,\text{m/s}^2)/(164) = -33.6$ N. From the x equation, $F_{Bx} = -F_{Ax} = 33.6$ N, and by assumption,

$F_{Ay} = F_{By} = \frac{1}{2}Mg = 73.5\,\text{N}$. Of course, the forces exerted *by* the door on the hinges (by Newton's third law) are the reactions to the forces, $\mathbf{F}_A$ and $\mathbf{F}_B$, just calculated.

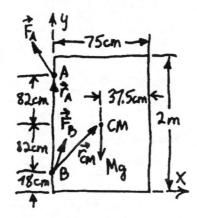

Problem 21 Solution.

Problem

23. Figure 14-39 shows a house designed to have high "cathedral" ceilings. Following a heavy snow, the total mass supported by each diagonal roof rafter is 170 kg, including building materials as well as snow. Under these conditions, what is the force in the horizontal tie beam near the roof peak? Is this force a compression or a tension? Neglect any horizontal component of force due to the vertical walls below the roof. Ignore the widths of the various structural components, treating contact forces as though they were concentrated at the roof peak and at the outside edge of the rafter/wall junction.

Solution

The forces on one of the diagonal rafters are drawn on Fig. 14-39. If the rafters are symmetrical (without internal stress), and we neglect the weight of the tie beam, $\mathbf{F}_B$ and $\mathbf{F}_C$ will be horizontal. $\mathbf{W}$ and $\mathbf{F}_A$ are vertical, the latter by assumption, and we suppose $\mathbf{W}$ acts at the center of the rafter. The equilibrium conditions needed to find F_B are: $\sum F_x = 0 = F_B - F_C$, and $(\sum \tau_z)_A = 0$, or $0 = (4\text{ m})F_C - (3.2\text{ m})F_B - (2.4\text{ m})W$. (Point A was chosen to eliminate $\mathbf{F}_A$ from the equation.) The solution is $F_B = 2.4W/(4 - 3.2) = 5.00\text{ kN}$. The force on the tie beam (the reaction to $\mathbf{F}_B$ is a tension. (The function of the tie beam is precisely to relieve any horizontal force that the roof may exert on the walls.)

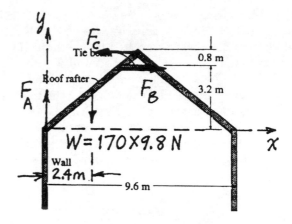

FIGURE 14-39 Problem 23 Solution.

Problem

25. A uniform sphere of radius R is supported by a rope attached to a vertical wall, as shown in Fig. 14-40. The point where the rope is attached to the sphere is located so a continuation of the rope would intersect a horizontal line through the sphere's center a distance $R/2$ beyond the center, as shown. What is the smallest possible value for the coefficient of friction between wall and sphere?

Solution

In equilibrium, the sum of the torques about the center of the sphere must be zero, so the frictional force is up, as shown. The lever arm of the tension in the rope is $\frac{1}{2}R\cos 30° = \sqrt{3}R = 4$, and the weight and normal force exert no torque about the center. Thus, $fR = T\sqrt{3}R/4$. The sum of the horizontal components of the forces is zero also, so $0 = N - T\sin 30°$, or $T = 2N$. Therefore $f = \frac{1}{2}\sqrt{3}$ N. Since $f \leq \mu_s N$, this implies $\mu_s \geq f/N = \frac{1}{2}\sqrt{3} = 0.87$.

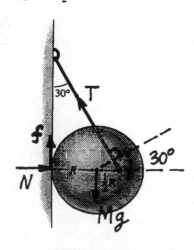

FIGURE 14-40 Problem 25 Solution.

Problem

27. A garden cart loaded with firewood is being pushed horizontally when it encounters a step 8.0 cm high, as shown in Fig. 14-41. The mass of the cart and its load is 55 kg, and the cart is balanced so that its center of mass is directly over the axle. The wheel diameter is 60 cm. What is the minimum horizontal force that will get the cart up the step?

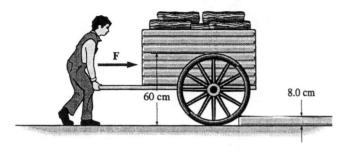

FIGURE 14-41 Problem 27.

Solution

We assume that a horizontal push on the cart results in a horizontal force exerted on the wheels by the axle, as shown. (We also suppose both wheels share the forces equally, so they can be treated together.) Also shown are the weight of the cart and the normal force of the ground, both acting through the center of the wheels, and the force of the step, F_s. If we consider the sum of the torques (positive CCW) about the step, the latter does not contribute, and the wheels (and cart) will remain stationary as long as $(\sum \tau)_{step} = MgR\sin\theta - NR\sin\theta - FR\cos\theta = 0$. When $N = 0$, however, the wheels begin to lose contact with the ground and go over the step. This occurs when $F = Mg\tan\theta$. From the geometry of the situation, $R(1 - \cos\theta) = h$, the height of the step, so $\theta =$

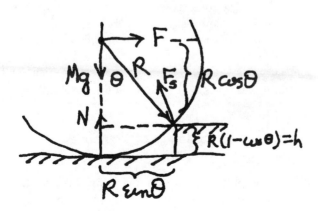

FIGURE 14-41 Problem 27 Solution.

$\cos^{-1}(1 - h/R) = \cos^{-1}(1 - 8/30)$. Then $F = (55\times 9.8\text{ N})\tan(\cos^{-1}(11/15)) = 500$ N is the minimum force.

Problem

29. The leaning Tower of Pisa (Fig. 14-43) currently leans at a 4.7° angle to the vertical. Treating the tower as a solid cylinder, what is the maximum angle at which it can lean before falling over? Treat the tower as a uniform cylinder 7.0 m in diameter and 55 m high, and assume the ground supports the tower's weight but does not provide any torque.

Solution

The center of mass of the tower must be somewhere over its footprint on the ground, or it will topple. This will be so, for a uniform cylindrical model of the tower, if the angle of tilt ϕ is less than the angle α that a diagonal makes with the length, as indicated on Fig. 14-43. Therefore $\phi \leq \alpha = \tan^{-1}(7\text{m}/55\text{m}) = 7.25°$.

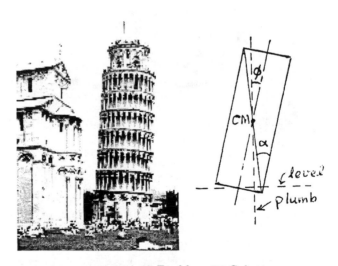

FIGURE 14-43 Problem 29 Solution.

Problem

31. The boom in the crane of Fig. 14-44 is free to pivot about point P and is supported by the cable that joins halfway along its 18-m total length. The cable passes over a pulley and is anchored at the back of the crane. The boom has mass 1700 kg, distributed uniformly along its length, and the mass hanging from the end of the boom is 2200 kg. The boom makes a 50° angle with the horizontal. What is the tension in the cable that supports the boom?

Solution

The forces on the boom are shown superposed on the figure. By assumption, T is horizontal and acts at the CM of the boom. To find T, we compute the torques about P, $(\sum \tau)_P = 0$, obtaining:

$$T\tfrac{1}{2}\ell \sin 50° - m_b g \tfrac{1}{2}\ell \cos 50° - mg\ell \cos 50° = 0,$$

or

$$T = (2m + m_b)g \cot 50° = (4400 + 1700)(9.8 \text{ N}) \cot 50°$$
$$= 50.2 \text{ kN}.$$

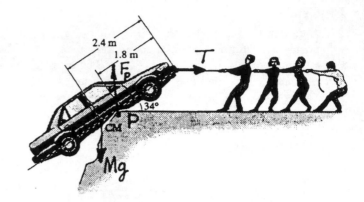

FIGURE 14-46 Problem 33 Solution.

Problem

35. Figure 14-47 shows a uniform board dangling over a *frictionless* edge, secured by a *horizontal* rope. If the angle θ in Fig. 14-47 were 30°, what fraction would the distance d shown in the figure be of the board length ℓ?

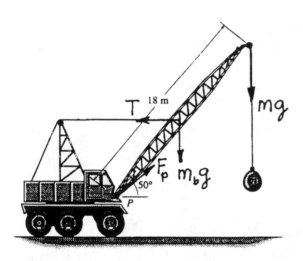

FIGURE 14-44 Problem 31 Solution.

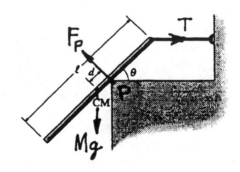

FIGURE 14-47 Problems 34, 35.

Problem

33. Figure 14-46 shows a 1250-kg car that has slipped over the edge of an embankment. A group of people are trying to hold the car in place by pulling on a horizontal rope, as shown. The car's bottom is pivoted on the edge of the embankment, and its center of mass lies further back, as shown. If the car makes a 34° angle with the horizontal, what force must the group apply to hold it in place?

Solution

Three forces act on the car, as shown added to Fig. 14-46. The unknown force, $\mathbf{F}_P$, exerted by the edge of the embankment, does not contribute to Equation 14-2 (positive torques CCW) if evaluated about point P, so the tension necessary to keep the car in equilibrium can be found directly. $(\sum \tau)_P = 0 = Mg(2.4 \text{ m} - 1.8 \text{ m}) \cos 34° - T(1.8 \text{ m}) \sin 34°$, or $T = (1250 \times 9.8 \text{ N})/3 \tan 34° = 6.05 \text{ kN}$.

Solution

The three forces acting on the board are in the same configuration as those in Problem 33, so Equation 14-2 about the edge gives $Mgd \cos \theta = T(\tfrac{1}{2}\ell - d) \sin \theta$. If the edge is frictionless, then $\mathbf{F}_P$ is perpendicular to the board, so Equation 14-1 requires $T = F_P \sin \theta$ and $Mg = F_P \cos \theta$. Substituting above, we find $F_P d \cos^2 \theta = F_P(\tfrac{1}{2}\ell - d) \sin^2 \theta$, or $d(\cos^2 \theta + \sin^2 \theta) = \tfrac{1}{2}\ell \sin^2 \theta$ and $\theta = \sin^{-1} \sqrt{2d/\ell}$. For $\theta = 30°$, $d/\ell = \tfrac{1}{2} \sin^2 30° = \tfrac{1}{8}$.

Section 14-4: Stability of Equilibria

Problem

37. A roly-poly toy clown is made from part of a sphere topped by a cone. The sphere is truncated at just the right point so that there is no discontinuity in angle as the surface changes from

sphere to cone (Fig. 14-48a). If the clown always returns to an upright position, what is the maximum possible height for its center of mass? Would your answer change if the continuity-of-angle condition were not met, as in Fig. 14-48b?

FIGURE 14-48 Problem 37.

Solution

Suppose the roly-poly rests on a flat horizontal surface. Its spherical surface is always tangent to the horizontal if the continuity-of-angle condition holds, as in sketch (a) (except when upside-down on the point of the cone). Gravity will always exert a restoring torque if the CM lies to the left of the vertical through the left-most point of contact, B, as shown. Since this

(a)

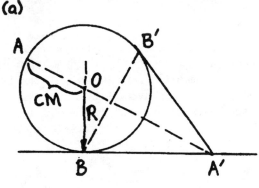

(b)

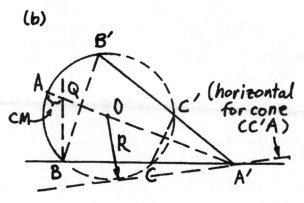

Problem 37 Solution.

vertical passes through O, the center of the sphere, the CM should lie "below" O, as measured from the bottom point, A, on the axis.

If the continuity-of-angle condition is not met, the cone, in general, intersects the sphere in one of two circles (through BB' or CC', as in sketch (b)). If CC' is the actual boundary, the reasoning in the first paragraph still applies. However, if BB' is the boundary, the maximum distance of the CM from point A should be $< AQ$ (considerably "lower" than O).

Problem

39. The potential energy as a function of position for a certain particle is given by

$$U(x) = U_0 \left(\frac{x^3}{x_0^3} + a\frac{x^2}{x_0^2} + 4\frac{x}{x_0} \right),$$

where U_0, x_0, and a are constants. For what values of a will there be two static equilibria? Comment on the stability of these equilibria.

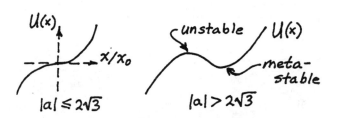

Problem 39 Solution.

Solution

The equilibrium condition, $dU/dx = 0$ (Equation 14-3), requires $3(x/x_0)^2 + 2a(x/x_0) + 4 = 0$. This quadratic has two real roots if the discriminant is positive, i.e., $a^2 - 12 > 0$, or $|a| > 2\sqrt{3}$. The roots are $(x/x_0)_\pm = \frac{1}{3}(-a \pm \sqrt{a^2 - 12})$. The second derivative of the potential energy, evaluated at these roots, is

$$\left(\frac{d^2U}{dx^2}\right)_\pm = \frac{U_0}{x_0^2}\left[6\left(\frac{x}{x_0}\right)_\pm + 2a\right]$$

$$= \pm 2\sqrt{a^2 - 12}\left(\frac{U_0}{x_0^2}\right).$$

Thus, the "plus" root is a position of metastable equilibrium (Equation 14-4), while the "minus" root represents unstable equilibrium, (Equation 14-5). A plot of the potential energy, which is a cubic, will clarify these remarks. For $|a| \le 2\sqrt{3}, U(x)$ has no wiggles, as shown. (U passes through the origin, but its position depends on the value of "a", and is not shown.)

Paired Problems

Problem

41. Figure 14-49 shows a 66-kg sign hung centered from a uniform rod of mass 8.2 kg and length 2.3 m. At one end the rod is attached to the wall by a pivot; at the other end it's supported by a cable that can withstand a maximum tension of 800 N. What is the minimum height h above the pivot for anchoring the cable to the wall?

Solution

Suppose that the sign is centered on the rod, so that its CM lies under the center of the rod. Then the total weight may be considered to act through the center of the rod, as shown. In equilibrium, Equation 14-2 calculated about the pivot (which does not contain the unknown pivot force) yields $0 = T\ell\sin\theta - Mg\frac{1}{2}\ell$, or $T = Mg/2\sin\theta$. But, $\tan\theta = h/\ell$, so $T = \frac{1}{2}Mg\times$ $\sqrt{1+\ell^2/h^2}$ (use the identity $1+\cot^2\theta = \csc^2\theta$). Therefore, the condition $T \le T_{\max}$ implies $1 + \ell^2 \div h^2 \le (2T_{\max}/Mg)^2$, or $h \ge \ell/\sqrt{(2T_{\max}/Mg)^2 - 1} = $ 2.3 m$/\sqrt{(1600/74.2\times9.8)^2 - 1} = 1.17$ m.

FIGURE 14-49 Problem 41 Solution.

Problem

43. A 4.2-kg plant hangs from the bracket shown in Fig. 14-51. The bracket has a mass of 0.85 kg, and its center of mass lies 9.0 cm from the wall. A single screw holds the bracket to the wall, as shown. Find the horizontal tension force in the screw. *Hint:* Imagine that the bracket is slightly loose and pivoting about its bottom end. Assume the wall is frictionless.

Solution

We assume that the screw provides the total support for the bracket, exerting a force with horizontal component F_x (the reaction to which is a tensile force

on the screw) and vertical component F_y (the reaction to which is a shearing force on the screw equal to the total weight) as shown. A normal contact force exerted by the wall could be distributed along the bracket (e.g., by tightening the screw), but if we only wish to estimate the minimum F_x, we may consider all the normal force to act at the lowest point of contact of the bracket, point O. Then Equation 14-2 about O gives $F_x(7.2 \text{ cm}) = [(4.2 \text{ kg})(28 \text{ cm}) + (0.85 \text{ kg})\times (9.0 \text{ cm})](9.8 \text{ m/s}^2)$ or $F_x = 170$ N.

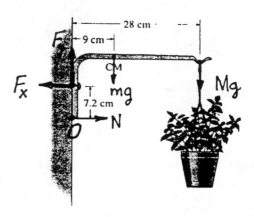

FIGURE 14-51 Problem 43 Solution.

Problem

45. A 5.0-m-long ladder has mass 9.5 kg and is leaning against a frictionless wall, making a 66° angle with the horizontal. If the coefficient of friction between the ladder and ground is 0.42, what is the mass of the heaviest person who can safely ascend to the top of the ladder? (The center of mass of the ladder is at its center.)

Solution

The forces on the uniform ladder are shown in the sketch, with the force exerted by the (frictionless) wall horizontal. The person is up the ladder a fraction α of its length. Equilibrium conditions require:

$$\sum F_x = 0 = f - F_{\text{wall}},$$
$$\sum F_y = 0 = N - (m_\ell + m)g,$$
$$(\sum\tau)_A = 0 = F_{\text{wall}}\ell\cos\theta - m_\ell g\frac{1}{2}\ell\sin\theta - mg\alpha\ell\sin\theta.$$

The ladder will not slip if $f \le \mu_s N$, which can be written as

$$f = F_{\text{wall}} = (\tfrac{1}{2}m_\ell + \alpha m)g\tan\theta \le \mu_s N = \mu_s(m_\ell + m)g,$$

or

$$\alpha \le [\mu_s(m_\ell + m)\cot\theta - \tfrac{1}{2}m_\ell]/m = $$
$$\mu_s\cot\theta + (m_\ell/m)(\mu_s\cot\theta - \tfrac{1}{2}).$$

(Here, we used the horizontal force equation to find f, the torque equation to find F_{wall}, and the vertical force equation to find N.) For a person at the top of the ladder, $\alpha = 1$ and the condition for no slipping becomes $m \leq m_\ell(\mu_s \cot\theta - \frac{1}{2})/(1 - \mu_s \cot\theta)$. With the data given for the ladder (note that $\cot\theta = \tan 66°$) $m \leq (9.5 \text{ kg})(0.42\tan 66° - 0.5)/(1 - 0.42\tan 66°) = 74.3$ kg.

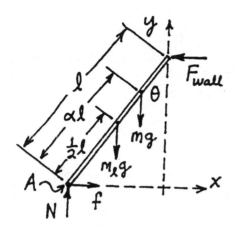

Problem 45 Solution.

Problem

47. A uniform, solid cube of mass m and side s is in stable equilibrium when sitting on a level tabletop. How much energy is required to bring it to an unstable equilibrium where it's resting on its corner?

Solution

When balancing on a corner, the CM of a uniform cube (i.e., its center) is a distance $\sqrt{(s/2)^2 + (s/2)^2 + (s/2)^2} = \sqrt{3}s/2$ above the corner resting on the tabletop. When in stable equilibrium, the CM is $s/2$ above the tabletop. Thus, the potential energy difference is $\Delta U = mg\,\Delta y_{cm} = mgs(\sqrt{3}-1)/2$.

Supplementary Problems
Problem

49. A uniform pole of mass M is at rest on an incline of angle θ, secured by a horizontal rope as shown in Fig. 14-52. What is the minimum coefficient of friction that will keep the pole from slipping?

Solution

The forces acting on the pole are the tension in the rope, gravity (acting at the CM at its center) and the contact force of the incline (perpendicular component N and parallel component f) as shown. Consideration of Equation 14-2 about the CM shows that a frictional

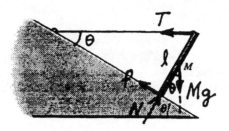

FIGURE 14-52 Problem 49 Solution.

force f must be acting up the plane if the rod is to remain in static equilibrium. (Since the weight of the rod, mg, and the normal force, N, contribute no torques about the CM, there must be a force to oppose the torque of the tension, T.) The equations for static equilibrium (parallel and perpendicular components of Equation 14-1, and CCW-positive component of Equation 14-2) are: $\sum F_\parallel = 0 = f + T\cos\theta - mg\sin\theta$, $\sum F_\perp = 0 = N - T\sin\theta - mg\cos\theta$, and $(\sum\tau)_{cm} = 0 = T\frac{1}{2}\ell\cos\theta - f\frac{1}{2}\ell$. The solutions for the forces are $f = \frac{1}{2}mg\sin\theta$, $T = \frac{1}{2}mg\tan\theta$, and $N = \frac{1}{2}mg(2\cos\theta + \sin^2\theta/\cos\theta)$, subject to the condition that $f \leq \mu N$. Therefore $\sin\theta \leq \mu(2\cos\theta + \sin^2\theta/\cos\theta)$ or $\mu \geq \tan\theta \div (2 + \tan^2\theta)$. (By use of the identities $\sin 2\theta = 2\sin\theta\cos\theta$, $\cos 2\theta = \cos^2\theta - \sin^2\theta$, and $\sin^2\theta = 1 - \cos^2\theta$, this may be rewritten as $\mu \geq \sin 2\theta \div (3 + \cos 2\theta)$.)

Problem

51. One end of a board of negligible mass is attached to a spring of spring constant k, while its other end rests on a frictionless surface, as shown in Fig. 14-53. If a mass m is placed on the middle of the board, by how much will the spring compress?

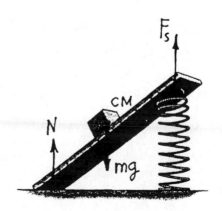

FIGURE 14-53 Problem 51 Solution.

Solution

If the frictionless surface is horizontal, the three forces acting on the board are vertical, as shown. For equilibrium, $N + F_s = mg$ and $(\sum \tau)_{cm} = 0$. The latter implies $N = F_s$, so $F_s = \frac{1}{2}mg = k\,\Delta x$, and the compression of the spring is $\Delta x = mg/2k$.

Problem

53. Figure 14-54 shows a wheel on a slope with inclination angle $\theta = 20°$, where the coefficient of friction is adequate to prevent the wheel from slipping; however, it might still roll. The wheel is a uniform disk of mass 1.5 kg, and it is weighted at one point on the rim with an additional 0.95-kg mass m. Find the angle ϕ shown in the figure such that the wheel will be in static equilibrium.

Solution

The wheel doesn't slide if $\sum F_{||} = 0$ and it doesn't roll if $(\sum \tau)_{center} = 0$. ("||" means parallel to the incline, and "center" is the center and CM of the wheel. These are the only equilibrium conditions needed for the solution of this problem.) With reference to the forces shown added to Fig. 14-52, these conditions are $f = (M + m)g \sin \theta$ and $fR = mgR \cos \phi$. Together, they imply $f = mg \cos \phi = (M + m)g \sin \theta$ or $\phi = \cos^{-1}[(1 + M/m) \sin \theta] = \cos^{-1}[(1 + 1.5/0.95) \times \sin 20°] = 28.1°$.

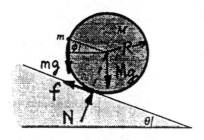

FIGURE 14-54 Problem 53 Solution.

Problem

55. A 2.0-m-long rod has a density described by $\lambda = a + bx$, where λ is the density in kilograms per meter of length, $a = 1.0$ kg/m, $b = 1.0$ kg/m^2, and x is the distance in meters from the left end of the rod. The rod rests horizontally with its ends each supported by a scale. What do the two scales read?

Solution

The rod is in static equilibrium under the three vertical forces shown in the sketch, so $\sum F_y = 0$ implies $F_{sr} + F_{sr} = Mg$, and $(\sum \tau)_{cm} = 0$ implies

$F_{s\ell}x_{cm} = F_{sr}(\ell - x_{cm})$. The solution for the left and right scale forces is $F_{s\ell} = Mg - F_{sr} = Mg \times (1 - x_{cm}/\ell)$. Equation 10-5 gives

$$x_{cm} = \int_0^\ell \lambda x\,dx \div \int_0^\ell \lambda\,dx$$

$$= \int_0^\ell (ax + bx^2)\,dx \div \int_0^\ell (a + bx)\,dx$$

$$= (a\tfrac{1}{2}\ell^2 + b\tfrac{1}{3}\ell^3)/(a\ell + b\tfrac{1}{2}\ell^2)$$

$$= \ell(3a + 2b\ell)/(6a + 3b\ell).$$

For the values given, $x_{cm}/\ell = \frac{7}{12}$ and note that $M = a\ell + \frac{1}{2}b\ell^2 = 4$ kg. Thus, $F_{sr} = Mgx_{cm}/\ell = 22.9$ N and $F_{s\ell} = 16.3$ N.

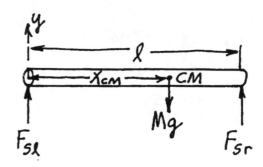

Problem 55 Solution.

Problem

57. A rectangular block twice as high as it is wide is resting on a board. The coefficient of static friction between board and block is 0.63. If the board is tilted as shown in Fig. 14-55, will the block first tip over or first begin sliding?

Solution

We suppose that the block is oriented with two sides parallel to the direction of the incline, and that its CM is at the center. The condition for sliding is that

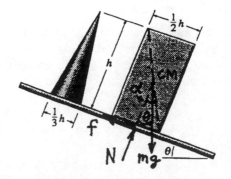

FIGURE 14-55 Problem 57 Solution.

$mg \sin \theta > f_s^{\max} = \mu_s N = \mu_s mg \cos \theta$, or $\tan \theta > \mu_s$. For $\mu_s = 0.63$, this condition is $\theta > \tan^{-1} 0.63 = 32.2°$. The condition for tipping over is that the CM lie to the left of the lower corner of the block (see sketch). Thus $\theta > \alpha$, where $\alpha = \tan^{-1}(w/h)$ is the diagonal angle of the block. For $h = 2w, \alpha = \tan^{-1} 0.5 = 26.6°$. Therefore, this block tips over before sliding.

Problem

59. A uniform solid cone of height h and base diameter $\frac{1}{3}h$ is placed on the board of Fig. 14-55. The coefficient of static friction between the cone and incline is 0.63. As the slope of the board is increased, will the cone first tip over or begin sliding? *Hint:* Begin with an integration to find the center of mass.

Solution

The analysis for Problem 57 applies to the cone, where α is the angle between the symmetry axis and a line from the CM to the edge of the base. The integration to find the CM is fastest when the cone is oriented like the aircraft wing in Example 10-3, for then, the equation of the sloping side is simple, as shown in the sketch. For mass elements, take thin disks parallel to the base, so $dm = \rho \pi y^2 dx = (3M/h^3)x^2\, dx$, where $\rho = M/\frac{1}{3}\pi R^2 h$ is the density (assumed constant) and M is the mass of the cone. Then $x_{cm} = M^{-1} \int x\, dm = (3/h^3) \int_0^h x^3\, dx = \frac{3}{4}h$, or the CM is $\frac{1}{4}h$ above the base. Since $\tan \alpha = (\frac{1}{6}h)/(\frac{1}{4}h) = \frac{2}{3} > 0.63 = \mu_s$, this cone will slide before tipping over.

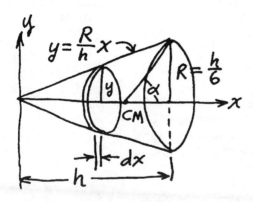

Problem 59 Solution.

Problem

61. An interstellar spacecraft from an advanced civilization is hovering above Earth, as shown in Fig. 14-57. The ship consists of two pods of mass m separated by a rigid shaft of negligible mass that is one Earth radius (R_E) long. Find (a) the

magnitude and direction of the net gravitational force on the ship and (b) the net torque about the center of mass. (c) Show that the ship's center of gravity is displaced approximately $0.083R_E$ from its center of mass.

Solution

(a) The force on each pod (pod #1 over the North pole) can be calculated from the law of universal gravitation, Equation 9-1 (vectors in the x-y Earth-centered frame shown on Fig. 14-57):

$$\mathbf{F}_1 = \frac{GM_E m}{(2R_E)^2}(-\hat{\jmath}), \mathbf{F}_2 = \frac{GM_E m}{(\sqrt{5}R_E)^2}\left(-\frac{1}{\sqrt{5}}\hat{\imath} - \frac{2}{\sqrt{5}}\hat{\jmath}\right).$$

When we replace GM_E by gR_E^2, the net force becomes:

$$\mathbf{F}_1 + \mathbf{F}_2 = -mg\left[\frac{\hat{\imath}}{5\sqrt{5}} + \left(\frac{1}{4} + \frac{2}{5\sqrt{5}}\right)\hat{\jmath}\right]$$
$$= -mg(0.894\hat{\imath} + 4.29\hat{\jmath}) \times 10^{-1}.$$

This has magnitude $(4.38 \times 10^{-1})mg$ and is directed 11.8° clockwise from the negative y axis. (b) The positions of the pods, relative to their CM midway between them, are $\mathbf{r}_1' = -\frac{1}{2}R_E\hat{\imath}$ and $\mathbf{r}_2' = \frac{1}{2}R_E\hat{\imath}$.

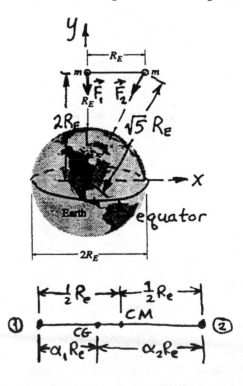

FIGURE 14-57 Problem 61 Solution.

Therefore, the net torque about the CM is:

$\mathbf{r}_1' \times \mathbf{F}_1 + \mathbf{r}_2' \times \mathbf{F}_2$

$= \left(-\frac{1}{2}R_E\hat{\mathbf{i}}\right) \times \left(-\frac{mg}{4}\hat{\mathbf{j}}\right) + \left(\frac{1}{2}R_E\hat{\mathbf{i}}\right) \times \frac{mg}{5\sqrt{5}}(-\hat{\mathbf{i}} - 2\hat{\mathbf{j}})$

$= \frac{mgR_E}{8}\hat{\mathbf{k}} - \frac{mgR_E}{5\sqrt{5}}\hat{\mathbf{k}}$

$= (3.56 \times 10^{-2})mgR_E\hat{\mathbf{k}}$ (out of page).

(c) The CG is positioned between the pods such that the net gravitational torque about it is zero. If the positions of the pods relative to the CG are $-\alpha_1 R_E\hat{\mathbf{i}}$ and $\alpha_2 R_E\hat{\mathbf{i}}$, respectively, where $\alpha_1 + \alpha_2 = 1$, then

$O = (-\alpha_1 R_E\hat{\mathbf{i}}) \times \left(-\frac{mg}{4}\hat{\mathbf{j}}\right) + (\alpha_2 R_E\hat{\mathbf{i}}) \times \frac{mg}{5\sqrt{5}}(-\hat{\mathbf{i}} - 2\hat{\mathbf{j}})$

$= mgR_E\left(\frac{\alpha_1}{4} - \frac{2\alpha_2}{5\sqrt{5}}\right)\hat{\mathbf{k}}.$

Solving for α_1 (or α_2), we find $\alpha_1 = 1 - \alpha_2 = 1 - 5\sqrt{5}\alpha_1/8$, or $\alpha_1 = (1 + 5\sqrt{5}/8)^{-1} = 0.417$ (and $\alpha_2 = 0.583$). Thus, the CG is $(0.5 - \alpha_1)R_E = 0.0829R_E$ closer to pod #1 than the CM.

PART 1   CUMULATIVE PROBLEMS

These problems combine material from chapters throughout the entire part or, in addition, from chapters in earlier parts, or they present special challenges.

Problem

1. A 170-g apple sits atop a 2.8-m-high post. A 45-g arrow moving horizontally at 130 m/s passes horizontally through the apple and strikes the ground 36 m from the base of the post, as shown in Fig. 1. Where does the apple hit the ground? Neglect the effect of air resistance on either object as well as any friction between apple and post.

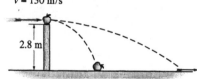

$v = 130$ m/s

2.8 m

FIGURE 1 Cumulative Problem 1.

Solution

We can assume that momentum is conserved during the inelastic collision (in a brief interval at $t = 0$) between the arrow (m_1) and the apple (m_2). The velocities of the arrow before and after are specified to be horizontal (in the x direction), therefore the velocity of the apple (which was at rest before) is also horizontal after the collision. Thus, $m_1v_{1i,x} = m_1v_{1f,x} + m_2v_{2f,x}$. Since both are moving horizontally after the collision, the arrow and the apple will each fall to the ground through the same vertical distance y (equal to the height of the post), in the same time

$t = \sqrt{2y/g}$. However, they strike the ground at different horizontal positions, which (in the absence of air resistance) are $x_1 = v_{1f,x}t$ and $x_2 = v_{2f,x}t$, relative to the base of the post. Since $x_1 = 36$ m, $y = 2.8$ m, and $v_{1i,x} = 130$ m/s are given, $v_{1f,x}$ and $v_{2f,x}$ can be eliminated and a solution for x_2 obtained: $x_2 = (m_1/m_2)(v_{1i,x} - v_{1f,x})t = (m_1/m_2)(v_{1i,x}\sqrt{2y/g} - x_1) = (45/170)[(130 \text{ m/s})\sqrt{2(2.8 \text{ m})/(9.8 \text{ m/s}^2)} - 36 \text{ m}] = 16.5$ m. (Alternatively, since external horizontal forces are neglected, the center of mass of the arrow/apple system moves horizontally at constant speed until it reaches ground level, $v_{cm,x} = \text{constant} = m_1v_{1i,x}/(m_1 + m_2)$ (its value before the collision). Then at ground level, $m_2x_2 = (m_1 + m_2)x_{cm} - m_1x_1 = (m_1 + m_2)v_{cm,x}t - m_1x_1 = m_1(v_{1i,x}t - x_1)$, as before.) Refer to relevant material in Chapters 4, 10, and 11 if necessary.

Problem

3. A block of mass m_1 is attached to the axle of a uniform solid cylinder of mass m_2 and radius R by massless strings. The two accelerate down a slope that makes an angle θ with the horizontal, as shown in Fig. 3. The cylinder rolls without slipping and the block slides with coefficient of kinetic friction μ between block and slope. The strings are attached to the cylinder's axle with frictionless loops so that the cylinder can roll freely without any torque from the string. Find an expression for the acceleration of the pair, assuming that the string remains taut.

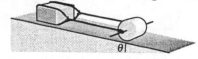

FIGURE 3 Cumulative Problem 3.

Solution

One must consider the forces on the block and the cylinder, exerted by gravity, the inclined surface, and the strings, as sketched (since the strings are assumed massless and other forces are neglected). If the strings remain taut, then the downslope acceleration, $a_{\parallel}$, of the block and the cylinder's center of mass are the same. If the string tension is parallel to the slope, the normal force on the block is $N_1 = m_1 g \cos\theta$, and the parallel component of its equation of motion is $F_{\text{net},\parallel} = m_1 g \sin\theta + T - \mu m_1 g \cos\theta = m_1 a_{\parallel}$. When rolling without slipping, the point of contact, P, of the cylinder with the slope, is instantaneously at rest, so the acceleration of its center of mass is $a_{\parallel} = \alpha R$ (this follows from $v = \omega R$ and $\omega = \omega_c = \omega_{cm}$; see Section 12-5). Since only the string tension and gravity exert torques about point P, the equation of motion of the cylinder is $\tau_P = (m_2 g \sin\theta - T)R = I_P \alpha = (\frac{3}{2}m_2 R^2)(a_{\parallel}/R)$, or $m_2 g \sin\theta - T = \frac{3}{2}m_2 a_{\parallel}$ (see the parallel-axis theorem in Chapter 12 for I_P). If T is eliminated by adding the two equations of motion, one finds $a_{\parallel} = 2g[(m_1 + m_2)\sin\theta - \mu m_1 \cos\theta] \div (2m_1 + 3m_2)$.

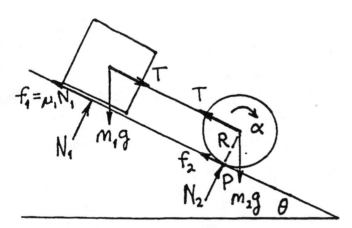

Cumulative Problem 3 Solution.

Problem

5. A solid ball of radius R is set spinning with angular speed ω about a horizontal axis. The ball is then lowered vertically with negligible speed until it just touches a horizontal surface and is released (Fig. 5). If the coefficient of kinetic friction between the ball and the surface is μ, find (a) the linear speed of the ball once it achieves pure rolling motion and (b) the distance it travels before its motion is pure rolling.

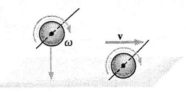

FIGURE 5 Cumulative Problem 5.

Solution

(a) While there is relative motion at the point of contact between the ball and the horizontal surface, the force of sliding friction ($f = \mu N = \mu mg$) slows the ball's rotation and accelerates its center of mass. The equation for the latter is $f = \mu mg = ma_{cm}$, or $a_{cm} = \mu g$ (positive to the right in the sketch and Fig. 5). The equation for the former is $\tau = -fR = -\mu mgR = I\alpha = (2mR^2/5)\alpha$, where $\alpha = -5\mu g/2R$ is the angular acceleration about the horizontal axis through the center of the ball (positive clockwise in the sketch and Fig. 5). The accelerations are constant, so the velocities are given by Equations 2-17 and 12-9 as $v_{cm} = a_{cm}t = \mu gt$ and $\omega = \omega_0 + \alpha t = \omega_0 - 5\mu gt/2R$, where the ball is released at $t = 0$ and the initial velocities, $v_0 = 0$ and ω_0, are given. The accelerated motion continues until the point of contact is instantaneously at rest (no more sliding friction). The ball rolls without slipping thereafter, at a constant velocity given by $v_{cm} = \omega R$. This occurs at a time t, when $(\omega_0 - 5\mu gt/2R)R = \mu gt$, or $t = 2\omega_0 R/7\mu g$. Thus, the final velocity is $v_{cm} = \mu gt = 2\omega_0 R/7$. (b) The distance traveled during this time is $\Delta x = \frac{1}{2}a_{cm}t^2 = \frac{1}{2}(\mu g)(2\omega_0 R/7\mu g)^2 = 2\omega_0^2 R^2/49\mu g$ (or $\Delta x = v_{cm}^2/2a_{cm}$ with the same result).

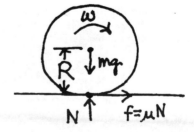

FIGURE 5 Cumulative Problem 5 Solution.

PART 2   OSCILLATIONS, WAVES, AND FLUIDS

CHAPTER 15   OSCILLATORY MOTION

ActivPhysics can help with these problems:
All Activities in Section 9

Sections 15-1 and 15-2:   Oscillations and
Simple Harmonic Motion

Problem

1. A doctor counts 77 heartbeats in one minute. What
   are the period and frequency of the heart's
   oscillations?

Solution

If 77 heartbeats take 1 min., then one heartbeat (one
cycle) takes $T = 1\,\text{min}\,/77 = 60\,\text{s}/77 = 0.77$ s, which is
the period. The frequency is $f = 77/\,\text{min} = 77/60\,\text{s} =$
1.28 Hz. (One can see that $T = 1/f$.)

Problem

3. The vibration frequency of a hydrogen chloride
   molecule is $8.66{\times}10^{13}$ Hz. How long does it take
   the molecule to complete one oscillation?

Solution

$T = 1/f = 1/(8.66{\times}10^{13}\,\text{Hz}) = 1.15{\times}10^{-14}$ s = 11.5 fs
(Equation 15-1).

Problem

5. Determine the amplitude, angular frequency, and
   phase constant for each of the simple harmonic
   motions shown in Fig. 15-33.

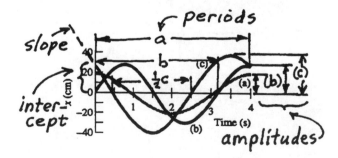

FIGURE 15-33 Problem 5 Solution.

Solution

The amplitude is the maximum displacement, read
along the x axis (ordinate) in Fig. 15-33. The angular
frequency is 2π times the reciprocal of the period,
which is the time interval between corresponding
points read along the t axis (abscissa). The phase
constant can be determined from the intercept and
slope (displacement and velocity) at $t = 0$. One sees
that (a) $A \simeq 20$ cm, $\omega \simeq 2\pi/4$ s $\simeq \frac{1}{2}\pi\,\text{s}^{-1}$, and $\phi \simeq 0$;
(b) $A \simeq 30$ cm, $\omega \simeq 2\pi/3.2$ s $\simeq 2\,\text{s}^{-1}$, and $\phi \simeq -90° \simeq$
$-\frac{1}{2}\pi$; (c) $A \simeq 40$ cm, $\omega \simeq 2\pi/(2{\times}2\,\text{s}) \simeq \frac{1}{2}\pi\,\text{s}^{-1}$, and
$\phi \simeq \cos^{-1}(27/40) \simeq 48° \simeq \frac{1}{4}\pi$.

Problem

7. An astronaut in an orbiting spacecraft is "weighed"
   by being strapped to a spring of constant $k =$
   400 N/m, and set into simple harmonic motion. If
   the oscillation period is 2.5 s, what is the
   astronaut's mass?

Solution

We suppose that the other end of the spring is
fastened to the spacecraft, whose mass is much greater
than the astronaut's, and that the orbiting system is
approximately inertial. Then Equation 15-8c gives
$m = k/\omega^2 = k(T/2\pi)^2 = (400\,\text{N/m})(2.5\,\text{s}/2\pi)^2 =$
63.3 kg.

Problem

9. A simple model of a carbon dioxide (CO_2) molecule
   consists of three mass points (the atoms) connected
   by two springs (electrical forces), as suggested in
   Fig. 15-34. One way this system can oscillate is if
   the carbon atom stays fixed and the two oxygens
   move symmetrically on either side of it. If the
   frequency of this oscillation is $4.0{\times}10^{13}$ Hz, what is
   the effective spring constant? The mass of an
   oxygen atom is 16 u.

FIGURE 15-34 Problem 9.

Solution

With the carbon atom end of either "spring" fixed, the frequency of either oxygen atom is $\omega = 2\pi f = \sqrt{k/m}$. Therefore $k = (2\pi \times 4 \times 10^{13} \text{ Hz})^2 (16 \times 1.66 \times 10^{-27} \text{ kg}) = 1.68 \times 10^3 \text{ N/m}$.

Problem

11. Two identical mass-spring systems consist of 430-g masses on springs of constant $k = 2.2$ N/m. Both are displaced from equilibrium and the first released at time $t = 0$. How much later should the second be released so the two oscillations differ in phase by $\pi/2$?

Solution

Suppose that both masses are released from their maximum positive displacements, the first at $t = 0$ and the second at $t = t_0$. Then, since the phase (the argument of the cosine in Equation 15-9) of each motion is zero at release, $\phi_1 = 0$ and $\omega_2 t_0 + \phi_2 = 0$. The difference in phase is $\frac{1}{2}\pi = (\omega_1 t + \phi_1) - (\omega_2 t + \phi_2) = (\omega_1 - \omega_2)t + (\phi_1 - \phi_2) = -\phi_2$, where $\omega_1 = \omega_2 = \sqrt{k/m}$ (for identical mass-spring systems). Thus, $t_0 = -\phi_2/\omega_2 = \frac{1}{2}\pi\sqrt{m/k} = \frac{1}{2}\pi\sqrt{(0.43 \text{ kg})/(2.2 \text{ N/m})} = 0.694$ s. (In terms of the period, $t_0 = -\phi_2/(2\pi/T) = \frac{1}{4}T$, which is intuitively more obvious.)

Problem

13. A mass m slides along a frictionless horizontal surface at speed v_0. It strikes a spring of constant k attached to a rigid wall, as shown in Fig. 15-35. After a completely elastic encounter with the spring, the mass heads back in the direction it came from. In terms of k, m, and v_0, determine (a) how long the mass is in contact with the spring and (b) the maximum compression of the spring.

FIGURE 15-35 Problem 13.

Solution

(a) While the mass is in contact with the spring, the net horizontal force on it is just the spring force, so it undergoes half a cycle of simple harmonic motion before leaving the spring with speed v_0 to the left. This takes time equal to half a period $\frac{1}{2}T = \pi/\omega = \pi\sqrt{m/k}$. (b) v_0 is the maximum speed, which is related to the maximum compression of the spring (the amplitude) by $v_0 = \omega A$. Thus $A = v_0/\omega = v_0\sqrt{m/k}$.

Problem

15. Show by substitution that $x(t) = A\sin\omega t$ is a solution to Equation 15-4.

Solution

If $x = A\sin\omega t$, $dx/dt = \omega A\cos\omega t$, and $d^2x/dt^2 = -\omega^2 A\sin\omega t = -\omega^2 x$. Substituting into Equation 15-4, we find $m(-\omega^2 x) = -kx$, which is satisfied if $\omega^2 = k/m$.

Section 15-3: Applications of Simple Harmonic Motion

Problem

17. How long should you make a simple pendulum so its period is (a) 200 ms; (b) 5.0 s; (c) 2.0 min?

Solution

The period and length of a simple pendulum (at the surface of the Earth) are related by Equation 15-18, therefore $\ell = (T/2\pi)^2 g = (0.248 \text{ m/s}^2)T^2 = 9.93$ mm, 6.21 m, and 3.57 km respectively, for the three values given.

Problem

19. A 640-g hollow ball 21 cm in diameter is suspended by a wire and is undergoing torsional oscillations at a frequency of 0.78 Hz. What is the torsional constant of the wire?

Solution

The rotational inertia of a hollow sphere about an axis through its center is $\frac{2}{3}MR^2$ (see Table 12-2). Equation 15-14 gives $\kappa = \omega^2 I = (2\pi \times 0.78 \text{ Hz})^2 \times (\frac{2}{3} \times 0.64 \text{ kg})(\frac{1}{2} \times 0.21 \text{ m})^2 = 0.113$ N·m/rad.

Problem

21. A pendulum of length ℓ is mounted in a rocket. What is its period if the rocket is (a) at rest on its launch pad; (b) accelerating upward with acceleration $a = \frac{1}{2}g$; (c) accelerating downward with acceleration $a = \frac{1}{2}g$; (d) in free fall?

Solution

(It may be helpful to think of an elevator instead of a rocket in this problem.) Let $\mathbf{a}$ be the acceleration of the pendulum relative to the rocket, and let $\mathbf{a_0}$ be the acceleration of the rocket relative to the ground (assumed to be an inertial system). Then Newton's second law is $\sum\mathbf{F} = m(\mathbf{a} + \mathbf{a_0})$, or $\sum\mathbf{F} - m\mathbf{a_0} = m\mathbf{a}$. The rotational analog of this equation is the appropriate generalization of Equation 15-15 for a simple pendulum in an accelerating frame. The "fictitious" torque (about the point of suspension), $\mathbf{r} \times (-m\mathbf{a_0})$, can be combined with the torque of gravity, $\mathbf{r} \times m\mathbf{g}$, if we replace g by $|\mathbf{g} - \mathbf{a_0}|$, while the right-hand side, $|\mathbf{r} \times m\mathbf{a}| = I\alpha$, is the same as Equation 15-15. For small oscillations about the equilibrium position (which depends on $\mathbf{a_0}$), the period is $T = 2\pi\sqrt{\ell/|\mathbf{g} - \mathbf{a_0}|}$. (a) If $\mathbf{a_0} = 0$, $T = 2\pi\sqrt{\ell/g}$, as before. (b) Take the y axis positive up. Then $\mathbf{a_0} = \frac{1}{2}g\hat{\mathbf{j}}$ and $\mathbf{g} = -g\hat{\mathbf{j}}$, so $T = 2\pi\sqrt{\ell/(g + \frac{1}{2}g)} = 2\pi\sqrt{2\ell/3g}$. (c) If $\mathbf{a_0} = -\frac{1}{2}g\hat{\mathbf{j}}$, $T = 2\pi\sqrt{\ell/(g - \frac{1}{2}g)} = 2\pi\sqrt{2\ell/g}$. (d) If $\mathbf{a_0} = \mathbf{g}$, $T = \infty$ (there is no restoring torque and the pendulum does not oscillate).

Problem

23. A mass is attached to a vertical spring, which then goes into oscillation. At the high point of the oscillation, the spring is in the original unstretched equilibrium position it had before the mass was attached; the low point is 5.8 cm below this. What is the period of oscillation?

Solution

At the highest point, there is no spring force (since the spring is unstretched), so the acceleration is just g (downward). This is also the maximum acceleration during the simple harmonic motion (since a_{max} occurs where the displacement is maximum), so $a_{max} = g = \omega^2 A$. The peak-to-peak displacement is $2A = 5.8$ cm, thus $T = 2\pi/\omega = 2\pi\sqrt{A/g} = 2\pi\sqrt{0.029 \text{ m}/(9.8 \text{ m/s}^2)} = 0.342$ s.

Problem

25. A solid disk of radius R is suspended from a spring of linear spring constant k and torsional constant κ, as shown in Fig. 15-36. In terms of k and κ, what value of R will give the same period for the vertical and torsional oscillations of this system?

FIGURE 15-36 Problem 25.

Solution

Equating the angular frequencies for vertical and torsional oscillations (Equations 15-8a and 15-14), we find $k/M = \kappa/I = \kappa/(\frac{1}{2}MR^2)$, or $R = \sqrt{2\kappa/k}$.

Problem

27. Geologists use an instrument called a **gravimeter** to measure the local acceleration of gravity. A particular gravimeter uses the period of a 1-m-long pendulum to determine g. If g is to be measured to within 1 mgal(1 gal $= 1$ cm/s^2) and if the period can be measured with arbitrary accuracy, how accurately must the length of the pendulum be known?

Solution

For a simple pendulum, $\ell = (T/2\pi)^2 g$. If the period is known precisely ($\Delta T \approx 0$), then the fractional errors in ℓ and g are the same, i.e., $\Delta\ell/\ell = \Delta g/g = 1$ mgal$\div 980$ gal $\approx 10^{-6}$.

Problem

29. A thin, uniform hoop of mass M and radius R is suspended from a thin horizontal rod and set oscillating with small amplitude, as shown in Fig. 15-38. Show that the period of the oscillations is $2\pi\sqrt{2R/g}$. *Hint:* You may find the parallel-axis theorem useful.

Solution

Equation 15-16 gives $T = 2\pi\sqrt{I/Mg\ell} = 2\pi\sqrt{(MR^2 + MR^2)/MgR} = 2\pi\sqrt{2R/g}$, where we used the parallel axis theorem for I, with $\ell = h = R$.

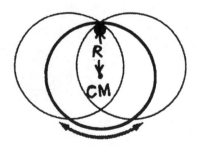

FIGURE 15-38 Problem 29 Solution.

Problem

31. A point mass m is attached to the rim of an otherwise uniform solid disk of mass M and radius R (Fig. 15-40). The disk is rolled slightly away from its equilibrium position and released. It rolls

back and forth without slipping. Show that the period of this motion is given by

$$T = 2\pi\sqrt{3MR/2mg}.$$

Solution

An exact solution of this problem requires advanced methods in analytical mechanics. However, there are two simple heuristic arguments showing that the period given is approximately correct, for small displacements from equilibrium, if the point mass is also assumed to be much smaller than the mass of the disk, i.e., $m \ll M$.

Consider the disk when it is displaced by a small angle, $\theta \ll 1$, from equilibrium, as shown in the sketch. The unbalanced torque about the point of contact, O (which is instantaneously at rest for rolling without slipping) is $\tau = I\alpha = -mgR\sin\theta \approx -mgR\theta$. This is a restoring torque because it is in the opposite direction to the angular displacement. If we neglect m compared to M, the rotational inertia is just that of the disk about O, or $I = \frac{1}{2}MR^2 + MR^2 = \frac{3}{2}MR^2$ (recall Equation 12-20), so $\alpha \approx -mgR\theta/I = -(2mg/3MR)\theta$. This is the equation for simple harmonic motion (see Equation 15-13) with period $T = 2\pi/\omega = 2\pi \times \sqrt{3MR/2mg}$ (see Equation 15-14).

Alternatively, the potential energy varies like that of just the point mass (since that of the disk is constant when it rolls on a horizontal surface), so $U = mgy = mgR(1 - \cos\theta) \approx \frac{1}{2}mgR\theta^2$, for small θ. For $m \ll M$, the kinetic energy is approximately that of just the disk, rolling without slipping, or $K = \frac{1}{2}I\omega^2 = \frac{1}{2}(\frac{3}{2}MR^2)(d\theta/dt)^2$. The total energy is $E = K + U \approx \frac{1}{2}(mgR)\theta^2 + \frac{1}{2}(\frac{3}{2}MR^2)(d\theta/dt)^2$. By analogy with a mass-spring system (where $E = \frac{1}{2}kx^2 + \frac{1}{2}m(dx/dt)^2$), this represents simple harmonic motion with angular frequency (analogous to $\sqrt{k/m}$) of $\sqrt{(mgR)/(\frac{3}{2}MR^2)} = \sqrt{2mg/3MR}$ (see also Problem 44).

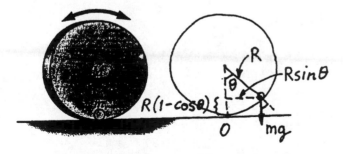

FIGURE 15-40 Problem 31 Solution.

Problem

33. A cyclist turns her bicycle upside down to tinker with it. After she gets it upside down, she notices the front wheel executing a slow, small-amplitude back-and-forth rotational motion with a period of 12 s. Considering the wheel to be a thin ring of mass 600 g and radius 30 cm, whose only irregularity is the presence of the tire valve stem, determine the mass of the valve stem.

Solution

The bicycle wheel may be regarded as a physical pendulum, with rotational inertia $I = MR^2 + mR^2$ about its central axle, where $M = 600$ g is the mass of the wheel (thin ring) and m is the mass of the valve stem (a circumferential point mass). The distance of the CM from the axle is given by $(M + m)\ell = mR$ (this is just Equation 10-4, with origin at the center of the wheel so $x_1 = 0$ for M, $x_2 = R$ for m, and $x_{cm} = \ell$). For small oscillations, the period is given by Equation 15-16 (where $M + m$ is the total mass), therefore $T = 2\pi/\omega = 2\pi\sqrt{I/(M+m)g\ell} = 2\pi \times \sqrt{(M+m)R^2/mgR}$. Solving for m, we find $m = M[(g/R)(T/2\pi)^2 - 1]^{-1} = (600 \text{ g})[(9.8/0.3)\times (12/2\pi)^2 - 1]^{-1} = 5.04$ g.

Problem

35. Repeat the previous problem for the case when the springs are connected as in Fig. 15-42.

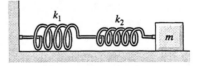

FIGURE 15-42 Problem 35.

Solution

The effective spring constant for this combination (see Chapter 5, Problem 73(b)) is $k = k_1 k_2/(k_1 + k_2)$, and $\omega^2 = k/m$.

Section 15-4: Circular Motion and Simple Harmonic Motion

Problem

37. The equation for an ellipse is

$$\frac{x^2}{a^2} + \frac{y^2}{b^2} = 1.$$

Show that two-dimensional simple harmonic motion whose two components have different amplitudes and are $\pi/2$ out of phase gives rise to elliptical motion. How are a and b related to the amplitudes?

Solution

Simple harmonic motions in the x and y directions, with different amplitudes and $\pi/2$ out of phase, are $x = a\cos(\omega t + \phi)$ and $y = b\cos(\omega t + \phi \pm \pi/2) = \mp b\sin(\omega t + \phi)$. These describe an elliptical path with semi-major or minor axis equal to the amplitudes, a or b, since $(x/a)^2 + (y/b)^2 = \cos^2(\omega t + \phi) + \sin^2(\omega t + \phi) = 1$.

Section 15-5: Energy in Simple Harmonic Motion

Problem

39. A 1400-kg car with poor shock absorbers is bouncing down the highway at 20 m/s, executing vertical harmonic motion at 0.67 Hz. If the amplitude of the oscillations is 18 cm, what is the total energy in the oscillations? What fraction of the car's kinetic energy is this? Neglect rotational energy of the wheels and the fact that not all the car's mass participates in the oscillation.

Solution

The total vibrational energy is $E_{\text{vib}} = \frac{1}{2}kA^2 = \frac{1}{2}m\omega^2 A^2 = \frac{1}{2}(1400\text{ kg})(2\pi\times 0.67\text{ Hz})^2(0.18\text{ m})^2 = 402$ J. The car's total energy is $E = E_{\text{trans}} + E_{\text{vib}} = \frac{1}{2}Mv_{cm}^2 + E_{\text{vib}} \approx \frac{1}{2}(1400\text{ kg})(20\text{ m/s})^2 = 280$ kJ, so E_{vib}/E is only about 0.144%.

Problem

41. The motion of a particle is described by

$$x = (45\text{ cm})[\sin(\pi t + \pi/6)],$$

with x in cm and t in seconds. At what time is the potential energy twice the kinetic energy? What is the position of the particle at this time?

Solution

The condition that the potential energy equal twice the kinetic energy implies that $U(t) = \frac{1}{2}kx(t)^2 = 2K(t) = mv(t)^2$, or $\omega x(t)/v(t) = \pm\sqrt{2}$, where $\omega = \sqrt{k/m}$. For $x(t) = (45\text{ cm})\sin(\pi t + \pi/6)$ as given (note that $\omega = \pi(\text{s}^{-1})$), $v(t) = dx/dt = \omega(45\text{ cm})\times \cos(\pi t + \pi/6)$, so the above condition becomes $\tan(\pi t + \pi/6) = \pm\sqrt{2}$. There are four angles in each cycle which satisfy this (since $\tan\theta = -\tan(\pi - \theta) = \tan(\pi + \theta) = -\tan(2\pi - \theta)$), which are $\pi(t + \frac{1}{6}) = 0.955$, 2.19, 4.10, and 5.33 radians. (We chose the cycle with phases between 0 and 2π radians; for any other cycle, an integer multiple of 2π can be added to these angles.) The times corresponding to these phases are $t = (0.955/\pi) - \frac{1}{6} = 0.137$ s, 0.529 s, 1.14 s, and 1.53 s, respectively. (An integer multiple of 2 can be added to get the times for any other cycle.) The positions of the particle corresponding to these phases are $x(0.137\text{ s}) = (45\text{ cm})\sin(0.955) = 36.7$ cm $= x(0.529\text{ s}) = -x(1.14\text{ s}) = -x(1.53\text{ s})$, since $\sin\theta = \sin(\pi - \theta) = -\sin(\pi + \theta) = -\sin(2\pi - \theta)$. (During each cycle, the particle passes each of the points ± 36.7 cm twice, traveling with the same speed, but in opposite directions.)

Problem

43. Show that the potential energy of a simple pendulum is proportional to the square of the angular displacement in the small-amplitude limit.

Solution

The potential energy of a simple pendulum (see Example 8-6 and Fig. 8-14) is $U = mgh = mg\ell(1 - \cos\theta)$. For small angles, $\cos\theta \approx 1 - \frac{1}{2}\theta^2$, so $U \approx \frac{1}{2}mg\ell\theta^2$.

Problem

45. A solid cylinder of mass M and radius R is mounted on an axle through its center. The axle is attached to a horizontal spring of constant k, and the cylinder rolls back and forth without slipping (Fig. 15-43). Write the statement of energy conservation for this system, and differentiate it to obtain an equation analogous to Equation 15-4 (see previous problem). Comparing your result with Equation 15-4, determine the angular frequency of the motion.

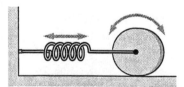

FIGURE 15-43 Problem 45.

Solution

With reference to Equation 12-24 (and the condition $v = \omega R$ for rolling without slipping) $K = \frac{1}{2}Mv^2 + \frac{1}{2}I_{cm}\omega^2 = \frac{1}{2}Mv^2 + \frac{1}{2}(\frac{1}{2}MR^2)(v/R)^2 = \frac{3}{4}Mv^2$. The potential energy of the spring is $U = \frac{1}{2}kx^2$, where $v = dx/dt$, so $E = K + U = \frac{3}{4}M(dx/dt)^2 + \frac{1}{2}kx^2$. Differentiating, we find:

$$\frac{dE}{dt} = 0 = \frac{3}{4}M\cdot 2\left(\frac{dx}{dt}\right)\left(\frac{d^2x}{dt^2}\right) + \frac{1}{2}k\cdot 2x\left(\frac{dx}{dt}\right), \quad \text{or}$$

$$\frac{d^2x}{dt^2} = -\frac{2k}{3M}x \equiv -\omega^2 x.$$

(The energy method is particularly convenient for analyzing small oscillations, since complicated details of the forces can be avoided.)

Section 15-6:   Damped Harmonic Motion

Problem

47. A 250-g mass is mounted on a spring of constant $k = 3.3$ N/m. The damping constant for this system is $b = 8.4 \times 10^{-3}$ kg/s. How many oscillations will the system undergo during the time the amplitude decays to $1/e$ of its original value?

Solution

Since the damping constant is small, the motion is under-damped and Equation 15-20 applies. The time for the amplitude to decay to $1/e$ of its original value is $t = 2m/b = 59.5$ s, while the period is $T = 2\pi\sqrt{m/k} = 1.73$ s, therefore the number of corresponding oscillations is $59.5/1.73 = 34.4$.

Section 15-7:   Driven Oscillations and Resonance

Problem

49. A mass-spring system has $b/m = \omega_0/5$, where b is the damping constant and ω_0 the natural frequency. How does its amplitude when driven at frequencies 10% above and below ω_0 compare with its amplitude at ω_0?

Solution

The amplitude at resonance ($\omega_d = \omega_0$) is $A_{\text{res}} = F_0 \div b\omega_0$, so that Equation 15-23 can be rewritten as:

$$\frac{A}{A_{\text{res}}} = \frac{A}{(F_0/b\omega_0)} = \frac{(b\omega_0/m)}{\sqrt{(\omega_d^2 - \omega_0^2)^2 + b^2\omega_d^2/m^2}}$$

$$= \left[\left(\frac{m\omega_0}{b}\right)^2 \left(\frac{\omega_d^2}{\omega_0^2} - 1\right)^2 + \frac{\omega_d^2}{\omega_0^2}\right]^{-1/2}$$

If $(m\omega_0/b) = 5$, and $(\omega_d/\omega_0) = 1.1$ (10% above resonance), then $A/A_{\text{res}} = 1/\sqrt{25(1.21-1)^2 + 1.21} = 65.8\%$, while for $\omega_d/\omega_0 = 0.9$ (10% below resonance), $A/A_{\text{res}} = 1/\sqrt{25(0.81-1)^2 + 0.81} = 76.4\%$.

Problem

51. Show by direct substitution that Equation 15-22 satisfies Equation 15-21 with A given by Equation 15-23.

Solution

When Equation 15-22 is substituted into Equation 15-21, one obtains $m[-\omega_d^2 A\cos(\omega_d t + \phi)] =$

$-kA\cos(\omega_d t + \phi) - b[-\omega_d A\sin(\omega_d t + \phi)] + F_0[\cos(\omega_d t + \phi)\cos\phi + \sin(\omega_d t + \phi)\sin\phi]$, where we let $\omega_d t = \omega_d t + \phi - \phi$ in the F_0-term, and used a trigonometric identity. This equation is true if the coefficients of the $\sin(\omega_d t + \phi)$ and $\cos(\omega_d t + \phi)$ terms on each side are equal, respectively, that is, $-m\omega_d^2 A = -kA + F_0\cos\phi$, and $0 = b\omega_d A + F_0\sin\phi$. Let $\omega_0^2 = k/m$, and these equations become $F_0\cos\phi = -m(\omega_d^2 - \omega_0^2)A$, and $F_0\sin\phi = -b\omega_d A$. Squaring and adding, we get Equation 15-23.

Paired Problems

Problem

53. A particle undergoes simple harmonic motion with amplitude 25 cm and maximum speed 4.8 m/s. Find (a) the angular frequency, (b) the period, and (c) the maximum acceleration.

Solution

(a) Since $v_{\text{max}} = \omega A$, $\omega = (4.8 \text{ m/s})/(0.25 \text{ m}) = 19.2$ s^{-1}. (b) $T = 2\pi/\omega = 0.327$ s. (c) $a_{\text{max}} = \omega v_{\text{max}} = 92.2$ m/s^2.

Problem

55. A massless spring of spring constant $k = 74$ N/m is hanging from the ceiling. A 490-g mass is hooked onto the unstretched spring and allowed to drop. Find (a) the amplitude and (b) the period of the resulting motion.

Solution

(a) The distance from the initial position of the mass on the unstretched spring, to the equilibrium position, where the net force is zero, is just the amplitude, since the initial velocity for a dropped mass is zero. Then at the equilibrium position, $mg = kA$, or $A = (0.49 \text{ kg}) \times (9.8 \text{ m/s}^2)/(74 \text{ N/m}) = 6.49$ cm. (Alternatively, when dropped at the unstretched position (zero spring force), the initial acceleration has its maximum magnitude which is just g, so $a_{\text{max}} = g = \omega^2 A$ gives the same result, since $\omega^2 = k/m$.) (b) $T = 2\pi\sqrt{m/k} = 0.511$ s.

Problem

57. A meter stick is suspended from one end and set swinging. What is the period of the resulting oscillations, assuming they have small amplitude?

Solution

The meter stick is a physical pendulum whose CM is $\ell = 0.5$ m below the point of suspension through one end. The rotational inertia of the stick about one end is $\frac{1}{3}M(1\text{ m})^2$, so Equation 15-16 gives the period as $T = 2\pi\sqrt{I/Mg\ell} = 2\pi\sqrt{2(1\text{ m})/3g} = 1.64$ s.

Problem

59. Two balls each of unknown mass m are mounted on opposite ends of a 1.5-m-long rod of mass 850 g. The system is suspended from a wire attached to the center of the rod and set into torsional oscillations. If the torsional constant of the wire is 0.63 N·m/rad and the period of the oscillations is 5.6 s, what is the unknown mass m?

Solution

The period of a torsional pendulum is given by Equation 15-14, $\omega = 2\pi/T = \sqrt{\kappa/I}$. The rotational inertia of the rod and two masses, about an axis perpendicular to the rod and through its center, is $I = (T/2\pi)^2\kappa = \frac{1}{12}M\ell^2 + 2m(\frac{1}{2}\ell)^2$ so $m = \frac{1}{2}[(T\div \pi\ell)^2\kappa - \frac{1}{3}M] = \frac{1}{2}[(5.6\text{ s}/\pi\times1.5\text{ m})^2(0.63\text{ N·m}) - \frac{1}{3}(0.85\text{ kg})] = 303$ g.

Problem

61. Two mass-spring systems with the same mass are undergoing oscillatory motion with the same amplitudes. System 1 has twice the frequency of system 2. How do (a) their energies and (b) their maximum accelerations compare?

Solution

(a) The energy of a mass-spring system is $E = \frac{1}{2}m\omega^2 A^2$ (see Section 15.5). If m and A are the same, but $\omega_1 = 2\omega_2$, then $E_1 = 4E_2$. (b) The maximum acceleration is $a_{max} = \omega^2 A$, so $a_{1,max} = 4a_{2,max}$ as well.

Supplementary Problems
Problem

63. While waiting for your plane to take off, you suspend your keys from a thread and set the resulting pendulum oscillating. It completes exactly 90 cycles in 1 minute. You repeat the experiment as the plane accelerates down the runway, and now find the pendulum completes exactly 91 cycles in 1 minute. Find the plane's acceleration.

Solution

The solution of Problem 21 shows that the period of a simple pendulum (point mass suspended in a constant gravitational field) in an accelerating reference frame is $T = 2\pi/\omega = 2\pi\sqrt{\ell/|\mathbf{g} - \mathbf{a_0}|}$. If $\mathbf{a_0}$ is perpendicular to $\mathbf{g}$ (as for the air plane in this problem) then $|\mathbf{g} - \mathbf{a_0}| =$

$\sqrt{g^2 + a_0^2}$, and the circular frequency is $\omega^2 = \sqrt{g^2 + a_0^2}/\ell$. Since $\omega_0^2 = g/\ell$ at rest, we can eliminate ℓ and solve for a_0: $(\omega/\omega_0)^2 = \sqrt{1 + (a_0/g)^2}$, or $(a_0/g) = \sqrt{(\omega/\omega_0)^4 - 1}$. The ratio of the frequencies is $\omega/\omega_0 = 91/90$, so numerically, $a_0 = (9.8\text{ m/s}^2)\times \sqrt{(91/90)^4 - 1} = 2.08$ m/s^2.

Problem

65. A 500-g block on a frictionless surface is connected to a rather limp spring of constant $k = 8.7$ N/m. A second block rests on the first, and the whole system executes simple harmonic motion with a period of 1.8 s. When the amplitude of the motion is increased to 35 cm, the upper block just begins to slip. What is the coefficient of static friction between the blocks?

Solution

If the surfaces of contact are horizontal, it is the frictional force which accelerates the upper block, hence $f_s = m_2 a(t) \leq \mu_s N = \mu_s m_2 g$, or $a(t) \leq \mu_s g$. In simple harmonic motion, $a_{max} = \omega^2 A$, so when the upper block begins to slip, $\omega^2 A = \mu_s g$, or $\mu_s = (2\pi/1.8\text{ s})^2(0.35\text{ m}/9.8\text{ m/s}^2) = 0.44$. [Note: the data given in the problem which were not used to find μ_s (i.e., k and m_1) can be used to calculate that $m_2 = 214$ g, since $\omega = 2\pi/T = \sqrt{k/(m_1 + m_2)}$.]

Problem 65 Solution.

Problem

67. Repeat Problem 46 for a small solid ball of mass M and radius R that rolls without slipping on the parabolic track.

Solution

The potential energy of the ball is $U = mgy = mgax^2$, and its kinetic energy (rolling without slipping) is (from Equation 12-24) $K = \frac{1}{2}Mv^2 + \frac{1}{2}I_{cm}\omega^2 = \frac{1}{2}Mv^2 + \frac{1}{2}(\frac{2}{5}MR^2)(v/R)^2 = \frac{7}{10}Mv^2$. Since the total mechanical energy, $E = U + K$, is constant,

$$\frac{dE}{dt} = 0 = \frac{d}{dt}\left(Mgax^2 + \frac{7}{10}Mv^2\right)$$

$$= 2Mgax\frac{dx}{dt} + \frac{7}{5}Mv\frac{dv}{dt}.$$

Note that in this problem (and in Problem 46) $v = \sqrt{(dx/dt)^2 + (dy/dt)^2}$, so the motion is not harmonic, in general. However, for small displacements, $v \approx dx/dt$, and $d^2x/dt^2 = -(10ga/7)x$, which is simple harmonic motion with period $T = 2\pi/\omega = 2\pi\sqrt{7/10ga}$.

Problem

69. A 1.2-kg block rests on a frictionless surface and is attached to a horizontal spring of constant $k = 23$ N/m (Fig. 15-46). The block is oscillating with amplitude 10 cm and with phase constant $\phi = -\pi/2$. A block of mass 0.80 kg is moving from the right at 1.7 m/s. It strikes the first block when the latter is at the rightmost point in its oscillation. The collision is completely inelastic, and the two blocks stick together. Determine the frequency, amplitude, and phase constant (relative to the *original* $t = 0$) of the resulting motion.

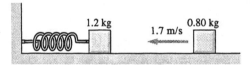

FIGURE 15-46 Problem 69.

Solution

The simple harmonic motion with just the first block on the spring can be described by Equation 15-9 and the given amplitude and phase constant; $x(t) = (10 \text{ cm})\cos(\omega_1 t - \pi/2) = (10 \text{ cm})\sin\omega_1 t$, where $\omega_1 = \sqrt{k/m_1} = \sqrt{(23 \text{ N/m})/(1.2 \text{ kg})} = 4.38 \text{ s}^{-1}$. This equation holds up to the time of the collision, i.e., for $t < t_c$, where $t_c = \pi/2\omega_1$, since for the rightmost point of oscillation, $\sin\omega_1 t_c = 1$, or $\omega_1 t_c = \pi/2$. (This specifies the original zero of time appropriate to the given phase constant of $-\pi/2$.)

Equation 15-9 also describes the simple harmonic motion after the collision; $x(t) = A\cos(\omega_2 t + \phi)$ for $t > t_c$, where $\omega_2 = \sqrt{k/(m_1 + m_2)} = 3.39 \text{ s}^{-1}$ is the angular frequency when both blocks oscillate on the spring (and $f_2 = \omega_2/2\pi = 0.540$ Hz, as asked in the problem.) It follows from this that $v(t) = -\omega_2 A \times \sin(\omega_2 t + \phi)$. The amplitude A and phase constant ϕ can be determined from these two equations evaluated just after the collision, essentially at t_c, if we assume that the collision takes place almost instantaneously; then conservation of momentum during the collision can be applied (see Equation 11-4). Just after the collision, $x(t_c) = 10$ cm (given) and $v(t_c) = (m_1v_1 + m_2v_2)/(m_1 + m_2)$, where just before the collision, $v_1 = 0$ (given m_1 at rightmost point of its original

motion) and $v_2 = -1.7$ m/s (also given). Numerically, $v(t_c) = (-1.7 \text{ m/s})(0.8)/(0.8 + 1.2) = -68$ cm/s. Thus, the two equations become $x(t_c) = 10 \text{ cm} = A \times \cos(\omega_2 t_c + \phi)$, and $v(t_c) = -68 \text{ cm/s} = -\omega_2 A \times \sin(\omega_2 t_c + \phi)$, where ω_2 and t_c are known. (In fact, $\omega_2 t_c = \omega_2 \pi/2\omega_1 = (\pi/2)\sqrt{m_1/(m_1 + m_2)} = \sqrt{0.6}\,(\pi/2)$ radians $= 69.7°$.)

Solving for A (using $\sin^2 + \cos^2 = 1$), we find $A = \sqrt{x(t_c)^2 + [-v(t_c)/\omega_2]^2} = \sqrt{(10 \text{ cm})^2 + (68 \text{ cm}/3.39)^2} = 22.4$ cm. Solving for ϕ (using $\sin/\cos = \tan$), we find $\phi = \tan^{-1}[-v(t_c) \div \omega_2 x(t_c)] - \omega_2 t_c = \tan^{-1}(68/3.39 \times 10) - 69.7° = -6.22° = -0.109$ radians.

(Note: The solution for A is equivalent to calculating the various energies in the second simple harmonic motion, since just after the collision, $K(t_c) = \frac{1}{2}(m_1 + m_2)v(t_c)^2 = \frac{1}{2}(2 \text{ kg})(-0.68 \text{ m/s})^2 = 0.462$ J, $U(t_c) = \frac{1}{2}kx(t_c)^2 = \frac{1}{2}(23 \text{ N/m})(0.1 \text{ m})^2 = 0.115$ J, $E = K(t_c) + U(t_c) = 0.577 \text{ J} = \frac{1}{2}kA^2$, or $A = \sqrt{2(0.577 \text{ J})/(23 \text{ N/m})}$. Once A is known, ϕ can also be found from either expression for $x(t_c)$ or $v(t_c)$, e.g., $\omega_2 t_c + \phi = \cos^{-1}(10/22.4) = \sin^{-1}(68/3.39 \times 22.4)$.)

Problem

71. A small object of mass m slides without friction in a circular bowl of radius R. Derive an expression for small-amplitude oscillations about equilibrium, and compare with that of a simple pendulum.

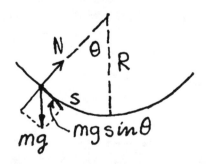

Problem 71 Solution.

Solution

What is intended in this problem is consideration of just motion in a vertical plane through the axis of the bowl. (This corresponds to the same initial conditions as for a simple pendulum. Other conditions lead to motion like various conical pendulums.) The tangential component of Newton's second law is $-mg\sin\theta = m(d^2s/dt^2)$, where $s = R\theta$. For small angles, $-g\sin\theta \approx -g\theta \approx R(d^2\theta/dt^2)$. This is simple harmonic motion, with the same equations as for a simple pendulum (see Problem 24), so $\omega = \sqrt{g/R}$ and $T = 2\pi\sqrt{R/g}$.

Problem

73. A mass m is connected between two springs of length L, as shown in Fig. 15-47. At equilibrium, the tension force in each spring is F_0. Find the period of oscillations *perpendicular* to the springs, assuming sufficiently small amplitude that the magnitude of the spring tension is essentially unchanged.

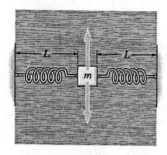

FIGURE 15-47 Problem 73.

Solution

Suppose that no forces with components in the direction of motion act on the mass other than the spring forces. If m is given a small displacement perpendicular to the springs (as sketched), the net force is $F_y = -2F_0 \sin\theta = -2F_0 y/\sqrt{L^2 + y^2} \approx -2F_0 y/L$, for $y \ll L$. Newton's second law gives $m d^2y/dt^2 = F_y$, or $d^2y/dt^2 \approx -(2F_0/mL)y$. This is the equation for simple harmonic motion with angular frequency $\omega = \sqrt{2F_0/mL}$ and period $T = 2\pi/\omega = 2\pi\sqrt{mL/2F_0}$.

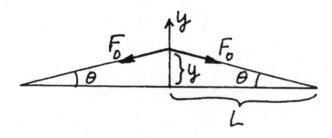

Problem 73 Solution.

Problem

75. A uniform piece of wire is bent into a V-shape with angle θ between two legs of length ℓ. The wire is placed over a pivot, as shown in Fig. 15-49. Show that the angular frequency of small-amplitude oscillations about this equilibrium is given by

$$\omega = \sqrt{\frac{3g\cos(\theta/2)}{2\ell}}.$$

Solution

The CM of the bent wire is a distance $h = (\ell/2)\times\cos(\theta/2)$ from the pivot. The rotational inertia of the bent wire (two thin rods) about the pivot is $I = 2[\frac{1}{3}(\frac{1}{2}M)\ell^2] = \frac{1}{3}M\ell^2$, where M is the mass of the whole wire. Equation 15-16 for the physical pendulum gives

$$\omega = \sqrt{\frac{Mgh}{I}} = \sqrt{\frac{Mg(\ell/2)\cos(\theta/2)}{M\ell^2/3}} = \sqrt{\frac{3g\cos(\theta/2)}{2\ell}}$$

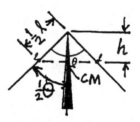

FIGURE 15-49 Problem 75 Solution.

CHAPTER 16 WAVE MOTION

ActivPhysics can help with these problems:
Activities 10.1, 10.2, 10.7, 10.10

Section 16-2: Wave Properties

Problem

1. Ocean waves with 18-m wavelength travel at 5.3 m/s. What is the time interval between wave crests passing under a boat moored at a fixed location?

Solution

Wave crests (adjacent wavefronts) take a time of one period to pass a fixed point, traveling at the wave speed (or phase velocity) for a distance of one wavelength. Thus $T = \lambda/v = 18$ m/(5.3 m/s) = 3.40 s.

Problem

3. An 88.7-MHz FM radio wave propagates at the speed of light. What is its wavelength?

Solution

From Equation 16-1, $\lambda = v/f = (3 \times 10^8$ m/s)$\div$ $(88.7 \times 10^6$ Hz) = 3.38 m.

Problem

5. A 145-MHz radio signal propagates along a cable. Measurement shows that the wave crests are spaced 1.25 m apart. What is the speed of the waves on the cable? Compare with the speed of light in vacuum.

Solution

The distance between adjacent wave crests is one wavelength, so the wave speed in the cable (Equation 16-1) is $v = f\lambda = (145 \times 10^6$ Hz)(1.25 m) = 1.81$\times$ 10^8 m/s = 0.604c, where $c = 3 \times 10^8$ m/s is the wave speed in vacuum.

Problem

7. Detecting objects by reflecting waves off them is effective only for objects larger than about one wavelength. (a) What is the smallest object that can be seen with visible light (maximum frequency 7.5×10^{14} Hz)? (b) What is the smallest object that can be detected with a medical ultrasound unit operating at 5 MHz? The speed of ultrasound waves in body tissue is about 1500 m/s.

Solution

(a) The wavelength of light corresponding to this maximum frequency is $\lambda = c/f = (3 \times 10^8$ m/s)$\div$ $(7.5 \times 10^{14}$ Hz) = 400 nm, violet in hue (see Equation 16-1). (b) The ultrasonic waves described have wavelength $\lambda = v/f = (1500$ m/s)/(5 MHz) = 0.3 mm.

Problem

9. In Fig. 16-28 two boats are anchored offshore and are bobbing up and down on the waves at the rate of six complete cycles each minute. When one boat is up the other is down. If the waves propagate at 2.2 m/s, what is the minimum distance between the boats?

Solution

The boats are 180° $= \pi$ rad out of phase, so the minimum distance separating them is half a wavelength. (In general, they could be an odd number of half-wavelengths apart.) The frequency is 6/60 s = 0.1 Hz, so $\frac{1}{2}\lambda = \frac{1}{2}v/f = \frac{1}{2}(2.2$ m/s)/(0.1/s) = 11 m. (Fig. 16-28 shows the answer, not the question.)

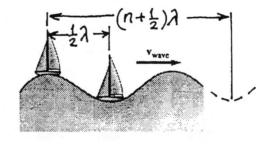

FIGURE 16-28 Problem 9 Solution.

Section 16-3: Mathematical Description of Wave Motion

Problem

11. An ocean wave has period 4.1 s and wavelength 10.8 m. Find (a) its wave number and (b) its angular frequency.

Solution

From Equations 16-3 and 4, (a) $k = 2\pi/10.8$ m = 0.582 m^{-1}, and (b) $\omega = 2\pi/(4.1$ s) = 1.53 s^{-1}.

Problem

13. A simple harmonic wave of wavelength 16 cm and amplitude 2.5 cm is propagating along a string in the negative x direction at 35 cm/s. Find (a) the angular frequency and (b) the wave number. (c) Write a mathematical expression describing the displacement y of this wave (in centimeters) as a function of position and time. Assume the displacement at $x = 0$ is a maximum when $t = 0$.

Solution

(b) Equation 16-4 gives $k = 2\pi/16$ cm $= 0.393$ cm^{-1}, and (a) Equation 16-6 gives $\omega = kv = (0.393$ cm$^{-1})\times(35$ cm/s$) = 13.7$ s^{-1}. (c) Equation 16-5, for a wave moving in the negative x direction, becomes
$$y(x,t) = (2.5 \text{ cm})\cos[(0.393 \text{ cm}^{-1})x + (13.7 \text{ s}^{-1})t].$$

Problem

15. What are (a) the amplitude, (b) the frequency in hertz, (c) the wavelength, and (d) the speed of a water wave whose displacement is $y = 0.25\sin(0.52x - 2.3t)$, where x and y are in meters and t in seconds?

Solution

Comparison of the given displacement with Equation 16-5 reveals that (a) $A = 0.25$ m, (b) $f = \omega/2\pi = (2.3$ s$^{-1})/2\pi = 0.366$ Hz, (c) $\lambda = 2\pi/k = 2\pi/(0.52$ m$^{-1}) = 12.1$ m, and $v = \omega/k = (2.3$ s$^{-1})/0.52$ m$^{-1} = 4.42$ m/s. (Note: The presence of a phase constant of $\phi = -\pi/2$ in the expression for $y(x,t) = A\sin(kx - \omega t) = A\cos(kx - \omega t + \phi)$ does not affect any of the quantities queried in this problem.)

Problem

17. At time $t = 0$, the displacement in a transverse wave pulse is described by $y = 2(x^4 + 1)^{-1}$, with both x and y in cm. Write an expression for the pulse as a function of position x and time t if it is propagating in the positive x direction at 3 cm/s.

Solution

From the shape of the pulse at $t = 0$, $y(x,0) = f(x)$, a pulse with the same waveform, traveling in the positive x direction with speed v, can be obtained by replacing x by $x - vt$, $y(x,t) = f(x - vt)$. For the given $f(x)$ and v, $y(x,t) = 2[(x - 3t)^4 + 1]^{-1}$, with x and y in cm and t in s.

Problem

19. Figure 16-30a shows a wave plotted as a function of position at time $t = 0$, while Fig. 16-30b shows the same wave plotted as a function of time at position $x = 0$. Find (a) the wavelength, (b) the period, (c) the wave speed, and (d) the direction of propagation.

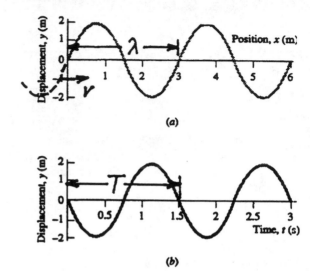

FIGURE 16-30 Problem 19.

Solution

(a) The wavelength is the distance between successive maxima at the same time (say $t = 0$), so Fig. 16-30a gives $\lambda = 3$ m. (b) The period is the time interval between successive maxima (or some other specific phases differing by 2π) at the same point (say $x = 0$), so Fig. 16-30b gives $T = 1.5$ s. (c) $v = \omega/k = \lambda/T = 2$ m/s. (d) Fig. 16-30b shows that as t increases from 0, the displacement at $x = 0$ first becomes negative. The waveform in Fig. 16-30a must therefore move to the right, in the positive x direction. [For the sinusoidal wave pictured, $y(x, 0) = A\sin kx$ and $y(0,t) = -A\sin\omega t = A\sin(-\omega t)$, so $y(x,t) = A\sin(kx - \omega t)$.]

Section 16-4: Waves on a String
Problem

21. The main cables supporting New York's George Washington Bridge have a mass per unit length of 4100 kg/m and are under tension of 250 MN. At what speed would a transverse wave propagate on these cables?

Solution

$v = \sqrt{F/\mu} = \sqrt{(2.5\times10^8 \text{ N})/(4100 \text{ kg/m})} = 247$ m/s (from Equation 16-7).

Problem

23. A transverse wave with 3.0-cm amplitude and 75-cm wavelength is propagating on a stretched spring whose mass per unit length is 170 g/m. If the wave speed is 6.7 m/s, find (a) the spring tension and (b) the maximum speed of any point on the spring.

Solution

(a) Equation 16-7 gives $F = \mu v^2 = (0.17 \text{ kg/m}) \times (6.7 \text{ m/s})^2 = 7.63$ N. (b) The unnumbered equation for the vertical velocity of the medium in Section 16.5 gives $u_{max} = (dy/dt)_{max} = \omega A = (2\pi v/\lambda)A = 2\pi(6.7 \text{ m/s})(3 \text{ cm})/(75 \text{ cm}) = 1.68$ m/s.

Problem

25. A 3.1-kg mass hangs from a 2.7-m-long string whose total mass is 0.62 g. What is the speed of transverse waves on the string? *Hint:* You can ignore the string mass in calculating the tension but not in calculating the wave speed. Why?

Solution

The tension in the string is approximately equal to the weight of the 3.1 kg mass (since the weight of the string is only 2% of this). Thus, $v = \sqrt{F/\mu} = \sqrt{(3.1 \text{ kg})(9.8 \text{ m/s}^2)(2.7 \text{ m})/(0.62 \text{ g})} = 364$ m/s. (0.62 g is small compared to 3.1 kg, but not small compared to zero!)

Problem

27. The density of copper is 8.29 g/cm^3. What is the tension in a 1.0-mm-diameter copper wire that propagates transverse waves at 120 m/s?

Solution

The linear mass density of copper wire with diameter d is $\mu = m/\ell = \rho\frac{1}{4}\pi d^2 = (8.29 \text{ g/cm}^3)\frac{1}{4}\pi(1 \text{ mm})^2 = 6.51\times10^{-3}$ kg/m, so $F = \mu v^2 = (6.51\times10^{-3} \text{ kg/m}) \times (120 \text{ m/s})^2 = 93.8$ N.

Problem

29. A 25-m-long piece of 1.0-mm-diameter wire is put under 85 N tension. If a transverse wave takes 0.21 s to travel the length of the wire, what is the density of the material comprising the wire?

Solution

From the length of wire, travel time, and Equation 16-7, $v = 25 \text{ m}/0.21 \text{ s} = \sqrt{85 \text{ N}/\mu}$, so $\mu = 6.00\times10^{-3}$ kg/m. But for a uniform wire of length ℓ and diameter d, $\rho = \mu/\frac{1}{4}\pi d^2 = (6.00\times10^{-3} \text{ kg/m}) \div \frac{1}{4}\pi(1 \text{ mm})^2 = 7.64$ g/cm^3 (see solution to Problem 27).

Problem

31. A steel wire can tolerate a maximum tension per unit cross-sectional area of 2.7 GN/m^2 before it undergoes permanent distortion. What is the maximum possible speed for transverse waves in a steel wire if it is to remain undistorted? Steel has a density of 7.9 g/cm^3.

Solution

The linear density is the (volume) density times the cross-sectional area (see solution to Problem 27), whereas the maximum tension is 2.7 GN/m^2 times the same cross-sectional area. Therefore $v_{max} = \sqrt{(2.7 \text{ GN/m}^2)/(7.9 \text{ g/cm}^3)} = 585$ m/s. (Recall that the prefix giga equals 10^9 and that 1 g/cm$^3 = 10^3$ kg/m^3.)

Section 16-5: Wave Power and Intensity

Problem

33. A rope with 280 g of mass per meter is under 550 N tension. A wave with frequency 3.3 Hz and amplitude 6.1 cm is propagating on the rope. What is the average power carried by the wave?

Solution

The average power transmitted by transverse traveling waves in a string is given by Equation 16-8, $\bar{P} = \frac{1}{2}\mu\omega^2 A^2 v = \frac{1}{2}(0.28 \text{ kg/m})(2\pi\times3.3 \text{ Hz})^2(0.061 \text{ m})^2 \times \sqrt{550 \text{ N}/(0.28 \text{ kg/m})} = 9.93$ W. (We used Equation 16-7 for v.)

Problem

35. A 600-g Slinky is stretched to a length of 10 m. You shake one end at the frequency of 1.8 Hz, applying a time-average power of 1.1 W. The resulting waves propagate along the Slinky at 2.3 m/s. What is the wave amplitude?

Solution

We assume that the elastic properties of a stretched string are shared by the Slinky, so Equation 16-8 applies. Then $A = \sqrt{2(1.1 \text{ W})/(0.06 \text{ kg/m})(2.3 \text{ m/s})}/(2\pi\times1.8 \text{ Hz}) = 35.3$ cm.

Problem

37. Figure 16-32 shows a wave train consisting of two cycles of a sine wave propagating along a string. Obtain an expression for the total energy in this wave train, in terms of the string tension F, the wave amplitude A, and the wavelength λ.

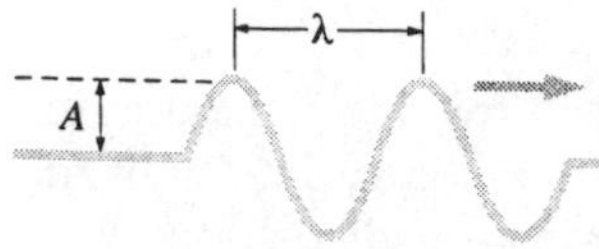

FIGURE 16-32 Problem 37.

Solution

The average wave energy, $d\bar{E}$, in a small element of string of length dx, is transmitted in time, dt, at the same speed as the waves, $v = dx/dt$. From Equation 16-8, $d\bar{E} = \bar{P}dt = \frac{1}{2}\mu\omega^2 A^2 v\, dt = \frac{1}{2}\mu\omega^2 A^2 dx$, so the average linear energy density is $d\bar{E}/dx = \frac{1}{2}\mu\omega^2 A^2$. The total average energy in a wave train of length $\ell = 2\lambda$ is $\bar{E} = (d\bar{E}/dx)\ell = \frac{1}{2}\mu\omega^2 A^2 (2\lambda)$. In terms of the quantities specified in this problem (see Equations 16-1 and 7) $\bar{E} = \frac{1}{2}(F/v^2)(2\pi v/\lambda)^2 A^2 (2\lambda) = 4\pi^2 F A^2/\lambda$. (Note: The relation derived can be written as $\bar{P} = (d\bar{E}/dx)v$. For a one-dimensional wave, $\bar{P}$ is the intensity, so the average intensity equals the average energy density times the speed of wave energy propagation. This is a general wave property, e.g., see the first unnumbered equation for S in Section 34-10.)

Problem

39. A loudspeaker emits energy at the rate of 50 W, spread in all directions. What is the intensity of sound 18 m from the speaker?

Solution

The wave power is spread out over a sphere of area $4\pi r^2$, so the intensity is $50\text{ W}/4\pi(18\text{ m})^2 = 12.3\text{ mW/m}^2$. (See Equation 16-9.)

Problem

41. Use data from Appendix E to determine the intensity of sunlight at (a) Mercury and (b) Pluto.

Solution

Equation 16-9 gives the ratio of intensities at two distances from an isotropic source of spherical waves as $I_2/I_1 = (r_1/r_2)^2$. If we use the average intensity of sunlight given in Table 16-1 and mean orbital distances to the sun from Appendix E, we obtain (a) $I_{\text{Merc}} = I_E (r_E/r_{\text{Merc}})^2 = (1368\text{ W/m}^2)(150\div 57.9)^2 = 9.18\text{ kW/m}^2$, and (b) $I_{\text{Pluto}} = (1368\text{ W/m}^2)\times (150/5.91\times10^3)^2 = 0.881\text{ W/m}^2$. (Alternatively, the luminosity of the sun, $\bar{P} = 3.85\times10^{26}$ W, from Appendix E, could be used directly in Equation 16-9, with only slightly different numerical results.)

Problem

43. Light emerges from a 5.0-mW laser in a beam 1.0 mm in diameter. The beam shines on a wall, producing a spot 3.6 cm in diameter. What are the beam intensities (a) at the laser and (b) at the wall?

Solution

If we assume that the power output of the laser is spread uniformly over the cross-sectional area of its beam, then $I = \bar{P}/\frac{1}{4}\pi d^2$. (a) When the beam emerges, $I = 5\text{ mW}/\frac{1}{4}\pi(1\text{ mm})^2 = 6.37\text{ kW/m}^2$, while (b) after its diameter has expanded by 36 times, at the wall, $I' = I(1/36)^2 = 4.91\text{ W/m}^2$.

Problem

45. Use Table 16-1 to determine how close to a rock band you should stand for it to sound as loud as a jet plane at 200 m. Treat the band and the plane as point sources. Is this assumption reasonable?

Solution

To have the same loudness, the soundwave intensities should be equal, i.e., $I_{\text{band}}(r) = I_{\text{jet}}(200\text{ m})$. Regarded as isotropic point sources, use of Equation 16-9 gives $\bar{P}_{\text{band}}/r^2 = \bar{P}_{\text{jet}}/(200\text{ m})^2$. The average power of each source can be found from Table 16-1 and a second application of Equation 16-9, $\bar{P}_{\text{band}} = 4\pi(4\text{ m})^2(1\text{ W/m}^2)$ and $\bar{P}_{\text{jet}} = 4\pi(50\text{ m})^2(10\text{ W/m}^2)$. Then $r^2 = (\bar{P}_{\text{band}}/\bar{P}_{\text{jet}})(200\text{ m})^2 = (200\text{ m})^2(4\text{ m})^2\times (1\text{ W/m}^2)/(50\text{ m})^2(10\text{ W/m}^2)$, or $r = 5.06$ m. The size of a rock band is several meters, nearly equal to this distance, so a point source is not a good approximation. Besides, the acoustical output of a rock band usually emanates from an array of speakers, which is not point-like. Moreover, the size of a jet plane is also not very small compared to 50 m.

Section 16-6: The Superposition Principle and Wave Interference

Problem

47. Two wave pulses are described by

$$y_1(x,t) = \frac{2}{(x-t)^2 + 1}, \quad y_2(x,t) = \frac{-2}{(x-5+t)^2 + 1},$$

where x and y are in cm and t in seconds.
(a) What is the amplitude of each pulse?
(b) At $t = 0$, where is the peak of each pulse, and in what direction is it moving? (c) At what time will the two pulses exactly cancel?

Solution

(a) The absolute value of the maximum displacement for each pulse is 2 cm, a value attained when the denominators are minimal ($x - t = 0$ for the first pulse and $x - 5 + t = 0$ for the second). (b) At $t = 0$, the peak of the first pulse is at $x = 0$ moving in the positive x direction. ($x - t = 0$ represents the peak, so if t increases so does x. This is why a wave traveling in the positive x direction is represented by a function of $x - vt$.) For the second pulse, the peak is at $x = 5$, moving in the negative x direction, when $t = 0$ ($x - 5 + t = 0$ implies $x = 5 - t$ and $dx/dt = -1 < 0$). (c) $y_1(x, t) + y_2(x, t) = 0$ for all values of x implies $(x - t)^2 = (x - 5 + t)^2$. This is true for all x, only if $(x - t) = +(x - 5 + t)$ or at $t = \frac{5}{2} = 2.5$ s. (The other root, $(x - t) = -(x - 5 + t)$, shows that $x = 2.5$ cm is always a node, i.e., the net displacement there is zero at all times.)

Problem

49. You're in an airplane whose two engines are running at 560 rpm and 570 rpm. How often do you hear the sound intensity increase as a result of wave interference?

Solution

As mentioned in the text, pilots of twin-engine airplanes use the beat frequency to synchronize the rpm's of their engines. The beat frequency is simply the difference of the two interfering frequencies, $f_{\text{beat}} = (570 - 560)/60$ s $= \frac{1}{6}$ s^{-1}, so you would hear one beat every six seconds.

Problem

51. What is the wavelength of the ocean waves in Example 16-5 if the calm water you encounter at 33 m is the *second* calm region on your voyage from the center line?

Solution

The second node occurs when the path difference is three half-wavelengths, or $AP - BP \equiv \Delta r = \frac{3}{2}\lambda_2$. (A phase difference of $k_2\Delta r = (2\pi/\lambda_2)\Delta r = 3\pi$, or an odd multiple of $\pi = 180°$ in general, insures complete destructive interference.) From Example 16-5, $2 \Delta r = 16.0$ m, so $\lambda_2 = 2 \Delta r/3 = 5.34$ m.

Section 16-7:   The Wave Equation
Problem

53. The following equation arises in analyzing the behavior of shallow water:

$$\frac{\partial^2 y}{dx^2} - \frac{1}{gh}\frac{\partial^2 y}{dt^2} = 0,$$

where h is the equilibrium depth and y the displacement from equilibrium. Give an expression for the speed of waves in shallow water. (Here *shallow* means the water depth is much less than the wavelength.)

Solution

The equation given is in the standard form for the one-dimensional linear wave equation (Equation 16-12), so the wave speed is the reciprocal of the square root of the quantity multiplying $\partial^2 y/\partial t^2$. Thus $v = \sqrt{gh}$.

Paired Problems
Problem

55. A wave on a taut wire is described by the equation $y = 1.5\sin(0.10x - 560t)$, where x and y are in cm and t is in seconds. If the wire tension is 28 N, what are (a) the amplitude, (b) the wavelength, (c) the period, (d) the wave speed, and (e) the power carried by the wave?

Solution

The wave has the form of Equation 16-5, with a phase constant of $-\frac{\pi}{2} = -90°$, $y(x, t) = A\sin(kx - \omega t) = A\cos(kx - \omega t - \frac{\pi}{2})$. Comparison reveals that $k = 0.1$ cm^{-1}, $\omega = 560$ s^{-1}, and (a) $A = 1.5$ cm (b) $\lambda = 2\pi/k = 2\pi/(0.1$ cm$^{-1}) = 62.8$ cm (Equation 16-4). (c) $T = 2\pi/\omega = 2\pi/(560$ s$^{-1}) = 11.2$ ms (Equation 16-3). (d) $v = \omega/k = 56$ m/s (Equation 16-6). (e) $\bar{P} = \frac{1}{2}\mu\omega^2 A^2 v = \frac{1}{2}(\omega A)^2(F/v) = \frac{1}{2}(560$ s$^{-1}\times 0.015$ m$)^2(28$ N$)/(56$ m/s$) = 17.6$ W (Equation 16-8, and Equation 16-7 to eliminate μ).

Problem

57. A spring of mass m and spring constant k has an unstretched length ℓ_0. Find an expression for the speed of transverse waves on this spring when it has been stretched to a length ℓ.

Solution

The spring may be regarded as a stretched string with tension, $F = k(\ell - \ell_0)$, and linear mass density $\mu = m/\ell$. Equation 16-7 gives the speed of transverse waves as $v = \sqrt{k\ell(\ell - \ell_0)/m}$.

Problem

59. At a point 15 m from a source of spherical sound waves, you measure a sound intensity of 750 mW/m^2. How far do you need to walk, directly away from the source, until the intensity is 270 mW/m^2?

Solution

The intensity of spherical waves from a point source is given by Equation 16-9. At a distance $r_1, I_1 = \bar{P}/4\pi r_1^2$, while after increasing the radial distance by d, $I_2 = \bar{P}/4\pi(r_1 + d)^2$. Dividing and solving for d, one finds $d = r_1(\sqrt{I_1/I_2} - 1) = (15\text{ m})(\sqrt{(750/270)} - 1) = 10.0$ m.

Problem

61. Two motors in a factory produce sound waves with the same frequency as their rotation rates. If one motor is running at 3600 rpm and the other at 3602 rpm, how often will workers hear a peak in the sound intensity?

Solution

The beat frequency equals the difference in the motors' rpm's, so the period of the beats is $T_{\text{beat}} = 1/f_{\text{beat}} = 1/(3602 - 3600)\text{ min}^{-1} = 30$ s. (See also Problem 49.)

Supplementary Problems

Problem

63. For a transverse wave on a stretched string, the requirement that the string be nearly horizontal is met if the amplitude is much less than the wavelength. (a) Show this by drawing an appropriate sketch. (b) Show that, under this approximation that $A \ll \lambda$, the maximum speed u of the string must be considerably less than the wave speed v. (c) If the amplitude is not to exceed 1% of the wavelength, how large can the string speed u be in relation to the wave speed v?

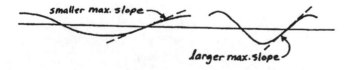

Problem 63 Solution.

Solution

(a) The relative "flatness" or "peakedness" of a sinusoidal waveform is determined by its maximum slope, $|dy/dx|_{\text{max}} = |\partial/\partial x[A\cos(kx - \omega t)]|_{\text{max}} =$

$kA = 2\pi(A/\lambda)$. If $A \ll \lambda$ (or $kA \ll 1$), the slope is nearly horizontal. (b) In terms of the speeds, $kA = \omega A/v = u_{\text{max}}/v$, so the string is nearly flat if $u_{\text{max}} \ll v$. (c) If $A/\lambda < 1\%$, then $u_{\text{max}}/v = 2\pi(A/\lambda) < 2\pi(1\%) = 6.3\%$.

Problem

65. An ideal spring is compressed until its total length is ℓ_1, and the speed of transverse waves on the spring is measured. When it's compressed further to a total length ℓ_2, waves propagate at the *same* speed. Show that the uncompressed spring length is just $\ell_1 + \ell_2$.

Solution

The tension in a compressed spring has magnitude $k(\ell_0 - \ell)$ while its linear mass-density is $\mu = m/\ell$. Therefore, the speed of transverse waves is $v = \sqrt{F/\mu} = \sqrt{k\ell(\ell_0 - \ell)/m}$ (as in Problem 57 for a stretched spring). If $v_1 = v_2$ for two different compressed lengths, then $\ell_1(\ell_0 - \ell_1) = \ell_2(\ell_0 - \ell_2)$ or $(\ell_1 - \ell_2)\ell_0 = \ell_1^2 - \ell_2^2 = (\ell_1 - \ell_2)(\ell_1 + \ell_2)$. Since $\ell_1 \neq \ell_2$, division by $\ell_1 - \ell_2$ gives $\ell_0 = \ell_1 + \ell_2$.

Problem

67. A 1-megaton nuclear explosion produces a shock wave whose amplitude, measured as excess air pressure above normal atmospheric pressure, is 1.4×10^5 Pa (1 Pa $= 1$ N/m^2) at a distance of 1.3 km from the explosion. An excess pressure of 3.5×10^4 Pa will destroy a typical woodframe house. At what distance from the explosion will such houses be destroyed? Assume the wavefront is spherical.

Solution

The intensity of a spherical wavefront varies inversely with the square of the distance from the central source (see Fig. 16-18b). In general, the intensity is proportional to the amplitude squared, so $A \sim 1/r$ for a spherical wave. (This can be proved rigorously by solution of the spherical wave equation, a generalization of Equation 16-12.) Therefore $A_1/A_2 = r_2/r_1$, or the overpressure reaches the stated limit at a distance $r_2 = (1.4\times10^5\text{ Pa}/3.5\times10^4\text{ Pa})(1.3\text{ km}) = 5.2$ km from the explosion.

Problem

69. In Example 16-5, how much farther would you have to row to reach a region of maximum wave amplitude?

Solution

In general, the interference condition for waves in the geometry of Example 16-5 is $AP - BP = n\lambda/2$, where

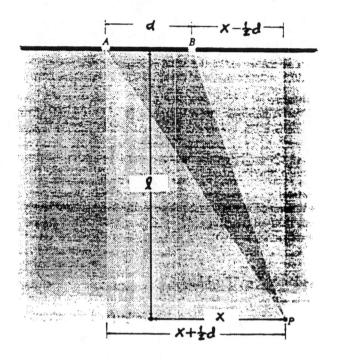

FIGURE 16-36 Problem 69 Solution.

n is an odd integer for destructive interference (a node) and n is an even integer for constructive interference (a maximum amplitude). (In Example 16-5, $n = 1$ gave the first node and in Problem 51, $n = 3$ gave the second node.) If $d = 20$ m is the distance between the openings, $\ell = 75$ m is the perpendicular distance from the breakwater, and x is the distance parallel to the breakwater measured from the midpoint of the openings, the interference condition is

$$\sqrt{\ell^2 + (x + \tfrac{1}{2}d)^2} - \sqrt{\ell^2 + (x - \tfrac{1}{2}d)^2} = n\lambda/2$$ (see

Fig. 16-36). In this problem, we wish to find x for the first maximum, $n = 2$, and the wavelength calculated in Example 16-5, $\lambda = 16.01$ m. Solving for x, we find:

$$
\begin{aligned}
x^2 &= \frac{[\ell^2 + (\tfrac{1}{2}d)^2 - (\tfrac{1}{4}n\lambda)^2]}{(2d/n\lambda)^2 - 1} \\
&= \frac{[(75)^2 + (10)^2 - (8.005)^2]\ \mathrm{m}^2}{(40/32.02)^2 - 1} = (100.5\ \mathrm{m})^2.
\end{aligned}
$$

This is 100.5 m $- 33$ m $= 67.5$ m farther than the first node in Example 16-5. (Note: We rounded off to three figures; if you round off to two figures, the answer is 67 m. Also, if $x = 33$ m is substituted into the general interference condition, one can recapture the wavelengths of the first and second nodes, for $n = 1$ and 3, calculated in Example 16-5 and Problem 51, respectively.)

CHAPTER 17   SOUND AND OTHER WAVE PHENOMENA

ActivPhysics can help with these problems:
Activities 10.3, 10.4, 10.5, 10.6, 10.8, 10.9

Sections 17-1 and 17-2:   Sound Waves and the Speed of Sound in Gases

Problem

1. Show that the quantity $\sqrt{P/\rho}$ has the units of speed.

Solution

The units of pressure (force per unit area) divided by density (mass per unit volume) are $(\text{N/m}^2)/(\text{kg/m}^3) = (\text{N/kg})(\text{m}^3/\text{m}^2) = (\text{m/s}^2)\text{m} = (\text{m/s})^2$, or those of speed squared.

Problem

3. Find the wavelength, period, angular frequency, and wave number of a 1.0-kHz sound wave in air under the conditions of Example 17-1.

Solution

The value of the speed of sound in air from Example 17-1 was 343 m/s. Therefore (a) $\lambda = v/f = (343 \text{ m/s})/(1 \text{ kHz}) = 34.3 \text{ cm}$; (b) $T = 1/f = 1 \text{ ms}$; (c) $\omega = 2\pi f = 6.28\times10^3 \text{ s}^{-1}$; and (d) $k = \omega/v = 2\pi/\lambda = 18.3 \text{ m}^{-1}$. (See Equations 16-1, 3, 4, and 6.)

Problem

5. Timers in sprint races start their watches when they see smoke from the starting gun, not when they hear the sound (Fig. 17-25). Why? How much error would be introduced by timing a 100-m race from the sound of the shot?

Solution

The sound of the starting gun takes $(100 \text{ m}) \div (340 \text{ m/s}) = 0.294 \text{ s}$ to reach the finish line. An error of this magnitude is significant in short races, where world records are measured in hundredths of a second. (This problem is almost the same as Problem 2, Chapter 2.)

Problem

7. At standard atmospheric pressure ($1.0\times10^5 \text{ N/m}^2$), what density of air would make the sound speed 1.0 km/s?

Solution

Solving for the density in Equation 17-1, we find: $\rho = \gamma P/v^2 = 1.4(1.0\times10^5 \text{ N/m}^2)/(10^3 \text{ m/s})^2 = 0.14 \text{ kg/m}^3$.

Problem

9. A gas with density 1.0 kg/m^3 and pressure $8.0\times10^4 \text{ N/m}^2$ has sound speed 365 m/s. Are the gas molecules monatomic or diatomic?

Solution

Solving for γ in Equation 17-1, we find $\gamma = \rho v^2/P = (1.0 \text{ kg/m}^3)(365 \text{ m/s})^2/(8.0\times10^4 \text{ N/m}^2) = 1.67$, very close to the value for an ideal monatomic gas. (Actually, $\gamma - 5/3 = -1.35\times10^{-3}$ for this gas.)

Problem

11. Saturn's moon Titan has one of the solar system's thickest atmospheres. Near Titan's surface, atmospheric pressure is 50% greater than standard atmospheric pressure on Earth, while the density in molecules per unit volume is one-third that of Earth's atmosphere. If Titan's atmosphere is essentially all nitrogen (N$_2$), what is the sound speed?

Solution

The data and method used in Example 17-1, combined with the numbers given for Titan's atmosphere, yield

$$v = \left[\frac{(1.4)(1.5\times1.01\times10^5 \text{ N/m}^2)}{(28 \text{ u})(1.66\times10^{-27} \text{ kg/u})(\frac{1}{3}\times2.51\times10^{25} \text{ m}^{-3})}\right]^{1/2} = 739 \text{ m/s}.$$

Problem

13. You see an airplane straight overhead at an altitude of 5.2 km. Sound from the plane, however, seems to be coming from a point back along the plane's path at a 35° angle to the

vertical (Fig. 17-26). What is the plane's speed, assuming an average 330 m/s sound speed?

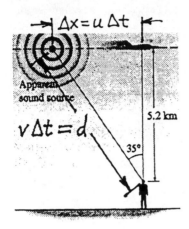

FIGURE 17-26 Problem 13.

Solution

The travel time of the sound from the airplane, reaching you along a line making an angle of 35° with the vertical (from the apparent sound source), is $\Delta t = d/v$. During this time, the airplane moved a horizontal distance $\Delta x = d\sin 35°$, so its speed is $u = \Delta x/\Delta t = d\sin 35°/(d/v) = v\sin 35° = (330 \text{ m/s})\sin 35° = 189$ m/s. (The airplane's altitude, 5.2 km $= d\cos 35°$, was not needed in this calculation.)

Section 17-3:   Sound Intensity

Problem

15. Sound intensity in normal conversation is about 1 μW/m^2. What is the displacement amplitude of air in a 2.5-kHz sound wave with this intensity?

Solution

As in Example 17-2, Equation 17-3c, combined with the atmospheric data in Example 17-1, can be used to calculate the displacement amplitude for sound waves of the specified frequency and intensity:

$$s_0 = \sqrt{2\bar{I}/\rho\omega^2 v} = \left(\frac{1}{2\pi \times 2.5 \times 10^3 \text{ Hz}}\right)$$
$$\times \sqrt{\frac{2(10^{-6} \text{ W/m}^2)}{(1.20 \text{ kg/m}^3)(343 \text{ m/s})}} = 4.44 \text{ nm}.$$

Problem

17. A speaker produces 440-Hz sound with total power 1.2 W, radiating equally in all directions. At a distance of 5.0 m, what are (a) the average intensity, (b) the decibel level, (c) the pressure amplitude, and (d) the displacement amplitude?

Solution

(a) Equation 16-9 gives the average intensity at a given distance from an isotropic point source, $\bar{I} = \bar{P}/4\pi r^2 = (1.2 \text{ W})/4\pi(5 \text{ m})^2 = 3.82$ mW/m^2. (b) From Equation 17-4, this corresponds to a sound level intensity of $\beta = (10 \text{ dB})\log(I/I_0) = (10 \text{ dB}) \times \log(3.82\times10^{-3}/10^{-12}) = 95.8$ dB $\simeq 96$ dB. (c) The pressure amplitude, for "normal air" of Example 17-1, follows from Equation 17-3b, $\Delta P_0 = \sqrt{2\rho v\bar{I}} = \sqrt{2(1.20 \text{ kg/m}^3)(343 \text{ m/s})(3.82\times10^{-3} \text{ W/m}^2)} = 1.77$ N/m^2. (d) The corresponding displacement amplitude is $s_0 = \Delta P_0/\rho\omega v = (1.77 \text{ Pa})/(1.20 \text{ kg/m}^2)(343 \text{ m/s})(2\pi\times440 \text{ Hz}) = 1.56$ μm.

Problem

19. What is the approximate frequency range over which sound with intensity 10^{-12} W/m^2 can be heard? Consult Fig. 17-3.

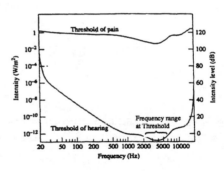

FIGURE 17-3 For reference.

Solution

Inspection of Fig. 17-3 shows that for frequencies approximately between 1 and 6.5 kHz, the threshold of hearing is at or below 10^{-12} W/m^2.

Problem

21. What are the intensity and pressure amplitudes in sound waves with intensity levels of (a) 65 dB and (b) −5 dB?

Solution

The exponentiation of Equation 17-4, to the base ten, relates the intensity to the decibel level, $I/I_0 = 10^{\beta/10 \text{ dB}}$ while Equation 17-3b gives the pressure amplitude, $\Delta P_0 = \sqrt{2\rho v I}$. Here, I is the average intensity, I_0 the threshold level, and we use values of ρ and v for air under the "normal" conditions in Example 17-1. (a) For $\beta = 65$ dB, $I = (10^{-12} \text{ W/m}^2)10^{6.5} = 3.16\times10^{-6}$ W/m^2, and

$\Delta P_0 = \sqrt{2(1.2 \text{ kg/m}^3)(343 \text{ m/s})(3.16 \times 10^{-6} \text{ W/m}^2)} = 5.10 \times 10^{-2} \text{ N/m}^2$. (b) For $\beta = -5$ dB, $I = I_0 10^{-5/10} = 3.16 \times 10^{-13} \text{ W/m}^2$ and $\Delta P_0 = 1.61 \times 10^{-5} \text{ N/m}^2$.

Problem

23. (a) What is the decibel level of a sound wave whose pressure amplitude is 2.9×10^{-4} N/m²? (b) Consult Fig. 17-3 to determine the approximate lowest frequency at which this sound would be audible.

Solution

(a) For a sound wave in air under "normal conditions" (see Example 17-1 for values of ρ and v), the average intensity and pressure amplitude are related by Equation 17-3b: $I = \frac{1}{2}\Delta P_0^2/\rho v = (2.9 \times 10^{-4} \text{ N/m}^2)^2 \div 2(1.20 \text{ kg/m}^3)(343 \text{ m/s}) = 1.02 \times 10^{-10} \text{ W/m}^2$. Compared to the "normal" threshold for human hearing at 1 kHz, $I_0 = 10^{-12} \text{ W/m}^2$, the sound corresponds to a decibel level (Equation 17-4) of $\beta = (10 \text{ dB})\log(102) = 20.1$ dB. (b) Fig. 17-3 shows approximately $f = 250$ Hz as the lowest audible frequency for this intensity.

Problem

25. Show that a doubling of sound intensity corresponds to very nearly a 3 dB increase in the decibel level.

Solution

If the sound intensity is doubled, $I' = 2I$, Equation 17-4 shows that $\beta' = (10 \text{ dB})\log(I'/I_0) = (10 \text{ dB})\log(2I/I_0) = (10 \text{ dB})\log(I/I_0) + (10 \text{ dB})\log 2 = \beta + 3.01$ dB, or the decibel level increases by about 3 dB.

Problem

27. At a distance 2.0 m from a localized sound source you measure the intensity level as 75 dB. How far away must you be for the perceived loudness to drop in half (i.e., to an intensity level of 65 dB)?

Solution

A change of -10 dB, (i.e., $\beta' - \beta = -10$ dB) corresponds to the intensity decreasing by a factor of one tenth (i.e., $I' = I/10$). To see this, note that Equation 17-4 may be written as $\beta' - \beta = (-10 \text{ dB})\log(I'/I)$, or $I'/I = 10^{(\beta'-\beta)/10 \text{ dB}}$. For an isotropic point source of sound, the intensity falls inversely with the square of the distance (Equation 16-9), so $r'^2 = 10r^2$, or $r' = \sqrt{10}(2 \text{ m}) = 6.32$ m.

Problem

29. Sound intensity from a certain extended source drops as $1/r^n$, where r is the distance from the source. If the intensity level drops by 3 dB every time the distance is doubled, what is n?

Solution

A 3 dB drop corresponds to a drop in intensity by a factor of one half (see Problem 25). Thus, if $I' = \frac{1}{2}I$ when $r' = 2r$, $I'/I = \frac{1}{2} = (1/2r)^n/(1/r)^n = 1/2^n$ implies $n = 1$.

Section 17-4: Sound Waves in Liquids and Solids

Problem

31. The bulk modulus for tungsten is 2.0×10^{11} N/m², and its density is 1.94×10^4 kg/m³. Find the sound speed in tungsten.

Solution

Substitution of the given values of bulk modulus and density for tungsten into Equation 17-6 gives $v = \sqrt{B/\rho} = \sqrt{(2.0 \times 10^{11} \text{ N/m}^2)/(1.94 \times 10^4 \text{ kg/m}^3)} = 3.21$ km/s.

Problem

33. The speed of sound in body tissues is essentially the same as in water. Find the wavelength of 2.0 MHz ultrasound used in medical diagnostics.

Solution

From Table 17-2, the speed of sound in water is 1497 m/s, so Equation 16-1 gives the wavelength of 2.0 MHz ultrasound as $\lambda = v/f = (1497 \text{ m/s}) \div (2.0 \text{ MHz}) = 0.749$ mm.

Problem

35. Mechanical vibration induces a sound wave in a mechanism consisting of a 12-cm-long steel rod attached to a 3.0-cm-long neoprene block. How long does it take the wave to propagate through this structure?

Solution

The travel time through each material is its length divided by the speed of sound. With reference to Table 17-2 for the speed of sound in steel and neoprene, the total travel time is (0.12 m ÷ 5940 m/s) + (0.03 m/1600 m/s) = $(20.2 + 18.8)\mu$s = 39.0 μs.

Problem

37. The bulk modulus of the steel whose sound speed is given in Table 17-2 is 1.6×10^{11} N/m². A 0.50-mm-diameter wire made from this steel can withstand a tension force of 50 N before it deforms permanently. Is it possible to put this wire under enough tension that the speed of transverse waves on the wire is the same as the sound speed in the wire? Answer by calculating the tension required.

Solution

If the speed of transverse waves (Equation 16-7) equals the speed of sound (Equation 17-6), then $F/\mu = B/\rho$. The ratio of the linear and volume densities of the wire is its cross-sectional area (see solution to Chapter 16, Problem 27), so the required tension would be $F = B(\mu/\rho) = B(\frac{1}{4}\pi d^2) =$ $(1.6 \times 10^{11}$ N/m²$)\frac{1}{4}\pi(0.50$ mm$)^2 = 31.4$ kN, far greater than the 50 N elastic limit.

Section 17-6: Standing Waves

Problem

39. When a stretched string is clamped at both ends, its fundamental standing-wave frequency is 140 Hz. (a) What is the next higher frequency? (b) If the same string, with the same tension, is now clamped at one end and free at the other, what is the fundamental frequency? (c) What is the next higher frequency in case (b)?

Solution

(a) The frequencies of the standing-wave modes of a string fixed at both ends are all the (positive) integer multiples of the fundamental frequency, $f_m = mf_1$, for $m = 1, 2, \ldots$. (This follows from Equation 17-8 if we use $f_m = v/\lambda_m = m(v/2L) = m(v/\lambda_1) = mf_1$.) Thus, $f_2 = 2f_1 = 2(140$ Hz$) = 280$ Hz. (b) The velocity of transverse waves is the same (for the same string under the same tension), but when one end is fixed and the other free, the standing-wave wavelengths are $\lambda_m = 4L, 4L/3, \ldots 4L/(2m-1), \ldots$, where $2m-1$ represents any odd integer for $m = 1, 2, \ldots$. (See Fig. 17-13 and its discussion in the text, or the solution to Problem 41.) Therefore, the fundamental frequency for the string fixed at one end is $f_1 = v/4L = \frac{1}{2}(v/2L) = \frac{1}{2}(140$ Hz$) = 70$ Hz, i.e., one half the fundamental frequency of the string fixed at both ends. (c) In this case, the standing-wave frequencies are only the odd multiples of the fundamental frequency, $f_m = (2m-1)f_1$, therefore the second standing-wave mode has frequency $f_2 = 3f_1 = 3(70$ Hz$) =$ 210 Hz for this string.

Problem

41. Show that only odd harmonics are allowed on a taut string with one end tight and the other free.

Solution

For a string free at one end, the amplitude factor in Equation 17-7 is a maximum for $x = L$, i.e., $2A\sin kL = \pm 2A$. Therefore, $kL = (2m-1)\pi/2$, where $2m-1$ is an odd integer for $m = 1, 2, \ldots$. In terms of standing-wave wavelengths, $kL = (2\pi/\lambda_m)L = (2m-1)\pi/2$, or $L = (2m-1)\lambda_m/4$, as stated on page 428. In terms of frequency, $f_m = v/\lambda_m = (2m-1)f_1$, where $f_1 = v/4L$ is the frequency of the fundamental. Thus, only odd harmonics occur.

Problem

43. Show that the standing-wave condition of Equation 17-8 is equivalent to the requirement that the time it takes a wave to make a round trip from one end of the medium to the other and back be an integer multiple of the wave period.

Solution

The round-trip time for waves on a string of length L, clamped at both ends, is $2L/v = 2L/(\lambda/T) = 2LT/(2L/m) = mT$ (a multiple of the wave period), where we used Equations 16-1 and 17-8.

Problem

45. "Vibrato" in a violin is produced by sliding the finger back and forth along the vibrating string. The G-string on a particular violin measures 30 cm between the bridge and its far end and is clamped rigidly at both points. Its fundamental frequency is 197 Hz. (a) How far from the end should the violinist place a finger so that the G-string plays the note A (440 Hz)? (b) If the violinist executes vibrato by moving the finger 0.50 cm to either side of the position in part (a), what range of frequencies results?

Solution

(a) The fundamental frequency of a string fixed at both ends is $f = v/2L$. Since fingering does not change the tension (and hence v) in a violin string appreciably, $f'/f = L/L'$, or $L' = (197$ Hz/440 Hz$)(30$ cm$) =$ 13.4 cm. This is the sounding length of the string, so the finger must be placed a distance $(30 - 13.4)$ cm = 16.6 cm from the ("nut") end. (b) Alteration of L' by ± 0.5 cm yields frequencies between:

$$f'' = (440 \text{ Hz})(13.4)/(13.4 \pm 0.5) = 424 \text{ to } 457 \text{ Hz}.$$

Problem

47. A bathtub 1.7 m long contains 13 cm of water. By sloshing water back and forth with your hand, you can build up large-amplitude oscillations. Determine the lowest frequency possible for such a resonant oscillation, using the fact that the speed of waves in shallow water of depth h is $v = \sqrt{gh}$. *Hint:* At resonance in this case, the wave has a crest at one end and a trough at the other.

Solution

The distance between a crest and a trough is an odd number of half-wavelengths, so the maximum resonant wavelength of the tub is $L = \frac{1}{2}\lambda$ or $\lambda = 2L = 2(1.7\text{ m}) = 3.4\text{ m}$. Thus, the minimum resonant frequency is $f = v/\lambda = \sqrt{gh}/\lambda = \sqrt{(9.8\text{ m/s}^2)(0.13\text{ m})}/3.4\text{ m} = 0.332$ Hz. (The period corresponding to this is $T = 1/f = 3.01$ s.)

Problem

49. What would be the fundamental frequency of the double bassoon of Example 17-4, if it were played in helium under conditions of Example 17-1?

Solution

The wavelength of the fundamental mode depends on the dimensions of the instrument, so the difference in fundamental frequency, for the bassoon played in helium versus air, is due to the change in the velocity of sound only, $f = v/\lambda$. Thus, if the speed of sound in helium from Example 17-1 is used in place of that in air in Example 17-4, one finds $f = (1000\text{ m/s}) \div (11\text{ m}) = 90.9$ Hz.

Problem

51. An astronaut smuggles a double bassoon (Example 17-4) to Mars and plays the instrument's fundamental note. If it sounds at 23 Hz, what is the sound speed on Mars?

Solution

Since the wavelength depends only on the dimensions of the bassoon, $v = f\lambda = (23\text{ Hz})(11\text{ m}) = 253$ m/s. (See Example 17-4 for the fundamental wavelength.)

Section 17-7: The Doppler Effect

Problem

53. A car horn emits 380-Hz sound. If the car moves at 17 m/s with its horn blasting, what frequency will a person standing in front of the car hear?

Solution

From Equation 17-10, with the minus sign in the denominator (car approaching the observer in front), $f' = f/(1 - u/v) = (380\text{ Hz})(1 - 17/343)^{-1} = 400$ Hz.

Problem

55. A fire truck's siren at rest wails at 1400 Hz; standing by the roadside as the truck approaches, you hear it at 1600 Hz. How fast is the truck going?

Solution

One can solve Equation 17-10 for u (with the minus sign appropriate to an approaching source) with the result: $u = v(1 - f/f') = (343\text{ m/s})(1 - 1400/1600) = 42.9\text{ m/s} = 154$ km/h. (We used the speed of sound in air from Example 17-1.)

Problem

57. The dominant frequency emitted by an airplane's engines is 1400 Hz. (a) What frequency will you measure if the plane approaches you at half the sound speed? (b) What frequency will you measure if the plane recedes at half the sound speed?

Solution

(a) For a 1400 Hz sound source approaching you at half the speed of sound, $u = \frac{1}{2}v$, Equation 17-10 gives a Doppler shifted observed frequency of $f' = f/(1 - u/v) = (1400\text{ Hz})/(1 - \frac{1}{2}) = 2800$ Hz. (b) The same equation for a receding source gives $f' = (1400\text{ Hz})/(1 + \frac{1}{2}) = 933$ Hz.

Problem

59. You're standing by the roadside as a truck approaches, and you measure the dominant frequency in the truck noise at 1100 Hz. As the truck passes the frequency drops to 950 Hz. What is the truck's speed?

Solution

The result of part (a) of the preceding problem gives $u/v = (1100 - 950)/(1100 + 950) = 0.0732$. For sound waves in "normal" air (Example 17-1), this implies a truck speed of $u = 0.0732(343\text{ m/s}) = 25.1\text{ m/s} = 90.4$ km/h. (From Equation 17-10, the frequency, emitted by the truck is $f = f_1(1 - u/v) = f_2(1 + u/v)$, where f_1 and f_2 are the observed frequencies when the truck is approaching or receding, respectively. The solution of this equation for the source's speed is $u/v = (f_1 - f_2)/(f_1 + f_2)$.)

Problem

61. Use the binomial approximation to show that Equations 17-10 and 17-11 give the same result in the limit $u \ll v$.

Solution

The binomial expansion for $(1 \pm u/v)^{-1}$ is $1 \mp u/v + \cdots$, so Equations 17-10 and 11 are the same, for small u/v. (Note the difference in sign convention for u in these two equations.)

Section 17-8:   Shock Waves

Problem

63. Figure 17-28 shows a projectile in supersonic flight, with shock waves clearly visible. By making appropriate measurements, determine the projectile's speed as compared with the sound speed.

Solution

The half-angle of the shock wave in Fig. 17-28, measured with a protractor, is about 45°, so (with reference to Fig. 17-23b) $u/v = 1/\sin 45° = 1.41$.

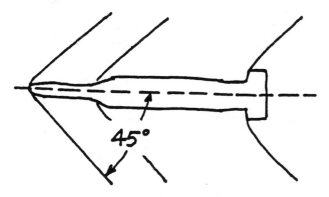

FIGURE 17-28 Problem 63 Solution.

Paired Problems

Problem

65. A 1.0-W sound source emits uniformly in all directions. Find (a) the intensity and (b) the decibel level 12 m from the source.

Solution

(a) The average intensity at any distance from an isotropic sound source is given by Equation 16-9, $\bar{I} = \bar{P}/4\pi r^2 = (1 \text{ W})/4\pi(12 \text{ m})^2 = 5.53 \times 10^{-4} \text{ W/m}^2$. (b) The corresponding decibel level (Equation 17-4) is $(10 \text{ dB})\log(5.53 \times 10^{-4}/10^{-12}) = 87.4 \text{ dB}$.

Problem

67. A pipe 80 cm long is open at both ends. When the pipe is immersed in a gas mixture, the frequency of a certain harmonic is 280 Hz and the next higher harmonic is 350 Hz. Determine (a) the sound speed and (b) the mode numbers of the two harmonics.

Solution

For an open pipe, the frequencies of the harmonics are integer multiples of the fundamental, $f_m = mf_1$, for $m = 1, 2, \ldots$ (b) It is given that $f_m/f_{m+1} = m/(m+1) = 280/350 = 4/5$, therefore $m = 4$ and $m + 1 = 5$ are the mode numbers. (a) Since the wavelengths of the harmonics are given by Equation 17-8, $\lambda_4 = 2L/m = 2(80 \text{ cm})/4 = 40 \text{ cm}$ and $\lambda_5 = 32 \text{ cm}$. From either, $v = f_4\lambda_4 = (280 \text{ Hz})(0.4 \text{ m}) = f_5\lambda_5 = (350 \text{ Hz})(0.32 \text{ m}) = 112 \text{ m/s}$.

Problem

69. Find the wave speed in a medium where a 28 m/s source speed causes a 3% increase in frequency measured by a stationary observer.

Solution

To cause an increase in frequency, the source must be approaching the stationary observer. Solving Equation 17-10 for the wave speed, we find $v = u/(1 - f/f')$. The given increase is 3%, so $f/f' = 1/1.03$ and $v = (28 \text{ m/s})/(1 - 1/1.03) = 961 \text{ m/s}$.

Supplementary Problems

Problem

71. The sound speed in air at 0°C is 331 m/s, and for temperatures within a few tens of degrees of 0°C it increases at the rate 0.590 m/s for every °C increase in temperature. How long would it take a sound wave to travel 150 m over a path where the temperature rises linearly from 5°C at one end to 15°C at the other end?

Solution

The speed of sound is given as a function of the Celsius temperature, $v(T) = 331 \text{ m/s} + (0.590 \text{ m/s/°C})T \equiv v_0 + aT$. The temperature itself varies linearly along the path, from $x_1 = 0$ to $x_2 = 150$ m, between the given values $T_1 = 5°C$ and $T_2 = 15°C$, so $T(x) = T_1 + (T_2 - T_1)(x/150 \text{ m}) = 5°C + (1°C/15 \text{ m})x \equiv T_1 + bx$. Thus, the speed, as a function of position along the path, is $v(x) = v(T(x)) = v_0 + aT(x) = v_0 + a(T_1 + bx) = v_0 + aT_1 + abx = dx/dt$. The time for a sound wave to travel an interval dx of path is $dt = dx/v(x)$, so the total time to travel the entire path is

$$t = \int_1^2 dt = \int_{x_1}^{x_2} \frac{dx}{v(x)} = \int_0^{x_2} \frac{dx}{v_0 + aT_1 + abx}$$

$$= \frac{1}{ab} \ln\left(\frac{v_0 + aT_2}{v_0 + aT_1}\right)$$

$$= \left(0.590 \frac{\text{m}}{\text{s} \cdot {}^\circ\text{C}}\right)^{-1} \left(\frac{1{}^\circ\text{C}}{15\ \text{m}}\right)^{-1} \ln\left(\frac{331 + 0.59 \times 15}{331 + 0.59 \times 5}\right)$$

$$= 0.445\ \text{s}.$$

(Note: $v_2 = v(x_2) = v_0 + aT_1 + abx_2 = v_0 + aT_2$, and $v_1 = v_0 + aT_1$.)

Problem

73. A rectangular trough is 2.5 m long and is much deeper than its length, so Equation 16-11 applies. Determine the wavelength and frequency of (a) the longest and (b) the next longest standing waves possible in this trough. Why isn't the higher frequency twice the lower?

Solution

Since the volume of water in the trough is constant, there must be as many "hills" as there are "valleys" in the standing-wave patterns of sinusoidal surface waves. Therefore, there are antinodes at each end, as shown. Since the distance between two antinodes is a multiple of half-wavelengths, $L = m\lambda/2$, so the two longest standing-wave wavelengths are $\lambda_1 = 2L = 5$ m and

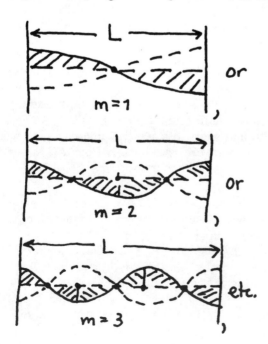

Problem 73 Solution.

$\lambda_2 = L = 2.5$ m. If we use Equation 16-11 for the speed of deep water surface waves, the corresponding standing-wave frequencies are $f_1 = v/\lambda_1 = \sqrt{g/2\pi\lambda_1} = \sqrt{(9.8\ \text{m/s}^2)/2\pi \times 5\ \text{m}} = 0.559$ Hz, and $f_2 = \sqrt{2} f_1 = 0.790$ Hz. These are not multiples of one another because of the way the wave speed depends on the wavelength, i.e., the dispersion relation for these waves is *not* $\omega = (\text{constant})k$.

Problem

75. A supersonic airplane flies directly over you at 6.5 km altitude. You hear its sonic boom 13 s later. What is the plane's Mach number?

Solution

The shock wave trails the airplane (P) with half angle $\theta = \sin^{-1}(v/u)$ as shown (see Fig. 17-23b also). The distance of the closest point on the wavefront (Q) from the observer (O) when the airplane is overhead is $OQ = h \cos\theta$. Since the wavefront travels with the speed of sound, v, the sonic boom reaches O after a time t, given by $OQ = vt = h \cos\theta$. Therefore, we can eliminate θ (from $\sin\theta = v/u$) and solve for u (the plane's speed): $(u/v) = 1/\sin\theta = 1/\sqrt{1 - \cos^2\theta} = 1/\sqrt{1 - (vt/h)^2}$. Suppose $h = 6.5$ km, $t = 13$ s, and $v = 340$ m/s. Then $(u/v) = 1/\sqrt{1 - (340 \times 13/6500)^2} = 1.36$, which is the Mach number.

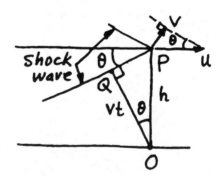

Problem 75 Solution.

Problem

77. Consider an object moving at speed u through a medium, and reflecting sound waves from a stationary source back toward the source. The object receives the waves at the shifted frequency given by Equation 17-11, and when it re-emits them they are shifted once again, this time according to Equation 17-10. Find an expression for the overall frequency shift that results, and show that, for $u \ll v$, this shift is approximately $2fu/v$.

Solution

The object receives waves at the frequency of an observer moving toward a stationary source, so $f' = f(1 + u/v)$. (See Equation 17-11.) The reflected waves are re-emitted by the moving object at this frequency, f', and so are received by the original stationary source at frequency $f'' = f'/(1 - u/v) = f(1 + u/v)/(1 - u/v)$. (See Equation 17-10.) The overall frequency shift is $f'' - f = \Delta f = f[(v + u) \times (v - u)^{-1} - 1] = 2uf/(v - u)$. If $u \ll v$, then $\Delta f \approx 2uf/v$. (Note: If the object is moving away from the stationary source, one replaces u with $-u$ in the above treatment.)

Problem

79. What is the frequency shift of a 70-GHz police radar signal when it reflects off a car moving at 120 km/h? (Radar waves travel at the speed of light.) *Hint:* See Problem 77.

Solution

Using the result of Problem 77 for $u \ll v = c$, one finds $\Delta f = 2uf/c = 2(120 \text{ m}/3.6 \text{ s})(7 \times 10^{10} \text{ Hz}) \div (3 \times 10^8 \text{ m/s}) = 15.6$ kHz. (Note that the formula for the Doppler shift for electromagnetic waves is different than for sound, but when $u \ll v$ or c, both formulas are the same.)

CHAPTER 18   FLUID MOTION

Problem

1. The density of molasses is 1600 kg/m^3. Find the mass of the molasses in a 0.75-liter jar.

Solution

The mass of molasses, which occupies a volume equal to the capacity of the jar, is $\Delta m = \rho \Delta V = (1600 \text{ kg/m}^3)(0.75 \times 10^{-3} \text{ m}^3) = 1.2$ kg.

Problem

3. The density of atomic nuclei is about 10^{17} kg/m^3, while the density of water is 10^3 kg/m^3. Roughly what fraction of the volume of water is *not* empty space?

Solution

The average density of a mixture of two substances, with definite volume fractions, is $\rho_{\text{av}} = \rho_1(V_1/V) + \rho_2(V_2/V)$, where $V_1 + V_2 = V$ is the total volume. (Try this formula in the preceding problem.) The density of water is approximately the average density ($\rho_{\text{av}} = 10^3$ kg/m^3) of the nuclei ($\rho_1 = 10^{17}$ kg/m^3) and empty space ($\rho_2 = 0$) provided we neglect the mass of the atomic electrons, so the volume fraction of nuclei in water is $(V_1/V) = \rho_{\text{av}}/\rho_1 = 10^3/10^{17} = 10^{-14}$.

Problem

5. A plant hangs from a 3.2-cm diameter suction cup affixed to a smooth horizontal surface (Fig. 18-42). What is the maximum weight that can be suspended (a) at sea level and (b) in Denver, where atmospheric pressure is about 0.80 atm?

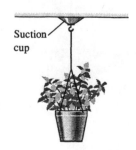

Suction cup

FIGURE 18-42 Problem 5.

Solution

(a) The force exerted on the suction cup by the atmosphere is $F = PA = P_{\text{atm}}(\pi d^2/4) = (1.013 \times 10^5 \text{ Pa})\pi(0.016 \text{ m})^2 = 81.5$ N (perfect vacuum inside cup assumed). This is equal to the maximum weight. (b) At Denver, $P = 0.8P_{\text{atm}}$, so the maximum weight is 80% of that in part (a), or 65.2 N (a slight variation in g with altitude is neglected).

Problem

7. Measurement of small pressure differences, for example, between the interior of a chimney and the ambient atmosphere, is often given in **inches of water**, where one inch of water is the pressure that will support a 1-in.-high water column. Express this unit in SI.

Solution

From Equation 18-2, the pressure of 1 in. of water is $\rho_{\text{H}_2\text{O}}g\,\Delta h = (10^3 \text{ kg/m}^3)(9.81 \text{ m/s}^2)(0.0254 \text{ m}) = 249$ Pa.

Problem

9. The fuselage of a 747 jumbo jet is roughly a cylinder 60 m long and 6 m in diameter. If the interior of the plane is pressurized to 0.75 atm, what is the net pressure force tending to separate half the cylinder from the other half when the plane is flying at 10 km, where air pressure is about 0.25 atm? (The earliest commercial jets suffered structural failure from just such forces; modern planes are better engineered.)

Solution

Consider the skin of the fuselage to be divided into infinitesimal strips, parallel to the cylinder's axis, of area dA (shown in cross-section in the sketch). Because of the pressure difference between the cabin interior and the outside, $\Delta P = (0.75 - 0.25)$ atm at 10 km altitude, there is a net force radially outward on dA of magnitude $dF = \Delta P\,dA$. These forces produce stresses in the skin, i.e., forces of one part of the cylinder on another part, that this problem asks us to estimate. The pressure force on one half of the cylinder is balanced by the stress force exerted by the other half. By symmetry, for every dA located at angle θ

shown, there is a dA' at angle $-\theta$ with opposite y component of pressure force, $dF_y + dF'_y = 0$, so only the x component of dF contributes to the net pressure force on the half cylinder. But $dF_x = \Delta P \, dA \cos\theta = \Delta P \, dA_y$, where dA_y is the projection of the area dA onto an axial plane parallel to the y axis, and the total projected area of the half-cylinder is just the diameter times the length, or $2RL$. Therefore, the net pressure force tending to separate two halves of the fuselage is $F_x = \Delta P(2RL) = (0.75 - 0.25)(101.3 \text{ kPa}) \times (6 \times 60 \text{ m}^2) = 1.82 \times 10^7 \text{ N} \approx 2050$ tons. Note: F_x can be expressed as a surface integral, which for the above area elements, $dA = RL \, d\theta$, reduces to

$$\int_{\text{half-cylinder}} dF_x = \int_{-\pi/2}^{\pi/2} \Delta P \cos\theta \, dA$$

$$= \Delta P \cdot LR \int_{-\pi/2}^{\pi/2} \cos\theta \, d\theta$$

$$= \Delta P \cdot LR \sin\theta \bigg|_{-\pi/2}^{\pi/2} = \Delta P \cdot 2RL.$$

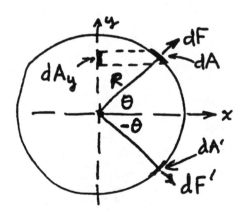

Problem 9 Solution.

Problem

11. A paper clip is made from wire 1.5 mm in diameter. You unbend a paper clip and push the end against the wall. What force must you exert to give a pressure of 120 atm?

Solution

An average pressure of 120 atm over the cross-sectional area of the wire, $\frac{1}{4}\pi d^2$, results in a force of $F = PA = (120 \times 101.3 \text{ kPa})\frac{1}{4}\pi(1.5 \times 10^{-3} \text{ m})^2 = 21.5 \text{ N}$.

Problem

13. When a couple with a total mass of 120 kg lies on a water bed, the pressure in the bed increases by

4700 Pa. What surface area of the two bodies is in contact with the bed?

Solution

The pressure increase times the average horizontal contact surface area equals the weight of the couple, or $A_{\text{av}} = mg/\Delta P = (120 \times 9.8 \text{ N})/(4700 \text{ Pa}) = 0.250 \text{ m}^2$.

Problem

15. The emergency escape window of a DC-9 jetliner measures 50 cm by 90 cm. The interior pressure is 0.75 atm, and the plane is at an altitude where atmospheric pressure is 0.25 atm. Is there any danger that a passenger could open the window? Answer by calculating the force needed to pull the window straight inward.

Solution

The pressure force on the (presumably flat) window is simply $F = \Delta P \cdot A = (0.75 - 0.25)(101.3 \text{ kPa}) \times (0.5 \times 0.9 \text{ m}^2) = 22.8 \text{ kN} = 2.56$ tons. It is unlikely any passenger is this strong.

Section 18-2:   Fluids at Rest: Hydrostatic Equilibrium

Problem

17. What is the density of a fluid whose pressure increases at the rate of 100 kPa for every 6.0 m of depth?

Solution

The increase in pressure with depth, for an incompressible fluid, is given by Equation 18-3. Thus, $\rho = \Delta P/gh = (100 \text{ kPa})/(9.8 \times 6 \text{ m}^2/\text{s}^2) = 1.70 \times 10^3 \text{ kg/m}^3$.

Problem

19. Scuba equipment provides the diver with air at the same pressure as the surrounding water. But at pressures greater than about 1 MPa, the nitrogen in air becomes dangerously narcotic. At what depth does nitrogen narcosis become a hazard?

Solution

In fresh water ($\rho \simeq 10^3 \text{ kg/m}^3$), the pressure is 1 MPa at a depth of $h = (P - P_0)/\rho g = (1 \text{ MPa} - 0.103 \text{ MPa})/(9.8 \times 10^3 \text{ N/m}^3) = 91.7 \text{ m}$, where P_0 is atmospheric pressure at the surface. (See Equation 18-3.) The depth is a little less in salt water since its density is slightly greater.

Problem

21. A vertical tube open at the top contains 5.0 cm of oil (density 0.82 g/cm^3) floating on 5.0 cm of water. Find the *gauge* pressure at the bottom of the tube.

Solution

The pressure at the top of the tube is atmospheric pressure, P_a. The absolute pressure at the interface of the oil and water is $P_i = P_a + \rho_{oil}gh_{oil}$, and at the bottom is $P = P_i + \rho_{water}gh_{water} = P_a + \rho_{oil}gh_{oil} + \rho_{water}gh_{water}$ (see Equation 18-3). Therefore, the gauge pressure at the bottom is $P - P_a = (\rho_{oil}h_{oil} + \rho_{water}h_{water})g = (0.82 + 1.00)(10^3 \text{ kg/m}^3)(0.05 \text{ m}) \times (9.8 \text{ m/s}^2) = 892$ Pa (gauge).

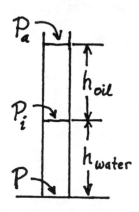

Problem 21 Solution.

Problem

23. A 1500-m-wide dam holds back a lake 95 m deep. What force does the water exert on the dam?

Solution

The pressure varies over the (assumed) vertical and rectangular surface of the dam in the same way as for the wall of the swimming pool in Example 18-2, so $F = P_a wH + \frac{1}{2}\rho gwH^2 = (1500 \text{ m})(95 \text{ m})[101.3 \text{ kPa} + \frac{1}{2}(9.8 \times 10^3 \text{ N/m}^3)(95 \text{ m})] = 80.8$ GN.

Problem

25. A U-shaped tube open at both ends contains water and a quantity of oil occupying a 2.0-cm length of the tube, as shown in Fig. 18-45. If the oil's density is 0.82 times that of water, what is the height difference h?

Solution

From Equation 18-3, the pressure at points at the same level in the water is the same, $P_1 = P_2$. Now,

$P_1 = P_{atm} + \rho_{H_2O}g(2 \text{ cm} - h)$ and $P_2 = P_{atm} + \rho_{oil}g(2 \text{ cm})$, so $h = (2 \text{ cm})(1 - \rho_{oil}/\rho_{H_2O}) = (2 \text{ cm}) \times (1 - 0.82) = 3.6$ mm (h is positive as shown).

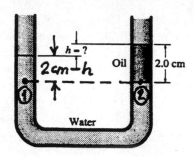

FIGURE 18-45 Problem 25 Solution.

Problem

27. Barometric pressure in the eye of a hurricane is 0.91 atm (27.2 inches of mercury). How does the level of the ocean surface under the eye compare with that under a distant fair-weather region where the pressure is 1.0 atm?

Solution

Equation 18-3 applied at points on the water surface under the hurricane eye and fair-weather region gives 1 atm = 0.91 atm + ρgy, therefore $y = (0.09 \text{ atm}) \times (1.013 \times 10^5 \text{ Pa/atm})/(9800 \text{ N/m}^3) = 93.0$ cm.

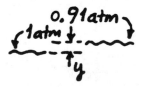

Problem 27 Solution.

Problem

29. A garage lift has a 45-cm diameter piston supporting the load. Compressed air with a maximum pressure of 500 kPa is applied to a small piston at the other end of the hydraulic system. What is the maximum mass the lift can support?

Solution

If we neglect the variation of pressure with height in the hydraulic system (which is usually small compared to the applied pressure), the fluid pressure is the same throughout, or $P_{appl} = F/A$ (for either the small or large cylinders). Thus, $F_{max} = (500 \text{ kPa})\frac{1}{4}\pi \times (0.45 \text{ m})^2 = 79.5$ kN, which corresponds to a mass-load of $F_{max}/g = 8.11$ tonnes (metric tons).

Section 18-3:   Archimedes' Principle and Buoyancy

Problem

31. On land, the most massive concrete block you can carry is 25 kg. How massive a block could you carry underwater, if the density of concrete is 2300 kg/m^3?

Solution

The 25×9.8 N force you exert underwater is equal to the apparent weight of the most massive block of concrete when submerged, $W_{app} = W - F_b = W(1 - \rho_w/\rho_c)$, where ρ_c/ρ_w is the ratio of the densities of concrete and water (also known as the specific gravity of concrete). (This relation is derived in Example 18-4, since the buoyant force on an object submerged in fluid is $F_b = W\rho_{fluid}/\rho$.) Thus, $W = W_{app}(1 - \rho_w/\rho_c)^{-1}$, or $m = W/g = (25\text{ kg})\times(1 - 1/2.3)^{-1} = 44.2$ kg.

Problem

33. The density of styrofoam is 160 kg/m^3. What per cent error is introduced by weighing a styrofoam block in air, which exerts an upward buoyancy force, rather than in vacuum? The density of air is 1.2 kg/m^3.

Solution

The fractorial error is $(W - W_{app})/W = \rho_{air}/\rho_{styro} = 1.2/160 = 0.75\%$ (see Example 18-4).

Problem

35. A partially full beer bottle with interior diameter 52 mm is floating upright in water, as shown in Fig. 18-47. A drinker takes a swig and replaces the bottle in the water, where it now floats 28 mm higher than before. How much beer did the drinker drink?

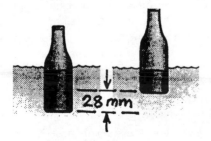

FIGURE 18-47 Problem 35 Solution.

Solution

Archimedes' principle implies that the weight of the beer swallowed equals the difference in the weight of

water displaced by the bottle, before and after. Therefore $\Delta m_{beer} = \rho_{H_2O}\,\Delta V$, where "$g$" was canceled from both sides. The difference in the volume of water displaced equals the cross-sectional area of the bottle times 28 mm. If we ignore the thickness of the walls of the bottle, $\Delta m_{beer} = (1\text{ g/cm}^3)\pi(\frac{1}{2}\times5.2\text{ cm})^2\times(2.8\text{ cm}) = 59.5$ g.

Problem

37. A typical supertanker has mass 2.0×10^6 kg and carries twice that much oil. If 9.0 m of the ship is submerged when it's empty, what is the minimum water depth needed for it to navigate when full? Assume the sides of the ship are vertical.

Solution

If the sides of the hull are vertical, and its bottom flat, the volume it displaces is proportional to its draft (depth in the water), i.e., $V = Ay$, where A is the cross-sectional area. Since the total mass of the full supertanker is 3× that when empty, the draft when full is simply $3\times(9\text{ m}) = 27$ m.

Problem 37 Solution.

Problem

39. (a) How much helium (density 0.18 kg/m^3) is needed to lift a balloon carrying two people in a basket, if the total mass of people, basket, and balloon (but not gas) is 280 kg? (b) Repeat for a hot air balloon, whose air density is 10% less than that of the surrounding atmosphere.

Solution

The buoyant force must exceed the weight of the load (mass M, including the balloon) plus the gas (mass m), $F_b \geq (M + m)g$. But $F_b = \rho_{air}gV$, and if we neglect the volume of the balloon's skin etc. compared to that of the gas it contains, $V = m/\rho_{gas}$, therefore $m = \rho_{gas}V = \rho_{gas}(F_b/\rho_{air}g) \geq (\rho_{gas}/\rho_{air})(M + m)$ or $m \geq M\rho_{gas}/(\rho_{air} - \rho_{gas})$. (a) When the gas is helium, $\rho_{air}/\rho_{He} = 1.2/0.18$, and $m \geq (280\text{ kg})(6.67 - 1)^{-1} = 49.4$ kg. (b) For hot air, $\rho_{gas} = 0.9\rho_{air}$, and $m \geq (280\text{ kg})(0.9/0.1) = 2520$ kg. (Note: these masses correspond to gas volumes of 275 m^3 for helium and 2330 m^3 for hot air.)

Sections 18-4 and 18-5: Fluid Dynamics and Applications

Problem

41. A fluid is flowing steadily, roughly from left to right. At left it is flowing rapidly; it then slows down, and finally speeds up again. Its final speed at right is not as great as its initial speed at left. Sketch a streamline pattern that could represent this flow.

Solution

In order to maintain a constant volume rate of flow, in an incompressible fluid, streamlines must be closer together (smaller cross-section of tube of flow) where the velocity is greater, as sketched.

Problem 41 Solution.

Problem

43. A typical mass flow rate for the Mississippi River is 1.8×10^7 kg/s. Find (a) the volume flow rate and (b) the flow speed in a region where the river is 2.0 km wide and an average of 6.1 m deep.

Solution

(a) The mass flow rate and the volume flow rate are related by Equations 18-4b and 5, namely $R_m = \rho v A = \rho R_V$. Therefore, $R_V = (1.8 \times 10^7 \text{ kg/s}) \div (10^3 \text{ kg/m}^3) = 1.8 \times 10^4$ m³/s for the Mississippi. (b) At a point in the river where the cross-sectional area is given, the average speed of flow is $v = R_V/A = (1.8 \times 10^4 \text{ m}^3/\text{s})/(2 \times 10^3 \times 6.1 \text{ m}^2) = 1.48$ m/s ($= 5.31$ km/h $= 3.30$ mph). The actual flow rate of any river varies with the season, local weather and vegetation conditions, and human water consumption.

Problem

45. A typical human aorta, or main artery from the heart, is 1.8 cm in diameter and carries blood at a speed of 35 cm/s. What will be the flow speed around a clot that reduces the flow area by 80%?

Solution

The continuity equation (Equation 18-5) is a reasonable approximation for blood circulation in an artery, so $v' = v(A/A')$. If the cross-sectional area is reduced by 80%, then $A/A' = 100\%/20\% = 5$, so $v' = 5(35 \text{ cm/s}) = 1.75$ m/s.

Problem

47. In Fig. 18-48 a horizontal pipe of cross-sectional area A is joined to a lower pipe of cross-sectional area $\frac{1}{2}A$. The entire pipe is full of liquid with density ρ, and the left end is at atmospheric pressure P_a. A small open tube extends upward from the lower pipe. Find the height h_2 of liquid in the small tube (a) when the right end of the lower pipe is closed, so the liquid is in hydrostatic equilibrium, and (b) when the liquid flows with speed v in the upper pipe.

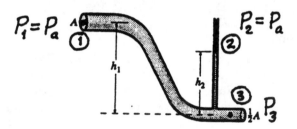

FIGURE 18-48 Problem 47.

Solution

The continuity equation (Equation 18-5) and Bernoulli's equation (Equation 18-6) can be applied to an incompressible fluid whether it is at rest or flowing steadily. (a) In hydrostatic equilibrium, the flow speed is zero everywhere. Since the pressure is P_a at the left end of the upper horizontal pipe and at the top of the liquid in the small vertical tube, Equation 18-6a for these points gives $P_a + 0 + \rho g h_1 = P_a + 0 + \rho g h_2$, or $h_1 = h_2$, where we measured the heights y from the lower horizontal tube. (b) In steady flow, Equation 18-5 gives the flow speed in the lower pipe as $v' = v(A/\frac{1}{2}A) = 2v$, where v is the speed in the upper pipe and the cross-sectional areas are given. Then Equation 18-6a gives $P_a + \frac{1}{2}\rho v^2 + \rho g h_1 = P_3 + \frac{1}{2}\rho(2v)^2 + 0$, where P_3 is the pressure anywhere in the lower pipe. (Since the lower pipe is horizontal, $y = 0$, and uniform in cross-section, $v = $ constant, hence $P_3 = $ constant.) Now, even when liquid flows in the pipes, the liquid in the small vertical tube is stagnant. If we assume the pressure is constant over the cross-section of the lower pipe, then Equation 18-3 gives $P_3 = P_a + \rho g h_2$. Combining these results, we find $P_3 - P_a = -\frac{3}{2}\rho v^2 + \rho g h_1 = \rho g h_2$, or $h_2 = h_1 - 3v^2/2g$.

Problem

49. The water in a garden hose is at a gauge pressure of 140 kPa and is moving at negligible speed. The hose terminates in a sprinkler consisting of many small holes. What is the maximum height reached by the water emerging from the holes?

Solution

The pressure, velocity, and height of the water in the hose (point 1) are $P_1 = P_{atm} + 140$ kPa, $v_1 \approx 0$, and $y_1 = 0$, while at the highest point of a jet of water from a hole (point 2), $P_2 = P_{atm}$, $v_2 \approx 0$, and $y_2 = h$. (We assume that the jets from the holes are the same.) Then Bernoulli's equation (Equation 18-6a) yields $P_{atm} + 140$ kPa $= P_{atm} + \rho gh$, or $h = 140$ kPa$\div$ $(9800$ N/m$^3) = 14.3$ m.

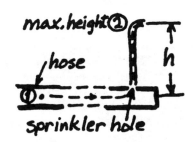

Problem 49 Solution.

Problem

51. The venturi flowmeter shown in Fig. 18-51 is used to measure the flow rate of water in a solar collector system. The flowmeter is inserted in a pipe with diameter 1.9 cm; at the venturi of the flowmeter the diameter is reduced to 0.64 cm. The manometer tube contains oil with density 0.82 times that of water. If the difference in oil levels on the two sides of the manometer tube is 1.4 cm, what is the volume flow rate?

Solution

If we apply Bernoulli's equation (Equation 18-6a) and the continuity equation (Equation 18-5) to points 1 and 2 in the flowmeter, we can calculate the volume rate of flow:

$$P_1 + \frac{1}{2}\rho v_1^2 = P_2 + \frac{1}{2}\rho v_2^2 \quad \text{and} \quad v_1 A_1 = v_2 A_2 \text{ imply}$$

$$P_1 - P_2 = \frac{1}{2}\rho(v_2^2 - v_1^2) = \frac{1}{2}\rho v_1^2 A_1^2 \left(\frac{1}{A_2^2} - \frac{1}{A_1^2}\right),$$

$$\text{or} \quad R_V = v_1 A_1 = \sqrt{\frac{2(P_1 - P_2)}{\rho(A_2^{-2} - A_1^{-2})}}.$$

(This is the same calculation as Example 18-9. Note that pressure variation with height in the flowmeter is assumed negligible.) The pressure difference is related to the difference in height and the density of oil in the manometer (where the fluid is assumed stagnant): $P_1 = P_3 + \rho gy_1$ and $P_2 = P_3 + \rho gy_2 + \rho_{oil}gh$ imply $P_1 - P_2 = (\rho - \rho_{oil})gh$, since $y_1 - y_2 = h$. If we use $A = \frac{1}{4}\pi d^2$ for each part of the flowmeter, we finally obtain $R_V = \frac{1}{4}\pi\sqrt{2gh(1 - \rho_{oil}/\rho)/(d_2^{-4} - d_1^{-4})} =$ 7.20 cm^3/s, when the given numerical values are substituted (we used h, d_1, d_2 in cm and $g = 980$ cm/s^2).

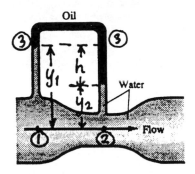

FIGURE 18-51 Problem 51 Solution.

Paired Problems

Problem

53. A steel drum has volume 0.23 m^3 and mass 16 kg. Will it float in water when filled with (a) water or (b) gasoline (density 860 kg/m^3)? Neglect the thickness of the steel.

Solution

An object will float in water if its average density is less than the density of water. (This follows from Archimedes' principle, since the volume of water displaced by an object floating on the surface is less than its total volume, i.e., $V_{dis} < V$. Because the buoyant force equals the weight of a floating object, $F_b = \rho_{H_2O}gV_{dis} = W = \rho_{av}gV$, this implies $\rho_{av} < \rho_{H_2O}$.) (a) Since $\rho_{steel} > \rho_{H_2O}$, when the drum is filled with water, $\rho_{av} > \rho_{H_2O}$ and the drum will sink. (The average density of a composite object is always greater than the smallest density of its components; see solution to Problem 3.) (b) When the drum is filled with gasoline, its average density is $\rho_{av} = (M_{steel} + M_{gas})/V$. If we neglect the volume occupied by the steel compared to the volume of the drum, then $M_{gas} = \rho_{gas}V$ and $\rho_{av} = (16$ kg/0.23 m$^3) +$ 860 kg/m$^3 = 930$ kg/m^3, which is less than ρ_{H_2O} so the drum floats.

Problem

55. A spherical rubber balloon with mass 0.85 g and diameter 30 cm is filled with helium (density 0.18 kg/m^3). How many 1.0-g paper clips can you hang from the balloon before it loses its buoyancy?

Solution

The buoyant force on the balloon will exceed its weight provided $\rho_{av} < \rho_{air}$ (see solution to Problem 53). If we neglect the volume occupied by the rubber and n paperclips (n is an integer) compared to the spherical volume of the balloon, $V = 4\pi R^3/3$, then $\rho_{av} = (M_{rubber} + nM_{clip} + \rho_{He}V)/V$, where $\rho_{He}V$ is (approximately) the mass of the helium. Thus, the balloon will have excess buoyancy in air of density 1.2 kg/m^3 if 1.2 kg/m^3 > $\rho_{av} = (0.18$ kg/m$^3) + (0.85+1.0n)\times10^{-3}$ kg/$\frac{4}{3}\pi(0.15$ m$)^3$, or $n < \frac{4}{3}\pi(1.5)^3\times (1.2 - 0.18) - 0.85 = 13.6$. (With $n = 13$ paper clips attached, the balloon will rise until its average density of 1.16 kg/m^3 equals the density of the surrounding air.)

Problem

57. Water at a pressure of 230 kPa is flowing at 1.5 m/s through a pipe, when it encounters an obstruction where the pressure drops by 5%. What fraction of the pipe's area is obstructed?

Solution

Assume horizontal flow in a narrow pipe (so there is no dependence on height). Then $v_1A_1 = v_2A_2$, and $P_1 + \frac{1}{2}\rho v_1^2 = P_2 + \frac{1}{2}\rho v_2^2$ (Equations 18-5 and 18-6a for steady incompressible fluid flow), where subscript 2 refers to the obstruction. Since the pressure at the obstruction is 5% less, $P_1 - P_2 = 0.05P_1 = \frac{1}{2}\rho v_1^2\times [(A_1^2/A_2^2) - 1]$, where we eliminated v_2. Then $(A_1/A_2)^2 = 1 + (0.1\times230$ kPa$)/(10^3$ kg/m$^3)\times (1.5$ m/s$)^2 = (3.35)^2$. The fraction of area obstructed is $(A_1 - A_2)/A_1 = 1 - (1/3.35) = 70.1\%$.

Problem

59. Find an expression for the volume flow rate from the siphon shown in Fig. 18-52, assuming the siphon area A is much less than the tank area.

Solution

Bernoulli's equation applied between points at the top surface of the water in the tank and at the mouth of the siphon (where both pressures are atmospheric pressure) gives $P_a + \frac{1}{2}\rho v_{top}^2 = P_a + \frac{1}{2}\rho v^2 - \rho gh$. (We find it convenient to measure heights from the water level in the tank.) As in Example 18-8, if the siphon cross-sectional area is much smaller than the tank

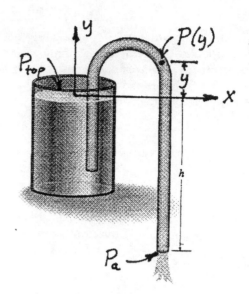

FIGURE 18-52 Problem 59.

area, the continuity equation, $v_{top}A_{tank} = vA$, implies $v_{top}/v = A/A_{tank} \ll 1$ so $v_{top} \approx 0$ can be neglected. Then $v = \sqrt{2gh}$. The volume rate of flow (Equation 18-5) is just $R_V = vA = A\sqrt{2gh}$. (Note: We assume the siphon tube has uniform cross-section and the water in it has constant flow speed, so the pressure at any other point in the siphon tube is $P(y) = P_a - \rho g(h + y)$, where y is the distance above the tank's water level. Since the absolute pressure is always positive, i.e., $P(y) \geq 0$, it follows that $y + h \leq P_a/\rho g = (101.3$ kPa$)/(9800$ N/m$^3) = 10.3$ m ≈ 34 ft. A siphon doesn't work if the highest point of its tube is more than this distance above the water surface. More precisely, h is the distance of the siphon's mouth below the water level, which we've assumed to be positive ($h \geq 0$, otherwise water doesn't flow out of the tank). Then $y < 10.3$ m $- h < 10.3$ m, which gives the previously stated limit for a siphon. On the other hand, when $h + y > 10.3$ m, the flow becomes unsteady. The assumption of steady incompressible flow, Equations 18-5 and 6, is quite restrictive.)

Supplementary Problems
Problem

61. A 1.0-m-diameter tank is filled with water to a depth of 2.0 m and is open to the atmosphere at the top. The water drains through a 1.0-cm-diameter pipe at the bottom; that pipe then joins a 1.5-cm-diameter pipe open to the atmosphere, as shown in Fig. 18-53. Find (a) the flow speed in the narrow section and (b) the water height in the *sealed* vertical tube shown.

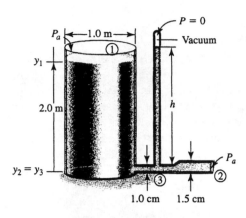

FIGURE 18-53 Problem 61.

Solution

(a) If we assume a steady incompressible flow (Equations 18-5 and 6), an argument similar to Example 18-8, comparing point 1 at the top of the tank with point 2 at the opening of the 1.5 cm pipe, gives $P_a' + \rho g y_1 \approx P_a + \rho g y_2 + \frac{1}{2}\rho v_2^2$ or $v_2 = \sqrt{2g(y_1 - y_2)}$. Here, we neglected the flow speed at the top, $v_1 \approx 0$, and $y_1 - y_2 = 2$ m. The continuity equation gives the speed in the narrower section of pipe, $v_3 = v_2(A_2/A_3) = (1.5/1.0)^2 \times \sqrt{2(9.8 \text{ m/s}^2)(2 \text{ m})} = 14.1$ m/s. (b) The pressure at the bottom of the stagnant column of water over the narrow section of pipe is $P_3 = \rho g h$, because there is no pressure exerted by a vacuum, and we assume this pressure is uniform over the cross-section of the narrow pipe. Another application of Bernoulli's equation gives $P_a + \rho g y_1 \approx P_3 + \rho g y_2 + \frac{1}{2}\rho v_3^2 = \rho g h + \rho g y_2 + (1.5)^4 \rho g(y_1 - y_2)$, where we have again neglected the flow speed at the top of the tank, and we used the expression for v_3 from part (a). Therefore, $h = (P_a/\rho g) - (y_1 - y_2)((1.5)^4 - 1) = (101.3 \text{ kPa} \div 9800 \text{ N/m}^3) - (2 \text{ m})(65/16) = 2.21$ m.

Problem

63. Figure 18-54 shows a simplified diagram of a Pitot tube, used for measuring aircraft speeds. The tube is mounted on the underside of the aircraft wing with opening A at right angles to the flow and opening B pointing into the flow. The gauge prevents airflow through the tube. Use Bernoulli's equation to show that the air speed relative to the wing is given by $v = \sqrt{2\Delta P/\rho}$, where ΔP is the pressure difference between the tubes and ρ is the density of air. *Hint:* The flow must be stopped at B, but continues past A with its normal speed.

Solution

Any difference in height between A and B is practically negligible ($y_A \approx y_B$), and $v_B = 0$, so Bernoulli's

equation gives $P_B = P_A + \frac{1}{2}\rho v_A^2$. Thus, $v_A = \sqrt{2(P_B - P_A)/\rho}$. (Note: Even though Equation 18-6 applies strictly to incompressible steady fluid flow, density variations in a gas are generally insignificant when the flow speed is much less than the speed of sound.)

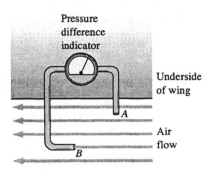

FIGURE 18-54 Problem 63 Solution.

Problem

65. With its throttle valve wide open, an automobile carburetor has a throat diameter of 2.4 cm. With each revolution, the engine draws 0.50 L of air through the carburetor. At an engine speed of 3000 rpm, what are (a) the volume flow rate, (b) the airflow speed, and (c) the difference between atmospheric pressure and air pressure in the carburetor throat? The density of air is 1.2 kg/m^3.

Solution

(a) The volume rate of flow of intake air is $R_V = (3000 \text{ rpm})(1 \text{ min}/60 \text{ s})(0.5 \text{ L/rev}) = 25$ L/s $= 0.025$ m^3/s. (b) This rate of flow, assumed constant over the cross-sectional area of the carburetor throat, implies a flow speed of $v = R_V/A = (0.025 \text{ m}^3/\text{s}) \div \frac{1}{4}\pi(0.024 \text{ m})^2 = 55.3$ m/s. (c) Since the flow speed is much smaller than the speed of sound in air at this density, we can use Equation 18-6 to calculate the pressure difference. We suppose that air enters the carburetor intake at a speed which is negligible compared to v, at essentially the same height as the throat. Then $\Delta P = P_a - P_{\text{throat}} = \frac{1}{2}\rho v^2 = \frac{1}{2} \times (1.2 \text{ kg/m}^3)(55.3 \text{ m/s})^2 = 1.83$ kPa.

Problem

67. A can of height h and cross-sectional area A_0 is initially full of water. A small hole of area $A_1 \ll A_0$ is cut in the bottom of the can. Find an expression for the time it takes all the water to drain from the can. *Hint:* Call the water depth y,

use the continuity equation to relate dy/dt to the outflow speed at the hole, then integrate.

Solution

If y is the height of the water above the bottom of the can, then $-dy/dt$ is the magnitude of the flow speed of the top surface of the water draining out (y decreases as a function of time). The continuity equation gives $-(dy/dt)A_0 = v_1 A_1$, where subscript 1 refers to the small hole in the bottom, so $dt = -(A_0/A_1)dy/v_1$. For most of the time, $v_1 \approx \sqrt{2gy}$ (see Example 18-8 and we assume the top of the can is open), thus

$$t = \int dt \approx -\int_h^0 \left(\frac{A_0}{A_1}\right)\frac{dy}{\sqrt{2gy}}$$

$$= \frac{A_0}{A_1\sqrt{2g}} \left.\frac{y^{1/2}}{(1/2)}\right|_0^h = \frac{A_0}{A_1}\sqrt{\frac{2h}{g}}.$$

This result is approximate since dy/dt cannot be neglected compared to v_1 when y is small. If we use Bernoulli's equation without this approximation, then $\frac{1}{2}\rho(dy/dt)^2 + \rho gy = \frac{1}{2}\rho v_1^2$, since the pressure is atmospheric pressure at both the top of the can and the hole. Combining with the continuity equation gives $v_1 = \sqrt{2gy + (dy/dt)^2} = -(A_0/A_1)\,dy/dt$, or $dy/dt = -\sqrt{2gy/[(A_0/A_1)^2 - 1]}$. Integration of this yields a more exact outflow time of $t = \sqrt{(2h/g)[(A_0/A_1)^2 - 1]}$.

Problem

69. A circular pan of liquid (density ρ) is centered on a horizontal turntable rotating with angular speed ω. Its axis coincides with the rotation axis, as shown in Fig. 18-56. Atmospheric pressure is P_a. Find expressions for (a) the pressure at the bottom of the pan and (b) the height of the liquid surface as functions of the distance r from the axis, given that the height at the center is h_0.

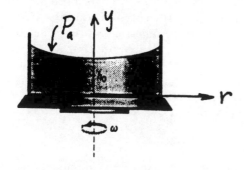

FIGURE 18-56 Problem 69.

Solution

When the water is in equilibrium at constant angular velocity, the vertical change in pressure balances the weight of the water, the radial change in pressure supplies the centripetal acceleration, and there is no change in pressure in the direction tangent to the rotation. Introduce vertical, radial, and tangential coordinates, y, r, and φ respectively, with origin at the bottom center of the pan and y axis positive upward. (These are cylindrical coordinates.) Consider a fluid element $dm = \rho\,dV = \rho\,dr(r\,d\varphi)\,dy$ as shown. (ρ is the density and dV is the volume element.)

The vertical pressure difference balances the gravitational force, as in Equation 18-2, $dF_{press} + dF_{grav} = 0 = -dP_y A_y - \rho g\,dV$, or $\partial P/\partial y = -\rho g$. Here, $A_y = r\,dr\,d\varphi$ is the area of the faces perpendicular to the y direction, and we wrote a partial derivative because the pressure varies with both y and r. Note that $\partial P/\partial y$ is negative because the pressure increases with depth (decreasing y).

Similarly, the pressure force in the radial direction equals the mass element times the centripetal acceleration, $dF_{press} = -dm\,\omega^2 r = -dP_r A_r = -\rho\,dV\,\omega^2 r$. (Recall that $a_r = -v^2/r = -\omega^2 r$.) In this equation, $A_r = r\,d\varphi\,dy$ is the area of the faces perpendicular to the radial direction. Since $dV = A_r\,dr$, after canceling $-A_r$, we find $dP_r = \rho\omega^2 r\,dr$, or $\partial P/\partial r = \rho\omega^2 r$. Here, $\partial P/\partial r$ is positive because the pressure increases with r.

For an incompressible fluid, ρ is a constant (not a function of r and y), thus $\partial P/\partial y = -\rho g$ and $\partial P/\partial r = \rho\omega^2 r$ require the presence of terms equal to $-\rho gy$ and $\frac{1}{2}\rho\omega^2 r^2$ in the expression for the pressure (then the partial derivatives have their specified values). Thus $P(r,y) = -\rho gy + \frac{1}{2}\rho\omega^2 r^2 + \text{constant}$. The constant term can be evaluated, since the pressure is atmospheric pressure at the surface above the center, i.e., $P(0, h_0) = P_a = -\rho gh_0 + \text{constant}$, or the constant $= P_a + \rho gh_0$. Then $P(r,y) = P_a - \rho g \times (y - h_0) + \frac{1}{2}\rho\omega^2 r^2$.

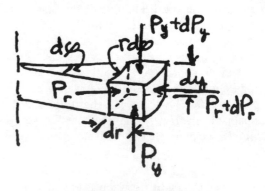

Problem 69 Solution.

(a) Along the bottom of the pan ($y = 0$), $P(r,\ 0) = P_a + \rho g h_0 + \frac{1}{2}\rho\omega^2 r^2$. (b) The pressure at the water's surface is P_a for all values of r, so the height of the surface, $y = h(r)$, is given by the equation $P(r, h(r)) = P_a$, or $-\rho g[h(r) - h_0] + \frac{1}{2}\rho\omega^2 r^2 = 0$. Thus, $h(r) = h_0 + \omega^2 r^2/2g$, i.e., parabolic. (Such a technique is used to shape large mirrors for astronomical telescopes by a process called spin casting.)

Problem

71. (a) Use the result of the preceding problem to express Earth's atmospheric density as a function of height (this is simple). (b) Use the result of (a) to find the height below which half of Earth's atmospheric mass lies (this will require integration).

Solution

(a) The variation of pressure with height in the Earth's atmosphere follows from Equation 18-2 (with h replaced by $-h$, since height is positive upward whereas depth is positive downward). Thus, $dP = -\rho g\, dh$. If pressure and density are proportional (as given in the previous problem), $dP = -P\, dh/h_0$. This equation can be integrated from the surface values, $h = 0$ and P_0, to yield $\int_{P_0}^{P} dP/P = \ln(P/P_0) = -\int_0^h dh/h_0 = -h/h_0$, or after exponentiation, $P = P_0 e^{-h/h_0}$. (This is called the law of atmospheres; it applies exactly if the temperature is constant.) In terms of density, $\rho = \rho_0 e^{-h/h_0}$, where $\rho_0 = P_0/h_0 g$. Both P and ρ fall to half their surface values at a height $h_0\ln 2 = (8.2\text{ km})(0.693) = 5.7$ km (since $e^{-h_0\ln2/h_0} = \frac{1}{2}$). (b) The mass of atmosphere contained in a thin spherical shell of thickness dh, at height h, is $dm = \rho\, dV = 4\pi\rho_0(R_E + h)^2 e^{-h/h_0}\, dh$, where R_E is the radius of the Earth and $R_E + h$ the radius of the shell. The mass of atmosphere below height h_1 is

$$M(h_1) = \int_0^{h_1} dm$$
$$= 4\pi\rho_0 R_E^2 \int_0^{h_1} \left(1 + 2\frac{h}{R_E} + \frac{h^2}{R_E^2}\right) e^{-h/h_0}\, dh.$$

The integrals can be evaluated easily enough with the use of the table of integrals in Appendix B, however, if $h_1/R_E \ll 1$, only the first term is important. (Even if h_1 is large, the exponential term is negligibly small for $h_1 \gg h_0$ and none of the terms contribute significantly for large h.) To a good approximation, therefore

$$M(h_1) \simeq 4\pi\rho_0 R_E^2 \int_0^{h_1} e^{-h/h_0}\, dh$$
$$= 4\pi\rho_0 R_E^2 h_0(1 - e^{-h_1/h_0}).$$

The total mass of the atmosphere is approximately $M(\infty) = 4\pi\rho_0 R_E^2 h_0$, so the height bounding half the total mass is given by the equation $\frac{1}{2}M = M(1 - e^{-h_1/h_0})$, or $h_1 = h_0\ln2$ as in part (a).

PART 2   CUMULATIVE PROBLEMS

Problem

1. A cylindrical log of total mass M and uniform diameter d has an uneven mass distribution that causes it to float in a vertical position, as shown in Figure 1. (a) Find an expression for the length ℓ of the submerged portion of the log when it is floating in equilibrium, in terms of M, d, and the water density ρ. (b) If the log is displaced vertically from its equilibrium position and released, it will undergo simple harmonic motion. Find an expression for the period of this motion, neglecting viscosity and other frictional effects.

Solution

(This problem is similar to Problem 18-68.)
(a) At equilibrium, the weight of the log is balanced by the buoyant force, as in Example 18-6. The former has magnitude Mg, while that of the latter is $F_b = \rho g V_{\text{sub}} = \rho g A\ell$, where $A = \frac{1}{4}\pi d^2$ is the cross-sectional

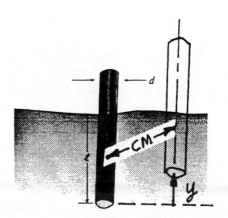

FIGURE 1 Cumulative Problem 1.

area and ℓ the equilibrium submerged length. Thus $\ell = M/\rho A = 4M/\rho\pi d^2$. (b) If the log is given a vertical displacement y (positive upwards as shown), the net force (neglecting frictional effects) is $F_b - Mg = \rho g A(\ell - y) - \rho g A\ell = -(Mg/\ell)y$, where we expressed the weight in terms of the equilibrium submerged length from part (a). Since y is also the displacement of the log's center of mass from its equilibrium position, the net force equals $M \, d^2y/dt^2$, or $d^2y/dt^2 = -(g/\ell)y$. This is the equation for simple harmonic motion with frequency $\omega^2 = g/\ell$ and period $T = 2\pi/\omega = 2\pi\sqrt{\ell/g} = 4\sqrt{\pi M/\rho g d^2}$.

Problem

3. Let P_0 and ρ_0 be the atmospheric pressure and density at Earth's surface. Assume that the ratio P/ρ is the same throughout the atmosphere (this implies that the temperature is uniform). Show that the pressure at vertical height z above the surface is given by $P(z) = P_0 e^{-\rho_0 g z/P_0}$, for z much less than Earth's radius (this amounts to neglecting Earth's curvature, and thus taking g to be constant).

Solution

(This problem is similar to Problems 18-70 and 71.) First, rewrite Equation 18-2 in terms of the height z above, instead of the depth h below, the Earth's surface ($z = -h$). Then the pressure variation with height in the atmosphere is given by $dP/dz = -\rho g$. If we assume $P/\rho = P_0/\rho_0$ is a constant (as for an ideal isothermal atmosphere—see Equation 20-1) then $dP/dz = -(\rho_0 g/P_0)P$. For $z \ll R_E, g$ is nearly constant, and the pressure equation can be integrated by separating variables:

$$\int_{P_0}^{P} \frac{dP}{P} = -\left(\frac{\rho_0 g}{P_0}\right) \int_0^z dz, \text{ or } \ln(P/P_0) = -(\rho_0 g/P_0)z.$$

Exponentiation gives the desired result, $P = P_0 e^{-z/H_0}$, which is called the barometric law. The constant $H_0 = P_0/\rho_0 g \approx 8.4$ km is known as the scale height.

Problem

5. A U-shaped tube containing liquid is mounted on a table that tilts back and forth through a slight angle, as shown in Fig. 4. The diameter of the tube is much less than either the height of it's arms or their separation. When the table is rocked very

slowly or very rapidly, nothing particularly dramatic happens. But when the rocking takes place at a few times per second, the liquid level in the tube oscillates violently, with maximum amplitude at a rocking frequency of 1.7 Hz. Explain what is going on, and find the total length of the liquid including both vertical and horizontal portions.

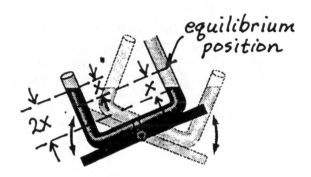

FIGURE 4 Cumulative Problem 5.

Solution

When the tube is rocked back and forth, the liquid in it is dragged along by viscous forces. We suppose that the dimensions of the tube (its small diameter compared to the height of the tube arms or their separation) allow us to treat the column of liquid, of total length ℓ, as a one-dimensional system which undergoes underdamped oscillations with weak damping (as in Section 15-7). The maximum amplitude occurs at a driving frequency very close to the undamped natural frequency, $\omega_d = 2\pi(1.7 \text{ Hz}) \approx \omega_0$. To find ω_0 in terms of ℓ, suppose one end of the liquid column is depressed a distance x from equilibrium, as shown. The net restoring force is the weight of a length $2x$ of liquid, or $F = -(\Delta m)g = -\rho g A \cdot 2x$, where A is the cross-sectional area of the tube and ρ is the density of the liquid. The mass of the entire column is $\rho A \ell$, so Newton's second law gives $-2\rho g A x = \rho A\ell \, d^2x/dt^2$. This is the equation for simple harmonic motion with natural frequency $\omega_0^2 = 2g/\ell$. In this case, $\ell = 2g/\omega_0^2 = 2(9.8 \text{ m/s}^2) \div (2\pi \times 1.7 \text{ Hz})^2 = 17.2$ cm.

PART 3   THERMODYNAMICS

CHAPTER 19   TEMPERATURE AND HEAT

Section 19-1:   Macroscopic and Microscopic Descriptions

Problem

1. The macroscopic state of a carton capable of holding a half-dozen eggs is specified by giving the number of eggs in the carton. The microscopic state is specified by telling where each egg is in the carton. How many microscopic states correspond to the macroscopic state of a full carton?

Solution

If we number the eggs (so that they are distinguishable), we could put the first egg in any one of six places, the second in any one of the remaining five places, etc. The total number of microscopic states is $6 \cdot 5 \cdot 4 \cdot 3 \cdot 2 \cdot 1 = 6! = 720$. (If the eggs are indistinguishable, there is only one microscopic state for a full carton.)

Section 19-3:   Measuring Temperature

Problem

3. Normal room temperature is 68°F. What is this in Celsius?

Solution

Equation 19-3, solved for the Celsius temperature, gives $T_C = \frac{5}{9}(T_F - 32) = 5(68 - 32)/9 = 20°C$.

Problem

5. At what temperature do the Fahrenheit and Celsius scales coincide?

Solution

In Equation 19-3, T_F and T_C are numerically equal when $T_F = \left(\frac{9}{5}\right)T_C + 32 = T_C$, or $T_C = -\left(\frac{5}{4}\right)(32) = -40 = T_F$.

Problem

7. The normal boiling point of nitrogen is 77.3 K. Express this in Celsius and Fahrenheit.

Solution

Equations 19-2 and 3 give $T_C = 77.3 - 273.15 \simeq -196°C$, and $T_F = \left(\frac{9}{5}\right)(-196) + 32 = -321°F$.

Problem

9. A constant-volume gas thermometer is filled with air whose pressure is 101 kPa at the normal melting point of ice. What would its pressure be at (a) the normal boiling point of water, (b) the normal boiling point of oxygen (90.2 K), and (c) the normal boiling point of mercury (630 K)?

Solution

The thermometric equation for an ideal constant-volume gas thermometer is $P/T = P_{\text{ref}}/T_{\text{ref}}$. (This is Equation 19-1 written for a reference point not necessarily equal to the triple point of water.) If we use the given values at the normal melting point of ice, $P = T(101 \text{ kPa})/(273.15 \text{ K})$. When the temperatures of the normal boiling points of water (100°C = 373.15 K), oxygen (90.2 K), and mercury (630 K) are substituted, pressures of (a) 138 kPa, (b) 33.4 kPa, and (c) 233 kPa are calculated.

Problem

11. The temperature of a constant-pressure gas thermometer is directly proportional to the gas volume. If the volume is 1.00 L at the triple point of water, what is it at water's normal boiling point?

Solution

The absolute temperature (for a given mass of gas at constant pressure) is proportional to the volume (this is known as the law of Charles and Gay-Lussac), therefore $T/T_3 = V/V_3$, where subscript 3 refers to the triple point values. Then $V = (T/T_3)V_3 = (373.15 \div 273.16)(1.00 \text{ L}) = 1.37 \text{ L}$, when $T = 100°C$.

Problem

13. A constant-volume gas thermometer supports a 72.5-mm-high mercury column when it's immersed

in liquid nitrogen at $-196°C$. What will be the column height when the thermometer is in molten lead at $350°C$?

Solution

For a constant-volume gas thermometer, P/T is constant (see Equation 19-1). Since pressure can be measured in mm of mercury ($P = \rho gh$) it is also true that h/T is constant. Under the conditions specified for liquid nitrogen and molten lead, $h/T = h'/T'$ implies $h' = (T'/T)h = (7.25 \text{ mm})(350 + 273) \div (-196 + 273) = 587$ mm. (Note: the temperature in Equation 19-1 is the absolute Kelvin temperature.)

Sections 19-4 and 19-5: Temperature and Heat, Heat Capacity and Specific Heat

Problem

15. If your mass is 60 kg, what is the minimum number of calories you would "burn off" climbing a 1700-m-high mountain? (The actual metabolic energy used would be much greater.)

Solution

The minimum energy burned-off is the work done against gravity, which equals the potential energy change: $-W_{grav} = \Delta U_{grav} = mg \, \Delta y = (60 \times 9.8 \text{ N})(1700 \text{ m})(1 \text{ kcal}/4184 \text{ J}) = 239$ kcal (recall Equations 8-2 and 3).

Problem

17. Typical fats contain about 9 kcal per gram. If the energy in body fat could be utilized with 100% efficiency, how much mass could a 78-kg person lose running a 26.2-mile marathon? The energy expenditure rate for that mass is 125 kcal/mile.

Solution

The energy expended in running a marathon for a person with the given mass is $(125 \text{ kcal/mi}) \times (26.2 \text{ mi}) = 3.28 \times 10^3$ kcal. This is equivalent to the energy content of $3.28 \times 10^3 \text{ kcal}/(9 \text{ kcal/g}) = 364$ g, or about 13 oz, of fat.

Problem

19. A circular lake 1.0 km in diameter averages 10 m deep. Solar energy is incident on the lake at an average rate of 200 W/m^2. If the lake absorbs all this energy and does not exchange heat with its surroundings, how long will it take to warm from $10°C$ to $20°C$?

Solution

Since the energy absorbed by the lake equals the solar power times the time, $\Delta t = \Delta Q/\mathcal{P} = mc\Delta T \div (200 \text{ W/m}^2)A$, where m/A is the mass per unit area of lake surface. Therefore:

$$\Delta t = (10^3 \text{ kg/m}^3)(10 \text{ m})(4184 \text{ J/kg·K})(10 \text{ K})/(200 \text{ W/m}^2)$$
$$= 2.09 \times 10^6 \text{ s} = 24.2 \text{ d}.$$

Problem

21. How much heat is required to raise an 800-g copper pan from $15°C$ to $90°C$ if (a) the pan is empty; (b) the pan contains 1.0 kg of water; (c) the pan contains 4.0 kg of mercury?

Solution

(a) When just the pan is heated, $\Delta Q = m_{Cu}c_{Cu} \Delta T = (0.8 \text{ kg})(386 \text{ J/kg·K})(90-15)\text{K} = 23.2$ kJ $= 5.54$ kcal. (b) If the pan contains water and both are heated between the same temperatures, $\Delta Q = (m_{Cu}c_{Cu} + m_w c_w) \Delta T = 23.2 \text{ kJ} + (1 \text{ kg})(4184 \text{ J/kg·K}) \times (75 \text{ K}) = 337$ kJ $= 80.5$ kcal. (c) With 4 kg of mercury replacing the water, $\Delta Q = 23.2 \text{ kJ} + (4 \text{ kg}) \times (140 \text{ J/kg·K})(75 \text{ K}) = 65.2$ kJ $= 15.6$ kcal. (See Table 19-1 for the specific heats.)

Problem

23. How much power does it take to raise the temperature of a 1.3-kg copper pipe by $15°C/s$?

Solution

Dividing Equation 19-5 by the time, we get $\mathcal{P} = \Delta Q/\Delta t = mc\Delta T/\Delta t = (1.3 \text{ kg})(386 \text{ J/kg·K}) \times (15 \text{ K/s}) = 7.53$ kW.

Problem

25. You insert your microwave oven's temperature probe in a roast and start it cooking. You notice that the temperature goes up $1°C$ every 20 s. If the roast has the same specific heat as water, and if the oven power is 500 W, what is the mass of the roast? Neglect heat loss.

Solution

With no losses, the heat absorbed by the roast per second (power) equals the mass times the specific heat of the roast times the temperature rise per second, or $\mathcal{P} = \Delta Q/\Delta t = mc\Delta T/\Delta t$. Thus $m = (500 \text{ W}) \div (4184 \text{ J/kg·K})(1 \text{ K}/20 \text{ s}) = 2.39$ kg, or about $5\frac{1}{4}$ lb.

Problem

27. A stove burner supplies heat at the rate of 1.0 kW, a microwave oven at 625 W. You can heat water in

the microwave in a paper cup of negligible heat capacity, but the stove requires a pan whose heat capacity is 1.4 kJ/K. (a) How much water do you need before it becomes quicker to heat on the stovetop? (b) What will be the rate at which the temperature of this much water rises?

Solution

The temperature rise per second is equal to the heat supplied per second (i.e., the power supplied if there are no losses) divided by the total heat capacity of the water and its container: $\Delta T/\Delta t = (\Delta Q/\Delta t)/C_{tot} = P/C_{tot}$ (Equation 19-4 divided by Δt). The total heat capacity is $C_{tot} = C_W + C_{cnt}$, provided the water and container both have the same instantaneous temperature. (This assumes that heat is supplied sufficiently slowly that the water and container share it and stay in instantaneous thermal equilibrium.) For the paper cup used in the microwave oven, $C_{cnt} \approx 0$, whereas for the pan used on the stove burner, $C_{cnt} = 1.4$ kJ/K. Thus, the rate of temperature rise is 625 W$/C_W$ for the microwave and 1.0 kW$/(C_W + 1.4$ kJ/K) for the stove burner. (a) When $C_W = m_W c_W$ is small, the microwave is faster, whereas when C_W is large, the stove burner is faster. (To see this, plot both rates as a function of C_W.) The rates of temperature rise are equal for $C_W = m_W \times (4.184$ kJ/kg·K$) = (1.4$ kJ/K$)(0.625)/(1 - 0.625) = 2.33$ kJ/K. Therefore, $m_W = (2.33$ kJ/K$) \div (4.184$ kJ/kg·K$) = 0.558$ kg. (b) For m_W in part (a), the rate of temperature rise is $\Delta T/\Delta t = (0.625$ kW$) \div (2.33$ kJ/K$) = 1.0$ kW$/(2.33 + 1.4)($kJ/K$) = 0.268$ K/s.

Problem

29. A 1.2-kg iron tea kettle sits on a 2.0-kW stove burner. If it takes 5.4 min to bring the kettle and the water in it from 20°C to the boiling point, how much water is in the kettle?

Solution

The energy supplied by the stove burner heats the kettle and water in it from 20°C to 100°C. If we neglect any losses of heat and the heat capacity of the burner, this energy is just the burner's power output times the time, so $\Delta Q = P \Delta t = (m_W c_W + m_K c_K) \times \Delta T$ (Equation 19-5 for water and kettle). Since all of these quantities are given except for the mass of the water, we can solve for m_W :

$$
\begin{aligned}
m_W &= [(P \Delta t/\Delta T) - m_K c_K]/c_W \\
&= \{[(2 \text{ kW})(5.4 \times 60 \text{ s})/80 \text{ K}] - (1.2 \text{ kg}) \\
&\quad \times (447 \text{ J/kg·K})\}/(4184 \text{ J/kg·K}) \\
&= 1.81 \text{ kg.}
\end{aligned}
$$

Problem

31. Two cars collide head-on at 90 km/h. If all their kinetic energy ended up as heat, what would be the temperature increase of the wrecks? The specific heat of the cars is essentially that of iron.

Solution

$\Delta Q = \Delta K = 2(\frac{1}{2}mv^2) = 2mc \, \Delta T$ (if there are no energy losses), therefore $\Delta T = v^2/2c_{Iron} = (90 \text{ m}/3.6 \text{ s})^2/(2 \times 447 \text{ J/kg·K}) = 0.699$ K.

Problem

33. A leaf absorbs sunlight with intensity 600 W/m². The leaf has a mass per unit area of 100 g/m², and its specific heat is 3800 J/kg·K. In the absence of any heat loss, at what rate would the leaf's temperature rise?

Solution

The derivative of Equation 19-5 with respect to time relates the rate of temperature rise to the rate of heat energy absorbed, $dQ/dt = mc(dT/dt)$. Using the values of dQ/dt and m given for unit areas of leaf, one finds $dT/dt = (600 \text{ W/m}^2)/(0.1 \text{ kg/m}^2) \times (3800 \text{ J/kg·K}) = 1.58$ K/s.

Problem

35. A piece of copper at 300°C is dropped into 1.0 kg of water at 20°C. If the equilibrium temperature is 25°C, what is the mass of the copper?

Solution

Let us assume that all the heat lost by the copper is gained by the water, with no heat transfer to the container or its surroundings. Then $-\Delta Q_{Cu} = \Delta Q_W$ (as in Example 19-3) or $-m_{Cu} c_{Cu}(T - T_{Cu}) = m_W c_W(T - T_W)$. Solving for m_{Cu} (in terms of the other quantities given in the problem or in Table 19-1) one finds $m_{Cu} = m_W c_W(T - T_W)/c_{Cu}(T_{Cu} - T) = (1 \text{ kg})(4184 \text{ J/kg·K})(25 - 20) \text{ K}/(386 \text{ J/kg·K}) \times (300 - 25) \text{ K} = 0.197$ kg.

Problem

37. A thermometer of mass 83.0 g is used to measure the temperature of a 150-g water sample. The thermometer's specific heat is 0.190 cal/g·°C, and it reads 20.0°C before immersion in the water. The water temperature is initially 60.0°C. What does the thermometer read after it comes to equilibrium with the water?

Solution

If we assume that the thermometer and water are thermally insulated from their surroundings, Equation 19-6 (and its solution from Example 19-4) gives:

$$T = \frac{m_t c_t T_t + m_W c_W T_W}{m_t c_t + m_W c_W}$$
$$= \frac{(83.0)(0.190)(20.0°C) + (150)(1)(60.0°C)}{(83.0)(0.190) + (150)(1)}$$
$$= 56.2°C,$$

(we omitted common units in the numerator and denominator).

Sections 19-6 and 19-7: Heat Transfer and Thermal Energy Balance

Problem

39. The top of a steel wood stove measures 90 cm×40 cm, and is 0.45 cm thick. The fire maintains the inside surface of the stove top at 310°C, while the outside surface is at 295°C. Find the rate of heat conduction through the stove top.

Solution

Assuming a steady flow of heat through the 90×40 cm^2 = 0.36 m^2 area face, with no flow through the edges, we can use Equation 19-7 and Table 19-2: $H = -kA \, \Delta T/\Delta x = -(46 \text{ W/m·K})(0.36 \text{ m}^2)(295°C - 310°C)/(0.45 \text{ cm}) = 55.2$ kW. (The heat flow is positive, for x going from the inside of the stove to the outside, because the temperature gradient, $\Delta T/\Delta x$, is negative.)

Problem

41. Building heat loss in the United States is usually expressed in Btu/h. What is 1 Btu/h in SI?

Solution

The conversion to SI units is (1 Btu/h)(1055 J/Btu)× (1 h/3600 s) = 0.293 W.

Problem

43. What is the R-factor of a wall that loses 0.040 Btu each hour through each square foot for each °F temperature difference?

Solution

Equation 19-7 for the rate of heat-flow through a slab, written in terms of the thermal resistance of the slab (Equation 19-9), is $H = -\Delta T/R$. Therefore, a heat loss of 0.040 Btu/h per Fahrenheit degree of temperature difference, corresponds to a thermal

resistance of $R = 1$ F°/(0.040 Btu/h) = 25 F°·h/Btu (the heat-flow is in the direction of decreasing temperature, so the thermal resistance is positive). Equation 19-11 gives the R-factor of $A = 1$ ft^2 of slab as $\mathcal{R} = 25$. (The units of the R-factor are such that the numerical value of $\mathcal{R}$ divided by the area in ft^2 is the thermal resistance in F°·h/Btu.)

Problem

45. A biology lab's walk-in cooler measures 3.0m× 2.0 m ×2.3 m and is insulated with 8.0-cm-thick styrofoam. If the surrounding building is at 20°C, at what average rate must the cooler's refrigeration unit remove heat in order to maintain 4.0°C in the cooler?

Solution

The total surface area (sides, top, and bottom) of the cooler is $A = 2(3 \times 2 + 3 \times 2.3 + 2 \times 2.3)$ m^2 = 35 m^2. A thickness of 8 cm of styrofoam of this area has a thermal resistance of $R = \Delta x/kA = (0.08 \text{ m}) \div (0.029 \text{ W/m·K})(35 \text{ m}^2) = 7.88 \times 10^{-2}$ K/W (see Equation 19-9). Therefore, a heat-flow of magnitude $|\Delta T|/R = (20°C - 4°C)/(7.88 \times 10^{-2}$ K/W$) = 203$ W (see Equation 19-7) must be balanced by the refrigeration unit to maintain the desired steady state temperatures.

Problem

47. (a) What is the R-factor for a wall consisting of $\frac{1}{4}$-in. pine paneling, $\mathcal{R}$-11 fiberglass insulation, $\frac{3}{4}$-in. pine sheathing, and 2.0-mm aluminum siding? (b) What is the heat-loss rate through a 20 ft×8 ft section of wall when the temperature difference across the wall is 55°F?

Solution

(a) The wall consists of conducting slabs of different materials and thickness, of the same area (connected in "series"), so the discussion following Equation 19-10 shows that the R-factor of the combination is the sum of the individual R-factors, $\mathcal{R} = \mathcal{R}_{\text{pine}} + \mathcal{R}_{\text{fiberglass}} + \mathcal{R}_{\text{Al}}$. The first and last terms we calculate from Equation 19-11; the middle term is given:

$$\mathcal{R} = (\tfrac{1}{4} + \tfrac{3}{4})/(0.78) + 11 + (0.2/2.54)/(1644) = 12.3.$$

(We remembered to express Δx in inches and k in Btu·in·/h·ft^2·F°. The contribution of the aluminum siding is negligible.) (b) The thermal resistance of a 20×8 ft^2 area of such a wall is $R = (12.3/20 \times 8) \times$ (h·F°/Btu) (see solution to Problem 43), so the heat-flow through the wall, for a steady temperature difference of −55 F° (i.e., $T_{\text{outside}} - T_{\text{inside}} = \Delta T$), is

$H = -\Delta T/R = (55\ \text{F}°)(20 \times 8/12.3)(\text{Btu/h·F}°) = 715\ \text{Btu/h}.$

Problem 47 Solution.

Problem

49. Repeat the preceding problem for a south-facing window where the average sunlight intensity is 180 W/m^2.

Solution

The difference in heat loss between R-factors of 19 and 2.1, for a window area of 40 ft^2 and given temperature difference, is $\Delta H = -A\ \Delta T(\mathcal{R}_2^{-1} - \mathcal{R}_1^{-1}) = -(40\ \text{ft}^2) \times (15°\text{F} - 68°\text{F})(2.1^{-1} - 19^{-1})(\text{ft}^2\text{·F}°\text{·h/Btu})^{-1} = 898\ \text{Btu/h}.$ (A positive ΔH represents a greater heat-flow from inside the house to outside, or a loss of energy.) Over a winter month, (898 Btu/h) (30 d)(24 h/d)(1 gal/10^5 Btu) = 6.47 gal of oil would have to be consumed to compensate for this loss.

On the other hand, if a southern window location resulted in a net gain from solar power of (180 W/m^2)(1 Btu/1055 J)(3600 s/h)(40 ft^2) × (0.3048 m/ft)^2 = 2284 Btu/h, this would be equivalent to a savings of (2284 Btu/h)(30 d)(24 h/d) × (1 gal/10^5 Btu) = 16.4 gal of oil over a month. The resulting net savings is 16.4 − 6.47 = 9.97 gal of oil for one winter month.

Problem

51. A house is insulated so its total heat loss is 370 W/°C. On a night when the outdoor temperature is 12°C the owner throws a party, and 40 people come. The average power output of the human body is 100 W. If there are no other heat sources in the house, what will be the house temperature during the party?

Solution

The average thermal resistance of the house is given, since $1/R = 370$ W/C°. (The thermal resistance is defined as the reciprocal of the rate of heat-flow per degree temperature difference, i.e., $-H/\Delta T = 1/R$; see Equations 19-7 and 9.) In thermal energy balance, the power released by the people (owner plus guests) is equal to the rate of heat loss from the house (otherwise the house would heat up or cool down). Therefore, $\mathcal{P} = 41 \times 100$ W $= |H| = |\Delta T|/R = |\Delta T|$ (370 W/C°), or $|\Delta T| = 11.1$ C°. Since the outside temperature is 12°C, the temperature inside the house is 12°C $+ |\Delta T| = 23.1°$C.

Problem

53. An electric stove burner has surface area 325 cm^2 and emissivity $e = 1.0$. The burner is at 900 K and the electric power input to the burner is 1500 W. If room temperature is 300 K, what fraction of the burner's heat loss is by radiation?

Solution

The net power radiated (emitted at T_1, absorbed at T_2) is $\mathcal{P} = e\sigma A(T_1^4 - T_2^4) = (1)(5.67 \times 10^{-8}\ \text{W/m}^2\text{·K}^4) \times (3.25 \times 10^{-2}\ \text{m}^2)(300\ \text{K})^4(3^4 - 1) = 1194$ W. This is 79.6% of the input power (1500 W).

Problem

55. The average human body produces heat at the rate of 100 W and has total surface area of about 1.5 m^2. What is the coldest outdoor temperature in which a down sleeping bag with 4.0-cm loft (thickness) can be used without the body temperature dropping below 37°C? Consider only conductive heat loss.

Solution

Assume that in thermal energy balance, the rate of heat generation by a human body equals the rate of heat-flow lost by conduction through the sleeping bag, i.e., 100 W $= -kA\ \Delta T/\Delta x$. The sleeping bag may be considered to be closefitting (same surface area as the body), consisting of goose down insulation (see Table 19-2) and a negligible fabric shell. Then 100 W $= -(0.043\ \text{W/m·K})(1.5\ \text{m}^2) \times (T - 37°\text{C}) \div (0.04\ \text{m})$, or $T = 37°\text{C} - 62.0°\text{C} = -25°\text{C} = -13°\text{F}.$

Problem

57. Scientists worry that a nuclear war could inject enough dust into the upper atmosphere to reduce significantly the amount of solar energy reaching Earth's surface. If an 8% reduction in solar input occurred, what would happen to Earth's 287-K average temperature?

Solution

If we assume that the Earth's average temperature is proportional to the one-fourth power of the effective solar intensity ($T_{av} \sim S^{1/4}$), as explained in the text's application to the greenhouse effect and global warming, then reducing the intensity to $0.92S$ alters the average temperature according to $T'_{av} = T_{av} \times (0.92)^{1/4}$. This would result in a decrease in the present $T_{av} = 287$ K of $\Delta T_{av} = T_{av} - T'_{av} = [1 - (0.92)^{1/4}](287 \text{ K}) = 5.92$ K.

Paired Problems

Problem

59. A blacksmith heats a 1.1-kg iron horseshoe to 550°C, then plunges it into a bucket containing 15 kg of water at 20°C. What is the final temperature?

Solution

If we assume that all of the heat lost by the horseshoe is transferred to the water (in reality, some heat is lost to the surroundings and bucket), then the analysis of Example 19-3 applies and the equilibrium temperature is

$$T = \frac{(1.1)(0.107)(550°C) + (15)(1)(20°C)}{(1.1)(0.107) + (15)(1)} = 24.1°C.$$

(Note: We used specific heats from Table 19-1 in cal/g·C°, since in those units, the numerical value for water is unity, and we canceled the common units from the numerator and denominator in the expression for T.)

Problem

61. What is the power output of a microwave oven that can heat 430 g of water from 20°C to the boiling point in 5.0 minutes? Neglect the heat capacity of the container.

Solution

The average power supplied to the water is $\mathcal{P} = \Delta Q / \Delta t = mc\,\Delta T / \Delta t = (430 \text{ g})(1 \text{ cal/g·C°})(100°C - 20°C) \times (4.184 \text{ J/cal})/(5 \times 60 \text{ s}) = 480$ W. This is also the output of the microwave, if we neglect the power absorbed by the container and any leakage in the unit.

Problem

63. A cylindrical log 15 cm in diameter and 65 cm long is glowing red hot in a fireplace. If it's emitting radiation at the rate of 34 kW, what is its temperature? The log's emissivity is essentially 1.

Solution

If we neglect the radiation absorbed by the log from its environment (the fireplace brick, for example, does radiate heat to the room, but is probably at a temperature far less than red hot), then the net power radiated by the log is just $\mathcal{P} = e\sigma A T^4$. The surface

area of the log is $\pi dL + 2\pi R^2 = \pi d(L + R) = \pi(0.15 \text{ m})(0.65 \text{ m} + 0.075 \text{ m}) = 0.342 \text{ m}^2$, so solving for T, we find $T = [(34 \times 10^3 \text{ W})/(5.67 \times 10^{-8} \text{ W/m}^2 \cdot \text{K}^4) \times (0.342 \text{ m}^2)]^{1/4} = 1.15$ kK.

Problem

65. An enclosed rabbit hutch has a thermal resistance of 0.25 K/W. If you put a 50-W heat lamp in the hutch on a day when the outside temperature is −15°C, what will be the hutch temperature? Neglect the rabbit's metabolism.

Solution

The rate of heat loss of the hutch by conduction is $H = -\Delta T/R = -[(-15°C) - T]/(0.25 \text{ K/W}) = (4 \text{ W/K})(T + 15°C)$ (see Equations 19-7 and 9). If we neglect any heat loss by radiation and convection, and any heat generated by the rabbit, this is equal to the power supplied by the 50 W lamp at the equilibrium temperature of the hutch. Thus, $T = [50 \text{ W} \div (4 \text{ W/K})] - 15°C = -2.5°C$. (Perhaps the rabbit would be more comfortable with a 100 W lamp.)

Supplementary Problems

Problem

67. Rework Example 19-5, now assuming that the house has 10 single-glazed windows, each measuring 2.5 ft × 5.0 ft. Four of the windows are on the south, and admit solar energy at the average rate of 30 Btu/h·ft². *All* the windows lose heat; their R-factor is 0.90. (a) What is the total heating cost for the month? (b) How much is the solar gain worth?

Solution

The window area is $10(2.5 \text{ ft} \times 5.0 \text{ ft}) = 125$ ft². The wall area is 125 ft² less than in Example 19-5, or $A_{walls} = 1506 \text{ ft}^2 - 125 \text{ ft}^2 = 1381$ ft². The heat losses through the various structural parts are:

$$|H|_{walls} = \left(\frac{1}{12.37} \frac{\text{Btu}}{\text{h·ft}^2 \cdot °\text{F}}\right)(1381 \text{ ft}^2)(50°\text{F})$$
$$= 5583 \text{ Btu/h},$$

$$|H|_{roof} = \left(\frac{1}{31.37} \frac{\text{Btu}}{\text{h·ft}^2 \cdot °\text{F}}\right)(1164 \text{ ft}^2)(50°\text{F})$$
$$= 1855 \text{ Btu/h},$$

$$|H|_{windows} = \left(\frac{1}{0.90} \frac{\text{Btu}}{\text{h·ft}^2 \cdot °\text{F}}\right)(125 \text{ ft}^3)(50°\text{F})$$
$$= 6944 \text{ Btu/h}.$$

If we include the gain from the south windows, $4(12.5 \text{ ft}^2)(30 \text{ Btu/h·ft}^2) = 1500$ Btu/h, the net rate of loss of energy from the entire house is $(5583 + 1855 + 6944 - 1500)$ Btu/h $= 12.88 \times 10^3$ Btu/h. The monthly fuel bill is $(12.88 \times 10^3$ Btu/h$)(24 \times 30$ h/mo$)(1$ gal$\div 10^5$ Btu$)(\$0.94$/gal$) = \$87.19$/mo. The solar gain from the south windows is worth $(1500)(24 \times 30) \times (\$0.94$/mo$)/10^5 = \$10.11$/mo.

Problem

69. My house currently burns 160 gallons of oil in a typical winter month when the outdoor temperature averages 15°F and the indoor temperature averages 66°F. Roof insulation consists of $\mathcal{R} = 19$ fiberglass, and the roof area is 770 ft^2. If I double the thickness of the roof insulation, by what percentage will my heating bills drop? A gallon of oil yields about 100,000 Btu of heat.

Solution

The rate of heat loss by conduction from the currently insulated house can be written as $H_0 = -\Delta T \times [(A/\mathcal{R})_{\text{roof}} + (A/\mathcal{R})_{\text{rest}}]$, where $\Delta T = 15°F - 66°F = -51$ F°, $A_{\text{roof}} = 770$ ft^2, $\mathcal{R}_{\text{roof}} = 19$ (ft$^2 \cdot$F°·h/Btu), and A_{rest} and $\mathcal{R}_{\text{rest}}$ are the effective area and R-factor for the rest of the house (i.e., walls, windows, floor, etc.). If the R-factor for the roof is doubled, the rate of heat loss will be changed by $\Delta H = \Delta T A_{\text{roof}}/2\,\mathcal{R}_{\text{roof}}$, which can be calculated from the given data: $\Delta H = (-51$ F°$)(770$ ft$^2)/2(19$ ft$^2 \cdot$F°·h/Btu$) = -1.03 \times 10^3$ Btu/h. (A negative change represents a reduction in heat loss; or a drop in heating costs.) The original rate of heat loss can be calculated from the given oil consumption: $H_0 = (160$ gal/mo$)(10^5$ Btu/gal$) \times (1$ mo/30×24 h$) = 2.22 \times 10^4$ Btu/h. Thus, the extra insulation would result in a savings of $|\Delta H/H_0| = 1.03/22.2 = 4.65\%$.

Problem

71. A copper pan 1.5 mm thick and a cast iron pan 4.0 mm thick are sitting on electric stove burners; the bottom area of each pan is 300 cm^2. Each contains 2.0 kg of water whose temperature is rising at the rate of 0.15 K/s. Find the temperature difference between the inside and outside bottom of each pan.

Solution

We assume that the heat-flow through the bottom of each pan by conduction, $H = -kA\,\Delta T_{\text{pan}}/\Delta x$ from Equation 19-7, raises the temperature of the water at the given rate, $H = \Delta Q/\Delta t = m_W c_W(\Delta T_W/\Delta t)$ from energy balance and Equation 19-5 divided by Δt. (Note that the heat-flow through the pan is from its

outside to its inside, so $\Delta T_{\text{pan}} = T_{\text{in}} - T_{\text{out}}$ is negative. We are also ignoring any heat transfer by convection and radiation.) Then, $T_{\text{out}} - T_{\text{in}} = -\Delta T_{\text{pan}} = m_W c_W(0.15$ K/s$)\,\Delta x/kA$. Using the given values for each pan and Tables 19-1 and 2, we find (a) $-\Delta T_{cu} = (2.0$ kg$)(4184$ J/kg·K$)(0.15$ K/s$)(1.5 \times 10^{-3}$ m$) \div (401$ W/m·K$)(0.03$ m$^2) = 0.157$ K, and (b) $-\Delta T_{Fe} = (4/1.5)/(80.4/401)(-\Delta T_{cu}) = 2.08$ K.

Problem

73. At low temperatures the specific heat of a solid is approximately proportional to the cube of the temperature; for copper the specific heat is given by $c = 31(T/343$ K$)^3$ J/g·K. When heat capacity is not constant, Equations 19-4 and 19-5 must be written in terms of the derivative dQ/dT and integrated to get the total heat involved in a temperature change. Find the heat required to bring a 40-g sample of copper from 10 K to 25 K.

Solution

If we write Equation 19-5 in the form $dQ = mc(T)\,dT$ and integrate, as suggested in the problem, we obtain

$$
\begin{aligned}
Q &= m \int_{T_1}^{T_2} c(T)\,dT \\
&= m \int_{10\ \text{K}}^{25\ \text{K}} (31\ \text{J/g·K})(T/343\ \text{K})^3\,dT \\
&= (40\ \text{g})(31\ \text{J/g·K})\frac{(25\ \text{K})^4 - (10\ \text{K})^4}{4(343\ \text{K})^3} \\
&= 2.92\ \text{J} = 0.699\ \text{cal.}
\end{aligned}
$$

Problem

75. A pipe of length ℓ and radius R_1 is surrounded by insulation of outer radius R_2 and thermal conductivity k. Use the methods of the preceding problem to show that the heat loss rate through the insulation is

$$
H = \frac{2\pi k\ell(T_1 - T_2)}{\ln(R_2/R_1)}.
$$

Hint: Consider the heat flow through a thin layer of thickness dr and temperature difference dT as shown in Fig. 19-28.

Solution

Although heat-flow is essentially a 3-dimensional process, the one-dimensional Equation 19-14 from the previous problem can be used if we assume a steady heat-flow, H, in the radial direction, through each thin cylindrical layer of insulating material of thickness dr and area $A = 2\pi r\ell$, so that $H = -kA \times dT/dr$. For steady flow, the temperature difference across any cylindrical layer is constant in time, so H is constant and the temperature must vary with radial distance, from T_1 at $r = R_1$ to T_2 at $r = R_2$, such that $dT = -(H/k)dr/2\pi r\ell$. Thus,

$$\int_{T_1}^{T_2} dT = T_2 - T_1 = -\frac{H}{2\pi k\ell} \int_{R_1}^{R_2} \frac{dr}{r}$$

$$= -\frac{H}{2\pi k\ell} \ln\left(\frac{R_2}{R_1}\right),$$

or $H = 2\pi k\ell(T_1 - T_2)/\ln(R_2/R_1)$. (Note that H is in the positive radial direction for $T_1 > T_2$ and we assumed no heat-flow along the length of the pipe.)

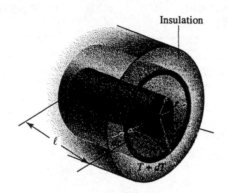

Insulation

FIGURE 19-28 Problem 75 Solution.

Problem

77. A house is at 20°C on a winter day when the outdoor temperature is −15°C. Suddenly the furnace fails. Use the result of the previous problem to determine how long it will take the house temperature to reach the freezing point. The heat capacity of the house is 6.5 MJ/K, and its thermal resistance is 6.67 mK/W.

Solution

In the previous problem, the heat flow from the object is proportional to the temperature difference between it and the surroundings, $H = dQ/dt = -(T - T_0)/R$. This can be transformed into the desired equation by using the chain rule and the definition of heat capacity: $dQ/dt = (dQ/dT)(dT/dt) = C(dT/dt) = -(T - T_0)/R$. If we separate variables and integrate from initial values $t = 0$ and T_1, to arbitrary final values t and T, we obtain:

$$\int_{T_1}^{T} \frac{dT'}{T' - T_0} = \ln\left(\frac{T - T_0}{T_1 - T_0}\right)$$

$$= -\int_{0}^{t} \frac{1}{RC} dt' = -\frac{t}{RC},$$

or $T - T_0 = (T_1 - T_0)\exp(-t/RC)$. The temperature at $t = 0$ was chosen to be T_1; as $t \to \infty$, $T - T_0 \to 0$, i.e., the body cools to the temperature of its surroundings.

This result can be applied to the house described in this problem: $(T - T_0)/(T_1 - T_0) = [0°C - (-15°C)]/[20°C - (-15°C)] = 3/7 = \exp(-t/RC)$ or $t = RC \ln(7/3) = (6.67 \text{ mK/W})(6.5 \text{ MJ/K}) \ln(7/3) = 3.67 \times 10^4$ s $= 10.2$ h (enough time for the emergency service to arrive?).

CHAPTER 20 THE THERMAL BEHAVIOR OF MATTER

ActivPhysics can help with these problems: Activities 8.1–8.4

Section 20-1: Gases

Problem

1. At Mars's surface, the planet's atmosphere has a pressure only 0.0070 times that of Earth, and an average temperature of 218 K. What is the volume of 1 mole of the Martian atmosphere?

Solution

The molar volume of an ideal gas at STP for the surface of Mars can be calculated as in Example 20-1. However, expressing the ideal gas law for 1 mole of gas at the surfaces of Mars and Earth as a ratio, $P_M V_M / T_M = P_E V_E / T_E$, and using the previous numerical result, we find $V_M = (P_E/P_M)(T_M/T_E) \times V_E = (1/0.0070)(218/273)(22.4\times10^{-3}$ m$^3) = 2.56$ m^3.

Problem

3. What is the pressure of an ideal gas if 3.5 moles occupy 2.0 L at a temperature of $-150°$C?

Solution

The ideal gas law in terms of the gas constant per mole, Equation 20-2, gives $P = nRT/V = (3.5$ mol$)(8.314$ J/K·mol$)(123$ K$)/(0.002$ m$^3) = 1.79\times10^6$ Pa. (The absolute temperature must be used, but any convenient units for the gas constant can be used, e.g., $R \simeq 0.0821$ L·atm/K·mol. Then $P = (3.5$ mol$)(0.0821$ L·atm/K·mol$)(123$ K$)/(2$ L$) = 17.7$ atm.)

Problem

5. If 2.0 mol of an ideal gas are at an initial temperature of 250 K and pressure of 1.5 atm, (a) what is the gas volume? (b) The pressure is now increased to 4.0 atm, and the gas volume drops to half its initial value. What is the new temperature?

Solution

(a) From Equation 20-2:

$$V = \frac{nRT}{P} = \frac{(2 \text{ mol})(8.314 \text{ J/mol·K})(250 \text{ K})}{(1.5 \text{ atm})(1.013\times10^5 \text{ Pa/atm})}$$

$$= 2.74\times10^{-2} \text{ m}^3 = 27.4 \text{ L}.$$

(b) The ideal gas law in ratio form (for a fixed quantity of gas, $N_1 = N_2$) gives:

$$\frac{T_2}{T_1} = \frac{P_2 V_2}{P_1 V_1}, \quad \text{or}$$

$$T_2 = \left(\frac{4.0 \text{ atm}}{1.5 \text{ atm}}\right)\left(\frac{0.5V_1}{V_1}\right)(250 \text{ K}) = 333 \text{ K}.$$

Problem

7. A pressure of 1.0×10^{-10} Pa is readily achievable with laboratory vacuum apparatus. If the residual air in this "vacuum" is at $0°$C, how many air molecules are in one liter?

Solution

$$N = PV/kT$$
$$= (10^{-10} \text{ Pa})(10^{-3} \text{ m}^3)/(1.38\times10^{-23} \text{ J/K})(273 \text{ K})$$
$$= 2.65\times10^7 \text{ (see Equation 20-1)}.$$

Problem

9. A helium balloon occupies 8.0 L at $20°$C and 1.0 atm pressure. The balloon rises to an altitude where air pressure is 0.65 atm and the temperature is $-10°$C. What is its volume when it reaches equilibrium at the new altitude? (Neglect tension forces in the material of the balloon.)

Solution

Use the ideal gas law (Equation 20-1) in ratio form, to compare two different states: $P_1 V_1/P_2 V_2 = N_1 T_1/N_2 T_2$. Since the balloon contains the same number of molecules of gas (if none escape), $N_1 = N_2$, and $V_2 = (P_1/P_2)(T_2/T_1)V_1 = (1/0.65)(263/293) \times (8.0$ L$) = 11.0$ L. (Note that absolute temperatures must be used, and that any consistent units for the ratio of the other quantities conveniently cancel.)

Problem

11. An aerosol can of whipped cream is pressurized at 440 kPa when it's refrigerated at $3°$C. The can warns against temperatures in excess of $50°$C. What is the maximum safe pressure for the can?

Solution

For a fixed amount of ideal gas at constant volume, the pressure is proportional to the absolute temperature, $P_1/T_1 = P_2/T_2 =$ constant. (See the solution to Problem 9.) Thus, the maximum safe aerosol pressure is $P_2 = (T_2/T_1)P_1 = (323/276)(440 \text{ kPa}) = 515 \text{ kPa}$, or about 5.1 atm.

Problem

13. A 3000-ml flask is initially open while in a room containing air at 1.00 atm and 20°C. The flask is then closed, and immersed in a bath of boiling water. When the air in the flask has reached thermodynamic equilibrium, the flask is opened and air allowed to escape. The flask is then closed and cooled back to 20°C. (a) What is the maximum pressure reached in the flask? (b) How many moles escape when air is released from the flask? (c) What is the final pressure in the flask?

Solution

The initial conditions of the gas are $P_1 = 1$ atm, $V_1 = 3$ L, $T_1 = 293$ K, and $n_1 = P_1V_1/RT_1 = (1 \text{ atm})(3 \text{ L})/(8.206 \times 10^{-2} \text{ L·atm/mol·K})(293 \text{ K}) = 0.125$ mol. (a) T_2 is the temperature of boiling water at 1 atm of pressure, or 373 K. Since the original quantity of gas was heated at constant volume, $P_2 = (T_2/T_1)P_1 = (373/293)(1 \text{ atm}) = 1.27$ atm, which is the maximum. (b) When the flask is opened at 373 K, the pressure decreases to 1 atm, so the quantity of gas remaining is $n_2 = P_1V_1/RT_2 = n_1(T_1/T_2) = (0.125 \text{ mol})(293/373) = 0.0980$ mol. Therefore, the amount which escaped was $n_1 - n_2 = 0.0268$ mol. (c) The pressure of the remaining gas is $P_3 = n_2RT_1/V_1 = (n_2/n_1)P_1 = (0.098/0.125)(1 \text{ atm}) = 0.786$ atm.

Problem

15. In which gas are the molecules moving faster: hydrogen (H_2) at 75 K or sulfur dioxide (SO_2) at 350 K?

Solution

Comparing the thermal speeds, $v_{th} = \sqrt{3kT/m}$, for H_2 (mass $\simeq 2$ u) and SO_2 (mass $\simeq 64$ u) at the given temperatures, we find $v_{th}(H_2)/v_{th}(SO_2) = \sqrt{(T_{H_2}/T_{SO_2})(m_{SO_2}/m_{H_2})} = \sqrt{(75/350)(64/2)} = 2.62$; the hydrogen is faster.

Problem

17. The van der Waals constants for helium gas (He) are $a = 0.0341$ L^2·atm/mol^2 and $b = 0.0237$ L/mol. What is the temperature of

3.00 mol of helium at 90.0 atm pressure if the gas volume is 0.800 L? How does this result differ from the ideal gas prediction?

Solution

From Equation 20-7,

$$T = \left(90 \text{ atm} + \frac{(3 \text{ mol})^2(0.0341 \text{ L}^2\text{·atm/mol}^2)}{(0.8 \text{ L})^2} \right)$$
$$\times \frac{[0.8 \text{ L} - (3 \text{ mol})(0.0237 \text{ L/mol})]}{(3 \text{ mol})(8.206 \times 10^{-2} \text{ L·atm/mol·K})} = 268 \text{ K}.$$

This differs from $(90 \text{ atm})(0.8 \text{ L})/(3 \text{ mol}) \times (8.206 \times 10^{-2} \text{ L·atm/mol·K}) = 292$ K, the ideal gas temperature, by about 8.4%.

Problem

19. Because the correction terms (n^2a/V^2 and $-nb$) in the van der Waals equation have opposite signs, there is a point at which the van der Waals and ideal gas equations predict the same temperature. For the gas of Example 20-3, at what pressure does that occur?

Solution

The van der Waals and ideal gas temperatures are the same when $(P + n^2a/V^2)(V - nb) = PV$, or $P = na(V - nb)/bV^2$. If $n = 1, V = 2$ L and the values of a and b from Example 20-3 are substituted, one finds $P = 1.76$ MPa.

Problem

21. In a sample of 10^{24} hydrogen (H_2) molecules, how many molecules have speeds between 900 and 901 m/s (a) at a temperature of 100 K and (b) at 450 K?

Solution

The right-hand side of Equation 20-6 is the number of molecules with speeds in the given range. If we substitute $N = 10^{24}$, $v = 900$ m/s, $\Delta v = 1$ m/s, $m = 2 \times 1.66 \times 10^{-27}$ kg (for H_2 molecules), and $k = 1.38 \times 10^{-23}$ J/K, we get $\Delta N = N(v) \Delta v = (2.41 \times 10^{27} \text{ K}^{3/2})T^{-3/2} \exp\{-97.4 \text{ K}/T\}$. (a) For $T = 100$ K, $\Delta N = 9.10 \times 10^{20}$, and (b) for $T = 450$ K, $\Delta N = 2.03 \times 10^{20}$.

Section 20-2: Phase Changes

Problem

23. How much energy does it take to melt a 65-g ice cube?

Solution

The energy required for a solid-liquid phase transition at the normal melting point of water (0°C) is (Equation 20-8) $Q = mL_f = (0.065 \text{ kg})(334 \text{ kJ/kg}) = 21.7 \text{ kJ}$, or 5.19 kcal. (See Table 20-1 for the heats of transformation.)

Problem

25. If it takes 840 kJ to vaporize a sample of liquid oxygen, how large is the sample?

Solution

Assuming the vaporization takes place at the normal boiling point for oxygen at atmospheric pressure, we may use Equation 20-8 and Table 20-1 to obtain $m = Q/L_v = 840 \text{ kJ}/(213 \text{ kJ/kg}) = 3.94 \text{ kg}$.

Problem

27. Find the energy needed to convert 28 kg of liquid oxygen at its boiling point into gas.

Solution

From Equation 20-8 and Table 20-1, $Q = mL_v = (28 \text{ kg})(213 \text{ kJ/kg}) = 5.96 \text{ MJ}$.

Problem

29. If a 1-megaton nuclear bomb were exploded deep in the Greenland ice cap, how much ice would it melt? Assume the ice is initially at about its freezing point, and consult Appendix C for the appropriate energy conversion.

Solution

A 1-megaton nuclear device releases about 4.16×10^{15} J. This amount of energy is capable of melting $m = Q/L_f = 4.16 \times 10^{15} \text{ J}/(334 \text{ kJ/kg}) = 1.25 \times 10^{10}$ kg of ice at the normal melting point of 0°C.

Problem

31. What is the power of a microwave oven that takes 20 min to boil dry a 300-g cup of water initially at its boiling point?

Solution

The energy required to vaporize the water at 100°C in 20 min is $Q = mL_v$ (see Equation 20-8 and Table 20-1), so the average power supplied by the microwave oven was $P = Q/t = (0.3 \text{ kg}) \times (2257 \text{ kJ/kg})/(20 \times 60 \text{ s}) = 564 \text{ W}$.

Problem

33. A refrigerator extracts energy from its contents at the rate of 95 W. How long will it take to freeze

750 g of water already at 0°C?

Solution

The refrigerator must extract $Q = mL_f = (0.75 \text{ kg}) \times (334 \text{ kJ/kg}) = 251 \text{ kJ}$ of heat energy in order to freeze the given quantity of water at 0°C. Since the rate of energy extraction is $P = Q/t = 95$ W, this requires a time $t = (251 \text{ kJ})/(95 \text{ W}) = 2.64 \times 10^3 \text{ s} = 43.9 \text{ min}$.

Problem

35. At its "thaw" setting a microwave oven delivers 210 W. How long will it take to thaw a frozen 1.8-kg roast, assuming the roast is essentially water and is initially at 0°C?

Solution

Using the same reasoning as in the solution to Problem 33, we find $t = mL_f/P = (1.8 \text{ kg}) \times (334 \text{ kJ/kg})/(210 \text{ W}) = 2.86 \times 10^3 \text{ s} = 47.7 \text{ min}$.

Problem

37. A 100-g block of ice, initially at −20°C, is placed in a 500-W microwave oven. (a) How long must the oven be on to produce water at 50°C? (b) Make a graph showing temperature versus time during this entire interval.

Solution

(a) To bring the ice to 0°C requires heat: $Q_1 = mc_{ice}\Delta T = (0.1 \text{ kg})(2.05 \text{ kJ/kg·K})[0°C - (-20°C)] = 4.10 \text{ kJ}$. To melt the ice requires $Q_2 = mL_f = (0.1 \text{ kg})(334 \text{ kJ/kg}) = 33.4 \text{ kJ}$. Finally, to bring the meltwater from 0°C to 50°C requires heat $Q_3 = mc_{water}\Delta T = (0.1 \text{ kg})(4.184 \text{ kJ/kg·K})(50°C - 0°C) = 20.9 \text{ kJ}$. Power must be supplied for a time $t = Q_{tot}/P$, or $t = (4.10 + 33.4 + 20.9) \text{ kJ}/0.5 \text{ kW} = 117 \text{ s}$. (b) It takes times $t_1 = Q_1/0.5 \text{ kW} = 8.2 \text{ s}, t_2 = 66.8 \text{ s}$, and $t_3 = 41.8 \text{ s}$ for the preceding steps, respectively. Since the power input is constant, the temperature is

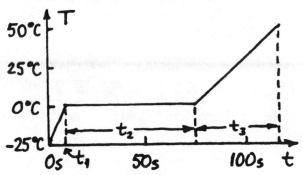

Problem 37 Solution.

linear for steps 1 and 3, and, of course, constant during melting, as shown.

Problem

39. How much energy does it take to melt 10 kg of ice initially at $-10°C$? Consult Table 19-1.

Solution

Energy must be supplied to first raise the ice temperature to the melting point, and then change its phase. Thus, $Q = mc_{ice}\Delta T + mL_f = (10\text{ kg}) \times$ $[(2.05\text{ kJ/kg·K})(10\text{ K}) + 334\text{ kJ/kg}] = 3.55$ MJ. (Combine Equations 19-5 and 20-8.)

Problem

41. A 250-g piece of ice at $0°C$ is placed in a 500-W microwave oven and the oven run for 5.0 min. What is the temperature at the end of this time?

Solution

The total energy supplied by the microwave oven is 500 W × 300 s = 150 kJ. The 0.25 kg piece of ice at $0°C$ absorbs $mL_f = 0.25\text{ kg}(334\text{ kJ/kg}) =$ 83.5 kJ while melting completely. Thus, there is $(150 - 83.5)$ kJ = 66.5 kJ of energy available to raise the temperature of 0.25 kg of water. The temperature rise is $\Delta T = \Delta Q/mc = 66.5\text{ kJ}/(0.25\text{ kg})(4.184\text{ kJ/kg·K}) =$ 63.6 K, which is also the numerical final Celsius temperature $T_f = T_i + \Delta T = 63.6°C$, since the initial temperature was $0°C$.

Problem

43. What is the minimum amount of ice in Example 20-6 that will ensure a final temperature of $0°C$?

Solution

In order for the final equilibrium temperature in Example 20-6 to be $0°C$, the original 1 kg of water must lose at least $Q_2 = 62.8$ kJ of heat energy. (It could lose more, if some or all of it froze, but this would clearly require a greater amount of original ice.) The minimum amount of original ice that could gain Q_2, without exceeding $0°C$, would be just completely melted, so $Q_2 = m_{ice}(c_{ice}\Delta T + L_f) = m_{ice}[(2.05\text{ kJ} \div$ kg·K$)(0°C - (-10°C)) + 334\text{ kJ/kg}]$. Therefore $m_{ice} = 62.8\text{ kJ}/(354.5\text{ kJ/kg}) = 177$ g. (Note: The maximum amount of original ice, which could produce a final temperature of $0°C$, while freezing all the original water, is $m_W(c_W\Delta T_W + L_f)/c_{ice}\Delta T_{ice} =$ $(62.8 + 334)\text{ kJ}/(20.5\text{ kJ/kg}) = 19.4$ kg. Between these limits, $m_{ice} = m_W c_W \Delta T_W/c_{ice}\Delta T_{ice} = (1\text{ kg}) \times$ $(4.184/2.05)(15/10) = 3.06$ kg gives a final mixture

with the original amounts of water and ice at $0°C$. For 177 g $< m_{ice} <$ 3.06 kg, some of the original ice melts, and for 3.06 kg $< m_{ice} <$ 19.4 kg, some of the original water freezes.)

Problem

45. A 500-g chunk of solid mercury at its 234 K melting point is added to 500 g of liquid mercury at room temperature (293 K). Determine the equilibrium mix and temperature.

Solution

Since the liquid mercury (Hg) is at a higher initial temperature than the solid chunk, it will lose heat energy (all of which is assumed to be absorbed by the solid chunk) and cool. The solid Hg will gain heat energy and start to melt. The heat energy that the liquid Hg could lose if it were to cool to the melting point is $m_{liq}c_{liq}\Delta T_{liq} = (0.5\text{ kg})(0.140\text{ kJ/kg·K}) \times$ $(293 - 234)$ K = 4.13 kJ. The heat energy that the solid Hg would gain if it were to melt completely at 234 K is $m_{solid}L_f = 0.5\text{ kg} \times 11.3\text{ kJ/kg} = 5.65$ kJ. Since the latter is greater than the former, we conclude that not all of the solid Hg melts, and that the final equilibrium temperature of the mixture is 234 K. The amount of solid Hg that does melt is determined by the heat energy lost by the liquid, $\Delta m = 4.13\text{ kJ}/L_f = 365$ g. Therefore, there is $500 - 365 = 135$ g of solid Hg and $500 + 365 = 865$ g of liquid Hg in the final mixture.

Problem

47. A bowl contains 16 kg of punch (essentially water) at a warm $25°C$. What is the minimum amount of ice at $0°C$ that will cool the punch to $0°C$?

Solution

Assume that the only heat transfer is between the punch and the ice. To cool to $0°C$, $\Delta Q =$ $(16\text{ kg})(4.184\text{ kJ/kg·K})(25°C - 0°C) = 1.67$ MJ of heat must be extracted from the punch. A minimum mass $m = \Delta Q/L_f = 1.67\text{ MJ}/(334\text{ kJ/kg}) = 5.01$ kg of ice at $0°C$ could do this, but the punch would be diluted with 5.01 kg of melt-water. (To reduce the dilution, sufficient ice at a temperature below $0°C$ is needed.)

Problem

49. A 50-g ice cube at $-10°C$ is placed in an equal mass of water. What must be the initial water temperature if the final mixture still contains equal amounts of ice and water?

Solution

Let us assume that all the heat gained by the ice was lost by the water, with no heat transfer to a container or the surroundings. An equilibrium mixture of ice and water (at atmospheric pressure) must be at 0°C, and if the masses of ice and water start out and remain equal, there is no net melting or freezing. Then $\Delta Q_{ice} = m_{ice}c_{ice}(0°C - T_{ice}) = -\Delta Q_W = m_W c_W \times (T_W - 0°C)$, or $T_W = m_{ice}c_{ice}(-T_{ice})/m_W c_W = (2.05 \div 4.184)(10°C) = 4.90°C$, where we canceled equal quantities and units.

Section 20-3: Thermal Expansion

Problem

51. A Pyrex glass marble is 1.00000 cm in diameter at 20°C. What will be its diameter at 85°C?

Solution

The linear expansion coefficient for Pyrex glass is given in Table 20-2, so we can calculate the diameter of the marble from Equation 20-10. $\Delta L = \alpha L \Delta T = (3.2 \times 10^{-6} \text{ K}^{-1})(1 \text{ cm})(85°C - 20°C) = 2.08 \times 10^{-4}$ cm, thus $L' = L + \Delta L = 1.00021$. (Note: We expressed the diameter at 85°C to the same accuracy as that given for 20°C.)

Problem

53. Suppose a single piece of welded steel railroad track stretched 5000 km across the continental United States. If the track were free to expand, by how much would its length change if the entire track went from a cold winter temperature of −25°C to a hot summer day of 40°C?

Solution

A naive application of Equation 20-10, with α for steel from Table 20-2 to two significant figures, gives $\Delta L = \alpha L \Delta T = (12 \times 10^{-6} \text{ K}^{-1})(5000 \text{ km})(40°C - (-25°C)) = 3.9$ km.

Problem

55. The tube in a mercury thermometer is 0.10 mm in diameter. What should be the volume of the thermometer bulb if a 1.0-mm rise is to correspond to a temperature change of 1.0°C? Neglect the expansion of the glass.

Solution

There should be a volume of mercury (V) in the reservoir bulb of the thermometer, such that the change in its volume $(\Delta V = \beta V \Delta T)$ over the full range of temperature equals the volume of the tube

into which it expands $(\frac{1}{4}\pi d^2 L)$. Here, we have neglected the expansion of the glass, as suggested, since $3\alpha_{glass} \ll \beta_{mercury}$ in Table 20-2. Thus $\frac{1}{4}\pi d^2 L = \beta V \Delta T$, or $V = (\frac{1}{4}\pi d^2/\beta)(L/\Delta T)$. Since the gradation of the thermometer, $L/\Delta T = 1$ mm/C°, is given, we find $V = \frac{1}{4}\pi(0.1 \text{ mm})^2(1 \text{ mm/K}) \div (18 \times 10^{-5} \text{ K}^{-1}) = 43.6 \text{ mm}^3$.

Problem

57. A steel ball bearing is encased in a Pyrex glass cube 1.0 cm on a side. At 330 K, the ball bearing fits tightly in the cube. At what temperature will it have a clearance of 1.0 μm all around?

Solution

Since the coefficient of linear expansion of steel is greater than that of Pyrex glass, the unit must be cooled to provide clearance. The difference in the contraction of steel and pyrex equals twice the given clearance on one side, so $|\Delta L_{steel}| - |\Delta L_{pyrex}| = 2 \mu\text{m} = (\alpha_{steel} - \alpha_{pyrex})L|\Delta T| = (12 - 3.2) \times 10^{-6} \text{ K}^{-1}(1 \text{ cm})(330 \text{ K} - T)$. Thus, $T = (330 - 2/0.088) \text{ K} = 307 \text{ K}$.

Problem

59. A rod of length L_0 is clamped rigidly at both ends. Its temperature increases by an amount ΔT, and in the ensuing expansion it cracks to form two straight pieces, as shown in Fig. 20-21. Find an expression for the distance d shown in the figure, in terms of $L_0, \Delta T$, and the coefficient of linear expansion, α.

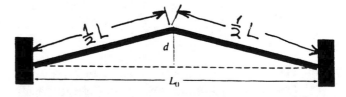

FIGURE 20-21 Problem 59.

Solution

If the two straight pieces in Fig. 20-21 are of equal length, the Pythagorean Theorem gives $d = \sqrt{(\frac{1}{2}L)^2 - (\frac{1}{2}L_0)^2}$, where $L = L_0(1 + \alpha\Delta T)$ is the total expanded length of the rod. Substitution gives $d = (\frac{1}{2}L_0)\sqrt{2\alpha\Delta T + \alpha^2\Delta T^2}$.

Paired Problems

Problem

61. What is the density, in moles per m³, of air in a tire whose absolute pressure is 300 kPa at 34°C?

Solution

The molar density implied by the ideal gas law (which is a good approximation for air under the conditions stated in the problem) is $n/V = P/RT = (300 \text{ kPa})/(8.314 \text{ J/K·mol})(273+34) \text{ K} = 118 \text{ mol/m}^3$.

Problem

63. What power is needed to melt 20 kg of ice in 6.0 min?

Solution

If the melting occurs at atmospheric pressure and the normal melting point, the heat of transformation from Table 20-1 requires $\mathcal{P} = Q/t = mL_f/t = (20 \text{ kg})(334 \text{ kJ/kg})/(6 \times 60 \text{ s}) = 18.6 \text{ kW}$.

Problem

65. You put 300 g of water into a 500-W microwave oven and accidentally set the time for 20 min instead of 2.0 min. If the water is initially at 20°C, how much is left at the end of 20 min?

Solution

In 20 min, $Q = (0.5 \text{ kW})(20 \times 60 \text{ s}) = 600 \text{ kJ}$ of heat energy is transferred to the water (if we ignore energy absorbed by a container or lost to the surroundings). The energy consumed in raising the water's temperature to the normal boiling point is $mc\Delta T = (0.3 \text{ kg})(4.184 \text{ kJ/kg·K})(100°C - 20°C) = 100 \text{ kJ}$, so 500 kJ is left to vaporize some of the water. Equation 20-6 gives the amount vaporized as $500 \text{ kJ}/(2257 \text{ kJ/kg}) = 221 \text{ g}$, therefore $300 - 221 = 78.7$ g of boiling water (or less than 3 oz) is all that remains.

Problem

67. Describe the composition and temperature of the equilibrium mixture after 1.0 kg of ice at −40°C is added to 1.0 kg of water at 5.0°C.

Solution

Assume that all the heat lost by the water is gained by the ice. The temperature of the water drops and that of the ice rises. If either reaches 0°C, a change of phase occurs, freezing or melting, depending on which reaches 0°C first. The water would lose $mc\Delta T = (1 \text{ kg})(4.184 \text{ kJ/kg·K})(5 \text{ K}) = 20.9 \text{ kJ}$ of heat cooling to 0°C, while the ice would gain $(1 \text{ kg})(2.05 \text{ kJ} \div \text{kg·K})(40 \text{ K}) = 82.0 \text{ kJ}$ of heat warming to 0°C. Evidently, the water reaches 0°C first, and can still lose 334 kJ of heat, more than enough to bring the ice to 0°C, if all of it were to freeze. In fact, only $82.0 - 20.9 = 61.1$ kJ of heat is transferred during the change of phase, therefore only $61.1 \text{ kJ}/(334 \text{ kJ/kg}) = 0.183$ kg of water freezes. The final mixture is at 0°C and contains 1.183 kg of ice and $1 - 0.183 = 0.818$ kg of water.

Supplementary Problems

Problem

69. How long will it take a 500-W microwave oven to vaporize completely a 500-g block of ice initially at 0°C?

Solution

The reasoning required here is similar to that in the Solutions to Problems 41 or 65: $Q = \mathcal{P}t = m(L_f + c_w(100 \text{ C°}) + L_v)$, or $t = 0.5 \text{ kg } [334 \text{ kJ/kg} + (4.184 \text{ kJ/kg·K})(100 \text{ K}) + 2257 \text{ kJ/kg}]/(500 \text{ W}) = 3.01 \times 10^3 \text{ s} = 50.2$ min.

Problem

71. A solar-heated house (Fig. 20-22) stores energy in 5.0 tons of Glauber salt ($Na_2SO_4 \cdot 10H_2O$), a substance that melts at 90°F. The heat of fusion of Glauber salt is 104 Btu/lb, and the specific heats of the solid and liquid are, respectively, 0.46 Btu/lb·°F and 0.68 Btu/lb·°F. After a week of sunny weather, the storage medium is all liquid at 95°F. Then a cool, cloudy period sets in during which the house loses heat at an average rate of 20,000 Btu/h. (a) How long is it before the temperature of the storage medium drops below 60°F? (b) How much of this time is spent at 90°F?

Solution

(a) In cooling from 95°F to 60°F (including the solidification at 90°F), the medium exhausts heat $Q = m[c_{liquid}(95°F-90°F)+L_f+c_{solid}(90°F-60°F)] = 1.21 \times 10^6$ Btu, where given values of m, the specific heats, and the heat of transformation were substituted. If all this heat were supplied to the house, which loses energy at the average rate of 2×10^4 Btu/h, it would take $(1.21 \times 10^6/2 \times 10^4) \text{ h} = 60.6$ h for this to occur. (b) The time spent during just the solidification at 90°F is $mL_f/H = (5 \times 2000 \text{ lb})(104 \text{ Btu/lb})/(2 \times 10^4 \text{ Btu/h}) = 52.0$ h.

Problem

73. Show that the coefficient of volume expansion of an ideal gas at constant pressure is just the reciprocal of its kelvin temperature.

Solution

As mentioned in the text following Equation 20-9, β is defined in general as $(dV/V)/dT = (dV/dT)/V$. For an ideal gas at constant pressure, $V = (nR/P)T$, so $dV/dT = nR/P$. Thus, $\beta = nR/PV = 1/T$.

Problem

75. Water's coefficient of volume expansion in the temperature range from 0°C to about 20°C is given approximately by $\beta = a + bT + cT^2$, where T is in Celsius and $a = -6.43\times10^{-5}$ °C^{-1}, $b = 1.70\times10^{-5}$ °C^{-2}, and $c = -2.02\times10^{-7}$ °C^{-3}. Show that water has its greatest density at approximately 4.0°C.

Solution

We do not actually need to differentiate the density or the volume [$\rho(T) = $ const. mass/$V(T)$], since Equation 20-9 shows that $dV/dT = \beta V = 0$ when $\beta(T) = 0$. Thus, the maximum density (or minimum volume) occurs for a temperature satisfying $a + bT + cT^2 = 0$. The quadratic formula gives $T = (-b \pm \sqrt{b^2 - 4ac})/2c$, or since both a and c are negative, $T = (b \mp \sqrt{b^2 - 4|a||c|})/2|c|$. Canceling a factor of 10^{-5} from the given coefficients, we find $T = (1.70 \mp \sqrt{(1.70)^2 - 4(6.43)(0.0202)})$°C/0.0404 = 3.97°C. (The other root, 80.2°C, can be discarded because it is outside the range of validity, $0 \le T \le 20$°C, of the original $\beta(T)$.) Thus, the maximum density of water occurs at a temperature close to 4°C. (That this represents a minimum volume can be verified by plotting $V(T)$, or from the second derivative, $d^2V/dT^2 = V(d\beta/dT) + \beta(dV/dT) = V(\beta^2 + d\beta/dT) = V(\beta^2 + b + 2cT) > 0$ for $T = 3.97$°C.)

Problem

77. Ignoring air resistance, find the height from which you must drop an ice cube at 0°C so it melts completely on impact. Assume no heat exchange with the environment.

Solution

The assumptions stated in the problem (no air resistance or heat exchange with the environment) imply that the change in the gravitational potential energy of the ice cube, per unit mass, is equal to the heat of transformation, $mgy = mL_f$, or $y = L_f/g = $ (334 kJ/kg)/(9.8 m/s^2) = 34.1 km! (This follows from the conservation of energy, since all of the initial mechanical energy of the ice cube goes into melting it on impact. Of course, the expression for potential energy difference, mgy, is not valid over such a large range, but $mgy\, R_E/(R_E + y)$ only changes this result to 34.3 km.) The thermal energies of ordinary macroscopic objects are very large compared to their mechanical energies.

Problem

79. Prove the relation $\beta = 3\alpha$ by considering a cube of side s and therefore volume $V = s^3$ that undergoes a small temperature change dT and corresponding length and volume changes ds and dV.

Solution

For a cubical volume $V = L^3$, the expansion coefficients are related by:

$$\beta = \frac{1}{V}\frac{dV}{dT} = \frac{1}{V}\frac{dV}{dL}\frac{dL}{dT} = \frac{1}{L^3}(3L^2)\frac{dL}{dT} = 3\frac{1}{L}\frac{dL}{dT} = 3\alpha.$$

(We used Equations 20-9 and 10 in differential form, with L in place of s, and the chain rule for differentiation.) Alternatively, use the binomial approximation for $\Delta V = (L + \Delta L)^3 - L^3 = 3L^2\,\Delta L$, keeping only the lowest order term in ΔL. Since $\Delta V = \beta V \Delta T$ and $\Delta L = \alpha L \Delta T$, one finds $3L^2(\alpha L\,\Delta T) = \beta L^3 \Delta T$, or $\beta = 3\alpha$.

CHAPTER 21 HEAT, WORK, AND THE FIRST LAW OF THERMODYNAMICS

ActivPhysics can help with these problems: Activities 8.5–8.13

Section 21-1: The First Law of Thermodynamics

Problem

1. In a perfectly insulated container, 1.0 kg of water is stirred vigorously until its temperature rises by 7.0°C. How much work was done on the water?

Solution

Since the container is perfectly insulated thermally, no heat enters or leaves the water in it. Thus, $Q = 0$ in Equation 21-1. The change in the internal energy of the water is determined from its temperature rise, $\Delta U = mc\,\Delta T$ (see comments in Section 19-4 on internal energy), so $W = -\Delta U = -(1\text{ kg}) \times (4.184\text{ kJ/kg·K})(7\text{ K}) = -29.3\text{ kJ}$. (The negative sign signifies that work was done on the water.)

Problem

3. A 40-W heat source is applied to a gas sample for 25 s, during which time the gas expands and does 750 J of work on its surroundings. By how much does the internal energy of the gas change?

Solution

$Q = 40\text{ W} \times 25\text{ s} = 1000\text{ J}$ of heat is added to the gas, which does $W = 750\text{ J}$ of work on its surroundings. Thus, the first law of thermodynamics requires that $\Delta U = Q - W = 1000\text{ J} - 750\text{ J} = 250\text{ J}$ (an increase in internal energy).

Problem

5. The most efficient large-scale electric power generating systems use high-temperature gas turbines and a so-called combined cycle system that maximizes the conversion of thermal energy into useful work. One such plant produces electrical energy at the rate of 360 MW, while extracting energy from its natural gas fuel at the rate of 670 MW. (a) At what rate does it reject waste heat to the environment? (b) Find its efficiency, defined

as the percent of the total energy extracted from the fuel that ends up as work.

Solution

(a) If we assume that the generating system operates in a cycle and choose it as "the system," then $dU/dt = 0$ and Equation 21-2 implies $dQ/dt = dW/dt$. Here, dW/dt is the rate that the generator supplies energy to its surroundings (360 MW in this problem) and dQ/dt is the net rate of heat flow into the generator from the surroundings. Since the system is just the generator, the net heat flow is the difference between the heat extracted from its fuel and the heat exhausted to the environment, i.e., $dQ/dt = (dQ/dt)_{\text{in}} - (dQ/dt)_{\text{out}} = 670\text{ MW} - (dQ/dt)_{\text{out}}$. Therefore, $(dQ/dt)_{\text{out}} = 670\text{ MW} - 360\text{ MW} = 310\text{ MW}$. (Note: If the system is assumed to be the generator and its fuel, as in Example 21-1, then dW/dt is still 360 MW, but the system's internal energy decreases because energy is extracted from the fuel, $dU/dt = -670\text{ MW}$, and there is no heat input. Then $dQ/dt = -670\text{ MW} + 360\text{ MW} = -310\text{ MW}$, representing the rate of heat rejected to the environment.) (b) The efficiency is $(dW/dt)/(dQ/dt)_{\text{in}} = 360\text{ MW}/670\text{ MW} = 53.7\%$ (see Section 22-2).

Problem

7. Water flows over Niagara Falls (height 50 m) at the rate of about 10^6 kg/s. Suppose that all the water passes through a turbine connected to an electric generator producing 400 MW of electric power. If the water has negligible kinetic energy after leaving the turbine, by how much has its temperature increased between the top of the falls and the outlet of the turbine?

Solution

Consider the electric generator to be the system. It operates in a cycle, so its internal energy doesn't change $dU/dt = 0$ (otherwise the generator would store energy). The rate of mechanical energy input to the generator from the gravitational potential energy of falling water is $-(dm/dt)gy = -(10^6\text{ kg/s}) \times (9.8\text{ m/s}^2)(50\text{ m}) = -490\text{ MW}$ (work done on the generator is negative), while the rate of work produced by the generator is 400 MW. Therefore, the first law of thermodynamics requires a heat flow of $dQ/dt = dW \div dt = -490$ MW $+ 400$ MW $= -90$ MW (negative for heat leaving the generator). This heat is absorbed by the water, causing a temperature rise satisfied by $(dm/dt)c\,\Delta T = 90$ MW. Thus $\Delta T = 90$ MW$\div (10^6$ kg/s$)(4.184$ kJ/kg·k$) = 2.15 \times 10^{-2}$ K ≈ 0.02 C°.

Section 21-2:   Thermodynamic Processes

Problem

9. Repeat the preceding problem for a process that follows the path ACB in Fig. 21-26.

Solution

AC is an isovolumic process, so $W_{AC} = 0$. CB is an isobaric process, so $W_{CB} = P_2(V_2 - V_1) = 2P_1 \times (2V_1 - V_1) = 2P_1V_1$. Of course, $W_{ACB} = W_{AC} + W_{CB}$. (In the PV diagram, Fig. 21-26, the area under AC is zero, and that under CB, a rectangle, is $2P_1V_1$. Equation 21-3 could also be used.)

Problem

11. A balloon contains 0.30 mol of helium. It rises, while maintaining a constant 300 K temperature, to an altitude where its volume has expanded 5 times. How much work is done by the gas in the balloon during this isothermal expansion? Neglect tension forces in the balloon.

Solution

During an isothermal expansion, the work done by a given amount of ideal gas is $W = nRT \ln(V_2/V_1) = (0.3$ mol$)(8.314$ J/mol·K$)(300$ K$)\ln(5) = 1.20$ kJ (see Equation 21-4).

Problem

13. How much work does it take to compress 2.5 mol of an ideal gas to half its original volume while maintaining a constant 300 K temperature?

Solution

In an isothermal compression of a fixed quantity of ideal gas, work is done on the gas so W is negative in Equation 21-4. For the values given, $W = nRT \ln(V_2/V_1) = (25$ mol$)(8.314$ J/mol·K$)(300$ K$) \times \ln(\frac{1}{2}) = -4.32$ kJ.

Problem

15. A 0.25 mol sample of an ideal gas initially occupies 3.5 L. If it takes 61 J of work to compress the gas isothermally to 3.0 L, what is the temperature?

Solution

In an isothermal compression, work is done on the gas, so W in Equation 21-4 is negative. Thus $W = -61$ J $= nRT \ln(V_2/V_1)$, or $T = -61$ J$/(0.25$ mol$) \times (8.314$ J/mol·K$)\ln(3.0/3.5) = 190$ K.

Problem

17. It takes 600 J to compress a gas isothermally to half its original volume. How much work would it take to compress it by a factor of 10 starting from its original volume?

Solution

For isothermal compressions starting from the same volume (and temperature), $W_{13}/W_{12} = \ln(V_3/V_1) \div \ln(V_2/V_1)$ (see Equation 20-4). If $V_2 = V_1/2, V_3 = V_1/10$, and $W_{12} = -600$ J, then $W_{13} = (-600$ J$) \times (\ln 10)/(\ln 2) = -1.99$ kJ. [Note: $\ln x = -\ln(1/x)$.]

Problem

19. A gas with $\gamma = 1.4$ is at 100 kPa pressure and occupies 5.00 L. (a) How much work does it take to compress the gas adiabatically to 2.50 L? (b) What is its final pressure?

Solution

The work done by an ideal gas undergoing an adiabatic process is $W_{12} = (P_1V_1 - P_2V_2)/(\gamma - 1)$ (see Equation 21-14). Since the compression is specified by given values of P_1, V_1, and V_2, we first find the final pressure from the adiabatic gas law. (b) $P_2 = P_1(V_1 \div V_2)^\gamma = (100$ kPa$)(5$ L$/2.5$ L$)^{1.4} = 264$ kPa. (a) Then the work done on the gas (which is $-W_{12}$) is $-W_{12} = (P_2V_2 - P_1V_1)/(\gamma - 1) = [(264$ kPa$)(2.5$ L$) - (100$ kPa$) \times (5$ L$)]/0.4 = 399$ J.

Problem

21. Repeat the preceding problem taking AB to be on an adiabat and using a specific heat ratio of $\gamma = 1.4$.

Solution

(a) If AB represents an adiabatic process for an ideal gas, then the adiabatic law and the given values yield $P_B = P_A(V_A/V_B)^\gamma = (60$ kPa$)(5)^{1.4} = 571$ kPa. (b) The work done by the gas over the adiabat AB is $W_{AB} = (P_AV_A - P_BV_B)/(\gamma - 1) = [(60$ kPa$)(5$ L$) - (571$ kPa$)(1$ L$)]/0.4 = -678$ J (see Equation 21-14). The process BC is isovolumic so $W_{BC} = 0$, and the process CA is isobaric so $W_{CA} = P_A(V_A - V_C) = (60$ kPa$)(5$ L $- 1$ L$) = 240$ J. The total work done by the gas is $W_{ABCA} = W_{AB} + W_{BC} + W_{CA} = -678$ J $+$

$0 + 240$ J $= -438$ J. The work done *on* the gas is the negative of this.

Problem

23. A gasoline engine has a compression ratio of 8.5. If the fuel-air mixture enters the engine at 30°C, what will be its temperature at maximum compression? Assume the compression is adiabatic and that the mixture has $\gamma = 1.4$.

Solution

Equation 21-13b gives $T = T_0(V_0/V)^{\gamma-1}$, where V_0/V is the compression ratio (for T and V at maximum compression.) Thus, $T = (303$ K$)(8.5)^{0.4} = 713$ K $= 440$°C. (Note: T appearing in the gas laws is the absolute temperature.)

Problem

25. By how much must the volume of a gas with $\gamma = 1.4$ be changed in an adiabatic process if the kelvin temperature is to double?

Solution

$V/V_0 = (T_0/T)^{1/(\gamma-1)} = (0.5)^{1/0.4} = 0.177$ (Equation 21-13b).

Problem

27. A gas expands isothermally from state A to state B, in the process absorbing 35 J of heat. It is then compressed isobarically to state C, where its volume equals that of state A. During the compression, 22 J of work are done on the gas. The gas is then heated at constant volume until it returns to state A. (a) Draw a PV diagram for this process. (b) How much work is done on or by the gas during the complete cycle? (c) How much heat is transferred to or from the gas as it goes from B to C to A?

Solution

(a) In the PV diagram shown, AB is an isotherm (given by $PV = $ constant $= P_AV_A = P_BV_B$, with $V_B > V_A$ for an expansion), BC is a straight horizontal line ($P = $ constant $= P_B = P_C$), and CA is a vertical line ($V = $ constant $= V_C = V_A$). (b) $W_{tot} = W_{AB} + W_{BC} + W_{CA}$. The first two terms are given, since for the isotherm, $Q_{AB} = W_{AB} = 35$ J, and $W_{BC} = -22$ J is the work done *by* the gas, while $W_{CA} = 0$ for an isovolumic process. Thus $W_{tot} = 35$ J $- 22$ J $+ 0 = 13$ J, positive for work done by the gas. (c) For the whole cycle, $\Delta U_{tot} = 0$, so $Q_{tot} = W_{tot}$. Since $Q_{tot} = Q_{AB} + Q_{BCA}$, we find that $Q_{BCA} = Q_{tot} - Q_{AB} = $

$W_{tot} - Q_{AB} = 13$ J $- 35$ J $= -22$ J. Negative Q_{BCA} means heat leaves the gas during the process BCA.

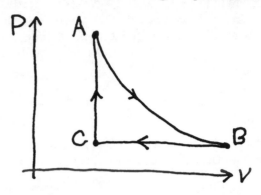

Problem 27 Solution.

Problem

29. A 2.0 mol sample of ideal gas with molar specific heat $C_V = \frac{5}{2}R$ is initially at 300 K and 100 kPa pressure. Determine the final temperature and the work done by the gas when 1.5 kJ of heat is added to the gas (a) isothermally, (b) at constant volume, and (c) isobarically.

Solution

(a) In an isothermal process, T is, of course, constant, so the final temperature is $T_2 = 300$ K. Since $\Delta U = 0$, $W = Q = 1.5$ kJ. (b) In an isovolumic process, $W = 0$ and $Q = nC_V\Delta T$. Therefore, $\Delta T = 1.5$ kJ/(2 mol)$\times$ $\frac{5}{2}R = 1.5$ kJ/(5×8.314 J/K) $= 36.1$ K, and $T_2 = 300$ K $+ \Delta T = 336$ K. (c) In an isobaric process, $Q = nC_P\,\Delta T = n(C_V + R)\,\Delta T = n(\frac{7}{2}R)\,\Delta T$, so $\Delta T = 2Q/7nR = 2(1.5$ kJ$)/7(2\times8.314$ J/K$) = 25.8$ K, and $T_2 = 326$ K. The work done is $W = P\Delta V = nR\,\Delta T = (R/C_P)Q = (\frac{2}{7})Q = 429$ J. (Refer to the relevant parts of Section 21-2 if necessary.)

Problem

31. An ideal gas with $\gamma = 1.67$ starts at point A in Fig. 21-29, where its volume and pressure are 1.00 m^3 and 250 kPa, respectively. It then undergoes an adiabatic expansion that triples its volume, ending at point B. It's then heated at constant volume to point C, then compressed isothermally back to A. Find (a) the pressure at B, (b) the pressure at C, and (c) the net work done on the gas.

Solution

(a) From the adiabatic law for an ideal gas (Equation 21-13a), $P_B = P_A(V_A/V_B)^{\gamma} = (250$ kPa$)\times$

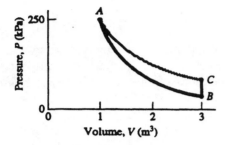

FIGURE 21-29 Problem 31.

$(\frac{1}{3})^{1.67} = 39.9$ kPa. (b) Point C lies on an isotherm with A, so the ideal gas law (Equation 20-2) yields $P_C = P_A V_A / V_C = (250 \text{ kPa})(\frac{1}{3}) = 83.3$ kPa. (c) $W_{net} = W_{AB} + W_{BC} + W_{CA}$. W_{AB} is for an adiabatic process (Equation 21-14) and equals $(P_A V_A - P_B V_B)/(\gamma - 1) = [(250 \text{ kPa})(1 \text{ m}^3) - (39.9 \text{ kPa})(3 \text{ m}^3)]/0.67 = 194$ kJ; W_{BC} is for an isovolumic process and equals zero; W_{CA} is for an isothermal process (Equation 21-4) and equals $nRT_A \ln(V_A/V_C) = P_A V_A \ln(V_A/V_C) = 250 \text{ kJ} \ln(\frac{1}{3}) = -275$ kJ. Thus, $W_{net} = -80.2$ kJ. The work done *on* the gas is the negative of this.

Problem

33. The gas of Example 21-5 starts at state A in Fig. 21-20 and is heated at constant volume until its pressure has doubled. It's then compressed adiabatically until its volume is one-fourth its original value, then cooled at constant volume to 300 K, and finally allowed to expand isothermally to its original state. Find the net work done on the gas.

Solution

The PV diagram for the cyclic process is shown. The work done *on* the gas, $-W_{ABCDA} = W_{ADCBA}$, can be

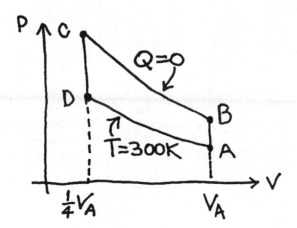

Problem 33 Solution.

calculated from the reversed cycle. AD is isothermal, so $W_{AD} = P_A V_A \ln(V_D/V_A) = (400 \text{ J})\ln(\frac{1}{4}) = -555$ J. DC and BA are isovolumic, so $W_{DC} = W_{BA} = 0$. CB is adiabatic, so $W_{CB} = (P_C V_C - P_B V_B)/(\gamma - 1)$. Since $V_B = V_A$, $P_B = 2P_A$, and $P_C = P_B(V_B/V_C)^\gamma$, $W_{CB} = P_B V_B[(V_B/V_C)^{\gamma-1} - 1]/(\gamma - 1) = 2P_A V_A \times [(V_A/V_C)^{\gamma-1} - 1]/(\gamma - 1) = 2(400 \text{ J})(4^{0.4} - 1)/0.4 = 1482$ J. Finally, $W_{ADCBA} = -555 \text{ J} + 0 + 1482 \text{ J} + 0 = 928$ J.

Problem

35. A 25 L sample of an ideal gas with $\gamma = 1.67$ is at 250 K and 50 kPa. The gas is compressed adiabatically until its pressure triples, then cooled at constant volume back to 250 K, and finally allowed to expand isothermally to its original state. (a) How much work is done on the gas? (b) What is the minimum volume reached? (c) Sketch this cyclic process in a PV diagram.

Solution

(c) The same individual processes are applied in the same order as in Example 21-5, so the PV diagram looks just like Fig. 21-20, except that $V_A = 25$ L and $P_B = 3P_A = 3(50 \text{ kPa})$. (b) The minimum volume attained can be found from the adiabatic law (Equation 21-13a): $V_C = V_B = V_A(P_A/P_B)^{1/\gamma} = (25 \text{ L})(\frac{1}{3})^{1/1.67} = 12.9$ L. (a) The work done *on* the gas is the negative of the work done *by* the gas: $-W_{AB} - W_{BC} - W_{CA} = (P_B V_B - P_A V_A)/(\gamma - 1) - 0 + P_A V_A \ln(V_C/V_A)$, since AB is adiabatic, BC is isovolumic, and CA is isothermal. Numerically, this is $[(150 \text{ kPa})(12.9 \text{ L}) - (50 \text{ kPa})(25 \text{ L})]/0.67 + (50 \text{ kPa}) \times (25 \text{ L})\ln(12.9/25) = 211$ J. (See Table 21-1 for the individual processes.)

Problem

37. A bicycle pump consists of a cylinder 30 cm long when the pump handle is all the way out. The pump contains air ($\gamma = 1.4$) at 20°C. If the pump outlet is blocked and the handle pushed until the internal length of the pump cylinder is 17 cm, by how much does the air temperature rise? Assume that no heat is lost.

Solution

If no heat is lost (or gained) by the gas, the compression is adiabatic and Equation 21-13b gives $TV^{\gamma-1} = T_0 V_0^{\gamma-1}$. Therefore, the temperature rise is $T - T_0 = \Delta T = T_0[(V_0/V)^{\gamma-1} - 1]$. Since $V_0/V = (30 \text{ cm}/17 \text{ cm})$, $\Delta T = [(30/17)^{0.4} - 1](293 \text{ K}) = 74.7$ C°.

Problem

39. A balloon contains 5.0 L of air at 0°C and 100 kPa pressure. How much heat is required to raise the air temperature to 20°C, assuming the gas stays in pressure equilibrium with its surroundings? Neglect tension forces in the balloon. The molar specific heat of air at constant volume is 2.5R.

Solution

The heat required, at constant pressure, is $Q = nC_P \, \Delta T$, where $C_P = C_V + R = 3.5R$ is the molar specific heat of air at constant pressure. The number of moles can be found from the ideal gas law and the initial conditions, $n = P_1V_1/RT_1$. Therefore: $Q = nC_P \, \Delta T = (P_1V_1/RT_1)(3.5R)(T_2 - T_1) = 3.5P_1V_1(T_2 - T_1)/T_1 = 3.5(100 \text{ kPa})(5 \text{ L})(20 \text{ C}°)/(273 \text{ K}) = 128 \text{ J}$. (The difference in Q between this isobaric process and the isovolumic process described in Problem 21-38 is just a factor of $C_P/C_V = \frac{7}{5}$.)

Problem

41. Problem 70 in Chapter 18 shows that pressure as a function of height in Earth's atmosphere is given approximately by $P = P_0 e^{-h/h_0}$, where P_0 is the surface pressure and $h_0 = 8.2$ km. A parcel of air, initially at the surface and at a temperature of 10°C, rises as described in Application: Smog Alert on page 529. (a) What will be its temperature when it reaches 2.0 km altitude? (b) If the temperature of the surrounding air decreases at the normal lapse rate of 6.5°C/km, is the atmosphere stable or unstable under the conditions of this problem?

Solution

(a) For an adiabatic expansion of the air parcel, the pressure and temperature are related to their surface values by $PT^{\gamma/(1-\gamma)} = P_0 T_0^{\gamma/(1-\gamma)}$ (see Problem 28). At a height of 2 km, where $P = P_0 e^{-2.0/8.2}$, and with $\gamma = \frac{7}{5}$ for air, $T = T_0(P/P_0)^{(\gamma-1)/\gamma} = (283 \text{ K}) \times (e^{-2.0/8.2})^{2/7} = 264 \text{ K} = -9.05°\text{C}$. (b) The temperature of the parcel of air drops about 19°C in 2 km, faster than the environmental lapse rate (13°C in 2 km), indicating stable atmospheric conditions.

Section 21-3: Specific Heats of an Ideal Gas

Problem

43. A mixture of monatomic and diatomic gases has specific heat ratio $\gamma = 1.52$. What fraction of the molecules are monatomic?

Solution

The internal energy of a mixture of two ideal gases is $U = f_1 N\bar{E}_1 + f_2 N\bar{E}_2$, where f_1 is the fraction of the total number of molecules, N, of type 1, and $\bar{E}_1$ is the average energy of a molecule of type 1, etc. Classically, $\bar{E} = g(\frac{1}{2}kT)$, where g is the number of degrees of freedom. The molar specific heat at constant volume is $C_V = (\frac{1}{n})(dU/dT) = (N_A/N)d/dT \times (f_1 N g_1 \frac{1}{2}kT + f_2 N g_2 \frac{1}{2}kT) = \frac{1}{2}R(f_1 g_1 + f_2 g_2)$. Suppose that the temperature range is such that $g_1 = 3$ for the monatomic gas, and $g_2 = 5$ for the diatomic gas, as discussed in Section 21-3. Then $C_V = R(1.5f_1 + 2.5f_2) = R(2.5 - f_1)$, where $f_2 = 1 - f_1$ since the sum of the fractions of the mixture is one. Now, C_V can also be specified by the ratio $\gamma = C_P/C_V = 1 + R/C_V$, or $C_V = R/(\gamma - 1)$, so in this problem, $2.5 - f_1 = 1/0.52$, or $f_1 = 57.7\%$.

Problem

45. A gas mixture contains monatomic argon and diatomic oxygen. An adiabatic expansion that doubles its volume results in the pressure dropping to one-third of its original value. What fraction of the molecules are argon?

Solution

From the pressures and volumes in the described adiabatic expansion, $P_0 V_0^\gamma = (\frac{1}{3}P_0)(2V_0)^\gamma$, we can calculate that $\gamma = \ln 3/\ln 2 = 1.58$. Then the result of Problem 43 gives $2.5 - f_{Ar} = 1/0.58$, or $f_{Ar} = 79.0\%$.

Problem

47. How much of a triatomic gas with $C_V = 3R$ would you have to add to 10 mol of monatomic gas to get a mixture whose thermodynamic behavior was like that of a diatomic gas?

Solution

Reference to the solution of Problem 43 shows that the specific heat of a mixture of two gases is $C_V = f_1 C_{V_1} + f_2 C_{V_2}$, where the f's are the number fractions of the gases. If gas 1 is monatomic ($C_V = \frac{3}{2}R$), gas 2 is triatomic (with $C_V = 3R$, as described in the text following Example 21-6), and we wish the mixture to have $C_V = \frac{5}{2}R$ appropriate to a diatomic gas, then $\frac{5}{2}R = \frac{3}{2}Rf_1 + 3Rf_2$, or $5 = 3f_1 + 6f_2$. Since $f_1 + f_2 = 1$, one finds $f_1 = \frac{1}{3}$ and $f_2 = \frac{2}{3}$. With 10 mol of gas 1, one needs 20 mol of gas 2.

Paired Problems

Problem

49. A 5.0 mol sample of ideal gas with $C_V = \frac{5}{2}R$ undergoes an expansion during which the gas does

5.1 kJ of work. If it absorbs 2.7 kJ of heat during the process, by how much does its temperature change? *Hint:* Remember that Equation 21-7 holds for *any* ideal gas process.

Solution

For any process connecting equilibrium states of an ideal gas, Equations 21-1 and 7 give $\Delta U = Q - W = nC_V\,\Delta T$, so $\Delta T = (2.7\text{ kJ} - 5.1\text{ kJ})/(5\text{ mol})(\frac{5}{2}\times 8.314\text{ J/mol·K}) = -23.1$ K.

Problem

51. A gas with $\gamma = \frac{5}{3}$ is at 450 K at the start of an expansion that triples its volume. The expansion is isothermal until the volume has doubled, then adiabatic the rest of the way. What is the final gas temperature?

Solution

For the isothermal part of the expansion, the temperature is constant at 450 K. The adiabatic part goes from $V_1 = 2V_0$ to $V_2 = 3V_0$, so Equation 21-13b gives $T_2 = T_1(V_1/V_2)^{\gamma-1} = (450\text{ K})(\frac{2}{3})^{(5/3)-1} = 343$ K.

Problem

53. An ideal gas with $\gamma = 1.4$ is initially at 273 K and 100 kPa. The gas expands adiabatically until its temperature drops to 190 K. What is its final pressure?

Solution

The ideal gas law can be used to eliminate V and write the adiabatic law in terms of P and T: $PV^\gamma = P^{1-\gamma}(PV)^\gamma = P^{1-\gamma}(nRT)^\gamma$, thus $P^{1-\gamma}T^\gamma$ is constant, or so is $PT^{\gamma/(1-\gamma)}$. For this problem, $P = P_0(T_0 \div T)^{\gamma/(1-\gamma)} = (100\text{ kPa})(273/190)^{-1.4/0.4} = 28.1$ kPa.

Problem

55. The curved path in Fig. 21-30 lies on the 350-K isotherm for an ideal gas with $\gamma = 1.4$. (a) Calculate the net work done on the gas as it goes around the cyclic path $ABCA$. (b) How much heat flows into or out of the gas on the segment AB?

Solution

(a) The work done *by* the gas in each segment of the cycle is summarized in Table 21-1. $-W_{AB} = 0$

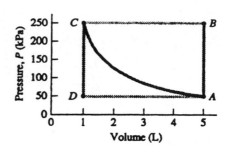

FIGURE 21-30 Problems 55 and 56.

(isovolumic), $-W_{BC} = P_B(V_B - V_C) = (250\text{ kPa})\times(5\text{ L} - 1\text{ L}) = 1000$ J (isobaric), $-W_{CA} = nRT_A\times \ln(V_C/V_A) = P_A V_A \ln(V_C/V_A) = (50\text{ kPa})(5\text{ L})\ln(\frac{1}{5}) = -402$ J (isothermal). The net work done *on* the gas is $-W_{ABCA} = 0 + 1000\text{ J} - 402\text{ J} = 598$ J. (b) Since V is constant, $Q_{AB} = nC_V\,\Delta T = nR(T_B - T_A)/(\gamma - 1) = (P_B V_B - P_A V_A)/(\gamma - 1) = (250\text{ kPa} - 50\text{ kPa})(5\text{ L}) \div 0.4 = 2.50$ kJ, positive for heat transferred into the gas (at constant volume, the gas must be heated in order to raise its pressure). (Note: Recall that $C_V = R/(\gamma - 1)$, as shown in the solution to Problem 43.)

Supplementary Problems
Problem

57. An 8.5-kg rock at 0°C is dropped into a well-insulated vat containing a mixture of ice and water at 0°C. When equilibrium is reached there are 6.3 g less ice. From what height was the rock dropped?

Solution

The mechanical energy of the rock (originally gravitational potential energy) melted the ice (changed its internal energy) and no heat energy was transferred ($Q = 0$). Therefore, $-W = m_{rock}gh = \Delta U = m_{ice}L_f$, or $h = m_{ice}L_f/m_{rock}g = (6.3\text{ g})(334\text{ J/g})/(8.5\times 9.8\text{ N}) = 25.3$ m. ($W < 0$ since the rock did work on the ice-water system.)

Problem

59. Repeat Problem 8 for the case when the gas expands along a path given by $P = P_1\left[1 + \left(\frac{V-V_1}{V_1}\right)^2\right]$. Sketch the path in the PV diagram, and determine the work done.

Solution

The path is a parabola in the PV plane, with vertex at point (V_1, P_1) and opening upwards. The work done is the area under the path between (V_1, P_1) and

$(V_2 = 2V_1, P_2 = 2P_1)$:

$$W = \int_1^2 P \, dV = \int_{V_1}^{V_2} P_1 \left[1 + \left(\frac{V - V_1}{V_1} \right)^2 \right] dV$$

$$= P_1 \left| V + \frac{V_1}{3} \left(\frac{V - V_1}{V_1} \right)^3 \right|_{V_1}^{V_2}$$

$$= P_1 \left[(V_2 - V_1) + \frac{V_1}{3} \left(\frac{V_2 - V_1}{V_1} \right)^3 \right] = \frac{4}{3} P_1 V_1.$$

Problem

61. Show that the application of Equation 21-3 to an adiabatic process results in Equation 21-14.

Solution

The work done by an ideal gas undergoing an adiabatic process from state 1 to state 2 can be found by integration of the adiabatic law, $P = P_1 V_1^\gamma / V^\gamma$.

$$W_{12} = \int_{V_1}^{V_2} P \, dV = \int_{V_1}^{V_2} (P_1 V_1^\gamma) \frac{dV}{V^\gamma}$$

$$= P_1 V_1^\gamma \left(\frac{V_2^{-\gamma+1} - V_1^{-\gamma+1}}{-\gamma + 1} \right)$$

$$= (P_1 V_1^\gamma V_1^{-\gamma+1} - P_2 V_2^\gamma V_2^{-\gamma+1})/(\gamma - 1),$$

which is Equation 21-14. (Note: $P_1 V_1^\gamma = P_2 V_2^\gamma$.)

Problem

63. An ideal gas is taken clockwise around the circular path shown in Fig. 21-31. (a) How much work does the gas do? (b) If there are 1.3 moles of gas, what is the maximum temperature reached?

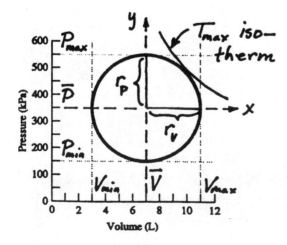

FIGURE 21-31 Problem 63 Solution.

Solution

(a) The circular path is most easily described by dimensionless variables:

$$x = (V - \bar{V})/r_V, \quad \text{and} \quad y = (P - \bar{P})/r_P,$$

where $\bar{V} = \frac{1}{2}(V_{min} + V_{max})$ and $\bar{P} = \frac{1}{2}(P_{min} + P_{max})$, and $r_V = \frac{1}{2}(V_{max} - V_{min})$ and $r_P = \frac{1}{2}(P_{max} - P_{min})$. (The "center" of the circle is $(\bar{V}, \bar{P})$, and the "radius," in volume and pressure units, is r_V or r_P, respectively.) Therefore, $x^2 + y^2 = 1$ for the path. The work done in one clockwise cycle is (see Equations 8-1 or 22-7)

$$W = \oint P \, dV = \oint (r_P y + \bar{P}) r_V \, dx$$

$$= r_P r_V \oint y \, dx + r_V \bar{P} \oint dx.$$

The first integral is the area of the unit circle, $x^2 + y^2 = 1$, which is π, and the second integral is zero. Therefore, $W = r_P r_V \pi = \frac{1}{4}\pi (P_{max} - P_{min})(V_{max} - V_{min}) = \frac{1}{4}\pi (550 - 150) \text{ kPa} (11 - 3) \text{ L} = 2.51 \text{ kJ}$. (For those unfamiliar with integrals over closed paths, they are the difference between the integrals over the upper and lower parts, as explained in Fig. 21-19. In this case, $\oint y \, dx = \int_{-1}^{1} y_+(x) \, dx - \int_{-1}^{1} y_-(x) \, dx$, where $y_\pm(x) = \pm\sqrt{1 - x^2}$ are the upper and lower semicircles. Then $\oint y \, dx = 2 \int_{-1}^{1} \sqrt{1 - x^2} \, dx = \pi$. The integral of a constant over a closed path is zero, since the upper and lower parts are the same. Then $\oint dx = \int_{-1}^{1} dx - \int_{-1}^{1} dx = 0$.)

(b) The maximum temperature reached is that of the isotherm ($PV = $ constant) which is tangent to the upper semicircle in the PV diagram. In terms of the dimensionless variables, the equation of an isotherm is $xy + (\bar{P}/r_P)x + (\bar{V}/r_V)y = $ constant, where $y = y_+(x) = \sqrt{1 - x^2}$ on the upper semicircle. For the particular numerical values given, $(\bar{P}/r_P = \bar{V}/r_V = \frac{7}{4})$, the equation simplifies to $xy + \frac{7}{4}(x + y) = $ constant. The condition for a maximum is:

$$0 = \frac{d}{dx}\left[xy + \frac{7}{4}(x + y) \right] = y - \frac{x^2}{y} + \frac{7}{4}\left(1 - \frac{x}{y}\right)$$

$$= \left(1 - \frac{x}{y}\right)\left(x + y + \frac{7}{4}\right),$$

where we used $dy/dx = d(\sqrt{1 - x^2})/dx = -x \div \sqrt{1 - x^2} = -x/y$. The second factor, $x + y + \frac{7}{4}$, is never zero on the unit circle, so the maximum occurs for $1 - x/y = 0$, or $x_m = y_m = \sqrt{1 - x_m^2} = 1/\sqrt{2}$. The corresponding values of volume and pressure are:

$$V_m = r_V x_m + \bar{V} = \frac{1}{\sqrt{2}}\left(\frac{11 - 3}{2}\right)\text{L} + \left(\frac{11 + 3}{2}\right)\text{L}$$

$$= 9.83 \text{ L},$$

and

$$P_m = r_P y_m + \bar{P}$$
$$= \frac{1}{\sqrt{2}}\left(\frac{550 - 150}{2}\right)\text{kPa} + \left(\frac{550 + 150}{2}\right)\text{kPa}$$
$$= 491 \text{ kPa}.$$

The maximum temperature follows from the ideal gas law:

$$T_m = \frac{P_m V_m}{nR} = \frac{(491 \text{ kPa})(9.83 \text{ L})}{1.3(8.314 \text{ J/K})} = 447 \text{ K}.$$

Problem

65. Show that the work done by a van der Waals gas undergoing isothermal expansion from volume V_1 to V_2 is

$$W - nRT \ln\left(\frac{V_2 - nb}{V_1 - nb}\right) + an^2\left(\frac{1}{V_2} - \frac{1}{V_1}\right),$$

where a and b are the constants in Equation 20-5.

Solution

If we solve for P in the van der Waals equation of state, Equation 20-5, and substitute into Equation 21-3, we find

$$W = \int_{V_1}^{V_2} P\,dV = \int_{V_1}^{V_2}\left[\frac{nRT}{(V - nb)} - \frac{n^2 a}{V}\right] dV$$
$$= nRT \int_{V_1}^{V_2}\frac{dV}{V - nb} - n^2 a \int_{V_1}^{V_2}\frac{dV}{V^2}$$
$$= nRT \ln\left(\frac{V_2 - nb}{V_1 - nb}\right) + an^2\left(\frac{1}{V_2} - \frac{1}{V_1}\right),$$

where, for an isothermal process, T is a constant and can be taken out of the integral.

Problem

67. A horizontal piston-cylinder system containing n mol of ideal gas is surrounded by air at temperature T_0 and pressure P_0. If the piston is displaced slightly from equilibrium, show that it executes simple harmonic motion with angular frequency $\omega = AP_0/\sqrt{MnRT_0}$, where A and M are the piston area and mass, respectively. Assume the gas temperature remains constant.

Solution

Since the piston-cylinder system is horizontal, we do not need to consider the force of gravity on the piston. At equilibrium, the pressure forces from inside and outside the piston are equal, so the gas pressure at the equilibrium position of the piston is P_0. We also assume that the gas temperature at equilibrium is T_0,

so $P_0 V_0 = nRT_0$, where $V_0 = Ax_0$ is the volume at equilibrium. When the piston is displaced from its equilibrium position by an amount Δx (positive to the right), the horizontal force on it is $PA - P_0 A$, and Newton's second law gives an acceleration of $d^2(\Delta x) \div dt^2 = (P - P_0)A/M$.

For isothermal expansions and compressions of the gas, $PV = P_0 V_0 = PA(x_0 + \Delta x) = P_0 Ax_0$, or $P/P_0 = (1 + \Delta x/x_0)^{-1}$. For small displacements ($\Delta x \ll x_0$), $P/P_0 \approx 1 - \Delta x/x_0$ (see the binomial approximation in Appendix A), so $d^2(\Delta x)/dt^2 = (P_0 A/M)[(P/P_0) - 1] \approx -(P_0 A/Mx_0)\,\Delta x$. This is the equation for simple harmonic motion of the piston, about its equilibrium position x_0, with angular frequency $\omega^2 = P_0 A/Mx_0$. Since $P_0 V_0 = P_0 Ax_0 = nRT_0$, we may eliminate x_0 to obtain $\omega = P_0 A/\sqrt{MnRT_0}$.

(Note: In order for the gas temperature to remain constant, as assumed above, heat must flow into and out of the gas. This requires time, so the motion of the piston must be very slow. If the motion is rapid (or if the cylinder is thermally insulated), there is no time for heat transfer in the gas and the expansions and compressions are adiabatic. In this case, $PV^\gamma = P_0 V_0^\gamma = PA^\gamma(x_0 + \Delta x)^\gamma = P_0 A^\gamma x_0^\gamma$, or $P/P_0 = (1 + \Delta x/x_0)^{-\gamma}$. For $\Delta x \ll x_0$, $d^2(\Delta x)/dt^2 = (P_0 A/M)\times [(P/P_0) - 1] \approx (P_0 A/M)[1 - \gamma \Delta x/x_0 + \cdots - 1] = -\gamma P_0 A\,\Delta x/M$. (This represents simple harmonic motion with $\omega = \sqrt{\gamma P_0 A/Mx_0} = P_0 A\sqrt{\gamma/MnRT_0}$.)

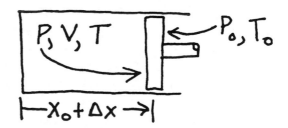

Problem 67 Solution.

Problem

69. A cylinder of cross-sectional area A is closed by a massless piston. The cylinder contains n mol of ideal gas with specific heat ratio γ, and is initially in equilibrium with the surrounding air at temperature T_0 and pressure P_0. The piston is initially at height h_1 above the bottom of the cylinder. Sand is gradually sprinkled onto the piston until it has moved downward to a final height h_2. Find the total mass of the sand if the process is (a) isothermal and (b) adiabatic.

Solution

Initially, $P_1 = P_0$ (since the piston is massless) and $T_1 = T_0$. Finally, in equilibrium with its load of sand, M, the net force on the piston is zero, or $P_2 A = Mg + P_0 A$. Therefore, $M = (P_2 - P_0)A/g$. (a) In an isothermal process, $T_2 = T_1 = T_0$, and $P_2 V_2 = P_1 V_1$. Therefore, $P_2 - P_1 = P_1[(V_1/V_2) - 1] = P_0[(h_1/h_2) - 1]$, and $M = (P_0 A/g)[(h_1/h_2) - 1]$. (b) In an adiabatic process, $P_2 V_2^\gamma = P_1 V_1^\gamma$, or $P_2 - P_1 = P_1[(V_1/V_2)^\gamma - 1]$. Thus $M = (P_0 A/g)[(h_1/h_2)^\gamma - 1]$. (Note: the given data are related by the ideal gas law, $P_0 A h_1 = nRT_0$.)

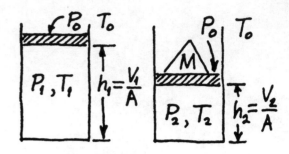

Problem 69 Solution.

CHAPTER 22   THE SECOND LAW OF THERMODYNAMICS

ActivPhysics can help with these problems: Activity 8.14.

Section 22-1:   Reversibility and Irreversibility

Problem

1. The egg carton shown in Fig. 22-27 has places for one dozen eggs. (a) How many distinct ways are there to arrange six eggs in the carton? (b) Of these, what fraction correspond to all six eggs being in the left half of the carton? Treat the eggs as distinguishable, so an interchange of two eggs gives rise to a new state.

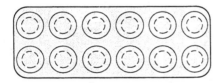

FIGURE 22-27 Problem 1.

Solution

(a) There are twelve places for first egg, eleven for the second, etc., so the number of arrangements (states) of six distinguishable eggs into twelve places is $12\times11\times10\times9\times8\times7 = 12!/6! \approx 6.65\times10^5$. (b) If limited to the left side of the carton only, there are 6! arrangements, so the fraction is $6!/(12!/6!) = 1/924$.

Problem

3. Estimate the energy that could be extracted by cooling the world's oceans by 1°C. How does your estimate compare with humanity's yearly energy consumption of about 2.5×10^{20} J?

Solution

The volume of the oceans is about 1.35×10^{18} m³ (the average depth is 3.73 km over 71% of the earth's surface). The heat extracted by cooling this volume of water by 1°C is $Q = \rho V c \,\Delta T = (10^3 \text{ kg/m}^3)\times$ $(1.35 \times 10^{18} \text{ m}^3)(4184 \text{ J/kg·K})(1 \text{ K}) \simeq 5.65\times10^{24}$ J, or about 23,000 times the world's annual energy consumption.

Sections 22-2 and 22-3:   The Second Law and its Applications

Problem

5. What are the efficiencies of reversible heat engines operating between (a) the normal freezing and boiling points of water, (b) the 25°C temperature at the surface of a tropical ocean and deep water at 4°C, and (c) a 1000°C candle flame and room temperature?

Solution

The efficiency of a reversible engine, operating between two absolute temperatures, $T_h > T_c$, is given by Equation 22-3. (a) $e = 1 - T_c/T_h = 1 - 273/373 = 26.8\%$. (b) $e = (T_h - T_c)/T_h = \Delta T/T_h = 21/298 = 7.05\%$. (c) With room temperature at $T_c = 20°C$, $e = 980/1273 = 77.0\%$.

Problem

7. A reversible Carnot engine operating between helium's melting point and its 4.25-K boiling point has efficiency of 77.7%. What is the melting point?

Solution

We can solve Equation 22-3 for the low temperature to find $T_c = (1 - e)T_h = (1 - 0.777)(4.25 \text{ K}) = 0.948 \text{ K}$.

Problem

9. The maximum temperature in a nuclear power plant is 570 K. The plant rejects heat to a river where the temperature is 0°C in the winter and 25°C in the summer. What are the maximum possible efficiencies for the plant in these seasons? Why might the plant not achieve these efficiencies?

Solution

The winter and summer thermodynamic efficiencies are $1 - (T_c/T_h)$, or $1 - (273/570) = 52.1\%$ and $1 - (298/570) = 47.7\%$, respectively. As explained in Section 22-3, irreversible processes, transmission losses, etc., make actual efficiencies less than the theoretical maxima.

Problem

11. A power plant's electrical output is 750 MW. Cooling water at 15°C flows through the plant at 2.8×10^4 kg/s, and its temperature rises by 8.5°C. Assuming the plant's only energy loss is to the cooling water and that the cooling water is effectively the low-temperature reservoir, find (a) the rate of energy extraction from the fuel, (b) the plant's efficiency, and (c) its highest temperature.

Solution

(a) In order to raise the temperature of the cooling water by 8.5 K, heat must be exhausted to it at a rate of $c(dm/dt) \, \Delta T = (4.184 \text{ kJ/kg·K})(2.8 \times 10^4 \text{ kg/s}) \times$ (8.5 K) = 996 MW (see Equation 19-4). If this is also all the heat rejected by the power plant, dQ_c/dt, then since the work output, dW/dt, is given, the heat input to the plant (extracted from its fuel) is $dQ_h/dt = dQ_c/dt + dW/dt = 996 \text{ MW} + 750 \text{ MW} = 1.75 \text{ GW}$. (b) The plant's actual efficiency (from the definition of efficiency in terms of rates) is $e = (dW/dt)/(dQ_h/dt) = 750 \text{ MW}/1.75 \text{ GW} = 43.0\%$. (c) If the power plant is considered to operate like an ideal Carnot engine, then $T_h/T_c = Q_h/Q_c = (dQ_h/dt)/(dQ_c/dt) = 1.75 \text{ GW} \div 996 \text{ MW} = 1.75$ (the energy rate per cycle and the energy rate per second are proportional). If $T_c = 15°C = 288 \text{ K}$, then $T_h = 1.75(288 \text{ K}) = 505 \text{ K} = 232°C$. The actual highest temperature would be somewhat greater than this, because the actual efficiency is always less than the Carnot efficiency.

Problem

13. The electric power output of all the thermal electric power plants in the United States is about 2×10^{11} W, and these plants operate at an average efficiency around 33%. What is the rate at which all these plants use cooling water, assuming an average 5°C rise in cooling-water temperature? Compare with the 1.8×10^7 kg/s average flow at the mouth of the Mississippi River.

Solution

The total rate at which heat is exhausted by all power plants is

$$\frac{dQ_c}{dt} = \frac{d}{dt}(Q_h - W) = \frac{dW}{dt}\left(\frac{1}{e} - 1\right)$$

$$= (2 \times 10^{11} \text{ W})\left(\frac{1}{33\%} - 1\right) = 4 \times 10^{11} \text{ W}.$$

The mass rate of flow at which water could absorb this amount of energy, with only a 5°C temperature rise, is

given by:

$$\frac{dQ_c}{dt} = \frac{dm}{dt} c_{\text{water}} \, \Delta T, \quad \text{or}$$

$$\frac{dm}{dt} = \frac{4 \times 10^{11} \text{ W}}{(4184 \text{ J/kg·K})(5°C)} = 1.91 \times 10^7 \text{ kg/s},$$

or about 1 Mississippi (a self-explanatory unit of river flow).

Problem

15. A Carnot engine absorbs 900 J of heat each cycle and provides 350 J of work. (a) What is its efficiency? (b) How much heat is rejected each cycle? (c) If the engine rejects heat at 10°C, what is its maximum temperature?

Solution

(a) The efficiency of the engine (by definition) is $e = W/Q_h = 350 \text{ J}/900 \text{ J} = 38.9\%$, where W and Q_h are the work done and heat absorbed per cycle. (b) These are related to the heat rejected per cycle by the first law of thermodynamics (since ΔU per cycle is zero), or $Q_c = Q_h - W = 900 \text{ J} - 350 \text{ J} = 550 \text{ J}$. (c) For a Carnot engine operating between two temperatures, $T_h/T_c = Q_h/Q_c$, so $T_h = (283 \text{ K}) \times (900/550) = 463 \text{ K} = 190°C$. (Carnot's theorem applies to the ratio of absolute temperatures.)

Problem

17. How much work does a refrigerator with a COP of 4.2 require to freeze 670 g of water already at its freezing point?

Solution

The amount of heat that must be extracted in order to freeze the water is $Q_c = mL_f = (0.67 \text{ kg}) \times (334 \text{ kJ/kg}) = 224 \text{ kJ}$. The work consumed by the refrigerator while extracting this heat is $W = Q_c/\text{COP} = 224 \text{ kJ}/4.2 = 53.3 \text{ kJ}$. (See Equation 22-4.)

Problem

19. A 4.0 L sample of water at 9.0°C is put into a refrigerator. The refrigerator's 130-W motor then runs for 4.0 min to cool the water to the refrigerator's low temperature of 1.0°C. (a) What is the COP of the refrigerator? (b) How does this compare with the maximum possible COP if the refrigerator exhausts heat at 25°C?

Solution

(a) The heat extracted from the water is $Q_c = mc \, \Delta T = \rho V c \, \Delta T = (10^3 \text{ kg/m}^3)(4 \times 10^{-3} \text{ m}^3) \times (4.184 \text{ kJ/kg·K})(9°C - 1°C) = 134 \text{ kJ}$, while the work input is $W = \mathcal{P}t = (130 \text{ W})(4 \times 60 \text{ s}) = 31.2 \text{ kJ}$. Therefore, the actual COP is (Equation 22-4) $Q_c/W = 134/31.2 = 4.29$. (b) This is about 38% of the maximum possible COP for a reversible refrigerator operating between $T_c = 274 \text{ K}$ (1°C) and $T_h = 298 \text{ K}$ (25°C), which is $T_c/(T_h - T_c) = 274/24 = 11.4$ (Equation 22-5).

Problem

21. A heat pump consumes electrical energy at the rate P_e. Show that it delivers heat at the rate $(\text{COP} + 1)P_e$.

Solution

For each cycle of operation of the heat pump, $Q_h = Q_c + W$ and $\text{COP} = Q_c/W = (Q_h/W) - 1$. Therefore, $Q_h = (1 + \text{COP})W$, which if written in terms of rates, is the relation stated in the question.

Problem

23. A heat pump transfers heat between the interior of a house and the outside air. In the summer the outside air averages 26°C, and the pump operates by chilling water to 5°C for circulation throughout the house. In winter the outside air averages 2°C, and the pump operates by heating water to 80°C for circulation throughout the house. (a) Find the coefficients of performance in summer and winter. How much work does the pump require (b) for each joule of heat removed from the house in summer and (c) for each joule supplied to the house in winter?

Solution

(a) If the heat pump is reversible, the summertime COP is $T_c/(T_h - T_c) = (273 + 5)/(26 - 5) = 13.2$, and the winter-time COP is $275/78 = 3.53$. (See Equation 22-5; if the heat pump is not reversible, then its COP cannot be found from just the hot and cold operating temperatures.) From the definition of COP and the first law, $W = Q_c/\text{COP} = Q_h/(1 + \text{COP})$. Therefore, (b) for each joule of heat removed in the summer, $W = 1 \text{ J}/13.2 = 7.55 \times 10^{-2} \text{ J}$ is required, while (c) for each joule of heat supplied in the winter, $W = 1 \text{ J}/(1 + 3.53) = 0.221 \text{ J}$ is required.

Problem

25. A 0.20-mol sample of an ideal gas goes through the Carnot cycle of Fig. 22-29. Calculate (a) the heat Q_h absorbed, (b) the heat Q_c rejected, and (c) the work done. (d) Use these quantities to determine efficiency. (e) Find the maximum and minimum temperatures, and show explicitly that

the efficiency as defined in Equation 22-1 is equal to the Carnot efficiency of Equation 22-3.

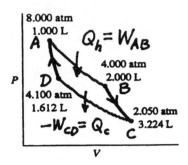

FIGURE 22-29 Problem 25 (Diagram is not to scale).

Solution

From the discussion of the efficiency of a Carnot engine in Section 22-2, (a) $Q_h = nRT_A \ln(V_B/V_A) = P_A V_A \ln(V_B/V_A) = (8)(1)(101.3 \text{ J})\ln 2 = 561.7 \text{ J}$, and (b) $Q_c = P_C V_C \ln(V_C/V_D) = (2.050)(3.224) \times (101.3 \text{ J})\ln(3.224/1.612) = 464.1 \text{ J}$ (we used 101.3 for the conversion factor of L·atm to J). (c) The work done in one cycle (from the first law) is $W = Q_h - Q_c = 97.66 \text{ J}$, resulting in an efficiency of (d) $e = W/Q_h = 0.1739$. (Note: For the Carnot cycle, $T_A = T_B$ and $T_C = T_D$, so for the adiabatic segments, $W_{BC} + W_{DA} = 0$. Thus, $W = W_{AB} + W_{BC} + W_{CD} + W_{DA} = Q_h - Q_c$, explicitly.) We find the maximum and minimum temperatures from the ideal gas law:

$$T_A = P_A V_A/nR$$
$$= (8)(1)(101.3 \text{ J})/(0.2)(8.314 \text{ J/K}) = 487.4 \text{ K},$$

and

$$T_C = (2.050)(3.224)(101.3 \text{ J})/(0.2)(8.314 \text{ J/K})$$
$$= 402.6 \text{ K}.$$

These imply a Carnot efficiency of $e = 1 - T_C/T_A = 0.1739$, exactly as before. Equations 22-1 and 22-3 are identical because $Q_c/Q_h = T_C/T_A = 0.8261$, explicitly. (We did not round off until after completing all the calculations, and we labeled the states as in Fig. 22-6.)

Problem

27. Use appropriate energy flow diagrams to show that the existence of a perfect heat engine would permit the construction of a perfect refrigerator, thus violating the Clausius statement of the second law.

Solution

If it were possible to construct a perfect heat engine (one which would extract heat and perform an equivalent amount of work), then it could be coupled to a real refrigerator in such a way that the work output of the engine equals the work input to the refrigerator, as shown. The net effect of this arrangement is to produce a perfect refrigerator (a cyclic device whose sole effect is the transfer of heat, $Q_c + Q_h - Q_c = Q_h$, from a cold reservoir to a hot one), in violation of the Clausius statement of the second law. This completes the proof of the equivalence of the Kelvin-Planck and Clausius statements in Section 22-2.

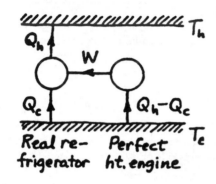

Problem 27 Solution.

Section 22-4: The Thermodynamic Temperature Scale

Problem

29. A Carnot engine operating between a vat of boiling sulfur and a bath of water at its triple point has an efficiency of 61.95%. What is the boiling point of sulfur?

Solution

From the efficiency of a reversible engine, $e = 1 - T_c \div T_h$, and the triple point temperature 273.16 K, we find $T_h = T_c/(1 - e) = 273.16 \text{ K}/(1 - 61.95\%) = 718 \text{ K}$ (in agreement with Table 20-1, to three-figure accuracy).

Problem

31. Calculate the entropy change associated with melting 1.0 kg of ice at 0°C.

Solution

Since the temperature is constant during a change of phase, Equation 22-8 gives $\Delta S = \Delta Q/T = mL_f/T = (1 \text{ kg})(334 \text{ kJ/kg})/273 \text{ K} = 1.22 \text{ kJ/K}$.

Problem

33. A 2.0-kg sample of water is heated to 35°C. If the entropy change is 740 J/K, what was the initial temperature?

Solution

Since the specific heat of water is approximately constant under normal conditions and no phase changes occur, Equation 22-9 can be solved for T_1 to yield $T_1 = T_2 e^{-\Delta S/mc} = (308 \text{ K})\exp\{-(740 \text{ J/K}) \div (2 \text{ kg})(4.184 \text{ kJ/kg·K})\} = 282 \text{ K} = 8.93°C$.

Problem

35. A shallow pond contains 94,000 kg of water. In winter it's entirely frozen. By how much does the entropy of the pond increase when the ice, already at 0°C, melts and then heats to its summer temperature of 15°C?

Solution

During the melting at $T_1 = 0°C$, $\Delta S_1 = \Delta Q/T_1 = mL_f/T_1$, and during the warming to $T_2 = 15°C$, $\Delta S_2 = mc \ln(T_2/T_1)$. The total change is $\Delta S = \Delta S_1 + \Delta S_2 = m[(L_f/T_1) + c\ln(T_2/T_1)] = (9.4 \times 10^4 \text{ kg})[(334 \text{ kJ/273 kg·K}) + (4.184 \text{ kJ/kg·K}) \ln(288/273)] = 136 \text{ MJ/K}$.

Problem

37. The temperature of n moles of ideal gas is changed from T_1 to T_2 while the gas volume is held constant. Show that the corresponding entropy change is $\Delta S = nC_V \ln(T_2/T_1)$.

Solution

From the first law of thermodynamics $(dQ = dU + dW)$ and the properties of an ideal gas $(dU = nC_V \, dT$ and $PV = nRT)$, an infinitesimal entropy change is $dS = \frac{dQ}{T} = nC_V \frac{dT}{T} + \frac{P}{T}dV = nC_V \frac{dT}{T} + nR\frac{dV}{V}$. If we integrate from state 1 (T_1, V_1) to state 2 (T_2, V_2), we obtain $\Delta S = nC_V \ln(T_2/T_1) + nR \ln(V_2/V_1)$. For an isovolumic process (in which the gas does no work), ΔS is as given in the problem. (Of course, we could have started with $dQ = nC_V \, dT$ at constant volume, but we wanted to display ΔS for a general ideal gas process, for use in other problems.)

Problem

39. A 5.0-mol sample of an ideal diatomic gas $(C_V = \frac{5}{2}R)$ is initially at 1.0 atm pressure and 300 K. What is the entropy change if the gas is heated to 500 K (a) at constant volume, (b) at constant pressure, and (c) adiabatically?

Solution

(a) From Problem 37, $\Delta S_V = nC_V \ln(T_2/T_1) = (5 \text{ mol})(2.5 \times 8.314 \text{ J/mol·K})\ln(500/300) = 53.1 \text{ J/K}$.

(b) From the general expression in the solution to Problem 37, with $V_2/V_1 = T_2/T_1$ at constant pressure and $C_P = C_V + R$, $\Delta S_P = nC_P \ln(T_2/T_1) = (C_P/C_V)\Delta S_V = 1.4(53.1 \text{ J/K}) = 74.3$ J/K. (c) For an adiabatic process, $dQ = T\,dS = 0$, so $\Delta S = 0$. (An adiabatic process is one at constant entropy.)

Problem

41. A 250-g sample of water at 80°C is mixed with 250 g of water at 10°C. Find the entropy changes for (a) the hot water, (b) the cool water, and (c) the system.

Solution

The equilibrium temperature for the mixture, assuming all the heat lost by the hot water is gained by the cold, is $T_{eq} = 45°C$. (a) $\Delta S_{\text{hot water}} = mc\ln(T_{eq}/T_{\text{hot}}) = (0.25 \text{ kg})(4.184 \text{ kJ/kg·K}) \times \ln(318/353) = -109$ J/K, (b) $\Delta S_{\text{cold water}} = (0.25 \text{ kg})(4.184 \text{ kJ/kg·K})\ln(318/283) = 122$ J/K, and (c) $\Delta S_{\text{tot}} = -109$ J/K $+ 122$ J/K $= 12.7$ J/K.

Problem

43. A 5.0-mol sample of ideal monatomic gas undergoes the cycle shown in Fig. 22-31, in which the process BC is isothermal. Calculate the entropy change associated with each of the three steps, and show explicity that there is zero net entropy change over the full cycle.

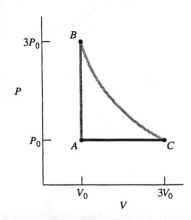

FIGURE 22-31 Problem 43.

Solution

For an ideal monatomic gas undergoing an isovolumic process $(dW = 0)$, $dQ = dU = nC_V\,dT = \frac{3}{2}nR\,dT$, and the entropy change is

$$\Delta S_{AB} = \int_A^B \frac{dQ}{T} = \int_A^B \frac{3}{2}nR\frac{dT}{T} = \frac{3}{2}nR\ln\left(\frac{T_B}{T_A}\right).$$

For the isothermal process $(dU = 0)$, $dQ = dW = P\,dV$, and

$$\Delta S_{BC} = \int_B^C \frac{dQ}{T} = \int_B^C \frac{P}{T}\,dV = \int_B^C nR\frac{dV}{V}$$
$$= nR\ln\left(\frac{V_C}{V_B}\right).$$

For the final isobaric process, $dQ = nC_P\,dT = \frac{5}{2}nR\,dT$, and $\Delta S_{CA} = \frac{5}{2}nR\ln(T_A/T_C)$. Therefore, the total entropy change for one cycle is $\Delta S = nR[\frac{3}{2}\ln(T_B/T_A) - \frac{5}{2}\ln(T_C/T_A) + \ln(V_C/V_B)]$. But $T_C = T_B$, while $P_BV_B = P_CV_C$ and $P_B/T_B = P_A/T_A$ imply $V_C/V_B = P_B/P_C = P_B/P_A = T_B/T_A$, so the logarithms are equal and $\Delta S = 0$ as expected. Numerically, $\Delta S_{AB} = (5 \text{ mol})(1.5)(8.314 \text{ J/mol·K}) \times \ln 3 = 68.5$ J/K, $\Delta S_{BC} = (5 \text{ mol})(8.314 \text{ J/mol·K}) \times \ln 3 = 45.7$ J/K, $\Delta S_{CA} = 2.5(5 \text{ mol}) \times (8.314 \text{ J/mol·K})\ln(\frac{1}{3}) = -114.2$ J/K, and $\Delta S_{\text{tot}} = 0$.

Problem

45. Ideal gas occupying 1.0 cm³ is placed in a 1.0-m³ vacuum chamber, where it expands adiabatically. If 6.5 J of energy become unavailable to do work, what was the initial gas pressure?

Solution

From Equations 22-10, 11 and the ideal gas law, $nRT = E_{\text{unavailable}}/\ln(V_2/V_1) = 6.5 \text{ J}/\ln(10^6) = 0.470$ J $= P_1V_1$. Therefore, $P_1 = 0.470 \text{ J}/(10^{-6} \text{ m}^3) = 470$ kPa.

Paired Problems

Problem

47. Cooling water circulates through a reversible Carnot engine at 3.2 kg/s. The water enters at 23°C and leaves at 28°C; the average temperature is essentially that of the engine's cool reservoir. If the engine's mechanical power output is 150 kW, what are (a) its efficiency and (b) its highest temperature?

Solution

(a) From the rate of flow and temperature rise of the cooling water, the rate of heat exhausted by the engine can be calculated: $(dQ_c/dt) = (dm/dt)c\,\Delta T = (3.2 \text{ kg/s})(4.184 \text{ kJ/kg·K})(28°C - 23°C) = 66.9$ kW. Then $(dQ_h/dt) = (dQ_c/dt) + (dW/dt) = 66.9$ kW $+ 150$ kW and $e = (dW/dt)/(dQ_h/dt) = 150$ kW $\div 216.9$ kW $= 69.1\%$. (b) For a Carnot engine with $T_c = \frac{1}{2}(23°C + 28°C) = 298.5$ K, $T_h = T_c/(1 - e) = (298.5 \text{ K})/(1 - 0.691) = 967$ K $= 694°C$.

Problem

49. Which would provide the greatest increase in efficiency of a Carnot engine, a 10 K increase in the maximum temperature or a 10 K decrease in the minimum temperature?

Solution

If we differentiate the efficiency with respect to T_c or T_h, respectively, we obtain: $de_c = -dT_c/T_h$, and $de_h = T_c\,dT_h/T_h^2$. If $dT_h = -dT_c$, then $de_h = (T_c/T_h)de_c < de_c$, since $T_c/T_h < 1$. The fact that a decrease in T_c produces a greater increase in efficiency than an increase of the same magnitude in T_h can also be demonstrated by direct substitution of $T_c - \Delta T$ or $T_h + \Delta T$ into Equation 22-3.

Problem

51. It costs \$180 to heat a house with electricity in a typical winter month. (Electric heat simply converts all the incoming electrical energy to heat.) What would be the monthly heating bill following conversion to an electrically powered heat pump system with COP = 2.1?

Solution

The same electrical energy W, as used for direct conversion in electric heating, would produce heat $Q_h = W + Q_c = (1 + \text{COP})W$ if used to run a heat pump. (A heat pump gives more heat for the money than electric heat.) Therefore, the cost of running a heat pump is a factor $(1 + \text{COP})^{-1}$ less, or $\$180(1 + 2.1)^{-1} = \$58.1$ for the house in this question.

Problem

53. A reversible engine contains 0.20 mol of ideal monatomic gas, initially at 600 K and confined to 2.0 L. The gas undergoes the following cycle:

- Isothermal expansion to 4.0 L.
- Isovolumic cooling to 300 K.
- Isothermal compression to 2.0 L.
- Isovolumic heating to 600 K.

(a) Calculate the net heat added during the cycle and the net work done. (b) Determine the engine's efficiency, defined as the ratio of the work done to only the heat *absorbed* during the cycle.

Solution

The P-V diagram for the cycle is as shown. For the isothermal expansion, $Q_{AB} = W_{AB} = nRT_A \ln(V_B/V_A) > 0$; for the isovolumic cooling, $W_{BC} = 0$, $Q_{BC} = \Delta U_{BC} = nC_V\,\Delta T_{BC} = \frac{3}{2}nR\times(T_C - T_B) < 0$; for the isothermal compression, $Q_{CD} =$

$W_{CD} = nRT_C \ln(V_D/V_C) < 0$; and for the final isovolumic heating $W_{DA} = 0$, $Q_{DA} = \Delta U_{DA} = \frac{3}{2}nR\times(T_A - T_D) > 0$. For these processes, it is given that $V_B = 2V_A = V_C = 2V_D = 4$ L, $T_A = T_B = 600$ K, and $T_C = T_D = 300$ K. (a) In a cyclic process ($\Delta U = 0$), the net heat added equals the net work done, $Q_{net} = W_{net} = nRT_A \ln(V_B/V_A) + nRT_C \ln(V_D/V_C) = nR(T_A - T_C)\ln(V_B/V_A) = (0.2\text{ mol})(8.314\text{ J/mol·K})\times(300\text{ K})\ln 2 = 346$ J. (Note explicitly, that since $\Delta T_{BC} = -\Delta T_{DA}$, $W_{BC} + W_{DA} = 0 = \Delta U_{BC} + \Delta U_{DA} = Q_{BC} + Q_{DA}$, and therefore $W_{net} = W_{AB} + W_{CD} = Q_{AB} + Q_{CD} = Q_{net}$.) (b) The heat absorbed during the cycle (just the positive values of heat) is $Q_+ = Q_{AB} + Q_{DA} = nRT_A \ln(V_B/V_A) + \frac{3}{2}nR(T_A - T_B) = (0.2\text{ mol})(3.314\text{ J/mol·K})[(600\text{ K})\ln 2 + 1.5(300\text{ K})] = 1.44$ kJ. (Note: $Q_- = -Q_{BC} - Q_{CD} = 1.09$ kJ is the heat exhausted per cycle, and $Q_{net} = Q_+ - Q_-$.) The efficiency, as defined in this problem, is $W_{net}/Q_+ = 346$ J/1.44 kJ = 24.0%. (Note: A Carnot engine operating between 600 K and 300 K has efficiency $1 - 300/600 = 50\%$. This is not a contradiction of Carnot's theorem, because the engine in this problem does *not* absorb and exhaust heat at constant temperatures.)

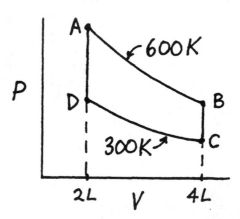

Problem 53 Solution.

Problem

55. You dump three 10-kg buckets of 10°C water into an empty tub, then add one 10-kg bucket of 70°C water. By how much does the entropy of the water increase?

Solution

Mixing 30 kg of water at 10°C = 283 K with 10 kg of water at 70°C = 343 K produces water at an equilibrium temperature given by $30(T_{eq} - 283\text{ K}) + 10(T_{eq} - 343\text{ K}) = 0$, or $T_{eq} = 298$ K. (This assumes there is no heat transfer to the tub or the surroundings.) The total entropy change of the water

is $\Delta S = m_1 c \ln(T_{eq}/T_1) + m_2 c \ln(T_{eq}/T_2) =$
$(4.184 \text{ kJ/kg·K})[(30 \text{ kg})\ln(298/283) + (10 \text{ kg})\ln \times$
$(298/343)] = 6.48 \text{ kJ/K} - 5.88 \text{ kJ/K} = 598 \text{ J/K}.$

Supplementary Problems

Problem

57. You're lying in a bathtub of water at 42°C. Suppose, in violation of the second law of thermodynamics, that the water spontaneously cooled to room temperature (20°C) and that the energy so released was transformed into your gravitational potential energy. Estimate the height to which you would rise above the bathtub.

Solution

A person taking a bath uses, on average, about 40 gal, or $(40 \text{ gal})(3.785 \times 10^{-3} \text{ m}^3/\text{gal})(10^3 \text{ kg/m}^3) - 151 \text{ kg}$, of water (see Appendix C). (In contrast, 5 min under a low-flow shower head consumes about 15 gal of water.) The amount of thermal energy that would be released by an average bathtub full of water, which cooled from 42°C to 20°C, is $\Delta Q = m c_w \Delta T = (151 \text{ kg}) \times (4184 \text{ J/kg·K})(22 \text{ K}) = 13.9 \text{ MJ}$ (see Equation 19-5 and Table 19-1). This amount is rather large compared to the increments of gravitational potential energy for ordinary objects near the Earth's surface, $\Delta U = mg \Delta y$ (see Equation 8-3). For a person of average mass 65 kg, it corresponds to a height difference of $\Delta y = (13.9 \text{ MJ})/(6 \times 9.81 \text{ N}) = 21.9 \text{ km}!$ (This energy is nearly half of the roughly 31.2 MJ required to launch 1 kg into a near Earth orbit (see Problem 9-45).)

Problem

59. A solar-thermal power plant is to be built in a desert location where the only source of cooling water is a small creek with average flow of 100 kg/s and an average temperature of 30°C. The plant is to cool itself by boiling away the entire creek. If the maximum temperature achieved in the plant is 500 K, what is the maximum electric power output it can sustain without running out of cooling water?

Solution

The maximum heatflow the creek could absorb is:
$$\frac{dQ_c}{dt} = \frac{dm}{dt}(c_W \Delta T + L_v)$$
$$= \left(10^2 \frac{\text{kg}}{\text{s}}\right)\left[\left(4.184 \frac{\text{kJ}}{\text{kg·K}}\right)\right.$$
$$\left. \times (100°C - 30°C) + 2257 \frac{\text{kJ}}{\text{kg}}\right] = 255 \text{ MW}.$$

If the plant had the maximum thermodynamic efficiency $(Q_c/Q_h = T_c/T_h)$, operating between $T_c = (273 + 30)$ K and $T_h = 500$ K, with the above rate of heat exhaust, its power output would be
$$\frac{dW}{dt} = \frac{d}{dt}(Q_h - Q_c) = \frac{dQ_c}{dt}\left(\frac{T_h}{T_c} - 1\right)$$
$$= 255 \text{ MW}\left(\frac{500}{303} - 1\right) = 166 \text{ MW}.$$

Problem

61. Gasoline engines operate approximately on the **Otto cycle**, consisting of two adiabatic and two constant-volume segments. The Otto cycle for a particular engine is shown in Fig. 22-33. (a) If the gas in the engine has specific heat ratio γ, find the engine's efficiency, assuming all processes are reversible. (b) Find the maximum temperature, in terms of the minimum temperature T_{min}. (c) How does the efficiency compare with that of a Carnot engine operating between the same two temperature extremes? *Note:* Fig. 22-33 neglects the intake of fuel-air and the exhaust of combustion products, which together involve essentially no net work.

Solution

It is convenient to solve this problem beginning with part (b), because we will need to express the temperatures of all the numbered points in Fig. 22-33 in terms of T_1, the minimum. This is accomplished by use of the adiabatic and ideal gas laws, Equations 21-13a and b, and 20-2, respectively. First,
$$\frac{T_3}{T_4} = \left(\frac{V_4}{V_3}\right)^{\gamma-1} = 5^{\gamma-1} = \left(\frac{V_1}{V_2}\right)^{\gamma-1} = \frac{T_2}{T_1}.$$

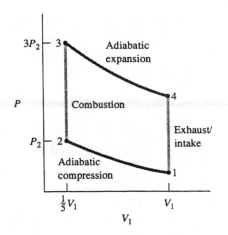

FIGURE 22-33 Problems 61 and 62.

Second,

$$\frac{P_3}{P_4} = \left(\frac{V_4}{V_3}\right)^\gamma = \left(\frac{V_1}{V_2}\right)^\gamma = \frac{P_2}{P_1}.$$

Last,

$$\frac{T_3}{T_2} = \frac{P_3}{P_2} = 3 = \frac{P_4}{P_1} = \frac{T_4}{T_1}.$$

Therefore, $T_2 = 5^{\gamma-1}T_1$, $T_4 = 3T_1$, and (the maximum temperature) $T_3 = 5^{\gamma-1}T_4 = 3\times5^{\gamma-1}T_1$. (a) Since no heat is transferred on the adiabatic segments, the heat input is $Q_h = Q_{23} = nC_V(T_3 - T_2) = nC_V5^{\gamma-1}\times(3-1)T_1 = 2\times5^{\gamma-1}nC_VT_1$, and the heat exhaust is $Q_c = -Q_{41} = -nC_V(T_1 - T_4) = 2nC_VT_1$. Therefore, the efficiency is $e = 1 - Q_c/Q_h = 1 - 5^{1-\gamma}$. (c) A Carnot engine operating between T_3 and T_1 has efficiency $1 - T_1/T_3 = 1 - \frac{1}{3}\times5^{1-\gamma}$, so $e_{\text{Otto}} < e_{\text{Carnot}}$.

Problem

63. A heat pump designed for southern climates extracts heat from outside air and delivers air at 40°C to the inside of a house. (a) If the average outside temperature is 5°C, what is the average COP of the heat pump? (b) Suppose the pump is used in a northern climate where the average winter temperature is −10°C. Now what is the COP? (c) Two *identical* houses, one in the north and one in the south, are heated by this pump. Both houses maintain indoor temperatures of 19°C. What is the ratio of electric power consumption in the two houses? *Hint:* Think about heat loss as well as COP!

Solution

(a) We assume the heat pump operates on a reversible Carnot cycle, so $\text{COP}_s = T_c/(T_h - T_c) = (273 + 5)\div(40 - 5) = 7.94$, where the subscript "s" is for southern. (b) Similarly, $\text{COP}_n = (273 - 10)/50 = 5.26$. (c) The heat flow needed from the pump (to balance the heat loss from the house) is proportional to the inside/outside temperature difference (see Equation 19-10), so for identical northern and southern houses, $H_n/H_s = [19°C - (-10°C)]\div(19°C - 5°C) = 2.07$. If we eliminate Q_c from Equation 22-4, we obtain $Q_h = W(1 + \text{COP})$. Now, $dQ_h/dt = H$ is the heat flow from the pump, while $dW/dt = P$ is the power consumption, so substitution into the ratio of heat losses yields $2.07 = P_n(1 + \text{COP}_n)/P_s(1 + \text{COP}_s)$, or $P_n/P_s = 2.07(1 + 7.94)/(1 + 5.26) = 2.96$.

Problem

65. A Carnot engine extracts heat from a block of mass m and specific heat c that is initially at temperature T_{h0} but which has no heat source to maintain that temperature. The engine rejects heat to a reservoir at a constant temperature T_c. The engine is operated so its mechanical power output is proportional to the temperature difference $T_h - T_c$:

$$P = P_0\frac{T_h - T_c}{T_{h0} - T_c},$$

where T_h is the instantaneous temperature of the hot block and P_0 is the initial power output. (a) Find an expression for T_h as a function of time, and (b) determine how long it takes for the engine's power output to reach zero.

Solution

(a) In time dt, the engine extracts heat $dQ_h = -mc\,dT_h$ from the block, and does work $dW = P\,dt$. Since it is a Carnot engine, $dW = e_{\max}dQ_h = [(T_h - T_c)/T_h](-mc\,dT_h) = P\,dt$. The power is also assumed to be proportional to $T_h - T_c$, so the equation becomes $-mc\,dT_h/T_h = P_0\,dt/(T_{h0} - T_c)$. Integrating from $t = 0$ and T_{h0} to t and T_h, we obtain

$$\int_{T_{h0}}^{T_h}\frac{dT_h'}{T_h'} = \ln\left(\frac{T_h}{T_{h0}}\right) = -\int_0^t\frac{P_0\,dt}{mc(T_{h0} - T_c)}$$

$$= -\frac{P_0 t}{mc(T_{h0} - T_c)},$$

or $T_h = T_{h0}\exp\{-P_0 t/mc(T_{h0} - T_c)\}$. (b) The power output is zero for $T_h = T_c$. This occurs at time $t_0 = (mc/P_0)(T_{h0} - T_c)\ln(T_{h0}/T_c)$. (Note: The expression for P was originally assumed valid for $T_h \geq T_c$, or for times $t \leq t_0$. If we allow $T_h < T_c$, or $t > t_0$, then $dW = P\,dt < 0$ becomes work input to an "engine" which acts like a refrigerator cooling the block.)

Problem

67. An ideal diatomic gas undergoes the cyclic process described in Fig. 22-34. Fill in the blank spaces in the table below:

	P	V	T	$U - U_A$	$S - S_A$
A	P_0	V_0	T_0	0	0
B	$3.4P_0$	V_0			
C					
D	P_0	$3.0V_0$			

$$V_C = \left(\frac{3^\gamma}{3.4}\right)^{1/(\gamma-1)} V_0 = \frac{3^{3.5}}{3.4^{2.5}} V_0 = 2.19V_0.$$

Then

$$P_C = P_B(V_B T_C / V_C T_B)$$
$$= (3.4P_0)(1/2.19)(1) = 1.55P_0.$$

The internal energies and entropies can be calculated from $\Delta U = nC_V \Delta T$ (Equation 21-7), and $\Delta S = nC_V \ln(T_2/T_1) + nR \ln(V_2/V_1)$ (see the solution to Problem 37, or Equation 22-9 for an isobaric or isovolumic process, and Equation 22-10 for an isothermal one), where $C_V = \frac{5}{2}R$. Thus, $U_B - U_A = nC_V(T_B - T_A) = \frac{5}{2}nR(3.4-1)T_0 = \frac{5}{2}(2.4)(nRT_0) = 6P_0V_0$, $U_C - U_B = 0$ (or $U_C - U_A = 6P_0V_0$) and, $U_D - U_A = \frac{5}{2}nR(3-1)T_0 = 5P_0V_0$, while $S_B - S_A = nC_V \ln(T_B/T_A) = \frac{5}{2}nR \ln(3.4) = 3.06nR$, $S_C - S_A = nC_V \ln(T_C/T_A) + nR \ln(V_C/V_A) = 3.06nR + nR \times \ln(2.19) = 3.85nR$, and $S_D - S_C = 0$ ($dS = dQ/T = 0$ on an adiabat), or alternatively, $S_D - S_A = nC_V \times \ln(T_D/T_A) + nR \ln(V_D/V_A) = nC_P \ln(T_D/T_A) = \frac{7}{2}nR \ln 3$. Of course, $nR = P_0V_0/T_0$, so all the tabulated results can be expressed in terms of the given parameters.

	P	V	T	$U - U_A$	$S - S_A$
A	P_0	V_0	T_0	0	0
B	$3.4P_0$	V_0	$3.4T_0$	$6P_0V_0$	$3.06nR$
C	$1.55P_0$	$2.19V_0$	$3.4T_0$	$6P_0V_0$	$3.85nR$
D	P_0	$3.0V_0$	$3T_0$	$5P_0V_0$	$3.85nR$

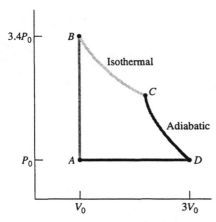

FIGURE 22-34 Problem 67.

Solution

We first calculate the blank P, V, and T's, using the adiabatic and ideal gas laws, with $\gamma = 1.4$. For states B and D:

$$T_B = (P_B V_B / P_A V_A)T_A = 3.4T_0, \quad \text{and}$$
$$T_D = (P_D V_D / P_A V_A)T_A = 3T_0.$$

For state C,

$$T_C = T_B,$$
$$P_C V_C^\gamma = P_D V_D^\gamma = P_0(3V_0)^\gamma, \quad \text{and}$$
$$P_C V_C = P_B V_B = 3.4P_0 V_0.$$

Division yields

$$V_C^{\gamma-1} = \frac{3^\gamma}{3.4} V_0^{\gamma-1}, \quad \text{or}$$

PART 3   CUMULATIVE PROBLEMS

Problem

1. Figure 1 shows the thermodynamic cycle of a diesel engine. Note that this cycle differs from that of a gasoline engine (see Fig. 22-33) in that combustion takes place isobarically. As with the gasoline engine, the compression ratio r is the ratio of maximum to minimum volume; $r = V_1/V_2$. In addition, the so-called *cutoff ratio* is defined by $r_c = V_3/V_2$. Find an expression for the engine's efficiency, in terms of the ratios r and r_c and the specific heat ratio γ. Although your expression suggests that the diesel engine might be less efficient than the gasoline engine (see Problem 61 of Chapter 22), the diesel's higher compression ratio more than compensates, giving it a higher efficiency.

Solution

From Table 21-1, the work done and heat absorbed during the four processes comprising the diesel cycle are:

$1 \rightarrow 2$ (adiabatic) $W_{12} = (P_1V_1 - P_2V_2)/(\gamma-1)$,
$$Q_{12} = 0,$$

$2 \rightarrow 3$ (isobaric) $W_{23} = P_2(V_3 - V_2)$,
$$Q_{23} = nC_P(T_3 - T_2) \equiv Q_h,$$

$3 \rightarrow 4$ (adiabatic) $W_{34} = (P_3V_3 - P_4V_4)/(\gamma-1)$,
$$Q_{34} = 0,$$

$4 \rightarrow 1$ (isovolumic) $W_{41} = 0$,
$$Q_{41} = nC_V(T_1 - T_4) \equiv -Q_c.$$

For the whole cycle, the work done is

$$W = \frac{P_1V_1 - P_2V_2 + P_3V_3 - P_4V_4}{(\gamma - 1)} + P_3V_3 - P_2V_2$$

$$= \frac{\gamma(P_3V_3 - P_2V_2) - P_4V_4 + P_1V_1}{(\gamma - 1)},$$

while the heat added can be written as $Q_h = nC_P(T_3 - T_2) = \gamma(P_3V_3 - P_2V_2)/(\gamma - 1)$, where we used the ideal gas law, $nT = PV/R$, and $C_P/R = C_P/(C_P - C_V) = \gamma/(\gamma - 1)$. Therefore,

$$W = Q_h - \frac{(P_4V_4 - P_1V_1)}{(\gamma - 1)} = Q_h - \frac{(P_4V_4 - P_1V_1)Q_h}{\gamma(P_3V_3 - P_2V_2)}$$

and the efficiency is $e = W/Q_h = 1 - (P_4V_4 - P_1V_1) \div \gamma(P_3V_3 - P_2V_2)$. The adiabatic law can now be used to express every product in terms of P_2V_2 and the compression and cutoff ratios: $P_2 = P_3$, $V_1 = V_4$, $V_1/V_2 = r$, $V_3/V_2 = r_c$, $P_1V_1^{\gamma} = P_2V_2^{\gamma}, P_3V_3^{\gamma} = P_4V_4^{\gamma}$, so $P_1V_1 = P_2V_2r^{1-\gamma}$, $P_3V_3 = P_2V_2r_c$, and $P_4V_4 = P_2V_2r^{1-\gamma}r_c^{\gamma}$. Finally, $e = 1 - r^{1-\gamma}(r_c^{\gamma} - 1)/\gamma(r_c - 1)$.

Problem

3. Equation 21-4 gives the work done by an ideal gas undergoing an isothermal expansion from volume V_1 to volume V_2. Find the analogous expression for a van der Waals gas described by Equation 20-7. Is the work done equal to the heat transferred to the gas in this case? Why or why not?

Solution

An expression for the work done by a van der Waals gas undergoing an isothermal expansion is derived in the solution to Problem 21-65. The van der Waals equation includes the effects of intermolecular forces, which contribute to the potential energy of a molecule, so we should not expect the internal energy, U, to be a function of T only (it depends on V and T). Therefore, $Q = \Delta U + W \neq W$ when a van der Waals gas undergoes an isothermal process.

Problem

5. The ideal Carnot engine shown in Fig. 4 operates between a heat reservoir and a block of ice with mass M. An external energy source maintains the reservoir at a constant temperature T_h. At time $t = 0$ the ice is at its melting point T_0, but it is insulated from everything except the engine, so it is free to change state and temperature. The engine is operated in such a way that it extracts heat from the reservoir at a constant rate P_h. (a) Find an expression for the time t_1 at which the ice is all melted, in terms of the quantities given and

any other appropriate thermodynamic parameters. (b) Find an expression for the mechanical power output of the engine as a function of time for times $t > t_1$. (c) Your expression in (b) holds only up to some maximum time t_2. Why? Find an expression for t_2.

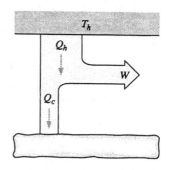

FIGURE 4 Cumulative Problem 5.

Solution

(a) The ice acts as a constant temperature cold reservoir at $T_c = T_0 = 273$ K, while it is melting for times $0 \leq t \leq t_1$. The Carnot engine exhausts heat at a constant rate, over this time interval, so the simplified form of Equation 22-2 gives $\mathcal{P}_c = (T_c/T_h)\mathcal{P}_h$, (where the first and second laws of thermodynamics have been written in terms of rate of heat flow or power, as in Example 22-2). The total heat exhausted is just sufficient to melt all the ice, therefore $\mathcal{P}_c t_1 = (T_0/T_h)\mathcal{P}_h t_1 = ML_f$, or $t_1 = ML_f T_h/\mathcal{P}_h T_0$. (Here, L_f is the heat of fusion of water and Equation 20-6 was used.) (b) After the ice is melted, the temperature of the cold reservoir increases with time. So does the rate of heat exhausted, while the mechanical power of the engine, $\mathcal{P} = \mathcal{P}_h - \mathcal{P}_c$, decreases. From Equation 19-5 in terms of rates, $\mathcal{P}_c dt = (T_c/T_h)\mathcal{P}_h dt = Mc\,dT_c$, or $dT_c/T_c = (\mathcal{P}_h/McT_h)\,dt$. This equation holds for a time interval $t_1 \leq t \leq t_2$ (see part (c) below), where $T_c = T_0$ at $t = t_1$, and can be integrated to give $\ln(T_c/T_0) = (\mathcal{P}_h/McT_h)(t - t_1)$, or $T_c = T_0 \exp\{\mathcal{P}_h(t - t_1)/McT_h\}$. The power output of the Carnot engine is therefore $\mathcal{P} = e\mathcal{P}_h = (1 - T_c/T_h)\mathcal{P}_h = [1 - (T_0/T_h)\exp\{\mathcal{P}_h(t - t_1)/McT_h\}]\mathcal{P}_h$, where e is the Carnot efficiency (Equation 22-3) and $\exp\{\ldots\}$ is the exponential function. (c) The Carnot engine operates as described above only as long as $T_c < T_h$. When $T_c = T_h$ at time t_2, the power output has dropped to zero. Then, from part (b), $\ln(T_h/T_0) = (\mathcal{P}_h/McT_h)(t_2 - t_1)$, or $t_2 = t_1 + (McT_h/\mathcal{P}_h) \times \ln(T_h/T_0)$.

PART 4   ELECTROMAGNETISM

CHAPTER 23   ELECTRIC CHARGE, FORCE, AND FIELD

ActivPhysics can help with these problems:
Activities 11.1–11.8

Section 23-2:   Electric Charge

Problem

1. Suppose the electron and proton charges differed by one part in one billion. Estimate the net charge you would carry.

Solution

Nearly all of the mass of an atom is in its nucleus, and about one half of the nuclear mass of the light elements in living matter (H, O, N, and C) is protons. Thus, the number of protons in a 65 kg average-sized person is approximately $\frac{1}{2}(65 \text{ kg})/(1.67\times10^{-27} \text{ kg}) \approx 2\times10^{28}$, which is also the number of electrons, since an average person is electrically neutral. If there were a charge imbalance of $|q_{\text{proton}} - q_{\text{electron}}| = 10^{-9}e$, a person's net charge would be about $\pm2\times10^{28}\times10^{-9}\times 1.6\times10^{-19}$ C $= \pm3.2$ C, or several coulombs (huge by ordinary standards).

Problem

3. Protons and neutrons are made from combinations of the two most common quarks, the u quark and the d quark. The u quark's charge is $+\frac{2}{3}e$ while the d quark carries $-\frac{1}{3}e$. How could three of these quarks combine to make (a) a proton and (b) a neutron?

Solution

(a) The proton's charge is $1e = \frac{2}{3}e + \frac{2}{3}e - \frac{1}{3}e$, corresponding to a combination of uud quarks; (b) for neutrons, $0 = \frac{2}{3} - \frac{1}{3} - \frac{1}{3}$ corresponds to udd. (See Chapter 39, or Chapter 45 in the extended version of the text.)

Section 23-3:   Coulomb's Law

Problem

5. If the charge imbalance of Problem 1 existed, what would be the approximate force between you and another person 10 m away? Treat the people as point charges, and compare the answer with your weight.

Solution

The magnitude of the Coulomb force between two point charges of 3.2 C (see solution to Problem 1), at a distance of 10 m, is $kq^2/r^2 = (9\times10^9 \text{ N·m}^2/\text{C}^2)\times (3.2 \text{ C}/10 \text{ m})^2 = 9.22\times10^8$ N. This is approximately 1.45 million times the weight of an average-sized 65 kg person.

Problem

7. The electron and proton in a hydrogen atom are 52.9 pm apart. What is the magnitude of the electric force between them?

Solution

$a_0 = 52.9$ pm is called the Bohr radius. For a proton and electron separated by a Bohr radius, $F_{\text{Coulomb}} = ke^2/a_0^2 \simeq (9\times10^9 \text{ N·m}^2/\text{C}^2)(1.6\times10^{-19} \text{ C} \div 5.29 \times 10^{-11} \text{ m})^2 = 8.23\times10^{-8}$ N.

Problem

9. Two charges, one twice as large as the other, are located 15 cm apart and experience a repulsive force of 95 N. What is the magnitude of the larger charge?

Solution

The product of the charges is $q_1 q_2 = r^2 F_{\text{Coulomb}}/k = (0.15 \text{ m})^2(95 \text{ N})/(9\times10^9 \text{ N·m}^2/\text{C}^2) = 2.38\times10^{-10}$ C^2. If one charge is twice the other, $q_1 = 2q_2$, then $\frac{1}{2}q_1^2 = 2.38\times10^{-10}$ C and $q_1 = \pm21.8$ μC.

Problem

11. A proton is on the x-axis at $x = 1.6$ nm. An electron is on the y-axis at $y = 0.85$ nm. Find the net force the two exert on a helium nucleus (charge $+2e$) at the origin.

Solution

A unit vector from the proton's position to the origin is $-\hat{\imath}$, so the Coulomb force of the proton on the helium nucleus is $\mathbf{F}_{p,He} = k(e)(2e)(-\hat{\imath})/(1.6 \text{ nm})^2 = -0.180\hat{\imath}$ nN. (Use Equation 23-1, with q_1 for the proton, q_2 for the helium nucleus, and the approximate values of k and e given.) A unit vector from the electron's position to the origin is $-\hat{\jmath}$, so its force on the helium nucleus is $\mathbf{F}_{e,He} = k(-e)(2e)(-\hat{\jmath}) \div (0.85 \text{ nm})^2 = 0.638\hat{\jmath}$ nN. The net Coulomb force on the helium nucleus is the sum of these. (The vector form of Coulomb's law and superposition, as explained in the solution to Problems 15 and 19, provides a more general approach.)

Problem

13. A charge q is at the point $x = 1$ m, $y = 0$. Write expressions for the unit vectors you would use in Coulomb's law if you were finding the force that q exerts on other charges located at (a) $x = 1$ m, $y = 1$ m; (b) the origin; (c) $x = 2$ m, $y = 3$ m. Note that you don't know the sign of q. Why doesn't this matter?

Solution

A unit vector from $\mathbf{r}_q = (1 \text{ m}, 0)$, the position of charge q, to any other point $\mathbf{r} = (x, y)$ is $\hat{\mathbf{n}} = (\mathbf{r} - \mathbf{r}_q) \div |\mathbf{r} - \mathbf{r}_q| = (x - 1 \text{ m}, y)/\sqrt{(x - 1 \text{ m})^2 + y^2}$. The sign of q doesn't affect this unit vector, but the signs of both charges do determine whether the force exerted by q is repulsive or attractive, i.e., in the direction of $+\hat{\mathbf{n}}$ or $-\hat{\mathbf{n}}$. (a) When the other charge is at position $\mathbf{r} = (1 \text{ m}, 1 \text{ m})$, $\hat{\mathbf{n}} = (0, 1 \text{ m})/\sqrt{0 + (1 \text{ m})^2} = (0, 1) = \hat{\jmath}$. (b) When $\mathbf{r} = (0, 0)$, $\hat{\mathbf{n}} = (-1 \text{ m}, 0)/\sqrt{(-1 \text{ m})^2 + 0} = (-1, 0) = -\hat{\imath}$. (c) Finally, when $\mathbf{r} = (2 \text{ m}, 3 \text{ m})$, $\hat{\mathbf{n}} = (1 \text{ m}, 3 \text{ m})/\sqrt{(1 \text{ m})^2 + (3 \text{ m})^2} = (1, 3)/\sqrt{10} = 0.316\hat{\imath} + 0.949\hat{\jmath}$.

Problem

15. A 9.5-μC charge is at $x = 16$ cm, $y = 5.0$ cm, and a -3.2-μC charge is at $x = 4.4$ cm, $y = 11$ cm. Find the force on the negative charge.

Solution

Denote the positions of the charges by $\mathbf{r}_1 = (16\hat{\imath} + 5\hat{\jmath})$ cm for $q_1 = 9.5$ μC, and $\mathbf{r}_2 = (4.4\hat{\imath} + 11\hat{\jmath})$ cm for $q_2 = -3.2$ μC. The vector from q_1 to q_2 is $\mathbf{r} = \mathbf{r}_2 - \mathbf{r}_1$, and a unit vector in this direction is $\hat{\mathbf{r}} = (\mathbf{r}_2 - \mathbf{r}_1) \div |\mathbf{r}_2 - \mathbf{r}_1|$. The vector form of Coulomb's law for the

electric force of q_1 on q_2 is $\mathbf{F}_{12} = kq_1 q_2 (\mathbf{r}_2 - \mathbf{r}_1) \div |\mathbf{r}_2 - \mathbf{r}_1|^3$. (This gives the Coulomb force between two point charges, as a function of their positions, and is a convenient form to memorize because of its direct applicability.) Substituting the given values for this problem, we find:

$$\mathbf{F}_{12} = \left(\frac{9 \times 10^9 \text{ N} \cdot \text{m}^2}{\text{C}^2}\right)(9.5 \ \mu\text{C})(-3.2 \ \mu\text{C})$$
$$\times \frac{(4.4\hat{\imath} + 11\hat{\jmath} - 16\hat{\imath} - 5\hat{\jmath}) \text{ cm}}{[(4.4 - 16)^2 + (11 - 5)^2]^{3/2} \text{ cm}^3}$$
$$= (14.2\hat{\imath} - 7.37\hat{\jmath}) \text{ N},$$

with magnitude 16.0 N and direction $\theta = -27.3°$ to the x-axis (negative angle measured CW).

Problem

17. A 60-μC charge is at the origin, and a second charge is on the positive x-axis at $x = 75$ cm. If a third charge placed at $x = 50$ cm experiences no net force, what is the second charge?

Solution

In order for the net force to be zero at a position between the first two charges, they must both have the same sign, i.e., $q_1 = 60$ μC at $x_1 = 0$ and $q_2 > 0$ at $x_2 = 75$ cm. (Then the separate forces of the first two charges on the third are in opposite directions.) Therefore, for the third charge q_3 at $x_3 = 50$ cm, $F_{3x} = kq_3[q_1(x_3 - x_1)^{-2} - q_2(x_2 - x_3)^{-2}] = kq_3 \times [60 \ \mu\text{C}(50 \text{ cm})^{-2} - q_2(25 \text{ cm})^{-2}] = 0$ implies $q_2 = 60 \ \mu\text{C}(25/50)^2 = 15 \ \mu\text{C}$.

Problem

19. In Fig. 23-39 take $q_1 = 68$ μC, $q_2 = -34$ μC, and $q_3 = 15$ μC. Find the electric force on q_3.

Solution

Denote the positions of the charges by $\mathbf{r}_1 = \hat{\jmath}$, $\mathbf{r}_2 = 2\hat{\imath}$, and $\mathbf{r}_3 = 2\hat{\imath} + 2\hat{\jmath}$ (distances in meters). The vector form of Coulomb's law (in the solution to Problem 15)

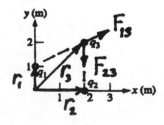

FIGURE 23-39 Problem 19 Solution.

and the superposition principle give the net electric force on q_3 as:

$$\mathbf{F}_3 = \mathbf{F}_{13} + \mathbf{F}_{23} = \frac{kq_1q_3(\mathbf{r}_3 - \mathbf{r}_1)}{|\mathbf{r}_3 - \mathbf{r}_1|^3} + \frac{kq_2q_3(\mathbf{r}_3 - \mathbf{r}_2)}{|\mathbf{r}_3 - \mathbf{r}_2|^3}$$

$$= (9 \times 10^9 \text{ N})(15 \times 10^{-6})[(68 \times 10^{-6})(2\hat{\mathbf{i}} + \hat{\mathbf{j}})/5\sqrt{5}$$
$$+ (-34 \times 10^{-6})2\hat{\mathbf{j}}/8]$$

$$= (1.64\hat{\mathbf{i}} - 0.326\hat{\mathbf{j}}) \text{ N},$$

or $F_3 = \sqrt{F_{3x}^2 + F_{3y}^2} = 1.67$ N at an angle of $\theta = \tan^{-1}(F_{3y}/F_{3x}) = -11.2°$ to the x-axis.

Problem

21. Four identical charges q form a square of side a. Find the magnitude of the electric force on any of the charges.

Solution

By symmetry, the magnitude of the force on any charge is the same. Let's find this for the charge at the lower left corner, which we take as the origin, as shown. Then $\mathbf{r}_1 = 0$, $\mathbf{r}_2 = a\hat{\mathbf{j}}$, $\mathbf{r}_3 = a(\hat{\mathbf{i}} + \hat{\mathbf{j}})$, $\mathbf{r}_4 = a\hat{\mathbf{i}}$, and

$$\mathbf{F}_1 = kq^2 \left[\frac{\mathbf{r}_1 - \mathbf{r}_2}{|\mathbf{r}_1 - \mathbf{r}_2|^3} + \frac{\mathbf{r}_1 - \mathbf{r}_3}{|\mathbf{r}_1 - \mathbf{r}_3|^3} + \frac{\mathbf{r}_1 - \mathbf{r}_4}{|\mathbf{r}_1 - \mathbf{r}_4|^3} \right]$$

$$= kq^2 \left[\frac{-a\hat{\mathbf{j}}}{a^3} - \frac{a(\hat{\mathbf{i}} + \hat{\mathbf{j}})}{2\sqrt{2}a^3} - \frac{a\hat{\mathbf{i}}}{a^3} \right]$$

$$= -\frac{kq^2}{a^2}(\hat{\mathbf{i}} + \hat{\mathbf{j}}) \left(1 + \frac{1}{2\sqrt{2}} \right),$$

(Use the vector form of Coulomb's law in the solution to Problem 15, and the superposition principle.) Since $|\hat{\mathbf{i}} + \hat{\mathbf{j}}| = \sqrt{2}$, $|\mathbf{F}_1| = (kq^2/a^2)\sqrt{2}(1 + 1/2\sqrt{2}) = (kq^2/a^2)(\sqrt{2} + \frac{1}{2}) = 1.91kq^2/a^2$.

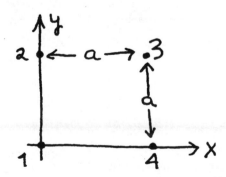

Problem 21 Solution.

Problem

23. Three charges lie in the x-y plane: $q_1 = 55$ μC at $x = 0$, $y = 2.0$ m; q_2 at $x = 3.0$ m, $y = 0$; and q_3

at $x = 4.0$ m, $y = 3.0$ m. If the force on q_3 is $8.0\hat{\mathbf{i}} + 15\hat{\mathbf{j}}$ N, find q_2 and q_3.

Solution

Using the vector form of Coulomb's law explained in the solution to Problem 15, and superposition, we can write the force on q_3 as $\mathbf{F}_3 = kq_3[q_1(\mathbf{r}_3 - \mathbf{r}_1) \times |r_3 - r_1|^{-3} + q_2(\mathbf{r}_3 - \mathbf{r}_2)|r_3 - r_2|^{-3}]$. Substituting the given values, $\mathbf{F}_3 = (8\hat{\mathbf{i}} + 15\hat{\mathbf{j}})$ N, $\mathbf{r}_1 = 2\hat{\mathbf{j}}$ m, $\mathbf{r}_2 = 3\hat{\mathbf{i}}$ m, and $\mathbf{r}_3 = (4\hat{\mathbf{i}} + 3\hat{\mathbf{j}})$ m, we find $(8\hat{\mathbf{i}} + 15\hat{\mathbf{j}})$ N·m$^2 = kq_3 \times [q_1(4\hat{\mathbf{i}} + \hat{\mathbf{j}})(4^2 + 1^2)^{-3/2} + q_2(\hat{\mathbf{i}} + 3\hat{\mathbf{j}})(1^2 + 3^2)^{-3/2}]$. Equating x and y components, we get 8 N·m$^2/kq_3 = 4q_1 17^{-3/2} + q_2 10^{-3/2}$ and 15 N·m$^2/kq_3 = q_1 17^{-3/2} + 3q_2 10^{-3/2}$. Dividing these equations and solving for q_2 in terms of $q_1 = 55$ μC, we find $q_2 = (10/17)^{3/2} \times (52/9)q_1 = 143$ μC. Substituting this into either component equation, we get $q_3 = (8 \text{ N·m}^2/k) \times [4q_1 17^{-3/2} + q_2 10^{-3/2}]^{-1} = (15 \text{ N·m}^2/k) [q_1 17^{-3/2} + 3q_2 10^{-3/2}]^{-1} = 116$ μC.

Section 23-4:　The Electric Field

Problem

25. An electron placed in an electric field experiences a 6.1×10^{-10} N electric force. What is the field strength?

Solution

From Equation 23-3a, $E = F/e = 6.1 \times 10^{-10}$ N $\div$ 1.6×10^{-19} C $= 3.81 \times 10^9$ N/C. (The field strength is the magnitude of the field.)

Problem

27. A 68-nC charge experiences a 150-mN force in a certain electric field. Find (a) the field strength and (b) the force that a 35-μC charge would experience in the same field.

Solution

Equations 23-3a and b give (a) $E = 150$ mN/68 nC $= 2.21$ MN/C, and (b) $F = (35$ μC$)(2.21$ MN/C$) = 77.2$ N.

Problem

29. The electron in a hydrogen atom is 0.0529 nm from the proton. What is the proton's electric field strength at this distance?

Solution

The proton in a hydrogen atom behaves like a point charge, for an electron one Bohr radius away (see solution to Problem 7), so Equation 23-4 gives $E = ke/a_0^2 = (9 \times 10^9 \text{ N·m}^2/\text{C}^2)(1.6 \times 10^{-19} \text{ C}) \div (5.29 \times 10^{-11} \text{ m})^2 = 5.15 \times 10^{11}$ N/C.

Section 23-5: Electric Fields of Charge Distributions

Problem

31. In Fig. 23-40, point P is midway between the two charges. Find the electric field in the plane of the page (a) 5.0 cm directly above P, (b) 5.0 cm directly to the right of P, and (c) at P.

Solution

Take the origin of x-y coordinates at the midpoint, as indicated, and use Equation 23-5. Let $\mathbf{r}_\pm = \pm(2.5 \text{ cm})\hat{\jmath}$ denote the positions of the charges, and $\mathbf{r}$ that of the field point. A unit vector from one charge to the field point is $(\mathbf{r} - \mathbf{r}_\pm)/|\mathbf{r} - \mathbf{r}_\pm|$, so the spacial factors in Coulomb's law are $\hat{\mathbf{r}}_i/r_i^2 = \mathbf{r}_i/r_i^3 = (\mathbf{r} - \mathbf{r}_\pm)/|\mathbf{r} - \mathbf{r}_\pm|^3$. (a) For $\mathbf{r} = (5.0 \text{ cm})\hat{\jmath}$, $\mathbf{r}_1 = \mathbf{r} - \mathbf{r}_+ = (5.0 \text{ cm})\hat{\jmath} - (2.5 \text{ cm})\hat{\jmath} = (2.5 \text{ cm})\hat{\jmath}$, and $\mathbf{r}_2 = \mathbf{r}_2 = \mathbf{r} - \mathbf{r}_- = (7.5 \text{ cm})\hat{\jmath}$. Then

$$\mathbf{E} = k\left(\frac{q_1 \mathbf{r}_1}{r_1^3} + \frac{q_2 \mathbf{r}_2}{r_2^3}\right)$$

$$= \left(9 \times 10^9 \frac{\text{N·m}^2}{\text{C}^2}\right)(2\ \mu\text{C})\left[\frac{\hat{\jmath}}{(2.5 \text{ cm})^2} - \frac{\hat{\jmath}}{(7.5 \text{ cm})^2}\right]$$

$$= (25.6 \text{ MN/C})\hat{\jmath}.$$

(b) For $\mathbf{r} = (5.0 \text{ cm})\hat{\imath}$,

$$\mathbf{E} = \left(9 \times 10^9 \frac{\text{N·m}^2}{\text{C}^2}\right)\left(\frac{2\ \mu\text{C}}{\text{cm}^2}\right)\left[\frac{(5.0\hat{\imath} - 2.5\hat{\jmath})}{(5.0^2 + (-2.5)^2)^{3/2}} - \frac{(5.0\hat{\imath} + 2.5\hat{\jmath})}{(5.0^2 + 2.5^2)^{3/2}}\right] = -(5.15 \text{ MN/C})\hat{\jmath}.$$

(c) For $\mathbf{r} = 0$,

$$\mathbf{E} = \left(9 \times 10^9 \frac{\text{N·m}^2}{\text{C}^2}\right)\left(\frac{2\ \mu\text{C}}{\text{cm}^2}\right)\left[\frac{-\hat{\jmath}}{(2.5)^2} - \frac{\hat{\jmath}}{(2.5)^2}\right]$$

$$= -(57.6 \text{ MN/C})\hat{\jmath}.$$

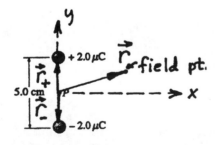

FIGURE 23-40 Problem 31 Solution.

Problem

33. A proton is at the origin and an ion is at $x = 5.0$ nm. If the electric field is zero at $x = -5$ nm, what is the charge on the ion?

Solution

The proton, charge e, is at $\mathbf{r}_p = 0$, and the ion, charge q, is at $\mathbf{r}_I = 5\hat{\imath}$ nm. The field at point $\mathbf{r} = -5\hat{\imath}$ nm is given by Equation 23-5, with spacial factors written as in the solutions to Problems 15 or 31:

$$\mathbf{E}(\mathbf{r}) = \sum_i kq_i \frac{(\mathbf{r} - \mathbf{r}_i)}{|\mathbf{r} - \mathbf{r}_i|^3}$$

$$= ke\frac{(-5\hat{\imath} \text{ nm})}{(5 \text{ nm})^3} + kq\frac{(-5\hat{\imath} \text{ nm} - 5\hat{\imath} \text{ nm})}{(10 \text{ nm})^3}.$$

Therefore, $\mathbf{E} = 0$ implies $2q/(10)^3 = -e/(5)^3$, or $q = -4e$. (Note how we used the general expression for the electric field, at position $\mathbf{r}$, due to a distribution of static point charges at positions $\mathbf{r}_i$.)

Problem

35. (a) Find an expression for the electric field on the y-axis due to the two charges q in Fig. 23-11.
    (b) At what point is the field on the y-axis a maximum?

Solution

(a) The electric field is the force per unit charge, so Example 23-2 shows that $\mathbf{E}(y) = 2kqy\hat{\jmath}(a^2 + y^2)^{-3/2}$.
(b) The magnitude of the field, a positive function, is zero for $y = 0$ and $y = \infty$, hence it has a maximum in between. Setting the derivative equal to zero, we find $0 = (a^2 + y^2)^{-3/2} - \frac{3}{2}y(a^2 + y^2)^{-5/2}(2y)$, or $a^2 + y^2 - 3y^2 = 0$. Thus, the field strength maxima are at $y = \pm a/\sqrt{2}$ (the directions at these points, of course, are opposite, by symmetry).

Problem

37. A dipole lies on the y-axis, and consists of an electron at $y = 0.60$ nm and a proton at $y = -0.60$ nm. Find the electric field (a) midway between the two charges, (b) at the point $x = 2.0$ nm, $y = 0$, and (c) at the point $x = -20$ nm, $y = 0$.

Solution

We can use the result of Example 23-6, with y replaced by x, and x by $-y$ (or equivalently, $\hat{\jmath}$ by $\hat{\imath}$, and $\hat{\imath}$ by $-\hat{\jmath}$). Then $\mathbf{E}(x) = 2kqa\ \hat{\jmath}(a^2 + x^2)^{-3/2}$, where $q = e = 1.6 \times 10^{-19}$ C and $a = 0.6$ nm. (Look at Fig. 23-18 rotated 90° CW.) The constant $2kq = 2(9 \times 10^9 \text{ N·m}^2/\text{C}^2)(1.6 \times 10^{-19} \text{ C}) = (2.88 \text{ GN/C}) \times (\text{nm})^2$. (a) At $x = 0$, $\mathbf{E}(0) = 2kq\hat{\jmath}/a^2 = (2.88 \text{ GN/C})\hat{\jmath}/(0.6)^2 = (8.00 \text{ GN/C})\hat{\jmath}$. (b) For $x = 2$ nm, $\mathbf{E} = (2.88 \text{ GN/C})\hat{\jmath}(0.6)(0.6^2 + 2^2)^{-3/2} = (190 \text{ MN/C})\hat{\jmath}$. (c) At $x = 20$ nm, $\mathbf{E} = (2.88 \text{ GN/C})\hat{\jmath} \times (0.6)(0.6^2 + 20^2)^{-3/2} = (216 \text{ kN/C})\hat{\jmath}$.

Problem

39. The dipole moment of the water molecule is 6.2×10^{-30} C·m. What would be the separation distance if the molecule consisted of charges $\pm e$? (The effective charge is actually less because electrons are shared by the oxygen and hydrogen atoms.)

Solution

The distance separating the charges of a dipole is $d = p/q = 6.2 \times 10^{-30}$ C·m$/1.6 \times 10^{-19}$ C $= 38.8$ pm.

Problem

41. Three charges form an equilateral triangle of side a. At one vertex is a charge $+2q$; at the other two vertices are charges $-q$. The triangle is oriented with the charge $2q$ on the positive x-axis and both charges $-q$ on the y-axis. (a) Find an expression for the electric field on the x-axis, in the approximation $x \gg a$. (b) Compare with Equation 23-7b to show that your result in (a) is a dipole field, and give an expression for the magnitude of the triangle's dipole moment.

Solution

(a) With the charges positioned as shown, the electric field on the positive x-axis, due to the two negative charges at $(0, \pm a/2)$, matches the field found in Example 23-2 (replace q with $-q$, a with $a/2$, y with x,

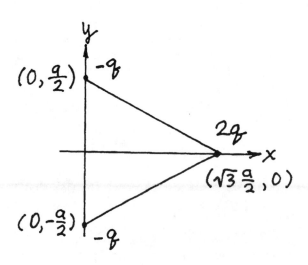

Problem 41 Solution.

and $\hat{\mathbf{j}}$ with $\hat{\mathbf{i}}$): $\mathbf{F}/Q = 2k(-q) \times (x^2 + a^2/4)^{-3/2}\hat{\mathbf{i}}$. For the positive charge at $(\sqrt{3}a/2, 0)$, the electric field on the x-axis to the right ($x > \sqrt{3}a/2$) is just $k(2q) \times (x - \sqrt{3}a/2)^{-2}\hat{\mathbf{i}}$ (the unit vector in Equation 23-5 is $\hat{\mathbf{i}}$ and the distance is $x - \sqrt{3}a/2$). The total field is the sum of these, $\mathbf{E}(x) = 2kq\hat{\mathbf{i}}[(x - \sqrt{3}a/2)^{-2} - x(x^2 + a^2/4)^{-3/2}]$. For $x \gg a$, one can use the binomial approximation (see Appendix A): $(x - \sqrt{3}a/2)^{-2} = x^{-2}(1 + \sqrt{3}a/x + \cdots)$ and $x(x^2 + a^2/4)^{-3/2} = x^{-2}(1 + \cdots)$, where $\cdots$ indicates terms of order a^2/x^2 or higher. Therefore, $\mathbf{E}(x) = 2kq\hat{\mathbf{i}}x^{-2}[1 + \sqrt{3}a/x + \cdots - (1 + \cdots)] \simeq 2\sqrt{3}kqax^{-3}\hat{\mathbf{i}}$. (b) This field is the same as Equation 23-7b, with $p = \sqrt{3}qa$. (In general, the dipole moment for a distribution of point charges is $|\sum \mathbf{r}_i q_i|$. Also note that the field in part (a) can also be found from Equation 23-5, with $\hat{\mathbf{r}}_i/r_i^2 = (\mathbf{r} - \mathbf{r}_i) \div |\mathbf{r} - \mathbf{r}_i|^3$ as in the solution to Problem 15.)

Problem

43. A 30-cm-long rod carries a charge of 80 μC spread uniformly over its length. Find the electric field strength on the rod axis, 45 cm from the end of the rod.

Solution

Applying the result of Example 23-7, at a distance $a = 0.45$ m from the near end of the rod, we get $E = kQ/a(a + \ell) = (9 \times 10^9$ N·m^2/C$^2)(80$ μC$) \div (0.45$ m$)(0.45$ m $+ 0.30$ m$) = 2.13$ MN/C.

Problem

45. A thin rod of length ℓ has its left end at the origin and its right end at the $x = \ell$. It carries a line charge density given by $\lambda = \lambda_0(x^2/\ell^2)\sin(\pi x/\ell)$, where λ_0 is a constant. Find the electric field strength at the origin.

Solution

The electric field at the origin, due to a small element of charge, $dq = \lambda \, dx$, located at position x, is $d\mathbf{E} =$

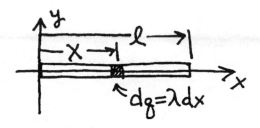

Problem 45 Solution.

$-\hat{\imath}k\lambda\ dx/x^2$. Using $\lambda/x^2 = (\lambda_0/\ell^2)\sin(\pi x/\ell)$ and integrating from $x = 0$ to $x = \ell$, we find

$$\mathbf{E} = \int_0^\ell -\hat{\imath}\left(\frac{k\lambda_0}{\ell^2}\right)\sin\left(\frac{\pi x}{\ell}\right)\ dx$$

$$= \hat{\imath}\frac{k\lambda_0}{\ell^2}\left.\frac{\ell}{\pi}\cos\left(\frac{\pi x}{\ell}\right)\right|_0^\ell$$

$$= \hat{\imath}(k\lambda_0/\ell\pi)(-1-1) = -2\hat{\imath}k\lambda_0/\pi\ell.$$

Problem

47. A uniformly charged ring is 1.0 cm in radius. The electric field on the axis 2.0 cm from the center of the ring has magnitude 2.2 MN/C and points toward the ring center. Find the charge on the ring.

Solution

From Example 23-8, the electric field on the axis of a uniformly charged ring is $kQx(x^2 + a^2)^{-3/2}$, where x is positive away from the center of the ring. For the given ring in this problem, -2.2 MN/C $= kQ(2\text{ cm})\times$ $(4\text{ cm}^2 + 1\text{ cm}^2)^{-3/2}$, or $Q = (-2.2\text{ MN/C})\times$ $(5.59\text{ cm}^2)/(9\times10^9\text{ N·m}^2/\text{C}) = -0.137\ \mu\text{C}$.

Problem

49. Use the result of the preceding problem to show that the field of an *infinite*, uniformly charged flat sheet is $2\pi k\sigma$, where σ is the surface charge density. Note that this result is independent of distance from the sheet.

Solution

An infinite flat sheet is the same as an infinite flat disk (as long as the dimensions are infinite in all directions, the shape is irrelevant). Thus, we can find the magnitude of the electric field from a uniformly changed infinite flat sheet by letting $R \to \infty$ in the result of the previous problem. Then, the limit of the second term is zero, and the magnitude is constant, $E = 2\pi k\sigma$. (The direction is perpendicularly away from (towards) the sheet for positive (negative) σ.)

Problem

51. The electric field 22 cm from a long wire carrying a uniform line charge density is 1.9 kN/C. What will be the field strength 38 cm from the wire?

Solution

For a very long wire ($\ell \gg 38$ cm), Example 23-9 shows that the magnitude of the radial electric field falls off like $1/r$. Therefore, $E(38\text{ cm})/E(22\text{ cm}) = 22\text{ cm}\div$ 38 cm; or $E(38\text{ cm}) = (22/38)1.9\text{ kN/C} = 1.10\text{ kN/C}$.

Problem

53. A straight wire 10 m long carries 25 μC distributed uniformly over its length. (a) What is the line charge density on the wire? Find the electric field strength (b) 15 cm from the wire axis, not near either end and (c) 350 m from the wire. Make suitable approximations in both cases.

Solution

(a) For a uniformly charged wire, $\lambda = Q/\ell =$ 2.5 μC/m. (b) Since $r = 15$ cm $\ll 10$ m $= \ell$ and the field point is far from either end, we may regard the wire as approximately infinite. Then Example 23-9 gives $E_r = 2k\lambda/r = (2\times 9\times 10^9\text{ N·m}^2/\text{C}^2)(2.5\ \mu\text{C/m})\div$ (0.15 m) $= 300$ kN/C. (c) At $r = 350$ m $\gg 10$ m $= \ell$, the wire behaves approximately like a point charge, so the field strength is $kQ/r^2 = (9\times 10^9\times$ $25\times 10^{-6}\text{ N·m}^2/\text{C})/(350\text{ m})^2 = 1.84$ N/C.

Section 23-6: Matter in Electric Fields

Problem

55. In this famous 1909 experiment that demonstrated quantization of electric charge, R. A. Millikan suspended small oil drops in an electric field. With a field strength of 20 MN/C, what mass drop can be suspended when the drop carries a net charge of 10 elementary charges?

Solution

In equilibrium under the gravitational and electro-static forces, $mg = qE$, or $m = (10\times 1.6\times 10^{-19}\text{ C})\times$ $(2\times 10^7\text{ N/C})/(9.8\text{ m/s}^2) = 3.27\times 10^{-12}$ kg. (Because this is so small, the size of such a drop may be better appreciated in terms of its radius, $R = (3m/4\pi\rho_{\text{oil}})^{1/3}$. Millikan used oil of density 0.9199 g/cm^3, so $R =$ 9.46 μm for this drop.)

Problem

57. A proton moving to the right at 3.8×10^5 m/s enters a region where a 56 kN/C electric field points to the left. (a) How far will the proton get before its speed reaches zero? (b) Describe its subsequent motion.

Solution

(a) Choose the x-axis to the right, in the direction of the proton, so that the electric field is negative to the left. If the Coulomb force on the proton is the only important one, the acceleration is $a_x = e(-E)/m$. Equation 2-11, with $v_{ox} = 3.8\times 10^5$ m/s and $v_x = 0$, gives a maximum penetration into the field region of

$$x - x_0 = -v_{ox}^2/2a_x = mv_{ox}^2/2eE =$$

$$\frac{(1.67{\times}10^{-27}\text{ kg})(3.8{\times}10^5\text{ m/s})^2}{2(1.6{\times}10^{-19}\text{ C})(56{\times}10^3\text{ N/C})} = 1.35\text{ cm}.$$

(b) The proton then moves to the left, with the same constant acceleration in the field region, until it exits with the initial velocity reversed.

Problem

59. An ink-jet printer works by "steering" charged ink drops to the right place on the page by passing moving drops through a uniform electric field that deflects them by the appropriate amount. Figure 23-47 shows an ink drop approaching the field region, which has length ℓ and width d between the charged plates that establish the field. Find an expression for the minimum speed a drop with mass m and charge q must have if it is to get through the region without hitting either plate.

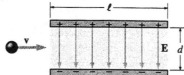

FIGURE 23-47 Problem 59.

Solution

If they enter the field region midway, moving horizontally, the maximum vertical deflection, during the transit time $t = \ell/v$, can be $d/2$, for the ink drops to pass through. Thus, $y = \frac{1}{2}at^2 = \frac{1}{2}(qE/m)(\ell/v)^2 < \frac{1}{2}d$, or $v > \ell\sqrt{qE/md}$.

Problem

61. An electron is moving in a circular path around a long, uniformly charged wire carrying 2.5 nC/m. What is the electron's speed?

Solution

The electric field of the wire is radial and falls off like $1/r$ (see Example 23-9). For an attractive force (negative electron encircling a positively charged wire), this is the same dependance as the centripetal acceleration. For circular motion around the wire, the Coulomb force provides the electron's centripetal acceleration, or $-eE/m = -2ke\lambda/mr = -v^2/r$. Thus, $v = \sqrt{2ke\lambda/m} = [2(9{\times}10^9\text{ N·m}^2/\text{C}^2)(1.6{\times}10^{-19}\text{ C}){\times}$ $(2.5{\times}10^{-9}\text{ C/m})/(9.11{\times}10^{-31}\text{ kg})]^{1/2} = 2.81\text{ Mm/s}.$

Problem

63. What is the line charge density on a long wire if a 6.8-μg particle carrying 2.1 nC describes a circular orbit about the wire with speed 280 m/s?

Solution

The solution to Problem 61 reveals that $\lambda = -mv^2 \div 2kq = -(6.8{\times}10^{-9}\text{ kg})(280\text{ m/s})^2/2(9{\times}10^9\text{ N·m}^2/\text{C}^2){\times}$ $(2.1{\times}10^{-9}\text{ C}) = -14.1\ \mu\text{C/m}.$ (In this case, the force on a positively charged orbiting particle is attractive for a wire with negative linear charge density.)

Problem

65. A dipole with dipole moment 1.5 nC·m is oriented at 30° to a 4.0-MN/C electric field. (a) What is the magnitude of the torque on the dipole? (b) How much work is required to rotate the dipole until it's antiparallel to the field?

Solution

(a) The torque on an electric dipole in an external electric field is given by Equation 23-11; $\tau = |\mathbf{p} \times \mathbf{E}| = pE\sin\theta = (1.5\text{ nC·m})(4.0\text{ MN/C})\sin 30° = 3.0\text{ mN·m}.$ (b) The work done against just the electric force is equal to the change in the dipole's potential energy (Equation 23-12); $W = \Delta U = (-\mathbf{p}{\cdot}\mathbf{E})_f - (-\mathbf{p}{\cdot}\mathbf{E})_i = pE(\cos 30° - \cos 180°) = (1.5\text{ nC·m})(4.0\text{ MN/C}){\times}$ $(1.866) = 11.2\text{ mJ}.$

Problem

67. Two identical dipoles, each of charge q and separation a, are a distance x apart as shown in Fig.23-49. By considering forces between pairs of charges in the different dipoles, calculate the net force between the dipoles. (a) Show that, in the limit $a \ll x$, the force has magnitude $6kp^2/x^4$, where $p = qa$ is the dipole moment. (b) Is the force attractive or repulsive?

Solution

All the forces are along the same line, so take the origin at the center of the left-hand dipole and the positive x-axis in the direction of the right-hand dipole in Fig. 23-49. The right-hand dipole has charges $+q$ at $x + a/2$, $-q$ at $x - a/2$, each of which experiences a force from both charges of the left-hand dipole, which are $+q$ at $a/2$ and $-q$ at $-a/2$. (There are forces between four pairs of changes.) The Coulomb force on a charge in the right-hand dipole, due to one in the left-hand one, is $kq_r q_\ell(x_r - x_\ell)\hat{\imath}/|x_r - x_\ell|^3$ (see solution to Problem 15), so the total force on the right-hand dipole is

$$F_x = kq^2\hat{\imath}\left[\frac{1}{x^2} - \frac{1}{(x+a)^2} - \frac{1}{(x-a)^2} + \frac{1}{x^2}\right]$$

$$= -\frac{2kq^2a^2(3x^2 - a^2)}{x^2(x^2 - a^2)^2}\hat{\imath}.$$

(a) In the limit $a \ll x$, $F_x \to -2kq^2a^2(3x^2)\hat{\mathbf{i}}/x^6 = -6kq^2a^2\hat{\mathbf{i}}/x^4 = -6kp^2\hat{\mathbf{i}}/x^4$, where $p = qa$ is the dipole moment of both dipoles. (b) The force on the right-hand dipole is in the negative x direction, indicating an attractive force.

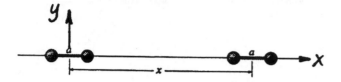

FIGURE 23-49 Problem 67 Solution.

Paired Problems

Problem

69. An electron is at the origin and an ion with charge $+5e$ is at $x = 10$ nm. Find a point where the electric field is zero.

Solution

The electron's field is directed toward the electron (a negative charge) and the ion's field is directed away from the ion (a positive charge). Therefore, the fields can cancel only at points on the negative x-axis ($x < 0$), since the directions are opposite there and the smaller charge is closer. The field from one point charge is $\mathbf{E}_q(x) = kq\hat{\mathbf{i}}(x - x_q)/|x - x_q|^3$, where $q = -e$, $x_q = 0$ for the electron, and $q = 5e$, $x_q = 10$ nm for the ion. The total field is zero when $0 = k[(-e)x \times |x|^{-3} + 5e(x - 10 \text{ nm})|x - 10 \text{ nm}|^{-3}]$. (See note to solution of Problem 33.) Since $x < 0$, $|x| = -x$ and $|x - 10 \text{ nm}| = 10 \text{ nm} - x$, so this implies $x^{-2} - 5 \times (10 \text{ nm} - x)^{-2} = 0$, or $4x^2 + 2(10 \text{ nm})x - (10 \text{ nm})^2 = 0$. The negative solution to this quadratic is

$$x = \frac{[-10 \text{ nm} - \sqrt{(10 \text{ nm})^2 + 4(10 \text{ nm})^2}]}{4}$$
$$= -2.5 \text{ nm}(1 + \sqrt{5}) = -8.09 \text{ nm}.$$

Problem

71. A thin rod of length ℓ has its left end at $x = -\ell$ and its right end at the origin. It carries a line charge density given by

$$\lambda = \lambda_0 \frac{x^2}{\ell^2},$$

where λ_0 is a constant. Find the electric field at the origin.

Solution

The electric field at the origin, due to an element of charge $dq = \lambda \, dx$, located at x, where $-\ell \le x \le 0$, is

$dE = k\lambda\hat{\mathbf{i}} \, dx/x^2 = k(\lambda_0/\ell^2)\hat{\mathbf{i}} \, dx$. The total field is the integral of this over the rod,

$$\mathbf{E}(0) = \int_{-\ell}^{0} (k\lambda_0/\ell^2)\hat{\mathbf{i}} \, dx$$
$$= (k\lambda_0/\ell^2)\hat{\mathbf{i}}[0 - (-\ell)] = (k\lambda_0/\ell)\hat{\mathbf{i}}.$$

Problem

73. A thin, flexible rod carrying charge Q spread uniformly over its length is bent into a quarter circle of radius a, as shown in Fig. 23-51a. Find the electric field strength at the point P, which is the center of the circle. *Hint:* Consult Problem 50.

Solution

It should be clear from the symmetry that the electric field is along the radius bisecting the arc, so take this as the x-axis, with P at the origin, and θ as defined in Fig. 23-45. The electric field at P, from each charge element, $dq = \lambda \, d\ell = \lambda a \, d\theta$, at θ, has the same magnitude, $dE = k \, dq/a^2 = k\lambda \, d\theta/a$, but direction $\hat{\mathbf{r}} = \hat{\mathbf{i}} \sin\theta - \hat{\mathbf{j}} \cos\theta$, as sketched. $\lambda = Q/\ell$ is constant, and the arc extends from $\theta_0 = 45°$ to $\pi - \theta_0 = 135°$, so the total field at P is the integral of $dE\hat{\mathbf{r}}$ from θ_0 to $\pi - \theta_0$:

$$\mathbf{E}(0) = (k\lambda/a) \int_{\theta_0}^{\pi-\theta_0} (\hat{\mathbf{i}} \sin\theta - \hat{\mathbf{j}} \cos\theta) \, d\theta$$
$$= (k\lambda/a)|-\hat{\mathbf{i}} \cos\theta - \hat{\mathbf{j}} \sin\theta|_{\theta_0}^{\pi-\theta_0}$$
$$= (k\lambda/a)2\hat{\mathbf{i}} \cos\theta_0 = \sqrt{2} \, k\lambda \, \hat{\mathbf{i}}/a.$$

Here, we used $\sin\theta_0 = \sin(\pi - \theta_0)$, $\cos\theta_0 = -\cos(\pi - \theta_0)$, and $\theta_0 = 45°$. In terms of the total charge, $\lambda = Q/\ell = Q/(\frac{1}{2}\pi a)$, so $\mathbf{E}(0) = 2\sqrt{2} \, kQ\hat{\mathbf{i}} \div \pi a^2$. [Note: in general, $\ell = (\pi - 2\theta_0)a$.]

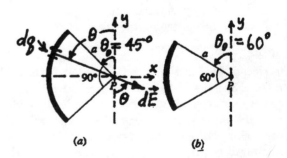

FIGURE 23-51 Problems 73 and 74 Solution.

Problem

75. Ink-jet printers work by deflecting moving ink droplets with an electric field so they hit the right place on the paper. Droplets in a particular printer have mass 1.1×10^{-10} kg, charge 2.1 pC, speed

12 m/s, and pass through a uniform 97-kN/C electric field in order to be deflected through a 10° angle. What is the length of the field region?

Solution

Suppose the ink droplets enter the field region perpendicular to the field, as in the geometry of Example 23-10. Then the analysis of that example shows that $v_y/v_x = \tan\theta = qE_y\,\Delta x/mv_x^2$, so $\Delta x = mv_x^2\tan\theta/qE_y = (0.11\;\mu\text{g})(12\text{ m/s})^2\tan 10°\div (2.1\text{ pC})(97\text{ kN/C}) = 1.37\text{ cm}.$

Supplementary Problems

Problem

77. A spring of spring constant 100 N/m is stretched 10 cm beyond its 90-cm equilibrium length. If you want to keep it stretched by attaching equal electric charges to the opposite ends, what magnitude of charge should you use?

Solution

The repulsive force between like charges, $kq^2/r^2\times$ ($r = 90$ cm $+ 10$ cm $= 1$ m), must balance the spring force, $k_s x$ ($x = 10$ cm is the stretch and k_s is the spring constant). Thus,

$$q = \pm\sqrt{r^2 k_s x/k}$$
$$= \pm[(100\text{ N/m})(1\text{ m})^2(0.1\text{ m})/(9\times 10^9\text{ N·m}^2/\text{C}^2)]^{1/2}$$
$$= \pm 33.3\;\mu\text{C}.$$

Problem

79. A charge $-q$ and a charge $\frac{4}{9}q$ are located a distance a apart, as shown in Fig. 23-53. Where would you place a third charge so that all three are in static equilibrium? What should be the sign and magnitude of the third charge?

Solution

Because of the vector nature of the forces, the third charge, Q, must be placed along the line joining the other two, as in Example 23-3. Q cannot go to the left of $-q$, since the magnitude of the force on it due to

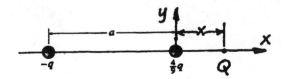

FIGURE 23-53 Problem 79 Solution.

$-q$ would always be greater than that due to $\frac{4}{9}q$. It cannot go between $-q$ and $\frac{4}{9}q$, since the forces on it would always be in the same direction. Thus, Q must go to the right of $\frac{4}{9}q$, as shown (x-axis to the right with origin at $\frac{4}{9}q$). The net force on each charge must be zero, so, for Q: $0 = kQ\left(\frac{4}{9}q\right)/x^2 + kQ(-q)/(x+a)^2$, or $\frac{4}{9}(x+a)^2 = x^2$. For $\frac{4}{9}q$: $0 = k\left(\frac{4}{9}q\right)(-q)/a^2 - k\times \left(\frac{4}{9}q\right)Q/x^2$, or $-qx^2 = Qa^2$. (The equation for the force on the third charge follows from the equations for the other two plus Newton's third law.) The solution of these two equations (for $x > 0$) is $x = 2a$ and $Q = -4q$. (The equilibrium is unstable. A slight displacement of the positive charge to the right, for example, would cause it to be attracted more strongly to the right.)

Problem

81. A 3.8-g particle with a 4.0-μC charge experiences a downward force of 0.24 N in a uniform electric field. Find the electric field, assuming that the gravitational force is *not* negligible.

Solution

If gravity and Coulomb forces both act, then $\mathbf{F}_{\text{net}} = -mg\hat{\mathbf{j}} + q\mathbf{E} = -(0.24\text{ N})\hat{\mathbf{j}}$, where $\hat{\mathbf{j}}$ is upward. Thus, $\mathbf{E} = (mg - 0.24\text{ N})\hat{\mathbf{j}}/q = [(3.8\times 10^{-3}\text{ kg})(9.8\text{ m/s}^2) - 0.24\text{ N}]\hat{\mathbf{j}}/(4.0\;\mu\text{C}) = -50.7\hat{\mathbf{j}}\text{ kN/C}.$

Problem

83. The electric field on the axis of a uniformly charged ring has magnitude 380 kN/C at a point 5.0 cm from the ring center. The magnitude 15 cm from the center is 160 kN/C; in both cases the field points away from the ring. Find the radius and charge of the ring.

Solution

The electric field on the axis of a uniformly charged ring is calculated in Example 23-8, so the data given in the question imply 380 kN/C $= kQ(5\text{ cm})\times [(5\text{ cm})^2 + a^2]^{-3/2}$, and 160 kN/C $= kQ(15\text{ cm})\times [(15\text{ cm})^2 + a^2]^{-3/2}$. Dividing these two equations and taking the $\frac{2}{3}$ root we get

$$\left(\frac{380\times 15}{160\times 5}\right)^{2/3} = 3.70 = \frac{(15\text{ cm})^2 + a^2}{(5\text{ cm})^2 + a^2},$$

which when solved for the radius, gives

$$a = \sqrt{[(15\text{ cm})^2 - (3.70)(5\text{ cm})^2]/2.70} = 7.00\text{ cm}.$$

Substituting for a in either of the field equations allows us to find

$$Q = \frac{(380\text{ kN/C})[(5\text{ cm})^2 + (7\text{ cm})^2]^{3/2}}{(9\times 10^9\text{ N·m}^2/\text{C}^3)(5\text{ cm})} = 538\text{ nC}.$$

Problem

85. A molecule with dipole moment p is located a distance r from a proton, oriented with its dipole moment vector **p** as shown in Fig. 23-54. (a) Use Equation 23-7b to find the force the molecule exerts on the proton. (b) Now find the net force on the molecule in the proton's nonuniform electric field by considering that the molecule consists of two opposite charges $\pm q$, separated by a distance d such that $qd = p$. Take the limit as d becomes very small compared with r, and show that the resulting force has the same magnitude as that of part (a), as required by Newton's third law.

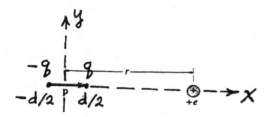

FIGURE 23-54 Problem 85.

Solution

Take the origin at the molecule, with the x-axis parallel to its dipole moment, as shown on Fig. 23-54. (a) Equation 23-7b gives the dipole's electric field at the position of the proton, so the force on the latter is $\mathbf{F}_p = e\mathbf{E}_{\text{dip}} = e(2kpr^{-3}\hat{\mathbf{i}}) = 2kepr^{-3}\hat{\mathbf{i}}$. (b) The electric field of the proton, for points on the x-axis to its left $(x < r)$ is $\mathbf{E}_p(x) = -ke(r - x)^{-2}\hat{\mathbf{i}}$ (see Equation 23-4). If the molecular dipole is considered to consist of charges $\pm q$ located at $x = \pm d/2$, where $p = qd$, then the force on it is $\mathbf{F}_{\text{dip}} = q\mathbf{E}_p(d/2) - q\mathbf{E}_p(-d/2) = -keq\,\hat{\mathbf{i}}[(r - d/2)^{-2} - (r + d/2)^{-2}] = -2kepr\times (r^2 - d^2/4)^{-2}\hat{\mathbf{i}}$. In the limit $d \ll r$, this becomes $\mathbf{F}_{\text{dip}} = -2kepr^{-3}\hat{\mathbf{i}} = -\mathbf{F}_p$, in agreement with the result of part (a) and Newton's third law.

Problem

87. Derive Equation 23-9 in Example 23-9 by making θ the integration variable, then evaluating the resulting integral.

Solution

In Fig. 23-23, $r = (x^2 + y^2)^{1/2} = y/\sin\theta$, and $\cot\theta = x/y$. Differentiating the latter with respect to x, we get $-\csc^2\theta\,d\theta = dx/y$, which relates dx to $d\theta$. (Extend the results in Appendix B, and recall that cosecant $= \sin e^{-1}$.) The limits $x = -\infty$ to ∞ correspond to $\theta = \pi$ to 0 (since when x is negative so is $\cot\theta$), thus the integral in Example 23-9 leading to Equation 23-9 becomes

$$E = k\lambda y \int_{-\infty}^{\infty} \frac{dx}{(x^2 + y^2)^{3/2}}$$
$$= k\lambda y \int_{\pi}^{0} \frac{-y\csc^2\theta\,d\theta}{(y/\sin\theta)^3} = \frac{k\lambda}{y} \int_{0}^{\pi} \sin\theta\,d\theta$$
$$= (k\lambda/y)[-\cos\pi + \cos 0] = 2k\lambda/y$$

as before.

CHAPTER 24 GAUSS'S LAW

ActivPhysics can help with these problems:
Activities 11.4–11.6

Section 24-1: Electric Field Lines

Problem

1. What is the net charge shown in Fig. 24-39? The magnitude of the middle charge is 3 μC.

Solution

The number of lines of force emanating from (or terminating on) the positive (or negative) charges is the same (14 in Fig. 24-39), so the middle charge is -3 μC and the outer ones are $+3$ μC. The net charge shown is therefore $3 + 3 - 3 = 3$ μC. This is reflected by the fact that 14 lines emerge from the boundary of the figure.

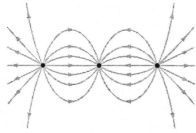

FIGURE 24-39 Problem 1 Solution.

Problem

3. Two charges $+q$ and a charge $-q$ are at the vertices of an equilateral triangle. Sketch some field lines for this charge distribution.

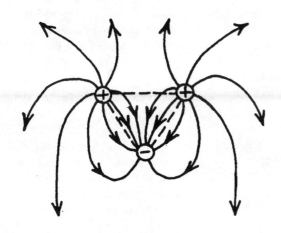

Problem 3 Solution.

Solution

(The sketch shown follows the text's convention of eight lines of force per charge magnitude q.)

Section 24-2: Electric Flux

Problem

5. A flat surface with area 2.0 m^2 is in a uniform electric field of 850 N/C. What is the electric flux through the surface when it is (a) at right angles to the field, (b) at 45° to the field, and (c) parallel to the field?

Solution

(a) When the surface is perpendicular to the field, its normal is either parallel or anti-parallel to $\mathbf{E}$. Then Equation 24-1 gives $\Phi = \mathbf{E} \cdot \mathbf{A} = EA\cos(0° \text{ or } 180°) = \pm(850 \text{ N/C})(2 \text{ m}^2) = \pm1.70 \text{ kN·m}^2/\text{C}$. (b) $\Phi = \mathbf{E} \cdot \mathbf{A} = EA\cos(45° \text{ or } 135°) = \pm(1.70 \text{ kN·m}^2/\text{C}) \times (0.866) = \pm1.20 \text{ kN·m}^2/\text{C}$. (c) $\Phi = EA\cos 90° = 0$.

Problem

7. A flat surface with area 0.14 m^2 lies in the x-y plane, in a uniform electric field given by $\mathbf{E} = 5.1\hat{\mathbf{i}} + 2.1\hat{\mathbf{j}} + 3.5\hat{\mathbf{k}}$ kN/C. Find the flux through this surface.

Solution

The surface can be represented by a vector area $\mathbf{A} = (0.14 \text{ m}^2)(\pm\hat{\mathbf{k}})$. (Since the surface is open, we have a choice of normal to the x-y plane.) Then $\Phi = \mathbf{E} \cdot \mathbf{A} = \pm\mathbf{E} \cdot \hat{\mathbf{k}} \times (0.14 \text{ m}^2) = \pm E_z(0.14 \text{ m}^2) = \pm(3.5 \text{ kN/C})(0.14 \text{ m}^2) = \pm490 \text{ N·m}^2/\text{C}$. (Only the z component of the field contributes to the flux through the x-y plane.)

Problem

9. What is the flux through the hemispherical open surface of radius R shown in Fig. 24-41? The uniform field has magnitude E. *Hint:* Don't do a messy integral! Imagine closing the surface with a flat, circular piece across the open end. What would be the flux through the entire closed surface? And what's the flux through the flat end? So what's the answer?

Solution

All of the lines of force going through the hemisphere also go through an equatorial disk covering its edge in Fig. 24-41. Therefore, the flux through the disk (normal in the direction of $\mathbf{E}$) equals the flux through the hemisphere. Since $\mathbf{E}$ is uniform, the flux through the disk is just $\pi R^2 E$. (Note: Gauss's Law gives the same result, since the flux through the closed surface, consisting of the hemisphere plus the disk, is zero. See Section 24-3.)

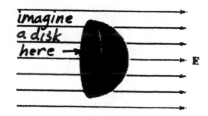

FIGURE 24-41 Problem 9 Solution.

Section 24-3: Gauss's Law

Problem

11. What is the electric flux through each closed surface shown in Fig. 24-43?

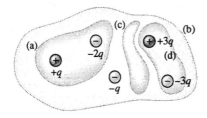

FIGURE 24-43 Problem 11.

Solution

From Gauss's law, $\Phi = q_{enclosed}/\varepsilon_0$. For the surfaces shown, this is (a) $(q - 2q)/\varepsilon_0 = -q/\varepsilon_0$, (b) $-2q/\varepsilon_0$, (c) and (d) 0.

Problem

13. A 2.6-μC charge is at the center of a cube 7.5 cm on each side. What is the electric flux through one face of the cube? *Hint:* Think about symmetry, and don't do an integral.

Solution

The symmetry of the situation guarantees that the flux through one face is $\frac{1}{6}$ the flux through the whole cubical surface, so $\Phi_{face} = \frac{1}{6} \oint_{cube} \mathbf{E} \cdot d\mathbf{A} = q_{enclosed}/6\varepsilon_0 = (2.6\ \mu C)/6(8.85 \times 10^{-12}\ C^2/N \cdot m^2) = 49.0\ kN \cdot m^2/C$.

Problem

15. A dipole consists of two charges $\pm 6.1\ \mu$C located 1.2 cm apart. What is the electric flux through each surface shown in Fig. 24-44?

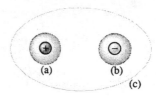

FIGURE 24-44 Problem 15.

Solution

It follows from Gauss's law that (a) $\Phi_a = +q/\varepsilon_0 = (6.1\ \mu C)/(8.85 \times 10^{-12}\ C^2/N \cdot m^2) = 689\ kN \cdot m^2/C$, (b) $\Phi_b = -\Phi_a$, and (c) $\Phi_c = 0$.

Section 24-4: Using Gauss's Law

Problem

17. The electric field at the surface of a uniformly charged sphere of radius 5.0 cm is 90 kN/C. What would be the field strength 10 cm from the surface?

Solution

The electric field due to a uniformly charged sphere is like the field of a point charge for points outside the sphere, i.e., $E(r) \sim 1/r^2$ for $r \geq R$. Thus, at 10 cm from the surface, $r = 15$ cm and $E(15\ cm) = (5/15)^2 E \times (5\ cm) = (90\ kN/C)/9 = 10\ kN/C$.

Problem

19. A crude model for the hydrogen atom treats it as a point charge $+e$ (the proton) surrounded by a uniform cloud of negative charge with total charge $-e$ and radius 0.0529 nm. What would be the electric field strength inside such an atom, halfway from the proton to the edge of the charge cloud?

Solution

At half the radius of the electron cloud, the field strength due to the cloud (a uniformly charged spherical volume) is given in Example 24-1: $E_e = k(-e)(\frac{1}{2}R)/R^3 = -ke/2R^2$. The field strength due to the proton (a point charge) at the same distance is $E_p = ke/(\frac{1}{2}R)^2 = 4ke/R^2$. (The electron's field is radially inward, negative, and the proton's is radially outward, positive.) The total field strength is $E = E_e + E_p = -ke/2R^2 + 4ke/R^2 = 7ke/2R^2 = 7(1.44 \times 10^{-9}\ N \cdot m^2/C)/2(5.29 \times 10^{-11}\ m)^2 = 1.80 \times 10^{12}\ N/C$.

Problem

21. A 10-nC point charge is located at the center of a thin spherical shell of radius 8.0 cm carrying -20 nC distributed uniformly over its surface. What are the magnitude and direction of the electric field (a) 2.0 cm, (b) 6.0 cm, and (c) 15 cm from the point charge?

Solution

The total electric field, the superposition of the fields due to the point charge and the spherical shell, is spherically symmetric about the center. Inside the shell ($r < R = 8$ cm), its field is zero, so the total field is just due to the 10 μC point charge. Outside ($r > R$), the shell's field is like that of a point charge of -20 μC at the same central location as the 10 μC charge. (This situation is described in Example 24-3.) (a) and (b) For $r = 2$ cm or 6 cm $< R$, $E = kq_{\text{pt.}}/r^2 = (9\times10^9 \text{ N·m}^2/\text{C}^2)(10 \text{ nC})/(2 \text{ cm or } 6 \text{ cm})^2 = 225$ kN/C or 25.0 kN/C, respectively, directed radially outward. (c) For $r = 15$ cm $> R$, $E = k(q_{\text{pt}} + q_{\text{shell}})/r^2 = (9\times 10^9 \text{ N·m}^2/\text{C}^2)(10 \text{ nC} - 20 \text{ nC})/(15 \text{ cm})^2 = -4.00$ kN/C, directed radially inward.

Problem

23. A point charge $-2Q$ is at the center of a spherical shell of radius R carrying charge Q spread uniformly over its surface. What is the electric field at (a) $r = \frac{1}{2}R$ and (b) $r = 2R$? (c) How would your answers change if the charge on the shell were doubled?

Solution

The situation is like that in Problem 21. (a) At $r = \frac{1}{2}R < R$ (inside shell), $E = E_{\text{pt}} + E_{\text{shell}} = k(-2Q)/(\frac{1}{2}R)^2 + 0 = -8 \ kQ/R^2$ (the minus sign means the direction is radially inward). (b) At $r = 2R > R$ (outside shell), $E = E_{\text{pt}} + E_{\text{shell}} = k(-2Q + Q)/(2R)^2 = -kQ/4R^2$ (also radially inward). (c) If $Q_{\text{shell}} = 2Q$, the field inside would be unchanged, but the field outside would be zero (since $q_{\text{shell}} + q_{\text{pt}} = 2Q - 2Q = 0$).

Problem

25. A spherical shell 30 cm in diameter carries a total charge 85 μC distributed uniformly over its surface. A 1.0-μC point charge is located at the center of the shell. What is the electric field strength (a) 5.0 cm from the center and (b) 45 cm from the center? (c) How would your answers change if the charge on the shell were doubled?

Solution

(a) The field due to the shell is zero inside, so at $r = 5$ cm, the field is due to the point charge only. Thus, $\mathbf{E} = kq\hat{\mathbf{r}}/r^2 = (9\times10^9 \text{ N·m}^2/\text{C}^2)(1 \ \mu\text{C})\hat{\mathbf{r}}/(0.05 \text{ m})^2 = (3.60\times10^6 \text{ N/C})\hat{\mathbf{r}}$. (b) Outside the shell, its field is like that of a point charge, so at $r = 45$ cm, $\mathbf{E} = k(q + Q)\hat{\mathbf{r}}/r^2 = (9\times10^9 \text{ N·m}^2/\text{C}^2)(86 \ \mu\text{C})\hat{\mathbf{r}}/(0.45 \text{ m})^2 = (3.82\times10^6 \text{ N/C})\hat{\mathbf{r}}$. (c) If the charge on the shell were doubled, the field inside would be unaffected, while the field outside would approximately double, $E = k(1.0 \ \mu\text{C} + 2\times85 \ \mu\text{C})/(45 \text{ cm})^2 = 7.60$ MN/C.

Problem

27. How should the charge density within a solid sphere vary with distance from the center in order that the magnitude of the electric field in the sphere be constant?

Solution

Assume that ρ is spherically symmetric, and divide the volume into thin shells with $dV = 4\pi r^2 \ dr$. From Gauss's law and Equation 24-5,

$$E = \frac{1}{4\pi\varepsilon_0 r^2} \int_V \rho \ dV = \frac{1}{4\pi\varepsilon_0 r^2} \int_0^r \rho(r')4\pi r'^2 \ dr'$$
$$= \frac{1}{\varepsilon_0 r^2} \int_0^r \rho r'^2 \ dr'.$$

It can be seen that if $\rho(r') \sim 1/r'$ then E is constant, but we can obtain the same result mathematically, by differentiation. If E is constant, $dE/dr = 0$. This implies

$$0 = \frac{d}{dr}\left(\frac{1}{r^2}\int_0^r \rho r'^2 dr'\right)$$
$$= \frac{1}{r^2}\frac{d}{dr}\left(\int_0^r \rho r'^2 \ dr'\right) + \left(\int_0^r \rho r'^2 \ dr'\right)\frac{d}{dr}\left(\frac{1}{r^2}\right)$$
$$= \frac{1}{r^2}\rho(r)r^2 - \frac{2}{r^3}\int_0^r \rho r'^2 \ dr',$$

or $\rho(r) = \frac{2}{r^3}\int_0^r \rho(r')r'^2 \ dr'.$

Since $r^{-2}\int_0^r \rho(r')r'^2 \ dr' = \varepsilon_0 E$ is a constant, by hypothesis, $\rho(r) = 2\varepsilon_0 E/r \sim 1/r$, as suspected. (Look up how to take the derivative of an integral in any calculus textbook.) Note that constant magnitude does not imply constant direction; $\mathbf{E} = E\hat{\mathbf{r}}$ is spherically symmetric, not uniform.

Problem

29. A long solid rod 4.5 cm in radius carries a uniform volume charge density. If the electric field strength at the surface of the rod (not near either end) is 16 kN/C, what is the volume charge density?

Solution

If the rod is long enough to approximate its field using line symmetry, we can equate the flux through a length ℓ of its surface (Equation 24-8) to the charge enclosed. The latter is the charge density (a constant) times the volume of a length ℓ of rod. Thus, $2\pi R\ell E = q_{enclosed}/\varepsilon_0 = \rho\pi R^2\ell/\varepsilon_0$, or $\rho = 2\varepsilon_0 E/R = 2(8.85\times 10^{-12}\ \text{C}^2/\text{N·m}^2)(16\ \text{kN/C})/(4.5\ \text{cm}) = 6.29\ \mu\text{C/m}^3$. (This is the magnitude of ρ, since the direction of the field at the surface, radially inward or outward, was not specified.)

Problem

31. An infinitely long rod of radius R carries a uniform volume charge density ρ. Show that the electric field strengths outside and inside the rod are given, respectively, by $E = \rho R^2/2\varepsilon_0 r$ and $E = \rho r/2\varepsilon_0$, where r is the distance from the rod axis.

Solution

The charge distribution has line symmetry (as in Problem 29) so the flux through a coaxial cylindrical surface of radius r (Equation 24-8) equals $q_{enclosed}/\varepsilon_0$, from Gauss's law. For $r > R$ (outside the rod), $q_{enclosed} = \rho\pi R^2\ell$, hence $E_{out} = \rho\pi R^2\ell/2\pi r\ell\varepsilon_0 = \rho R^2/2\varepsilon_0 r$. For $r < R$ (inside the rod), $q_{enclosed} = \rho\pi r^2\ell$, hence $E_{in} = \rho\pi r^2\ell/2\pi r\ell\varepsilon_0 = \rho r/2\varepsilon_0$. (The field direction is radially away from the symmetry axis if $\rho > 0$, and radially inward if $\rho < 0$.)

Problem

33. A long, thin wire carries a uniform line charge density $\lambda = -6.8\ \mu\text{C/m}$. It is surrounded by a thick concentric cylindrical shell of inner radius 2.5 cm and outer radius 3.5 cm. What uniform volume charge density in the shell will result in zero electric field outside the shell?

Solution

In order to have $\mathbf{E} = 0$ outside the shell, it is only necessary for the charge per unit length of shell to cancel that of the wire, i.e., $\lambda_{shell} = +6.8\ \mu\text{C/m}$ (see Gauss's law and Equation 24-8, with $q_{enclosed} = 0$). A uniform charge density which guarantees this is $\rho = \lambda_{shell}\ell/V$, where V is the volume of length ℓ of shell. Thus, $\rho = \lambda_{shell}\ell/\pi(r_2^2 - r_1^2)\ell = (6.8\ \mu\text{C/m})/\pi(3.5^2 - 2.5^2)\times 10^{-4}\ \text{m}^2 = 3.61\times 10^{-3}\ \text{C/m}^3$.

Problem

35. If you "painted" positive charge on the floor, what surface charge density would be necessary in order to suspend a 15-μC, 5.0-g particle above the floor?

Solution

A positive surface charge density σ, on the floor, would produce an approximately uniform electric field upward of $E = \sigma/2\varepsilon_0$, at points near the floor and not near an edge. The field needed to balance the weight of a particle, of mass m and charge q, is given by $mg = qE$, therefore $\sigma = 2\varepsilon_0 E = 2\varepsilon_0 mg/q = 2(8.85\times 10^{-12}\ \text{C}^2/\text{N·m}^2)(5\times 10^{-3}\ \text{kg})(9.8\ \text{m/s}^2)/(15\times 10^{-6}\ \text{C}) = 57.8\ \text{nC/m}^2$.

Problem

37. Figure 24-47 shows sections of three infinite flat sheets of charge, each carrying surface charge density with the same magnitude σ. Find the magnitude and direction of the electric field in each of the four regions shown.

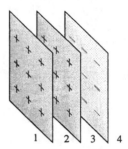

FIGURE 24-47 Problem 37.

Solution

The field from each sheet has magnitude $\sigma/2\varepsilon_0$ and points away from the positive sheets and toward the negative sheet. Take the x axis perpendicular to the sheets, to the right in Fig. 24-47. Superposition gives the field in each of the four regions, as shown.

	1	2	3	4
First sheet:	$-\sigma\hat{\imath}/2\varepsilon_0$ (+) $\leftarrow$	$\sigma\hat{\imath}/2\varepsilon_0$ (+) $\rightarrow$	$\sigma\hat{\imath}/2\varepsilon_0$ (−) $\rightarrow$	$\sigma\hat{\imath}/2\varepsilon_0$ $\rightarrow$
	(+)	(+)	(−)	
Second sheet:	$-\sigma\hat{\imath}/2\varepsilon_0$ (+) $\leftarrow$	$-\sigma\hat{\imath}/2\varepsilon_0$ (+) $\leftarrow$	$\sigma\hat{\imath}/2\varepsilon_0$ (−) $\rightarrow$	$\sigma\hat{\imath}/2\varepsilon_0$ $\rightarrow$
	(+)	(+)	(−)	
Third sheet:	$-\sigma\hat{\imath}/2\varepsilon_0$ (+) $\leftarrow$	$\sigma\hat{\imath}/2\varepsilon_0$ (+) $\rightarrow$	$\sigma\hat{\imath}/2\varepsilon_0$ (−) $\rightarrow$	$-\sigma\hat{\imath}/2\varepsilon_0$ $\leftarrow$
	(+)	(+)	(−)	
Sum:	$-\sigma\hat{\imath}/2\varepsilon_0$ (+) $\leftarrow$	$\sigma\hat{\imath}/2\varepsilon_0$ (+) $\rightarrow$	$3\sigma\hat{\imath}/2\varepsilon_0$ (−) $\rightarrow$	$\sigma\hat{\imath}/2\varepsilon_0$ $\rightarrow$
	(+)	(+)	(−)	

Section 24-5: Fields of Arbitrary Charge Distributions

Problem

39. A nonconducting square plate 75 cm on a side carries a uniform surface charge density. The electric field strength 1 cm from the plate, not near an edge, is 45 kN/C. What is the approximate field strength 15 m from the plate?

Solution

The electric field strength close to the plate $(1 \text{ cm} \ll 75 \text{ cm})$ has approximate plane symmetry $(E = \sigma/2\varepsilon_0)$, so the charge on the plate is $q = \sigma A = 2\varepsilon_0 EA = 2(8.85 \times 10^{-12} \text{ C}^2/\text{N·m}^2)(45 \text{ kN/C}) \times (0.75 \text{ m})^2 = 448 \text{ nC}$. Very far from the plate $(15 \text{ m} \gg 0.75 \text{ m})$, the field strength is like that from a point charge, $E = kq/r^2 = (9 \times 10^9 \text{ N·m}^2/\text{C}^2) \times (448 \text{ nC})(15 \text{ m})^{-2} = 17.9 \text{ N/C}$.

Problem

41. The electric field strength on the axis of a uniformly charged disk is given by $E = 2\pi k\sigma(1 - x/\sqrt{x^2 + a^2})$, with σ the surface charge density, a the disk radius, and x the distance from the disk center. If $a = 20 \text{ cm}$, (a) for what range of x values does treating the disk as an infinite sheet give an approximation to the field that is good to within 10%? (b) For what range of x values is the point-charge approximation good to 10%?

Solution

(Note: The expression given, for the field strength on the axis of a uniformly charged disk, holds only for positive values of x.) (a) For small x, using the field strength of an infinite sheet, $E_{\text{sheet}} = \sigma/2\varepsilon_0 = 2\pi k\sigma$, produces a fractional error less than 10% if $|E_{\text{sheet}} - E|/E < 0.1$. Since $E_{\text{sheet}} > E$, this implies that $E_{\text{sheet}}/E < 1.1$ or $2\pi k\sigma/2\pi k\sigma(1 - x/\sqrt{x^2 + a^2}) < 1.1$. The steps in the solution of this inequality are: $1.1x < 0.1\sqrt{x^2 + a^2}$, $1.21x^2 < 0.01(x^2 + a^2)$, $x < a\sqrt{0.01/1.20} = 9.13 \times 10^{-2}a$. For $a = 20 \text{ cm}, x < 1.83 \text{ cm}$. (b) For large x, the point charge field, $E_{\text{pt}} = kq/x^2 = k\pi\sigma a^2/x^2$, is good to 10% for $|E_{\text{pt}} - E|/E < 0.1$. The solution of this inequality is simplified by defining an angle ϕ, such that $\cos\phi = x/\sqrt{x^2 + a^2}$ and $\tan\phi = a/x$. In terms of ϕ, one finds $E = 2\pi k\sigma(1 - \cos\phi)$, $E_{\text{pt}} = k\pi\sigma \tan^2\phi$, and $E_{\text{pt}}/E = \tan^2\phi/2(1 - \cos\phi)$. Furthermore, $\tan^2\phi = \sin^2\phi/\cos^2\phi = (1 - \cos\phi)(1 + \cos\phi)/\cos^2\phi$, so $E_{\text{pt}}/E = (1 + \cos\phi)/2\cos^2\phi$. The range $0 \leq x < \infty$ corresponds to $0 < \phi \leq \pi/2$, so $E_{\text{pt}}/E > 1$ and the inequality becomes $E_{\text{pt}}/E = (1 + \cos\phi)/2\cos^2\phi < 1.1$, or $2.2 \cos^2\phi - \cos\phi - 1 > 0$. The quadratic formula for the positive root gives $\cos\phi > (1 + \sqrt{1 + 8.8})/4.4 = 0.939$, or $\phi < 20.2°$. This implies $x = a/\tan\phi > a/\tan 20.2° = 2.72 a$. For $a = 20 \text{ cm}$, $x > 54.5 \text{ cm}$.

Section 24-6: Gauss's Law and Conductors

Problem

43. What is the electric field strength just outside the surface of a conducting sphere carrying surface charge density $1.4 \text{ }\mu\text{C/m}^2$?

Solution

At the surface of a conductor, $E = \sigma/\varepsilon_0$ (positive away from the surface), or $(1.4 \text{ }\mu\text{C/m}^2)(8.85 \times 10^{-12} \text{ C}^2/\text{N·m}^2)^{-1} = 158 \text{ kN/C}$ in this problem.

Problem

45. A net charge of $5.0 \text{ }\mu\text{C}$ is applied on one side of a solid metal sphere 2.0 cm in diameter. After electrostatic equilibrium is reached, what are (a) the volume charge density inside the sphere and (b) the surface charge density on the sphere? Assume there are no other charges or conductors nearby. (c) Which of your answers depends on this assumption, and why?

Solution

(a) The electric field within a conducting medium, in electrostatic equilibrium, is zero. Therefore, Gauss's law implies that the net charge contained in any closed surface, lying within the metal, is zero. (b) If the volume charge density is zero within the metal, all of the net charge must reside on the surface of the sphere. If the sphere is electrically isolated, the charge will be uniformly distributed (i.e., spherically symmetric), so $\sigma = Q/4\pi R^2 = (5 \text{ }\mu\text{C})/4\pi(1 \text{ cm})^2 = 3.98 \times 10^{-3} \text{ C/m}^2$. (c) Spherical symmetry for σ depends on the proximity of other charges and conductors.

Problem

47. A 250-nC point charge is placed at the center of an uncharged spherical conducting shell 20 cm in radius. (a) What is the surface charge density on the outer surface of the shell? (b) What is the electric field strength at the shell's outer surface?

Solution

(a) There is a non-zero field outside the shell, because the net charge within is not zero. Therefore, there is a surface charge density $\sigma = \varepsilon_0 E$ on the outer surface of the shell, which is uniform, if we ignore the possible presence of other charges and conducting surfaces outside the shell. Gauss's law (with reasoning similar to Example 24-7) requires that the charge on the shell's outer surface is equal to the point charge within, so $\sigma = q/4\pi R^2 = 250 \text{ nC}/4\pi(0.20 \text{ m})^2 = 497 \text{ nC/m}^2$. (b) Then the field strength at the outer surface is $E = \sigma/\varepsilon_0 = 56.2 \text{ kN/C}$.

Problem

49. An irregular conductor containing an irregular, empty cavity carries a net charge Q. (a) Show that the electric field inside the cavity must be zero.

(b) If you put a point charge inside the cavity, what value must it have in order to make the surface charge density on the outer surface of the conductor everywhere zero?

Solution

(a) When there is no charge inside the cavity, the flux through any closed surface within the cavity (S_1) is zero, hence so is the field. (b) If the surface charge density on the outer surface (and also the electric field there) is to vanish, then the net charge inside a gaussian surface containing the conductor (S_2) is zero. Thus, the point charge in the cavity must equal $-Q$. (Note: The argument in part (a) depends on the conservative nature of the electrostatic field (see Section 25-1), for then positive flux on one part of S_1 canceling negative flux on another part is ruled out.)

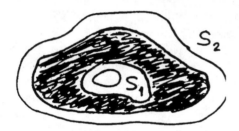

Problem 49 Solution.

Problem

51. A total charge of 18 μC is applied to a thin, square metal plate 75 cm on a side. Find the electric field strength near the plate's surface.

Solution

The net charge of 18 μC must distribute itself over the outer surface of the plate, in accordance with Gauss's law for conductors. The outer surface consists of two plane square surfaces on each face, plus the edges and corners. Symmetry arguments imply that for an isolated plate, the charge density on the faces is the same, but not necessarily uniform because the edges and corners also have charge. If the plate is thin, we could assume that the edges and corners have negligible charge and that the density on the faces is approximately uniform. Then the surface charge density is the total charge divided by the area of both faces, $\sigma = 18 \mu C/2(75 \text{ cm})^2 = 16.0 \mu C$, and the field strength near the plate (but not near an edge) is $E = \sigma/\varepsilon_0 = 1.81$ MN/C.

Problem

53. A conducting sphere 2.0 cm in radius is concentric with a spherical conducting shell with inner radius

8.0 cm and outer radius 10 cm. The small sphere carries 50 nC charge and the shell has no net charge. Find the electric field strength (a) 1.0 cm, (b) 5.0 cm, (c) 9.0 cm, and (d) 15 cm from the center.

Solution

If we assume the two-conductor system is isolated and in electrostatic equilibrium, then the field has spherical symmetry. Gauss's law requires that the field inside the conducting material be zero (for $0 \leq r < 2$ cm and 8 cm $< r < 10$ cm in this problem), and that, since the shell is neutral, the field elsewhere is like that from a point charge of 50 μC located at the center of symmetry ($r = 0$). Thus, (a) $E(1 \text{ cm}) = 0$, (b) $E(5 \text{ cm}) = kq/r^2 = (9 \times 10^9 \text{ N·m}^2/\text{C}^2) \times (50 \mu C)/(5 \text{ cm})^2 = 180$ kN/C, (c) $E(9 \text{ cm}) = 0$, and (d) $E(15 \text{ cm}) = kq/r^2 = (\frac{1}{9})E(5 \text{ cm}) = 20$ kN/C.

Paired Problems

Problem

55. A point charge $-q$ is at the center of a spherical shell carrying charge $+2q$. That shell, in turn, is concentric with a larger shell carrying charge $-\frac{3}{2}q$. Draw a cross section of this structure, and sketch the electric field lines using the convention that 8 lines correspond to a charge of magnitude q.

Solution

The field from the given charges is spherically symmetric, so (from Gauss's law) is like that of a point charge, located at the center, with magnitude equal to the net charge enclosed by a sphere of radius

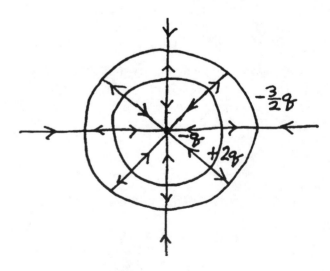

Problem 55 Solution.

equal to the distance to the field point. Thus, $E = -kq/r^2$ inside the first shell (8 lines radially inward), $E = +kq/r^2$ between the first and second shells (8 lines radially outward), and $E = -kq/2r^2$ outside the second shell (4 lines radially inward).

Problem

57. A point charge q is at the center of a spherical shell of radius R carrying charge $2q$ spread uniformly over its surface. Write expressions for the electric field strength at (a) $\frac{1}{2}R$ and (b) $2R$.

Solution

As explained in Example 24-3, (a) for $r = \frac{1}{2}R < R$, $q_{enclosed} = q$ and $E = kq/(\frac{1}{2}R)^2 = 4kq/R^2$, and (b) for $r = 2R > R$, $q_{enclosed} = q + 2q$ and $E = 3kq/(2R)^2 = 3\ kq/4R^2$.

Problem

59. A long, thin hollow pipe 4.0 cm in diameter carries charge at a density of $-2.6\ \mu C/m$, uniformly distributed over the pipe. It is concentric with 10-cm diameter pipe carrying $+2.6\ \mu C/m$, also uniformly distributed. Find the magnitude of the electric field at (a) 0.50 cm, (b) 3.5 cm, and (c) 12 cm from the axis of the pipes.

Solution

Assume the electric field has line symmetry, and apply Gauss's law to a coaxial cylindrical surface of radius r. The result is Equation 24-8 set equal to $\lambda_{enclosed}\ell/\varepsilon_0$, so $E(r) = \lambda_{enclosed}/2\pi\varepsilon_0 r$. (a) At $r = 0.5$ cm < 2 cm (inside inner pipe), $\lambda_{enclosed}$ is zero and so is E. (b) At 2 cm $< r = 3.5$ cm < 5 cm (between the pipes), $\lambda_{enclosed}$ is just the inner pipe, so $E = (-2.6\ \mu C/m) \div 2\pi\varepsilon_0(3.5$ cm$) = -1.34$ MN/C. (The minus sign means the direction of E is radially inward toward the axis of the pipes; the magnitude is the absolute value of E.) (c) At $r = 12$ cm > 5 cm (outside outer pipe), $\lambda_{enclosed}$ and E are again zero, since the pipes have opposite linear charge densities.

Problem

61. An early (and incorrect) model for the atom pictured its positive charge as spread uniformly throughout the spherical atomic volume. For a hydrogen atom of radius 0.0529 nm, what would be the electric field due to such a distribution of positive charge (a) 0.020 nm from the center and (b) 0.20 nm from the center?

Solution

(a) Inside a uniformly charged spherical volume, $E = kQr/R^3 = (9\ GN\cdot m^2/C^2)(1.6\times10^{-19}\ C)\times (0.02$ nm$)/(0.0529$ nm$)^3 = 195$ GN/C (see Equation 24-7). (b) Outside, the field is like that of a point charge, $E = kQ/r^2 = (9\ GN\cdot m^2/C^2)(1.6\times10^{-19}\ C) \div (0.2$ nm$)^2 = 36.0$ GN/C (see Equation 24-6).

Problem

63. A sphere of radius $2a$ has a hole of radius a, as shown in Fig. 24-50. The solid portion carries a uniform volume charge density ρ. Find an expression for the electric field strength within the solid portion, as a function of the distance r from the center.

Solution

From Gauss's law and Equation 24-5, $E = q_{enclosed} \div 4\pi\varepsilon_0 r^2$, where $q_{enclosed}$ is the charge within a spherical gaussian surface of radius r about the center of symmetry. For $a < r < 2a$, $q_{enclosed} = \rho V = \frac{4}{3}\pi\rho(r^3 - a^3)$, so $E = (\rho/3\varepsilon_0)(r - a^3/r^2)$.

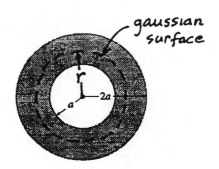

FIGURE 24-50 Problem 63 Solution.

Supplementary Problems
Problem

65. Repeat Problem 10 for the case $\mathbf{E} = E_0 \left(\frac{y}{a}\right)^2 \hat{\mathbf{k}}$.

Solution

Since the electric field depends only on y, break up the square in Fig. 24-42 (see Problem 10) into strips of area $d\mathbf{A} = \pm a\ dy\ \hat{\mathbf{k}}$, of length a parallel to the x axis and width dy, the normal to which could be $\pm\hat{\mathbf{k}}$. The electric flux through the square is

$$\Phi = \int_{square} \mathbf{E}\cdot d\mathbf{A} = \pm\int_0^a E_0 \left(\frac{y}{a}\right)^2 a\ dy$$

$$= \pm\left(\frac{E_0}{a}\right)\int_0^a y^2\ dy = \pm\frac{1}{3}E_0 a^2.$$

Problem

67. A proton is released from rest 1.0 cm from a large sheet carrying a surface charge density of -24 nC/m^2. How much later does it strike the sheet?

Solution

The proton is accelerated toward the sheet by an electric field in that direction. For the field, we can use that from an infinite plane sheet (assuming we are not near an edge) so $E = \sigma/2\varepsilon_0$ and the acceleration $a = eE/m_p$ is uniform. Starting from rest, the proton travels to the sheet in time

$$t = \sqrt{2(x - x_0)/a} = \sqrt{4\varepsilon_0 m_p (x - x_0)/e\sigma}$$

$$= \sqrt{\frac{4(8.85 \times 10^{-12} \text{ C}^2/\text{N·m}^2)(1.67 \times 10^{-27} \text{ kg})(1 \text{ cm})}{(1.6 \times 10^{-19} \text{ C})(24 \text{ nC/m}^2)}}$$

$$= 392 \text{ ns}.$$

Problem

69. Repeat Problem 36 for the case when the charge density in the slab is given by $\rho = \rho_0 |x/d|$, where ρ_0 is a constant.

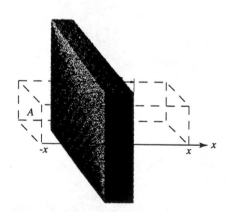

FIGURE 24-45 Problem 69 Solution.

Solution

Gauss's law, plane symmetry, and Equation 24-9 can be used to find the electric field strength, but we must integrate to get the charge enclosed by the gaussian surface. We use charge elements that are thin parallel sheets of the same area as the face of the gaussian surface and of thickness dx, as shown. (a) Inside the slab ($|x| < d/2$), $2EA = \varepsilon_0^{-1} \int_{-x}^{x} \rho A \, dx = (\rho_0 A/\varepsilon_0 d) \times$ $(\int_{-x}^{0}(-x)\,dx + \int_{0}^{x} x\,dx) = \rho_0 A x^2/\varepsilon_0 d$, hence $E = \rho_0 x^2/2\varepsilon_0 d$. (b) Outside, $2EA = \varepsilon_0^{-1} \int_{-d/2}^{d/2} \rho A \, dx =$

$(\rho_0 A/\varepsilon_0 d)(\int_{-d/2}^{0}(-x)\,dx + \int_{0}^{d/2} x\,dx) =$ $(\rho_0 A/\varepsilon_0 d)(d/2)^2$, hence $E = \rho_0 d/8\varepsilon_0$. (This is equivalent, of course, to the field strength outside an infinite sheet with $\sigma = \rho_0 d/4$.)

Problem

71. A small object of mass m and charge q is attached by a thread of length ℓ to a large, flat, nonconducting plate carrying a uniform surface charge density σ with the same sign as q (Fig. 24-52). If the object is displaced slightly sideways from its equilibrium, show that it undergoes simple harmonic motion with period $T = 2\pi\sqrt{2\varepsilon_0 m\ell/q\sigma}$. Assume the gravitational force is negligible.

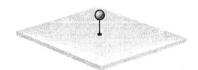

FIGURE 24-52 Problem 71.

Solution

The Coulomb force on the particle is approximately uniform and greater than its weight. Therefore, the net upward force field on the particle is $F_{\text{net}}/m = (q/m)E - g$, and since it is tethered by the string, it oscillates like an upside-down pendulum with period $T = 2\pi\sqrt{\ell/(qE/m - g)}$. If we neglect gravity and use $E = \sigma/2\varepsilon_0$ for an infinite sheet as an approximation, then $T = 2\pi\sqrt{2\varepsilon_0 m\ell/q\sigma}$.

Problem

73. A thick spherical shell of inner radius a and outer radius b carries a charge density given by $\rho = \dfrac{ce^{-r/a}}{r^2}$, where a and c are constants. Find expressions for the electric field strength for (a) $r < a$, (b) $a < r < b$, and (c) $r > b$.

Solution

Spherical symmetry, Equation 24-5 and Gauss's law give a field strength of $E(r) = q_{\text{enclosed}}/4\pi\varepsilon_0 r^2$, where $q_{\text{enclosed}} = \int_{0}^{r} \rho \, dV$ is the charge within a concentric spherical surface of radius r, and $dV = 4\pi r^2 \, dr$ is the volume element for a thin shell with this surface. (a) For $r < a$, $q_{\text{enclosed}} = 0$ hence $E(r) = 0$. (b) For $a \leq r \leq b$, $q_{\text{enclosed}} = 4\pi c \int_{a}^{r} e^{-r/a} \, dr = 4\pi ac(e^{-1} - e^{-r/a})$ hence $E(r) = ac(e^{-1} - e^{-r/a})/\varepsilon_0 r^2$. (c) For $r > b$, $q_{\text{enclosed}} = 4\pi c \int_{a}^{b} e^{-r/a} \, dr$ hence $E(r) = ac(e^{-1} - e^{-b/a})/\varepsilon_0 r^2$.

Problem

75. A solid sphere of radius R carries a uniform volume charge density ρ. A hole of radius $R/2$ occupies a region from the center to the edge of the sphere, as shown in Fig. 24-53. Show that the electric field everywhere in the hole points horizontally and has magnitude $\rho R/6\varepsilon_0$. *Hint: Treat the hole as a superposition of two charged spheres of opposite charge.*

Solution

A large solid sphere can be considered to be the superposition of the sphere with a cavity plus a small solid sphere filling the cavity, both with uniform charge density ρ. The electric field inside the solid spheres is $\rho\mathbf{r}/3\varepsilon_0$, where $\mathbf{r}$ is a vector from the center of each sphere to the field point P, in both (see Equation 24-7). For the large sphere, whose center we take at the origin, $\mathbf{r} = \mathbf{r}_P$, and for the small sphere, whose center is at $\frac{1}{2}R\hat{\imath}$, $\mathbf{r} = \mathbf{r}_P - \frac{1}{2}R\hat{\imath}$. Therefore, $\mathbf{E}(\text{large sphere}) = \mathbf{E}(\text{sphere with cavity}) + \mathbf{E}(\text{small sphere})$, or $\rho\mathbf{r}_P/3\varepsilon_0 = \mathbf{E} + \rho(\mathbf{r}_P - \frac{1}{2}R\hat{\imath})/3\varepsilon_0$. Thus, $\mathbf{E} = \rho R\hat{\imath}/6\varepsilon_0$, that is, for any point inside the cavity, the electric field of the sphere with the cavity is uniform (with direction parallel to the line between the centers of the sphere and cavity). (Note that this result holds for any size spherical cavity if one replaces $\frac{1}{2}R\hat{\imath}$ with the vector to the center of the cavity.)

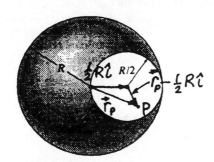

FIGURE 24-53 Problem 75 Solution.

Problem

77. Two flat, parallel, closely spaced metal plates of area 0.080 m² carry total charges of −2.1 μC and +3.8 μC. Find the surface charge densities on the inner and outer faces of each plate.

Solution

If the thickness and separation of the plates is small compared to their lateral dimensions ($\sqrt{0.08 \text{ m}^2} \approx 28$ cm), we can assume that the electric field near the plates (edge effects neglected) is uniform and normal to the plates. (Of course, far away, the field goes like $1/r^2$.) The general case, where the plates carry arbitrary charges q_1 and q_2, can be viewed as the superposition of three simpler cases: a neutral plate, a charged plate, and two oppositely charged plates (see last paragraph in Section 24-6). In each of the three regions, left of, between, and right of the plates (the thickness, assumed negligible, does not add to any region), the electric field is the sum of three contributions, as shown. The final diagram, together with Equation 24-11 (which gives $\sigma = \varepsilon_0 E$ at each surface), shows that the charge densities on the outer surfaces are equal, $\sigma_{\text{out}} = \frac{1}{2}(q_1 + q_2)/A$, and that the charge densities on the inner surfaces are equal and opposite, $\sigma_{\text{in}} = \pm\frac{1}{2}(q_2 - q_1)/A$. Numerically, $\sigma_{\text{out}} = \frac{1}{2}(-2.1 + 3.8) \text{ } \mu\text{C}/0.08 \text{ m}^2 = 10.6 \text{ } \mu\text{C/m}^2$, and $\sigma_{\text{in}} = \pm\frac{1}{2}[3.8 - (-2.1)] \text{ } \mu\text{C}/0.08 \text{ m}^2 = \pm36.9 \text{ } \mu\text{C/m}^2$. (Note: the direction of the fields is shown for positive total charge, with the more positive plate on the right.)

Problem 77 Solution.

CHAPTER 25   ELECTRIC POTENTIAL

ActivPhysics can help with these problems:
Activities 11.9, 11.10

Section 25-2:   Potential Difference

Problem

1. How much work does it take to move a 50-μC charge against a 12-V potential difference?

Solution

The potential difference and the work per unit charge, done by an external agent, are equal in magnitude, so $W = q\,\Delta V = (50\ \mu C)(12\ V) = 600\ \mu J$. (Note: Since only magnitudes are needed in this problem, we omitted the subscripts A and B.)

Problem

3. It takes 45 J to move a 15-mC charge from point A to point B . What is the potential difference ΔV_{AB}?

Solution

The work done by an external agent equals the potential energy change, $\Delta U_{AB} = 45\ J = q\,\Delta V_{AB}$, hence $\Delta V_{AB} = 45\ J/15\ mC = 3\ kV$. (Since the work required to move the charge from A to B is positive, $V_B > V_A$ and ΔV_{AB} is positive.)

Problem

5. Find the magnitude of the potential difference between two points located 1.4 m apart in a uniform 650 N/C electric field, if a line between the points is parallel to the field.

Solution

For ℓ in the direction of a uniform electric field, Equation 25-2b gives $|\Delta V| = E\ell = (650\ N/C) \times$ $(1.4\ m) = 910\ V$. (See note in solution to Problem 1. Since $dV = -\mathbf{E}\cdot d\boldsymbol{\ell}$, the potential always decreases in the direction of the electric field.)

Problem

7. Two points A and B lie 15 cm apart in a uniform electric field, with the path AB parallel to the field. If the potential difference ΔV_{AB} is 840 V, what is the field strength?

Solution

Equation 25-2b for a uniform field gives $E = |\Delta V/\ell| =$ 840 V/0.15 m = 5.60 kV/m. (See notes in solutions to Problems 1 and 5.)

Problem

9. A proton, an alpha particle (a bare helium nucleus), and a singly ionized helium atom are accelerated through a potential difference of 100 V. Find the energy each gains.

Solution

The energy gained is $q\,\Delta V$ (see Example 25-1). The proton and singly-ionized helium atom have charge e, so they gain 100 eV = $(1.6\times10^{-19}\ C)(100\ V) =$ 1.6×10^{-17} J, while the α-particle has charge $2e$ and gains twice this energy.

Problem

11. What is the potential difference between the terminals of a battery that can impart 7.2×10^{-19} J to each electron that moves between the terminals?

Solution

The magnitude of the potential difference is $|W/q| =$ 7.2×10^{-19} J/1.6×10^{-19} C = 4.5 V. (The energy imparted per electron is 4.5 eV.)

Problem

13. A 12-V car battery stores 2.8 MJ of energy. How much charge can move between the battery terminals before it is totally discharged? Assume the potential difference remains at 12 V, an assumption that is not realistic.

Solution

A charge q, moving through a potential difference ΔV, is equivalent to electrostatic potential energy $\Delta U =$ $q\,\Delta V$, stored in the battery. Thus, $q = 2.8$ MJ/12 V = 2.33×10^5 C.

Problem

15. Two large, flat metal plates are a distance d apart, where d is small compared with the plate size. If

the plates carry surface charge densities $\pm\sigma$, show that the potential difference between them is $V = \sigma d/\varepsilon_0$.

Solution

The electric field between the plates is uniform, with $E = \sigma/\varepsilon_0$, directed from the positive to the negative plate (see last paragraph of Section 24-6 and Fig. 24-35). Then Equation 25-2b gives $V = V_+ - V_- = -(\sigma/\varepsilon_0)(-d) = \sigma d/\varepsilon_0$ (the displacement from the negative to the positive plate is opposite to the field direction).

Problem

17. A 5.0-g object carries a net charge of 3.8 μC. It acquires a speed v when accelerated from rest through a potential difference V. A 2.0-g object acquires twice the speed under the same circumstances. What is its charge?

Solution

The speed acquired by a charge q, starting from rest at point A and moving through a potential difference of V, is given by $\frac{1}{2}mv_B^2 = q(V_A - V_B) = qV$, or $v_B = \sqrt{2V(q/m)}$. (This is the work-energy theorem for the electric force. A positive charge is accelerated in the direction of decreasing potential.) If the second object acquires twice the speed of the first object, moving through the same potential difference, it must have four times the charge to mass ratio, q/m. Thus, $q_2 = 4(q_1/m_1)m_2 = 4(3.8\ \mu\text{C})(2\text{g}/5\text{g}) = 6.08\ \mu\text{C}$.

Section 25-3:   Calculating Potential Difference

Problem

19. The classical picture of the hydrogen atom has a single electron in orbit a distance 0.0529 nm from the proton. Calculate the electric potential associated with the proton's electric field at this distance.

Solution

The potential of the proton, at the position of the electron (both of which may be regarded as point-charge atomic constituents) is (Equation 25-4) $V = ke/a_0$, where a_0 is the Bohr radius. Numerically, $V = (9\times10^9\ \text{N}\cdot\text{m}^2/\text{C}^2)(1.6\times10^{-19}\ \text{C})/(5.29\times10^{-11}\ \text{m}) = 27.2$ V. (The energy of an electron in a classical, circular orbit, around a stationary proton, is one half its potential energy, or $\frac{1}{2}U = \frac{1}{2}(-e)V = -13.6$ eV. The excellent agreement with the ionization energy of hydrogen was one of the successes of the Bohr model.)

Problem

21. Points A and B lie 20 cm apart on a line extending radially from a point charge Q, and the potentials at these points are $V_A = 280$ V, $V_B = 130$ V. Find Q and the distance r between A and the charge.

Solution

Since $V_A = kQ/r_A$ and $V_B = kQ/r_B$, division yields $r_B = (V_A/V_B)r_A = (280/130)r_A = 2.15r_A$. But $r_B - r_A = 20$ cm, so $r_A = (20\ \text{cm})(2.15 - 1)^{-1} = 17.3$ cm. Then $Q = V_A r_A/k = (280\ \text{V})(17.3\ \text{cm})\div (9\times10^9\ \text{N}\cdot\text{m}^2/\text{C}^2) = 5.39$ nC.

Problem

23. A 3.5-cm-diameter isolated metal sphere carries a net charge of 0.86 μC. (a) What is the potential at the sphere's surface? (b) If a proton were released from rest at the sphere's surface, what would be its speed far from the sphere?

Solution

(a) An isolated metal sphere has a uniform surface charge density, so Equation 25-4 gives the potential at its surface, $V_{\text{surf}} = kQ/R = (9\ \text{GN}\cdot\text{m}^2/\text{C}^2)(0.86\ \mu\text{C})\div (\frac{1}{2}\times3.5\ \text{cm}) = 442$ kV. (b) The work done by the repulsive electrostatic field (the negative of the change in the proton's potential energy) equals the proton's kinetic energy at infinity, $W = -e(V_\infty - V_{\text{surf}}) = eV_{\text{surf}} = \frac{1}{2}mv^2$. Then $v = \sqrt{2eV_{\text{surf}}/m} = [2(1.6\times10^{-19}\ \text{C})(442\ \text{kV})/(1.67\times10^{-27}\ \text{kg})]^{1/2} = 9.21$ Mm/s.

Problem

25. A thin spherical shell of charge has radius R and total charge Q distributed uniformly over its surface. What is the potential at its center?

Solution

The potential at the surface of the shell is kQ/R (as in Example 25-3). The electric field inside a uniformly charged shell is zero, so the potential anywhere inside is a constant, equal, therefore, to its value at the surface.

Problem

27. Find the potential as a function of position in an electric field given by $\mathbf{E} = ax\,\hat{\imath}$, where a is a constant and where $V = 0$ at $x = 0$.

Solution

Since $V(0) = 0$, $V(\mathbf{r}) = -\int_0^r \mathbf{E}\cdot d\mathbf{r} = -\int_0^r ax\hat{\imath}\cdot d\mathbf{r} = -\int_0^x ax\,dx = -\frac{1}{2}ax^2$.

Problem

29. The potential difference between the surface of a 3.0-cm-diameter power line and a point 1.0 m distant is 3.9 kV. What is the line charge density on the power line?

Solution

If we approximate the potential from the line by that from an infinitely long charged wire, Equation 25-5 can be used to find λ: $\lambda = 2\pi\varepsilon_0 \, \Delta V_{AB}/\ln(r_A/r_B) = $ (3.9 kV)$[2(9\times10^9 \text{ N·m}^2/\text{C}^2)\ln(100/1.5)]^{-1} = $ 51.6 nC/m. (Note: $\Delta V_{AB} = V_B - V_A$ so B is at the surface of the wire and A is 100 cm distant.)

Problem

31. A charge $+Q$ lies at the origin, and $-3Q$ at $x = a$. Find two points on the x-axis where $V = 0$.

Solution

The potential on the x-axis, $\Sigma \, kq_i/r_i = kQ/|x| + k(-3Q)/|x - a|$ (from Equation 25-6), is zero when $3|x| = |x - a|$. For $x < 0$, this implies $-3x = a - x$, or $x = -a/2$. Note that the absolute value of a negative number is minus the number, and we assume that a is a positive constant. For $0 < x < a$, the condition is $3x = a - x$, or $x = a/4$. For $x > a$, there are no solutions. (The same results follow from the quadratic $8x^2 + 2ax - a^2 = 0$, which results from the square of the above condition.)

Problem

33. Find the potential 10 cm from a dipole of moment $p = 2.9$ nC·m (a) on the dipole axis, (b) at 45° to the axis, and (c) on the perpendicular bisector. The dipole separation is much less than 10 cm.

Solution

Equation 25-7 gives the potential from a point dipole as a function of distance and angle from the dipole axis. For the dipole moment and distance given, $V(r, \theta) = kp \cos\theta/r^2 = (9 \text{ GN·m}^2/\text{C}^2)(2.9 \text{ nC·m})\times \cos\theta/(10 \text{ cm})^2 = (2.61 \text{ kV}) \cos\theta$. For the three given angles,
(a) $V = (2.61 \text{ kV}) \cos 0° = 2.61$ kV;
(b) $V = (2.61 \text{ kV}) \cos 45° = 1.85$ kV; and
(c) $V = (2.61 \text{ kV}) \cos 90° = 0$.

Problem

35. A hollow, spherical conducting shell of inner radius b and outer radius c surrounds, and is concentric with, a solid conducting sphere of radius a, as shown in Fig. 25-39. The sphere carries a net charge $-Q$ and the shell carries a net charge $+3Q$. Both conductors are in electrostatic equilibrium. Find an expression for the potential difference from infinity to the surface of the sphere.

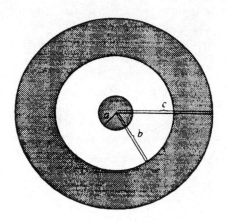

FIGURE 25-39 Problem 35.

Solution

The electric field between the solid sphere and the shell is like that due to a point charge $-Q$ located at their common center (the origin), so the potential difference between the sphere and the shell is $V_{\text{sph}} - V_{\text{shell}} = -kQ(a^{-1} - b^{-1})$. The electric field outside the shell is like that due to a point charge $2Q$ at the origin (the total charge is $3Q - Q$), so the potential difference between the shell and infinity is $V_{\text{shell}} - V_\infty = 2kQc^{-1}$. (See Examples 24-3 and 25-3.) The entire shell is at one potential, as is the entire sphere, because each is a conductor in electrostatic equilibrium. Therefore, the potential difference between the sphere and infinity is $V_{\text{sph}} - V_\infty = V_{\text{sph}} - V_{\text{shell}} + V_{\text{shell}} - V_\infty = kQ(2c^{-1} + b^{-1} - a^{-1})$.

Problem

37. A thin ring of radius R carries a charge $3Q$ distributed uniformly over three-fourths of its circumference, and $-Q$ over the rest. What is the potential at the center of the ring?

Solution

The result in Example 25-6 did not depend on the ring being uniformly charged. For a point on the axis of the ring, the geometrical factors are the same, and $\int_{\text{ring}} dq = Q_{\text{tot}}$ for any arbitrary charge distribution, so $V = kQ_{\text{tot}}(x^2 + a^2)^{-1/2}$ still holds. Thus, at the center ($x = 0$) of a ring of total charge $Q_{\text{tot}} = 3Q - Q = 2Q$, and radius $a = R$, the potential is $V = 2kQ/R$.

Problem

39. The annulus shown in Fig. 25-40 carries a uniform surface charge density σ. Find an expression for the potential at an arbitrary point P on its axis.

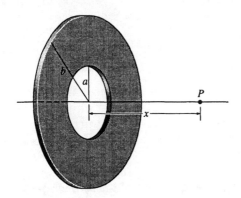

FIGURE 25-40 Problem 39.

Solution

The annulus can be considered to be composed of thin rings of radius $r(a \le r \le b)$ and charge $dq = 2\pi\sigma r\, dr$ (see Example 25-7 and Figs. 25-15 and 16). The element of potential from a ring on its axis, a distance x from the center, is $dV = k\, dq/\sqrt{x^2 + r^2}$ (see Example 25-6) so the potential from the whole annulus is:

$$V = \int dV = 2\pi\sigma k \int_a^b \frac{r\, dr}{\sqrt{x^2 + r^2}} = 2\pi k\sigma \left. \left|\sqrt{x^2 + r^2}\right.\right|_a^b$$
$$= 2\pi k\sigma (\sqrt{x^2 + b^2} - \sqrt{x^2 + a^2}).$$

(Note: This reduces to the potential on the axis of a uniformly charged disk if $a \to 0$.)

Problem

41. (a) Find the potential as a function of position in the electric field $\mathbf{E} = E_0(\hat{\mathbf{i}} + \hat{\mathbf{j}})$, where $E_0 = 150$ V/m. Take the zero of potential at the origin. (b) Find the potential difference from the point $x = 2.0$ m, $y = 1.0$ m to the point $x = 3.5$ m, $y = -1.5$ m.

Solution

(a) Equation 25-2b gives the potential for a uniform field. Take the zero of potential at the origin (point A in Equation 25-2b) and let $\boldsymbol{\ell} = \mathbf{r} = x\hat{\mathbf{i}} + y\hat{\mathbf{j}} + z\hat{\mathbf{k}}$ be the vector from the origin to the field point (point B in Equation 25-2b). Then $\Delta V_{AB} = V_B - V_A = V(\mathbf{r}) - 0 = V(x, y) = -E_0(\hat{\mathbf{i}} + \hat{\mathbf{j}}) \cdot \mathbf{r} = -E_0(x + y)$. (The potential is independent of z, so we wrote $V(\mathbf{r}) = V(x, y)$.)
(b) $V(3.5$ m, -1.5 m$) - V(2.0$ m, 1.0 m$) = -(150$ V/m$)(3.5$ m $- 1.5$ m $- 2.0$ m $- 1.0$ m$) = 150$ V.

Section 25-4:   Potential Difference and the Electric Field

Problem

43. Figure 25-41 shows a plot of potential versus position along the x-axis. Make a plot of the x component of the electric field for this situation.

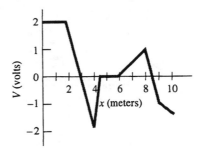

FIGURE 25-41 Problem 43.

Solution

Figure 25-24 illustrates the relation $E_x = -dV/dx$, which may be used to estimate E_x for the seven straight-line segments shown in Fig. 25-41. Thus, for $x = 0$ to 2 m, $E_x = 0$, for $x = 2$ m to 4 m, $E_x = -(-2$ V $- 2$ V$)/(4$ m $- 2$ m$) = 2$ V/m; etc., as sketched below.

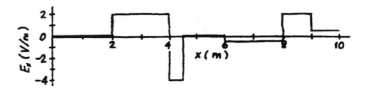

Problem 43 Solution.

Problem

45. The potential in a certain region is given by $V = axy$, where a is a constant. (a) Determine the electric field in the region. (b) Sketch some equipotentials and field lines.

Solution

(a) The x and y components of the electric field can be found from Equation 25-10: $E_x = -\partial V/\partial x = -\partial/\partial x(axy) = -ay$, and $E_y = -\partial V/\partial y = -ax$. Thus $\mathbf{E} = -a(y\hat{\mathbf{i}} + x\hat{\mathbf{j}})$. (The field has no z component.)
(b) See sketch below. The field lines (dashed) are perpendicular to the equipotentials (solid) in the direction of decreasing potential (arrow for $a > 0$, in this case). These equipotentials and field lines are confocal hyperbolas, proportional to xy and $\frac{1}{2}(x^2 - y^2)$ respectively, and are sketched only for x and y in the first quadrant.

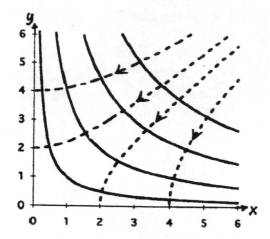

Problem 45 Solution.

Problem

47. The electric potential in a region of space is given by $V = 2xy - 3zx + 5y^2$, with V in volts and the coordinates in meters. If point P is at $x = 1$ m, $y = 1$ m, $z = 1$ m, find (a) the potential at P and (b) the x, y, and z components of the electric field at P.

Solution

(a) Direct substitution gives $V(P) = 2(1)(1) - 3(1)(1) + 5(1)^2 = 4$ V. (b) Use of Equation 25-10 gives $E_x = -\partial V/\partial x = -2y + 3z$, $E_y = -\partial V/\partial y = -2x - 10y$, and $E_z = -\partial V/\partial z = 3x$. At $P(x = y = z = 1)$, $E_x = 1$ V/m, $E_y = -12$ V/m, and $E_z = 3$ V/m.

Problem

49. Use the result of Example 25-6 to determine the on-axis field of a charged ring, and verify that your answer agrees with the result of Example 23-8.

Solution

On the axis of a uniformly charged ring (the x-axis), $V = kQ/\sqrt{x^2 + a^2}$ (Equation 25-9), and the electric field only has an x component (by symmetry). Then $E = (-dV/dx)\hat{\imath} = kQx(x^2 + a^2)^{-3/2}\hat{\imath}$, in accord with Example 23-8. (In general, one needs to know the potential in a 3-dimensional region in order to calculate the field from its partial derivatives.)

Problem

51. The electric potential in a region is given by $V = -V_0(r/R)$, where V_0 and R are constants, r is the radial distance from the origin, and where the zero of potential is taken at $r = 0$. Find the magnitude and direction of the electric field in this region.

Solution

Since $V = -(V_0/R)r$ depends only on r, the field is spherically symmetric and its direction is radial. From Equation 25-10 (which applies unmodified for the radial coordinate), $E_r = -dV/dr = V_0/R$, and $\mathbf{E} = (V_0/R)\hat{\mathbf{r}}$ ($\hat{\mathbf{r}}$ is a unit vector radially outward).

Section 25-5: Potentials of Charged Conductors

Problem

53. The spark plug in an automobile engine has a center electrode made from wire 2.0 mm in diameter. The electrode is worn to a hemispherical shape, so it behaves approximately like a charged sphere. What is the minimum potential on this electrode that will ensure the plug sparks in air? Neglect the presence of the second electrode.

Solution

If we can treat the field from the central electrode as that from an isolated sphere, then $E = kq/R^2$ and $V = kq/R$, so that $V = RE$. Dielectric breakdown in air would occur for potentials exceeding $V = (1$ mm$)\times(3\times10^6$ V/m$) = 3$ kV.

Problem

55. Two metal spheres each 1.0 cm in radius are far apart. One sphere carries 38 nC of charge, the other -10 nC. (a) What is the potential on each? (b) If the spheres are connected by a thin wire, what will be the potential on each once equilibrium is reached? (c) How much charge must move between the spheres in order to achieve equilibrium?

Solution

(a) Since the spheres are far apart (approximately isolated), we can use Equation 25-11 to find their potentials: $V_1 = kQ_1/R_1 = (9$ GN·m^2/C$^2)(38$ nC$)\div(1$ cm$) = 34.2$ kV and $V_2 = kQ_2/R_2 = -9$ kV. (b) When connected by a thin wire, the spheres reach electrostatic equilibrium with the same potential, so $V = kQ_1'/R_1 = kQ_2'/R_2$. Since the radii are equal, so must be the charges, $Q_1' = Q_2'$. The total charge is 38 nC $- 10$ nC $= 28$ nC $= Q_1' + Q_2' = 2Q_1'$ (if we assume that the wire is so thin that it has a negligible charge), so $Q_1' = Q_2' = 14$ nC. Then $V' = k(14$ nC$)\div(1$ cm$) = 12.6$ kV. (c) In this process, the first sphere loses $38 - 14 = 24$ nC to the second.

Problem

57. Two conducting spheres are each 5.0 cm in diameter and each carries 0.12 μC. They are 8.0 m

apart. Determine (a) the potential on each sphere; (b) the field strength at the surface of each sphere; (c) the potential midway between the spheres; (d) the potential difference between the spheres.

Solution

Since the spheres are small and widely separated, at small distances ($\ll 8$ m) they behave like isolated spheres, and at large distances ($\gg 5$ cm) they behave like two point charges. (a) At the surface of either sphere, Equation 25-11 gives $V_{surf} = kq/R = (9 \times 10^9$ N·m^2/C$^2)(1.2\times10^{-7}$ C$)/(0.025$ m$) = 43.2$ kV. (b) Then $E_{surf} = \sigma/\varepsilon_0 = kq/R^2 = V_{surf}/R = (43.2$ kV$)/(2.5$ cm$) = 1.73\times10^6$ V/m. (c) Midway between the spheres, the potential from each one is the same, so $V_{\text{mid-pt.}} = 2kq/r = 2\times (9 \times 10^9$ V·m/C$)(1.2\times10^{-7}$ C$)/(4$ m$) = 540$ V. (d) The spheres are at the same potential, so the difference is zero.

Paired Problems

Problem

59. Three 50-pC charges sit at the vertices of an equilateral triangle 1.5 mm on a side. How much work would it take to bring a proton from very far away to the midpoint of one of the triangle's sides?

Solution

Two of the charges are at distances of $\frac{1}{2}(1.5$ mm$) = 0.75$ mm, and the third is at $\sqrt{3}\,(0.75$ mm$)$ from the midpoint of one side. Therefore, the potential at this point (Equation 25-6) is $V(P) = k(50$ pC$)(1 + 1 + 1/\sqrt{3})/(0.75$ mm$) = 1.55$ kV. The work it would take to move a proton of charge e from infinity (where the potential is zero) to this point is $eV = 1.55$ keV $= (1.6\times10^{-19}$ C$)(1.55$ kV$) = 2.47\times10^{-16}$ J.

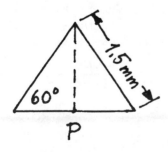

Problem 59 Solution.

Problem

61. A pair of equal charges q lies on the x-axis at $x = \pm a$. (a) Find expressions for the potential at

points on the x-axis for which $x > a$ and (b) show that your result reduces to a point-charge potential for $x \gg a$.

Solution

(a) From Equation 25-6, the potential on the x-axis is $V(x) = kq(|x - a|^{-1} + |x + a|^{-1})$. For $x > a$, this becomes $V(x) = kq[(x - a)^{-1} + (x + a)^{-1}] = 2kqx(x^2 - a^2)^{-1}$. (b) For $x \gg a$, $V(x) \to 2kq/x$, the potential of a point charge $2q$ at a distance x.

Problem

63. A 2.0-cm-radius metal sphere carries 75 nC and is surrounded by a concentric spherical conducting shell of radius 10 cm carrying -75 nC. (a) Find the potential difference between the shell and the sphere. (b) How would your answer change if the shell charge were changed to $+150$ nC?

Solution

(a) The electric field outside the sphere (radius $R_1 = 2$ cm), but inside the shell inner radius $R_2 = 10$ cm), is only due to the charge q_1 on the sphere (recall Gauss's law), and equals kq_1/r^2 radially outward. The potential difference is $V_1 - V_2 = -\int_{R_2}^{R_1} kq_1\,dr/r^2 = kq_1(R_1^{-1} - R_2^{-1}) = (9$ GN·m^2/C$^2)(75$ nC$)[(2$ cm$)^{-1} - (10$ cm$)^{-1}] = 27.0$ kV. (We used Equation 25-2a, with A at R_2 and B at R_1. Note that $\mathbf{E} \cdot d\mathbf{r} = kq_1 dr/r^2$ regardless of the choice of points A and B.) (b) Adding more charge to the conducting shell does not affect the field inside, nor the potential difference between points inside, like $V_1 - V_2$ in part (a). The field outside the shell and the potential relative to infinity would change.

Problem

65. On the x-axis, the electric field of a certain charge distribution is given by $\mathbf{E} = a/x^4\hat{\imath}$, where $a = 55$ V·m^3. Find the potential difference from the point $x = 1.3$ m to the point $x = 2.8$ m.

Solution

Direct application of Equation 25-2a yields:

$$V(2.8\text{ m}) - V(1.3\text{ m})$$
$$= -\int_{1.3\,\text{m}}^{2.8\,\text{m}} E_x\,dx = -\int_{1.3\,\text{m}}^{2.8\,\text{m}} \frac{a}{x^4}\,dx$$
$$= \left(\frac{55\text{ V·m}^3}{3}\right)\left[\frac{1}{(2.8\text{ m})^3} - \frac{1}{(1.3\text{ m})^3}\right]$$
$$= -7.51\text{ V}.$$

Problem

67. The potential as a function of position in a certain region is given by $V(x) = 3x - 2x^2 - x^3$, with x in meters and V in volts. Find (a) all points on the x-axis where $V = 0$, (b) an expression for the electric field, and (c) all points on the x-axis where $\mathbf{E} = 0$.

Solution

(a) The expression for the potential can be factored into $V(x) = x(x+3)(1-x)$, so $V(x) = 0$ at $x = 0$, 1 m, and -3 m. (b) V is independent of y and z, hence $\mathbf{E} = E_x\hat{\imath}$ has only an x component, and $E_x = -dV/dx = 3x^2 + 4x - 3$. (c) $E_x = 0$ for $x = (-2 \pm \sqrt{4+9})/3 = 0.535$ m and -1.87 m.

Supplementary Problem

Problem

69. A conducting sphere 5.0 cm in radius carries 60 nC. It is surrounded by a concentric spherical conducting shell of radius 15 cm carrying -60 nC. (a) Find the potential at the sphere's surface, taking the zero of potential at infinity. (b) Repeat for the case when the shell also carries $+60$ nC.

Solution

The potential difference between the sphere and the shell depends only on the electric field inside the shell (which is due to q_{sphere} only). For a spherically symmetric configuration, this was found to be $V_{\text{sphere}} - V_{\text{shell}} = kq_{\text{sphere}}(R_{\text{sphere}}^{-1} - R_{\text{shell}}^{-1})$ in Problem 63. (Here, R_{shell} is the inner radius of the shell.) Outside the shell, the potential depends on the total charge, which, at the surface of a spherical distribution, was found to be $V_{\text{shell}} - V_\infty = k(q_{\text{sphere}} + q_{\text{shell}})/R_{\text{shell}}$ in Example 25-3. (Here, R_{shell} is the outer radius of the shell, the conducting material of which is assumed to be an equipotential region in electrostatic equilibrium, and $V_\infty = 0$.) Thus, $V_{\text{sphere}} - V_\infty = V_{\text{sphere}} - V_{\text{shell}} + V_{\text{shell}} - V_\infty = kq_{\text{sphere}}(R_{\text{sphere}}^{-1} - R_{\text{shell}}^{-1}) + k(q_{\text{sphere}} + q_{\text{shell}})R_{\text{shell}}^{-1} = kq_{\text{sphere}}R_{\text{sphere}}^{-1} + kq_{\text{shell}}R_{\text{shell}}^{-1}$. (For a thin shell, the inner and outer radii are approximately equal.) (a) In this problem, $q_{\text{sphere}} = -q_{\text{shell}} = 60$ nC, and $R_{\text{sphere}} = \frac{1}{3}R_{\text{shell}} = 5$ cm, so $V_{\text{sphere}} = (9 \times \frac{6}{5}) \times (1 - \frac{1}{3})$ kV $= 7.2$ kV. (b) If q_{shell} is changed to equal $+q_{\text{sphere}}$, $V_{\text{sphere}} = (9 \times \frac{6}{5})(1 + \frac{1}{3})$ kV $= 14.4$ kV.

Problem

71. The potential on the axis of a uniformly charged disk at 5.0 cm from the disk center is 150 V; the potential 10 cm from the disk center is 110 V. Find the disk radius and its total charge.

Solution

Combining the given data with the potential in Exercise 25-7, we find 150 V $= 2kQ/a^2(\sqrt{(5\text{ cm})^2 + a^2} - 5$ cm$)$ and 110 V $= 2kQ/a^2(\sqrt{(10\text{ cm})^2 + a^2} - 10$ cm$)$. The charge can be eliminated by division,

$$\left(\frac{150}{110}\right) = \frac{\sqrt{1 + (a/5\text{ cm})^2} - 1}{\sqrt{4 + (a/5\text{ cm})^2} - 2}.$$

Several lines of algebra to remove the square roots finally yields $a = (5\text{ cm})\sqrt{105 \times 209}/52 = 14.2$ cm. We can now solve for Q from either of the first two equations, $Q = (Va^2/2k)[\sqrt{x^2 + a^2} - x]^{-1} = 1.67$ nC.

Problem

73. A power line consists of two parallel wires 3.0 cm in diameter spaced 2.0 m apart. If the potential difference between the wires is 4.0 kV, what is the charge per unit length on each wire? The wires carry equal but opposite charges. *Hint:* The wires are far enough apart that they don't greatly affect each other's fields.

Solution

Since the radius of the wires ($a = 1.5$ cm) is much smaller than their separation ($b = 200$ cm), their potential difference can be found from the superposition of the potentials of two (approximately infinite) line charges $\pm\lambda$ (as calculated in Example 25-4). Thus, for any two points between the wires,

$$V_2 - V_1 = (V_2^+ - V_1^+) + (V_2^- - V_1^-)$$
$$= \frac{\lambda}{2\pi\varepsilon_0}\ln\left(\frac{r_1}{r_2}\right) + \frac{(-\lambda)}{2\pi\varepsilon_0}\ln\left(\frac{b - r_1}{b - r_2}\right)$$
$$= \frac{\lambda}{2\pi\varepsilon_0}\ln\left(\frac{r_1(b - r_2)}{r_2(b - r_1)}\right).$$

To obtain the potential difference between the wires, we let $r_2 = a$ (the positive wire is at a higher potential) and $r_1 = b - a$. Then

$$\Delta V = \frac{\lambda}{2\pi\varepsilon_0}\ln\frac{(b-a)(b-a)}{a(b-b+a)} = \frac{\lambda}{\pi\varepsilon_0}\ln\left(\frac{b-a}{a}\right).$$

From the given numerical values, we can solve for λ:

$$\lambda = \frac{\pi\varepsilon_0\,\Delta V}{\ln\left(\frac{b-a}{a}\right)} = \frac{4000\text{ V}}{(36 \times 10^9\text{ V·m/C})\ln\left(\frac{200}{1.5} - 1\right)}$$
$$= 22.7\text{ nC/m}.$$

(Incidentally, we have found the capacitance per unit length of a bifilar transmission line, $C/\ell = \pi\varepsilon_0 \div \ln\left(\frac{b}{a} - 1\right)$, in the limit $a \ll b$; see Section 26-4.)

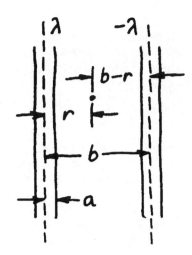

Problem 73 Solution.

Problem

75. A thin rod of length ℓ lies on the x-axis with its center at the origin. It carries a line charge density given by $\lambda = \lambda_0(x/\ell)^2$, where λ_0 is a constant. (a) Find an expression for the potential on the x-axis for $x > \ell/2$. (b) Integrate the charge density to find the total charge on the rod. (c) Show that your answer for (a) reduces to the potential of a point charge whose charge is the answer to (b), for $x \gg \ell$.

Solution

(a) For points P on the x-axis at $x > \ell/2$, the potential from a charge element $dq = \lambda\,dx'$ at x' along the rod $(-\ell/2 \le x' \le \ell/2)$ is $dV = k\,dq/|x - x'| = k\lambda\,dx' \div (x - x')$. (We used x' for the variable position of dq, and took the potential relative to zero at infinity.) The potential due to the entire rod, for $\lambda = \lambda_0(x'/\ell)^2$, is

$$V(x) = \int_{-\ell/2}^{\ell/2} dV = \frac{k\lambda_0}{\ell^2}\int_{-\ell/2}^{\ell/2}\frac{x'^2\,dx'}{(x - x')}$$

$$= \frac{k\lambda_0}{\ell^2}\left|-x^2\ln(x - x') - xx' - \frac{x'^2}{2}\right|_{-\ell/2}^{\ell/2}$$

$$= \frac{k\lambda_0}{\ell^2}\left[x^2\ln\!\left(\frac{x + \ell/2}{x - \ell/2}\right) - x\ell\right].$$

(Use partial fractions, $x'^2/(x - x') = -x - x' + x^2 \div (x - x')$, or standard tables to evaluate the integral.)
(b) The total charge on the rod is $Q = \int dq = \int_{-\ell/2}^{\ell/2}(\lambda_0/\ell^2)x^2\,dx = (\lambda_0/\ell^2)\left(\frac{2}{3}\right)(\ell/2)^3 = \lambda_0\ell/12$.
(c) For $x \gg \ell$, we can expand the logarithms, $\ln(1 \pm \ell/2x) = \pm(\ell/2x) - \frac{1}{2}(\ell/2x)^2 \pm\frac{1}{3}(\ell/2x)^3 - \dots$.

Therefore:

$$V(x) = (k\lambda_0/\ell^2)[x^2\ln(1 + \ell/2x)$$
$$- x^2\ln(1 - \ell/2x) - x\ell]$$

$$= \left(\frac{k\lambda_0}{\ell^2}\right)\left[x^2\left(\frac{\ell}{2x} - \frac{1}{2}\left(\frac{\ell}{2x}\right)^2\right.\right.$$
$$\left.+ \frac{1}{3}\left(\frac{\ell}{2x}\right)^3 - \cdots\right)$$
$$- x^2\left(-\frac{\ell}{2x} - \frac{1}{2}\left(\frac{\ell}{2x}\right)^2\right.$$
$$\left.\left.- \frac{1}{3}\left(\frac{\ell}{2x}\right)^3 - \cdots\right) - x\ell\right]$$

$$= \left(\frac{k\lambda_0}{\ell^2}\right)\left[x^2\left(\frac{\ell}{x} + \frac{2}{3}\left(\frac{\ell}{2x}\right)^3 + \cdots\right) - x\ell\right]$$

$$= \frac{k\lambda_0\ell}{12x} = kQ/x,$$

as expected.

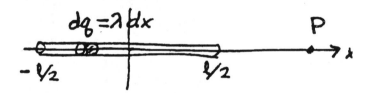

Problem 75 Solution.

Problem

77. For the situation of Example 25-10, find an equation for the equipotential with $V = 0$ in the x-y plane. Plot the equipotential, and show that it passes through the points described in Example 25-10 and its exercise.

Solution

The potential in the x-y plane, for the charges in Example 25-10, is

$$V(x,y) = \frac{k(2q)}{r_+} + \frac{k(-q)}{r_-}$$

$$= kq\left[\frac{2}{\sqrt{(x + a)^2 + y^2}} - \frac{1}{\sqrt{(x - a)^2 + y^2}}\right],$$

where r_+, r_- are the distances from a point (x, y) to the charges $+2q$ at $(-a, 0)$ and $-q$ at $(a, 0)$ respectively. The potential is zero when $r_+ = 2r_-$, or $(x + a)^2 + y^2 = 4[(x - a)^2 + y^2]$. This equation simplifies to $y^2 + x^2 - 10ax/3 + a^2 = 0$, or after completion of the square in x, to $y^2 + (x - 5a/3)^3 = (4a/3)^2$. This is the equation of a circle centered at $(5a/3, 0)$ with radius $4a/3$. The intercepts of the circle on the x-axis are at $5a/3 \pm 4a/3 = 3a$ or $a/3$, as found in Example 25-10 and the following exercise.

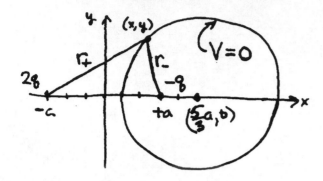

Problem 77 Solution.

Problem

79. An open-ended cylinder of radius a and length $2a$ carries charge q spread uniformly over its surface.

Find the potential on the cylinder axis at its center. *Hint:* Treat the cylinder as a stack of charged rings, and integrate.

Solution

The cylinder can be considered to be composed of rings of radius a, width dx, and charge $dq = (q/2a)\, dx$. The potential at the center of the cylinder (which we take as the origin, with x-axis along the cylinder axis) due to a ring at $x\,(-a \le x \le a)$ is $dV = k\, dq/\sqrt{x^2 + a^2}$ (see Example 25-6). The whole potential at the center follows from integration:

$$V = \int_{-a}^{a} \left(\frac{kq}{2a}\right) \frac{dx}{\sqrt{x^2 + a^2}} = \frac{kq}{2a} \ln\left(\frac{a + \sqrt{2}a}{-a + \sqrt{2}a}\right)$$

$$= \frac{kq}{a} \ln(1 + \sqrt{2}) = 0.881 \frac{kq}{a}.$$

CHAPTER 26   ELECTROSTATIC ENERGY AND CAPACITORS

Section 26-1:   Energy of a Charge Distribution

Problem

1. Three point charges, each of $+q$, are moved from infinity to the vertices of an equilateral triangle of side ℓ. How much work is required?

Solution

The sentence preceding Example 26-1 allows us to rewrite Equation 26-1 (for the electrostatic energy of a distribution of point charges) as $W = \Sigma_{\text{pairs}}\, kq_iq_j/r_{ij}$. For three equal charges (three different pairs) at the corners of an equilateral triangle ($r_{ij} = \ell$ for each pair) $W = 3kq^2/\ell$.

Problem

3. Four 50-μC charges are brought from far apart onto a line where they are spaced at 2.0-cm intervals. How much work does it take to assemble this charge distribution?

Solution

Number the charges $q_i = 50\ \mu$C, $i = 1, 2, 3, 4$, as they are spaced along the line at $a = 2$ cm intervals. There are six pairs, so $W = \Sigma_{\text{pairs}}\, kq_iq_j/r_{ij} = k(q_1q_2/a + q_1q_3/2a + q_1q_4/3a + q_2q_3/a + q_2q_4/2a + q_3q_4/a) = (kq^2/a)(1 + \frac{1}{2} + \frac{1}{3} + 1 + \frac{1}{2} + 1) = 13kq^2/3a = 13\times(9\times10^9\ \text{m/F})(50\ \mu\text{C})^2/(3\times2\ \text{cm}) = 4.88$ kJ. (See solution to Problem 1.)

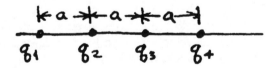

Problem 3 Solution.

Problem

5. Suppose two of the charges in Problem 1 are held in place, while the third is allowed to move freely. If this third charge has mass m, what will be its speed when it's far from the other two charges?

Solution

With one charge removed to infinity, the potential energy is reduced to that of just one pair of charges, $W_f = kq^2/\ell$. The initial potential energy was $W_i = 3kq^2/\ell$ (see Problem 1), so the kinetic energy of the charge at infinity (from the conservation of energy) is $K = W_i - W_f = 2kq^2/\ell$. Thus, $\nu = \sqrt{2K/m} = q/\sqrt{\pi\varepsilon_0 m\ell}$.

Problem

7. Four identical charges q, initially widely separated, are brought to the vertices of a tetrahedron of side a (Fig. 26-26). Find the electrostatic energy of this configuration.

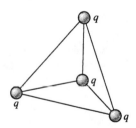

FIGURE 26-26 Problem 7.

Solution

There are six different pairs of equal charges and the separation of any pair is a. Thus, $W = \Sigma_{\text{pairs}}\, kq_iq_j \div a = 6kq^2/a$. (See Problem 1.)

Section 26-2:   Two Isolated Conductors

Problem

9. Two square conducting plates 25 cm on a side and 5.0 mm apart carry charges $\pm1.1\ \mu$C. Find (a) the electric field between the plates, (b) the potential difference between the plates, and (c) the stored energy.

Solution

(a) The electric field between two closely spaced, oppositely charged, parallel conducting plates is approximately uniform (directed from the positive to the negative plate), with strength $E = \sigma/\varepsilon_0 = q \div \varepsilon_0 A = (1.1\ \mu\text{C})/(8.85\ \text{pF/m})(0.25\ \text{m})^2 = 1.99$ MV/m.

(See the last paragraph of Section 24-6.) (b) Since E is uniform, $V = Ed = (1.99 \text{ MV/m})(5 \text{ mm}) = 9.94 \text{ kV}$. (See Section 26-2.) (c) The energy stored is $U = \frac{1}{2}q^2 d/\varepsilon_0 A = \frac{1}{2}qV = \frac{1}{2}(1.1 \ \mu\text{C})(9.94 \text{ kV}) = 5.47 \text{ mJ}$. (See Equation 26-2, and note that $U = \frac{1}{2}\varepsilon_0 E^2 Ad$.)

Problem

11. (a) How much charge must be transferred between the initially uncharged plates of the preceding problem in order to store 15 mJ of energy? (b) What will be the potential difference between the plates?

Solution

(a) From Equation 26-2, $Q = \sqrt{2\varepsilon_0 AU/d} = [2(8.85 \text{ pF/m})(5 \text{ cm})^2(15 \text{ mJ})/(1.2 \text{ mm})]^{1/2} = 0.744 \ \mu\text{C}$. (b) The argument leading to Equation 26-2 shows that $V = Qd/\varepsilon_0 A = 40.3 \text{ kV}$. Alternatively, the second expression for U in the solution to Problem 9 part (c) gives the same results, $V = 2U/Q = 2(15 \text{ mJ})/(0.744 \ \mu\text{C})$.

Problem

13. A conducting sphere of radius a is surrounded by a concentric spherical shell of radius b. Both are initially uncharged. How much work does it take to transfer charge from one to the other until they carry charges $\pm Q$?

Solution

When a charge q (assumed positive) is on the inner sphere, the potential difference between the spheres is $V = kq(a^{-1} - b^{-1})$. (See the solution to Problem 25-63(a).) To transfer an additional charge dq from the outer sphere requires work $dW = V \, dq$, so the total work required to transfer charge Q (leaving the spheres oppositely charged) is $W = \int_0^Q V \, dq = \int_0^Q kq \, dq(a^{-1} - b^{-1}) = \frac{1}{2}kQ^2(a^{-1} - b^{-1})$. (Incidentally, this shows that the capacitance of this spherical capacitor is $1/k(a^{-1} - b^{-1}) = ab/k(b - a)$; see Equation 26-8a.)

Problem

15. Two conducting spheres of radius a are separated by a distance $\ell \gg a$; since the distance is large, neither sphere affects the other's electric field significantly, and the fields remain spherically symmetric. (a) If the spheres carry equal but opposite charges $\pm q$, show that the potential difference between them is $2kq/a$. (b) Write an expression for the work dW involved in moving an infinitesimal charge dq from the negative to the

positive sphere. (c) Integrate your expression to find the work involved in transferring a charge Q from one sphere to the other, assuming both are initially uncharged.

Solution

(a) The potential difference between the two (essentially isolated) spheres is $\Delta V = kq/a - k(-q) \div a = 2kq/a$ (see Equation 25-12). (b) ΔV is the work per unit positive charge transferred between the spheres, so $dW = dq \, \Delta V = 2kq \, dq/a$. (c) The integration yields $W = \int dW = \int_0^Q 2kq \, dq/a = kQ^2/a$.

Problem

17. A car battery stores about 4 MJ of energy. If all this energy were used to create a uniform electric field of 30 kV/m, what volume would it occupy?

Solution

In a uniform field, Equation 26-4 can be written as $U = \frac{1}{2}\varepsilon_0 E^2 \times$ (Volume of field region). Therefore, the volume is $2(4 \text{ MJ})/(8.85 \text{ pF/m})(30 \text{ kV/m})^2 = 1.00 \times 10^9 \text{ m}^3 = 1 \text{ km}^3$.

Problem

19. Find the electric field energy density at the surface of a proton, taken to be a uniformly charged sphere 1 fm in radius.

Solution

For this model of the proton, the field strength at the surface is $E = ke/R^2$ (from spherical symmetry and Gauss's law). Thus, the energy density in the surface electric field is $u = \frac{1}{2}\varepsilon_0 E^2 = ke^2/8\pi R^4 \simeq (9 \times 10^9 \text{ m/F})(1.6 \times 10^{-19} \text{ C})^2/8\pi \ (1 \text{ fm})^4 = 9.17 \times 10^{30} \text{ J/m}^3 = 57.3 \text{ keV/fm}^3$.

Problem

21. The electric field strength as a function of position x in a certain region is given by $E = E_0(x/x_0)$, where $E_0 = 24 \text{ kV/m}$ and $x_0 = 6.0 \text{ m}$. Find the total energy stored in a cube 1.0 m on a side, located between $x = 0$ and $x = 1.0 \text{ m}$. (The field strength is independent of y and z.)

Solution

Since there is no y or z dependence, the volume element of the cube can be written as $dV = \ell^2 \, dx$, where $\ell = 1 \text{ m}$ is the cube's edge. Then $U = \int u \, dV = \int_0^\ell \frac{1}{2}\varepsilon_0(E_0/x_0)^2 x^2 \ell^2 \, dx = \frac{1}{2}\varepsilon_0(E_0/x_0)^2 \ell^5/3$. Numerically, $U = \frac{1}{6}(8.85 \text{ pF/m})(24 \text{ kV/m})^2(1 \text{ m})^5 \div (6 \text{ m})^2 = 23.6 \ \mu\text{J}$.

Problem

23. A sphere of radius R carries a total charge Q distributed over its surface. Show that the total energy stored in its electric field is $U = kQ^2/2R$.

Solution

The calculation of the electrostatic energy for a sphere with uniform surface charge density is, in fact, given in Example 26-3. We simply set $R_2 = R$, the radius of the sphere, and $R_1 = \infty$ (so the integral covers all the space where the field is non-zero).

Problem

25. Two 4.0-mm-diameter water drops each carry 15 nC. They are initially separated by a great distance. Find the change in the electrostatic potential energy if they are brought together to form a single spherical drop. Assume all charge resides on the drops' surfaces.

Solution

The initial electrostatic energy of two isolated spherical drops, with charge Q on their surfaces and radii R, is $U_i = 2(\frac{1}{2}kQ^2/R)$ (see Problem 23 and Example 26-3). Together, a drop of charge $2Q$, radius $2^{1/3}R$, and energy $U_f = \frac{1}{2}k(2Q)^2/(2^{1/3}R) = 2^{2/3}kQ^2/R$, is created. The work required is the difference in energy, $W = U_f - U_i = (2^{2/3} - 1)kQ^2 \div R = (0.587)(9\times10^9 \text{ m/F})(1.5\times10^{-8} \text{ C})^2 \div (2\times10^{-3} \text{ m}) = 5.95\times10^{-4}$ J.

Problem

27. A long, solid rod of radius a carries uniform volume charge density ρ. Find an expression for the electrostatic energy per unit length contained *within* the rod. *Hint:* See Problem 24-31.

Solution

The electric field within such a rod (assumed to have line symmetry) is radially away from the axis with magnitude $E_r = \rho r/2\varepsilon_0$ (see Problem 24-31). The energy density on a cylindrical shell of radius r, length ℓ, and volume $dV = 2\pi r\ell\, dr$, is $u = \frac{1}{2}\varepsilon_0 E_r^2 = \rho^2 r^2/8\varepsilon_0$. Hence, the energy per unit length inside the rod is $U/\ell = \ell^{-1}\int_{\text{rod}} u\, dV = (\pi\rho^2/4\varepsilon_0)\int_0^a r^3\, dr = \pi\rho^2 a^4/16\varepsilon_0$.

Problem

29. The "memory" capacitor in a VCR has a capacitance of 4.0 F and is charged to 3.5 V. What is the charge on its plates?

Solution

The definition of capacitance (Equation 26-5) gives the magnitude of the charge on either plate, $Q = CV = 4.0 \text{ F} \times 3.5 \text{ V} = 14.0$ C. (This is a very large capacitor.)

Problem

31. Figure 26-27 shows data from an experiment in which known amounts of charge are placed on a capacitor and the resulting voltage measured. Fit a line to the data, and use it to determine the capacitance.

Solution

Since $V = Q/C$, the slope of the best straight line through the data points is the inverse of the capacitance, or $C = (Q_2 - Q_1)/(V_2 - V_1)$. An "eyeball fit" to Fig. 26-27 passes through the origin and (12 mC, 1.85 V), so $C \approx 12 \text{ mC}/1.85 \text{ V} \approx 6.5$ mF.

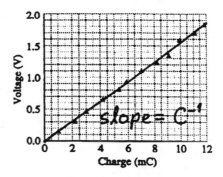

FIGURE 26-27 Problem 31 Solution.

Problem

33. Find the capacitance of a parallel-plate capacitor consisting of circular plates 20 cm in radius separated by 1.5 mm.

Solution

For a (closely spaced) parallel plate capacitor, with circular plates, Example 26-4 shows that $C = \varepsilon_0\pi r^2 \div d = (8.85 \text{ pF/m})\pi(20 \text{ cm})^2/(1.5 \text{ mm}) = 741$ pF.

Problem

35. Find the capacitance of a 1.0-m-long piece of coaxial cable whose inner conductor radius is 0.80 mm and whose outer conductor radius is 2.2 mm, with air in between.

Solution

The capacitance of air-filled ($\kappa = 1$) cylindrical capacitor was found in Example 26-5: $C = 2\pi\varepsilon_0\ell \div \ln(b/a) = 2\pi(8.85 \text{ pF/m})(1 \text{ m})/\ln(2.2/0.8) = 55.0$ pF.

Problem

37. Figure 26-28 shows a capacitor consisting of two electrically connected plates with a third plate between them, spaced so its surfaces are a distance d from the other plates. The plates have area A. Neglecting edge effects, show that the capacitance is $2\varepsilon_0 A/d$.

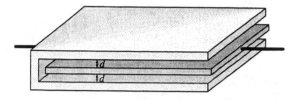

FIGURE 26-28 Problem 37.

Solution

When the third (middle) plate is positively charged, the electric field (not near an edge) is approximately uniform and away from the plate, with magnitude $E = \sigma/\varepsilon_0$. Since half of the total charge Q is on either side (by symmetry), $\sigma = Q/2A$. The potential difference between the third plate and the outer two plates (which are both at the same potential and carry charges of $-Q/2$ on their inner surfaces) is $V = Ed = \sigma d/\varepsilon_0 = Qd/2\varepsilon_0 A$. Therefore the capacitance is $C = Q/V = 2\varepsilon_0 A/d$. (The arrangement is like two capacitors in parallel.)

Problem

39. Find the capacitance of a capacitor that stores 350 μJ when the potential difference across its plates is 100 V.

Solution

Equation 26-8b relates the capacitance, voltage, and energy stored in a capacitor, so $C = 2U_C/V^2 = 2(350\ \mu\text{J})/(100\ \text{V})^2 = 70$ nF.

Problem

41. Which can store more energy, a 1-μF capacitor rated at 250 V or a 470 pF capacitor rated at 3 kV?

Solution

The first capacitor stores $U_C = \frac{1}{2}(1\ \mu\text{F})(250\ \text{V})^2 = 31.3$ m of energy, while the second only $\frac{1}{2}(470\ \text{pF}) \times (3\ \text{kV})^2 = 2.12$ mJ, about 14.8 times less. (See Equation 26-8b.)

Problem

43. A 0.01-μF, 300-V capacitor costs 25¢, a 0.1-μF, 100-V capacitor costs 35¢, and a 30-μF, 5-V capacitor costs 88¢. (a) Which can store the most charge? (b) Which can store the most energy? (c) Which is the most effective energy storage device, as measured by energy stored per unit cost?

Solution

(a) $Q = CV = (0.01\ \mu\text{F})(300\ \text{V}) = 3\ \mu$C for the first capacitor, 10 μC for the second, and 150 μC for the third. (b) $U = \frac{1}{2}QV$ (or $\frac{1}{2}CV^2$) $= \frac{1}{2}(3\ \mu\text{C})(300\ \text{V}) = 450\ \mu$J for the first, 500 μJ for the second, and 375 μJ for the third. (c) The cost effectiveness, measured in J/¢, is 18.0, 14.3, and 4.26 for these capacitors, respectively.

Problem

45. A camera flashtube requires 5.0 J of energy per flash. The flash duration is 1.0 ms. (a) What is the power used by the flashtube *while it is actually flashing*? (b) If the flashtube operates at 200 V, what size capacitor is needed to supply the flash energy? (c) If the flashtube is fired once every 10 s, what is its *average* power consumption?

Solution

(a) $\mathcal{P}_{\text{flash}} = W/t = 5\ \text{J}/1\ \text{ms} = 5$ kW. (b) $U = \frac{1}{2}CV^2$, so $C = 2U/V^2 = 2(5\ \text{J})/(200\ \text{V})^2 = 250\ \mu$F. (c) $\mathcal{P}_{\text{av}} = 5\ \text{J}/10\ \text{s} = 0.5$ W, only 10^{-4} times $\mathcal{P}_{\text{flash}}$.

Problem

47. A solid conducting slab is inserted between the plates of a charged capacitor, as shown in Fig. 26-29. The slab thickness is 60% of the plate spacing, and its area is the same as the plates. (a) What happens to the capacitance? (b) What happens to the stored energy, assuming the capacitor is not connected to anything?

Solution

(a) The charge on the plates remains the same, and so does the electric field ($E = \sigma/\varepsilon_0$) in the gaps between either plate and the slab. However, the separation (i.e., the thickness of the field region) between the plates is reduced to 40% of its original value $d' = d_1 + d_2 = 0.4d$, therefore the capacitance is increased, $C' = \varepsilon_0 A/d' = \varepsilon_0 A/0.4d = 2.5\ C$. (The equations $V = E\ell$ and $C = Q/V$ lead to the same result.) In fact, the configuration behaves like a series combination of two parallel plate capacitors, $1/C' = C_1^{-1} + C_2^{-1} = (d_1/\varepsilon_0 A) + (d_2/\varepsilon_0 A) = (d_1 + d_2)/\varepsilon_0 A = 0.4d/\varepsilon_0 A =$

1/2.5 C. (b) When the charge is constant (no connections to anything isolates the system), the energy stored is inversely proportional to the capacitance, $U = Q^2/2C$. Thus $U' = Q^2/2C' = Q^2 \div 2(2.5C) = 0.4U$, or the energy decreases to 40% of its original value. (With the slab inserted, there is less field region and less energy stored. While the slab is being inserted, work is done by electrical forces to conserve energy.)

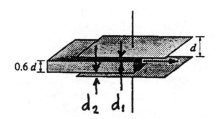

FIGURE 26-29 Problem 47 Solution.

Problem

49. The cylindrical capacitor of Example 26-5 is charged to a voltage V. Obtain an expression for the energy density as a function of radial position in the capacitor, and integrate to show explicitly that the stored energy is $\frac{1}{2}CV^2$.

Solution

The electric field in the capacitor is approximately $\mathbf{E} = \lambda\hat{\mathbf{r}}/2\pi\varepsilon_0 r$, where $\hat{\mathbf{r}}$ is the radial unit vector in cylindrical coordinates (see Example 25-4). (The assumption of line symmetry neglects fringing fields at the ends of the capacitor.) The energy density is $u = \frac{1}{2}\varepsilon_0 E^2 = \lambda^2/8\pi^2\varepsilon_0 r^2$. We can take the volume element to be a cylindrical shell of radius r, thickness dr, and length L, so $dV = 2\pi r L\, dr$. Then the stored energy is

$$U = \int u\, dV = \int_a^b \frac{\lambda^2 2\pi r L\, dr}{8\pi^2\varepsilon_0 r^2} = \frac{\lambda^2 L}{4\pi\varepsilon_0}\int_a^b \frac{dr}{r}$$
$$= \frac{\lambda^2 L}{4\pi\varepsilon_0}\ln\left(\frac{b}{a}\right).$$

Reference to Example 26-5 shows that this is precisely $\frac{1}{2}CV^2$.

Problem

51. Two capacitors are connected in series and the combination charged to 100 V. If the voltage across each capacitor is 50 V, how do their capacitances compare?

Solution

For capacitors in series, the total voltage is the sum of the voltages across each one, $V = V_1 + V_2$, whereas the charge on each capacitor is the same, $Q_1 = Q_2 = C_1 V_1 = C_2 V_2$. Thus, $V = V_1 + (C_1/C_2)V_1$, or $V_1 = VC_2/(C_1 + C_2)$, and similarly $V_2 = VC_1/(C_1 + C_2)$ (a general result). If $V_1 = V_2 = \frac{1}{2}V$ as in this problem, either equation implies $C_1 = C_2$.

Problem

53. You're given three capacitors: $1.0\ \mu F, 2.0\ \mu F$, and $3.0\ \mu F$. Find (a) the maximum, (b) the minimum, and (c) two intermediate values of capacitance you could achieve with various combinations of all three capacitors.

Solution

The capacitors can be connected (a) all in parallel: $1 + 2 + 3 = 6\ \mu F$; (b) all in series: $1/1 + 1/2 + 1/3 = 11/6$, or $6/11 = 0.545\ \mu F$; (c) one in parallel with the other two in series:

$$1 + \frac{2\times3}{2+3} = \frac{11}{5} = 2.20\ \mu F,$$
$$2 + \frac{1\times3}{1+3} = \frac{11}{4} = 2.75\ \mu F,$$
$$3 + \frac{1\times2}{1+2} = \frac{11}{3} = 3.67\ \mu F,$$

or one in series with the other two in parallel:

$$\frac{1(2+3)}{1+2+3} = \frac{5}{6} = 0.933\ \mu F,$$
$$\frac{2(1+3)}{2+1+3} = \frac{4}{3} = 1.33\ \mu F,$$
$$\frac{3(1+2)}{3+1+2} = \frac{3}{2} = 1.50\ \mu F.$$

Problem

55. You have an unlimited supply of 2.0-μF, 50-V capacitors. Describe combinations that would be equivalent to (a) a 2.0-μF, 100-V capacitor and (b) a 0.50-μF, 200-V capacitor.

Solution

In parallel, the voltage across each element is the same, so to increase the voltage rating of a combination of equal capacitors, series connections

Problem 55 Solution.

must be considered. The general result of Problem 51 shows that for two equal capacitors in series, the voltage across each is one half the total, so the voltage rating of a series combination is doubled. Thus, in part (a) we must use two capacitors in series (rating 50 V + 50 V = 100 V, and capacitance $C = (2\ \mu\text{F}) \times (2\ \mu\text{F})/(2\ \mu\text{F} + 2\ \mu\text{F}) = 1\ \mu\text{F}$), while in part (b), four capacitors in series are required (rating 200 V, and capacitance $C^{-1} = 4(2\ \mu\text{F})^{-1}$ or $C = \frac{1}{2}\ \mu\text{F}$). In part (b), one series combination of four capacitors is sufficient, but in part (a), we need to increase the total capacitance to twice that of just two in series, without altering the voltage rating. This can be accomplished with a parallel combination of two pairs in series, i.e., a parallel combination of two 1 μF, 100 V series pairs. (Note that for equal capacitor elements, a parallel combination of two pairs in series has the same properties as a series combination of two pairs in parallel.) Schematically the connections described look like the following.

Problem

57. What is the equivalent capacitance in Fig. 26-32?

Solution

Number the capacitors as shown. Relative to points A and B, C_1, C_4 and the combination of C_2 and C_3 are in series, so the capacitance is given by $C_{AB}^{-1} = C_1^{-1} + C_4^{-1} + C_{23}^{-1}$. C_{23} is a parallel combination, hence $C_{23} = C_2 + C_3$, therefore $C_{AB}^{-1} = (3\ \mu\text{F})^{-1} + (2\ \mu\text{F})^{-1} + (2\ \mu\text{F} + 1\ \mu\text{F})^{-1}$, or $C_{AB} = \frac{6}{7}\ \mu\text{F} = 0.857\ \mu\text{F}$.

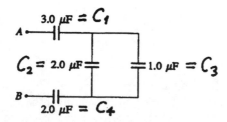

FIGURE 26-32 Problem 57 Solution.

Problem

59. Two capacitors C_1 and C_2 are in series, with a voltage V across the combination. Show that the voltages across the individual capacitors are

$$V_1 = \frac{C_2 V}{C_1 + C_2} \quad \text{and} \quad V_2 = \frac{C_1 V}{C_1 + C_2}.$$

Solution

This is shown in the solution to Problem 51.

Problem

61. A variable "trimmer" capacitor used to make fine adjustments has a capacitance range from 10 to 30 pF. The trimmer is in parallel with a capacitor of about 0.001 μF. Over what percentage range can the capacitance of the combination be varied?

Solution

"Capacitors in parallel add" (Equation 26-9a), so the combination covers a range from 1010 to 1030 pF, or about $\pm 10/1020 \approx \pm 1\%$ from the central value.

Problem

63. A 5.0-μF capacitor is charged to 50 V, and a 2.0-μF capacitor is charged to 100 V. The two are disconnected from their charging batteries and connected in parallel, positive to positive.
(a) What is the common voltage across each after they are connected? *Hint:* Charge is conserved.
(b) Compare the total electrostatic energy before and after the capacitors are connected. Speculate on the discrepancy.

Solution

(a) The charge on the parallel combination is the sum of the original charges, $Q_{\parallel} = Q_1 + Q_2 = C_1 V_1 + C_2 V_2 = (5\ \mu\text{F})(50\ \text{V}) + (2\ \mu\text{F})(100\ \text{V}) = 450\ \mu\text{C}$, while the capacitance is $C_{\parallel} = C_1 + C_2 = 7\ \mu\text{F}$. Thus, the voltage is $V_{\parallel} = Q_{\parallel}/C_{\parallel} = 450\ \mu\text{C}/7\ \mu\text{F} = 64.3\ \text{V}$.
(b) The total energy stored in both capacitors before they are connected is $\frac{1}{2}C_1 V_1^2, +\frac{1}{2}C_2 V_1^2 = \frac{1}{2}(5\ \mu\text{F})(50\ \text{V})^2 + \frac{1}{2}(2\ \mu\text{F})(100\ \text{V})^2 = 16.3\ \text{mJ}$. After the connection, $U_{\parallel} = \frac{1}{2}C_{\parallel} V_{\parallel}^2 = \frac{1}{2}(7\ \mu\text{F})(64.3\ \text{V})^2 = 14.5\ \text{mJ}$, a difference of 1.79 mJ. It takes work to redistribute the original charges when the capacitors are connected. (The new charges are $Q_1' = (5\ \mu\text{F})(64.3\ \text{V}) = 321\ \mu\text{C}$, and $Q_2' = 129\ \mu\text{C}$, respectively.)

Problem

65. A 470-pF capacitor consists of two circular plates 15 cm in radius, separated by a sheet of polystyrene. (a) What is the thickness of the sheet? (b) What is the working voltage?

Solution

(a) With reference to Equations 26-6, 26-11, and Table 26-1, one finds that $C = \kappa C_0 = \kappa \varepsilon_0 A/d$, or $d = \kappa \varepsilon_0 A/C = (2.6)(8.85\ \text{pF/m})\pi(0.15\ \text{m})^2/470\ \text{pF} = 3.46\ \text{mm}$. (Since this is much less than the radius of the plates, the parallel plate approximation (plane symmetry) is a good one.) (b) The dielectric breakdown field for polystyrene is $E_{\max} = 25\ \text{kV/mm}$,

so the maximum voltage for this capacitor is
$V_{max} = E_{max}d = (25 \text{ kV/mm})(3.46 \text{ mm}) = 86.5 \text{ kV}$.
(Note: in practice, the working voltage would be less
than this by a comfortable safety margin.)

Problem

67. Repeat Problem 35 for the more realistic case of a
    cable insulated with polyethylene.

Solution

With polyethylene insulation occupying the space
between the conductors, instead of air, the capacitance
is increased by a factor of $\kappa = 2.3$ (see Equation 26-11
and Table 26-1). Thus, $C = \kappa C_0 = (2.3)(55.0 \text{ pF}) =$
126 pF, where we used the value of C_0 from the
solution to Problem 35.

Problem

69. The capacitor of the preceding problem is
    connected to its 900-V charging battery and left
    connected as the plexiglass sheet is inserted, so the
    potential difference remains at 900 V. What are
    (a) the charge on the plates and (b) the stored
    energy both before and after the plexiglass is
    inserted?

Solution

(a) The capacitances before and after the insertion of
the plexiglass insulation are $C_0 = \varepsilon_0 A/d =$
$(8.85 \text{ pF/m})(76 \text{ cm}^2)/(1.2 \text{ mm}) = 56.1 \text{ pF}$, and $C =$
$\kappa C_0 = (3.4)(56.1 \text{ pF}) = 191 \text{ pF}$, as found previously.
Therefore, since the voltage stays at 900 V in this case
(due to the battery), $Q_0 = C_0(900 \text{ V}) = 50.4 \text{ nC}$, and
$Q = C(900 \text{ V}) = \kappa Q_0 = 172 \text{ nC}$, before and after
insertion, respectively. (b) The stored energy is $U_0 =$
$\frac{1}{2}C_0(900 \text{ V})^2 = 22.7 \text{ μJ}$ before, and $U = \frac{1}{2}C(900 \text{ V})^2 =$
$\kappa U_0 = 77.2 \text{ μJ}$ after. (The difference between this
situation and the one in the previous problem is that
the battery does additional work moving more charge
to the capacitor plates, while maintaining the constant
voltage. Equation 26-12 applies to an isolated
capacitor only.)

Paired Problems

Problem

71. A pair of parallel conducting plates of area
    0.025 m^2 carrying equal but opposite charges
    stores 1.6 J in its electric field. When the
    magnitude of the charge on both plates is
    increased by 5.0 μC, the stored energy increases
    to 2.4 J. Find the plate separation.

Solution

Using Equation 26-8a, we can write $U_1 = Q_1^2/2C =$
1.6 J, and $U_2 = (Q_1 + 5 \text{ μC})^2/2C = 2.4$ J, from which
Q_1 can be eliminated and C can be found: $\sqrt{2CU_2} -$
$\sqrt{2CU_1} = 5 \text{ μC}$, or $C = [5 \text{ μC}/(\sqrt{2 \times 2.4 \text{ J}} -$
$\sqrt{2 \times 1.6 \text{ J}})]^2 = 155$ pF. For an air-insulated parallel
plate capacitor, Equation 26-6 then gives $d = \varepsilon_0 A/C =$
$(8.55 \text{ pF/m})(0.025 \text{ m}^2)/155 \text{ pF} = 1.43$ mm.

Problem

73. A 20-μF air-insulated parallel-plate capacitor is
    charged to 300 V. The capacitor is then
    disconnected from the charging battery, and its
    plate separation is doubled. Find the stored
    energy (a) before and (b) after the plate
    separation increases. Where does the extra energy
    come from?

Solution

(a) Initially, the stored energy is $U_0 = \frac{1}{2}C_0 V_0^2 =$
$\frac{1}{2}(20 \text{ μF})(300 \text{ V})^2 = 0.9$ J. (b) Disconnected from the
battery, the charge stays constant, but the capacitance
is halved when the separation is doubled ($C = \varepsilon_0 A \div$
$2d = C_0/2$). Therefore, the stored energy is doubled,
since $U = Q^2/2C = Q^2/2(C_0/2) = 2U_0 = 1.8$ J. Work
must be done, against the attractive force between the
oppositely charged plates, to increase their separation.

Problem

75. In the capacitor network of Fig. 26-33, take
    $C = 6.0$ μF. Find (a) the equivalent capacitance
    between A and B and (b) the charge on C when
    30 V is applied between A and B.

Solution

(a) The 2 μF capacitor is in series with the parallel
combination of the 1 μF capacitor and the series
combination of the 3 μF and 6 μF capacitors (see the
numbering added to the figure). Therefore, the total
capacitance between A and B is

$$C_{tot} = \frac{C_1 C_{\parallel}}{C_1 + C_{\parallel}} = \frac{C_1(C_2 + C_{34})}{C_1 + C_2 + C_{34}}$$

$$= \frac{C_1\left(C_2 + \dfrac{C_3 C_4}{C_3 + C_4}\right)}{C_1 + C_2 + \dfrac{C_3 C_4}{C_3 + C_4}}$$

$$= \frac{2 \text{ μF}\left(1 \text{ μF} + \dfrac{3 \text{ μF} \times 6 \text{ μF}}{3 \text{ μF} + 6 \text{ μF}}\right)}{2 \text{ μF} + \left(1 \text{ μF} + \dfrac{3 \times 6 \text{ μF}}{3 + 6}\right)} = \frac{2 \times 3}{5} \text{ μF}$$

$$= 1.2 \text{ μF}.$$

(b) When 30 V is applied across A and B, the voltage across the parallel combination (whose capacitance is $1\ \mu\text{F} + 3\times6\ \mu\text{F}/(3+6) = 3\ \mu\text{F}$) is $(30\ \text{V})2/(2+3) = 12\ \text{V}$ (since this is in series with the $2\ \mu\text{F}$ capacitor—see the result of Problem 59). A second application of this result (i.e., Problem 59) to the series combination of the $3\ \mu\text{F}$ and $6\ \mu\text{F}$ capacitor gives $V = (12\ \text{V})3/(3+6) = 4\ \text{V}$ for the voltage across C. Then $Q = CV = (6\ \mu\text{F})(4\ \text{V}) = 24\ \mu\text{C}$.

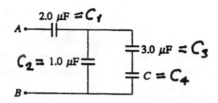

FIGURE 26-33 Problem 75 Solution.

Supplementary Problems

Problem

77. A typical lightning flash transfers 30 C across a potential difference of 30 MV. Assuming such flashes occur every 5 s in the thunderstorm of Example 26-2, roughly how long could the storm continue if its electrical energy were not replenished?

Solution

The energy in the thunderstorm of Example 26-2 was about 1.4×10^{11} J, while the energy in a lightning flash is $qV = (30\ \text{C})(30\ \text{MV}) = 9\times10^{8}$ J. Thus, there is energy for about $1.4\times10^{11}/9\times10^{8} = 156$ flashes, which at a rate of one flash in 5 s, would last for $156\times5\ \text{s} = 13\ \text{min}$.

Problem

79. Six charges $\pm q$, initially widely separated, are positioned to form a hexagon of side a, as shown in Fig. 26-35. What is the electrostatic energy of this configuration?

Solution

The electrostatic energy is $U = \Sigma_{\text{pairs}}\ kq_iq_j/r_{ij}$ (see solution to Problem 1). For the six charges at the corners of a regular hexagon, of side a, shown in Fig. 26-32, there are a total of 15 different pairs: 6 pairs of opposite charges separated by distance a, 6 pairs of equal charges separated by $\sqrt{3}a$, and 3 pairs of opposite charges separated by $2a$ (see geometry

added to Fig. 26-32). Thus,

$$U = kq^2\left(-\frac{6}{a} + \frac{6}{\sqrt{3}a} - \frac{3}{2a}\right) = \frac{kq^2}{a}\left(2\sqrt{3} - \frac{15}{2}\right)$$

$$= -4.04\frac{kq^2}{a}.$$

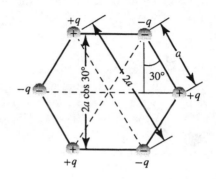

FIGURE 26-35 Problem 79 Solution.

Problem

81. An air-insulated parallel-plate capacitor of capacitance C_0 is charged to voltage V_0 and then disconnected from the charging battery. A slab of material with dielectric constant κ, whose thickness is essentially equal to the capacitor spacing, is then inserted halfway into the capacitor (Fig. 26-36). Determine (a) the new capacitance, (b) the stored energy, and (c) the force on the slab in terms of C_0, V_0, κ, and the capacitor plate length L.

Solution

(a) In so far as fringing fields can be neglected, the electric field between the plates is uniform, $E = V/d$ (but when the dielectric is inserted, $V \neq V_0$ and E depends on x). In fact, on the left side, where the slab has penetrated, $E = (1/\kappa)(\sigma_\ell/\varepsilon_0)$, and on the right, $E = \sigma_r/\varepsilon_0$, where σ_ℓ and σ_r are the charge densities on the left and right sides. Thus, $\sigma_\ell = \kappa\varepsilon_0 E$ and $\sigma_r = \varepsilon_0 E$, and the charge can be written (in terms of geometrical variables superposed on Fig. 26-36) as $q = \sigma_\ell wx + \sigma_r w(L - x) = \varepsilon_0 Ew(\kappa x + L - x) = \varepsilon_0(V/d)w\times(\kappa x + L - x)$. From Equation 26-5, $C = q/V = C_0(\kappa x + L - x)/L$, where $C_0 = \varepsilon_0 A/d$ and $A = Lw$. Although the question specifies $x = \frac{1}{2}L$, for which value the capacitance is $\frac{1}{2}C_0(\kappa + 1)$, we give C as a function of x, because we will need to differentiate with respect to x in part (c). (b) When the battery is disconnected, the capacitor is isolated and the charge on it is a constant, $q = q_0$. The stored energy is (Equation 26-8a) $U = q^2/2C = q_0^2 L/2C_0(\kappa x + L - x) = U_0 L/(\kappa x + L - x)$, where $U_0 = \frac{1}{2}q_0^2/C_0 = \frac{1}{2}C_0 V_0^2$. For

$x = \frac{1}{2}L$, the energy is $C_0V_0^2/(\kappa + 1)$. (c) The force on a part of an isolated system is related to the potential energy of the system by Equation 8-9. The force on the slab is therefore

$$F_x = -\frac{dU}{dx} = -\frac{d}{dx}\left(\frac{U_0L}{\kappa x + L - x}\right) = \frac{U_0L(\kappa - 1)}{(\kappa x + L - x)^2},$$

in the direction of increasing x (so as to pull the slab into the capacitor). For $x = \frac{1}{2}L$, the magnitude of the force is $2C_0V_0^2(\kappa - 1)/L(\kappa + 1)^2$. It turns out that if we rewrite the force, for any value of x, in terms of the voltage for that x, using $q_0 = C_0V_0 = CV = C_0V(\kappa x + L - x)/L$, the expression can be used in the succeeding problem. Thus,

$$F_x = \frac{C_0V_0^2L(\kappa - 1)}{2(\kappa x + L - x)^2} = \frac{C_0}{2}\left(\frac{V}{L}\right)^2 L(\kappa - 1)$$
$$= \frac{C_0V^2(\kappa - 1)}{2L}.$$

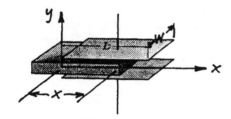

FIGURE 26-36 Problems 81 and 82 Solution.

Problem

83. We live inside a giant capacitor! Its plates are Earth's surface and the ionosphere, a conducting layer of the atmosphere beginning at about 60 km altitude. (a) What is its capacitance? *Hint:* You can treat it as either a spherical or a parallel-plate capacitor. Why? (b) The potential difference between Earth and ionosphere is about 6 MV. Find the total energy stored in this planetary capacitor.

Solution

(a) Since the radius of the Earth, $R_E = 6370$ km $= a$, is much larger than the altitude of the ionosphere, 60 km $= b - a = d$ (which is the separation of the plates in this planetary capacitor), the result of Problem 80 shows that either the spherical or parallel plate expressions for the capacitance are approximately the same. Thus, $C \approx \varepsilon_0 4\pi(6370 \text{ km})^2 \div$ 60 km $= 75.2$ mF. (b) Then $U = \frac{1}{2}CV^2 =$ $\frac{1}{2}(75.2 \text{ mF})(6 \text{ MV})^2 = 1.35 \times 10^{12}$ J.

Problem

85. Equation 26-2 gives the potential energy of a pair of oppositely charged plates. (a) Differentiate this expression with respect to the plate spacing to find the magnitude of the attractive force between the plates. (b) Compare with the answer you would get by multiplying one plate's charge by the electric field between the plates. Why do your answers differ? Which is right?

Solution

(a) Equation 26-2 gives the potential energy of two isolated oppositely charged plates, $U(x) = Q^2x/2\varepsilon_0A$, where x is their separation. Equation 8-9, $F_x = -dU/dx$, implies an attractive force of $F_x = -Q^2 \div 2\varepsilon_0A$ acting between the plates. (b) Multiplying the charge by the total electric field between the plates gives one $Q(\sigma/\varepsilon_0) = Q^2/\varepsilon_0A$, an expression equal to twice the magnitude of the force. The total field includes the field of both plates, whereas the force on one plate depends on only the field of the other plate. (In general, the force per unit area on the surface charge distribution on a conductor is $\frac{1}{2}\sigma E = \sigma^2/2\varepsilon_0$ for the same reason.)

Problem

87. A small dipole lies on the x axis, centered at the origin. Find an expression for the total electrostatic energy contained in a thin cylindrical volume of diameter d and length ℓ, with its left end a distance ℓ from the dipole center, as shown in Fig. 26-37. Assume that ℓ is much greater than the dipole spacing. *Hint:* Since the cylinder is very thin, you can use the on-axis dipole field (Equation 23-5b) for the field throughout the cylinder.

Solution

The field on the x axis from the dipole is approximately $E_x = 2kp/x^3$, so the energy density is approximately $U = \frac{1}{2}\varepsilon_0E^2 = kp^2/2\pi x^6$. The volume element of the cylinder can be taken to be a disk of area $\pi d^2/4$ and thickness dx. Thus,

$$U = \int u\, dV = \int_\ell^{2\ell}\left(\frac{\pi d^2}{4}\right)\left(\frac{kp^2}{2\pi}\right)\frac{dx}{x^6}$$
$$= \frac{kp^2d^2}{8}\left.\left|-\frac{1}{5x^5}\right|\right._\ell^{2\ell} = \frac{kp^2d^2}{40\ell^5}\left(1 - \frac{1}{2^5}\right)$$
$$= 31kp^2d^2/1280\ell^5.$$

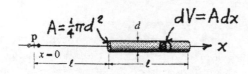

FIGURE 26-37 Problem 87 Solution.

Problem

89. A TV antenna cable consists of two 0.50-mm-diameter wires spaced 12 mm apart. Estimate the capacitance per unit length of this cable, neglecting dielectric effects of the insulation.

Solution

The capacitance per unit length for a bifilar cable, in air, when the diameter of the wires is small compared to their separation, is calculated in the solution to Problem 25-73:

$$\frac{C}{L} = \frac{\lambda}{\Delta V} = \frac{\pi \varepsilon_0}{\ln[(b/a) - 1]}$$
$$= \frac{\pi (8.85 \text{ pF/m})}{\ln[(12/0.25) - 1]} = 7.22 \text{ pF/m}.$$

Problem

91. Use the fact that the static electric field is conservative to argue that there *must* be fringing field at the edges of a parallel plate capacitor. *Hint:* Remember that the plates are equipotentials, and consider the potential

differences V_{AB} and V_{CD} in Fig. 26-38. What does your argument say about the strength of the fringing field relative to the field between the plates?

Solution

The potential difference between the plates, via path $A \rightarrow B$ is $V_{AB} = -\int_{A \rightarrow B} \mathbf{E} \cdot d\ell \neq 0$, since the field is non-zero and parallel to $d\ell$. If there were no fringing field, then the integral of $\mathbf{E}$ over path $C \rightarrow D$ would vanish, $V_{CD} = -\int_{C \rightarrow D} \mathbf{E} \cdot d\ell = 0$, in contradiction to the path-independence of a conservative field. Since the plates are equipotentials, V_{AB} must equal V_{CD}. If we choose a path along an electric field line, we can define an average field strength by $\int \mathbf{E} \cdot d\ell = \int E d\ell = E_{av}\ell$. It is clear that the average field strength is weaker along the longer field line between the same two points.

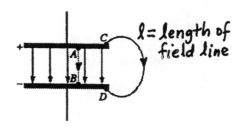

FIGURE 26-38 Problem 91 Solution.

CHAPTER 27 ELECTRIC CURRENT

Section 27-1: Electric Current

Problem

1. A wire carries 1.5 A. How many electrons pass through the wire in each second?

Solution

The current is the amount of charge passing a given point in the wire, per unit time, so in one second, $\Delta q = I\Delta t = (1.5 \text{ A})(1 \text{ s}) = 1.5$ C. The number of electrons in this amount of charge is $1.5 \text{ C}/1.6\times10^{-19}$ C $= 9.38\times10^{18}$.

Problem

3. A 12-V car battery is rated at 80 ampere-hours, meaning it can supply 80 A of current for 1 hour before it becomes discharged. If you accidentally leave the headlights on until the battery discharges, how much charge moves through the lights?

Solution

A battery rated at 80 A·h can supply a net charge of $\Delta q = I\Delta t = (80 \text{ C/s})(3600 \text{ s}) = 2.88\times10^5$ C.

Problem

5. Microbiologists measure total current due to potassium ions (K^+) moving through a cell membrane of a rock crab neuron cell to be 30 nA. How many ions pass through the membrane each second?

Solution

The charge moving through the membrane each second is 30 nC. Since singly-charged ions carry one elementary charge (about 160 zC), this corresponds to 30 nC/(160 zC/ion) $= (1.88\times10^{11}$ ion)/(6.02×10^{23} ion/mol) $= 0.311$ pmol. (See Example 27-2(a), Appendix B, and Table 1-1.) Chemical drug-testing instrumentation can detect amounts of substances this low.

Problem

7. The National Electrical Code specifies a maximum current of 10 A in 16-gauge (0.129 cm diameter) copper wire. What is the corresponding current density?

Solution

The cross-section of a wire is uniform, so Equation 27-3a gives $J = I/\frac{1}{4}\pi d^2 = 10$ A$/\frac{1}{4}\pi(0.129 \text{ cm})^2 = 7.65$ MA/m^2.

Problem

9. What is the drift speed in a silver wire carrying a current density of 150 A/mm^2? Each silver atom contributes 1.3 free electrons.

Solution

Calculating the density of conduction electrons as in Example 27-1, and using Equation 27-3a for the drift speed, we find $n = (1.3)(10.5\times10^3 \text{ kg/m}^3)\div (107.87 \text{ u/ion})(1.66\times10^{-27} \text{ kg/u}) = 7.62\times10^{28}$ m^{-3}, and $v_d = J/ne = (150 \text{ A/mm}^2)/(7.62\times10^{28} \text{ m}^{-3})\times (1.6\times10^{-19} \text{ C}) = 1.23$ cm/s.

Problem

11. A gold film in an integrated circuit measures 2.5 μm thick by 0.18 mm wide. It carries a current density of 6.8×10^5 A/m^2. What is the total current?

Solution

(Assuming the current density is perpendicular to the cross-sectional area of the film) we find, from Equation 27-3a, $I = JA = (6.8\times10^5 \text{ A/m}^2)(2.5 \mu\text{m}\times 0.18 \text{ mm}) = 306 \mu$A.

Problem

13. A plasma used in fusion research contains 5.0×10^{18} electrons and an equal number of protons per cubic meter. Under the influence of an electric field the electrons drift in one direction at 40 m/s, while the protons drift in the opposite direction at 6.5 m/s. (a) What is the current density? (b) What fraction of the current is carried by the electrons?

Solution

The proton current density is $J_p = nev_{d,p} = (5.0\times10^{18} \text{ m}^{-3})(1.6\times10^{-19} \text{ C})(6.5 \text{ m/s}) = 5.20$ A/m^2 (positive in the direction of $v_{d,p}$), and the electron current density is $J_e = n(-e)v_{d,e} = (5.0\times10^{18} \text{ m}^{-3})\times$

$(-1.6\times10^{-19}$ C)$(-40$ m/s$) = 32.0$ A/m^2. (a) The total current density is $J_p + J_e = 37.2$ A/m^2, and (b) the fraction carried by electrons is $32.0/37.2 = 86.0\%$.

Problem

15. In a study of proteins mediating cell membrane transport, microbiologists measure current versus time through the cell membranes of oocytes (nearly mature egg cells) taken from the African clawed frog, *Xenopus*. The measured current versus time is given approximately by $I = 60t + 200t^2 + 4.0t^3$, with t in seconds and I in nA. Find the total charge that flows through the cell membrane in the interval from $t = 0$ to $t = 5.0$ s.

Solution

(a) If we use Equation 27-1b for the charge, then
$$q = \int_0^{5\,s} I\,dt = \left|(60\text{ nA/s})\tfrac{1}{2}t^2 + (200\text{ nA/s}^2)\tfrac{1}{3}t^3 +\right.$$
$$(4\text{ nA/s}^3)\tfrac{1}{4}t^4\Big|_0^{5\,s} = (30 + \tfrac{1}{3}\times1000 + 25)(25\text{ nC}) =$$
$9.71\ \mu$C.

Section 27-2: Conduction Mechanisms

Problem

17. What electric field is necessary to drive a 7.5-A current through a silver wire 0.95 mm in diameter?

Solution

From Ohm's law (which applies to silver) and the definition of current density (which we assume is uniform in the wire) one finds $E = \rho J = \rho I/\tfrac{1}{4}\pi d^2 = (1.59\times10^{-8}\ \Omega\cdot\text{m})(7.5\text{ A})/\tfrac{1}{4}\pi(0.95\text{ mm})^2 = 0.168$ V/m.

Problem

19. A 1.0-cm-diameter rod carries a 50-A current when the electric field in the rod is 1.4 V/m. What is the resistivity of the rod material?

Solution

If the rod has a uniform current density and obey's Ohm's law (Equations 27-3a and 4b), then its resistivity is $\rho = E/J = E/(I/\tfrac{1}{4}\pi d^2) = \tfrac{1}{4}\pi(10^{-2}\text{ m})^2\times (1.4\text{ V/m})/(50\text{ A}) = 2.20\times10^{-6}\ \Omega\cdot\text{m}$.

Problem

21. Use Table 27-1 to determine the conductivity of (a) copper and (b) sea water.

Solution

Equations 27-4a and b show that the conductivity and the resistivity are reciprocals of one another.

Thus, (a) $\rho^{-1} = \sigma = (1.68\times10^{-8}\ \Omega\cdot\text{m})^{-1} = 5.95\times10^7(\Omega\cdot\text{m})^{-1}$ for copper, and (b) $\sigma = (0.22\ \Omega\cdot\text{m})^{-1} = 4.55(\Omega\cdot\text{m})^{-1}$ for typical seawater. (The salinity of open-ocean water varies between 33 and 37‰, but can vary from 1 to 80‰ in shallow coastal waters.)

Problem

23. The free-electron density in aluminum is 2.1×10^{29} m^{-3}. What is the collision time in aluminum?

Solution

We can estimate τ from Equation 27-5 and Table 27-1: $\tau = m_e/\rho n e^2 = (9.11\times10^{-31}$ kg$)(2.65\times10^{-8}\ \Omega\cdot\text{m})^{-1}\times (2.1\times10^{29}$ m$^{-3})^{-1}(1.6\times10^{-19}$ C$)^{-2} = 6.39\times10^{-15}$ s. (An explicit confirmation of the units is:

$$\frac{\text{kg}}{(\Omega\cdot\text{m})(\text{m}^{-3})(\text{C}^2)} = \frac{\text{kg}\cdot\text{m}^2}{(\text{V/A})\text{C}^2}$$
$$= \frac{\text{kg}\cdot\text{m}^2(\text{C/s})}{(\text{J/C})\text{C}^2} = \frac{\text{kg}\cdot\text{m}^2/\text{s}}{\text{kg}\cdot\text{m}^2/\text{s}^2} = \text{s}.\Big)$$

Problem

25. The resistivity of copper as a function of temperature is given approximately by $\rho = \rho_0[1 + \alpha(T - T_0)]$, where ρ_0 is the value listed in Table 27-1 for 20°C, $T_0 = 20$°C, and $\alpha = 4.3\times10^{-3}$°C^{-1}. Find the temperature at which copper's resistivity is twice its room-temperature value.

Solution

Taking room temperature to be 20°C, we find that the resistivity doubles when $2 = 1 + \alpha(T - T_0)$, or $T = T_0 + 1/\alpha = 20$°C $+ 1$°C$/4.3\times10^{-3} = 253$°C.

Section 27-3: Resistance and Ohm's Law

Problem

27. What is the resistance of a heating coil that draws 4.8 A when the voltage across it is 120 V?

Solution

The macroscopic form of Ohm's Law is probably applicable to the heating coil, which is typically a coil of wire. Equation 27-6 gives $R = V/I = 120$ V $\div 4.8$ A $= 25\ \Omega$.

Problem

29. What is the current in a 47-kΩ resistor with 110 V across it?

Solution

Provided the resistor obeys Ohm's law, $I = V/R = 110 \text{ V}/47 \text{ k}\Omega = 2.34 \text{ mA}$.

Problem

31. What current flows when a 45-V potential difference is imposed across a 1.8-kΩ resistor?

Solution

If the resistor obeys Ohm's law, $I = V/R = 45 \text{ V}/1.8 \text{ k}\Omega = 25 \text{ mA}$.

Problem

33. The presence of a few ions makes air a conductor, albeit a poor one. If the total resistance between the ionosphere and Earth is 200 Ω, how much current flows as a result of a 300-kV potential difference between Earth and ionosphere?

Solution

If the total atmospheric resistance is 200 Ω at 300 kV, Equation 27-6 gives a current of $I = V/R = 300 \text{ kV}/200 \text{ }\Omega = 1.5 \text{ kA}$.

Problem

35. A cylindrical iron rod measures 88 cm long and 0.25 cm in diameter. (a) Find its resistance. If a 1.5-V potential difference is applied between the ends of the rod, find (b) the current, (c) the current density, and (d) the electric field in the rod.

Solution

(a) Equation 27-7 gives the resistance of a uniform object of Ohmic material, $R = \rho\ell/A = (9.71 \times 10^{-8} \text{ }\Omega\text{·m})(88 \text{ cm})/\frac{1}{4}\pi(0.25 \text{ cm})^2 = 17.4 \text{ m}\Omega$ (see Table 27-1 for the resistivity of iron). (b) Equation 27-6 (Ohm's law) gives $I = V/R = 1.5 \text{ V} \div 17.4 \text{ m}\Omega = 86.2 \text{ A}$. (c) Equation 27-3a gives $J = I/\frac{1}{4}\pi d^2 = 86.2 \text{ A}/\frac{1}{4}\pi(0.25 \text{ cm})^2 = 17.6 \text{ MA/m}^2$. (d) Equation 27-4b gives $E = \rho J = (9.71 \times 10^{-8} \text{ }\Omega\text{·m})(17.6 \text{ MA/m}^2) = 1.70 \text{ V/m}$. (The quantities were calculated in the order queried; alternatively, in reverse order, $E = V/\ell$, $J = E/\rho$, $I = JA$, and $R = V/I$.)

Problem

37. How must the diameters of copper and aluminum wire be related if they are to have the same resistance per unit length?

Solution

From Equation 27-7, the resistance per unit length (of a uniform wire of Ohmic material) is $R/\ell = \rho/\frac{1}{4}\pi d^2$, so equal values for copper and aluminum wires imply that $\rho_{Cu}/d_{Cu}^2 = \rho_{Al}/d_{Al}^2$, or $d_{Al}/d_{Cu} = \sqrt{\rho_{Al}/\rho_{Cu}} = \sqrt{2.65/1.68} = 1.26$ (see Table 27-1 for the resistivities).

Problem

39. Engineers call for a power line with a resistance per unit length of 50 mΩ/km. What wire diameter is required if the line is made of (a) copper or (b) aluminum? (c) If the costs of copper and aluminum wire are \$1.53/kg and \$1.34/kg, which material is more economical? The densities of copper and aluminum are 8.9 g/cm^3 and 2.7 g/cm^3, respectively.

Solution

From Equation 27-7, $R/\ell = \rho/(\frac{1}{4}\pi d^2)$, so $d = 2\sqrt{\rho/\pi(R/\ell)}$. With resistivities from Table 27-1, (a) $d_{Cu} = 2\sqrt{(1.68 \times 10^{-8} \text{ }\Omega\text{·m})/(50\pi \text{ m}\ell/\text{km})} = 2.07 \text{ cm}$, and (b) $d_{Al} = \sqrt{2.65/1.68} \, d_{Cu} = 2.60 \text{ cm}$ (see Problem 37). (c) With these diameters, the cost of one meter of wire is $\frac{1}{4}\pi d_{Cu}^2(1 \text{ m})(8.9 \text{ g/cm}^3)(\$1.53/\text{kg}) = \$4.58$ for copper, and \$1.92 for aluminum.

Problem

41. Corrosion at battery terminals results in increased resistance, and is a frequent cause of hard starting in cars. In an effort to diagnose hard starting, a mechanic measures the voltage between the battery terminal and the wire carrying current to the starter motor. While the motor is cranking, this voltage is 4.2 V. If the motor draws 125 A, what is the resistance at the battery terminal?

Solution

From Ohm's law, $R = V/I = 4.2 \text{ V}/125 \text{ A} = 33.6 \text{ m}\Omega$. (The resistance of the battery cable in Example 27-5 was 0.60 mΩ, so most of the resistance just calculated was in the connection.)

Section 27-4:   Electric Power

Problem

43. A car's starter motor draws 125 A with 11 V across its terminals. What is its power consumption?

Solution

Equation 27-8 gives the power supplied to the motor, $\mathcal{P} = VI = (11 \text{ V})(125 \text{ A}) = 1.38 \text{ kW}$.

Problem

45. A watch uses energy at the rate of 240 μW. How much current does it draw from its 1.5-V battery?

Solution

Rearranging Equation 27-8, we find $I = \mathcal{P}/V = 240$ μW/1.5 V $= 160$ μA.

Problem

47. What is the resistance of a standard 120-V, 60-W light bulb?

Solution

The bulb's resistance, from Equation 27-9b, is $R = V^2/\mathcal{P} = (120$ V$)^2/60$ W $= 240$ Ω, at its operating temperature. (This equation, with average values of voltage and power, can be used for an ac-resistor.)

Problem

49. If the electrons of Problem 4 are accelerated through a potential difference of 10 kV, how much power must be supplied to produce the electron beam?

Solution

The electron beam in Problem 4 carries a current of 4.8 mA, and so must be supplied with a power of $\mathcal{P} = VI = (10$ kV$)(4.8$ mA$) = 48$ W. (Recall Equation 27-2, $I = (nA)ev_d = (5.0\times10^6$ electrons/cm$)\times (1.6\times10^{-19}$ C/electron$)(6.0\times10^7$ m/s), where nA is the electron density per length of beam in Problem 4.)

Problem

51. How much total energy could the 12-V battery of Problem 3 supply?

Solution

If the battery is a typical 12 V automotive model, it can supply a total energy of $(12$ V$)(80$ A$)(1$ h$) = (0.960$ kW·h$)(3600$ s/h$) = 3.46$ MJ. (Note that the total energy can be thought of as either the power times 1 h of operating time, $\mathcal{P}\,\Delta t = VI\,\Delta t$, or the total charge times the potential difference, $\Delta qV = (I\,\Delta t)V$ as in Problem 3.)

Problem

53. Two cylindrical resistors are made from the same material and have the same length. When connected across the same battery, one dissipates twice as much power as the other. How do their diameters compare?

Solution

At the same voltage, the ratio of the power dissipated is the inverse of the ratio of the resistances, which in turn, goes as the inverse of the square of the ratio of the diameters: $\mathcal{P}_1/\mathcal{P}_2 = (V^2/R_1)/(V^2/R_2) = R_2/R_1 = (\rho\ell/\frac{1}{4}\pi d_2^2)/(\rho\ell/\frac{1}{4}\pi d_1^2) = (d_1/d_2)^2$. We used Equation 27-9b for the power, and Equation 27-7 for the resistance. Thus, if $\mathcal{P}_1 = 2\mathcal{P}_2$, then $d_1 = \sqrt{2}\,d_2$.

Problem

55. A 2000-horsepower electric railroad locomotive gets its power from an overhead wire with 0.20 Ω/km. The potential difference between wire and track is 10 kV. Current returns through the track, whose resistance is negligible. (a) How much current does the locomotive draw? (b) How far from the power plant can the train go before 1% of the energy is lost in the wire?

Solution

(a) If we neglect possible energy losses in the locomotive's engine, $\mathcal{P}_{\text{in}} = \mathcal{P}_{\text{out}}$, or $VI = 2000$ hp. Thus, $I = (2000$ hp$)(746$ W/hp$)/(10^4$ V$) = 149$ A.
(b) The power loss in getting current to and from the locomotive is $\mathcal{P}_{\text{loss}} = I^2R$, where I is the current from part (a) and R is the resistance of the feed/return circuit. The high voltage wire has resistance $(0.2$ Ω/km$)\ell$, where its length, ℓ, is also the distance from the power station, and the resistance of the return rail is negligible (see Problem 30). Then $\mathcal{P}_{\text{loss}} = 1\%$, $\mathcal{P}_{\text{in}}$ implies $I^2R = 0.01VI$, or $\ell = (0.01)(10^4$ V$)/(149$ A$)(0.2$ Ω/km$) = 3.35$ km.

Paired Problems

Problem

57. Electrons in a fine silver wire 20 μm in diameter drift at 0.14 mm/s. What is the current in the wire? Each silver atom contributes 1.3 free electrons.

Solution

For a uniform wire, Equation 27-2 gives $I = ne(\frac{1}{4}\pi d^2)v_d = (7.62\times10^{28}$ m$^{-3})(1.6\times10^{-19}$ C$)\times \frac{1}{4}\pi(20$ μm$)^2(0.14$ mm/s$) = 536$ μA. We used the density of conduction electrons for silver from Problem 9.

Problem

59. What is the resistance of a column of mercury 0.75 m long and 1.0 mm in diameter?

Solution

At ordinary temperatures, metallic mercury obeys Ohm's law, so Equation 27-7 and Table 27-1 give a resistance of $R = \rho\ell/A = (9.84\times10^{-7}\ \Omega\cdot\text{m})\times (0.75\ \text{m})/\frac{1}{4}\pi(10^{-3}\ \text{m})^2 = 0.940\ \Omega$.

Problem

61. A power plant produces 1000 MW to supply a city 40 km away. Current flows from the power plant on a single wire of resistance 0.050 Ω/km, through the city, and returns via the ground, assumed to have negligible resistance. At the power plant the voltage between the wire and ground is 115 kV. (a) What is the current in the wire? (b) What fraction of the power is lost in transmission?

Solution

(a) If the power supplied by the plant is 1 GW at 115 kV, then the current supplied is $I = \mathcal{P}_{\text{out}}/V = 10^9\ \text{W}/115\ \text{kV} = 8.70\ \text{kA}$. (b) The power loss in the transmission line is $\mathcal{P}_{\text{loss}} = I^2 R_{\text{line}} = (8.70\ \text{kA})^2 \times (0.050\ \Omega/\text{km} \times 40\ \text{km}) = 151\ \text{MW}$, or 15.1% of the plants 1 GW output. (Note: The voltage drop along the transmission line is $IR_{\text{line}} = 17.4\ \text{kV}$, not 115 kV.)

Problem

63. A 240-V electric motor is 90% efficient, meaning that 90% of the energy supplied to it ends up as mechanical work. If the motor lifts a 200-N weight at 3.1 m/s, how much current does it draw?

Solution

The electrical power input to the motor is $\mathcal{P}_{\text{in}} = VI$, while the mechanical power output (lifting the weight) is $\mathcal{P}_{\text{out}} = Fv = 90\%\ \mathcal{P}_{\text{in}}$. Thus, $I = Fv/0.9\ V = (200\ \text{N})(3.1\ \text{m/s})/(0.9\times240\ \text{V}) = 2.87\ \text{A}$.

Supplementary Problems
Problem

65. A metal bar has a rectangular cross section 5.0 cm by 10 cm, as shown in Fig. 27-25. The bar has a nonuniform conductivity, ranging from zero at the bottom to a maximum at the top. As a result, the current density increases linearly from zero at the bottom to 0.10 A/cm² at the top. What is the total current in the bar?

Solution

The current density is $\mathbf{J} = (0.1\ \text{A/cm}^2)(x/10\ \text{cm})\hat{\mathbf{k}}$, in the coordinate system superposed on Fig. 27-25. The cross section can be divided into strips of area $d\mathbf{A} = (5\ \text{cm})\,dx\,\hat{\mathbf{k}}$ (over which $\mathbf{J}$ is constant), so the total current in the bar (Equation 27-3a in differential form) is:

$$I = \int_{x\text{-sect}} \mathbf{J}\cdot d\mathbf{A} = \int_0^{10\ \text{cm}} (10^{-2}\ \text{A/cm}^3)(5\ \text{cm})x\ dx$$
$$= (0.05\ \text{A/cm}^2)\frac{1}{2}(10\ \text{cm})^2 = 2.5\ \text{A}.$$

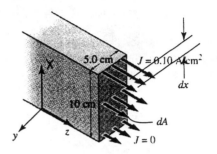

FIGURE 27-25 Problem 65 Solution.

Problem

67. General Motors' EV1 electric car has a mass of 1500 kg and is powered by 26 12-V batteries connected in series, for a total of 312 V. About 85% of the electrical energy from the batteries ends up as mechanical energy at the drive wheels. How much current do the batteries supply when the car is climbing a 10° slope at 45 km/h? Neglect frictional losses and air resistance.

Solution

If frictional losses and air resistance are neglected, the motor only supplies power to work against gravity, $\mathcal{P}_{\text{out}} = F_\parallel v = (mg\sin 10°)v$, which equals 85% of the input electrical power from the batteries, $\mathcal{P}_{\text{in}} = VI$. Therefore, $I = mg\sin 10° v/0.85V = (1500\times9.8\ \text{N})\times (45\ \text{m}/3.6\ \text{s})\sin 10°/(0.85\times312\ \text{V}) = 120\ \text{A}$.

Problem

69. A 100-Ω resistor of negligible mass is mounted inside a calorimeter. When a 12-V battery is connected for 5.0 min, the temperature inside the calorimeter rises by 26°C. What is the heat capacity of the calorimeter contents?

Solution

The electrical energy dissipated in the resistor raises the temperature inside the calorimeter, $\mathcal{P}t = (V^2/R)t = C\ \Delta T$, where C is the heat capacity of the contents (the heat capacity of a resistor of negligible mass is approximately zero). Therefore, $C =$

$V^2t/R \, \Delta T = (12 \text{ V})^2(5 \times 60 \text{ s})/(100 \text{ }\Omega)(26 \text{ C}°) =$
16.6 J/C°.

Problem

71. Figure 27-27 shows a resistor made from a truncated cone of material with uniform resistivity ρ. Consider the cone to be made of thin slices of thickness dx, like the one shown; Equation 27-7 shows that the resistance of each slab is $dR = \rho \, dx/A$. By integrating over all such slices, shows that the resistance between the two flat faces is $R = \rho\ell/\pi ab$. (This method assumes the equipotentials are planes, which is only approximately true.)

Solution

As suggested, consider that the total resistance R equals $\int_{\text{cone}} dR$, where $dR = \rho \, dx/A$ is the resistance of a thin disk as shown on Fig. 27-27. The area of

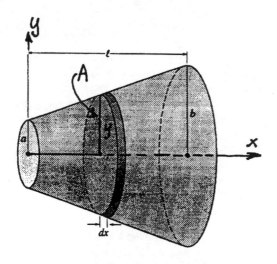

FIGURE 27-27 Problem 71 Solution.

such a disk is $A = \pi y^2$, where $y = a + (b-a)x/\ell$ is the radius and $0 \le x \le \ell$ as shown. Then

$$
\begin{aligned}
R &= \int_0^\ell \frac{\rho \, dx}{\pi[a + (b-a)x/\ell]^2} \\
&= \frac{\rho}{\pi} \left| -\left(\frac{\ell}{b-a}\right) \frac{1}{[a+(b-a)x/\ell]} \right|_0^\ell \\
&= \frac{\rho}{\pi}\left(\frac{\ell}{b-a}\right)\left(\frac{1}{a} - \frac{1}{b}\right) = \frac{\rho\ell}{\pi ab}.
\end{aligned}
$$

(This result depends on the condition that the flat faces and parallel circular cross-sections of the cone are equipotential surfaces.)

Problem

73. At some point in a material of resistivity ρ the current density is J. Show that the power per unit volume dissipated at that point is $J^2\rho$.

Solution

Consider a small volume in the material, ΔV, where the current density is $\mathbf{J} = nq\,\mathbf{v}_d$ (for simplicity, we assume only one type of charge carrier). The electric field does work on each charge carrier at a rate equal to $\mathbf{F}_{\text{el}}\cdot\mathbf{v}_d = q\,\mathbf{E}\cdot\mathbf{v}_d$, so the total power is this times the number of charge carriers in ΔV, or $\Delta \mathcal{P} = (n\,\Delta V)(q\,\mathbf{E}\cdot\mathbf{v}_d) = \mathbf{J}\cdot\mathbf{E}\,\Delta V$. If the electric field in the material is given by Ohm's law (such a material is called a passive region; if other effective fields are present, the material is an active region), then $\mathbf{E} = \rho\mathbf{J}$, where ρ is the resistivity, and $\Delta\mathcal{P}/\Delta V = \mathbf{J}\cdot(\rho\mathbf{J}) = \rho J^2$. This power is transformed into random thermal motion of the charge carriers and ions of the material, and is referred to as Joule heat; $J^2\rho$ is the rate of Joule heating per unit volume of material.

CHAPTER 28   ELECTRIC CIRCUITS

ActivPhysics can help with these problems:
Section 12, "DC Circuits"

Section 28-1:   Circuits and Symbols

Problem

1. Sketch a circuit diagram for a circuit that includes a resistor R_1 connected to the positive terminal of a battery, a pair of parallel resistors R_2 and R_3 connected to the lower-voltage end of R_1, then returned to the battery's negative terminal, and a capacitor across R_2.

Solution

A literal reading of the circuit specifications results in connections like those in sketch (a). Because the connecting wires are assumed to have no resistance (a real wire is represented by a separate resistor), a topologically equivalent circuit diagram is shown in sketch (b).

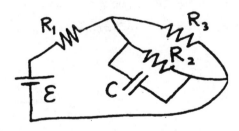

Problem 1 Solution (a).

Problem 1 Solution (b).

Problem

3. Resistors R_1 and R_2 are connected in series, and this series combination is in parallel with R_3. This parallel combination is connected across a battery whose internal resistance is R_{int}. Draw a diagram representing this circuit.

Solution

The circuit has three parallel branches: one with R_1 and R_2 in series; one with just R_3; and one with the battery (an ideal emf in series with the internal resistance).

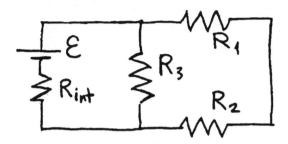

Problem 3 Solution.

Section 28-2:   Electromotive Force

Problem

5. A 1.5-V battery stores 4.5 kJ of energy. How long can it light a flashlight bulb that draws 0.60 A?

Solution

The average power, supplied by the battery to the bulb, multiplied by the time equals the energy capacity of the battery. For an ideal battery, $\mathcal{P} = \mathcal{E}I$, therefore $\mathcal{E}It = 4.5$ kJ, or $t = 4.5$ kJ$/(1.5$ V$)\times$ $(0.60$ A$) = 5\times10^3$ s $= 1.39$ h.

Problem

7. A battery stores 50 W·h of chemical energy. If it uses up this energy moving 3.0×10^4 C through a circuit, what is its voltage?

Solution

The emf is the energy (work done going through the source from the negative to the positive terminal) per unit charge: $\mathcal{E} = (50$ W·h$)(3600$ s/h$)/(3\times10^4$ C$) =$ 6 V. (This is the average emf; the actual emf may vary with time.)

Section 28-3:   Simple Circuits: Series and Parallel Resistors

Problem

9. What resistance should be placed in parallel with a 56-kΩ resistor to make an equivalent resistance of 45 kΩ?

Solution

The solution for R_2 in Equation 28-3a is $R_2 = R_1 R_{\text{parallel}}/(R_1 - R_{\text{parallel}}) = (56 \text{ k}\Omega)(45)/(56 - 45) = 229 \text{ k}\Omega$.

Problem

11. In Fig. 28-49, take all resistors to be 1.0 Ω. If a 6.0-V battery is connected between points A and B, what will be the current in the vertical resistor?

Solution

The circuit in Fig. 28-49, with a battery connected across points A and B, is similar to the circuit analysed in Example 28-4. In this case, $R_{||} = (1 \text{ }\Omega) \times (2)/(1 + 2) = (\frac{2}{3}) \text{ }\Omega$, and $R_{\text{tot}} = 1 \text{ }\Omega + 1 \text{ }\Omega + \frac{2}{3} \text{ }\Omega = \frac{8}{3} \text{ }\Omega$. The total current (that through the battery) is $I_{\text{tot}} = \mathcal{E}/R_{\text{tot}} = 6 \text{ V}/(\frac{8}{3} \text{ }\Omega) = (\frac{9}{4}) \text{ A}$. The voltage across the parallel combination is $I_{\text{tot}} R_{||} = (\frac{9}{4} \text{ A}) \times (\frac{2}{3} \text{ }\Omega) = \frac{3}{2} \text{ V}$, which is the voltage across the vertical 1 Ω resistor. The current through this resistor is then $(\frac{3}{2} \text{ V})/(1 \text{ }\Omega) = 1.5 \text{ A}$.

Problem

13. What is the internal resistance of the battery in the preceding problem?

Solution

The solution of the preceding problem (or the reasoning of Example 28-2) gives $R_{\text{int}} = 0.02 \text{ }\Omega$ (i.e., (12 V − 6 V)/300 A).

Problem

15. When a 9-V battery is temporarily short-circuited, a 200-mA current flows. What is the internal resistance of the battery?

Solution

From the equation for a battery short-circuited, in the subsection "Real Batteries," $R_{\text{int}} = \mathcal{E}/I = 9 \text{ V}/0.2 \text{ A} = 45 \text{ }\Omega$.

Problem

17. A partially discharged car battery can be modeled as a 9-V emf in series with an internal resistance

of 0.08 Ω. Jumper cables are used to connect this battery to a fully charged battery, modeled as a 12-V emf in series with a 0.02-Ω internal resistance. How much current flows through the discharged battery?

Solution

Terminals of like polarity are connected with jumpers of negligible resistance. Kirchhoff's voltage law gives $\mathcal{E}_1 - \mathcal{E}_2 - IR_1 - IR_2 = 0$, or $I = (\mathcal{E}_1 - \mathcal{E}_2)/(R_1 + R_2) = (12 - 9) \text{ V}/(0.02 + 0.08) \text{ }\Omega = 30 \text{ A}$.

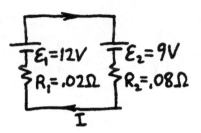

Problem 17 Solution.

Problem

19. What is the equivalent resistance between A and B in each of the circuits shown in Fig. 28-50? *Hint:* In (c), think about symmetry and the current that would flow through R_2.

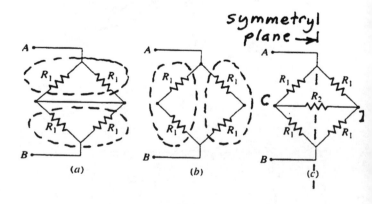

FIGURE 28-50 Problem 19 Solution.

Solution

(a) There are two parallel pairs ($\frac{1}{2}R_1$) in series, so $R_{AB} = \frac{1}{2}R_1 + \frac{1}{2}R_1 = R_1$. (b) Here, there are two series pairs ($2R_1$) in parallel, so $R_{AB} = (2R_1)(2R_1) \div (2R_1 + 2R_1) = R_1$. (c) Symmetry requires that the current divides equally on the right and left sides, so points C and D are at the same potential. Thus, no current flows through R_2, and the circuit is equivalent to (b). (Note that the reasoning in parts (a) and (b) is easily generalized to resistances of different values; the generalization in part (c) requires the equality of ratios of resistances which are mirror images in the plane of symmetry.)

Problem

21. How many 100-W, 120-V light bulbs can be connected in parallel before they below a 20-A circuit breaker?

Solution

The circuit breaker is activated if $I = 120$ V$\div R_{min} > 20$ A, or if $R_{min} < 6$ Ω. The resistance of each light bulb is $R = V^2/\mathcal{P} = (120$ V$)^2/100$ W $= 144$ Ω, and n bulbs in parallel have resistance $R_{||} = R/n$. Therefore $R_{||} \geq R_{min}$ implies $n \leq 144/6 = 24$, so more than 24 bulbs would blow the circuit.

Problem

23. Take $\mathcal{E} = 12$ V and $R_1 = 270$ Ω in the voltage divider of Fig. 28-5. (a) What should be the value of R_2 in order that 4.5 V appear across R_2? (b) What will be the power dissipation in R_2?

Solution

(a) For this voltage divider, Equation 28-2b gives $V_2 = R_2\mathcal{E}/(R_1 + R_2)$, or $R_2 = R_1V_2/(\mathcal{E} - V_2) = (270$ $\Omega)(4.5)/(12 - 4.5) = 162$ Ω. (b) The power dissipated (Equation 27-9b) is $\mathcal{P}_2 = V_2^2/R_2 = (4.5$ V$)^2/162$ $\Omega = 125$ mW.

Problem

25. In the circuit of Fig. 28-52, R_1 is a variable resistor, and the other two resistors have equal resistances R. (a) Find an expression for the voltage across R_1, and (b) sketch a graph of this quantity as a function of R_1 as R_1 varies from 0 to $10R$. (c) What is the limiting value as $R_1 \to \infty$?

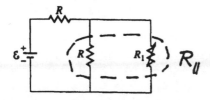

FIGURE 28-52 Problem 25.

Solution

(a) The resistors in parallel have an equivalent resistance of $R_{||} = RR_1/(R + R_1)$. The other R, and

$R_{||}$, is a voltage divider in series with $\mathcal{E}$, so Equation 28-2 gives $V_{||} = \mathcal{E}R_{||}/(R + R_{||}) = \mathcal{E}R_1/(R + 2R_1)$. (b) and (c) If $R_1 = 0$ (the second resistor shorted out), $V_{||} = 0$, while if $R_1 = \infty$ (open circuit), $V_{||} = \frac{1}{2}\mathcal{E}$ (the value when R_1 is removed). If $R_1 = 10R$, $V_{||} = (10/21)\mathcal{E}$ (as in Problem 24).

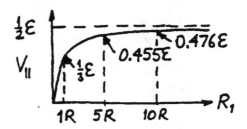

Problem 25 Solution.

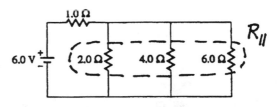

FIGURE 28-53 Problem 26 Solution, and Problem 27.

Problem

27. In the circuit of Fig. 28-53, how much power is being dissipated in the 4-Ω resistor?

Solution

The three resistors in parallel have an effective resistance of $1/R_{||} = (1/2 + 1/4 + 1/6)$ Ω^{-1}, or $R_{||} = (12/11)$ Ω. Equation 28-2 gives the voltage across them as $V_{||} = (6$ V$)(12/11)/(1 + 12/11) = (72/23)$ V. Thus, $\mathcal{P}_4 = V_{||}^2/R_4 = (72/23)^2$ V$^2/4$ $\Omega = 2.45$ W.

Section 28-4: Kirchhoff's Laws and Multiloop Circuits

Problem

29. In the circuit of Fig. 28-54 it makes no difference whether the switch is open or closed. What is $\mathcal{E}_3$ in terms of the other quantities shown?

Solution

If the switch is irrelevant, then there is no current through its branch of the circuit. Thus, points A and B must be at the same potential, and the same current flows through R_1 and R_2. Kirchhoff's voltage law applied to the outer loop, and to the left-hand loop, gives $\mathcal{E}_1 - IR_1 - IR_2 + \mathcal{E}_2 = 0$, and $\mathcal{E}_1 - IR_1 + \mathcal{E}_3 = 0$, respectively. Therefore,

$$\mathcal{E}_3 = IR_1 - \mathcal{E}_1 = \left(\frac{\mathcal{E}_1 + \mathcal{E}_2}{R_1 + R_2}\right)R_1 - \mathcal{E}_1 = \frac{\mathcal{E}_2 R_1 - \mathcal{E}_1 R_2}{R_1 + R_2}.$$

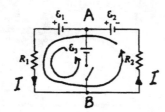

FIGURE 28-54 Problem 29 Solution.

Problem

31. In Fig. 28-56, what is the equivalent resistance measured between points A and B?

Solution

The effective resistance is determined by the current which would flow through a pure emf if it were connected between A and B: $R_{AB} = \mathcal{E}/I$. Since I is but one of six branch currents, the direct solution of Kirchhoff's circuit laws is tedious (6×6 determinants). (The method of loop currents, not mentioned in the text, involves more tractable 3×3 determinants.) However, because of the special values of the resistors in Fig. 28-56, a symmetry argument greatly simplifies the calculation.

The equality of the resistors on opposite sides of the square implies that the potential difference between A and C equals that between D and B, i.e., $V_A - V_C = V_D - V_B$. Equivalently, $V_A - V_D = V_C - V_B$. Since $V_A - V_C = I_1 R$, $V_A - V_D = I_2(2R)$, etc., the symmetry argument requires that both R-resistors on the perimeter carry the same current, I_1, and both $2R$-resistors carry current I_2. Then Kirchhoff's current law implies that the current through $\mathcal{E}$ is $I_1 + I_2$, and the current through the central resistor is $I_1 - I_2$ (as added to Fig. 28-56). Now there are only two independent branch currents, which can be found from Kirchhoff's voltage law, applied, for example, to loops $ACBA$, $\mathcal{E} - I_1 R - I_2(2R) = 0$, and $ACDA$, $-I_1 R - (I_1 - I_2)R + I_2(2R) = 0$. These equations may be rewritten as $I_1 + 2I_2 = \mathcal{E}/R$ and $-2I_1 + 3I_2 = 0$, with solution $I_1 = 3\mathcal{E}/7R$ and $I_2 = 2\mathcal{E}/7R$. Therefore, $I = I_1 + I_2 = 5\mathcal{E}/7R$, and $R_{AB} = \mathcal{E}/I = 7R/5$. (The configuration of resistors in Fig. 28-56 is called a Wheatstone bridge.)

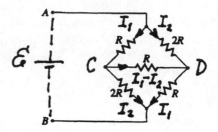

FIGURE 28-56 Problem 31 Solution.

Problem

33. Find all three currents in the circuit of Fig. 28-18 with the values given, but with battery $\mathcal{E}_2$ reversed.

Solution

The general solution of the two loop equations and one node equation given in Example 28-5 can be found using determinants (or I_1 and I_2 can be found in terms of I_3, as in Example 28-5). The equations and the solution are:

$$I_1 R_1 + 0 + I_3 R_3 = \mathcal{E}_1 \qquad \text{(loop 1)},$$
$$0 - I_2 R_2 + I_3 R_3 = \mathcal{E}_2 \qquad \text{(loop 2)},$$
$$I_1 - I_2 - I_3 = 0 \qquad \text{(node A)};$$

$$\Delta \equiv \begin{vmatrix} R_1 & 0 & R_3 \\ 0 & -R_2 & R_3 \\ 1 & -1 & -1 \end{vmatrix} = R_1 R_2 + R_2 R_3 + R_3 R_1,$$

$$I_1 = \frac{1}{\Delta} \begin{vmatrix} \mathcal{E}_1 & 0 & R_3 \\ \mathcal{E}_2 & -R_2 & R_3 \\ 0 & -1 & -1 \end{vmatrix} = \frac{\mathcal{E}_1(R_2 + R_3) - \mathcal{E}_2 R_3}{\Delta},$$

$$I_2 = \frac{1}{\Delta} \begin{vmatrix} R_1 & \mathcal{E}_1 & R_3 \\ 0 & \mathcal{E}_2 & R_3 \\ 1 & 0 & -1 \end{vmatrix} = \frac{\mathcal{E}_1 R_3 - \mathcal{E}_2(R_1 + R_3)}{\Delta},$$

$$I_3 = \frac{1}{\Delta} \begin{vmatrix} R_1 & 0 & \mathcal{E}_1 \\ 0 & -R_2 & \mathcal{E}_2 \\ 1 & -1 & 0 \end{vmatrix} = \frac{\mathcal{E}_2 R_1 + \mathcal{E}_1 R_2}{\Delta}.$$

With the particular values of emf's and resistors in this problem, we find currents of $I_1 = [(4 + 1)6 - 1(-9)]$ A$/14 = 2.79$ A, $I_2 = [1 \times 6 - (2 + 1)(-9)]$ A$/14 = 2.36$ A, and $I_3 = [4 \times 6 + 2 \times (-9)]$ A$/14 = 0.429$ A. Or, one could retrace the reasoning of Example 28-5, with $\mathcal{E}_2 = -9$ V replacing the original value in loop 2. Then, everything is the same until the equation $-9 + 2(6 - 3I_3) - I_3 = 0$, or $I_3 = (\frac{3}{7})$ A, $I_2 = \frac{1}{2}(6 - 3 \times \frac{3}{7})$ A $= (33/14)$ A, and $I_1 = I_2 + I_3 = (39/14)$ A.

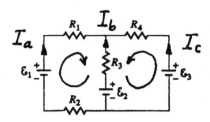

FIGURE 28-57 Problem 34 Solution, and Problems 35 and 36.

Problem

35. With all the values except $\mathcal{E}_2$ in Fig. 28-57 as given in the preceding problem, find the condition on $\mathcal{E}_2$ that will make the current in R_3 flow upward.

Solution

Let us choose the positive sense for each of the three branch currents in Fig. 28-57 as upward through their respective emf's (at least one must be negative, of course), and consider the two smaller loops shown. Kirchhoff's circuit laws give:

$$I_a + I_b + I_c = 0 \qquad \text{(top node)}$$
$$(R_1 + R_2)I_a - R_3 I_b = \mathcal{E}_1 - \mathcal{E}_2 \qquad \text{(left loop)}$$
$$-R_3 I_b + R_4 I_c = \mathcal{E}_3 - \mathcal{E}_2 \qquad \text{(right loop)}.$$

Solve for I_a and I_c from the loop equations and substitute into the node equation:

$$\frac{(\mathcal{E}_1 - \mathcal{E}_2) + R_3 I_b}{R_1 + R_2} + I_b + \frac{(\mathcal{E}_3 - \mathcal{E}_2) + R_3 I_b}{R_4} = 0.$$

Then

$$I_b = \frac{R_4(\mathcal{E}_2 - \mathcal{E}_1) + (R_1 + R_2)(\mathcal{E}_2 - \mathcal{E}_3)}{R_3 R_4 + (R_1 + R_2)(R_3 + R_4)},$$

with similar expressions for I_a and I_c. One can see that I_b is positive if $R_4(\mathcal{E}_2 - \mathcal{E}_1) + (R_1 + R_2) \times (\mathcal{E}_2 - \mathcal{E}_3) > 0$, or

$$\mathcal{E}_2 > \frac{R_4 \mathcal{E}_1 + (R_1 + R_2)\mathcal{E}_3}{R_1 + R_2 + R_4}$$
$$= \frac{(820\ \Omega)(6\ \text{V}) + (420\ \Omega)(4.5\ \text{V})}{(1240\ \Omega)} = 5.49\ \text{V}.$$

Problem

37. Figure 28-58 shows a portion of a circuit used to model the electrical behavior of long, cylindrical biological cells such as muscle cells or the axons of neurons. Find the current through the emf $\mathcal{E}_3$, given that all resistors have the same value $R = 1.5\ \text{M}\Omega$ and that $\mathcal{E}_1 = 75\ \text{mV}$, $\mathcal{E}_2 = 45\ \text{mV}$, and $\mathcal{E}_3 = 20\ \text{mV}$. Be sure to specify the direction of the current.

Solution

The circuits in Figs. 28-57 and 58 each have three branches (with emf's and total resistances $\mathcal{E}_a, R_a, \mathcal{E}_b, R_b,$ and $\mathcal{E}_c, R_c$) connected in parallel between two nodes. (The emf's and currents in each branch are taken as positive upward in the figures.) The expression for I_b in the solution to Problem 35 can be used for any permutation of indices a, b, and c (any order of parallel branches is equivalent). For example, in Fig. 28-58, take b to be the third branch etc., so that $\mathcal{E}_b = \mathcal{E}_3 = 20\ \text{mV}$, $R_b = 2R$, $\mathcal{E}_a = 75\ \text{mV}$, $R_a = 2R$, $\mathcal{E}_c = 45\ \text{mV}$, and $R_c = R$, with $R = 1.5\ \text{M}\Omega$. Then

$$I_b = \frac{(\mathcal{E}_b - \mathcal{E}_a)R_c + (\mathcal{E}_b - \mathcal{E}_c)R_a}{R_a R_b + R_b R_c + R_c R_a}$$
$$= \frac{(\mathcal{E}_b - \mathcal{E}_a)R + (\mathcal{E}_b - \mathcal{E}_c)2R}{4R^2 + 2R^2 + 2R^2}$$
$$= (3\mathcal{E}_b - 2\mathcal{E}_c - \mathcal{E}_a)/8R = (60 - 90 - 75)\ \text{mV}/12\ \text{M}\Omega$$
$$= -8.75\ \text{nA}.$$

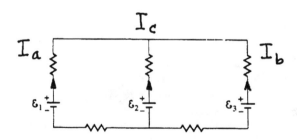

FIGURE 28-58 Problems 37, 38.

Section 28-5:   Electrical Measuring Instruments

Problem

39. A voltmeter with 200-kΩ resistance is used to measure the voltage across the 10-kΩ resistor in Fig. 28-59. By what percentage is the measurement in error because of the finite meter resistance?

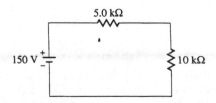

FIGURE 28-59 Problems 39 and 40.

Solution

The voltage across the 10 kΩ resistor in Fig. 28-59 is $(150\ \text{V})(10)/(10 + 5) = 100\ \text{V}$ (the circuit is just a voltage divider as described by Equations 28-2a and b), as would be measured by an ideal voltmeter with infinite resistance. With the real voltmeter connected in parallel across the 10 kΩ resistor, its effective resistance is changed to $R_{||} = (10 \text{ k}\Omega)(200 \text{ k}\Omega) \div (210 \text{ k}\Omega) = 9.52$ kΩ, and the voltage reading is only $(150 \text{ V})(9.52)/(9.52 + 5) = 98.4$ V, or about 1.64% lower.

Problem

41. A neophyte mechanic foolishly connects an ammeter with 0.1-Ω resistance directly across a 12-V car battery whose internal resistance is 0.01 Ω. What is the power dissipation in the meter? No wonder it gets destroyed!

Solution

The current through the misconnected ammeter is $I = \mathcal{E}/(R_{int} + R_m)$, so the power dissipated in it is $\mathcal{P} = I^2 R_m = \mathcal{E}^2 R_m/(R_{int} + R_m)^2 = (12 \text{ V}/0.11 \text{ }\Omega)^2(0.1 \text{ }\Omega) = 1.19$ kW (comparable to a small toaster-oven).

Problem

43. In Fig. 28-61 what are the meter readings when (a) an ideal voltmeter or (b) an ideal ammeter is connected between points A and B?

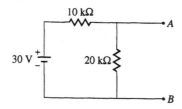

FIGURE 28-61 Problem 43.

Solution

(a) An ideal voltmeter has infinite resistance, so AB is still an open circuit (as shown on Fig. 28-61) when such a voltmeter is connected. The meter reads the voltage across the 20 kΩ resistor (part of a voltage divider), or $(30 \text{ V})20/(20 + 10) = 20$ V (see Equation 28-2a or b). (b) An ideal ammeter has zero resistance, and thus measures the current through the points A and B when short-circuited (i.e., no current flows through the 20 kΩ resistor). In Fig. 28-61, this would be $I_{AB} = 30 \text{ V}/10 \text{ }\Omega = 3$ mA. (Such a connection does not measure the current in the original circuit, since an ammeter should be connected in series with the current to be measured.)

Section 28-6: Circuits with Capacitors

Problem

45. Show that the quantity RC has the units of time (seconds).

Solution

The SI units for the time constant, RC, are $(\Omega)(\text{F}) = (\text{V}/\text{A})(\text{C}/\text{V}) = (\text{s}/\text{C})(\text{C}) = $ s, as stated.

Problem

47. Show that a capacitor is charged to approximately 99% of the applied voltage in five time constants.

Solution

After five time constants, Equation 28-6 gives a voltage of $V_C/\mathcal{E} = 1 - e^{-5} = 1 - 6.74 \times 10^{-3} \simeq 99.3\%$ of the applied voltage.

Problem

49. Figure 28-62 shows the voltage across a capacitor that is charging through a 4700-Ω resistor in the circuit of Fig. 28-29. Use the graph to determine (a) the battery voltage, (b) the time constant, and (c) the capacitance.

Solution

(a) For the circuit considered, the voltage across the capacitor asymptotically approaches the battery voltage after a long time (compared to the time constant). In Fig. 28-62, this is about 9 V. (b) The time constant is the time it takes the capacitor voltage to reach $1 - e^{-1} = 63.2\%$ of its asymptotic value, or 5.69 V in this case. From the graph, $\tau \simeq 1.5$ ms. (c) The time constant is RC, so $C = 1.5 \text{ ms}/4700 \text{ }\Omega = 0.319$ μF.

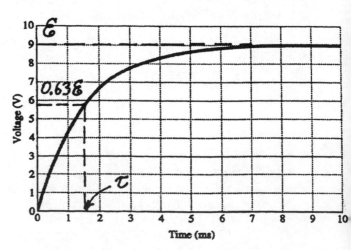

FIGURE 28-62 Problem 49 Solution.

Problem

51. A 1.0-μF capacitor is charged to 10.0 V. It is then connected across a 500-kΩ resistor. How long does it take (a) for the capacitor voltage to reach 5.0 V and (b) for the energy stored in the capacitor to decrease to half its initial value?

Solution

A capacitor discharging through a resistor is described by exponential decay, with time constant RC (see Equation 28-8), and, of course, $U_C(t) = \frac{1}{2}CV(t)^2 = \frac{1}{2}CV_0^2 e^{-2t/RC} = U_C(0)e^{-2t/RC}$ is the energy stored (see Equation 26-8b). (a) $V(t)/V(0) = 1/2$ implies $t = RC \ln 2 = (500 \text{ k}\Omega)(1 \mu F)(0.693) = 347$ ms. (b) $U_c(t)/U_c(0) = 1/2$ implies $t = \frac{1}{2}RC \ln 2 = 173$ ms.

Problem

53. A capacitor is charged until it holds 5.0 J of energy. It is then connected across a 10-kΩ resistor. In 8.6 ms, the resistor dissipates 2.0 J. What is the capacitance?

Solution

Equation 28-8 gives a voltage $V = V_0 e^{-t/RC}$ for a capacitor discharging through a resistor. If 2 J is dissipated in time t, the energy stored in the capacitor drops from $U_0 = 5$ J to $U = 3$ J (assuming there are no losses due to radiation, etc.). Since $U = \frac{1}{2}CV^2$, $U_0/U = (V_0/V)^2 = e^{2t/RC}$ and we may solve for C: $C = 2t/R\ln(U_0/U) = 2(8.6 \text{ ms})/(10 \text{ k}\Omega)\times \ln(5/3) = 3.37 \mu F$.

Problem

55. For the circuit of Example 28-9, take $\mathcal{E} = 100$ V, $R_1 = 4.0$ kΩ, and $R_2 = 6.0$ kΩ, and assume the capacitor is initially uncharged. What are the currents in both resistors and the voltage across the capacitor (a) just after the switch is closed and (b) a long time after the switch is closed? Long after the switch is closed it is again opened. What are I_1, I_2, and V_C (c) just after this switch opening and (d) a long time later?

Solution

In addition to the explanation in Example 28-9, we note that when the switch is in the closed position, Kirchhoff's voltage law applied to the loop containing both resistors yields $\mathcal{E} = I_1 R_1 + I_2 R_2$, and to the loop containing just R_2 and C, $V_C = I_2 R_2$. (a) If the switch is closed at $t = 0$, Example 28-9 shows that $V_C(0) = 0$, $I_2(0) = 0$, and $I_1(0) = \mathcal{E}/R_1 = 100$ V/4 k$\Omega = 25$ mA. (b) After a long time, $t = \infty$, Example 28-9 also shows

that $I_1(\infty) = I_2(\infty) = \mathcal{E}/(R_1 + R_2) = 100$ V/10 k$\Omega = 10$ mA, and $V_C(\infty) = I_2(\infty)R_2 = (10 \text{ mA})(6 \text{ k}\Omega) = 60$ V. (c) Under the conditions stated, the fully charged capacitor ($V_C = 60$ V) simply discharges through R_2. (R_1 is in an open-circuit branch, so $I_1 = 0$ for the entire discharging process.) The initial discharging current is $I_2 = V_C/R_2 = 60$ V/6 k$\Omega = 10$ mA. (d) I_2 and V_C decay exponentially to zero.

Problem

57. In the circuit for Fig. 28-65 the switch is initially open and the capacitor is uncharged. Find expressions for the current I supplied by the battery (a) just after the switch is closed and (b) a long time after the switch is closed.

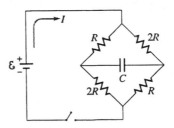

FIGURE 28-65 Problem 57.

Solution

(a) Just after the switch is closed, the uncharged capacitor acts instantaneously like a short circuit and the resistors act like two parallel pairs in series. The effective resistance of the combination is $2\times(R)(2R)/(R+2R) = 4R/3$, and the current supplied by the battery is $I(0) = 3\mathcal{E}/4R$. (b) A long time after the switch is closed, the capacitor is fully charged and acts like an open circuit. Then the resistors act like two series pairs in parallel, with an effective resistance of $(\frac{1}{2})(R + 2R) = 3R/2$. The battery current is $I(\infty) = 2\mathcal{E}/3R$.

Paired Problems
Problem

59. A 3.3-kΩ resistor and a 4.7-kΩ resistor are connected in parallel, and the pair is in series with a 1.5-kΩ resistor. What is the resistance of the combination?

Solution

Equation 28-3c for the resistance of the parallel pair, combined with Equation 28-1 for the resistance of this in series with the third resistor gives an effective resistance of $R_{\text{eff}} = 1.5$ kΩ + (3.3)(4.7) kΩ/(3.3 + 4.7) = 3.44 kΩ.

Problem

61. A battery's voltage is measured with a voltmeter whose resistance is 1000 Ω; the result is 4.36 V. When the measurement is repeated with a 1500-Ω meter the result is 4.41 V. What are (a) the battery voltage and (b) its internal resistance?

Solution

The internal resistance of the battery (R_i) and the resistance of the voltmeter (R_m) are in series with the battery's emf, so the current is $I = \mathcal{E}/(R_i + R_m)$. The potential drop across the meter (its reading) is $V_m = IR_m = \mathcal{E}R_m/(R_i + R_m)$. From the given data, 4.36 V = $\mathcal{E}(1 \text{ k}\Omega)/(R_i + 1 \text{ k}\Omega)$ and 4.41 V = $\mathcal{E}(1.5 \text{ k}\Omega)/(R_i + 1.5 \text{ k}\Omega)$, which can be solved simultaneously for $\mathcal{E}$ and R_i. One obtains $R_i + 1 \text{ k}\Omega = \mathcal{E}(1 \text{ k}\Omega/4.36 \text{ V})$ and $R_i + 1.5 \text{ k}\Omega = \mathcal{E}(1.5 \text{ k}\Omega/4.41 \text{ V})$, or

$$\mathcal{E} = (1.5 \text{ k}\Omega - 1 \text{ k}\Omega)\left(\frac{1.5 \text{ k}\Omega}{4.41 \text{ V}} - \frac{1 \text{ k}\Omega}{4.36 \text{ V}}\right)^{-1} = 4.51 \text{ V}$$

and $R_i = (4.51 \text{ V})(1 \text{ k}\Omega/4.36 \text{ V}) - 1 \text{ k}\Omega = 35.2 \text{ }\Omega$.

Problem

63. In Fig. 28-67, take $\mathcal{E}_1 = 12$ V, $\mathcal{E}_2 = 6.0$ V, $\mathcal{E}_3 = 3.0$ V, $R_1 = 1.0 \text{ }\Omega$, $R_2 = 2.0 \text{ }\Omega$, and $R_3 = 4.0 \text{ }\Omega$. Find the current in R_2 and give its direction.

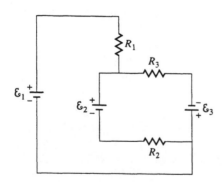

FIGURE 28-67 Problem 63.

Solution

An obvious reconfiguration of the circuit in Fig. 28-67 results in a circuit like that in Fig. 28-57, with R_1 replacing ($R_1 + R_2$), R_2 for R_3, R_3 for R_4, and $-\mathcal{E}_3$ for $\mathcal{E}_3$. Thus, the solution to Problem 35, properly altered, gives

$$I_b = \frac{(\mathcal{E}_2 - \mathcal{E}_1)R_3 + (\mathcal{E}_2 + \mathcal{E}_3)R_1}{R_1R_2 + R_1R_3 + R_2R_3}$$

$$= \frac{(6 - 12)(4) + (6 + 3)(1)}{(1)(2 + 4) + (2)(4)} \text{ A} = -1.07 \text{ A}.$$

(A negative current is opposite to the direction of the emf $\mathcal{E}_2$.)

Problem

65. In Fig. 28-68 what are the meter readings when (a) an ideal voltmeter or (b) an ideal ammeter is connected between points A and B?

Solution

(a) An ideal voltmeter has $R_m = \infty$ (AB open circuited), so V_{AB} is just the voltage across the 5.6 kΩ resistor. This is part of a voltage divider (in series with the 4.7 kΩ resistor), so Equation 28-2 gives $V_{AB} = (24 \text{ V})(5.6)/(5.6 + 4.7) = 13.0$ V. (b) An ideal ammeter has $R_m = 0$ (AB short circuited), so I_{AB} is just the current through the 3.3 kΩ resistor. This is part of parallel combination, $R_\parallel = (3.3 \text{ k}\Omega)(5.6) \div (3.3 + 5.6) = 2.08 \text{ k}\Omega$, which, in series with the 4.7 kΩ resistor, draws a total current of $I_{tot} = 24 \text{ V} \div (2.08 + 4.7) \text{ k}\Omega = 3.54$ mA. Now, two resistors in parallel form a current divider, each one taking a fraction of the total current, given by $I_1 = (R_\parallel/R_1)I_{tot}$, and $I_2 = (R_\parallel/R_2)I_{tot}$, respectively. (This follows directly from Ohm's law: $V_\parallel = I_{tot}R_\parallel = I_1R_1 = I_2R_2$.) Therefore, in the case of the ideal ammeter, $I_{AB} = (R_\parallel/3.3 \text{ k}\Omega)I_{tot} = (2.08/3.3) \times (3.54 \text{ mA}) = 2.23$ mA.

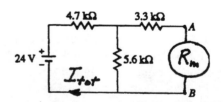

FIGURE 28-68 Problem 65 Solution.

Problem

67. An initially uncharged capacitor in an RC circuit reaches 75% of its full charge in 22.0 ms. What is the time constant?

Solution

From Equation 28-6, $V_C/\mathcal{E} = 75\% = 1 - e^{-t/\tau}$, which implies $e^{t/\tau} = 4$, or $\tau = t/\ln 4 = 22 \text{ ms}/\ln 4 = 15.9$ ms.

Supplementary Problems

Problem

69. Suppose the currents into and out of a circuit node differed by 1 μA. If the node consists of a small metal sphere with diameter 1 mm, how long would it take for the electric field around the node to reach the breakdown field in air (3 MV/m)?

Solution

The charge on the node (whether positive or negative) accumulates at a rate of 1 μA = 1 μC per second, so $|q(t)| = (1\ \mu\text{A})t$ (where we assume that $q(0) = 0$). If the node is treated approximately as an isolated sphere, the electric field strength at its surface, $k\,|q|\,/r^2 = k(1\ \mu\text{A})t/r^2$, equals the breakdown field for air, when

$$t = (3\ \text{MV/m})(0.5\ \text{mm})^2/(9{\times}10^9\ \text{m/F})(1\ \mu\text{A})$$

$$= 83.3\ \mu\text{s}.$$

Problem

71. In Fig. 28-70, what is the current in the 4-Ω resistor when each of the following circuit elements is connected between points A and B. (a) an ideal ammeter; (b) an ideal voltmeter; (c) another 4.0-Ω resistor; (d) an uncharged capacitor, right after it's connected; (e) long after the capacitor of part (d) is connected; (f) an ideal 12-V battery, with its positive terminal at A; (g) a capacitor initially charged to 12 V, right after it's connected with its positive plate at A; (h) long after the capacitor in part (g) is connected?

Solution

(a) An ideal ammeter connected across AB short-circuits the 4 Ω resistor, so $I_4 = 0$. (This is a dangerous connection unless the ammeter can handle 3 A.) (b) An ideal voltmeter across AB maintains the open circuit, so $I_4 = 6$ V/(2 + 4) Ω = 1 A. (c) Another 4 Ω resistor in parallel across AB divides the total current equally; $I_{\text{tot}} = 6$ V/[2 + 4×4/(4 + 4)] Ω =

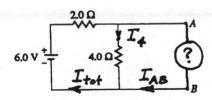

FIGURE 28-70 Problem 71 Solution.

1.5 A, and $I_4 = 0.75$ A. (d) The instantaneous voltage across an uncharged capacitor is zero, so AB is momentarily short-circuited and $I_4 = 0$, as in (a). (e) When the capacitor is fully charged $(dq/dt = 0)$ it behaves like an open circuit, so $I_4 = 1$ A, as in (b). (f) Kirchhoff's voltage law applied to the loop containing just AB (with the ideal 12 V emf) and the 4 Ω resistor, gives 12 V = $I_4(4\ \Omega)$, or $I_4 = 3$ A. (g) Instantaneously, a capacitor charged to 12 V appears like the battery in (f) when first connected, so $I_4 = 3$ A. (h) Charge leaves the capacitor until, after a long time, $dq/dt = 0 = I_C$ and it appears like an open circuit. Then $I_4 = 1$ A, as in (b) and (e).

Problem

73. A parallel-plate capacitor is insulated with a material of dielectric constant κ and resistivity ρ. Since the resistivity is finite, the capacitor "leaks" charge and can be modeled as an ideal capacitor in parallel with a resistor. (a) Show that the time constant of the capacitor is independent of its dimensions (provided the spacing is small enough that the usual parallel-plate approximation applies) and is given by $\varepsilon_0\kappa\rho$. (b) If the insulating material is polystyrene ($\kappa = 2.6$, $\rho = 10^{16}$ Ω·m), how long will it take for the stored energy in the capacitor to decrease by a factor of 2?

Solution

(a) In the parallel plate approximation, the electric field between the plates is constant, $E = V/d$, the leakage current density is uniform, $J = E/\rho$, and the capacitance is $C = \kappa\varepsilon_0 A/d$. These relations imply that the resistance of the dielectric slab between the plates is $R = V/I = V/JA = V/(EA/\rho) = \rho d/A$, and the time constant for discharging through the dielectric is $\tau = RC = (\rho d/A)(\kappa\varepsilon_0 A/d) = \kappa\varepsilon_0\rho$. (b) The stored energy in the capacitor, $U_C = \frac{1}{2}CV_C^2$, decays with half the voltage time constant, i.e., $U_C = \frac{1}{2}C(V_0 e^{-t/RC})^2 = \frac{1}{2}CV_0^2 e^{-2t/RC} = U_0 e^{-t/(RC/2)}$. To decay by 50% takes time $t = (RC/2)\ln 2 = (\kappa\varepsilon_0\rho/2)\ln 2$. With the values given for polystyrene, $t = (2.6)(8.85\ \text{pF/m})(10^{16}\ \Omega\text{·m})\frac{1}{2}\ln 2 = 7.97{\times}10^4$ s = 22.2 h.

Problem

75. Write the loop and node laws for the circuit of Fig. 28-71, and show that the time constant for this circuit is $R_1 R_2 C/(R_1 + R_2)$.

Solution

Consider the loops and node added to Fig. 28-71. Kirchhoff's laws are $\mathcal{E} = I_1 R_1 + I_2 R_2$, $V_C = I_2 R_2$, and $I_C = I_1 - I_2$. Since $V_C = q/C$ and $I_C = dq/dt$, the equations can be combined to yield

$$\mathcal{E} - I_1 R_1 - I_2 R_2 = \mathcal{E} - (I_C + I_2) R_1 - I_2 R_2$$

$$= \mathcal{E} - I_C R_1 - \left(\frac{V_C}{R_2}\right)(R_1 + R_2)$$

$$= \mathcal{E} - I_C R_1 - \frac{q}{CR_2/(R_1 + R_2)} = 0.$$

This is exactly in the same form as the first equation, solved in the text, in the section "The RC Circuit:

Charging" (with $I \to I_C$, $R \to R_1$ and $C \to CR_2 \div (R_1 + R_2)$), so the time constant for the circuit is $\tau = CR_1 R_2/(R_1 + R_2)$ (the ratio of the coefficients of I_C and q).

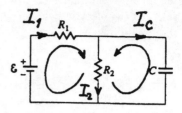

FIGURE 28-71 Problem 75 Solution.

CHAPTER 29   THE MAGNETIC FIELD

ActivPhysics can help with these problems:
Activities 13.4, 13.6, 13.7, 13.8

Section 29-2:   The Magnetic Force and Moving Charge

Problem

1. (a) What is the minimum magnetic field needed to exert a 5.4×10^{-15}-N force on an electron moving at 2.1×10^7 m/s? (b) What magnetic field strength would be required if the field were at 45° to the electron's velocity?

Solution

(a) From Equation 29-1b, $B = F/ev\sin\theta$, which is a minimum when $\sin\theta = 1$ (the magnetic field perpendicular to the velocity). Thus, $B_{min} = (5.4\times10^{-15}$ N$)/(1.6\times10^{-19}$ C$)(2.1\times10^7$ m/s$) = 1.61\times10^{-3}$ T $= 16.1$ G. (b) For $\theta = 45°$, $B = B_{min} \div \sin 45° = \sqrt{2}\,B_{min} = 22.7$ G.

Problem

3. What is the magnitude of the magnetic force on a proton moving at 2.5×10^5 m/s (a) at right angles; (b) at 30°; (c) parallel to a magnetic field of 0.50 T?

Solution

From Equation 29-1b, $F = evB\sin\theta$, so (a) when $\theta = 90°$, $F = (1.6\times10^{-19}$ C$)(2.5\times10^5$ m/s$)(0.5$ T$) = 2.0\times10^{-14}$ N, (b) $F = (2.0\times10^{-14}$ N$)\sin 30° = 1.0\times10^{-14}$ N, and (c) $F = evB\sin 0° = 0$.

Problem

5. A particle carrying a 50-μC charge moves with velocity $\mathbf{v} = 5.0\hat{\imath} + 3.2\hat{k}$ m/s through a uniform magnetic field $\mathbf{B} = 9.4\hat{\imath} + 6.7\hat{\jmath}$ T. (a) What is the force on the particle? (b) Form the dot products $\mathbf{F}\cdot\mathbf{v}$ and $\mathbf{F}\cdot\mathbf{B}$ to show explicitly that the force is perpendicular to both $\mathbf{v}$ and $\mathbf{B}$.

Solution

(a) From Equation 27-2, $\mathbf{F} = q\mathbf{v}\times\mathbf{B} = (50~\mu$C$)\times(5\hat{\imath} + 3.2\hat{k}$ m/s$)(9.4\hat{\imath} + 6.7\hat{\jmath}$ T$) = (50\times10^{-6}$ N$)\times(5\times6.7\hat{k} + 3.2\times9.4\hat{\jmath} - 3.2\times6.7\hat{\imath}) = (-1.072\hat{\imath} + 1.504\hat{\jmath} + 1.675\hat{k})\times10^{-3}$ N. (The magnitude and direction can be found from the components, if

desired.) (b) The dot products $\mathbf{F}\cdot\mathbf{v}$ and $\mathbf{F}\cdot\mathbf{B}$ are, respectively, proportional to $(-1.072)(5) + (1.675)\times(3.2) = 0$, and $(-1.072)(9.4) + (1.504)(6.7) = 0$, since the cross product of two vectors is perpendicular to each factor. (We did not round off the components of $\mathbf{F}$, so that the vanishing of the dot products could be exactly confirmed.)

Problem

7. A proton moving with velocity $\mathbf{v}_1 = 3.6\times10^4~\hat{\jmath}$m/s experiences a magnetic force of $7.4\times10^{-16}~\hat{\imath}$N. A second proton moving on the x axis experiences a magnetic force of $2.8\times10^{-16}~\hat{\jmath}$N. Find the magnitude and direction of the magnetic field, and the velocity of the second proton.

Solution

The magnetic force on the first proton is $(7.4\times10^{-16}$ N$)\hat{\imath} = e(v_1\hat{\jmath})(B_x\hat{\imath} + B_y\hat{\jmath} + B_z\hat{k}) = ev_1(-B_x\hat{k} + B_z\hat{\imath})$, so $B_x = 0$ and 7.4×10^{-16} N $= ev_1B_z$. The force on the second proton is $(2.8\times10^{-16}$ N$)\hat{\jmath} = e(v_2\hat{\imath})\times(B_y\hat{\jmath} + B_z\hat{k}) = ev_2(B_y\hat{k} - B_z\hat{\jmath})$, so $B_y = 0$ and 2.8×10^{-16}N $= -ev_2B_z$. Therefore, $\mathbf{B} = B_z\hat{k} = \hat{k}(7.4\times10^{-16}$ N$)\div(1.6\times10^{-19}$ C$)(3.6\times10^4$ m/s$) = (0.128$ T$)\hat{k}$, and $\mathbf{v}_2 = v_2\hat{\imath} = \hat{\imath}(-2.8/7.4)(3.6\times10^4$ m/s$) = (-1.36\times10^4$ m/s$)\hat{\imath}$.

Problem

9. An alpha particle (2 protons, 2 neutrons) is moving with velocity $\mathbf{v} = 150\hat{\imath} + 320\hat{\jmath} - 190\hat{k}$ km/s in a magnetic field is $\mathbf{B} = 0.66\hat{\imath} - 0.41\hat{\jmath}$ T. Find the magnitude of the force on the particle.

Solution

The magnetic force on the alpha particle is (Equation 29-1a): $\mathbf{F}_B = 2e\mathbf{v}\times\mathbf{B} = 2(1.6\times10^{-19}$ C$)\times(150\hat{\imath} + 320\hat{\jmath} - 190\hat{k})(10^3$ m/s$)\times(0.66\hat{\imath} - 0.41\hat{\jmath})$ T $= (3.2\times10^{-16}$ N$)(-150\times0.41\hat{k} - 320\times0.66\hat{k} - 190\times0.66\hat{\jmath} - 190\times0.41\hat{\imath}) = -(24.9\hat{\imath} + 40.1\hat{\jmath} + 87.3\hat{k})$ fN.

Problem

11. A 1.4-μC charge moving at 185 m/s experiences a magnetic force $\mathbf{F}_B = 2.5\hat{\imath} + 7.0\hat{\jmath}~\mu$N in a magnetic field $\mathbf{B} = 4.2\hat{\imath} - 15\hat{\jmath}$ mT. What is the angle

between the particle's velocity and the magnetic field?

Solution

Equation 29-1b gives $\sin\theta = F/qvB = \sqrt{(2.5^2) + (7.0)^2}$ μN $\div$ (1.4 μC)(185 m/s)$\times$ $\sqrt{(42^2) + (-15)^2}$ mT = 0.644. Then $\theta = 40.1°$ or $140°$ (both are possible since $\sin\theta = \sin(180° - \theta)$).

Problem

13. A region contains an electric field
$\mathbf{E} = 7.4\hat{\imath} + 2.8\hat{\jmath}$ kN/C and a magnetic field
$\mathbf{B} = 15\hat{\jmath} + 36\hat{k}$ mT. Find the electromagnetic force on (a) a stationary proton, (b) an electron moving with velocity $\mathbf{v} = 6.1\hat{\imath}$ Mm/s.

Solution

The force on a moving charge is given by Equation 29-2 (called the Lorentz force) $\mathbf{F} = q(\mathbf{E} + \mathbf{v}\times\mathbf{B})$. (a) For a stationary proton, $q = e$ and $\mathbf{v} = 0$, so $\mathbf{F} = e\mathbf{E} = (1.6\times10^{-19}$ C)(7.4$\hat{\imath}$ + 2.8$\hat{\jmath}$) kN/C = (1.18$\hat{\imath}$ + 0.448$\hat{\jmath}$) fN. (b) For the electron, $q = -e$ and $\mathbf{v} = 6.1\hat{\imath}$ Mm/s, so the electric force is the negative of the force in part (a) and the magnetic force is $-e\mathbf{v}\times\mathbf{B} = (-1.6\times10^{-19}$ C)(6.1$\hat{\imath}$ Mm/s)$\times$(15$\hat{\jmath}$ + 36$\hat{k}$) mT = $(-14.6\hat{k} + 35.1\hat{\jmath})$ fN. The total Lorentz force is the sum of these, or $(-1.18\hat{\imath} + 34.7\hat{\jmath} - 14.6\hat{k})$ fN.

Section 29-3: The Motion of Charged Particles in Magnetic Fields

Problem

15. What is the radius of the circular path described by a proton moving at 15 km/s in a plane perpendicular to a 400-G magnetic field?

Solution

From Equation 29-3, the radius of the orbit is $r = mv/eB = (1.67\times10^{-27}$ kg)(15 km/s)/(1.6$\times10^{-19}$ C)$\times$ $(4\times10^{-2}$ T) = 3.91 mm. (SI units and data for the proton are summarized in the appendices and inside front cover.)

Problem

17. Radio astronomers detect electromagnetic radiation at a frequency of 42 MHz from an interstellar gas cloud. If this radiation is caused by electrons spiraling in a magnetic field, what is the field strength in the gas cloud?

Solution

If this is electromagnetic radiation at the electron's cyclotron frequency, Equation 29-5 implies a field

strength of $B = 2\pi f(m/e) = 2\pi(42$ MHz)$\times$ $(9.11\times10^{-31}$ kg/1.6$\times10^{-19}$ C) = 1.50$\times10^{-3}$ T = 15.0 G.

Problem

19. Electrons and protons with the same kinetic energy are moving at right angles to a uniform magnetic field. How do their orbital radii compare?

Solution

It is convenient to anticipate the result of Problem 22 for the orbital radius of a non-relativistic charged particle in a plane perpendicular to a uniform magnetic field. From Equation 29-3, $r = mv/qB$. For a non-relativistic particle, $K = \frac{1}{2}mv^2$, or $v = \sqrt{2K/m}$, therefore $r = \sqrt{2Km}/qB$. (Note: All quantities can, of course, be expressed in standard SI units, but in many applications, atomic units are more convenient. The conversion factor for electron volts to joules is just the numerical magnitude of the electronic charge, so if K is expressed in MeV, m in MeV/c^2, q in multiples of e, and B in teslas, we obtain

$$r = \frac{\sqrt{2K(e\times10^6)m(e\times10^6/c^2)}}{(qe)B}$$

$$= \frac{10^6\sqrt{2Km}}{(3\times10^8)qB} = \frac{\sqrt{2Km}}{300qB}.)$$

From this expression, it follows that protons and electrons with the same kinetic energy have radii in the ratio $r_p/r_e = \sqrt{m_p/m_e} = \sqrt{1836} \approx 43$, in the same magnetic field. Heavier particles are more difficult to bend.

Problem

21. Microwaves in a microwave oven are produced by electrons circling in a magnetic field at a frequency of 2.4 GHz. (a) What is the magnetic field strength? (b) The electrons' motion takes place inside a special tube called a magnetron. If the magnetron can accommodate electron orbits with a maximum diameter of 2.5 mm, what is the maximum electron energy?

Solution

(a) A cyclotron frequency of 2.4 GHz for electrons implies a magnetic field strength of $B = 2\pi f(m/e) = 2\pi(2.4$ GHz)(9.11$\times10^{-31}$ kg/1.6$\times10^{-19}$ C) = 85.9 mT (see Equation 29-5). (b) The kinetic energy of an electron, with the maximum orbital radius allowed for this magnetron tube (half the diameter), in the field found in part (a), is $K = (reB)^2/2m = (1.25$ mm $\times$

1.6×10^{-19} C $\times$ 85.9 mT)$^2/(2 \times 9.11 \times 10^{-31}$ kg) $=$ $(1.62 \times 10^{-16}$ J)$/(1.6 \times 10^{-19}$ J/eV) $= 1.01$ keV (see Problem 22). The same calculation in atomic units, explained in the solution to Problem 19, is $(1.25 \times 10^{-3} \times 300 \times 85.9 \times 10^{-3})^2 (2 \times 0.511)^{-1}$ MeV. The electron's kinetic energy could also be expressed in terms of the cyclotron frequency directly, $K = (2\pi f rm)^2/2m = 2m(\pi r f)^2$, with the same result.

Problem

23. Two protons, moving in a plane perpendicular to a uniform magnetic field of 500 G, undergo an elastic head-on collision. How much time elapses before they collide again? *Hint:* Draw a picture.

Solution

In an elastic head on collision between particles of equal mass, the particles exchange velocities $(v_{1f} = v_{2i}$ and $v_{2f} = v_{1i}$, see Section 11-4). Moving in a plane perpendicular to **B**, each proton describes a different circle of radius $r = mv/eB$, with period (which is independent of r and v) of $T = 2\pi m/eB = 2\pi(1.67 \times 10^{-27}$ kg)$/(1.6 \times 10^{-19}$ C)$(5 \times 10^{-2}$ T) $=$ 1.31 μs. After one period, each proton would be back at the site of the collision, and could collide again.

Problem

25. A cyclotron is designed to accelerate deuterium nuclei. (Deuterium has one proton and one neutron in its nucleus.) (a) If the cyclotron uses a 2.0-T magnetic field, at what frequency should the dee voltage be alternated? (b) If the vacuum chamber has a diameter of 0.90 m, what is the maximum kinetic energy of the deuterons? (c) If the magnitude of the potential difference between the dees is 1500 V, how many orbits do the deuterons complete before achieving the energy of part (b)?

Solution

(a) The frequency of the accelerating voltage is the cyclotron frequency for deuterons (Equation 29-5), $f = eB/2\pi m \simeq (1.6 \times 10^{-19}$ C)$(2$ T)$/2\pi(2 \times 1.67 \times 10^{-27}$ kg) $= 15.2$ MHz. (b) We can use the result of Problem 19 (expressed in atomic units), with the maximum orbital radius equal to the radius of the dees. Thus, $K_{max} = (300qBr_{max})^2/2m \simeq (300 \times 1 \times 2 \times 0.45)^2/2(2 \times 938) = 19.4$ MeV. (c) If the deuterons start with essentially zero kinetic energy, and gain 1500 eV each half-orbit, they will make 19.4 MeV/2(1500 eV) $= 6.48 \times 10^3$ orbits. (Of course, the same results follow in standard SI units.)

Problem

27. Figure 29-38 shows a simple mass spectrometer, designed to analyze and separate atomic and molecular ions with different charge-to-mass ratios. In the design shown, ions are accelerated through a potential difference V, after which they enter a region containing a uniform magnetic field. They describe semicircular paths in the magnetic field, and land on a detector a lateral distance x from where they entered the field region, as shown. Show that x is given by

$$x = \frac{2}{B}\sqrt{\frac{2V}{(q/m)}},$$

where B is the magnetic field strength, V the accelerating potential, and q/m the charge-to-mass ratio of the ion. By counting the number of ions accumulated at different positions x, one can determine the relative abundances of different atomic or molecular species in a sample.

Solution

The positive ions enter the field region with speed (determined from the work-energy theorem) of $\frac{1}{2}mv^2 = qV$, or $v = \sqrt{2V(q/m)}$. They are bent into a semicircle with diameter $x = 2r = 2mv/qB = 2(m/qB)\sqrt{2V(q/m)} = 2\sqrt{2(m/q)V}/B$, as shown in Fig. 29-38 (see Equation 29-3).

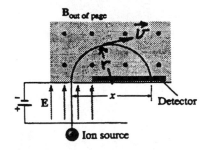

FIGURE 29-38 Problem 27 Solution.

Problem

29. A mass spectrometer is used to separate the fissionable uranium isotope U-235 from the much more abundant isotope U-238. To within what percentage must the magnetic field be held constant if there is to be no overlap of these two isotopes? Both isotopes appear as constituents of uranium hexafluoride gas (UF_6), and the gas molecules are all singly ionized.

Solution

The separation of different uranium isotopes in UF_6 molecules can be found by differentiation of the result of Problem 27. Keeping the spectrometer parameters fixed, $x \sim m^{1/2}$, $dx \sim \frac{1}{2}m^{-1/2}dm$, and $dx/x = \frac{1}{2}(dm/m)$. The molecular masses of the two species are approximately 235 or 238 plus 6×19 which equals 349 or 352, respectively, so $\frac{1}{2}(dm/m) \approx 3/2 \times 350 = 0.43\%$. For a particular ion, $x \sim B^{-1}$ and $dx \sim B^{-2}dB$, therefore $dx/x = -dB/B$. Thus, variations in B should be less than 0.43% to separate these isotopes.

Problem

31. An electron moving at 3.8×10^6 m/s enters a region containing a uniform magnetic field $\mathbf{B} = 18\hat{\mathbf{k}}$ mT. The electron is moving at 70° to the field direction, as shown in Fig. 29-39. Find the radius r and pitch p of its spiral path, as indicated in the figure.

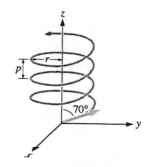

FIGURE 29-39 Problem 31.

Solution

The parallel and perpendicular components of the electron's velocity (relative to the field direction, or z axis) are $v_{\parallel} = v \cos 70° = (3.8 \times 10^6$ m/s$) \cos 70° = 1.30$ Mm/s, and $v_{\perp} = v \sin 70° = 3.57$ Mm/s. Then $r = mv_{\perp}/eB = 1.13$ mm and the pitch $p = v_{\parallel}T = v_{\parallel}(2\pi m/eB) = 2.58$ mm.

Section 29-4: The Magnetic Force on a Current

Problem

33. What is the magnitude of the force on a 50-cm-long wire carrying 15 A at right angles to a 500-G magnetic field?

Solution

The force on a straight current-carrying wire in a uniform magnetic field is (Equation 29-6) $\mathbf{F} = I\boldsymbol{\ell} \times \mathbf{B}$. Thus, $F = I\ell B \sin \theta = (15$ A$)(0.5$ m$)(0.05$ T$) \sin 90° = 0.375$ N. (The direction is given by the right-hand rule.)

Problem

35. A wire carrying 15 A makes a 25° angle with a uniform magnetic field. The magnetic force per unit length of wire is 0.31 N/m. (a) What is the magnetic field strength? (b) What is the maximum force per unit length that could be achieved by reorienting the wire in this field?

Solution

Equation 29-6 gives the magnetic force on a straight current carrying wire in a uniform magnetic field, $\mathbf{F} = I\boldsymbol{\ell} \times \mathbf{B}$. (a) From the magnitude of $\mathbf{F}$ and the given data, we find $B = F/I\ell \sin \theta = (0.31$ N/m$) \div (15$ A$) \sin 25° = 48.9$ mT. (b) By placing the wire perpendicular to the field ($\sin \theta = 1$) a maximum force per unit length of $IB = (15$ A$)(48.9$ mT$) = 0.734$ N/m could be attained.

Problem

37. In a high-magnetic-field experiment, a conducting bar carrying 7.5 kA passes through a 30-cm-long region containing a 22-T magnetic field. If the bar makes a 60° angle with the field direction, what force is necessary to hold it in place?

Solution

The magnitude of the force necessary to balance the magnetic force on the bar (Equation 29-6) is $F = |I\boldsymbol{\ell} \times \mathbf{B}| = I\ell B \sin \theta = (7.5$ kA$)(0.3$ m$)(22$ T$) \times \sin 60° = 42.9$ kN (nearly 5 tons). The direction of this force is perpendicular to the plane of $\boldsymbol{\ell} \times \mathbf{B}$ in the opposite sense as the magnetic force.

Problem

39. A piece of wire with mass per unit length 75 g/m runs horizontally at right angles to a horizontal magnetic field. A 6.2-A current in the wire results in its being suspended against gravity. What is the magnetic field strength?

Solution

A magnetic force equal in magnitude to the weight of the wire requires that $I\ell B = mg$ (since the wire is perpendicular to the field), or $B = (m/\ell)(g/I) = (75$ g/m$)(9.8$ m/s$^2)/(6.2$ A$) = 0.119$ T.

Problem

41. A wire carrying 1.5 A passes through a region containing a 48-mT magnetic field. The wire is perpendicular to the field and makes a quarter-circle turn of radius 21 cm as it passes through the field region, as shown in Fig. 29-43.

Find the magnitude and direction of the magnetic force on this section of wire.

Solution

Take the x-y axes as shown on Fig. 29-43, with the z axis out of the page and the origin at the center of the quarter-circle arc. With θ measured clockwise from the y axis, $d\boldsymbol{\ell} = R\,d\theta(\hat{\mathbf{i}}\cos\theta - \hat{\mathbf{j}}\sin\theta)$ as shown, and $\mathbf{B} = B(-\hat{\mathbf{k}})$. The magnetic force on the arc of wire is found from Equation 29-8.

$$\mathbf{F} = I\int_{arc} d\boldsymbol{\ell}\times\mathbf{B}$$
$$= I\int_0^{90°} R\,d\theta(\hat{\mathbf{i}}\cos\theta - \hat{\mathbf{j}}\sin\theta)\times B(-\hat{\mathbf{k}})$$
$$= IRB\int_0^{90°}(\hat{\mathbf{j}}\cos\theta + \hat{\mathbf{i}}\sin\theta)\,d\theta$$
$$= IRB|-\hat{\mathbf{i}}\cos\theta + \hat{\mathbf{j}}\sin\theta|_0^{90°} = IRB(\hat{\mathbf{i}}+\hat{\mathbf{j}}).$$

This has magnitude $\sqrt{2}\,IRB = \sqrt{2}(1.5\text{ A})(0.21\text{ m})\times(48\text{ mT}) = 21.4$ mN and direction 45° between the positive x and y axes.

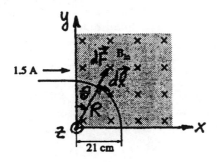

FIGURE 29-43 Problem 41 Solution.

Problem

43. Apply Equation 29-8 to a closed current loop of arbitrary shape in a *uniform* magnetic field, and show that the net force on the loop is zero. *Hint:* Both I and $\mathbf{B}$ are constant as you go around the loop, so you can take them out of the integral. What is the remaining vector integral?

Solution

If both I and $\mathbf{B}$ are constants, Equation 29-8, for a closed loop, may be written as $\mathbf{F} = \oint I\,d\boldsymbol{\ell}\times\mathbf{B} = I(\oint d\boldsymbol{\ell})\times\mathbf{B}$. But the integral is the sum of vectors, $d\boldsymbol{\ell}$, beginning and ending at the same point, hence $\oint d\boldsymbol{\ell} = 0$. (This integral relation is a special case of Equation 8-1 for a constant $\mathbf{F}$.)

Problem

45. The probe in a Hall-effect magnetometer uses a semiconductor doped to a charge-carrier density of 7.5×10^{20} m^{-3}. The probe measures 0.35 mm thick in the direction of the magnetic field being measured, and carries a 2.5-mA current perpendicular to the field. If its Hall potential is 4.5 mV, what is the magnetic field strength?

Solution

If we assume the charge-carriers are of one type, with charge of magnitude e, then Equation 29-7 and the given data require $B = nqV_H t/I = (7.5\times10^{20}\text{ m}^{-3})\times(1.6\times10^{-19}\text{ C})(4.5\text{ V})(0.35\text{ mm})/(2.5\text{ mA}) = 75.6$ mT.

Section 29-5:   A Current Loop in a Magnetic Field

Problem

47. A single-turn square wire loop 5.0 cm on a side carries a 450-mA current. (a) What is the magnetic moment of the loop? (b) If the loop is in a uniform 1.4-T magnetic field with its dipole moment vector at 40° to the field direction, what is the magnitude of the torque it experiences?

Solution

(a) Equation 29-10 for the magnetic moment of a loop gives $\mu = NIA = (1)(450\text{ mA})(5\text{ cm})^2 = 1.13\times10^{-3}$ A·m^2. (b) Equation 29-11 gives the torque on a magnetic dipole moment in a uniform magnetic field, $\tau = |\boldsymbol{\mu}\times\mathbf{B}| = \mu B\sin\theta = (1.13\times10^{-3}\text{ A·m}^2)(1.4\text{ T})\times\sin40° = 1.01\times10^{-3}$ N·m.

Problem

49. A bar magnet experiences a 12-mN·m torque when it is oriented at 55° to a 100-mT magnetic field. What is the magnitude of its magnetic dipole moment?

Solution

Equation 29-11, solved for the magnitude of the dipole moment, gives $\mu = \tau/B\sin\theta = (12\times10^{-3}\text{ N·m})\div(0.1\text{ T})\sin55° = 0.146$ A·m^2.

Problem

51. A simple electric motor like that of Fig. 29-36 consists of a 100-turn coil 3.0 cm in diameter, mounted between the poles of a magnet that produces a 0.12-T field. When a 5.0-A current flows in the coil, what are (a) its magnetic dipole moment and (b) the maximum torque developed by the motor?

Solution

(a) From Equation 29-10, the magnetic moment of the coil has magnitude $\mu = NIA = 100(5\text{ A})\frac{1}{4}\pi(0.03\text{ m})^2 = 0.353\text{ A·m}^2$. The direction of μ, determined from the right-hand rule (see Fig. 29-33), rotates with the coil. (b) The maximum torque (from Equation 29-11, with $\sin\theta = 1$) is $\tau_{max} = \mu B = (0.353\text{ A·m}^2)(0.12\text{ T}) = 4.24\times10^{-2}\text{ N·m}$.

Problem

53. Nuclear magnetic resonance (NMR) is a technique for analyzing chemical structures and is also the basis of magnetic resonance imaging used for medical diagnosis. The NMR technique relies on sensitive measurements of the energy needed to flip atomic nuclei upside-down in a given magnetic field. In an NMR apparatus with a 7.0-T magnetic field, how much energy is needed to flip a proton ($\mu = 1.41\times10^{-26}$ A·m^2) from parallel to antiparallel to the field?

Solution

From Equation 29-12, the energy required to reverse the orientation of a proton's magnetic moment from parallel to antiparallel to the applied magnetic field is $\Delta U = 2\mu B = 2(1.41\times10^{-26}\text{ A·m}^2)(7.0\text{ T}) = 1.97\times10^{-25}\text{ J} = 1.23\times10^{-6}$ eV. (This amount of energy is characteristic of radio waves of frequency 298 MHz, see Chapter 39.)

Paired Problems

Problem

55. Find the magnetic force on an electron moving with velocity $\mathbf{v} = 8.6\times10^5\hat{\mathbf{i}} - 4.1\times10^5\hat{\mathbf{j}}$ m/s in a magnetic field $\mathbf{B} = 0.18\hat{\mathbf{j}} + 0.64\hat{\mathbf{k}}$ T.

Solution

From Equation 29-1a, $\mathbf{F} = -e\mathbf{v}\times\mathbf{B} = -(1.6\times10^{-19}\text{ C})(8.6\hat{\mathbf{i}} - 4.1\hat{\mathbf{j}})(10^5\text{ m/s})\times(0.18\hat{\mathbf{j}} + 0.64\hat{\mathbf{k}})\text{ T} = (16\text{ fN})(4.1\times0.64\hat{\mathbf{i}} + 8.6\times0.64\hat{\mathbf{j}} - 8.6\times0.18\hat{\mathbf{k}}) = (42.0\hat{\mathbf{i}} + 88.1\hat{\mathbf{j}} - 24.8\hat{\mathbf{k}})$ fN. (If necessary, review the cross product of unit vectors; see Fig. 13-9 and use the right hand rule.)

Problem

57. Proponents of space-based particle-beam weapons have to confront the effect of Earth's magnetic field on their beams. If a beam of protons with kinetic energy 100 MeV is aimed in a straight line perpendicular to Earth's magnetic field in a region where the field strength is 48 μT, what will be the radius of the protons' circular path?

Solution

It is simplest to use the result of Problem 22, in atomic units, to find the radius (see solution to Problem 19), $r = \sqrt{2Km}/300qB$, where r is in meters, K is in MeV, m in MeV/c^2, q in units of e, and B in teslas. Then $r = \sqrt{2(100)(938)}/300(0.48\times10^{-4}) = 30.1$ km. (The mass of a proton in atomic units is approximately 938 MeV/c^2.)

Problem

59. A 170-mT magnetic field points into the page, confined to a square region as shown in Fig. 29-44. A square conducting loop 32 cm on a side carrying a 5.0-A current in the clockwise sense extends partly into the field region, as shown. Find the magnetic force on the loop.

Solution

The forces on the upper and lower horizontal parts of the loop in the field region cancel one-another, so the net force is just due to the magnetic force on the right vertical side, which is directed to the right in Fig. 29-44, with magnitude $I\ell B = (5\text{ A})(0.32\text{ m})\times(170\text{ mT}) = 0.272$ N. (See Equation 29-6.)

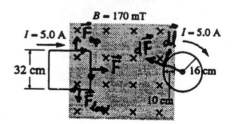

FIGURE 29-44 Problem 59 Solution.

Problem

61. An old-fashioned analog meter uses a wire coil in a magnetic field to deflect the meter needle. If the coil is 2.0 cm in diameter and consists of 500 turns of wire, what should be the magnetic field strength if the maximum torque is to be 1.6 μN·m when the current in the coil is 1.0 mA?

Solution

The maximum torque on a flat coil, with magnetic moment $\mu = N\pi R^2 I$, is $\tau_{max} = \mu B$ (see Equations 29-10 and 11). Thus, $B = \tau_{max}/NI\pi R^2 = (1.6\ \mu\text{N·m})/(500)(1\text{ mA})\pi(1\text{ cm})^2 = 102$ G.

Supplementary Problems

Problem

63. Electrons in a TV picture tube are accelerated through a 30-kV potential difference and head straight for the center of the tube, 40 cm away. If the electrons are moving at right angles to Earth's 0.50-G magnetic field, by how much do they miss the screen's exact center?

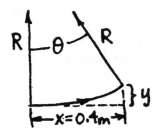

Problem 63 Solution.

Solution

The electrons suffer the maximum deflection when moving perpendicularly to the magnetic field. They are bent into a circle of radius

$$R = \sqrt{2Km}/eB$$

$$= \frac{\sqrt{2(3 \times 10^4 \text{ eV})(1.6 \times 10^{-19} \text{ J/eV})(9.11 \times 10^{-31} \text{ kg})}}{(1.6 \times 10^{-19} \text{ C})(5 \times 10^{-5} \text{ T})}$$

$$= 11.7 \text{ m}.$$

Geometry gives the maximum deflection from screen center: $y = R(1 - \cos\theta)$, where $\sin\theta = x/R$. Numerically, $y = (11.7 \text{ m})\{1 - \cos[\sin^{-1}(0.4/11.7)]\} = 6.85 \text{ mm}$.

Problem

65. A conducting bar with mass 15.0 g and length 22.0 cm is suspended from a spring in a region where a 0.350-T magnetic field points into the page, as shown in Fig. 29-45. With no current in the bar, the spring length is 26.0 cm. The bar is supplied with current from outside the field region, using wires of negligible mass. When a 2.00-A current flows from left to right in the bar, it rises 1.2 cm from its equilibrium position. Find (a) the spring constant and (b) the unstretched length of the spring.

Solution

In equilibrium, the vertical forces on the bar must sum to zero. These include the weight of the bar, $-mg$ (negative downward), the upward spring force

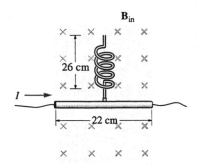

FIGURE 29-45 Problem 65.

$k\,\Delta\ell = k(\ell - \ell_0)$ (where ℓ is the length and ℓ_0 the unstretched length of the spring), and when current flows from left to right in the bar, an upward magnetic force of $F_B = ILB$ (L is the length of the bar). The conditions stated require that $k(26.0 \text{ cm} - \ell_0) - mg = 0$, and $k(24.8 \text{ cm} - \ell_0) + ILB - mg = 0$ or $k(26 \text{ cm} - \ell_0) = mg$ and $k(26 \text{ cm} - \ell_0 - 1.2 \text{ cm}) = mg - ILB$. These equations can be solved for k and ℓ_0 (subtract to eliminate ℓ_0, and divide to eliminate k) with the result $k = ILB/(1.2 \text{ cm}) = (2 \text{ A})(22 \text{ cm}) \times (0.35 \text{ T})/(1.2 \text{ cm}) = 12.8 \text{ N/m}$, and $\ell_0 = 26 \text{ cm} - (mg/ILB)(1.2 \text{ cm}) = 26 \text{ cm} - (0.015 \times 9.8 \text{ N}) \div (12.8 \text{ N/m}) = 24.9 \text{ cm}$.

Problem

67. A solid disk of mass M and thickness d sits on an incline, as shown in Fig. 29-46. A loop of wire is wrapped around the disk, running along a diameter and oriented so the loop is parallel to the incline. A uniform magnetic field **B** points

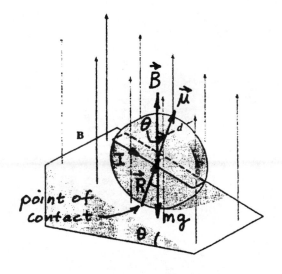

FIGURE 29-46 Problem 67.

vertically upward. Find an expression for the current I in the loop that will keep it from rolling down the incline.

Solution

The torque of gravity causing the disk to roll down the incline (about the point of contact) is $\tau_{\text{grav}} = mgR \times \sin(180° - \theta) = mgR \sin \theta$ into the page in Fig. 29-46. The magnetic torque on the magnetic moment of the loop, which would cancel this, is $\tau_{\text{mag}} = \mu B \sin \theta$ out of the page for the current shown, where $\mu = IA$ and $A = 2Rd$ is the area of the loop. (The magnetic torque is due to a couple and is the same about any point.) In equilibrium, $mg \sin \theta = I(2Rd)B \sin \theta$, or $I = mg/2Bd$. (We assume that static friction between the disk and the incline is sufficient to satisfy the other conditions for equilibrium.)

Problem

69. A 10-turn wire loop measuring 8.0 cm by 16 cm carrying 2.0 A lies in a horizontal plane but is free to rotate about the axis shown in Fig. 29-47. A 50-g mass hangs from one side of the loop, and a uniform magnetic field points horizontally, as shown. What magnetic field strength is required to hold the loop in its horizontal position?

Solution

For the direction of current shown in Fig. 29-47, the magnetic moment of the loop is downward, and the magnetic torque $\boldsymbol{\mu} \times \mathbf{B}$ is along the axis, out of the page. The gravitational torque $\mathbf{r} \times m\mathbf{g}$ is along the axis, into the page. The two torques cancel when $\mu B = mgr$, or $B = mgr/NIA = (0.05 \text{ kg})(9.8 \text{ m/s}^2) \times (0.04 \text{ m})/(10)(2.0 \text{ A})(0.8 \times 0.16 \text{ m}^2) = 76.6 \text{ mT}$.

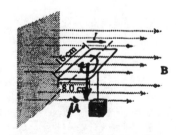

FIGURE 29-47 Problem 69 Solution.

Problem

71. A circular wire loop of mass m and radius R carries a current I. The loop is hanging horizontally below a cylindrical bar magnet,

suspended by the magnetic force, as shown in Fig. 29-49. If the field lines crossing the loop make an angle θ with the vertical, show that the strength of the magnet's field at the loop's position is $B = mg/2\pi RI \sin \theta$.

Solution

We assume that the magnetic field has axial symmetry, with vertical axis through the center of the loop's horizontal plane area. The magnetic field lines intersect the loop at right angles, and if the current circulates clockwise (as seen from the magnet), the magnetic force on an element $I \, d\ell$ has an upward component $dF_y = dF \sin \theta = I \, d\ell \, B \sin \theta$. This is the same at any position on the loop, so the net upward force is $F_y = \int IB \sin \theta \, d\ell = 2\pi RIB \sin \theta$. (The horizontal forces on diametrically opposite elements of loop cancel, so $F_x = 0$.) When the loop is suspended, the upward magnetic force balances the loop's weight, so $F_y = mg$, or $B = mg/2\pi RI \sin \theta$.

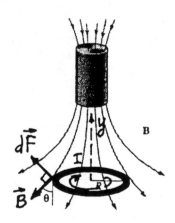

FIGURE 29-49 Problem 71 Solution.

Problem

73. Early models pictured the electron in a hydrogen atom as being in a circular orbit of radius 5.29×10^{-11} m about the stationary proton, held in orbit by the electric force. Find the magnetic dipole moment of such an atom. This quantity is called the *Bohr magneton* and is typical of atomic-sized magnetic moments. *Hint:* The full electron charge passes any given point in the orbit once per orbital period. Use this fact to calculate the average current.

Solution

One electronic charge passes a given point on the orbit every period of revolution, so the magnitude of the average current corresponding to the electron's orbital motion is $I = \Delta q/\Delta t = e/(2\pi r/v)$. (The current circulates opposite to the orbital motion, since the electron is negatively charged.) In the simplest version of the Bohr model for the hydrogen atom, the electron moves in a circular orbit, around a fixed proton, under the influence of the Coulomb force, so that $mv^2/r = ke^2/r^2$, or $v/r = \sqrt{ke^2/mr^3}$. Thus,

$I = (e/2\pi)(v/r) = (e^2/2\pi)\sqrt{k/mr^3}$. The magnetic dipole moment associated with this orbital atomic current (called a Bohr magneton) has magnitude $\mu_B = I\pi r^2 = \frac{1}{2}e^2\sqrt{kr/m} \approx \frac{1}{2}(1.6\times10^{-19}\text{ C})^2\times$ $\sqrt{(9\times10^9\text{ N}\cdot\text{m}^2/\text{C}^2)(5.29\times10^{-11}\text{ m})/(9.11\times10^{-31}\text{ kg})} \approx$ $9.25\times10^{-24}\text{ A}\cdot\text{m}^2$, and is typical of the size of atomic magnetic dipole moments in general.

CHAPTER 30   SOURCES OF THE MAGNETIC FIELD

ActivPhysics can help with these problems:
Activities 13.1–13.3, 13.5

Section 30-1:   The Biot-Savart Law

Problem

1. A wire carries 15 A. You form the wire into a single-turn circular loop with magnetic field 80 μT at the loop center. What is the loop radius?

Solution

Equation 30-3, with $x = 0$, gives the magnetic field at the center of a circular loop, $B = \mu_0 I/2a$ (with direction along the axis of the loop, consistent with the sense of circulation of the current and the right-hand rule). Thus, the radius is $a = \mu_0 I/2B = (2\pi \times 10^{-7}\ \text{N/A}^2)(15\ \text{A})/(80\ \mu\text{T}) = 11.8$ cm.

Problem

3. A 2.2-m-long wire carrying 3.5 A is wound into a tight, loop-shaped coil 5.0 cm in diameter. What is the magnetic field at its center?

Solution

The magnetic field at the center of each of the tightly-wound turns is the same (Equation 30-3 with $x = 0$), so the net field is $B = N(\mu_0 I/2a)$, where N is the number of turns. If the wire has length L and the coil has diameter D, then $N = L/\pi D$ and $B = (L/\pi D)(\mu_0 I/D) = \mu_0 IL/\pi D^2 = (4\times 10^{-7}\ \text{N/A}^2)\times (3.5\ \text{A})(2.2\ \text{m})/(5\ \text{cm})^2 = 1.23$ mT.

Problem

5. Suppose Earth's magnetic field arose from a single loop of current at the outer edge of the planet's liquid core (core radius 3000 km), concentric with Earth's center. What current would be necessary to give the observed field strength of 62 μT at the north pole? (The currents responsible for Earth's field are more complicated than this problem suggests.)

Solution

The north pole is on the axis of the hypothetical circular current loop, with radius $a = 3.0$ Mm, at a distance $x = R_E = 6.37$ Mm from the earth's center.

Equation 30-3 gives a current of $I = 2B(x^2 + a^2)^{3/2} \div \mu_0 a^2 = 2(62\ \mu\text{T})[(6.37\ \text{Mm})^2 + (3.0\ \text{Mm})^2]^{3/2} \div (4\pi \times 10^{-7}\ \text{T·m/A})(3.0\ \text{Mm})^2 = 3.83$ GA.

Problem

7. A single-turn current loop carrying 25 A produces a magnetic field of 3.5 nT at a point on its axis 50 cm from the loop center. What is the loop area, assuming the loop diameter is much less than 50 cm?

Solution

If the radius of the loop is assumed to be much smaller than the distance to the field point ($a \ll x = 50$ cm), then Equation 30-4 for the field on the axis of a magnetic dipole can be used to find $\mu = 2\pi x^3 B/\mu_0$. The magnetic moment of a single-turn loop is $\mu = IA$, therefore $A = \mu/I = 2\pi x^3 B/\mu_0 I = (3.5\ \text{nT})(50\ \text{cm})^3 \div (2\times 10^{-7}\ \text{N/A}^2)(25\ \text{A}) = 0.875$ cm^2.

Problem

9. You have a spool of thin wire that can handle a maximum current of 0.50 A. If you wind the wire into a loop-like coil 20 cm in diameter, how many turns should the coil have if the magnetic field at its center is to be 2.3 mT at this maximum current?

Solution

Equation 30-3 can be modified for N turns of wire (as in the solution to Problem 3), so at the center of a flat circular coil, $B = N\mu_0 I/2a$. Thus, $N = 2aB/\mu_0 I = (0.2\ \text{m})(2.3\ \text{mT})/(4\pi \times 10^{-7}\ \text{N/A}^2)(0.5\ \text{A}) = 732$.

Problem

11. Two long, parallel wires are 6.0 cm apart. One carries 5.0 A and the other 10 A, with both currents in the same direction. Where on a line perpendicular to both wires is the magnetic field zero?

Solution

The magnetic fields from each parallel wire are in opposite directions at a point between the wires, with magnitudes given by Equation 30-5 (or Equation 30-8). If d is the distance between the wires and x the distance of the field point from the first wire, then $B = B_1 - B_2 = (\mu_0/2\pi)[(I_1/x) - I_2/(d-x)]$. Evidently, $B = 0$ when $I_1/x = I_2/(d-x)$, or $x = d/(1 + I_2/I_1)$. For the case $I_2 = 2I_1$ and $d = 6$ cm, this gives a null field at $x = 2$ cm from the first (smaller current) wire.

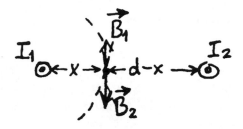

Problem 11 Solution.

Problem

13. A power line carries a 500-A current toward magnetic north and is suspended 10 m above the ground. The horizontal component of Earth's magnetic field at the power line's latitude is 0.24 G. If a magnetic compass is placed on the ground directly below the power line, in what direction will it point?

Solution

A compass needle (small dipole magnet) is free to rotate in a horizontal plane until it is aligned with the direction of the total horizontal magnetic field (see Equation 29-11). A long, straight wire (the power line) carrying a current of 500 A parallel to the ground, in the direction of magnetic north, produces a magnetic field, at a distance 10 m below, to the west, with magnitude given by Equation 30-8 (see diagram): $B_y = \mu_0 I/2\pi r = (2\times10^{-7} \text{ N/A}^2)(500 \text{ A})/10 \text{ m} = 10^{-5}$ T $= 0.1$ G. The horizontal component of the Earth's magnetic field is $B_x = 0.24$ G, so the compass

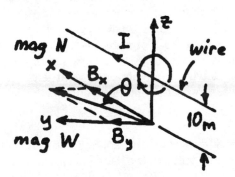

Problem 13 Solution.

needle will point $\theta = \tan^{-1} B_y/B_x = \tan^{-1}(0.1/0.24) = 22.6°$ west of magnetic north.

Problem

15. Part of a long wire is bent into a semicircle of radius a, as shown in Fig. 30-50. A current I flows in the direction shown. Use the Biot-Savart law to find the magnetic field at the center of the semicircle (point P).

Solution

The Biot-Savart law (Equation 30-2) written in a coordinate system with origin at P, gives $\mathbf{B}(P) = (\mu_0 I/4\pi) \int_{\text{wire}} d\boldsymbol{\ell} \times \hat{\mathbf{r}}/r^2$, where $\hat{\mathbf{r}}$ is a unit vector from an element $d\boldsymbol{\ell}$ on the wire to the field point P. On the straight segments to the left and right of the semicircle, $d\boldsymbol{\ell}$ is parallel to $\hat{\mathbf{r}}$ or $-\hat{\mathbf{r}}$, respectively, so $d\boldsymbol{\ell} \times \hat{\mathbf{r}} = 0$. On the semicircle, $d\boldsymbol{\ell}$ is perpendicular to $\hat{\mathbf{r}}$ and the radius is constant, $r = a$. Thus,

$$B(P) = \frac{\mu_0 I}{4\pi} \int_{\text{semicircle}} \frac{d\ell}{a^2} = \frac{\mu_0 I}{4\pi}\cdot\frac{\pi a}{a^2} = \frac{\mu_0 I}{4a}.$$

The direction of $\mathbf{B}(P)$, from the cross product, is into the page.

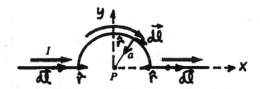

FIGURE 30-50 Problem 15 Solution.

Problem

17. Figure 30-51 shows a conducting loop formed from concentric semicircles of radii a and b. If the loop carries a current I as shown, find the magnetic field at point P, the common center.

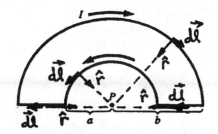

FIGURE 30-51 Problem 17 Solution.

Solution

The Biot-Savart law gives $\mathbf{B}(P) = (\mu_0/4\pi)\times \int_{\text{loop}} I d\boldsymbol{\ell} \times \hat{\mathbf{r}}/r^2$. $I d\boldsymbol{\ell} \times \hat{\mathbf{r}}/r^2$ on the inner semicircle has magnitude $Id\ell/a^2$ and direction out of the page, while on the outer semicircle, the magnitude is $Id\ell/b^2$ and the direction is into the page. On the straight segments, $d\ell \times \hat{r} = 0$, so the total field at P is $(\mu_0/4\pi)[(I\pi a/a^2) - (I\pi b/b^2)] = \mu_0 I(b-a)/4ab$ out of the page. (Note: the length of each semicircle is $\int d\ell = \pi r$.)

Section 30-2:   The Magnetic Force Between Two Conductors

Problem

19. It would take a rather large apparatus to implement the definition of the ampere given at the end of Section 30-2. Suppose you wanted to use a smaller apparatus, with wires 50 cm long separated by 2.0 cm. What force would correspond to a current of 1 A?

Solution

Since 50 cm $= \ell \gg d = 2$ cm, the wires can be considered approximately infinitely long and Equation 30-6 gives $F = \mu_0 I_1 I_2 \ell/2\pi d = (2\times10^{-7} \text{ N/A}^2)\times$ $(1 \text{ A})^2(50 \text{ cm})/(2 \text{ cm}) = 5 \text{ }\mu\text{N}$ (about half the weight of a raindrop, see Table 1-4).

Problem

21. The structure shown in Fig. 30-52 is made from conducting rods. The upper horizontal rod is free to slide vertically on the uprights, while maintaining electrical contact with them. The upper rod has mass 22 g and length 95 cm. A battery connected across the insulating gap at the bottom of the left-hand upright drives a 66-A current through the structure. At what height h will the upper wire be in equilibrium?

Solution

If h is small compared to the length of the rods, we can use Equation 30-6 for the repulsive magnetic force between the horizontal rods (upward on the top rod) $F = \mu_0 I^2 \ell/2\pi h$. The rod is in equilibrium when this equals its weight, $F = mg$, hence $h = \mu_0 I^2 \ell/2\pi mg =$

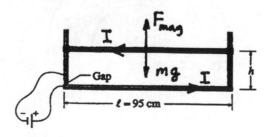

FIGURE 30-52 Problem 21 Solution.

$(2\times10^{-7} \text{ N/A}^2)(66 \text{ A})^2(0.95 \text{ m})/(0.022\times9.8 \text{ N}) =$ 3.84 mm. (This is indeed small compared to 95 cm, as assumed.)

Problem

23. The wires in Fig. 30-49 carry 2.5-A currents in the directions indicated. Find the net force per unit length on the wire at lower left.

Solution

The force per unit length between any pair of wires has magnitude given by Equation 30-6, and is attractive for parallel and repulsive for antiparallel currents. For the bottom pair, for example, the force per unit length on the lower left wire is $F/\ell = (\mu_0 I^2/2\pi d)$ to the right, where d is the side of the square. The other forces on this wire are (F/ℓ) down and $(F/\ell)/\sqrt{2}$ diagonally down and left, as shown. The resultant force is $(F/\ell)\hat{i} + (F/\ell)(-\hat{j}) + (F/\ell)(-\hat{i}\cos 45° - \hat{j}\sin 45°)/\sqrt{2} = (F/\ell)(\frac{1}{2}\hat{i} - \frac{3}{2}\hat{j})$. This has magnitude $\frac{1}{2}\sqrt{10}(F/\ell) = \frac{1}{2}\sqrt{10}(2\times10^{-7} \text{ N/A}^2)(25 \text{ A})^2/(0.15 \text{ m}) = 13.2 \text{ }\mu\text{N/m}$, and direction $\theta = \tan^{-1}(-3) = -71.6°$ below x axis shown (lower right quadrant).

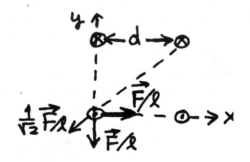

Problem 23 Solution.

Problem

25. A solenoid 10 cm in diameter is made with 2.1-mm-diameter copper wire wound so tightly that adjacent turns touch, separated only by enamel insulation of negligible thickness. The solenoid carries a 28-A current. In the long, straight wire approximation, what is the net force between two adjacent turns of the solenoid?

Solution

Equation 30-6 provides an approximate value of the force per unit length between adjacent turns of $F/\ell = \mu_0 I^2/2\pi d = (2\times10^{-7} \text{ N/A}^2)(28 \text{ A})^2 \div$ (2.1 mm) $= 74.7 \text{ mN/m}$. The length of one turn is $2\pi r = 10\pi$ cm $= 0.314$ m, so the force on one turn is approximately $(74.7 \text{ mN/m})(0.314 \text{ m}) = 23.5 \text{ mN}$.

Section 30-3:   Ampère's Law

Problem

27. The line integral of the magnetic field on a closed path surrounding a wire has the value 8.8 μT·m. What is the current in the wire?

Solution

If the only current encircled by the path is that flowing in the wire, Ampère's law (Equation 30-7) gives 8.8 μT·m = $\oint \mathbf{B} \cdot d\boldsymbol{\ell} = \mu_0 I_{\text{wire}}$, or $I_{\text{wire}} = $ (8.8 μT·m)/$(4\pi \times 10^{-7}$ N/A^2) = 7.00 A.

Problem

29. The magnetic field shown in Fig. 30-56 has uniform magnitude 75 μT, but its direction reverses abruptly. How much current is encircled by the rectangular loop shown?

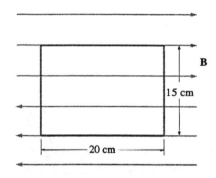

FIGURE 30-56 Problem 29.

Solution

Ampère's law applied to the loop shown in Fig. 30-56 (going clockwise) gives $\oint \mathbf{B} \cdot d\boldsymbol{\ell} = 2B\ell = 2(75\ \mu\text{T}) \times$ (0.2 m) = $\mu_0 I_{\text{encircled}} = (0.4\pi \mu\text{T·m/A}) I_{\text{encircled}}$ (since the sides of the loop perpendicular to **B** give no contribution to the line integral). Thus, $I_{\text{encircled}} = (75/\pi)$ A = 23.9 A. As explained in the text, the current flows along the boundary surface between the regions of oppositely directed **B**, positive into the page in Fig. 30-56, for clockwise circulation around the loop.

Problem

31. Figure 30-58 shows a magnetic field pointing in the x direction. Its strength, however, varies with position in the y direction. At the top and bottom of the rectangular loop shown the field strengths are 3.4 μT and 1.2 μT, respectively. How much current flows through the area encircled by the loop?

Solution

Equation 30-7 applied to the loop shown (going around in the direction of **B** at the top, i.e., clockwise) gives $\oint \mathbf{B} \cdot d\boldsymbol{\ell} = B_{\text{top}}\ell - B_{\text{bot}}\ell = (3.4 - 1.2)\ \mu\text{T}(7\ \text{cm}) = \mu_0 I_{\text{encircled}}$, so $I_{\text{encircled}} = (2.2\ \mu\text{T})(0.07\ \text{m}) \div (0.4\pi\ \mu\text{T·m/A}) = 123$ mA (positive current into the page in Fig. 30-58). (Note: $\mathbf{B} \cdot d\boldsymbol{\ell} = 0$ on the 4 cm sides of the amperian loop and a right-hand screw turned clockwise advances into the page.)

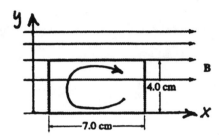

FIGURE 30-58 Problem 31 Solution.

Section 30-4:   Using Ampère's Law

Problem

33. A solid wire 2.1 mm in diameter carries a 10-A current with uniform current density. What is the magnetic field strength (a) at the axis of the wire, (b) 0.20 mm from the axis, (c) at the surface of the wire, and (d) 4.0 mm from the wire axis?

Solution

The magnetic field strength is given by Equation 30-9 inside the wire ($r \le R$) and Equation 30-8 outside ($r \ge R$) as shown in Fig. 30-24. (a) For $r = 0, B = 0$. (b) For $r = 0.2$ mm $< R = \frac{1}{2} \times 2.1$ mm = 1.05 mm, $B = \mu_0 Ir/2\pi R^2 = (2 \times 10^{-7}$ N/A$^2)(10$ A$) \times$ (0.2 mm)/(1.05 mm)^2 = 3.63 G. (c) For $r = R$, $B = \mu_0 I/2\pi R = (2 \times 10^{-7}$ N/A$^2)(10$ A$)/(1.05$ mm$) = $ 19.0 G. (d) For $r = 4$ mm $> R$, $B = \mu_0 I/2\pi r = (2 \times 10^{-7}$ N/A$^2)(10$ A$)/(4$ mm$) = 5$ G.

Problem

35. A long conducting rod of radius R carries a nonuniform current density given by $J = J_0 r/R$, where J_0 is constant and r is the radial distance from the rod's axis. Find expressions for the magnetic field strength (a) inside and (b) outside the rod.

Solution

The magnetic field from a long conducting rod is approximately cylindrically symmetric, as discussed in Section 30-4, so (b) the field outside ($r \ge R$) is given by Equation 30-8, and has direction circling the rod according to the right-hand rule. The total current can be related to the current density by integrating over the cross-sectional area of the rod, $I = \int \mathbf{J} \cdot d\mathbf{A} = \int_0^R J_0(r/R)2\pi r \, dr = 2\pi J_0 R^2/3$. Here, area elements were chosen to be circular rings of radius r, thickness dr, and area $dA = 2\pi r \, dr$. Equation 30-8 can then be written as $B = \mu_0 I/2\pi r = \mu_0 J_0 R^2/3r$, for $r \geq R$. (a) Inside the rod, Ampère's law can be used to find the field, as in Example 30-4, by integrating the current density over a smaller cross-sectional area, corresponding to $I_{\text{encircled}}$ for an amperian loop with $r \leq R$. Then $I_{\text{encircled}} = \int_0^r J_0(r/R)2\pi r \, dr = 2\pi J_0 r^3/3R$, and Ampère's law gives $2\pi r B = \mu_0 I_{\text{encircled}}$, or $B = \mu_0 J_0 r^2/3R$, for $r \leq R$.

Problem

37. Typically, cylindrical wires made from yttrium-barium-copper-oxide superconductor can carry a maximum current density of 6.0 MA/m^2 at a temperature of 77 K, as long as the magnetic field at the conductor surface does not exceed 10 mT. Suppose such a wire is to carry the maximum current density. (a) At what wire diameter would the surface magnetic field equal the 10-mT limit? (b) Is this a maximum or minimum value for the diameter if the field is not to exceed the limit? (c) What current would a wire with this diameter carry?

Solution

The magnetic field strength at the surface of a wire with axial symmetry is $B = \mu_0 I/2\pi R$ (see Problem 34). If the current density in the wire is uniform over its circular cross-section, $I = J\pi R^2$ and $B = \frac{1}{2}\mu_0 J R$. (a) and (b) If $J = J_{\max}$, then $B \leq B_{\max}$ implies $2R \leq 4B_{\max}/\mu_0 J_{\max} = (10 \text{ mT}) + (0.1\pi \ \mu\text{T·m/A})/(6 \text{ MA/m}^2) = 5.31$ mm, which is the maximum diameter. (c) With the diameter above, the total current is $I = J_{\max}\pi R^2 = (6 \text{ MA/m}^2)\pi \times (5.31 \text{ mm}/2)^2 = 133$ A, not excessive for a superconductor.

Problem

39. Two large, flat conducting plates lie parallel to the x-y plane. They carry equal currents, one in the $+x$ and the other in the $-x$ direction. In each plate the current per meter of width in the y direction is J_s. Find the magnetic field strength (a) between and (b) outside the plates.

Solution

The total field is the superposition of fields due to two (approximately infinite) flat parallel current sheets. Equation 30-10 and Fig. 30-25 give the magnitude and direction of the individual fields. (a) Between the plates, both fields are in the negative y direction, thus, $\mathbf{B}_{\text{btw}} = -\frac{1}{2}\mu_0 J_{s,1}\hat{\jmath} - \frac{1}{2}\mu_0 J_{s,2}\hat{\jmath} = -\mu_0 J_s\hat{\jmath}$. (b) Outside the plates, the fields are in opposite directions, and thus cancel, $\mathbf{B}_{\text{out}} = 0$.

Problem 39 Solution.

Problem

41. A hollow conducting pipe of inner radius a and outer radius b carries a current I parallel to its axis and distributed uniformly through the pipe material (Fig. 30-61). Find expressions for the magnetic field for (a) $r < a$, (b) $a < r < b$, and (c) $r > b$, where r is the radial distance from the pipe axis.

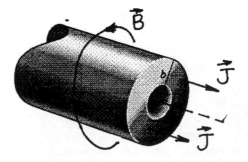

FIGURE 30-61 Problem 41.

Solution

The symmetry argument in the text, for the field of a straight wire, shows that the magnetic field lines (from a very long pipe) are concentric circles, counterclockwise for current out of the page. Ampère's law, for loops along the field lines, gives $2\pi r B = \mu_0 I_{\text{encircled}}$. For uniform current density, $I_{\text{encircled}}$ is proportional to the cross-sectional area of conducting material. Therefore,

$$B(r) = \frac{\mu_0}{2\pi r} \begin{cases} 0, & r < a \\ I\dfrac{(r^2 - a^2)}{(b^2 - a^2)}, & a \leq r \leq b \\ I, & r > b \end{cases}$$

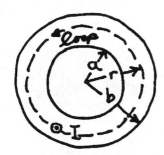

Problem 41 Solution.

Section 30-5: Solenoids and Toroids

Problem

43. A superconducting solenoid has 3300 turns per meter and can carry a maximum current of 4.1 kA. What is the magnetic field strength in the solenoid?

Solution

Equation 30-11 for a long thin solenoid (if applicable) gives $B = \mu_0 nI = (4\pi \times 10^{-7} \text{ N/A}^2)(3300/\text{m}) \times (4.1 \text{ kA}) = 17.0$ T.

Problem

45. You have 10 m of 0.50-mm-diameter copper wire and a battery capable of passing 15 A through the wire. What magnetic field strengths could you obtain (a) inside a 2.0-cm-diameter solenoid wound with the wire as closely spaced as possible and (b) at the center of a single circular loop made from the wire?

Solution

(a) The length of a solenoid, with one layer of N turns of closely spaced wire of diameter d, is $\ell = Nd$, so the number of turns per unit length is $n = N/\ell = d^{-1} = (0.5 \text{ mm})^{-1} = 2000 \text{ m}^{-1}$. (Although not needed, the value of $N = 10 \text{ m}/2\pi$ cm $= 159$ turns can be used to check that $\ell = 159 \times 0.5$ mm $= 7.96$ cm is barely long enough, compared to 2 cm, to justify the use of Equation 30-11 as an approximation to B at the solenoid's center.) Then $B = \mu_0 nI = (4\pi \times 10^{-7} \text{ N/A}^2) \times (2000 \text{ m}^{-1})(15 \text{ A}) = 3.77 \times 10^{-2}$ T. (b) The magnetic field at the center of a flat, circular current loop can

be found from Equation 30-3, $B = \mu_0 I/2a$. Here, $2\pi a = 10$ m, so $B = \mu_0 \pi I/10 \text{ m} = (4\pi^2 \times 10^{-7} \text{ N/A}^2) \times (15 \text{ A})/10 \text{ m} = 5.92 \times 10^{-6}$ T.

Problem

47. A toroidal fusion reactor requires a magnetic field that varies by no more than 10% from its central value of 1.5 T. If the minor radius of the toroidal coil producing this field is 30 cm, what is the minimum value for the major radius of the device?

Solution

The central value of the toroidal field is given by Equation 30-12 with $r = R_{\text{maj}}$. At other values of r inside the toroid, the percent difference in field strength is $100(B - B_{\text{maj}})/B_{\text{maj}} = 100(R_{\text{maj}} - r)/r$. The extremes of r are $R_{\text{maj}} \pm R_{\text{min}}$, so it is required that $10 \geq 100 \left| [R_{\text{maj}} - (R_{\text{maj}} \pm R_{\text{min}})] \div (R_{\text{maj}} \pm R_{\text{min}}) \right|$, or $R_{\text{maj}} \geq (10 \mp 1)R_{\text{min}}$. The minimum major radius (corresponding to the limit using the smallest value of r, or the lower sign above) is $R_{\text{maj}} \geq 11R_{\text{min}} = 11 \times 30$ cm $= 3.30$ m. (Since 10% is not infinitesimal, differentiation w.r.t. r gives an alternative approximate limit: $|dB/B| = |-dr/r| = R_{\text{min}}/R_{\text{maj}} \leq 10\%$, or $R_{\text{maj}} \geq 3$ m.)

Problem

49. We noted that there is a nonzero magnetic field component outside a solenoid, encircling the device, associated with the component of current flow parallel to the solenoid axis. For a long solenoid of radius R, find an expression for the ratio of this external encircling field just outside the solenoid to the field inside, and show explicitly that this ratio tends to zero as the number of turns per unit length becomes large.

Solution

If the current I to the solenoid enters at one end and leaves at the other, this constitutes a net flow along the solenoid axis. Treating this axial current as a long straight wire, one finds that a magnetic field of $B' = \mu_0 I/2\pi R$, encircling the solenoid just outside its surface, is generated. The field inside the solenoid is $B = \mu_0 nI$, so the ratio is $B'/B = 1/2\pi nR$, which becomes small when the number of turns per unit length, n, is large. (Note: There must be another field outside the solenoid due to its finite length and the fact that the magnetic field lines passing inside are parts of closed loops.)

Section 30-6: Magnetic Matter

Problem

51. When a sample of a certain substance is placed in a 250.0-mT magnetic field, the field inside the sample is 249.6 mT. Find the magnetic susceptibility of the substance. Is it ferromagnetic, paramagnetic, or diamagnetic?

Solution

Equation 30-13 gives the relation between the internal and applied magnetic fields in terms of the relative permeability or the magnetic susceptibility. For the sample described in this problem, the latter is $\chi_M = (B_{\text{int}} - B_{\text{app}})/B_{\text{app}} = (249.6 - 250.0)/250.0 = -1.6 \times 10^{-3}$. Since $\chi_M < 0$ (or $B_{\text{int}} < B_{\text{app}}$) the material is diamagnetic.

Problem

53. A ferromagnetic material is placed in a 2.5-G magnetic field and the field within the material is determined to be 1.8 T. What is the magnetic susceptibility of this material?

Solution

From Equation 30-13, $\chi_M = (B_{\text{int}}/B_{\text{app}}) - 1 = (1.8 \text{ T}/2.5 \text{ G}) - 1 \simeq 7.20 \times 10^3$. (In ferromagnetic materials, $\chi_M \approx \kappa_M$ and both are functions of the applied field and the past history of the sample.)

Paired Problems

Problem

55. Two concentric, coplanar circular current loops have radii a and $2a$. If the magnetic field is zero at their common center, how does the current in the outer loop compare with that in the inner loop?

Solution

The magnetic field strength at the center of a circular current loop is $\mu_0 I/2R$ (Equation 30-3). In order for the net field to cancel, the currents must be in opposite directions and have magnitudes such that $\mu_0 I_{\text{outer}}/2(2a) = \mu_0 I_{\text{inner}}/2a$. Then $I_{\text{outer}} = 2I_{\text{inner}}$.

Problem

57. Figure 30-64 shows a wire of length ℓ carrying current fed by other wires that are not shown. Point A lies on the perpendicular bisector, a distance y from the wire. Adapt the calculation of Example 30-2 to show that the magnetic field at A due to the straight wire alone has magnitude $\frac{\mu_0 I \ell}{2\pi y \sqrt{\ell^2 + 4y^2}}$. What is the field direction?

Solution

To find the magnetic field at point A, one may use the same argument as in Example 30-2, except that for a wire of finite length, one integrates from $x = -\ell/2$ to $x = \ell/2$. For current flowing in the positive x direction, **B** is out of the page at point A, with strength

$$B(A) = \int_{-\ell/2}^{\ell/2} \frac{\mu_0 I}{4\pi} \frac{y\,dx}{(x^2 + y^2)^{3/2}}$$

$$= \frac{\mu_0 I y}{4\pi} \left. \frac{x}{y^2 \sqrt{x^2 + y^2}} \right|_{-\ell/2}^{\ell/2} = \frac{\mu_0 I}{2\pi y} \frac{\ell}{\sqrt{\ell^2 + 4y^2}}.$$

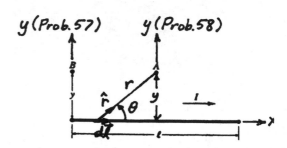

FIGURE 30-64 Problem 57 Solution.

Problem

59. The largest lightning strikes have peak currents around 250 kA, flowing in essentially cylindrical channels of ionized air. How far from such a flash would the resulting magnetic field be equal to Earth's magnetic field strength, about 50 μT?

Solution

Supposing that the cylindrical channel of ionized air acts like a long straight wire, we can use Equation 30-5 to estimate the distance: $y = \mu_0 I/2\pi B = (2 \times 10^{-7} \text{ N/A}^2)(250 \text{ kA})/(50 \ \mu\text{T}) = 1 \text{ km}$.

Problem

61. A coaxial cable like that shown in Fig. 30-60 consists of a 1.0-mm-diameter inner conductor and an outer conductor of inner diameter 1.0 cm and 0.20 mm thickness. A 100-mA current flows down the center conductor and back along the outer conductor. Find the magnetic field strength (a) 0.10 mm, (b) 5.0 mm, and (c) 2.0 cm from the cable axis.

Solution

For a long, straight cable, the magnetic field can be found from Ampere's law. The field lines are cylindrically symmetric and form closed loops, hence must be concentric circles, which we also choose as amperian loops. Take positive circulation counterclockwise so that positive current is out of the page. Then $\oint_c \mathbf{B} \cdot d\boldsymbol{\ell} = 2\pi r B = \mu_0 I_{\text{encircled}}$. Assume that the current density in each conductor is uniform; i.e., the current is proportional to the cross-sectional area. We may calculate $I_{\text{encircled}}$ in four regions of space. (a) For $r \leq R_a$, $I_{\text{encircled}} = I(\pi r^2 / \pi R_a^2) = Ir^2 / R_a^2$, so $B = \mu_0 Ir / 2\pi R_a^2$. (b) For $R_a \leq r \leq R_b$, $I_{\text{encircled}} = I$, so $B = \mu_0 I / 2\pi r$. (Although not asked for, for $R_b \leq r \leq R_c$,

$$I_{\text{encircled}} = I - I \frac{\pi(r^2 - R_b^2)}{\pi(R_c^2 - R_b^2)} = I \left(\frac{R_c^2 - r^2}{R_c^2 - R_b^2} \right), \text{so}$$

$$B = \frac{\mu_0 I}{2\pi(R_c^2 - R_b^2)} \left(\frac{R_c^2}{r} - r \right).$$

The outer radius R_c is the inner radius plus the thickness of the outer conductor.) (c) Finally, for $r \geq R_c$, $I_{\text{encircled}} = 0$, so $B = 0$. Numerically, $R_a = 0.5$ mm, $R_b = 5$ mm, $R_c = 5.2$ mm, and $I = 0.1$ A, so inside the inner conductor (a) when $r = 0.1$ mm $\leq R_a$, $B = (2 \times 10^{-7} \text{ T·m/A}) \times (0.1 \text{ A})(0.1 \text{ mm})/(0.5 \text{ mm})^2 = 8 \ \mu\text{T}$, between the two conductors (b) when $R_a \leq r = 5$ mm $\leq R_b$, $B = (2 \times 10^{-7} \text{ T·m/A})(0.1 \text{ A})/(5 \text{ mm}) = 4 \ \mu\text{T}$; and outside the outer conductor (c) when $r = 2$ cm $\geq R_c$, $B = 0$.

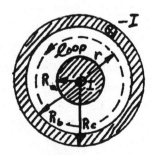

Problem 61 Solution.

Supplementary Problems
Problem

63. A circular wire loop of radius 15 cm and negligible thickness carries a 2.0-A current. Use suitable approximations to find the magnetic field of this loop (a) in the loop plane, 1.0 mm outside the loop, and (b) on the loop axis, 3.0 m from the loop center.

Solution

(a) As mentioned in the paragraph following Equation 30-4, the calculation of the magnetic field from a circular current loop, at points not on its axis, is difficult. Fortunately, at a distance of 1 mm from a loop of radius 15 cm, the field is approximately that of a long, straight wire, Equation 30-5 gives $B \approx \mu_0 I / 2\pi y = (2 \times 10^{-7} \text{ N/A}^2)(2 \text{ A})/(10^{-3} \text{ m}) = 4$ G. (b) Since 3 m $\gg$ 15 cm, the approximation $x \gg a$ in Equation 30-4 is justified. Then $B \approx \mu_0 I a^2 / 2 |x|^3 = (2\pi \times 10^{-7} \text{ N/A}^2)(2 \text{ A})(0.15 \text{ m})^2 / (3 \text{ m})^3 = 1.05 \times 10^{-5}$ G.

Problem

65. A long, hollow conducting pipe of radius R and length ℓ carries a uniform current I flowing around the pipe, as shown in Fig. 30-66. Find expressions for the magnetic field (a) inside and (b) outside the pipe. *Hint:* What configuration does this pipe resemble?

Solution

The current distribution is similar to a solenoid, where the number of turns per unit length and the current in each turn are related to the total current in the pipe by $n\ell I_t = I$. Therefore (see Section 30-5), the field is approximately that of an infinite solenoid: $B = \mu_0 n I_t = \mu_0 I / \ell$ inside, and $B = 0$ outside, directed parallel to the axis to the left in Fig. 30-66.

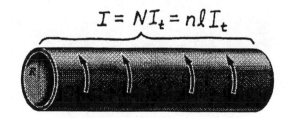

FIGURE 30-66 Problem 65 Solution.

Problem

67. A wide, flat conducting spring of spring constant $k = 20$ N/m and negligible mass consists of two 6.0-cm-diameter turns, as shown in Fig. 30-67. In

FIGURE 30-67 Problem 67.

its unstretched configuration the coils are nearly touching. A 10-g mass is hung from the spring, and at the same time a current I is passed through it. The spring stretches 2.0 mm. Find I, assuming the coils remain close enough to be treated as parallel wires.

Solution

The length of one turn, $\ell = 6\pi$ cm, is large compared to the separation of the two turns, $d = 2$ mm, so that Equation 30-6 can be used to find the magnetic force in the spring. $F_{mag} = \mu_0 I^2 \ell / 2\pi d = (2 \times 10^{-7} \text{ N/A}^2) \times I^2 (6\pi \text{ cm})/2 \text{ mm} = 1.88 I^2 \times 10^{-5} \text{ N/A}^2$. The elastic force in the spring, in the same direction as F_{mag}, is $F_{el} = kd = (20 \text{ N/m})(2 \text{ mm}) = 4 \times 10^{-2}$ N. At equilibrium, $F_{el} + F_{mag} = mg = (10^{-2} \text{ kg}) \times (9.8 \text{ m/s}^2) = 9.8 \times 10^{-2}$ N. Thus, $1.88 I^2 \times 10^{-5} \text{ N/A}^2 = 5.8 \times 10^{-2}$ N, or $I = 55.5$ A.

Problem

69. A disk of radius a carries a uniform surface charge density σ, and is rotating with angular speed ω about the central axis perpendicular to the disk. Show that the magnetic field at the disk's center is $\frac{1}{2}\mu_0 \sigma \omega a$.

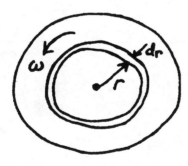

Problem 69 Solution.

Solution

The disk may be considered to be composed of rings of radius r, thickness dr, and charge $dq = 2\pi r \, dr \cdot \sigma$. Each ring represents a circular current loop (see hint in problem 29-73) $dI = dq/T = dq/(2\pi/\omega) = \omega \sigma r \, dr$, which produces a magnetic field strength $dB = \mu_0 \, dI/2r = \frac{1}{2}\mu_0 \omega \sigma \, dr$ at the center of the disk, directed out of the page, as sketched for positive charge density. The total field strength is $B = \int_0^a dB = \frac{1}{2}\mu_0 \omega \sigma \int_0^a dr = \frac{1}{2}\mu_0 \omega \sigma a$.

Problem

71. Work Example 30-2 by expressing all variables in terms of the angle θ and integrating over the appropriate range in θ.

Solution

In Example 30-2 and Fig. 30-9, $\cos \theta = -x/r$, $\tan \theta = -y/x$, so $d(\tan \theta) = d\theta/\cos^2 \theta = y \, dx/x^2$. Then $dx = (x^2/\cos^2 \theta) \, d\theta/y = r^2 \, d\theta/y$. Thus, we can write the field element (out of the page) as

$$dB = \frac{\mu_0 I}{4\pi} \frac{dx \sin \theta}{r^2} = \frac{\mu_0 I}{4\pi} \left(\frac{r^2 d\theta}{y} \right) \frac{\sin \theta}{r^2} = \frac{\mu_0 I}{4\pi y} \sin \theta \, d\theta.$$

The limits of integration $x = -\infty$ to $+\infty$ correspond to $\theta = 0$ to π (or 180°), hence

$$B = \frac{\mu_0 I}{4\pi y} \int_0^\pi \sin \theta \, d\theta = \frac{\mu_0 I}{4\pi y} \left. -\cos \theta \right|_0^\pi = \frac{\mu_0 I}{2\pi y},$$

which is the same as Equation 30-5.

CHAPTER 31   ELECTROMAGNETIC INDUCTION

ActivPhysics can help with these problems:
Activities 13.9, 13.10

Sections 31-2 and 31-3:   Faraday's Law and Induction and the Conservation of Energy

Problem

1. Show that the volt is the correct SI unit for the rate of change of magnetic flux, making Faraday's law dimensionally correct.

Solution

The units of $d\phi_B/dt$ are $\text{T·m}^2/\text{s} = (\text{N/A·m})(\text{m}^2/\text{s}) = (\text{N·m/A·s}) = \text{J/C} = \text{V}$.

Problem

3. Find the magnetic flux through a circular loop 5.0 cm in diameter oriented with the loop normal at 30° to a uniform 80-mT magnetic field.

Solution

For a stationary plane loop in a uniform magnetic field, the integral for the flux in Equation 31-1 is just $\phi_B = \mathbf{B} \cdot \mathbf{A} = BA \cos\theta = (80 \text{ mT})\pi(2.5 \text{ cm})^2 \cos 30° = 1.36 \times 10^{-4}$ Wb. (The SI unit of flux, T·m², is also called a weber, Wb.)

Problem

5. A conducting loop of area 240 cm² and resistance 12 Ω lies at right angles to a spatially uniform magnetic field. The loop carries an induced current of 320 mA. At what rate is the magnetic field changing?

Solution

The flux through a stationary loop perpendicular to a magnetic field is $\phi_B = BA$ (see Problem 3), so Faraday's law (Equation 31-2) and Ohm's law (Equation 27-6) relate this to the magnitude of the induced current: $I = |\mathcal{E}/R| = |d\phi_B/dt|/R = A|dB/dt|/R$. Therefore $|dB/dt| = IR/A = (320 \text{ mA})(12 \text{ } \Omega) \div (240 \text{ cm}^2) = 160$ T/s.

Problem

7. A conducting loop with area 0.15 m² and resistance 6.0 Ω lies in the x-y plane. A spatially uniform magnetic field points in the z direction. The field varies with time according to $B_z = at^2 - b$, where $a = 2.0$ T/s² and $b = 8.0$ T. Find the loop current (a) when $t = 3.0$ s and (b) when $B_z = 0$.

Solution

The reasoning in the solution to Problem 5 shows that $|I| = A|dB/dt|/R = A(2at)/R = (0.15 \text{ m}^2) \times (2 \times 2 \text{ T/s}^2)t/(6 \text{ } \Omega) = 10^{-1}t$ (A/s). (a) For $t = 3$ s, $|I| = 0.3$ A, and (b) for $t = \sqrt{b/a} = \sqrt{8 \text{ T}/(2 \text{ T/s}^2)} = 2$ s, $|I| = 0.2$ A. (Note: In this problem, there is enough information to also specify the direction of I. Choose the x-y axes as shown and the z-axis out of the page. For positive normal to the loop along the z-axis, a positive sense of circulation for the induced current is CCW. Then $I = \mathcal{E}/R = -(1/R)d(\mathbf{B} \cdot \mathbf{A})/dt = -(A/R)(dB_z/dt) = -(A/R)(2at)$. Negative currents are CW around the z-axis.)

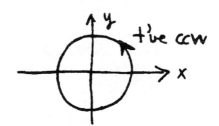

Problem 7 Solution.

Problem

9. A square wire loop of side ℓ and resistance R is pulled with constant speed v from a region of no magnetic field until it is fully inside a region of constant, uniform magnetic field $\mathbf{B}$ perpendicular to the loop plane. The boundary of the field region is parallel to one side of the loop. Find an expression for the total work done by the agent pulling the loop.

Solution

The loop can be treated analogously to the situation analyzed in Section 31-3, under the heading "Motional EMF and Lenz's Law"; instead of exiting, the loop is entering the field region at constant velocity. All quantities have the same magnitudes, except the current in the loop is CCW instead of CW, as in Fig. 31-13. Since the applied force acts over a displacement equal to the side-length of the loop, the work done can be calculated directly: $W_{app} = \mathbf{F}_{app} \cdot \ell = (I\ell B)\ell = I\ell^2 B$. But, $I = \mathcal{E}/R = |d\phi_B/dt|/R = d/dt(B\ell x)/R = B\ell v/R$, as before, so $W_{app} = B^2\ell^3 v/R$. [Alternatively, the work can be calculated from the conservation of energy: $I = B\ell v/R$, $\mathcal{P}_{diss} = I^2 R = (B\ell v)^2/R$, and $W_{app} = \mathcal{P}_{diss}t = [(B\ell v)^2/R](\ell/v)$.]

Problem

11. In Fig. 31-26 the loop radius is 15 cm, and the magnetic field is decreasing at the rate of 550 T/s. If the gap width is small compared with the loop circumference, what is the voltage across the gap?

Solution

If the gap is small, essentially the entire induced emf around the loop appears across it (as explained after Example 31-7 in the sub section of the text on "Induction in Open Circuits"), so $|\mathcal{E}| = |d\phi_B/dt| = \pi R^2 |dB/dt| = \pi(0.15\text{ m})^2(550\text{ T/s}) = 38.9$ V. (The polarity is as mentioned in the text.)

Problem

13. The wingspan of a 747 jetliner is 60 m. If the plane is flying at 960 km/h in a region where the vertical component of Earth's magnetic field is 0.20 G, what emf develops between the plane's wingtips?

Solution

A motional emf causes electrons in the wing to drift until equilibrium with the electrostatic field from accumulated wingtip charges is achieved. The magnetic and electric forces on an electron have magnitudes $F_{mag} = |-e\mathbf{v} \times \mathbf{B}| = evB_\perp$ (we suppose that the 747 is flying horizontally so only the vertical

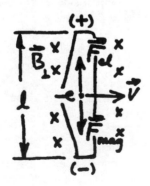

Problem 13 Solution.

magnetic field gives a force parallel to the wingspan), and $F_{el} = |-e\mathbf{E}_s| = eV/\ell$, where V is the potential difference between the wingtips. At equilibrium, $evB = eV/\ell$, or $V = B\ell v = (2\times10^{-5}\text{ T})(60\text{ m})\times(960\text{ m}/3.6\text{ s}) = 320$ mV. (Motional emf's like this need to be considered in rocket measurements of ionospheric electric fields.)

Problem

15. In Example 31-2 take $a = 1.0$ cm, $w = 3.5$ cm, and $\ell = 6.0$ cm. Suppose the rectangular loop is a conductor with resistance 50 mΩ and that the current I in the long wire is increasing at the rate of 25 A/s. Find the induced current in the loop. In what direction does it flow?

Solution

The normal to the loop in Fig. 31-6 was taken to be in the direction of the magnetic field of the wire, or into the page, so the positive sense of circulation around the loop is clockwise (from the right-hand rule). Faraday's and Ohm's laws, together with the result of Example 31-2, give an induced current the loop of $\mathcal{E}/R = (-d\phi_B/dt)/R = -(\mu_0\ell/2\pi R)\ln[(a + w)/a]\times(dI/dt) = -(2\times10^{-7}\text{ N/A}^2)(6\text{ cm})(25\text{ A/s})\times\ln[4.5/1]/(50\text{ m}\Omega) = -9.02$ μA. A negative current is counterclockwise in the loop.

Problem

17. A square conducting loop of side $s = 0.50$ m and resistance $R = 5.0$ Ω moves to the right with speed $v = 0.25$ m/s. At time $t = 0$ its rightmost edge enters a uniform magnetic field B=1.0 T pointing into the page, as shown in Fig. 31-46. The magnetic field covers a region of width $\omega = 0.75$ m. Plot (a) the current and (b) the power dissipation in the loop as functions of time, taking a clockwise current as positive and covering the time until the entire loop has exited the field region.

Solution

Let x be the distance between the right side of the loop and the left edge of the field region. Take $t = 0$ when $x = 0$, so that $x = vt$. The loop enters the field region at $t = 0$, is completely within the region for t between $\ell/v = 2$ s and $w/v = 3$ s, and is out of the region for $t \geq (w + \ell)/v = 5$ s. The area of loop overlapping the field region increases linearly from 0 to ℓ^2, stays constant at ℓ^2, then decreases to 0 between these times. (We use ℓ for side

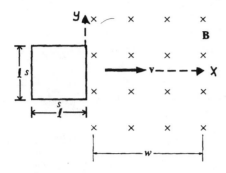

FIGURE 31-46 Problem 17 Solution.

length to avoid confusion with time units.). Thus,

$$\phi_B = BA = B\ell^2 \begin{cases} 0 \\ vt/\ell \\ 1 \\ (w+\ell-vt)/\ell \\ 0 \end{cases}$$

$$= 0.25 \text{ Wb} \begin{cases} 0, & t \le 0 \\ 0.5t, & 0 \le t \le 2 \\ 1, & 2 \le t \le 3 \\ 0.5(5-t), & 3 \le t \le 5 \\ 0, & 5 \le t \end{cases}$$

(We substituted the given numerical values and used SI units for flux, with time t in seconds, see solution to Problem 3.)

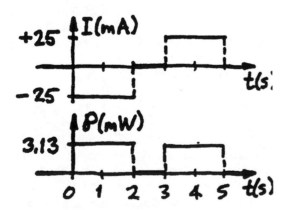

Problem 17 Solution.

(a) The induced current (positive clockwise) is given by Faraday's and Ohm's laws:

$$I = -\frac{1}{R}\frac{d\phi_B}{dt} = 25 \text{ mA} \begin{cases} 0, & t \le 0 \\ -1, & 0 \le t \le 2 \\ 0, & 2 \le t \le 3 \\ +1, & 3 \le t \le 5 \\ 0, & 5 \le t \end{cases}$$

(b) The power dissipated, I^2R, is $(\pm 25 \text{ mA})^2(5 \ \Omega) = 3.13$ mW when the current is not zero.

Problem

19. A solenoid 2.0 m long and 30 cm in diameter consists of 5000 turns of wire. A 5-turn coil with negligible resistance is wrapped around the solenoid and connected to a 180-Ω resistor, as shown in Fig. 31-47. The direction of the current in the solenoid is such that the solenoid's magnetic field points to the right. At time $t = 0$ the solenoid current begins to decay exponentially, being given by $I = I_0 e^{-t/\tau}$, where $I_0 = 85$ A, $\tau = 2.5$ s, and t is the time in seconds. (a) What is the direction of the current in the resistor as the solenoid current decays? What is the value of the resistor current at (b) $t = 1.0$ s and (c) $t = 5.0$ s?

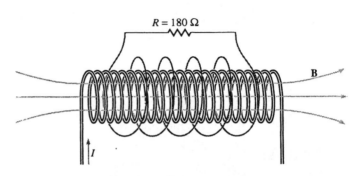

FIGURE 31-47 Problem 19.

Solution

(a) From Lenz's law, the direction of the induced current in the coil, I_c, must oppose the decrease in the solenoid's field, B_s, to the right, so the induced field due to I_c must also be to the right. Thus, the right-hand rule for the field of the coil requires that I_c flow from left to right in the 180 Ω resistor in Fig. 31-47. (b) The field inside the solenoid ($B_s = \mu_0(N_s/\ell)I_s$) links each turn of the coil, so the total flux through the latter is $\phi_B = N_c B_s A_s$. (Recall that approximately none of the solenoid's field lines go through the coil outside the solenoid's cross-sectional area.) From Faraday's and Ohm's laws, the induced current in the coil has magnitude $I_c = |\mathcal{E}|/R = (1/R)|-d\phi_B/dt| = (\mu_0 N_c N_s/\ell)\frac{1}{4}\pi D_s^2(1/R)|-dI_s/dt|$. With $I_s = I_0 e^{-t/\tau}$, $|dI_s/dt| = |-(I_0/\tau)e^{-t/\tau}|$, and the given numerical data,

$$I_c = \left(\pi^2 \times 10^{-7} \frac{\text{N}}{\text{A}^2}\right)\left(\frac{5000 \times 5}{2 \text{ m}}\right)\frac{(0.3 \text{ m})^2}{(180 \ \Omega)}\left(\frac{85 \text{ A}}{2.5 \text{ s}}\right)e^{-t/2.5 \text{ s}}$$
$$= (210 \ \mu A)e^{-t/2.5 \text{ s}}.$$

At $t = 1$ s, $I_c = (210 \ \mu A)e^{-0.4} = 141 \ \mu A$, and (c) at $t = 5$ s, $I_c = (210 \ \mu A)e^{-2} = 28.4 \ \mu A$.

Problem

21. (a) Find an expression for the resistor current in Problem 19 if the solenoid current is given by $I = I_0 \sin \omega t$, where $I_0 = 85$ A and $\omega = 210$ s^{-1}.
(b) What is the peak current in the resistor?
(c) What is the resistor current when the solenoid current is a maximum?

Solution

(a) In Problem 19, $I_c = \mu_0(N_cN_s/\ell)\frac{1}{4}\pi(D_s^2/R) \times (-dI_s/dt)$ was the expression for the current in the coil, taken as positive from left to right in the resistor. When the current in the solenoid is $I_s = I_0 \sin \omega t$, $dI_s/dt = \omega I_0 \cos \omega t$, so $I_c = -(\mu_0\pi/4)(N_cN_s/\ell) \times (D_s^2/R)\omega I_0 \cos \omega t \equiv -I_{\text{peak}}\cos \omega t$. (b) Numerically, $I_{\text{peak}} = (\pi^2 \times 10^{-7}\text{N/A}^2)(5 \times 5000/2 \text{ m})(0.3 \text{ m})^2 \times (210 \text{ s}^{-1})(85 \text{ A})/(180 \ \Omega) = 110$ mA. (c) When $\sin \omega t$ is a maximum, $\cos \omega t$ is zero.

Problem

23. In the preceding problem, what is the first time after $t = 0$ when the loop current will be zero?

Solution

Take the normal to the loop along the z-axis, so that the positive sense of circulation for the induced emf and current is CCW (looking down on the x-y plane) as shown. The flux through the loop is $\phi_B = \mathbf{B} \cdot \mathbf{A} = (B_0 \sin \omega t \ \hat{\mathbf{k}})(A\hat{\mathbf{k}}) = B_0 A \sin \omega t$, so the induced current is $I = \mathcal{E}/R = (-d\phi_B/dt)/R = -(\omega B_0 A/R) \times \cos \omega t$. The first time $\cos \omega t$ is zero, after $t = 0$ is when $\omega t = \pi/2$, or $t = \pi/2\omega = \pi/20$ s^{-1} = 157 ms.

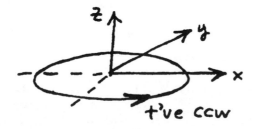

Problem 23 Solution.

Problem

25. A credit-card reader extracts information from the card's magnetic stripe as it is pulled past the reader's head. At some instant the card motion results in a magnetic field at the head that is changing at the rate of 450 μT/ms. If this field passes perpendicularly through a 5000-turn head coil 2.0 mm in diameter, what will be the induced emf?

Solution

The magnetic flux through the coil in the reader's head is changing at a rate of $NA\ dB/dt = (5000)\pi(1 \text{ mm})^2(450 \ \mu\text{T/ms}) = 7.07$ mV. According to Faraday's law, this is equal to the magnitude of the induced emf.

Problem

27. Figure 31-49 shows a pair of parallel conducting rails a distance ℓ apart in a uniform magnetic field **B**. A resistance R is connected across the rails, and a conducting bar of negligible resistance is being pulled along the rails with velocity **v** to the right. (a) What is the direction of the current in the resistor? (b) At what rate must work be done by the agent pulling the bar?

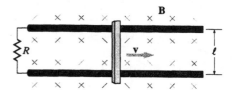

FIGURE 31-49 Problem 27.

Solution

(a) The force on a (hypothetical) positive charge carrier in the bar, $q\mathbf{v} \times \mathbf{B}$, is upward in Fig. 31-49, so current will circulate CCW around the loop containing the bar, the resistor, and the rails (i.e., downward in the resistor). (The force per unit positive charge is the motional emf in the bar.) Alternatively, since the area enclosed by the circuit, and the magnetic flux through it, are increasing, Lenz's law requires that the induced current oppose this with an upward induced magnetic field. Thus, from the right-hand rule, the induced current must circulate CCW. (Take the positive sense of circulation around the circuit CW, so that the normal to the area is in the direction of **B**, into the page.) (b) In Example 31-4, which analyzed the same situation, the current in the bar was found to be $I = |\mathcal{E}|/R = B\ell v/R$. Since this is perpendicular to the magnetic field, the magnetic force on the bar is $F_{\text{mag}} = I\ell B$ (to the left in Fig. 31-49). The agent pulling the bar at constant velocity must exert an equal force in the direction of **v**, and therefore does work at the rate $\mathbf{F} \cdot \mathbf{v} = I\ell Bv = (B\ell v)^2/R$. (Note: The conservation of energy requires that this equal the rate energy is dissipated in the resistor (we neglected the resistance of the bar and the rails), $I^2R = (B\ell v/R)^2R$.)

Problem

29. A battery of emf $\mathcal{E}$ is inserted in series with the resistor in Fig. 31-49, with its positive terminal toward the top rail. The bar is initially at rest, and now no agent pulls it. (a) Describe the bar's subsequent motion. (b) The bar eventually reaches a constant speed. Why? (c) What is that constant speed? Express in terms of the magnetic field, the battery emf, and the rail spacing ℓ. Does the resistance R affect the final speed? If not, what role does it play?

Solution

(a) The battery causes a CW current (downward in the bar) to flow in the circuit composed of the bar, resistor, and rails. (For positive circulation CW, the right-hand rule gives a positive normal to the area bounded by the circuit into the page, so that the flux $\phi_B = \mathbf{B} \cdot \mathbf{A} = B\ell x$ is positive. The length of the circuit is x, as in Example 31-4.) Thus, there is a magnetic force $\mathbf{F}_{mag} = I\boldsymbol{\ell} \times \mathbf{B} = I\ell B$ to the right, which accelerates the bar in that direction. (Any other forces on the bar are assumed to cancel, or be negligible.) An induced emf opposes the battery ($\mathcal{E}_i = -d\phi_B/dt = -B\ell v$, as in Example 31-4, the negative sign indicating a CCW sense in the circuit) so the instantaneous current is $I(t) = (\mathcal{E} + \mathcal{E}_i)/R = (\mathcal{E} - B\ell v)/R$. Thus, as v increases, I (and the accelerating force) decreases. (b) Eventually ($t \to \infty$), $I(\infty) = 0$, $F_{mag} = I(\infty)\ell B = 0$, and the velocity v_∞ stays constant. (c) When $I(\infty) = 0$, $\mathcal{E} - B\ell v_\infty = 0$, so $v_\infty = \mathcal{E}/B\ell$. Although v_∞ doesn't depend on the resistance, the value of R does affect how rapidly v approaches v_∞. For large R, I charges slowly and v takes a long time to reach v_∞. (The equation of motion of the bar (mass m) is $m(dv/dt) = I\ell B = (\mathcal{E} - B\ell v)\ell B/R$, which can be separated: $dv/(v_\infty - v) = (\ell^2 B^2/mR)dt$. For $v_0 = 0$, this integrates to $\ln(1 - v/v_\infty) = -\ell^2 B^2 t/mR$, or $v = v_\infty(1 - e^{-\ell^2 B^2 t/mR})$. The time constant, $\tau = mR/\ell^2 B^2$, depends on the resistance.)

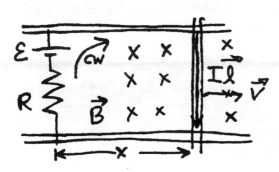

Problem 29 Solution.

Problem

31. A pair of parallel conducting rails 10 cm apart lie at right angles to a uniform magnetic field $\mathbf{B}$ of magnitude 2.0 T, as shown in Fig. 31-51. A 5.0-Ω and a 10-Ω resistor lie across the rails and are free to slide along them. (a) The 5-Ω resistor is held fixed, and the 10-Ω resistor is pulled to the right at 50 cm/s. What are the direction and magnitude of the induced current? (b) Now the 10-Ω resistor is held fixed, and the 5-Ω resistor is pulled to the left at 50 cm/s. What are the direction and magnitude of the induced current? (c) What is the power dissipation in the 10-Ω resistor in both cases?

Solution

(a) Let x be the distance between the resistors. If we take the normal to the area bounded by the resistors and the rails into the page (positive circulation clockwise), then the flux linking the circuit is $\phi_B = BA = B\ell x$, so Faraday's law gives the magnitude of the induced emf as $\mathcal{E}_i = -d\phi_B/dt = -B\ell(dx/dt) = -B\ell v = -(2 \text{ T})(0.1 \text{ m})(0.5 \text{ m/s}) = -0.1 \text{ V}$. The current, therefore, is $I = \mathcal{E}_i/R_{tot} = -0.1 \text{ V} \div (5 + 10) \ \Omega = -6.67 \text{ mA}$. (Lenz's law also gives the direction of I, which opposes the increase in flux, as counterclockwise.) (b) The relative velocity between the loop and the field is the same as in part (a), thus the induced current is the same. (c) $\mathcal{P}_{10\,\Omega} = I^2 R_{10\,\Omega} = (6.67 \text{ mA})^2(10 \ \Omega) = 0.444 \text{ mW}$. (We assumed the rails have negligible resistance in calculating I.)

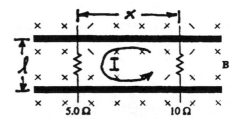

FIGURE 31-51 Problem 31 Solution.

Problem

33. In Fig. 31-49, take $\ell = 10$ cm, $B = 0.50$ T, $R = 4.0 \ \Omega$, and $v = 2.0$ m/s. Find (a) the current in the resistor, (b) the magnetic force on the bar, (c) the power dissipation in the resistor, and (d) the mechanical work done by the agent pulling the bar. Compare your answers to (c) and (d).

Solution

The situation is like that described in Example 31-4 and the solution to Problem 27. (a) $I = \mathcal{E}/R = B\ell v/R = (0.5 \text{ T})(0.1 \text{ m})(2 \text{ m/s})/4 \ \Omega = 25$ mA.

(Neglect the resistance of the bar and rails.)
(b) $F_{mag} = I\ell B = (25 \text{ mA})(0.1 \text{ m})(0.5 \text{ T}) =$
1.25×10^{-3} N. (c) $\mathcal{P}_J = I^2 R = (25 \text{ mA})^2 (4 \text{ }\Omega) =$
2.5 mW. (d) The agent pulling the bar must exert a
force equal in magnitude to F_{mag} and parallel to **v**.
Therefore, it does work at a rate $Fv =$
$(1.25 \times 10^{-3} \text{ N})(2 \text{ m/s}) = 2.5$ mW. The conservation of
energy requires the answers to parts (c) and (d) to be
equal.

Problem

35. A circular loop 40 cm in diameter is made from a
flexible conductor and lies at right angles to a
uniform 12-T magnetic field. At time $t = 0$ the
loop starts to expand, its radius increasing at the
rate of 5.0 mm/s. Find the induced emf in the
loop (a) at $t = 1.0$ s and (b) at $t = 10$ s.

Solution

The flux through the loop is $\phi_B = BA = B\pi r^2$ (if we
take the normal to the loop's area in the direction of **B**),
so Faraday's law gives $\mathcal{E}_i = -d\phi_B/dt = -2\pi Br(dr/dt)$.
For $r(t) = r_0 + (5 \text{ mm/s})t$, $\mathcal{E}_i = -2\pi(12 \text{ T}) \times$
$(5 \text{ mm/s})(20 \text{ cm} + 5t \text{ mm/s})$. At $t = 1$ s,
$\mathcal{E}_i = -77.3$ mV, and (b) at $t = 10$ s, $\mathcal{E}_i = -94.2$ mV.
The minus sign means that $\mathcal{E}_i$ opposes the increase in
flux; a right-hand screw rotated in the sense of $\mathcal{E}_i$
would advance in a direction opposite to **B**.

Section 31-4: Induced Electric Fields

Problem

37. Find the electric force on a 50-μC charge inside
the solenoid of Problem 18, if the charge is 5.0 cm
from the solenoid axis.

Solution

The induced electric field inside a long thin solenoid is
the subject of the exercise accompanying Example 31-9,
where a similar argument gives $E = -\frac{1}{2}r(dB/dt)$.
Such a field would produce a force on a point charge q,
of magnitude $\frac{1}{2}qr(dB/dt) = \frac{1}{2}\mu_0 nqr(dI/dt)$, since
$B = \mu_0 nI$. For the solenoid in Problem 18, this is
$\frac{1}{2}(4\pi \times 10^{-7} \text{ N/A}^2)(2000/2 \text{ m})(50 \text{ }\mu\text{C})(5 \text{ cm}) \times$
$(10^3 \text{ A/s}) = 1.57 \text{ }\mu\text{N}$.

Problem

39. A uniform magnetic field points into the page in
Fig. 31-54. In the same region an electric field
points straight up, but increases with position at
the rate of 10 V/m^2 as you move to the right.
Apply Faraday's law to a rectangular loop to show
that the magnetic field must be changing with
time, and calculate the rate of change.

Solution

Assume that there are no sources for **E** other than the
changing **B**. Choose coordinate axes and a rectangular
loop as shown superposed on Fig. 31-54. The
clockwise direction of circulation is such that $d\mathbf{A}$ is
parallel to **B** (into the page). The electric field is
$\mathbf{E} = [E_0 + (10 \text{ V/m}^2)x]\hat{\jmath}$. This is perpendicular to the
sides of width w, and parallel or antiparallel to the
sides of length ℓ, located at x_1 and x_2 respectively.
Thus, $\oint \mathbf{E} \cdot d\boldsymbol{\ell} = E_1\ell - E_2\ell = [E_0 + (10 \text{ V/m}^2)x_1 -$
$E_0 - (10 \text{ V/m}^2)x_2]\ell = -(10 \text{ V/m}^2)w\ell$, since
$x_2 - x_1 = w$. The flux through the loop is
$\int \mathbf{B} \cdot d\mathbf{A} = BA = Bw\ell$, so Faraday's law gives
$\oint \mathbf{E} \cdot d\boldsymbol{\ell} = -(10 \text{ V/m}^2)w\ell = -(d/dt) \int \mathbf{B} \cdot d\mathbf{A} =$
$-(dB/dt)w\ell$, or $dB/dt = 10$ T/s.

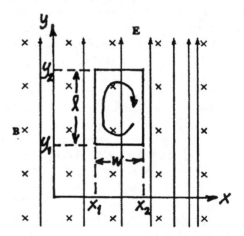

FIGURE 31-54 Problem 39 Solution.

Problem

41. Figure 31-56 shows a magnetic field pointing into
the page; the field is confined to a layer of
thickness h in the vertical direction but extends
infinitely to the left and right. The field strength
is increasing with time: $B = bt$, where b is a
constant. Find an expression for the electric field
at all points outside the field region. *Hint:*
Consult Example 30-5.

FIGURE 31-56 Problem 41.

Solution

A changing magnetic field acts as a source for an
induced electric field, just like a current density is a

source for a magnetic field. In fact, Faraday's law (for a loop fixed in space), $\oint_{\text{loop}} E \cdot d\ell = -d\phi_B/dt = \int_{\text{surface}}(\partial \mathbf{B}/\partial t) \cdot d\mathbf{A}$, is analogous to Ampère's law, $\oint_{\text{loop}} \mathbf{B} \cdot d\ell = \mu_0 I_{\text{encircled}} = \int_{\text{surface}} \mu_0 \mathbf{J} \cdot d\mathbf{A}$ (compare $\mathbf{E}$ and $\partial \mathbf{B}/\partial t$ with $\mathbf{B}$ and $\mu_0 \mathbf{J}$). The geometry of the source and symmetry of the field in Fig. 31-56 is similar to that in Fig. 30-25 for an infinite current sheet. The induced electric field, $\mathbf{E}$, should have the same magnitude above and below the source region for $\partial \mathbf{B}/\partial t$, and should circulate in a CCW sense so as to oppose the increase of flux into the page (i.e., CCW circulation is out of the page, opposite to the normal to an area into the page). For the rectangular loop shown added to Fig. 31-56, $\oint_{\text{loop}} \mathbf{E} \cdot d\ell = 2E\ell = \int_{\text{area}}(\partial B/\partial t) \, dA = (\partial B/\partial t)\ell h$, or $E = \frac{1}{2} |\partial B/\partial t| \, h = \frac{1}{2} bh$.

Paired Problems

Problem

43. A magnetic field is given by $\mathbf{B} = B_0(x/x_0)^2 \hat{\mathbf{k}}$, where B_0 and x_0 are constants. Find an expression for the magnetic flux through a square of side $2x_0$ that lies in the x-y plane with one corner at the origin and two sides coinciding with the positive x and y axes.

Solution

Take elements of area, $d\mathbf{A} = 2x_0 \, dx \hat{\mathbf{k}}$, which are rectangular strips parallel to the y-axis. Then

$$\phi_B = \int_{\text{square}} \mathbf{B} \cdot d\mathbf{A} = \left(\frac{B_0}{x_0^2}\right) 2x_0 \int_0^{2x_0} x^2 \, dx$$
$$= (2B_0/x_0)(2x_0)^3/3 = 16B_0 x_0^2/3.$$

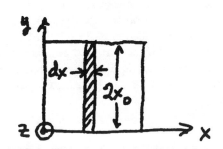

Problem 43 Solution.

Problem

45. A uniform magnetic field is given by $\mathbf{B} = bt\hat{\mathbf{k}}$, where $b = 0.35$ T/s. Find the current in a conducting loop with area 240 cm^2 and resistance 0.20 Ω that lies in the x-y plane. In what direction is the current, as viewed from the positive z-axis?

Solution

A normal to the loop parallel to the z-axis corresponds to CCW positive circulation (via the right-hand rule), when viewed *from* the positive z-axis. Faraday's and Ohm's law give the current in the loop: $I = \mathcal{E}/R = -(d\phi_B/dt)/R = -A(dB/dt)/R = -Ab/R = -(240 \text{ cm}^2)(0.35 \text{ T/s})/(0.2 \, \Omega) = -42.0$ mA, the minus sign indicating a CW circulation viewed from positive z-axis.

Problem

47. A pair of vertical conducting rods are a distance ℓ apart and are connected at the bottom by a resistance R. A conducting bar of mass m runs horizontally between the rods and can slide freely down them while maintaining electrical contact. The whole apparatus is in a uniform magnetic field $\mathbf{B}$ pointing horizontally and perpendicular to the bar. When the bar is released from rest it soon reaches a constant speed. Find this speed.

Solution

When the bar is falling, a motional emf causes an induced current to flow in the bar (in the direction of $\mathbf{v} \times \mathbf{B}$) as shown. When we are looking horizontally in the direction of $\mathbf{B}$ (into the page), the forces on the bar are gravity, mg downward, and the magnetic force, $I\ell B$ upward, where the induced current opposes the decrease in flux. The velocity is constant when $I\ell B = mg$. Now, $\phi_B = B\ell y$ and $I = \mathcal{E}/R = -(d\phi_B/dt)/R = -(B\ell/R)(dy/dt) = B\ell v/R$ (where $dy/dt = -v$ is the speed downward). Therefore, $v = IR/B\ell = mgR/B^2\ell^2$.

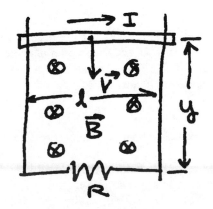

Problem 47 Solution.

Problem

49. Figure 31-58 shows an unusual design for a generator, consisting of a conducting bar that

rotates about a central axis while making contact with a conducting ring of radius R. A uniform magnetic field is perpendicular to the ring. Wires from the axis and ring carry power to a load. Find an expression for the emf induced in this generator when the bar rotates with angular speed ω.

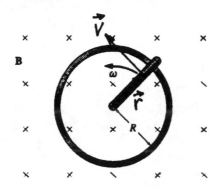

FIGURE 31-58 Problem 49.

Solution

Each rotation, the bar sweeps through an area of πR^2 perpendicular to the magnetic field B, so the flux changes by $\Delta\phi_B = \pi R^2 B$ in one period of rotation $\Delta t = 2\pi/\omega$. Then the magnitude of the induced emf is $\mathcal{E} = |-\Delta\phi_B/\Delta t| = \pi R^2 B/(2\pi/\omega) = \frac{1}{2}\omega R^2 B$. Alternatively, at a point r on the bar, there is a motional emf resulting from an equivalent electric field of $\mathbf{v} \times \mathbf{B} = -vB\hat{\mathbf{r}} = -\omega rB\hat{\mathbf{r}}$ (minus $\hat{\mathbf{r}}$ is toward the center). The emf developed across the length of the bar is $\mathcal{E} = \int_0^R \mathbf{E} \cdot d\mathbf{r} = -\omega B \int_0^R r\, dr = -\frac{1}{2}\omega BR^2$ (the axis is positive relative to the rim).

Problem

51. An electron is inside a solenoid, 28 cm from the solenoid axis. It experiences an electric force of magnitude 1.3 fN. At what rate is the solenoid's magnetic field changing?

Solution

The electric field has magnitude $E = F/e$. If we suppose that this is the electric field induced by the changing magnetic field in the solenoid, and use the axial symmetric approximation in Example 31-9 (modified as in the accompanying exercise, $\phi_B = \pi r^2 B$ inside the solenoid, and $2\pi r E = -\pi r^2(dB/dt)$), then $E = \frac{1}{2}r\,|dB/dt| = F/e$, or $|dB/dt| = 2F/re = 2(1.3\text{ fN})/(28\text{ cm}\times 1.6\times 10^{-19}\text{ C}) = 58.0$ T/ms.

Supplementary Problems

Problem

53. At time prior to $t = 0$, there is no current in either the solenoid or the small coil of Problem 19. Subsequently, the current in the small coil is observed to increase at 10 μA/s. What is the solenoid current as a function of time?

Solution

If we assume that the current in the small coil is only the current induced by the solenoid, then differentiation of the result of Problem 19 and substitution of numerical data gives

$$10\frac{\mu A}{s} = \frac{dI_C}{dt} = \mu_0\left(\frac{N_S N_C}{\ell}\right)\frac{1}{4}\pi D_S^2 \frac{1}{R}\frac{d^2 I_S}{dt^2},$$

or

$$\frac{d^2 I_S}{dt^2} = 1.62\frac{A}{s^2}.$$

(We only need to find the magnitude of I_S, since its direction was discussed in Problem 19.) The solution of this differential equation, which satisfies the initial conditions ($I_S = 0$ and $dI_S/dt \sim I_C = 0$ at $t = 0$), is simply $I_S = \frac{1}{2}(1.62$ A/s$^2)t^2$.

Problem

55. So-called magnetohydrodynamic generators have been proposed as a means of extracting electrical energy from charged particles released in fusion reactions; they've also been suggested as a way to generate electricity from flowing water. An MHD generator consists of two metal plates on either side of a channel carrying conducting fluid in a magnetic field, as shown in Fig. 31-60. The magnetic force on free charges in the fluid drives positive charge to one plate, negative to the other. If there's no electrical load connected across the plates, the electric field that develops eventually halts any further charge motion. (a) Show in this

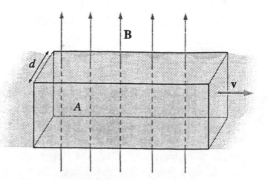

FIGURE 31-60 Problem 55.

case that the voltage between the plates is $V = vBd$, where v is the fluid velocity and d the plate spacing. (b) Now suppose a resistance R is connected between the plates. Show that the current through R is $I = \frac{vABd}{\rho d + AR}$, where A is the plate area and ρ is the resistivity of the fluid.

Solution

(a) The equivalent motional electric field on charge carriers in the fluid channel, $\mathbf{E}_i = \mathbf{F}_{mag}/q = \mathbf{v} \times \mathbf{B}$, with magnitude $E_i = vB$, results in an emf of $\mathcal{E}_i = \int \mathbf{E}_i \cdot d\boldsymbol{\ell} = vBd$ between the plates. When no circuit is present, the static field of accumulated charge cancels the motional emf, and the potential difference between the plates has the same magnitude. (b) When current is free to flow in a circuit containing the plates, its resistance outside the fluid channel is R, and for the volume of conducting fluid inside the channel, is $\rho d/A$ (assuming equipotential surfaces parallel to the plates, no fringing fields, etc.). Then the magnitude of the current generated by the motional emf in part (a) is $I = \mathcal{E}_i/R_{tot} = vBd/(R + \rho d/A)$, similar to a battery of internal resistance $\rho d/A$, and emf vBd, connected to a resistance R.

Problem

57. Clever farmers whose lands are crossed by large power lines have been known to steal power by stringing wire near the power line and making use of the induced current—a practice that has been ruled legally to be theft. The scene of a particular crime is shown in Fig. 31-61. The power line carries 60-Hz alternating current with a peak current of 10 kA (that is, the current is given by $I = I_0 \sin \omega t$, where $I_0 = 10$ kA and $\omega = 2\pi f$, with $f = 60$ Hz). (a) If the farmer wants a peak voltage of 170 V, what should be the length ℓ of the loop shown in Fig. 31-61? (170 V is the peak of standard 120-V AC power.) (b) If all the equipment the farmer connects to the loop has an equivalent resistance of 5.0 Ω, what is the farmer's average power consumption? *Note:* The *average* power consumption is half the product of the peak

voltage and peak current. (c) If the power company charges 10¢ per kWh, what is the monetary value of the energy stolen each day? (d) Without examining the farmer's lands, how, in principle, could the power company know that a crime is being committed?

Solution

(a) The induced emf $\mathcal{E}$ in the farmer's loop (which depends on the dimensions of the loop) can be calculated from the flux found in Example 31-2 for a similar configuration of wire and loop, $\phi_B = (\mu_0/2\pi) \times I\ell \ln(1 + w/a)$. (The dimensions of the loop and its distance from the power line are shown in Fig. 31-61.) For $I = I_0 \sin \omega t$ (and $dI/dt = \omega I_0 \cos \omega t$), Faraday's law gives $\mathcal{E} = -d\phi_B/dt = -(\mu_0/2\pi)\omega I_0 \ell \cos \omega t \ln \times (1 + w/a) \equiv -\mathcal{E}_P \cos \omega t$, where $\mathcal{E}_P$ is the peak voltage in the farmer's loop. Using the numerical data given, we find:

$$\ell = \frac{170 \text{ V}}{(2 \times 10^{-7} \text{ N/A}^2)(10^4 \text{ A})(2\pi \times 60 \text{ Hz}) \ln(5.5/5)}$$
$$= 2.37 \text{ km}.$$

(b) For sinusoidal current, the average power is one half the peak power (see Chapter 33), so the average power stolen is $\mathcal{P}_{av} = \frac{1}{2}\mathcal{E}_P^2/R = \frac{1}{2}(170 \text{ V})^2/5 \text{ Ω} = 2.89$ kW. (c) For continuous consumption, the farmer's dishonesty costs $(2.89 \text{ kW})(24 \text{ h/d})(\$0.10/\text{kW·h}) = \$6.94/\text{d}$. (d) The company could measure the power on either side of the farmer's property to detect the difference, but, the stolen power is such a small fraction of the total power output, very accurate measurements would be necessary. Certainly, the company is producing more power than it's being paid for, and this is, in principle, detectable.

Problem

59. A conducting disk with radius a, thickness h, and resistivity ρ is inside a solenoid of circular cross section. The disk axis coincides with the solenoid axis. The magnetic field in the solenoid is given by $B = bt$, with b a constant. Find expressions for (a) the current density in the disk as a function of the distance r from the disk center and (b) the rate of power dissipation in the entire disk. *Hint:* Consider the disk to be made up of infinitesimal conducting loops.

Solution

The changing magnetic field in the solenoid will induce a circular current density J in the disk. (a) From the exercise accompanying Example 31-9 (as mentioned in the solutions to Problems 37, 42, and 51) the induced

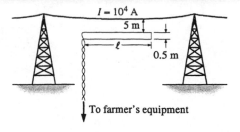

FIGURE 31-61 Problem 57.

electric field in the disk has magnitude $E = \frac{1}{2}rb$ (when $dB/dt = b$), so Ohm's law (in point form) gives $J = E/\rho = rb/2\rho$. (b) The result of Problem 27-73 gives the power dissipated per unit volume of disk (the joule heat). For volume elements, take rings of radius r, thickness dr, and width h, so

$$\mathcal{P}_J = \int J^2 \, \rho dV = \frac{b^2}{4\rho} \int_0^a r^2 2\pi r h \, dr = \frac{b^2}{4\rho} \cdot 2\pi h \cdot \frac{a^4}{4}$$

$$= \frac{\pi b^2 a^4 h}{8\rho}.$$

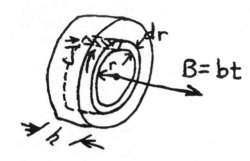

Problem 59 Solution.

Problem

61. Find an expression for the speed of the left-hand resistor in Problem 32 as a function of time, in terms of its mass m, the field strength B, the speed v of the right-hand bar, the time t, and the resistance R_{left} and R_{right}.

Solution

When the right-hand resistor has constant speed v, the left-hand resistor (initially at rest at $t = 0$) will begin moving to the right, as explained in the solution to Problem 32. We assume that the only horizontal force acting on the resistor is the magnetic force, $F_B = I\ell B$ (to the right), so its horizontal equation of motion is $m \, dv_{\text{left}}/dt = I\ell B$. The current has magnitude $I = (1/R_{\text{tot}})(d\phi_B/dt) = (\ell B/R_{\text{tot}}) \, d(x_{\text{right}} - x_{\text{left}})/dt = (\ell B/R_{\text{tot}})(v - v_{\text{left}})$, where $R_{\text{tot}} = R_{\text{left}} + R_{\text{right}}$, and x and $v = dx/dt$ are the position and speed of a resistor. Newton's second law becomes $m \, dv_{\text{left}}/dt = (B^2\ell^2/R_{\text{tot}})(v - v_{\text{left}})$. This equation is easily solved by separation of variables and integration:

$$\int \frac{dv_{\text{left}}}{v - v_{\text{left}}} = \int \left(\frac{B^2\ell^2}{mR_{\text{tot}}} \right) dt = -\ln\left(1 - \frac{v_{\text{left}}}{v} \right)$$

$$= \left(\frac{B^2\ell^2}{mR_{\text{tot}}} \right) t, \text{ or } v_{\text{left}} = v(1 - e^{B^2\ell^2 t/mR_{\text{tot}}})$$

Problem

63. A *flip coil* consists of a small coil used to measure magnetic fields. The flip coil is placed in a magnetic field with its plane perpendicular to the field, and then rotated abruptly through 180° about an axis in the plane of the coil. The coil is connected to instrumentation to measure the total charge Q that flows during this process. If the coil has N turns of area A and if its rotation axis is perpendicular to the magnetic field, show that the field strength is given by $B = QR/2NA$, where R is the coil resistance.

Solution

Initially, the flux through the flip coil is $\phi_B = NBA$, but is reversed to $-NBA$ when the coil is rotated 180°, so $\Delta\phi_B = -2NBA$. The total charge which flows is $\Delta Q = I_{\text{av}}\Delta t$, where I_{av} is the average induced current and Δt the time for the rotation. From Faraday's and Ohm's laws, $I_{\text{av}} = (-\Delta\phi_B/\Delta t)/R = 2NBA/R \, \Delta t$, hence $\Delta Q = 2NBA/R$. If the properties of the coil are known and the total charge is measured, one can find the magnetic field strength, $B = R \, \Delta Q/2NA$.

CHAPTER 32   INDUCTANCE AND MAGNETIC ENERGY

ActivPhysics can help with these problems:
Activity 14.1

Section 32-1:   Mutual Inductance

Problem

1. Two coils have a mutual inductance of 2.0 H. If current in the first coil is changing at the rate of 60 A/s, what is the emf in the second coil?

Solution

From Equation 32-2, $\mathcal{E}_2 = -M(dI_1/dt) = -(2\text{ H})\times(60\text{ A/s}) = -120$ V. (The minus sign, Lenz's law, signifies that an induced emf opposes the process which creates it.)

Problem

3. The current in one coil is given by $I = I_p \sin 2\pi ft$, where $I_p = 75$ mA, $f = 60$ Hz, and $t =$ time. Find the peak emf in a second coil if the mutual inductance between the coils is 440 mH.

Solution

Suppose $I_1 = I_p \sin 2\pi ft$ in Equation 32-2. Then $\mathcal{E}_2 = -M\,dI_1/dt = -2\pi fMI_p \cos 2\pi ft$, and the peak emf (when $\cos 2\pi ft = \pm1$) is $2\pi fMI_p = (2\pi \times 60\text{ Hz})(440\text{ mH})(75\text{ mA}) = 12.4$ V.

Problem

5. An alternating current given by $I_p \sin 2\pi ft$ is supplied to one of two coils whose mutual inductance is M. (a) Find an expression for the emf in the second coil. (b) When $I_p = 1.0$ A and $f = 60$ Hz, the peak emf in the second coil is measured at 50 V. What is the mutual inductance?

Solution

(a) From Equation 32-2, $\mathcal{E}_2 = -M\,dI_1/dt = -M2\pi fI_p \cos(2\pi ft)$. (b) The peak value of the cosine is 1, so $|M| = \mathcal{E}_{2p}/2\pi fI_p = 50\text{ V}/(2\pi \times 60\text{ Hz})(1\text{ A}) = 133$ mH. (From the information given, only the magnitude of M can be determined; its sign depends on how the coils are coupled.)

Problem

7. Two long solenoids of length ℓ both have n turns per unit length. They have circular cross sections with radii R and $2R$, respectively. The smaller solenoid is mounted inside the larger one, with their axes coinciding. Find the mutual inductance of this arrangement, neglecting any nonuniformity in the magnetic field near the ends.

Solution

All of the flux from the smaller solenoid (number one) links the larger solenoid (number two), so $\phi_{B,2} = N_2\mathbf{B}_1 \cdot \mathbf{A}_1 = N_2(\mu_0 n_1 I_1)(\pm\pi R^2) = \pm\mu_0 n^2 \pi R^2 \ell I_1$, since both solenoids have the same number of turns and length. Dividing by I_1 gives M (see Equation 32-1). Note that the sign of M depends on the relative direction of the windings in the two solenoids. (Alternatively, only a fraction A_1/A_2 of the flux from the larger solenoid links the smaller solenoid, so $\phi_{B,1} = \pm N_1 B_2 A_1$ and $M = \phi_{B,1}/I_2$ is the same.)

Problem

9. A rectangular loop of length ℓ and width w is located a distance a from a long, straight, wire, as shown in Fig. 32-20. What is the mutual inductance of this arrangement?

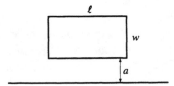

FIGURE 32-20 Problem 9.

Solution

When current I_1 flows to the left in the wire, the flux through the loop is $\phi_{B,2} = (\mu_0 I_1/2\pi) \int_a^{a+w} \ell\, dr/r = (\mu_0 I_1 \ell/2\pi)\ln(1 + w/a)$ (see Example 31-2). Then Equation 32-1 gives $M = \phi_{B,2}/I_1 = (\mu_0 \ell/2\pi) \times \ln(1 + w/a)$. (In calculating the flux, the normal to the loop area was taken into the page, so the positive sense of circulation around the loop is CW. This determines the direction of the induced emf $\mathcal{E}_2$ in Equation 32-2.)

Section 32-2: Self-Inductance

Problem

11. What is the self-inductance of a solenoid 50 cm long and 4.0 cm in diameter that contains 1,000 turns of wire?

Solution

Equation 32-4 gives $L = \mu_0 N^2 A/\ell = (4\pi \times 10^{-7}\,\text{H/m}) \times (10^3)^2 \pi (2\,\text{cm})^2/(50\,\text{cm}) = 3.16\,\text{mH}$. (The long thin solenoid approximation is valid here.)

Problem

13. A 2.0-A current is flowing in a 20-H inductor. A switch is opened, interrupting the current in 1.0 ms. What emf is induced in the inductor?

Solution

Assume that the current changes uniformly from 2 A to zero in 1 ms (or consider average values). Then $dI/dt = -2\,\text{A/ms}$, and Equation 32-5 gives $\mathcal{E} = -(20\,\text{H})(-20\,\text{A/ms}) = 40\,\text{kV}$. (The emf opposes the decreasing current.)

Problem

15. A cardboard tube measures 15 cm long by 2.2 cm in diameter. How many turns of wire must be wound on the full length of the tube to make a 5.8-mH inductor?

Solution

From Equation 32-4, $N = \sqrt{L\ell/\mu_0 A} = [(5.8\,\text{mH}) \times (15\,\text{cm})/(4\pi \times 10^{-7}\,\text{H/m})\pi(1.1\,\text{cm})^2]^{1/2} = 1.35 \times 10^3$ turns.

Problem

17. The emf in a 50-mH inductor has magnitude $|\mathcal{E}| = 0.020t$, with t in seconds and $\mathcal{E}$ in volts. At $t = 0$ the inductor current is 300 mA. (a) If the current is increasing, what will be its value at $t = 3.0$ s? (b) Repeat for the case when the current is decreasing.

Solution

(a) $\mathcal{E}$ has the opposite sign to dI/dt in Equation 32-5. When I is increasing, $dI/dt > 0$, $\mathcal{E}$ is negative, $\mathcal{E} = -|\mathcal{E}|$. Thus,

$$\frac{dI}{dt} = -\frac{\mathcal{E}}{L} = \frac{(0.02\,\text{V/s})t}{(0.05\,\text{H})} = \left(0.4\frac{\text{A}}{\text{s}^2}\right)t,$$

and

$$I = \frac{1}{2}\left(0.4\frac{\text{A}}{\text{s}^2}\right)t^2 + I_0.$$

At $t = 3$ s, $I = (0.2\,\text{A/s}^2)(3\,\text{s})^2 + 0.3\,\text{A} = 2.1\,\text{A}$. (b) For $dI/dt < 0$, $\mathcal{E} = |\mathcal{E}| > 0$, so $dI/dt = -(0.4\,\text{A/s}^2)t$, and $I = -(0.2\,\text{A/s}^2)t^2 + I_0$. At $t = 3$ s in this case, $I = -1.5\,\text{A}$. (After $t = \sqrt{3/2}$ s, the current reverses direction and begins increasing in absolute value.)

Problem

19. A 2,000-turn solenoid is 65 cm long and has cross-sectional area 30 cm². What rate of change of current will produce a 600-V emf in this solenoid?

Solution

The self-inductance (of this long, thin solenoid) is $L = \mu_0 N^2 A/\ell = (4\pi \times 10^{-7}\,\text{H/m})(2000)^2(30\,\text{cm}^2) \div (65\,\text{cm}) = 23.2\,\text{mH}$ (see Equation 32-4), so Equation 32-5 gives $|dI/dt| = |\mathcal{E}|/L = (600\,\text{V}) \div (23.2\,\text{mH}) = 25.9\,\text{A/ms}$.

Problem

21. The emf in a 50-mH inductor is given by $\mathcal{E} = \mathcal{E}_p \sin \omega t$, where $\mathcal{E}_p = 75$ V and $\omega = 140$ s^{-1}. What is the peak current in the inductor? (Assume the current swings symmetrically about zero.)

Solution

From Equation 32-5, $dI/dt = -(\mathcal{E}_p/L)\sin \omega t$, so integration yields $I(t) = (\mathcal{E}_p/\omega L)\cos \omega t$. (Since $I(t)$ is symmetric about $I = 0$, the constant of integration is zero.) The peak current is $I_p = \mathcal{E}_p/\omega L = 75\,\text{V}/(140\,\text{s}^{-1} \times 50\,\text{mH}) = 10.7\,\text{A}$.

Section 32-3: Inductors in Circuits

Problem

23. Show that the inductive time constant has the units of seconds.

Solution

A henry is a volt·second/ampere (see Equation 32-5), so the units of $\tau_L = L/R$ are $\text{H}/\Omega = \text{V·s}/\Omega\text{·A} = \text{s}$.

Problem

25. The current in a series RL circuit rises to 20% of its final value in 3.1 μs. If $L = 1.8$ mH, what is the resistance R?

Solution

The buildup of current in an RL circuit with a battery is given by Equation 32-8, $I(t) = I_\infty(1 - e^{-Rt/L})$, where $I_\infty = \mathcal{E}_0/R$ is the final current. Solving for R, one finds $R = -(L/t)\ln(1 - I/I_\infty) = -(1.8\,\text{mH} \div 3.1\,\mu\text{s})\ln(1 - 20\%) = 130\,\Omega$.

Problem

27. A 10-H inductor is wound of wire with resistance 2.0 Ω. If the inductor is connected across an ideal 12-V battery, how long will it take the current to reach 95% of its final value?

Solution

Reference to the solution to Problem 25 shows that $t = -(L/R)\ln(1 - I/I_\infty) = -(10 \text{ H}/2 \ \Omega)\ln(1 - 0.95) = 15.0$ s. (The percentage I/I_∞ is independent of $\mathcal{E}_0$.)

Problem

29. In Fig. 32-8a, take $R = 2.5$ kΩ and $\mathcal{E}_0 = 50$ V. When the switch is closed, the current through the inductor rises to 10 mA in 30 μs. (a) What is the inductance? (b) What will be the current in the circuit after many time constants?

Solution

(b) After a long time ($t \to \infty$), the exponential term in Equation 32-8 is negligible. Thus, $I_\infty = \mathcal{E}_0/R = 50$ V/2.5 kΩ = 20 mA. (a) The current has risen to half its final value in 30 μs. Thus (Equation 32-8 again), $\frac{1}{2} = 1 - e^{-Rt/L}$, or $L = Rt/\ln 2 = (2.5 \text{ k}\Omega) \times (30 \ \mu\text{s})/\ln 2 = 108$ mH.

Problem

31. In Fig. 32-8a, take $R = 100 \ \Omega$, $L = 2.0$ H, and $\mathcal{E}_0 = 12$ V. At 20 ms after the switch is closed, what are (a) the circuit current, (b) the inductor emf, (c) the resistor voltage, (d) the rate of change of the circuit current, and (e) the power dissipation in the resistor?

Solution

(a) The time constant is $L/R = 2$ H/100 Ω = 20 ms, and the final current is $\mathcal{E}_0/R = 12$ V/100 Ω = 120 mA. After one time constant, Equation 32-8 gives $I = 120 \text{ mA}(1 - e^{-1}) = 75.9$ mA. (b) The voltage drop across the inductor is (from Equation 32-7) $V_L = \mathcal{E}_0 - IR = -\mathcal{E}_L = \mathcal{E}_0 e^{-Rt/L}$. After one time constant, $V_L = (12 \text{ V})e^{-1} = 4.41$ V. (Note that $V_L + V_R = \mathcal{E}_0$ is Kirchhoff's loop law.) (c) $V_R = IR = (75.9 \text{ mA}) \times (100 \ \Omega) = 7.59$ V. (Alternatively, $V_R = \mathcal{E}_0 - V_L = 12$ V $- 4.41$ V.) (d) From Equation 32-5 and the loop law, $V_L = -\mathcal{E}_L = L \ dI/dt$. After one time constant, $dI/dt = V_L/L = 4.41$ V/2 H $= 2.21$ A/s. (e) $P_R = I^2R = (75.9 \text{ mA})^2(100 \ \Omega) = 575$ mW.

Problem

33. Resistor R_2 in Fig. 32-22 is to limit the emf that develops when the switch is opened. What should

be its value in order that the inductor emf not exceed 100 V?

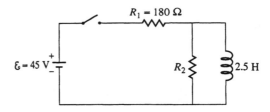

FIGURE 32-22 Problem 33.

Solution

As explained in Example 32-6, when the switch is opened (after having been closed a long time), the voltage across R_2 (which equals the inductor emf) is $V_2 = I_2R_2 = \mathcal{E}_0R_2/R_1$. If we choose to limit this to no more than 100 V, then $R_2 \leq (100 \text{ V})(180 \ \Omega)/45$ V $= 400 \ \Omega$.

Problem

35. A 5.0-A current is flowing through a nonideal inductor with $L = 500$ mH. If the inductor is suddenly short-circuited, the inductor current drops to 2.5 A in 6.9 ms. What is the resistance of the inductor?

Solution

A real inductor can be represented by a resistance, R, in series with an inductance, L. When short-circuited, the inductor constitutes an LR circuit without a battery, and the current decays exponentially with time constant L/R, $I = I_0 e^{-Rt/L}$ (see Equation 32-9). When the given data is substituted, we can solve for the resistance,

$$R = \left(\frac{L}{t}\right)\ln\left(\frac{I_0}{I}\right) = \left(\frac{500 \text{ mH}}{6.9 \text{ ms}}\right)\ln 2 = 50.2 \ \Omega.$$

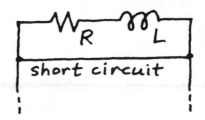

Problem 35 Solution.

Problem

37. In Fig. 32-24, take $\mathcal{E}_0 = 20$ V, $R_1 = 10 \ \Omega$, $R_2 = 5.0 \ \Omega$, and assume the switch has been open

for a long time. (a) What is the inductor current immediately after the switch is closed? (b) What is the inductor current a long time after the switch is closed? (c) If after a long time the switch is again opened, what will be the voltage across R_1 immediately afterward?

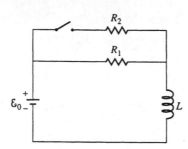

FIGURE 32-24 Problem 37.

Solution

(a) If the switch has been open a long time, a steady current flows through the inductance ($dI_L/dt = 0$). When the switch is closed (at $t = 0$), I_L cannot change instantaneously, so $I_L(0) = \mathcal{E}/R_1 = 20$ V/10 Ω = 2 A. (Of course, $I_1(0) = I_L(0)$, and $I_2(0) = 0$.) (b) After another long time ($t \to \infty$), the currents are steady again and $\mathcal{E}_L = 0$ (the inductance behaves like a short circuit). The resistors are in parallel; therefore $I_L(\infty) = \mathcal{E}(1/R_1 + 1/R_2) = 20$ V$(\frac{1}{5} + \frac{1}{10})$ $\Omega^{-1} = 6$ A. (c) When the switch is again opened, the current through R_2 is zero, but I_L cannot change instantly, so $I_L = I_1 = I_L(\infty) = 6$ A. Thus, the voltage across R_1 is $V_1 = I_1 R_1 = (6$ A$)(10$ $\Omega) = 60$ V.

Section 32-4: Magnetic Energy

Problem

39. What is the current in a 10-mH inductor when the stored energy is 50 μJ?

Solution

From Equation 32-10, $I = \sqrt{2U/L} = \sqrt{2(50\ \mu\text{J})/10\ \text{mH}} = 0.1$ A.

Problem

41. A 12-V battery, 5.0-Ω resistor, and 18-H inductor are connected in series and allowed to reach a steady state. (a) What is the energy stored in the inductor? (b) Once in the steady state, over what time interval is the energy dissipated in the resistor equal to that stored in the inductor?

Solution

(a) The steady state (i.e., final) current in an RL circuit with emf $\mathcal{E}_0$ is $I = \mathcal{E}_0/R$, so the energy stored in the inductor (Equation 32-10) is $U_L = \frac{1}{2}LI^2 = \frac{1}{2}(18$ H$)(12$ V/5 $\Omega)^2 = 51.8$ J. (b) Energy is dissipated in the resistor at the rate $\mathcal{P}_R = I^2R$ (the joule heat), so the time interval queried is $\Delta t = U_L/\mathcal{P}_R = \frac{1}{2}LI^2/I^2R = L/2R = 18$ H$/(2\times5$ $\Omega) = 1.8$ s.

Problem

43. The current in a 2.0-H inductor is decreased linearly from 5.0 A to zero over 10 ms. (a) What is the average rate at which energy is being extracted from the inductor during this time? (b) Is the instantaneous rate constant?

Solution

(a) The energy falls from $U_i = \frac{1}{2}LI^2 = \frac{1}{2}(2$ H$)\times(5$ A$)^2 = 25$ J to $U_f = 0$ in $\Delta t = 10$ ms, so the rate of decrease is $\Delta U/\Delta t = -25$ J/10 ms $= -2.5$ kW. (b) The discussion in the text leading to Equation 32-10 shows that the instantaneous power is $\mathcal{P}_L = LI(dI/dt)$, so even if dI/dt is constant, I and $\mathcal{P}_L$ are not.

Problem

45. The current in a 2.0-H inductor is increasing. At some instant, the current is 3.0 A and the inductor emf is 5.0 V. At what rate is the inductor's magnetic energy increasing at this instant?

Solution

The rate at which energy is stored in an inductor is $\mathcal{P}_L = LI(dI/dt)$ (see the discussion of "Magnetic Energy in an Inductor" leading to Equation 32-10). When the current is increasing, as for this inductor, $L(dI/dt) = |\mathcal{E}_L|$, and $\mathcal{P}_L = I|\mathcal{E}_L| = (3$ A$)(5$ V$) = 15$ W.

Problem

47. A superconducting solenoid with inductance $L = 3.5$ H carries 1.8 kA. Copper is embedded in the coils to carry the current in the event of a quench (see Example 32-7). (a) What is the magnetic energy in the solenoid? (b) What is the maximum resistance of the copper that will limit the power dissipation to 100 kW immediately after a loss of superconductivity? (c) With this resistance, how long will it take the power to drop to 50 kW?

Solution

(a) In its superconducting state, the solenoid's stored energy (Equation 32-10) is $U = \frac{1}{2}LI^2 = \frac{1}{2}(3.5\text{ H})\times (1.8\text{ kA})^2 = 5.67$ MJ. (b) The current in an inductor cannot change instantaneously, so if the power dissipated in the copper (I^2R) just after a sudden loss of superconductivity must be less than 100 kW, then $R \leq 100$ kW$/(1.8$ kA$)^2 = 30.9$ mΩ. (c) The power drops to one half its maximum original value, when the current drops to $1/\sqrt{2}$ times its initial value. From Equation 32-9, the decay time is $t = (L/R)\ln(I_0/I) = (3.5\text{ H}/30.9\text{ m}\Omega)\ln\sqrt{2} = 39.3$ s.

Problem

49. The Alcator fusion experiment at MIT has a 50-T magnetic field. What is the magnetic energy density in Alcator?

Solution

From Equation 32-11, $u_B = (50\text{ T})^2/(8\pi\times 10^{-7}\text{ N/A}^2) = 995$ MJ/m^3. (This is about 2.8% of the energy density content of gasoline; see Appendix C.)

Problem

51. The magnetic field of a neutron star is about 10^8 T. How does the energy density in this field compare with the energy density stored in (a) gasoline and (b) pure uranium-235 (mass density 19×10^3 kg/m^3)? Consult Appendix C.

Solution

The energy density in a field of this strength is $u_B = B^2/2\mu_0 = (10^8\text{ T})^2/(8\pi\times10^{-7}\text{ H/m}) = 3.98\times10^{21}$ J/m^3 (see Equation 32-11). This is about (a) 1.1×10^{11} times the energy density content of gasoline (44 MJ/kg × 800 kg/m$^3 = 3.52\times10^{10}$ J/m^3), and (b) 2600 times that of pure U^{235}(8×10^{13} J/kg × 19×10^3 kg/m$^3 = 1.52\times10^{18}$ J/m^3).

Problem

53. A single-turn loop of radius R carries current I. How does the magnetic energy density at the loop center compare with that of a long solenoid of the same radius, carrying the same current, and consisting of n turns per unit length?

Solution

The energy density at the center of the loop is $u_B^{(\text{loop})} = B^2/2\mu_0 = (\mu_0 I/2R)^2/2\mu_0 = \mu_0 I^2/8R^2$ (see Equations 30-3 and 32-11). In a long thin solenoid of the same radius, $u_B^{(\text{solenoid})} = (\mu_0 nI)^2/2\mu_0 = \mu_0 n^2 I^2/2$, so the ratio of $u_B^{(\text{loop})}$ to $u_B^{(\text{solenoid})}$ is $1/4n^2R^2$.

Problem

55. A toroidal coil has inner radius R and a square cross section of side ℓ (Fig. 32-25). It is wound with N turns of wire, and carries a current I. Show that the magnetic energy in the toroid is given by

$$U = \frac{\mu_0 N^2 I^2 \ell}{4\pi}\ln\left(\frac{R+\ell}{R}\right).$$

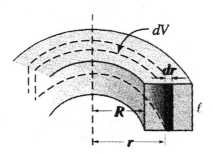

FIGURE 32-25 Problem 55.

Solution

The magnetic field inside a cylindrically symmetric toroid was found by using Ampere's law, $B = \mu_0 NI \div 2\pi r$, where r is the distance from the central axis (see Equation 30-12). The energy in this magnetic field can be found by integrating the energy density, $u_B = B^2/2\mu_0$, over the volume of the toroid, with volume elements consisting of cylindrical shells of radius r, appropriate height, and thickness dr. For a toroid of square cross section of side ℓ, and inner radius R, $dV = 2\pi r\ell\,dr$ (see figure), and

$$U = \int_{\text{toroid}} u_B\,dV = \int_R^{R+\ell}\frac{1}{2\mu_0}\left(\frac{\mu_0 NI}{2\pi r}\right)^2 2\pi r\ell\,dr$$

$$= \frac{\mu_0 N^2\ell I^2}{4\pi}\int_R^{R+\ell}\frac{dr}{r} = \frac{\mu_0 N^2\ell I^2}{4\pi}\ln\left(1 + \frac{\ell}{R}\right).$$

An alternative approach is to use the flux linkage of the toroid from the solution to Problem 31-30, $N\phi_B = \phi_{\text{toroid}} = LI$ (recall that $a = R$, $b = R + \ell$, and ϕ_B was the flux through one turn), and $U = \frac{1}{2}I\phi_{\text{toroid}}$.

Paired Problems

Problem

57. Two coils have mutual inductance M. The current supplied to coil A is given by $I = bt^2$. Find an expression for the magnitude of the induced emf in coil B.

Solution

From Equation 32-2, $\mathcal{E}_B = -M(dI_A/dt) = -M(2bt)$, where the negative value means the emf opposes the change in current.

Problem

59. In the circuit of Fig. 32-8a, take $\mathcal{E}_0 = 5.0$ V and $R = 1.8$ Ω. At 2.5 s after the switch is closed, the circuit current is 250 mA. Find the inductance.

Solution

Equation 32-8 for the build-up of current in the RL circuit gives $L = -Rt/\ln(1 - IR/\mathcal{E}_0) = -(1.8 \text{ Ω}) \times (2.5 \text{ s})/\ln(1 - 0.25 \times 1.8/5) = 47.7$ H.

Problem

61. In Fig. 32-13a, take $\mathcal{E}_0 = 25$ V, $R_1 = 1.5$ Ω, and $R_2 = 4.2$ Ω. What is the voltage across R_2 (a) immediately after the switch is first closed and (b) a long time after the switch is closed? (c) Long after the switch is closed it is again opened. Now what is the voltage across R_2?

Solution

The circuit is analyzed in Example 32-6 at the instants of time mentioned in this problem, so all that is necessary here is to find numerical values. (a) At the moment the switch is first closed, $I = \mathcal{E}_0/(R_1 + R_2) = 25 \text{ V}/(1.5 + 4.2) \text{ Ω} = 4.39$ A, and $V_2 = (4.39 \text{ A})(4.2 \text{ Ω}) = 18.4$ V. (b) When the current is steady, the current through R_2 is zero (since the inductor is an ideal one). (c) When the switch is reopened, $V_2 = \mathcal{E}_0 R_2/R_1 = (25 \text{ V})(4.2/1.5) = 70.0$ V, momentarily.

Problem

63. A wire of radius R carries a current I distributed uniformly over its cross section. Find an expression for the magnetic energy per unit length in the region from R to $100R$.

Solution

The magnetic field from a long straight wire has cylindrical symmetry. A thin coaxial cylindrical shell of length ℓ, radius r, and thickness dr, has volume $dV = 2\pi r\ell \, dr$ and magnetic energy density $u_B = B^2/2\mu_0$. Outside the wire, $B = \mu_0 I/2\pi r$ (see Equation 30-8), so the energy per unit length in a volume extending between $r = R$ and $100R$ is

$$\frac{U}{\ell} = \int \frac{u_B dV}{\ell} = \int_R^{100R} \frac{\mu_0 I^2}{8\pi^2 r^2} 2\pi r \, dr = \frac{\mu_0 I^2}{4\pi} \ln 100.$$

Supplementary Problems

Problem

65. (a) Use the result of Problem 55 to determine the inductance of a toroid. (b) Show that your result

reduces to the inductance of a long solenoid when $R \gg \ell$.

Solution

(a) Since $U = \frac{1}{2}LI^2$, dividing the expression for U in Problem 55 by $\frac{1}{2}I^2$, we find $L = (\mu_0 N^2/2\pi)\ell \times \ln(1 + \ell/R)$. (b) For $\ell \ll R$, $\ln(1 + \ell/R) \approx \ell/R$, so $L \approx \mu_0 N^2 \ell^2/2\pi R$. Since $\ell^2 = A$ is the cross-sectional area, $2\pi R = \ell_{\text{Solenoid}}$ is the length, and $n = N/\ell_{\text{Solenoid}}$ is the number of turns per unit length of the toroidal solenoid, this result is approximately the same as Equation 32-4.

Problem

67. (a) Use Equation 32-9 to write an expression for the power dissipation in the resistor as a function of time, and (b) integrate from $t = 0$ to $t = \infty$ to show that the total energy dissipated is equal to the energy initially stored in the inductor, namely, $\frac{1}{2}LI_0^2$.

Solution

Equation 32-9 gives the current decaying through a resistor connected to an inductor carrying an initial current I_0. (a) The instantaneous power dissipated in the resistor is $\mathcal{P}_R = I^2 R = I_0^2 \, Re^{-2Rt/L}$. (b) In a time interval dt, the energy dissipated is $dU = \mathcal{P}_R \, dt$, so the total energy dissipated is

$$U = \int_0^\infty I_0^2 Re^{-2Rt/L} dt = I_0^2 R \left.\frac{e^{-2Rt/L}}{(-2R/L)}\right|_0^\infty$$

$$= \frac{I_0^2 RL}{2R} = \frac{1}{2}LI_0^2.$$

This is precisely the energy initially stored in the inductor.

Problem

69. An electric field and a magnetic field have the same energy density. Obtain an expression for the ratio E/B, and evaluate this ratio numerically. What are its units? Is your answer close to any of the fundamental constants listed inside the front cover?

Solution

The combination of Equations 26-3 and 32-11 implies that if $u_E = \frac{1}{2}\varepsilon_0 E^2 = u_B = B^2/2\mu_0$, then $E/B = 1/\sqrt{\mu_0\varepsilon_0}$. Numerically, $\mu_0 = 4\pi \times 10^{-7}$ N/A^2 and $(1/4\pi\varepsilon_0) \approx 9 \times 10^9$ N·m^2/ C^2, so $1/\sqrt{\mu_0\varepsilon_0} \approx$

$$\sqrt{(9 \times 10^9 \text{ N·m}^2/\text{C}^2)/(10^{-7} \text{ N/A}^2)} = 3 \times 10^8 \text{ m/s},$$

which is, in fact, the speed of light (see Section 34-5).

Problem

71. The switch in the circuit of Fig. 32-27 is closed at time $t = 0$, at which instant the inductor current is zero. Write the loop and node laws for this circuit, and show that they are satisfied if the inductor current is given by $I = (\mathcal{E}_0/R_1)(1 - e^{-R_{\parallel}t/L})$, where $R_{\parallel}$ is the resistance of R_1 and R_2 were they connected in parallel.

Solution

The node law (applied to either node, with currents as shown superimposed on Fig. 32-27) gives $I_1 - I_2 - I_L = 0$. Two independent loop equations (for both loops containing $\mathcal{E}_0$) are $\mathcal{E}_0 - I_1 R_1 - I_2 R_2 = 0$ and $\mathcal{E}_0 - I_1 R_1 + \mathcal{E}_L = 0$. Here, $\mathcal{E}_L = -L(dI_L/dt)$ is the inductor's induced emf. Note that I_1 and I_2 can be determined from I_L, since $I_1 = (\mathcal{E}_0 + \mathcal{E}_L)/R_1$ and $I_2 = I_1 - I_L$. These currents automatically satisfy the node equation and the second loop equation, so we need only check that the expression given for I_L

satisfies the first loop equation. The equation can be written entirely in terms of I_L: $0 = \mathcal{E}_0 - I_1 R_1 - I_2 R_2 = \mathcal{E}_0 - (\mathcal{E}_0 + \mathcal{E}_L) - (I_1 - I_L)R_2 = -\mathcal{E}_L - (\mathcal{E}_0 + \mathcal{E}_L)(R_2/R_1) + I_L R_2 = -\mathcal{E}_L(1 + R_2/R_1) + I_L R_2 - \mathcal{E}_0 R_2/R_1$. If we divide by R_2, substitute for $\mathcal{E}_L$, and use $R_{\parallel} = R_1 R_2/(R_1 + R_2)$, the equation becomes $(L/R_{\parallel})(dI_L/dt) + I_L - \mathcal{E}_0/R_1 = 0$. Substituting the given expression for I_L, we, indeed, find that

$$\frac{L}{R_{\parallel}}\left(-\frac{\mathcal{E}_0}{R_1}\right)\left(-\frac{R_{\parallel}}{L}e^{-R_{\parallel}t/L}\right) + \frac{\mathcal{E}_0}{R_1}(1 - e^{-R_{\parallel}t/L}) - \frac{\mathcal{E}_0}{R_1} = 0.$$

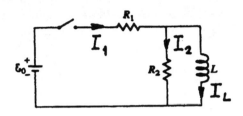

FIGURE 32-27 Problem 71 Solution.

CHAPTER 33   ALTERNATING-CURRENT CIRCUITS

ActivPhysics can help with these problems:
Activities 14.2, 14.3

Section 33-1:   Alternating Current

Problem

1. Much of Europe uses AC power at 230 V rms and 50 Hz. Express this AC voltage in the form of Equation 33-3, taking $\phi = 0$.

Solution

Use of Equations 33-1 and 2 allows us to write $V_p = \sqrt{2} \, V_{rms} = \sqrt{2}(230 \text{ V}) = 325 \text{ V}$, and $\omega = 2\pi f = 2\pi(50 \text{ Hz}) = 314 \text{ s}^{-1}$. Then the voltage expressed in the form of Equation 33-3 is $V(t) = (325 \text{ V}) \times \sin[(314 \text{ s}^{-1})t]$.

Problem

3. An oscilloscope displays a sinusoidal signal whose peak-to-peak voltage (see Fig. 33-1) is 28 V. What is the rms voltage?

Solution

As shown in Fig. 33-1, the peak-to-peak voltage is twice the peak voltage, so Equation 33-1 gives $V_{rms} = V_p/\sqrt{2} = V_{p\text{-}p}/2\sqrt{2} = 28 \text{ V}/2\sqrt{2} = 9.90 \text{ V}$.

Problem

5. An AC current is given by $I = 495\sin(9.43t)$, with I in milliamperes and t in milliseconds. Find (a) the rms current and (b) the frequency in Hz.

Solution

Comparison of the current with Equation 33-3 shows that its amplitude and angular frequency are $I_p = 495 \text{ mA}$ and $\omega = 9.43 \text{ (ms)}^{-1}$. Application of Equations 33-1 and 2 give (a) $I_{rms} = 495 \text{ mA}/\sqrt{2} = 350 \text{ mA}$, and (b) $f = 9.43/2\pi \text{(ms)} = 1.50 \text{ kHz}$.

Problem

7. The rms amplitude is defined as the square root of the average of the square of the signal. For a periodic function, the time average is the integral over one period, divided by the period. For a sinusoidal voltage given by $V = V_p \sin \omega t$, show explicitly that $V_{rms} = V_p/\sqrt{2}$.

Solution

The period of a sinusoidal signal is $T = 2\pi/\omega$ (Equation 15-6), so

$$V_{rms}^2 = \langle V^2 \rangle = \frac{1}{T}\int_0^T (V_p^2 \sin^2 \omega t)\, dt$$

$$= \frac{V_p^2}{2T}\int_0^T (1 - \cos 2\omega t)\, dt$$

$$= \frac{V_p^2}{2T}\left[T - \left| \frac{1}{2\omega}\sin 2\omega t \right|_0^T \right] = \frac{V_p^2}{2}.$$

(We used an identity from Appendix A.) A graphical argument that $\langle \sin^2 \omega t \rangle = \frac{1}{2}$ is suggested in Fig. 16-16.

Problem

9. How are the rms and peak voltages related for the triangle wave in Fig. 33-29? See Problem 7.

Solution

Define the zero of time to coincide with a positive apex of the waveform, which has period T, as drawn upon Fig. 33-29. The analytic form of the voltage signal (for $0 \le t \le T$) is

$$V(t) = V_p \begin{cases} -(4t/T) + 1, & 0 \le t \le \frac{1}{2}T \\ (4t/T) - 3, & \frac{1}{2}T \le t \le T. \end{cases}$$

Because the negative part of the waveform is the reflection of the positive part, the square of the waveform has period $\frac{1}{2}T$. Thus, the average over T equals the average over $\frac{1}{2}T$, or

$$\langle V^2 \rangle = \frac{1}{T}\int_0^T V^2 dt = \frac{1}{\frac{1}{2}T}\int_0^{T/2} V^2 dt$$

$$= \frac{2V_p^2}{T}\int_0^{T/2}\left(\frac{16t^2}{T^2} - \frac{8t}{T} + 1 \right) dt$$

$$= \frac{2V_p^2}{T}\left[\frac{16}{3T^2}\left(\frac{T}{2}\right)^3 - \frac{8}{2T}\left(\frac{T}{2}\right)^2 + \frac{T}{2} \right]$$

$$= 2V_p^2\left(\frac{2}{3} - 1 + \frac{1}{2} \right) = \frac{V_p^2}{3}.$$

Consequently, $V_{rms} = \sqrt{\langle V^2 \rangle} = V_p/\sqrt{3}$. (We used a symmetry of the waveform to reduce the number of integrations by half.)

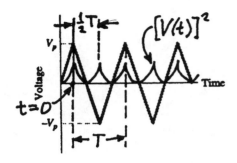

FIGURE 33-29 Problem 9 Solution.

Problem

11. The most general expression for a sinusoidal AC current may be written either $I = I_1 \sin\omega t + I_2 \cos\omega t$ or $I = I_p \sin(\omega t + \phi)$. Find relations between I_1, I_2, I_p, and ϕ that make these expressions equivalent. (See Appendix A for trig identities.)

Solution

Write $I_p \sin(\omega t + \phi) = I_p(\sin\omega t \cos\phi + \cos\omega t \sin\phi) = I_1 \sin\omega t + I_2 \cos\omega t$. The sine and cosine functions are independent, so their coefficients are equal (or, since the equality holds for all t, consider $\omega t = 0$ and $\omega t = \pi/2$.) Thus, $I_1 = I_p \cos\phi$ and $I_2 = I_p \sin\phi$. Inverting these relations, $I_p = \sqrt{I_1^2 + I_2^2}$ and $\phi = \tan^{-1}(I_2/I_1)$, where one must consider the signs of I_1 and I_2 to decide in which quadrant ϕ belongs.

Section 33-2:   Circuit Elements in AC Circuits

Problem

13. What is the rms current in a 1.0-μF capacitor connected across the 120-V rms, 60-Hz AC line?

Solution

Equation 33-5 can be used with the rms current and voltage, since both are $1/\sqrt{2}$ times their peak values. Thus, $I_{\text{rms}} = \omega C V_{\text{rms}} = (2\pi \times 60 \text{ Hz})(1\ \mu\text{F})(120 \text{ V}) = 45.2$ mA.

Problem

15. Find the reactance of a 3.3-μF capacitor at (a) 60 Hz, (b) 1.0 kHz, and (c) 20 kHz.

Solution

Equation 33-5 gives $X_C = 1/\omega C = 1/2\pi f(3.3\ \mu\text{F}) = 48.2$ kΩ·Hz$/f$. For the given frequencies, $X_C =$ (a) 804 Ω, (b) 48.2 Ω, (c) 2.41 Ω respectively. (One can see that a capacitor has the greatest effect at low frequency.)

Problem

17. A capacitor and a 1.8-kΩ resistor pass the same current when each is separately connected across a 60-Hz power line. What is the capacitance?

Solution

The currents (rms or peak) are the same if $X_C = R = 1/\omega C$, so $C = 1/\omega R = [2\pi(60 \text{ Hz})(1.8 \text{ k}\Omega)]^{-1} = 1.47\ \mu\text{F}$.

Problem

19. A 50-mH inductor is connected across a 10-V rms AC generator, and an rms current of 2.0 mA flows. What is the generator frequency?

Solution

From Equation 33-7, $f = \omega/2\pi = V_p/2\pi I_p L$. Since the ratio of the peak values of voltage and current is the same as that of the rms values, $f = 10 \text{ V}/2\pi(2 \text{ mA}) \times (50 \text{ mH}) = 15.9$ kHz.

Problem

21. A 1.2-μF capacitor is connected across a generator whose output is given by $V = V_p \sin 2\pi ft$, where $V_p = 22$ V, $f = 60$ Hz, and t is in seconds. (a) What is the peak current? (b) What are the magnitudes of the voltage and (c) the current at $t = 6.5$ ms?

Solution

(a) $I_p = V_p \omega C = (22 \text{ V})(377 \text{ s}^{-1})(1.2\ \mu\text{F}) = 9.95$ mA (see Equation 33-5 and Example 33-1). (b) $V(6.5 \text{ ms}) = V_p \sin\omega t = (22 \text{ V})\sin[(377 \text{ s}^{-1})(6.5 \text{ ms})] = 14.0$ V. (Remember that ωt is in radians.) (c) In a capacitor, the current leads the voltage by 90°, so $I(6.5 \text{ ms}) = I_p \sin(\omega t + \pi/2) = (9.95 \text{ mA})\cos[(377 \text{ s}^{-1})(6.5 \text{ ms})] = -7.67$ mA (the magnitude is 7.67 mA).

Problem

23. What is the maximum charge on the plates of a 16-μF capacitor connected across the 120-V rms, 60-Hz AC power line?

Solution

The charge on the capacitor is $q(t) = CV_C(t)$, so the maximum charge (for sinusoidal voltage) is $q_{\text{max}} = CV_{C,p} = \sqrt{2}\ CV_{C,\text{rms}} = \sqrt{2}\ (16\ \mu\text{F})(120 \text{ V}) = 2.72$ mC (independent of frequency).

Problem

25. A 0.75-H inductor is in series with a flourescent lamp, and the series combination is across the

120-V rms, 60-Hz power line. If the rms inductor voltage is 90 V, what is the rms lamp current?

Solution

In a series circuit, the same current flows through the inductor and lamp. Therefore, since the ratio of the rms quantities for a given circuit element equals that of the peak values, Equation 33-7 gives $I_{rms} = V_{rms}/\omega L = 90 \text{ V}/(2\pi \times 60 \text{ Hz})(0.75 \text{ H}) = 318 \text{ mA}$.

Section 33-3: LC Circuits

Problem

27. Find the resonant frequency of an LC circuit consisting of a 0.22-μF capacitor and a 1.7-mH inductor.

Solution

Equations 33-2 and 11 give $f = 1/2\pi\sqrt{LC} = 1/2\pi\sqrt{(0.22 \ \mu\text{F})(1.7 \ \text{mH})} = 8.23 \text{ kHz}$.

Problem

29. You have a 2.0-mH inductor and wish to make an LC circuit whose resonant frequency spans the AM radio band (550 kHz to 1600 kHz). What range of capacitance should your variable capacitor cover?

Solution

The resonant frequency of an LC circuit is (Equation 33-11) $\omega = 1/\sqrt{LC}$, so the capacitance should cover a range from $C = 1/\omega^2 L = 1/(2\pi \times 550 \text{ kHz})^2(2 \text{ mH}) = 41.9 \text{ pF}$ down to $C = 1/(2\pi \times 1.6 \text{ MHz})^2(2 \text{ mH}) = 4.95 \text{ pF}$.

Problem

31. You want to use an LC circuit in a timing application. The circuit is to start with the capacitor fully charged, and the voltage should drop to zero in 15 s. You have available a 25-H inductor. What capacitance should you use?

Solution

The capacitor voltage drops to zero in one quarter of a period, so $\omega = 2\pi/T = 2\pi/(4 \times 15 \text{ s})$. From Equation 33-11, $C = 1/\omega^2 L = (60 \text{ s}/2\pi)^2/(25 \text{ H}) = 3.65 \text{ F}$.

Problem

33. An LC circuit includes a 0.025-μF capacitor and a 340-μH inductor. (a) If the peak voltage on the capacitor is 190 V, what is the peak current in the inductor? (b) How long after the voltage peak does the current peak occur?

Solution

(a) In an LC circuit, the peak current and voltage are related by $I_p = \omega q_p = \omega C V_p = C V_p/\sqrt{LC} = V_p\sqrt{C/L}$ (see Example 33-3). Thus, $I_p = (190 \text{ V})\times \sqrt{0.025 \ \mu\text{F}/340 \ \mu\text{H}} = 1.63 \text{ A}$. (b) The current peaks one quarter of a period after the voltage, or $\Delta t = \frac{1}{4}T = \frac{1}{4}(2\pi/\omega) = \frac{1}{2}\pi\sqrt{LC} = (\pi/2)\sqrt{0.025 \ \mu\text{F} \times 340 \ \mu\text{H}} = 4.58 \ \mu\text{s}$.

Problem

35. At the instant when the electric and magnetic energies are equal in the LC circuit of Problem 33, the current is 540 mA. (a) What is the instantaneous voltage? Find (b) the peak voltage, (c) the peak current, and (d) the total energy.

Solution

(a) The energies are instantaneously equal when $\frac{1}{2}CV(t)^2 = \frac{1}{2}LI(t)^2$, or $V(t) = I(t)\sqrt{L/C} = (540 \text{ mA})\times \sqrt{340 \ \mu\text{H}/0.025 \ \mu\text{F}} = 63.0 \text{ V}$. (b) The times when the energies are equal are given by $\sin^2 \omega t = \cos^2 \omega t$, so $|\sin \omega t| = |\cos \omega t| = 1/\sqrt{2}$ at those times (recall that $\sin^2 + \cos^2 = 1$). From $V(t) = V_p \cos \omega t$, we get $V_p = (63.0 \text{ V})\sqrt{2} = 89.1 \text{ V}$. (c) Similarly, $I_p = (540 \text{ mA})\sqrt{2} = 764 \text{ mA}$. (d) The total energy is $\frac{1}{2}CV_p^2 = \frac{1}{2}LI_p^2 = \frac{1}{2}CV(t)^2 + \frac{1}{2}LI(t)^2 = LI(t)^2 = (340 \ \mu\text{H})(540 \text{ mA})^2 = 99.1 \ \mu\text{J}$. (We chose to evaluate the total energy from twice the inductor's energy, at the instant the capacitor and inductor energies are equal, because these numerical values were given.)

Problem

37. One-eighth of a cycle after the capacitor in an LC circuit is fully charged, what are each of the following as fractions of their peak values: (a) capacitor charge, (b) energy in the capacitor, (c) inductor current, (d) energy in the inductor?

Solution

The equations in Section 33-3 give the desired quantities, which we evaluate when $\omega t = \omega(T/8) = 2\pi/8 = \frac{1}{4}\pi = 45°$ (i.e., $\frac{1}{8}$ cycle). (Note that phase constant zero corresponds to a fully charged capacitor at $t = 0$.) (a) From Equation 33-10, $q/q_p = \cos 45° = 1/\sqrt{2}$. (b) From the equation for electric energy, $U_E/U_{E,p} = \cos^2 45° = 1/2$. (c) From Equation 33-12, $I/I_p = -\sin 45° = -1/\sqrt{2}$. (The direction of the current is away from the positive capacitor plate at $t = 0$.) (d) From the equation for magnetic energy, $U_B/U_{B,p} = \sin^2 45° = 1/2$.

Problem

39. The 2000-μF capacitor in Fig. 33-30 is initially charged to 200 V. (a) Describe how you would manipulate switches A and B to transfer all the energy from the 2000-μF capacitor to the 500-μF capacitor. Include the times you would throw the switches. (b) What will be the voltage across the 500-μF capacitor once you've finished?

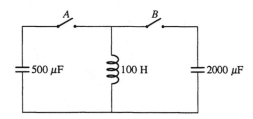

FIGURE 33-30 Problem 39.

Solution

(a) The energy initially stored in the first capacitor is $\frac{1}{2}(2 \text{ mF})(200 \text{ V})^2 = 40$ J. First close switch B for one quarter of a period of the LC circuit containing the 2000 μF capacitor, or $t_B = \frac{1}{4}T_B = \frac{1}{4}(2\pi/\omega_B) = \frac{1}{2}\pi\sqrt{LC_B} = \frac{1}{2}\pi\sqrt{(100 \text{ H})(2 \text{ mF})} = 702$ ms. This transfers 40 J to the inductor. Then open switch B and close switch A for one quarter of a period of the LC circuit containing the 500 μF capacitor, or $t_A = \frac{1}{2}\pi\sqrt{(100 \text{ H})(0.5 \text{ mF})} = \frac{1}{2}t_B = 351$ ms. This transfers 40 J to the second capacitor from the inductor. Finally, open switch A. (b) When the second capacitor has 40 J of stored energy, its voltage is $\sqrt{2(40 \text{ J})/(0.5 \text{ mF})} = 400$ V.

Problem

41. A damped RLC circuit includes a 5.0-Ω resistor and a 100-mH inductor. If half the initial energy is lost after 15 cycles, what is the capacitance?

Solution

If only half the energy is lost after 15 cycles, the damping is small and the energy varies like the square of Equation 33-13, namely $U_{\text{tot}} = U_p e^{-Rt/L} \cos^2 \omega t$. (The energy time constant is L/R, one half the charge time constant.) After 15 cycles, $t = 15\,T = 15(2\pi/\omega)$, the fraction of energy remaining is $\frac{1}{2} = e^{-15RT/L} \times \cos^2 30\pi = e^{-15RT/L}$. Take logarithms and use $\omega = 1/\sqrt{LC}$ to get $L \ln 2 = 15RT = 30\pi R\sqrt{LC}$, from which we find

$$C = \left(\frac{\ln 2}{30\pi R}\right)^2 L = \left(\frac{\ln 2}{30\pi \times 5 \text{ }\Omega}\right)^2 (100 \text{ mH})$$
$$= 0.216 \text{ }\mu\text{F}.$$

Section 33-4: Driven RLC Circuits and Resonance

Problem

43. If the speaker system of Example 33-4 is driven by a 10-V peak, 1.0-kHz sine wave, what will be the peak voltage across the capacitor?

Solution

The peak capacitor voltage in a series RLC circuit at resonance ($\omega_0 = 1$ kHz in Example 33-4) is $V_{C,p} = I_p X_C = V_p/\omega_0 RC = 10$ V$/[2\pi(1 \text{ kHz})(8.51 \text{ }\Omega) \times (11.5 \text{ }\mu\text{F})] = 16.2$ V. (Alternatively, $V_{C,p} = V_p \omega_0 L/R = (10 \text{ V})2\pi(1 \text{ kH})(2.2 \text{ mH})/(8.51 \text{ }\Omega)$, since $X_C = X_L$ at resonance. Finally, since $\omega_0 = 1/\sqrt{LC}$, $V_{C,p} = (V_p/R)\sqrt{L/C} = (10 \text{ V}/8.51 \text{ }\Omega)\sqrt{2.2 \text{ mH}/11.5 \text{ }\mu\text{F}}$. We calculated a value of $R = 8.51$ Ω without the round-off errors in Example 33-4.)

Problem

45. TV channel 2 occupies the frequency range from 54 MHz to 60 MHz. A series RLC tuning circuit in a TV receiver includes an 18-pF capacitor and resonates in the middle of the channel 2 band. (a) What is the inductance? (b) To let the whole signal in, the resonance curve must be broad enough that the current throughout the band be no less than 70% of the current at the resonant frequency. What constraint does this place on the circuit resistance?

Solution

(a) The condition for resonance in a series RLC circuit requires $L = 1/\omega_0^2 C = 1/(2\pi \times 57 \text{ MHz})^2(18 \text{ pF}) = 0.433$ μH. (b) The peak current at any frequency is $I_p = V_p/Z$, while at resonance, $I_{\text{res}} = V_p/R$. Thus, $I_p/I_{\text{res}} = R/Z = 1/\sqrt{1 + (X_L - X_C)^2/R^2}$ (see Equation 33-14). If this ratio is required to be not less than 70%, then

$$\left(\frac{X_L - X_C}{R}\right)^2 \leq \left(\frac{1}{0.7}\right)^2 - 1, \text{ or } R \geq \sqrt{\frac{49}{51}} |X_L - X_C|.$$

The reactance can be expressed in several equivalent forms, one being in terms of the given data and the resonant frequency:

$$|X_L - X_C| = \left|\omega L - \frac{1}{\omega C}\right| = \sqrt{\frac{L}{C}} \left|\omega\sqrt{LC} - \frac{1}{\omega\sqrt{LC}}\right|$$
$$= \frac{1}{\omega_0 C} \left|\frac{\omega}{\omega_0} - \frac{\omega_0}{\omega}\right|.$$

If this is evaluated at the lower band edge (54 MHz), one gets

$$R \geq \frac{\sqrt{49/51}}{(2\pi \times 57 \text{ MHz})(18 \text{ pF})} \left(\frac{57}{54} - \frac{54}{57}\right) = 16.4 \ \Omega.$$

(The upper band edge, 60 MHz, gives a weaker limit, $R \geq 15.6 \ \Omega$.)

Problem

47. A 2.0-H inductor and a 3.5-μF capacitor are connected in series with a 50-Ω resistor, and the combination is connected to an AC generator supplying 24 V peak at 60 Hz. (a) At the instant the generator voltage is at its peak, what is the instantaneous voltage across each circuit element? Show explicitly that these sum to the generator voltage. (b) If rms voltmeters are connected across each of the three components, what will they read? Do their readings sum to the rms generator voltage? Does this contradict the loop law?

Solution

Before calculating numerical results, we display the general time-dependent voltages in an RLC series circuit, which can be derived from the phasor diagram in Fig. 33-17b (the angle between V_p and the horizontal axis is ωt). Note that the phase constants in Equation 33-3 (for the applied voltage and current, i.e., the resistor voltage) are related to the phase difference in Equation 33-16 by $\phi_V = 0$ and $\phi_I = -\phi$. (See paragraphs following Equation 33-16.)

$$V = V_p \sin \omega t \qquad V_{R,p} = I_p R$$
$$Z = \sqrt{R^2 + (X_L - X_C)^2}$$
$$V_R = V_{R,p} \sin(\omega t - \phi) \qquad V_{C,p} = I_p X_C$$
$$\phi = \tan^{-1}\left(\frac{X_L - X_C}{R}\right)$$
$$V_C = V_{C,p} \sin(\omega t - \phi - \tfrac{1}{2}\pi) \quad V_{L,p} = I_p X_L$$
$$-\tfrac{1}{2}\pi \leq \phi \leq \tfrac{1}{2}\pi$$
$$V_L = V_{L,p} \sin(\omega t - \phi + \tfrac{1}{2}\pi) \quad I_p = V_p/Z$$

(a) At the instant when $V = V_p = 24$ V, $\omega t = 90°$. From the given values of R, L, C, ω, and V_p, we can calculate ϕ and the instantaneous voltages: $V_R = (23.93 \text{ V}) \sin(90° + 4.458°) = 23.86$ V; $V_C = (362.7 \text{ V}) \sin 4.458° = 28.19$ V; $V_L = (360.8 \text{ V}) \times \sin 184.458° = -28.05$ V. Four significant figures are necessary to verify the loop law, $V_R + V_C + V_L = 24.00$ V, in this case. (b) The rms voltages are $1/\sqrt{2}$ times the peak values found in part (a) (approximately 16.9, 256, and 255 volts for R, C, and L respectively). Of course, since the instantaneous voltages have different phases, the sum of the rms voltages does not equal $V_{\text{rms}} = V_p/\sqrt{2} = 17.0$ V.

Problem

49. Figure 33-31 shows the phasor diagram for an RLC circuit. (a) Is the driving frequency above or below resonance? (b) Complete the diagram by adding the applied voltage phasor, and from your diagram determine the phase difference between applied voltage and current.

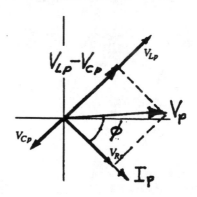

FIGURE 33-31 Problem 49 Solution.

Solution

(a) From the observation that $V_{L,p} = I_p \omega L > V_{C,p} = I_p/\omega C$, we conclude that the frequency is above resonance, $\omega^2 > 1/LC$. (b) The applied voltage phasor is the vector sum of the resistor, capacitor, and inductor voltage phasors, as drawn on Fig. 33-31. The current is in phase with the voltage across the resistor, in this case lagging the applied voltage (since $\phi = \tan^{-1}[(V_{L,p} - V_{C,p})/V_{R,p}] > 0$) by approximately 50° (as estimated from Fig. 33-31).

Problem

51. For the circuit of Problem 46, find the phase relation between applied voltage and current at frequencies of (a) 550 Hz and (b) 700 Hz.

Solution

The phase constant (relative phase of the current and the applied voltage) for a series RLC circuit is given by $\phi = \tan^{-1}[(\omega L - 1/\omega C)/R]$ (Equation 33-16). For the given values of R, L, and C in Problem 46, and the lower frequency $\phi = \tan^{-1}[(2\pi(550 \text{ Hz})(50 \text{ mH}) - (2\pi)^{-1}(550 \text{ Hz})^{-1}(1.5 \ \mu\text{F})^{-1})/10 \ \Omega] = -63.6°$. As expected, for a frequency below resonance $(\omega_0 = 2\pi(581 \text{ Hz})$, the current leads the voltage $(X_C > X_L$ so $\phi < 0)$. A similar calculation at 700 Hz gives $\phi = +81.7°$, or the current lags the voltage at a frequency above resonance $(X_L > X_C$ so $\phi > 0)$.

Problem

53. An electric drill draws 4.6 A rms at 120 V rms. If the current lags the voltage by 25°, what is the drill's power consumption?

Solution

The average power consumed by an AC circuit is given by Equation 33-17, $P_{av} = V_{rms}I_{rms}\cos\phi = (120\text{ V})\times(4.6\text{ A})\cos(+25°) = 500\text{ W}$.

Problem

55. A series RLC circuit has power factor 0.80 and impedance 100 Ω at 60 Hz. (a) What is the circuit resistance? (b) If the inductance is 0.10 H, what is the resonant frequency?

Solution

(a) The geometry of Fig. 33-17b, with the peak voltages expressed in terms of the peak current, shows that $\cos\phi = V_{R,p}/V_p = (I_pR)/(I_pZ) = R/Z$. (Alternatively, use Equations 33-15 and 16 and the trigonometric identity $\sec^2\phi = 1 + \tan^2\phi$.) Therefore, $R = Z\cos\phi = (100\text{ Ω})(0.8) = 80\text{ Ω}$. (b) The reactance of the circuit can be expressed in terms of the inductance, the resonant frequency, and the given values by the use of Equation 33-15:

$$X_L - X_C = \pm\sqrt{Z^2 - R^2} = \pm\sqrt{100^2 - 80^2}\text{ Ω} = \pm60\text{ Ω}$$
$$= \omega L - 1/\omega C = \omega L(1 - 1/\omega^2 LC)$$
$$= \omega L(1 - \omega_0^2/\omega^2), \text{ or } \omega_0^2/\omega^2 = 1 \pm (60\text{ Ω}/\omega L).$$

Since $\omega L = (2\pi \times 60\text{ Hz})(0.1\text{ H}) = 37.7\text{ Ω} < 60\text{ Ω}$, we can discard the unphysical solution (with $\omega_0^2/\omega^2 < 0$) to find $\omega_0 = \omega\sqrt{1 + (60/37.7)}$, or $f_0 = 60\text{ Hz}\sqrt{2.59} = 96.6\text{ Hz}$. (If the value of ωL has been >60 Ω, there would have been two physically acceptable solutions.)

Problem

57. A power plant produces 60-Hz power at 365 kV rms and 200 A rms. The plant is connected to a small city by a transmission line with total resistance 100 Ω. What fraction of the power is lost in transmission if the city's power factor is (a) 1.0 or (b) 0.60? (c) Is it more economical for the power company if the load has a large power factor or a small one? Explain.

Solution

(a) We assume that the average power supplied to the city is $P_{av} = I_{rms}V_{rms}\cos\phi = (200\text{ A})(365\text{ kV})(1) = 73.0\text{ MW}$, and that the average power lost in the transmission line is $\Delta P = I_{rms}^2R = (200\text{ A})^2(100\text{ Ω}) = 4\text{ MW}$. Thus, the percent lost is $(4/73)\times100 \simeq 5.5\%$.

(b) If the power factor in part (a) were 0.6 instead of 1.0, the percent lost would be 5.5%/0.6 = 9.1%.
(c) Since ΔP is a constant, the larger P_{av} (which is proportional to $\cos\phi$), the smaller the fraction of power lost, $\Delta P/P_{av}$. A large power factor is better for the power plant owners.

Section 33-6:   Transformers and Power Supplies

Problem

59. A transformer steps up the 120-V rms AC power line voltage to 23 kV rms for a TV picture tube. If the rms current in the primary is 1.0 A, and the transformer is 95% efficient, what is the secondary current?

Solution

Only 95% of the power in the primary is transformed to power in the secondary, so Equation 33-19 can be modified to yield $(0.95)(120\text{ V})(1\text{ A}) = (23\text{ kV})I_{sec}$, or $I_{sec} = 4.96$ mA (rms).

Problem

61. The transformer in the power supply of Fig. 33-26a has an output voltage of 6.3 V rms at 60 Hz, and the capacitance is 1200 μF. (a) With an infinite load resistance, what would be the output voltage of the power supply? (b) What is the minimum load resistance for which the output would not drop more than 1% from this value? Assume that the discharge time in Fig. 33-26b is essentially a full cycle.

Solution

(a) With no load present, the capacitor cannot discharge through the transformer because the diode blocks such a current. Therefore, it remains charged to the peak AC voltage, $V_p = \sqrt{2}\,V_{rms} = \sqrt{2}(6.3\text{ V}) = 8.91\text{ V}$. (b) With the load resistance present, the capacitor voltage must not decay to less than 99% of its maximum value over one period of AC ($T = \frac{1}{60}$ s) (see Fig. 33-26b). Therefore, $e^{-T/RC} \geq 0.99$, or $R \geq -T/C\ln(0.99) = [-(60\text{ Hz})(1200\text{ μF})\ln(0.99)]^{-1} = 1.38\text{ kΩ}$. (Note: The voltage for a discharging RC circuit is given by Equation 28-8.)

Paired Problems
Problem

63. A sine-wave generator delivers a signal whose peak voltage is independent of frequency. Two identical capacitors are connected in parallel across the generator, and the generator supplies a peak current I_p at frequency f_1. The capacitors are

then connected in series across the generator. To what frequency should the generator be tuned to bring the current back to I_p?

Solution

For constant $V_p = I_p X_C$ (i.e., independent of frequency), the same peak current will be supplied if the capacitive reactances for the two connections are equal, i.e., $2\pi f_1 C_1 = 2\pi f_2 C_2$. Thus, for parallel and series combinations of two equal capacitors, $f_2 = f_1(C_{\text{parallel}}/C_{\text{series}}) = f_1(C_a + C_b)^2/C_a C_b = 4f_1$. (We left the capacitors general in the intermediate step for application to the next problem.)

Problem

65. The peak current in an oscillating LC circuit is 850 mA. If $L = 1.2$ mH and $C = 5.0\ \mu$F, what is the peak voltage?

Solution

In an LC circuit, the peak electric and magnetic energies are equal (see the analysis in the text), so $\frac{1}{2}CV_p^2 = \frac{1}{2}LI_p^2$. Thus, $V_p = \sqrt{L/C}\,I_p = \sqrt{1.2\text{ mH}/5.0\ \mu\text{F}}(850\text{ mA}) = 13.2$ V.

Problem

67. An RLC circuit includes a 3.3-μF capacitor and a 27-mH inductor. The capacitor is charged to 35 V, and the circuit begins oscillating. Ten full cycles later the capacitor voltage peaks at 28 V. What is the resistance?

Solution

For the damped oscillations of an RLC circuit, the voltage decays according to Equation 33-13, $V(t) = q(t)/C = V_p e^{-Rt/2L} \cos \omega t$, with frequency given by Equation 33-11. If in ten cycles ($t = 10\ T = 10(2\pi/\omega) = 20\pi\sqrt{LC}$) the peak voltage has decayed from 35 V to 28 V, then $\ln(35/28) = Rt/2L = 10\pi R\sqrt{C/L}$, or $R = (10\pi)^{-1}\sqrt{27\text{ mH}/3.3\ \mu\text{F}}\times \ln(35/28) = 642$ mΩ.

Problem

69. A series RLC circuit with $R = 5.5\ \Omega$, $L = 180$ mH, and $C = 0.12\ \mu$F is connected across a sine-wave generator. If the inductor can handle a maximum current of 1.5 A, what is the maximum safe value for the generator's peak output voltage when it is tuned to resonance?

Solution

At resonance, the impedance of a series RLC circuit is $Z = R$, so $V_p = ZI_p = RI_p = (5.5\ \Omega)(1.5\text{ A}) = 8.25$ V at the maximum safe peak current.

Supplementary Problems

Problem

71. Two capacitors are connected in parallel across a 10-V rms, 10-kHz sine-wave generator, and the generator supplies a total rms current of 30 mA. When the capacitors are rewired in series, the rms generator current drops to 5.5 mA. Find the values of the two capacitances.

Solution

Equation 33-5 gives the rms current when capacitors are connected to an AC generator, $I_{\text{rms}} = V_{\text{rms}}/X_C = \omega C V_{\text{rms}}$. For the parallel connection, 30 mA $= (2\pi\times10^5$ V/s$)(C_1 + C_2)$, while for the series connection, 5.5 mA $= (2\pi\times10^5$ V/s$)C_1 C_2/(C_1 + C_2)$. Thus, $C_1 + C_2 = 47.7$ nF and $C_1 C_2 = (20.4\text{ nF})^2$. Eliminate one capacitance from the second equation and substitute into the first equation to obtain a quadratic, $C^2 - (47.7\text{ nF})C + (20.4\text{ nF})^2 = 0$, whose two solutions are C_1 and C_2, i.e., $\frac{1}{2}[(47.7\text{ nF})\pm \sqrt{(47.7\text{ nF})^2 - 4(20.4\text{ nF})^2}] = 11.5$ nF and 36.2 nF. (The equations are symmetric in C_1 and C_2.)

Problem

73. An undriven RLC circuit with inductance L and resistance R starts oscillating with total energy U_0. After N cycles the energy is U_1. Find an expression for the capacitance, assuming the circuit is not heavily damped.

Solution

The total energy in an underdamped oscillating RLC circuit decays as the square of Equation 33-13 ($U = q^2/2C$), or $U_{\text{tot}} = U_0 e^{-Rt/L}$. After N cycles, $t = N(2\pi/\omega) = 2\pi N\sqrt{LC}$ (see Equation 33-11), so $U_1 = U_0 e^{-2\pi RN\sqrt{C/L}}$, and $C = L[\ln(U_0/U_1)/2\pi RN]^2$.

Problem

75. You wish to make a "black box" with two input connections and two output connections, as shown in Fig. 33-33. When you put a 12-V rms, 60-Hz sine wave across the input, a 6.0-V, 60-Hz signal should appear at the output, with the output voltage leading the input voltage by 45°. Design a circuit that could be used in the "black box."

FIGURE 33-33 Problem 75.

Solution

Since the voltage across a resistor leads the voltage across a capacitor in series with it (see Problem 82), and the voltage across an inductor leads the voltage across a resistor in series with it (see Problem 83), either circuit can be adapted to the criteria of the "black box" in this problem. (We include the solution to these three problems below.)

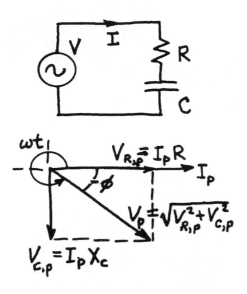

Problem 75 Solution (see Problem 82).

In the circuit diagram for Problem 82, it can be seen that $V = V_R + V_C$ and $I = I_C = I_R$. In the corresponding phasor diagram, V_C lags I by 90°, V_R and I are in phase, and V is the vector sum of these (see Table 33-1). (We drew I horizontally for convenience.) The impedance is

$$Z = \frac{V_p}{I_p} = \frac{\sqrt{V_{R,p}^2 + V_{C,p}^2}}{I_p}$$

$$= \sqrt{R^2 + X_C^2} = \sqrt{R^2 + \frac{1}{\omega^2 C^2}},$$

and the phase angle is $\tan\phi = -V_{C,p}/V_{R,p} = -X_C/R = -1/\omega RC$. I always leads V, because $\phi < 0$. (Recall that ϕ is defined by $I = I_p \sin(\omega t - \phi)$ when $V = V_p \sin\omega t$.)

When an inductor in Problem 83 replaces the capacitor of Problem 82, the phasors for V_R and I are still parallel, but V_L leads I by 90°. V is the vector sum of V_R and V_L, so the impedance is

$$Z = \frac{V_p}{I_p} = \frac{\sqrt{V_{R,p}^2 + V_{L,p}^2}}{I_p}$$

$$= \sqrt{R^2 + X_L^2} = \sqrt{R^2 + \omega^2 L^2},$$

and the phase angle is $\tan\phi = V_{L,p}/V_{R,p} = X_L/R = \omega L/R$. In this case, I always lags V, because $\phi > 0$. (Negative ϕ is in the same sense as ωt, measured from V.)

The solution to Problem 82 shows that, in a series RC circuit, V_R leads the applied voltage, V, by an angle $\tan^{-1}(1/\omega RC)$, which may be adjusted to 45° if $\omega RC = 1$. The peak voltage across the entire resistance is $V_{R,p} = V_p \cos 45° = V_p/\sqrt{2}$, so if we divide the resistance into two parts, $R_1 + R_2 = R$, with $R_2/R = 1/\sqrt{2}$, then the peak voltage across R_2 will be $(1/\sqrt{2})V_{R,p} = \frac{1}{2}V_p$, as desired (rms voltages have the same ratio as peak voltages).

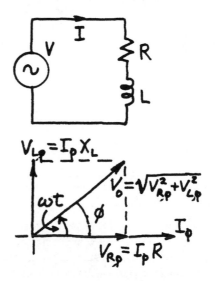

Problem 75 Solution (see Problem 83).

Alternatively, the solution to Problem 83 shows that, in a series RL circuit, V_L leads V by 90° − $\tan^{-1}(\omega L/R)$, which equals 45° if $\omega L = R$. Again, $V_{L,p} = V_p/\sqrt{2}$, so if we divide L into $L_1 + L_2$, with $L_2 = L/\sqrt{2}$, the peak voltage across L_2 is $\frac{1}{2}V_p$. Both circuits are sketched below.

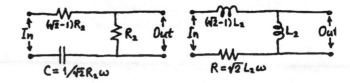

Problem 75 Solution (3).

(Note that V_{out} is the open-circuit output voltage. If a load is connected across the output terminals, the magnitude and the phase of the voltage will be changed accordingly.)

Problem

77. A sine-wave generator with peak output voltage of 20 V is applied across a series RLC circuit. At the resonant frequency of 2.0 kHz the peak current is 50 mA, while at 1.0 kHz it is 15 mA. Find R, L, and C.

Solution

At resonance, $I_p = V_p/R$, so $R = 20$ V/50 mA = 400 Ω. The impedance at resonance is $Z = R$ [i.e., $X = X_L - X_C = 0$], while at half the resonant frequency (1 kHz = $\frac{1}{2}$(2 kHz)], $Z = V_p/I_p = 20$ V/15 mA = 1.33 kΩ = (10/3)R (i.e., $|X| = \sqrt{Z^2 - R^2} = \sqrt{(10/3)^2 - 1}\,R = \sqrt{91}\,R/3$.) Therefore,

$$\frac{1}{\omega_0 C} - \omega_0 L = 0, \text{ and } \frac{1}{\frac{1}{2}\omega_0 C} - \frac{1}{2}\omega_0 L = \sqrt{91}\,R/3.$$

These equations can be solved for C and L, with the following result:

$$L = \frac{2\sqrt{91}\,R}{9\omega_0} = \frac{2\sqrt{91}(400\ \Omega)}{9 \times 2\pi \times 2\text{ kHz}} = 67.5 \text{ mH, and}$$
$$C = (\omega_0^2 L)^{-1} = (2\pi \times 2 \text{ kHz})^{-2}(67.5 \text{ mH})^{-1} = 93.8 \text{ nF.}$$

(Note: Below resonance, $X_C > X_L$.)

Problem

79. A 2.5-H inductor is connected across a 1500-μF capacitor. A 5.0-kg mass is connected to a spring. What should be the spring constant if the mechanical and electrical systems have the same resonant frequency?

Solution

The comparison of Equations 15-4 and 33-9 in the text shows that the resonant frequency of each system is $\omega_0^2 = 1/LC = k/m$ (see Equations 15-12 and 33-11), so $k = m/LC = (5 \text{ kg})/(2.5 \text{ H})(1.5 \text{ mF}) = 1.33 \text{ kN/m}$.

Problem

81. For RLC circuits in which the resistance is not too large, the Q factor may be defined as the ratio of the resonant frequency to the difference between the two frequencies where the power dissipated in the circuit is half that dissipated at resonance. Show, using suitable approximations, that this definition leads to the expression $Q = \omega_0 L/R$, with ω_0 the resonant frequency.

Solution

From Equations 33-14 (with rms values), 33-17, and the result in the solution to Problem 55(a), the average power in a series RLC circuit becomes $\langle \mathcal{P} \rangle = I_{rms} V_{rms} \cos\phi = (V_{rms}/Z)V_{rms}(R/Z) = V_{rms}^2 R/Z^2$. From Equation 33-15, one sees that the power falls to half its resonance value (V_{rms}^2/R) when $Z = \sqrt{2}R$, or when $|X_L - X_C| = R$. In terms of the resonant frequency, $\omega_0 = 1/\sqrt{LC}$, this condition becomes

$$\left|\omega L - \frac{1}{\omega C}\right| = L\left|\omega - \frac{\omega_0^2}{\omega}\right| = R, \text{ or } \omega^2 - \omega_0^2 = \pm\frac{R}{L}\omega.$$

The solutions of these quadratics, with $\omega > 0$, are

$$\omega = \frac{1}{2}\left[\pm\frac{R}{L} + \sqrt{\frac{R^2}{L^2} + 4\omega_0^2}\right].$$

If $R/L \ll \omega_0$ (equivalent to $R \ll \sqrt{L/C}$), we can neglect the first term under the square root sign compared to the second, obtaining $\omega \approx \omega_0 \pm R/2L$. The difference between these two values of ω is $\Delta\omega = R/L$, from which the desired expression for Q follows.

Problem

83. Consider a series circuit containing an AC generator, a resistor, and an inductor. Construct a phasor diagram, and derive expressions for the circuit impedance and the phase angle between the applied voltage and the current. Show that the voltage always leads the current.

Solution

See solution to Problem 75.

CHAPTER 34   MAXWELL'S EQUATIONS AND ELECTROMAGNETIC WAVES

Section 34-2:   Ambiguity in Ampère's Law

Problem

1. A uniform electric field is increasing at the rate of 1.5 V/m·µs. What is the displacement current through an area of 1.0 cm^2 at right angles to the field?

Solution

Maxwell's displacement current is $\varepsilon_0 \partial\phi_E/\partial t = (8.85\times10^{-12}$ F/m$)(1.5$ V/m·µs$)(1$ cm$^2) = 1.33$ nA. (See Equations 34-1 and 24-2.)

Problem

3. A parallel-plate capacitor of plate area A and spacing d is charging at the rate dV/dt. Show that the displacement current in the capacitor is equal to the conduction current flowing in the wires feeding the capacitor.

Solution

The displacement current is $I_D = \varepsilon_0 \partial\phi_E/\partial t$. For a parallel-plate capacitor, $E = q/\varepsilon_0 A$, so $I_D = \varepsilon_0\, \partial(EA)/\partial t = \varepsilon_0 \partial(q/\varepsilon_0)/\partial t = dq/dt$. But dq/dt is just the conduction current (the rate at which charge is flowing onto the capacitor plates); hence $I_D = I$.

Problem

5. A parallel-plate capacitor has circular plates with radius 50 cm and spacing 1.0 mm. A uniform electric field between the plates is changing at the rate 1.0 MV/m·s. What is the magnetic field between the plates (a) on the symmetry axis, (b) 15 cm from the axis, and (c) 150 cm from the axis?

Solution

(a) As explained in Example 34-1, cylindrical symmetry and Gauss's law for magnetism require that the **B**-field lines be circles around the symmetry axis, as in Fig. 34-5. For a radius, r, less than the radius of the plates, R, the displacement current is $I_D = \varepsilon_0 d\phi_E/dt = \varepsilon_0(d/dt)\int \mathbf{E}\cdot d\mathbf{A} = \varepsilon_0\pi r^2(dE/dt)$, where the integral is over a disk of radius r centered between

the plates. Maxwell's form of Ampère's law gives $\oint \mathbf{B}\cdot d\boldsymbol{\ell} = 2\pi r B = \mu_0 I_D$, where the line integral is around the circumference of the disk. Thus, $B = \frac{1}{2}\mu_0\varepsilon_0 r(dE/dt) = r(dE/dt)/2c^2$, where c is the speed of light (Equation 34-16). On the symmetry axis, $r = 0$, so $B = 0$. (b) For $r = 15$ cm $< R$, $B = \frac{1}{2}(0.15$ m$)(10^6$ V/m·s$)/(3\times10^8$ m/s$)^2 = 8.33\times10^{-13}$ T. (c) For $r > R$, the displacement current is $I_D = \varepsilon_0\pi R^2(dE/dt)$, so $B = (dE/dt)R^2/2c^2 r$. At $r = 150$ cm, $B = (10^6$ V/m·s$)(50$ cm$)^2/2(3\times10^8$ m/s$)^2\times(150$ cm$) = 9.26\times10^{-13}$ T.

Section 34-4:   Electromagnetic Waves

Problem

7. At a particular point the instantaneous electric field of an electromagnetic wave points in the $+y$ direction, while the magnetic field points in the $-z$ direction. In what direction is the wave propagating?

Solution

For electromagnetic waves in vacuum, the directions of the electric and magnetic fields, and of wave propagation, form a right-handed coordinate system, as shown. (The vector relationship is summarized in Equation 34-20b.) Therefore, the given wave is headed in the $-x$-direction.

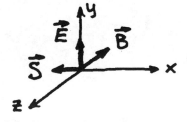

Problem 7 Solution.

Problem

9. The electric field of a radio wave is given by $\mathbf{E} = E\sin(kz - \omega t)(\hat{\mathbf{i}} + \hat{\mathbf{j}})$. (a) What is the peak amplitude of the electric field? (b) Give a unit vector in the direction of the magnetic field at a place and time where $\sin(kz - \omega t)$ is positive.

Solution

(a) The peak amplitude is the magnitude of $E(\hat{\imath}+\hat{\jmath})$, which is $E\sqrt{2}$. Note that $\hat{\imath}+\hat{\jmath}=\sqrt{2}\hat{n}$, where $\hat{n}$ is a unit vector $45°$ between the positive x and y axes.
(b) When $\mathbf{E}$ is parallel to $\hat{n}$ (for $\sin(kz-\omega t)$ positive) $\mathbf{B}$ points $45°$ into the second quadrant (so that $\mathbf{E}\perp\mathbf{B}$, and $\mathbf{E}\times\mathbf{B}$ is in the $+z$ direction). Thus, $\mathbf{B}$ is parallel to the unit vector $(-\hat{\imath}+\hat{\jmath})/\sqrt{2}$.

Problem

11. Show that it is impossible for an electromagnetic wave in a vacuum to have a time-varying component of its electric field in the direction of its magnetic field. *Hint:* Assume $\mathbf{E}$ does have such a component, and show that you cannot satisfy both Gauss and Faraday.

Solution

Consider a wave propagating in the x direction through a vacuum (no charges or currents present). Gauss's laws for electricity and magnetism require the field lines to continue forever in the y-z plane (no E_x or B_x). We may choose the z direction parallel to the magnetic field, $\mathbf{B}=B_z\hat{k}$. Suppose $\mathbf{E}=E_y\hat{\jmath}+E_z\hat{k}$. The discussion of Faraday's law leading to Equation 34-12b shows that $\partial E_y/\partial x=-\partial B_z/\partial t$. But consider a corresponding loop in the x-z plane. Then $-\partial E_z/\partial x=-\partial B_y/\partial t=0$, since there is no B_y by assumption. Because all the space-time dependence in the wave occurs in the combination $kx-\omega t$ (see Problem 10), $\omega\partial E_z/\partial t=-k\partial E_z/\partial x=0$. Then E_z must be a constant and is therefore not a part of the wave. (Similar consideration of Ampère's law over the loop in the x-y plane gives $\varepsilon_0\mu_0\partial E_z/\partial t=0$ directly, with the same conclusion regarding E_z.)

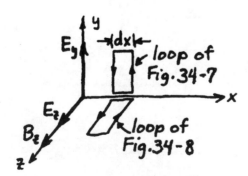

Problem 11 Solution.

Section 34-5: The Speed of Electromagnetic Waves

Problem

13. Your intercontinental telephone call is carried by electromagnetic waves routed via a satellite in geosynchronous orbit at an altitude of 36,000 km. Approximately how long does it take before your voice is heard at the other end?

Solution

Assuming the satellite is approximately overhead, we can estimate the round-trip travel time by $\Delta t = \Delta r/c \doteq (2\times36{,}000 \text{ km})/(3\times10^5 \text{ km/s}) = 0.24 \text{ s}$.

Problem

15. Roughly how long does it take light to go 1 foot?

Solution

In $\Delta t = 1$ ns, light travels about $\Delta r = c\,\Delta t = (3\times10^8 \text{ m/s})(10^{-9} \text{ s}) = 30$ cm, or about one foot.

Problem

17. "Ghosts" on a TV screen occur when part of the signal goes directly from transmitter to receiver, while part takes a longer route, reflecting off mountains or buildings (Fig. 34-31). The electron beam in a 50-cm-wide TV tube "paints" the picture by scanning the beam from left to right across the screen in about 10^{-4} s. If a "ghost" image appears displaced about 1 cm from the main image, what is the difference in path lengths of the direct and indirect signals?

Solution

The time it takes for the electron beam to sweep across 1 cm of the TV screen is 1 cm/$(50 \text{ cm}/10^{-4} \text{ s}) = 2\times10^{-6}$ s, which equals the time delay of the ghost signal. Therefore, the path difference is $\Delta r = c\,\Delta t = (3\times10^8 \text{ m/s})(2\times10^{-6} \text{ s}) = 600$ m.

Problem

19. Problem 69 shows that the speed of electromagnetic waves in a transparent dielectric is given by $1/\sqrt{\kappa\varepsilon_0\mu_0}$, where κ is the dielectric constant described in Chapter 26. An experimental measurement gives 1.97×10^8 m/s for the speed of light in a piece of glass. What is the dielectric constant of this glass at optical frequencies?

Solution

Since $c = 1/\sqrt{\varepsilon_0\mu_0}$ (speed of light in vacuum), we can write $v = c/\sqrt{\kappa}$ for the speed of light in a dielectric.

Then $\kappa = (c/v)^2 = (3/1.97)^2 = 2.32$. (Note: $\sqrt{\kappa}$ at optical frequencies is called the index of refraction; see Section 35-3.)

Section 34-6:   Properties of Electromagnetic Waves

Problem

21. A 60-Hz power line emits electromagnetic radiation. What is the wavelength?

Solution

The wavelength in a vacuum (or air) is $\lambda = c/f = (3 \times 10^8$ m/s$)/(60$ Hz$) = 5 \times 10^6$ m, almost as large as the radius of the Earth.

Problem

23. A CB radio antenna is a vertical rod 2.75 m high. If this length is one-fourth of the CB wavelength, what is the CB frequency?

Solution

From Equation 34-17b, $f = c/\lambda = (3 \times 10^8$ m/s$) \div (4 \times 2.75$ m$) = 27.3$ MHz.

Problem

25. What would be the electric field strength in an electromagnetic wave whose magnetic field equaled that of Earth, about 50 μT?

Solution

For a wave in free space, Equation 34-18 gives $E = cB = (3 \times 10^8$ m/s$)(0.5 \times 10^{-4}$ T$) = 15$ kV/m.

Problem

27. A radio receiver can detect signals with electric fields as low as 320 μV/m. What is the corresponding magnetic field?

Solution

From Equation 34-18 for waves in free space, $B = E/c = (320$ μV/m$)/(3 \times 10^8$ m/s$) = 1.07$ pT.

Section 34-8:   Polarization

Problem

29. Polarized light is incident on a sheet of polarizing material, and only 20% of the light gets through. What is the angle between the electric field and the polarization axis of the material?

Solution

From the law of Malus (Equation 34-19), $S/S_0 = \cos^2 \theta = 20\%$, or $\theta = \cos^{-1}(\sqrt{0.2}) = 63.4°$.

Problem

31. A polarizer blocks 75% of a polarized light beam. What is the angle between the beam's polarization and the polarizer's axis?

Solution

Equation 34-19 gives $\theta = \cos^{-1}\sqrt{S/S_0} = \cos^{-1}\sqrt{1 - 75\%} = \cos^{-1}\sqrt{\frac{1}{4}} = 60°$.

Problem

33. Unpolarized light of intensity S_0 passes first through a polarizer with its polarization axis vertical, then through one with its axis at 35° to the vertical. What is the light intensity after the second polarizer?

Solution

Only 50% (one half the intensity) of the unpolarized light is transmitted through the first polarizer, and the second cuts this down by $\cos^2 35°$. Therefore $\frac{1}{2} \cos^2 35° = 33.6\%$ of the unpolarized intensity gets through both polarizers.

Problem

35. Unpolarized light with intensity S_0 passes through a stack of five polarizing sheets, each with its axis rotated 20° with respect to the previous one. What is the intensity of the light emerging from the stack?

Solution

Only $\frac{1}{2}$ (or 50%) of the incident unpolarized intensity gets through the first polarizing sheet, while $\cos^2 20°$ (or 88.3%) is passed through each of the succeeding four sheets. The net percentage emerging is $\frac{1}{2}(\cos^2 20°)^4 = 30.4\%$.

Problem

37. Polarized light with average intensity S_0 passes through a sheet of polarizing material which is rotating at 10 rev/s. At time $t = 0$ the polarization axis is aligned with the incident polarization. Write an expression for the transmitted intensity as a function of time.

Solution

Because the frequency of light is much greater than that of the rotating polarizer (5×10^{14} Hz $\gg 10$ Hz), the law of Malus relates the average light intensities (see discussion leading to Equation 34-21a). Thus, $S = S_0 \cos^2 \theta$. For $\theta = \omega t$, where $\omega = 2\pi \times 10$ s^{-1}, $S = S_0 \cos^2(20\pi$ s$^{-1})t = \frac{1}{2}S_0[1 + \cos(40\pi$ s$^{-1})t]$.

Section 34-10: Energy in Electromagnetic Waves

Problem

39. What would be the average intensity of a laser beam so strong that its electric field produced dielectric breakdown of air (which requires $E_p = 3 \times 10^6$ V/m)?

Solution

Equation 34-21b for the average intensity of electromagnetic waves gives $\bar{S} = E_p^2/2\mu_0 c = (3 \times 10^6 \text{ V/m})^2 (8\pi \times 10^{-7} \text{ H/m})^{-1}(3 \times 10^8 \text{ m/s})^{-1} = 11.9 \text{ GW/m}^2$.

Problem

41. A radio receiver can pick up signals with peak electric fields as low as 450 μV/m. What is the average intensity of such a signal?

Solution

From Equation 34-21b, $\bar{S} = E_p^2/2\mu_0 c = (450 \ \mu\text{V/m})^2/(8\pi \times 10^{-7} \text{ H/m})(3 \times 10^8 \text{ m/s}) = 2.69 \times 10^{-10} \text{ W/m}^2$.

Problem

43. A laser blackboard pointer delivers 0.10 mW average power in a beam 0.90 mm in diameter. Find (a) the average intensity, (b) the peak electric field, and (c) the peak magnetic field.

Solution

(a) If the average power is spread uniformly over the beam area, $\bar{S} = 0.1 \text{ mW}/\frac{1}{4}\pi(0.9 \text{ mm})^2 = 157 \text{ W/m}^2$.
(b) Equation 34-21b gives $E_p = \sqrt{2\mu_0 c\bar{S}} = 344$ V/m, and (c) Equation 34-18 gives $B_p = E_p/c = 1.15 \ \mu$T.

Problem

45. The United States' safety standard for continuous exposure to microwave radiation is 10 mW/cm^2. The glass door of a microwave oven measures 40 cm by 17 cm and is covered with a metal screen that blocks microwaves. What fraction of the oven's 625-W microwave power can leak through the door window without exceeding the safe exposure to someone right outside the door? Assume the power leaks uniformly through the window area.

Solution

The power corresponding to the safety standard of intensity, uniformly distributed over the window area,

is $(10 \text{ mW/cm}^2)(40 \times 17 \text{ cm}^2) = 6.8$ W, which is 1.09% of the microwave's 625 W power output.

Problem

47. Use the fact that sunlight intensity at Earth's orbit is 1368 W/m^2 to calculate the Sun's total power output.

Solution

If the Sun is emitting isotropically, its power output is $\mathcal{P} = 4\pi r^2 \bar{S}$ (from Equation 34-22). Using values of comparable accuracy for the Earth's average orbital distance, we find $\mathcal{P} = 4\pi(1.496 \times 10^{11} \text{ m})^2 \times (1368 \text{ W/m}^2) = 3.85 \times 10^{26}$ W.

Problem

49. During its 1989 encounter with Neptune, the Voyager II spacecraft was 4.5×10^9 km from Earth. Its images of Neptune were broadcast by a radio transmitter with a mere 21-W average power output. What would be (a) the average intensity and (b) the peak electric field received at Earth if the transmitter broadcast equally in all directions? (The received signal was actually somewhat stronger because Voyager used a directional antenna.)

Solution

(a) The average intensity at a distance r from an isotropic emitter is (Equation 34-22) $\bar{S} = \mathcal{P}/4\pi r^2 = 21 \text{ W}/4\pi(4.5 \times 10^{12} \text{ m})^2 = 8.25 \times 10^{-26} \text{ W/m}^2$.
(b) This corresponds to a peak electric field of only (Equation 34-21b) $E_p = \sqrt{2\mu_0 c\bar{S}} = 7.89 \times 10^{-12}$ V/m. (The one-way travel time queried in the caption of Fig. 34-32 is $r/c = 4$ h, 10 min.)

Problem

51. At 1.5 km from the transmitter, the peak electric field of a radio wave is 350 mV/m. (a) What is the transmitter's power output, assuming it broadcasts uniformly in all directions? (b) What is the peak electric field 10 km from the transmitter?

Solution

(a) Equations 34-22 and 21b can be combined to express the average power output of an isotropic transmitter in terms of the peak electric field at a distance r, $\mathcal{P} = 4\pi r^2(E_p^2/2\mu_0 c) = (1.5 \text{ km})^2 \times (350 \text{ mV/m})^2/(2 \times 10^{-7} \times 3 \times 10^8 \text{ H/s}) = 4.59$ kW.
(b) Since $r^2 E_p^2$ is a constant, $E_p' = (r/r')E_p = (1.5/10)(350 \text{ mV/m}) = 52.5$ mV/m at a distance of 10 km.

Problem

53. A typical fluorescent lamp is a little over 1 m long and a few cm in diameter. How do you expect the light intensity to vary with distance (a) near the lamp but not near either end and (b) far from the lamp?

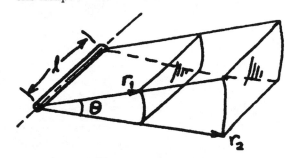

Problem 53 Solution.

Solution

(a) Near the lamp, but far from its ends, light waves travel approximately radially outwards from the tube axis. The power crossing two co-axial cylindrical patches is the same, but the area of each patch is proportional to the radius. Therefore, the intensity varies as $1/r$. ($S_1 A_1 = S_2 A_2 = S_1 \theta r_1 \ell = S_2 \theta r_2 \ell$.) This is the same relation as depicted in the solution to Problem 16-44. (b) Very far away, the lamp appears as a point source, and Equation 34-22 holds, so the intensity varies like $1/r^2$.

Section 34-11:   Wave Momentum and Radiation Pressure

Problem

55. What is the radiation pressure exerted on a light-absorbing surface by a laser beam whose intensity is 180 W/cm²?

Solution

The radiation pressure generated by a totally absorbed electromagnetic wave of given average intensity (from Equation 34-24) is $P_{rad} = \bar{S}/c = (180 \text{ W/cm}^2)/(3\times10^8 \text{ m/s}) = 6$ mPa.

Problem

57. The average intensity of noonday sunlight is about 1 kW/m². What is the radiation force on a solar collector measuring 60 cm by 2.5 m if it is oriented at right angles to the incident light and absorbs all the light?

Solution

For sunlight incident normally on a perfect absorber, $P_{rad} = \bar{S}/c$ (Equation 34-24). Therefore, the force on

the solar collector is $P_{rad}A = (1 \text{ kW/m}^2)\times (0.6\times2.5 \text{ m}^2)/(3\times10^8 \text{ m/s}) = 5 \ \mu N$.

Problem

59. A 65-kg astronaut is floating in empty space. If the astronaut shines a 1.0-W flashlight in a fixed direction, how long will it take the astronaut to accelerate to a speed of 10 m/s?

Solution

The reaction force of the light emitted on the flashlight equals the rate at which momentum is carried away by the beam, or $F = dp/dt = (dU/dt)/c = \mathcal{P}/c$. Such a force could accelerate a mass m from rest to a speed v in time $t = v/a = mv/F = mcv/\mathcal{P}$. For the values given for the astronaut and flashlight, $t = (65 \text{ kg})\times (3\times10^8 \text{ m/s})(10 \text{ m/s})/(1 \text{ W}) = 1.95\times10^{11} \text{ s} = 6.18\times10^3$ y (impractically long).

Paired Problems

Problem

61. Find the peak electric and magnetic fields 1.5 m from a 60-W light bulb that radiates equally in all directions.

Solution

For an isotropic source of electromagnetic waves (in a medium with vacuum permittivity and permeability) Equations 34-21b and 22 give $\bar{S} = \mathcal{P}/4\pi r^2 = E_p^2/2\mu_0 c$, therefore $E_p = \sqrt{2\mu_0 c \mathcal{P}/4\pi r^2} = (2\times10^{-7}\times 3\times10^8 \times 60)^{1/2}(\text{ V/m})/(1.5 \text{ m}) = 40$ V/m. Then Equation 34-18 gives $B_p = E_p/c = 133$ nT.

Problem

63. Unpolarized light is incident on two polarizers with their axes at 45°. What fraction of the incident light gets through?

Solution

50% of the incident unpolarized light (a random mixture of all polarizations) is transmitted by the first polarizer, and $\cos^2 45° = 50\%$ of that (see Equation 34-19) by the second, for a total of $0.5 \times 0.5 = 25\%$ for both.

Problem

65. What is the radiation force on the door of a microwave oven if 625 W of microwave power hits the door at right angles and is reflected?

Solution

The radiation pressure for normally incident, perfectly reflected electromagnetic waves is $2\bar{S}/c$. We suppose that the microwave power is uniformly spread over the area of the door (as for plane waves), so $\bar{S} = \mathcal{P}/A$. Then the force on the door is simply $\mathcal{P}_{rad} A = (2\bar{S}/c)A = 2\mathcal{P}/c = 2(625 \text{ W})/(3\times10^8 \text{ m/s}) = 4.17 \ \mu\text{N}$.

Problem

67. A 60-W light bulb is 6.0 cm in diameter. What is the radiation pressure on an opaque object at the bulb's surface?

Solution

Suppose that the filament in the bulb is small enough to be considered as an isotropic point source. Then the radiation pressure on an opaque (perfectly absorbing) object is $P_{rad} = \bar{S}/c$ and $\bar{S} = \mathcal{P}/4\pi r^2$. Thus $P_{rad} = (60 \text{ W}/4\pi)(3 \text{ cm})^{-2}(3\times10^8 \text{ m/s})^{-1} = 17.7 \ \mu\text{Pa}$.

Supplementary Problems
Problem

69. Maxwell's equations in a dielectric resemble those in vacuum (Equations 34-6 through 34-9), but with ϕ_E in Ampère's law replaced by $\kappa\phi_E$, where κ is the dielectric constant introduced in Chapter 26. Show that the speed of electromagnetic waves in such a dielectric is $c/\sqrt{\kappa}$.

Solution

The effect of a linear, isotropic, homogeneous dielectric medium in Gauss's law is to replace ε_0 by $\kappa\varepsilon_0$. Maxwell defined the displacement current in a dielectric analogously, as $\kappa\varepsilon_0 d\phi_E/dt$. Therefore, Maxwell's equations in a dielectric medium (containing no free charge or conduction currents) are just those in Table 34-2 with $\kappa\varepsilon_0$ replacing ε_0. The discussion in Sections 34-4 through 6 applies to waves in a dielectric medium, with the same replacement. In particular, the wave speed (Equation 34-16) becomes $1/\sqrt{\kappa\varepsilon_0\mu_0} = c/\sqrt{\kappa}$. (In Section 35-3, $\sqrt{\kappa}$ is defined as the index of refraction of the medium.)

Problem

71. A radar system produces pulses consisting of 100 full cycles of a sinusoidal 70-GHz electromagnetic wave. The average power while the transmitter is on is 45 MW, and the waves are confined to a beam 20 cm in diameter. Find (a) the peak electric field, (b) the wavelength, (c) the total energy in a pulse, and (d) the total momentum in a pulse. (e) If the transmitter

produces 1000 pulses per second, what is its average power output?

Solution

(a) The average intensity of a pulse is the average power during a pulse divided by the beam area, $\bar{S} = \mathcal{P}/\pi R^2$, and the peak electric field is therefore $E_p = \sqrt{2\mu_0 c\bar{S}} = (1/R)\sqrt{2\mu_0 c\mathcal{P}/\pi} = [(8\times10^{-7} \text{ H/m})\times (3\times10^8 \text{ m/s})(45 \text{ MW})]^{1/2}/(0.1 \text{ m}) = 1.04 \text{ MV/m}$. (b) The wavelength is $\lambda = c/f = (3\times10^8 \text{ m/s})\div (70 \text{ GHz}) = 4.29 \text{ mm}$. (c) 100 full cycles has a duration of $100/f = 100/(70 \text{ GHz}) = 1.43 \text{ ns}$, so the total energy in a pulse is $\mathcal{P}t = (45 \text{ MW})(1.43 \text{ ns}) = 64.3 \text{ mJ}$. (d) From Equation 34-23, the momentum per pulse is $(64.3 \text{ mJ})/(3\times10^8 \text{ m/s}) = 2.14\times10^{-10} \text{ kg}\cdot\text{m/s}$. (e) The average power output (which includes the time between pulses) is $(1000 \text{ pulses/s})(64.3 \text{ mJ/pulse}) = 64.3 \text{ W}$. (This is, of course, much less than the average power during a pulse, since pulses are emitted for only $10^3\times1.43 \text{ ns} = 1.43 \ \mu\text{s}$ each second. Thus $\mathcal{P}_{out} = \mathcal{P}_{pulse} \times 1.43 \ \mu\text{s/s}$, as above.)

Problem

73. The peak electric field measured at 8.0 cm from a light source is 150 W/m^2, while at 12 cm it measures 122 W/m^2. Describe the shape of the source.

Solution

The intensity is proportional to the square of the peak electric field (Equation 34-21b), so the given data implies that the ratio of intensities is proportional to that of the inverse distances, $(150/122)^2 = 1.51 \approx (12/8)$. Thus, $\bar{S}r$ is roughly constant. The intensity near a long, cylindrically symmetric source, where r is the axial distance, has this space dependence (see Problem 53).

Problem

75. Studies of the origin of the solar system suggest that sufficiently small particles might be blown out of the solar system by the force of sunlight. To see how small such particles must be, compare the force of sunlight with the force of gravity, and solve for the particle radius at which the two are equal. Assume the particles are spherical and have density 2 g/cm^3. Why do you not need to worry about the distance from the Sun.

Solution

The force due to radiation pressure (away from the Sun) on a totally absorbing spherical particle (radius R, cross-sectional area πR^2) is $P_{rad}\pi R^2$, where

$P_{rad} = \bar{S}/c$, and $\bar{S} = \mathcal{P}_{\odot}/4\pi r^2$ is the intensity of Sunlight at a distance r from the Sun. The force of the Sun's gravity on the particle (toward the Sun) is $GM_{\odot}(\frac{4}{3}\pi R^3 \rho)/r^2$, where ρ is the particle's density. The two forces are equal in magnitude for particles with

$$R\rho = 3\mathcal{P}_{\odot}/16\pi GM_{\odot}c$$
$$= \frac{3(3.85\times10^{26}\ \text{W}/16\pi)}{(6.67\times10^{-11}\ \text{N·m}^2/\text{kg}^2)(1.99\times10^{30}\ \text{kg})(3\times10^8\ \text{m/s})}$$
$$= 5.77\times10^{-5}\ \text{g/cm}^2.$$

(Note that since both forces are proportional to r^{-2}, the result is independent of r.) For particles with density 2 g/cm^3, $R = 2.89\times10^{-5}$ cm = 0.289 μm. Particles of the same density but smaller radius would be swept away by the solar radiation pressure.

PART 4   CUMULATIVE PROBLEMS

Problem

1. An air-insulated parallel-plate capacitor has plate area 100 cm^2 and spacing 0.50 cm. The capacitor is charged and then disconnected from the charging battery. A thin-walled, nonconducting box of the same dimensions as the capacitor is filled with water at 20.00°C. The box is released at the edge of the capacitor and moves without friction into the capacitor (Fig. 1). When it reaches equilibrium the water temperature is 21.50°C. What was the original voltage on the capacitor?

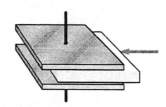

FIGURE 1 Cumulative Problem 1.

Solution

We assume that the entire difference between the initial and final values of the electrostatic energy stored in the capacitor is eventually dissipated as heat in the water (the dielectric medium), as mentioned following Equation 26-12. Thus, $U_0 - U = mc\,\Delta T$. From Equation 26-12, $U_0 - U = U_0 - U_0/\kappa = \frac{1}{2}C_0 V_0^2(\kappa - 1)/\kappa$, where $C_0 = \varepsilon_0 A/d = (8.85\ \text{pF/m})\times (10^{-2}\ \text{m}^2)/(5\times10^{-3}\ \text{m}) = 17.7$ pF, and $\kappa = 78$ for water; see Table 26-1. The amount of water is $m = (1\ \text{g/cm}^3)(100\ \text{cm}^2)(0.5\ \text{cm}) = 50$ g, so putting

everything together and solving for V_0, we find:

$$V_0 = \sqrt{\frac{2mc\,\Delta T}{C_0(\kappa - 1)/\kappa}}$$
$$= \sqrt{\frac{2(50\ \text{g})(4.184\ \text{J/g·°C})(1.5\text{°C})}{(17.7\ \text{pF})(77/78)}} = 5.99\ \text{MV}.$$

Problem

3. Five wires of equal length 25 cm and resistance 10 Ω are connected, as shown in Fig. 3. Two solenoids, each 10 cm in diameter, extend a long way perpendicular to the page. The magnetic fields of both solenoids point out of the page; the field strength in the left-hand solenoid is increasing at 50 T/s while that in the right-hand solenoid is decreasing at 30 T/s. Find the current in the resistance wire shared by both triangles. Which way does the current flow?

Solution

The increasing (decreasing) magnetic flux in the left-hand (right-hand) triangle in Fig. 3 gives rise to a clockwise (counterclockwise) induced emf in loop 1 (2), so the circuit behaves like the one shown in Fig. 28-18, but with both batteries reversed. Then the solution to

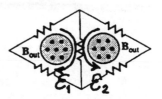

FIGURE 3 Cumulative Problem 3 Solution.

Problem 28-33 gives the current in the middle wire as

$$I_3 = -(\mathcal{E}_2 R_1 + \mathcal{E}_1 R_2)/(R_1 R_2 + R_2 R_3 + R_3 R_1).$$

The minus sign means that the direction of the current is opposite to that shown in Fig. 28-18, i.e., downward. There are two wires in the left and right branches and one in the middle, so the corresponding resistances are $R_1 = R_2 = 2R_3 = 2(10\ \Omega)$. Then I_3 becomes $-(\mathcal{E}_1 + \mathcal{E}_2)/4(10\ \Omega)$. The magnitudes of the emf's are given by Faraday's law, $\mathcal{E} = |d\phi/dt| = \pi r^2 |dB/dt|$, and both solenoids have the same cross-sectional area, so $I_3 = -\pi(5\text{ cm})^2 (50\text{ T/s} + 30\text{ T/s})/4(10\ \Omega) = -15.7$ mA.

Problem

5. A coaxial cable consists of an inner conductor of radius a and an outer conductor of radius b; the space between the conductors is filled with insulation of dielectric constant κ (Fig. 4). The cable's axis is the z axis. The cable is used to carry electromagnetic energy from a radio transmitter to a broadcasting antenna. The electric field between the conductors points radially from the axis, and is given by $E = E_0(a/r)\cos(kz - \omega t)$. The magnetic field encircles the axis, and is given by $B = B_0(a/r)\cos(kz - \omega t)$. Here E_0, B_0, k, and ω are constants. (a) Show, using appropriate closed surfaces and loops, that these fields satisfy Maxwell's equations. Your result shows that the cable acts as a "waveguide," confining an electromagnetic wave to the space between the conductors. (b) Find an expression for the speed at which the wave propagates along the cable.

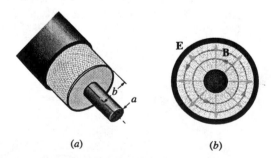

(a)   (b)

FIGURE 4 Cumulative Problem 5.

Solution

The method for (a) demonstrating that the given fields in a coaxial cable (so-called transverse electric and magnetic waves, or TEM-waves) satisfy Maxwell's equations and for (b) finding their speed is the same as that used in Sections 34-4 and 5, except that cylindrical coordinates r, θ, z, and Maxwell's equations in the dielectric (see solution to Problem 34-69) must be used. Consider an infinitesimal volume of dielectric at a point (r, θ, z) bounded by mutually orthogonal coordinate displacements $dr, r\,d\theta, dz$. The dielectric contains no free charges or conduction currents, so the given radial electric field and circulating magnetic field satisfy Gauss's laws for electricity and magnetism in the dielectric.

In Faraday's law, take a CCW loop and sides dr and dz, at fixed θ, so that the normal to the area it bounds is parallel to $\mathbf{B}$, as shown. Then, since $\mathbf{E} \perp d\mathbf{z}$ and $\mathbf{B} \perp d\mathbf{r}$ and $d\mathbf{z}$,

$$\oint \mathbf{E} \cdot d\boldsymbol{\ell} = -E\,dr + \left(E + \frac{\partial E}{\partial z}\,dz\right) dr = \frac{\partial E}{\partial z}\,dz\,dr$$

$$= -\frac{d\phi_B}{dt} = -\frac{d}{dt}(B\,dz\,dr) = -\frac{\partial B}{\partial t}\,dz\,dr,$$

or $\partial E/\partial z = -(E_0 ak/r)\sin(kz - \omega t) = -\partial B/\partial t = -(B_0 a\omega/r)\sin(kz - \omega t)$. Thus, $E_0 k = B_0 \omega$ (same as Equation 34-14).

In Ampere's law, take a CCW loop with sides $r\,d\theta$ and dz, at fixed r, so that the normal to its area is parallel to $\mathbf{E}$. Then, since $\mathbf{B} \perp d\mathbf{z}$ and $\mathbf{E} \perp r\,d\theta$ and $d\mathbf{z}$,

$$\oint \mathbf{B} \cdot d\boldsymbol{\ell} = Br\,d\theta - \left(B + \frac{\partial B}{\partial z}\,dz\right) r\,d\theta = -\frac{\partial B}{\partial z} r\,d\theta\,dz$$

$$= \kappa\varepsilon_0\mu_0 \frac{d\phi_E}{dt} = \kappa\varepsilon_0\mu_0 \frac{d}{dt}(Er\,d\theta\,dz)$$

$$= \kappa\varepsilon_0\mu_0 \frac{\partial E}{\partial t} r\,d\theta\,dz,$$

or $-\partial B/\partial z = (B_0 ak/r)\sin(kz - \omega t) = \kappa\varepsilon_0\mu_0 \times (\partial E/dt) = \kappa\varepsilon_0\mu_0(E_0 a\omega/r)\sin(kz - \omega t)$. Thus, $B_0 k = \kappa\varepsilon_0\mu_0 E_0 \omega$ (same as Equation 34-15 except for κ).

Therefore, Maxwell's equations are satisfied provided E_0, B_0, k and ω are related by $B_0\omega = E_0 k$ and $B_0 k = E_0\omega\kappa\varepsilon_0\mu_0$. The wave speed follows by dividing these equations to eliminate the amplitudes. Then $\omega/k = k/\omega\kappa\varepsilon_0\mu_0$, or $(\omega/k)^2 = 1/\kappa\varepsilon_0\mu_0 = v^2$, and $v = 1/\sqrt{\kappa\varepsilon_0\mu_0} = c/\sqrt{\kappa}$.

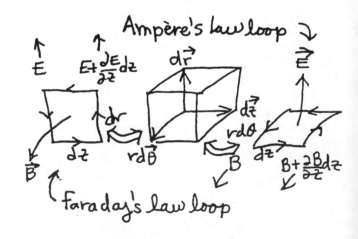

Cumulative Problem 5 Solution.

PART 5   OPTICS

CHAPTER 35   REFLECTION AND REFRACTION

ActivPhysics can help with these problems:
Activities 15.1, 15.2

Section 35-2:   Reflection

Problem

1. Through what angle should you rotate a mirror in order that a reflected ray rotate through 30°?

Solution

Since $\theta_1 = \theta_1'$ for specular reflection, (Equation 35-1) a reflected ray is deviated by $\phi = 180° - 2\theta_1$ from the incident direction. If rotating the mirror changes θ_1 by $\Delta\theta_1$, then the reflected ray is deviated by $\Delta\phi = |-2\Delta\theta_1|$ or twice this amount. Thus, if $\Delta\phi = 30°$, $|\Delta\theta_1| = 15°$.

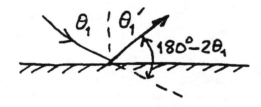

Problem 1 Solution.

Problem

3. To what angular accuracy must two ostensibly perpendicular mirrors be aligned in order that an incident ray returns on a path within 1° of its incident direction?

Solution

A ray incident on the first mirror at a grazing angle α is deflected through an angle 2α (this follows from the law of reflection). It strikes the second mirror at a grazing angle β, and is deflected by an additional angle 2β. The total deflection is $2\alpha + 2\beta = 2(180° - \theta)$. If this is to be within $180° \pm 1°$, then the angle between the mirrors, θ, must be within $90° \pm \frac{1}{2}°$.

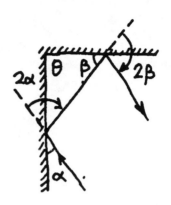

Problem 3 Solution.

Problem

5. Suppose the angle in Fig. 35-33 is changed to 75°. A ray enters the mirror system parallel to the axis. (a) How many reflections does it make? (b) Through what angle is it turned when it exits the system?

Solution

Now, after the first reflection, the ray leaves the top mirror at a grazing angle of $37\frac{1}{2}°$, and so makes a grazing angle of $180° - 75° - 37\frac{1}{2}° = 67\frac{1}{2}°$ with the bottom mirror. It is therefore deflected through an angle of $2(37\frac{1}{2}°) + 2(67\frac{1}{2})° = 210°$ CW, as it exits the system, after being reflected once from each mirror.

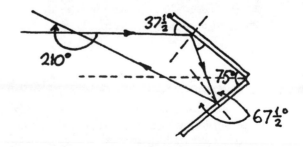

FIGURE 35-33 Problem 5 Solution.

Problem

7. Two plane mirrors make an angle ϕ. A light ray enters the system and is reflected once off each mirror. Show that the ray is turned through an angle $360° - 2\phi$.

Solution

The diagram in the solution to Problem 3 shows that the deflection of a ray, reflected once from each of two mirrors making an angle ϕ, is $2(180° - \phi) = 360° - 2\phi$. (Of course, there are conditions on ϕ and α, the grazing angle for the first mirror, if there is to be just one reflection from each mirror.)

Section 35-3: Refraction

Problem

9. Information in a compact disc is stored in "pits" whose depth is essentially one-fourth of the wavelength of the laser light used to "read" the information. That wavelength is 780 nm in air, but the wavelength on which the pit depth is based is measured in the $n = 1.55$ plastic that makes up most of the disc. Find the pit depth.

Solution

Equation 35-4 and the reasoning in Example 35-4 show that the wavelength in the plastic is $\lambda = \lambda_{air}/n = 780$ nm$/1.55 = 503$ nm. The pit depth is one quarter of this, or 126 nm.

Problem

11. A light ray propagates in a transparent material at 15° to the normal to the surface. When it emerges into the surrounding air, it makes a 24° angle with the normal. What is the refractive index of the material?

Solution

Snell's law (Equation 35-3), with air as medium 1, gives $n_2 = n_1 \sin\theta_1/\sin\theta_2 = 1 \times \sin 24°/\sin 15° = 1.57$.

Problem

13. A block of glass with $n = 1.52$ is submerged in one of the liquids listed in Table 35-1. For a ray striking the glass with incidence angle 31.5°, the angle of refraction is 27.9°. What is the liquid?

Solution

With the unknown liquid as medium 1, and the glass as medium 2, Snell's law gives $n_1 = n_2 \sin\theta_2/\sin\theta_1 = 1.52 \times \sin 27.9°/\sin 31.5° = 1.361$. This is the same as ethyl alcohol.

Problem

15. You look at the center of one face of a solid cube of glass, on a line of sight making a 55° angle with the normal to the cube face. What is the minimum refractive index of the glass for which you will see through the opposite face of the cube?

Solution

The angle of refraction in the glass, given by $\sin\theta_2 = \sin 55°/n_2$, must be less than $\tan^{-1}(\frac{1}{2}) = 26.6°$, for the ray to emerge from the opposite face (see diagram). Therefore, $n_2 \geq \sin 55°/\sin 26.6° = 1.83$.

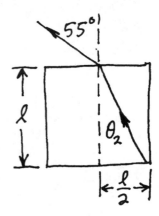

Problem 15 Solution.

Problem

17. You're standing 2.3 m horizontally from the edge of a 4.5-m-deep lake, with your eyes 1.7 m above the water surface. A diver holding a flashlight at the lake bottom shines the light so you can see it. If the light in the water makes a 42° angle with the vertical, at what horizontal distance is the diver from the edge of the lake?

Solution

Snell's law gives the angle of refraction (θ_1) in terms of the angle of incidence ($\theta_2 = 42°$) for the light path from the flashlight to your eye. These can be related to the other given distances by means of a carefully drawn diagram. Thus, $\theta_1 = \sin^{-1}(n_2 \sin\theta_2/n_1) = \sin^{-1}(1.333 \sin 42°) = 63.1°$, where we used indices of refraction from Table 35-1, with $n_1 \approx 1$ for air. The geometry of the diagram makes the horizontal distances apparent: $\tan\theta_1 = (2.3$ m $+ x_1)/(1.7$ m$)$, or $x_1 = (1.7$ m$)\tan 63.1° - 2.3$ m $= 1.05$ m, and $\tan\theta_2 = x_2/(4.5$ m$)$, or $x_2 = (4.5$ m$)\tan 42° = 4.05$ m. The total horizontal distance from the edge is $x_1 + x_2 = 5.11$ m.

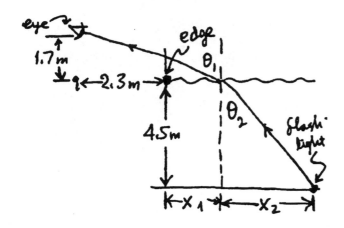

Problem 17 Solution.

Problem

19. A light ray is propagating in a crystal where its wavelength is 540 nm. It strikes the interior surface of the crystal with an incidence angle of 34° and emerges into the surrounding air at 76° to the surface normal. Find (a) the light's frequency and (b) its wavelength in air.

Solution

The index of refraction of the crystal (Equation 35-3) is $n_2 = \sin 76°/\sin 34° = 1.74$ (relative to air, with $n_1 \approx 1$). (a) The frequency of the light is $f = v_2/\lambda_2 = c/n_2\lambda_2 = (3\times10^8 \text{ m/s})/(1.74\times540 \text{ nm}) = c/(937 \text{ nm}) = 3.20\times10^{14}$ Hz, where we used the velocity (Equation 35-2) and wavelength of light in the crystal. (b) Of course, $\lambda_1 = n_2\lambda_2 = 937$ nm is the wavelength of the light in air (Equation 35-4). (Obviously, $f = v_1/\lambda_1$ is an alternative approach for part (a), since $v_1 \approx c$ in air.)

Section 35-4: Total Internal Reflection

Problem

21. Find the critical angle for total internal reflection in (a) ice, (b) polystyrene, and (c) rutile. Assume the surrounding medium is air.

Solution

For $n_{\text{air}} \approx 1$, the critical angle for total internal reflection in a medium of refractive index n is $\theta_c = \sin^{-1}(1/n)$. (Air is medium-2 in Equation 35-5.) From Table 35-1, $n = 1.309$ (ice), 1.49 (polystyrene), and 2.62 (rutile), so $\theta_c = \sin^{-1}(1/1.309) = 49.8°$, 42.2° and 22.4°, respectively, for these media.

Problem

23. What is the critical angle for light propagating in glass with $n = 1.52$ when the glass is immersed in (a) water, (b) benzene, and (c) diiodomethane?

Solution

The critical angle in medium-1, at an interface with medium-2, is $\theta_c = \sin^{-1}(n_2/n_1)$, where $n_1 > n_2$ (Equation 35-5). (a) For glass ($n_1 = 1.52$) immersed in water ($n_2 = 1.333$), $\theta_c = \sin^{-1}(1.333/1.52) = 61.3°$. (b) The same glass immersed in benzene has $\theta_c = \sin^{-1}(1.501/1.52) = 80.9°$. (c) Since the index of refraction of diiodomethane ($n_2 = 1.738$) is not smaller than that for this glass, there is no total internal reflection for light propagating in the glass. (However, for light originating in the liquid, $\theta'_c = \sin^{-1}(1.52 \div 1.738) = 61.0°$ at the glass interface.)

Problem

25. Light propagating in a medium with refractive index n_1 encounters a parallel-sided slab with index n_2. On the other side is a third medium with index $n_3 < n_1$. Show that the condition for avoiding internal reflection at *both* interfaces is that the incidence angle at the n_1-n_2 interface be less than the critical angle for an n_1-n_3 interface. In other words, the index of the intermediate material doesn't matter.

Solution

Since medium-2 has parallel interfaces with media-1 and 3, the angle of refraction at the 1-2 interface equals the angle of incidence at the 2-3 interface, as shown. (The normals to the interfaces are also parallel, so the alternate angles, marked θ_2, are equal.) Thus Snell's law implies $n_1 \sin \theta_1 = n_2 \sin \theta_2 = n_3 \sin \theta_3$, so that the angles in media-1 and 3 are related as if media-2 were not present. Of course, there are conditions on the intensity of the light transmitted

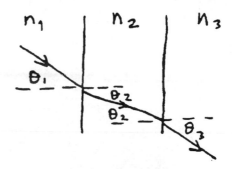

Problem 25 Solution.

through medium-2 which do depend on n_2 and the critical angle, if any, at the 1-2 interface. If $n_2 < n_1$, the phenomenon of frustrated total reflection (i.e., transmission of light for angles greater than the critical angle) may occur if the thickness of the slab is on the order of a few wavelengths of the incident light.

Problem

27. What is the minimum refractive index for which total internal reflection will occur as shown in Fig. 35-15a? Assume the surrounding medium is air and that the prism is an isosceles right triangle.

Solution

Figure 35-15a shows a 45° right-triangle prism with critical angle less than 45°. Thus, $\theta_c = \sin^{-1}(\frac{1}{n}) < 45°$, or $n > 1/\sin 45° = \sqrt{2}$. (We used $n_2 = 1$ for air, and $n_1 = n$ for the prism, in Equation 35-5.)

Problem

29. What is the speed of light in a material for which the critical angle at an interface with air is 61°?

Solution

From Equations 35-5 and 2, $\sin\theta_c = n_{air}/n \approx \frac{1}{n} = v/c$, so $v = c\sin\theta_c = (3\times10^8 \text{ m/s}) \sin 61° = 2.62\times10^8$ m/s. (The critical angle and the speed of light in a material are both related to the index of refraction.)

Problem

31. A compound lens is made from crown glass $(n = 1.52)$ bonded to flint glass $(n = 1.89)$. What is the critical angle for light incident on the flint-crown interface?

Solution

The critical angle for light incident in the glass of higher index of refraction (flint glass) is $\theta_c = \sin^{-1}(n_{crown}/n_{flint}) = \sin^{-1}(1.52/1.89) = 53.5°$.

Problem

33. A scuba diver sets off a camera flash a distance h below the surface of water with refractive index n. Show that light emerges from the water surface through a circle of diameter $2h/\sqrt{n^2 - 1}$.

Solution

Light from the flash will strike the water surface at the critical angle for a distance $r = h\tan\theta_c$ from a point directly over the flash. Therefore, the diameter of the circle through which light emerges is $2r = 2h\tan\theta_c$.

But $\sin\theta_c = 1/n$ (Equation 33-5 at the water-air interface), and $\tan^2\theta_c = (\csc^2\theta_c - 1)^{-1}$ (a trigonometric identity), so we can write $2r = 2h \div \sqrt{n^2 - 1}$.

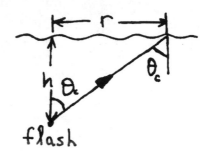

Problem 33 Solution.

Section 35-5: Dispersion

Problem

35. Suppose the red and blue beams of the preceding problem are now propagating in the same direction *inside* the glass. For what range of incidence angles on the glass-air interface will one beam be totally reflected and the other not?

Solution

The critical angle for blue light is less than for red light, $\theta_{c,blue} = \sin^{-1}(1/1.680) = 36.5°$, and $\theta_{c,red} = \sin^{-1}(1/1.621) = 38.1°$. For incidence angles between these values, blue light will be totally reflected, while some red light is refracted at the glass-air interface.

Problem

37. Two of the prominent spectral lines—discrete wavelengths of light—emitted by glowing hydrogen are hydrogen-α at 656.3 nm and hydrogen-β at 486.1 nm. Light from glowing hydrogen passes through a prism like that of Fig. 35-22, then falls on a screen 1.0 m from the prism. How far apart will these two spectral lines be? Use Fig. 35-20 for the refractive index.

Solution

The angular dispersion of H_α and H_β light in the prism of Fig. 35-22 can be found from the analysis in Example 35-6. For normal incidence on the prism, rays emerge with refraction angles of $\sin^{-1}(n \sin 40°)$. From Fig. 35-20, we estimate that $n_\alpha = 1.517$ and $n_\beta = 1.528$, so the angular dispersion is $\gamma = 79.2° - 77.2° = 1.98°$. We can assume that the size of the prism is small compared to the distance, r, to the screen, so the separation on the screen corresponding to γ is $\Delta x = \gamma\cdot r = (1.98°)(\pi/180°)(1 \text{ m}) = 3.45$ cm.

Section 35-6:   Reflection and Polarization

Problem

39. Find the polarizing angle for diamond when light is incident from air.

Solution

Equation 35-6 gives the polarizing angle, for light in air reflected from diamond; $\theta_p = \tan^{-1}(2.419/1) = 67.5°$.

Problem

41. What is the polarizing angle for light incident from below on the surface of a pond?

Solution

In Equation 35-6, n_2 is the index of refraction of the reflecting medium, so $\theta_p = \tan^{-1}(n_{air}/n_{water}) = \tan^{-1}(1/1.333) = 36.9°$.

Paired Problems

Problem

43. Light propagating in air strikes a transparent crystal at incidence angle 35°. If the angle of refraction is 22°, what is the speed of light in the crystal?

Solution

Combining Equations 35-2 and 3 (or using the form of Snell's law preceding Equation 35-2), we find $v = c/n = c \sin\theta_2 / \sin\theta_1 = (3\times10^8 \text{ m/s}) \sin 22°/\sin 35° = 1.96\times10^8$ m/s. (Note: $n_{air} \approx 1$.)

Problem

45. A cylindrical tank 2.4 m deep is full to the brim with water. Sunlight first hits part of the tank bottom when the rising Sun makes a 22° angle with the horizon. Find the tank's diameter.

Solution

The rays of sunlight which first hit the bottom of the tank just skim the opposite edge of the rim, as sketched. The diameter and depth of the tank (d and h) are related to the angle of refraction as shown, $\tan\theta_2 = d/h$. Combining this with Snell's law (Equation 35-3), we find $d = h \tan\theta_2 = h \tan[\sin^{-1} \times (n_1 \sin\theta_1/n_2)] = (2.4 \text{ m}) \tan[\sin^{-1}(\sin 68°/1.333)] = 2.32$ m. (Note: $n_1 = 1$ for air and $n_2 = 1.333$ for water; see Table 35-1.)

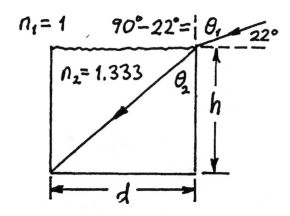

Problem 45 Solution.

Problem

47. Light is incident from air on the flat wall of a polystyrene water tank. If the incidence angle is 40°, what angle does the light make with the tank normal in the water?

Solution

If the plastic wall of the tank has parallel faces, it does not affect the angle of refraction in the water (see solution to Problem 25). Then $n_1 \sin\theta_1 = n_3 \sin\theta_3$, or $\theta_3 = \sin^{-1}(n_1 \sin\theta_1/n_3) = \sin^{-1}(\sin 40°/1.333) = 28.8°$. (Media-1, 2, and 3 are air, polystyrene, and water, respectively, in the solution to Problem 25.)

Problem

49. Light strikes a right-angled glass prism ($n = 1.52$) in a direction parallel to the prism's base, as shown in Fig. 35-39. The point of incidence is high enough that the refracted ray hits the opposite sloping side. (a) Through which side of the prism does the beam emerge? (b) Through what angle has it been deflected?

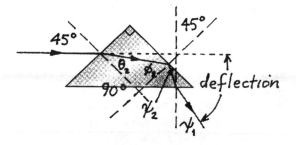

FIGURE 35-39 Problem 49 Solution.

Solution

The prism geometry of Fig. 35-39 and Snell's law imply $\phi_2 = 90° - \theta_2 = 90° - \sin^{-1}(\sin 45°/1.52) = 62.3°$, which is greater than the critical angle for total reflection from the glass-air interface, $\theta_c = \sin^{-1}(1 \div 1.52) = 41.1°$. Therefore, the incident light is totally reflected in the glass, as shown superposed on Fig. 35-39, and hits the base of the prism at an incidence angle of $\psi_2 = \phi_2 - 45° = 17.3°$. Its angle of refraction in air, after emerging, is $\psi_1 = \sin^{-1} \times (1.52 \sin 17.3°) = 26.8°$, giving it a net deflection of $90° - 26.8° = 63.2°$, measured clockwise from its original direction.

Problem

51. Repeat Problem 20 for the case $n = 1.75$, $\alpha = 40°$, and $\theta_1 = 25°$.

Solution

A general treatment of refraction through a prism of index of refraction $n_2 = n$, surrounded by air of index $n_1 = 1$, for the geometry of Fig. 35-36, is given in the solution to Problem 55. For $n = 1.75$, $\alpha = 40°$, and $\theta_1 = 25°$, the other angles defined there are:

$\theta_2 = \sin^{-1}(\sin \theta_1/n) = \sin^{-1}(\sin 25°/1.75) = 14.0°$,
$\phi_2 = \alpha - \theta_2 = 40° - 14.0° = 26.0°$,
$\phi_1 = \sin^{-1}(n \sin \phi_2) = \sin^{-1}(1.75 \sin 26.0°) = 50.2°$,
and $\delta = \theta_1 + \phi_1 - \alpha = 35.2°$.

(Note that ϕ_2 is less than the critical angle for this prism, which is $\sin^{-1}(1/1.75) = 34.8°$.)

Supplementary Problems
Problem

53. A cubical block is made from two equal-size slabs of materials with different refractive indices, as shown in Fig. 35-40. Find the index of the right-hand slab if a light ray is incident on the center of the left-hand slab and then describes the path shown.

Solution

In Fig. 35-40, the incident ray appears to hit the cube at the center of a side and the thickness of each material is the same, so $x_1 = \ell \tan \theta_1$, $x_2 = \ell \tan \theta_2$, and $x_1 + x_2 = \ell$. Thus $\tan \theta_1 + \tan \theta_2 = 1$, where θ_1 and θ_2 are the angles of refraction in the two materials, as shown. From Snell's law, $\sin 35° = n_1 \times \sin \theta_1 = n_2 \sin \theta_2$, so θ_1 and θ_2 can be eliminated in terms of the indices of refraction of the two materials, and since n_1 is given, n_2 can be easily determined. With the aid of a calculator, the intermediate steps are $\theta_1 = \sin^{-1}(\sin 35°/1.43) = 23.6°$, $\theta_2 = \tan^{-1}(1 - $

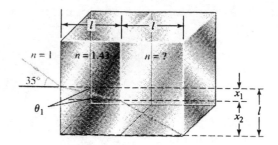

FIGURE 35-40 Problem 53 Solution.

$\tan \theta_1) = 29.3°$, and $n_2 = \sin 35°/\sin \theta_2 = 1.17$. In order to write the solution compactly for general values of n_1 and n_2, first notice that $\tan \theta = \sin \theta \div \cos \theta = \sin \theta/\sqrt{1 - \sin^2 \theta} = 1/\sqrt{\csc^2 \theta - 1}$, and that $n_1 \csc 35° = \csc \theta_1$ and $n_2 \csc 35° = \csc \theta_2$ (recall that the cosecant is the reciprocal of the sine). Then

$$\tan \theta_1 + \tan \theta_2 = 1 = \left(1/\sqrt{n_1^2 \csc^2 35° - 1}\right)$$
$$+ \left(1/\sqrt{n_2^2 \csc^2 35° - 1}\right).$$

Since $n_1 = 1.43$, one finds $\sqrt{n_2^2 \csc^2 35° - 1} = 1.78$, and $n_2 = 1.17$.

Problem

55. Light is incident with incidence angle θ_1 on a prism with apex angle α and refractive index n, as shown in Fig. 35-36. Show that the angle δ through which the outgoing beam deviates from the incident beam is given by

$$\delta = \theta_1 - \alpha + \sin^{-1}\left\{n \sin\left[\alpha - \sin^{-1}\left(\frac{\sin \theta_1}{n}\right)\right]\right\}.$$

Assume the surrounding medium has $n = 1$.

Solution

It is an exercise in ray tracing to determine the angles shown superposed on Fig. 35-36; from Snell's law and plane geometry: $\theta_2 = \sin^{-1}(\sin \theta_1/n)$ (Snell's law for the first refraction, with $n_1 = 1$ and $n_2 = n$), $\phi_2 = \alpha - \theta_2$ (α is the exterior angle to the triangle formed by the ray segment in the prism and the normals to the surfaces), $\phi_1 = \sin^{-1}(n \sin \phi_2)$ (Snell's law for the second refraction), and finally, $\delta = \theta_1 - \theta_2 + \phi_1 - \phi_2 = \theta_1 + \phi_1 - \alpha$ (the total deflection is the sum of the deflections at each refraction, clockwise deflection positive in Fig. 35-36). Writing down these steps in reverse order, substituting for each angle, one gets

$$\delta = \theta_1 + \sin^{-1}\left\{n \sin\left[\alpha - \sin^{-1}\left(\frac{\sin \theta_1}{n}\right)\right]\right\} - \alpha.$$

(Note: Problems 20, 36, 49–52, and also Problems 27, 28, 30, 37 and 38, involve ray tracing through prisms, in which a similar, but not identical, analysis is useful. Of course, ϕ_1 must be a real angle, i.e., less than 90°, or total internal reflection occurs instead of the second refraction. This is determined by the given values of n, α and θ_1.)

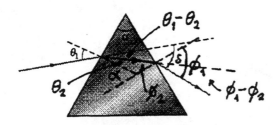

FIGURE 35-36 Problem 55 Solution.

Problem

57. Show that a three-dimensional corner reflector (three mirrors in three mutually perpendicular planes, or a solid cube in which total internal reflection occurs) turns an incident light ray through 180°, so it returns in the direction from which it came. *Hint:* Let $\mathbf{q} = q_x\hat{\mathbf{i}} + q_y\hat{\mathbf{j}} + q_z\hat{\mathbf{k}}$ be a vector in the direction of propagation. How does this vector get changed on reflection by a mirror in a plane defined by two of the coordinate axes?

Solution

A single plane mirror reverses the direction of just the normal component of a ray striking its surface. For example, a ray incident in the direction $\hat{\mathbf{q}} = \hat{\mathbf{i}}\cos\alpha_x + \hat{\mathbf{j}}\cos\alpha_y + \hat{\mathbf{k}}\cos\alpha_z$, on a mirror normal to the x-axis, is reflected into the direction $-\hat{\mathbf{i}}\cos\alpha_x + \hat{\mathbf{j}}\cos\alpha_y + \hat{\mathbf{k}}\cos\alpha_z$. (In our notation, $\hat{\mathbf{q}}$ is a unit vector, and $\cos^2\alpha_x + \cos^2\alpha_y + \cos^2\alpha_z = 1$.) If the ray also strikes mirrors which are normal to the y- and z-axes, as in a corner reflector, it emerges in the direction $\hat{\mathbf{q}}' = -\hat{\mathbf{q}}$, or opposite to the initial direction. In order to strike all three mirrors, the direction cosines of the incident ray must have magnitudes greater than some minimum non-zero value, depending on the size of the reflector (i.e., $|\cos\alpha_i| > 0$ for $i = x$, y, and z).

Problem

59. (a) Differentiate the result of the preceding problem to show that the maximum value of ϕ occurs when the incidence angle θ is given by $\cos^2\theta = \frac{1}{3}(n^2 - 1)$. (b) Use this result and that of the preceding problem to find $\phi_{\max}$ in water with $n = 1.333$.

Solution

One can differentiate ϕ, with respect to θ, directly, by using $d[\sin^{-1}(x/a)]/dx = 1/\sqrt{a^2 - x^2}$ (see the integral table in Appendix A.) Then

$$\frac{d\phi}{d\theta} = \frac{4}{\sqrt{n^2 - \sin^2\theta}}\frac{d(\sin\theta)}{d\theta} - 2 = \frac{4\cos\theta}{\sqrt{n^2 - \sin^2\theta}} - 2.$$

The condition for a maximum, $d\phi/d\theta = 0$, implies that $2\cos\theta_m = \sqrt{n^2 - \sin^2\theta_m}$, or $4\cos^2\theta_m = n^2 - (1 - \cos^2\theta_m)$, so $\cos^2\theta_m = \frac{1}{3}(n^2 - 1)$. If this value of θ is substituted into the expression for ϕ, after noting that $\sin^2\theta_m = \frac{1}{3}(4 - n^2)$, one gets $\phi_{\max} = 4\sin^{-1} \times (\sqrt{(4 - n^2)/3n^2}) - 2\cos^{-1}\sqrt{(n^2 - 1)/3}$, which equals 42.1° for $n = 1.333$. (This is the average angle, above the anti-solar direction, that an observer sees a rainbow, because n is the average index of refraction for visible wavelengths.)

Problem

61. *Fermat's principle* states that the path of a light ray between two points is such that the time to traverse that path is an extremum (either a minimum or maximum) when compared with the times for nearby paths. Consider two points A and B on the same side of a reflecting surface, and show that a light ray traveling from A to B via a point on the reflecting surface will take the least time if its path obeys the law of reflection. Thus, the law of reflection (Equation 35-1) follows from Fermat's principle.

Solution

Take the x-z plane to be the reflecting surface, and the x-y plane to contain the points A and B, with the y-axis through A. Suppose a ray traveling from A to B

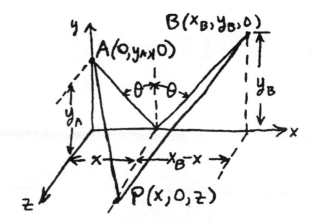

Problem 61 Solution.

via the surface is reflected at point $P(x, 0, z)$, as shown. (It is assumed that the path of a ray between two points in the same medium, with no intervening reflections or refractions, is a straight line, which also follows from Fermat's principle.) Since the ray propagates in only one medium, the time to traverse the path APB is proportional to the total distance $(t = D/v)$, where $D = AP + PB = \sqrt{x^2 + y_A^2 + z^2} + \sqrt{(x_B - x)^2 + y_B^2 + z^2}$. The conditions for D to be an extremum (a minimum in this case) are $\partial D/\partial z = 0$ and $\partial D/\partial x = 0$. The first condition requires that $(z/\sqrt{x^2 + y_A^2 + z^2}) + (z/\sqrt{(x_B - x)^2 + y_B^2 + z^2}) = 0$,

which is satisfied only for $z = 0$. (Thus, the incident ray, the normal to the surface, and the reflected ray all lie in the same plane, called the plane of incidence.) With $z = 0$, the second condition requires that $(x/\sqrt{x^2 + y_A^2}) - (x_B - x)/\sqrt{(x_B - x)^2 + y_B^2} = 0$. But $\sin \theta = x/\sqrt{x^2 + y_A^2}$, and $\sin \theta' = (x_B - x) \div \sqrt{(x_B - x)^2 + y_B^2}$, therefore $\sin \theta - \sin \theta' = 0$ which is the law of reflection. (Note: If A and B are on opposite sides of the x-z plane in media where the velocity of light is different, differentiation of $t = (AP/v) + (PB/v')$ leads to $(\sin \theta/v) - (\sin \theta'/v') = 0$, which is the subject of the next problem.)

CHAPTER 36   IMAGE FORMATION AND OPTICAL INSTRUMENTS

ActivPhysics can help with these problems:
All activities in Section 15

Sections 36-1 and 36-2:   Plane and Curved Mirrors

Problem

1. A shoe store uses small floor-level mirrors to let customers view prospective purchases. At what angle should such a mirror be inclined so that a person standing 50 cm from the mirror with eyes 140 cm off the floor can see her feet?

Solution

A small mirror (M) on the floor intercepts rays coming from a customer's shoes (O), which are traveling nearly parallel to the floor. The angle to the customer's eye (E) from the mirror is twice the angle of reflection, so $\tan 2\alpha = h/d$, or $\alpha = \frac{1}{2}\tan^{-1} \times (140/50) = 35.2°$, for the given distances. Therefore, the plane of the mirror should be tilted by 35.2° from the vertical to provide the customer with a floor-level view of her shoes.

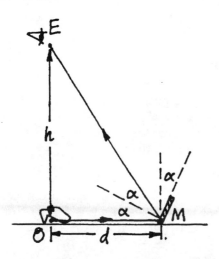

Problem 1 Solution.

Problem

3. (a) What is the focal length of a concave mirror if an object placed 50 cm in front of the mirror has a real image 75 cm from the mirror? (b) Where and what type will the image be if the object is moved to a point 20 cm from the mirror?

Solution

(a) The mirror equation relates the given distances (both positive for a real object and image) to the focal length: $f^{-1} = (50 \text{ cm})^{-1} + (75 \text{ cm})^{-1}$, or $f = 30$ cm. (See Equation 36-2.) (b) A second application of the mirror equation yields $(\ell')^{-1} = (30 \text{ cm})^{-1} - (20 \text{ cm})^{-1}$, or $\ell' = -60$ cm. A negative distance indicates a virtual, erect image located behind the mirror. (Fig. 36-8c and Table 36-1 confirm these results.)

Problem

5. An object is five focal lengths from a concave mirror. (a) How do the object and image heights compare? (b) Is the image upright or inverted?

Solution

(a) One can solve Equation 36-2 for ℓ', and substitute into Equation 36-1, to yield $h'/h = -\ell'/\ell = -f/(\ell - f) = -f/(5f - f) = -\frac{1}{4}$. (b) A negative magnification applies to a real, inverted image.

Problem

7. A virtual image is located 40 cm behind a concave mirror with focal length 18 cm. (a) Where is the object? (b) By how much is the image magnified?

Solution

(a) The mirror equation (Equation 36-2) gives $\ell = f\ell'/(\ell' - f) = (18 \text{ cm})(-40 \text{ cm})/(-58 \text{ cm}) = 12.4$ cm (positive distances are in front of the mirror, negative distances behind). (b) Equation 36-1 gives $M = -\ell'/\ell = +40 \text{ cm}/12.4 \text{ cm} = 3.22$.

Problem

9. A 12-mm-high object is 10 cm from a concave mirror with focal length 17 cm. (a) Where, (b) how high, and (c) what type is its image?

Solution

For an object on the mirror's axis, Equations 36-2 and 1 give (a) $\ell' = f\ell/(\ell - f) = (17 \times 10 \text{ cm})/(10 - 17) =$

−24.3 cm (i.e., behind the mirror). (c) A negative image distance indicates a virtual image. (b) $M = -\ell'/\ell = 24.3/10 = 2.43 = h'/h$, so the image is upright and 2.43×12 mm $= 29.1$ mm high.

Problem

11. An object's image in a 27-cm-focal-length concave mirror is upright and magnified by a factor of 3. Where is the object?

Solution

An upright image in a concave mirror must be virtual, so $M = +3 = -\ell'/\ell$. The mirror equation gives $(1/\ell) + (1/\ell') = (1/\ell) - (1/3\ell) = 1/f$, or $\ell = (\frac{2}{3})f = (\frac{2}{3})(27$ cm$) = 18$ cm (positive in front of the mirror).

Problem

13. When viewed from Earth, the moon subtends an angle of 0.5° in the sky. How large an image of the moon will be formed by the 3.6-m-diameter mirror of the Canada-France-Hawaii telescope, which has a focal length of 8.5 m?

Solution

The main mirror of a telescope is concave, since only such a mirror collects light from a distant object into a real image. Inspection of Fig. 36-8a (partially redrawn below) shows that the angular size of the object and image are equal ($h/\ell = h'/\ell' \approx \theta =$ size ÷ distance). Since the object distance is astronomical, $1/\ell \approx 0 \approx 1/f - 1/\ell'$, or $\ell' \approx f$. Thus, the image distance equals the focal length, and its size is $\theta \times f = \frac{1}{2}^{\circ} \times (\pi/180°) \times 8.5$ m $= 7.42$ cm. (See Problem 6.)

Problem 13 Solution.

Problem

15. You look into a reflecting sphere 80 cm in diameter and see an image of your face at one-third its normal size (Fig. 36-46). How far are you from the sphere's surface?

Solution

The sphere reflects like a convex mirror of focal length $f = R/2 = -40$ cm$/2 = -20$ cm. (Only a convex mirror produces a reduced, upright, virtual image.)

The equation in the solution to Problem 5(a) can be used to find the object distance in terms of f and the magnification, $\ell = f(1 - h/h') = f(1 - 1/M) = -20$ cm$(1 - 3) = 40$ cm.

Section 36-3: Lenses

Problem

17. A light bulb is 56 cm from a convex lens, and its image appears on a screen located 31 cm on the other side of the lens. (a) What is the focal length of the lens? (b) By how much is the image enlarged or reduced?

Solution

(a) The object and image distances are both positive, for a real image formed by a single lens (recall that only a real image can appear on a screen), so the lens equation gives $f^{-1} = (56$ cm$)^{-1} + (31$ cm$)^{-1} = (20.0$ cm$)^{-1}$. (b) Equation 36-4 gives a magnification of $M = -31/56 = -0.554$, so the inverted image is reduced to nearly 55% of the actual size of the bulb.

Problem

19. A lens with 50-cm focal length produces a real image the same size as the object. How far from the lens are image and object?

Solution

For a real image the same size as the object, $h' = -h$, so $M = -1 = -\ell'/\ell$, or $\ell' = \ell$. The lens equation (Equation 36-5) then gives $(1/\ell) + (1/\ell') = 2/\ell = 1/f$ or $\ell = \ell' = 2f = 100$ cm.

Problem

21. A simple camera uses a single converging lens to focus an image on its film. If the focal length of the lens is 45 mm, what should be the lens-to-film distance for the camera to focus on an object 80 cm from the lens?

Solution

Set $\ell = 80$ cm and $f = 45$ mm in the lens equation and solve for ℓ'. The result is $\ell' = \ell f/(\ell - f) = (80\times45$ cm$)/(800 - 45) = 4.77$ cm.

Problem

23. How far from a page should you hold a lens with 32-cm focal length in order to see the print magnified 1.6 times?

Solution

When a virtual, upright image is formed by a converging lens (a diverging lens always produces a reduced image), the magnification is positive, $M = 1.6 = -f/(\ell - f)$. (Use Equations 36-5 and 4.) Therefore, $\ell = (M - 1)f/M = 0.6 (32 \text{ cm})/1.6 = 12$ cm.

Problem

25. The largest refracting telescope in the world, at Yerkes Observatory, has a 1-m-diameter lens with focal length 12 m (Fig. 36-47). If an airplane flew 1 km above the telescope, where would its image occur in relation to the images of the very distant stars?

Solution

Very distant stars ($\ell = \infty$) produce images at the focus, $\ell' = f = 12$ m. For an airplane 1 km away, $\ell' = \ell f/(\ell - f) = (1 \text{ km})(12 \text{ m})/(988 \text{ m}) = 12.1$ m, or 14.6 cm farther than the focal point. ($\ell' - f = f^2 \div (\ell - f) = (144/988)$ m can be calculated more accurately.)

Problem

27. A lens has focal length $f = 35$ cm. Find the type and height of the image produced when a 2.2-cm-high object is placed at distances (a) $f + 10$ cm and (b) $f - 10$ cm.

Solution

The lens equation and magnification for a thin (converging, i.e., positive f) lens, Equations 36-4 and 5, give $M = -\ell'/\ell = -f/(\ell - f)$, so $h' = Mh = -fh \div (\ell - f)$. (a) If $f = 35$ cm and $\ell = f + 10$ cm, then $h' = -(35 \text{ cm})(2.2/10) = -7.7$ cm. A negative image height signifies a real, inverted image. (b) If $\ell = f - 10$ cm, then $h' = -(35 \text{ cm})(2.2)/(-10) = +7.7$ cm, which represents a virtual, erect image of the same size.

Problem

29. A candle and a screen are 70 cm apart. Find two points between candle and screen where you could put a convex lens with 17 cm focal length to give a sharp image of the candle on the screen.

Solution

Since ℓ and ℓ' are both positive for a real image, and the distance $a \equiv \ell + \ell' = 70$ cm is fixed, the lens equation can be rewritten as a quadratic, $af = (\ell + \ell')f = \ell\ell' = \ell(a - \ell) = \ell'(a - \ell')$. The solutions for ℓ or ℓ' are $\frac{1}{2}a(1 \pm \sqrt{1 - 4f/a}) = (35 \text{ cm})(1 \pm \sqrt{1 - (4 \times 17)/70}) = 29.1$ cm or 40.9 cm, which are the desired lens locations. (Note that this situation has a real solution only if $0 < 4f \leq a$.)

Section 36-4   Refraction in Lenses: The Details

Problem

31. You're standing in a wading pool and your feet appear to be 30 cm below the surface. How deep is the pool?

Solution

The image formed by a single refracting interface between two media, for paraxial rays, is described by Equation 36-6, with sign conventions for distances defined in the paragraph following. For the flat surface of the wading pool, $R = \infty$, $n_1 = 1.333$ for water, ℓ is the depth of the pool (your feet, the object, are on the bottom), $n_2 = 1$ for air, and $\ell' = -30$ cm (for a virtual image at the apparent depth). Thus, $(1.333/\ell) + (1/(-30 \text{ cm})) = (n_2 - n_1)/\infty = 0$, or $\ell = 40$ cm. (This problem could also be solved directly from Snell's law, as in the previous chapter, without the paraxial ray approximation.)

Problem

33. Use Equation 36-6 to show that an object at the center of a glass sphere will appear to be its actual distance—one radius—from the edge. Draw a ray diagram showing why this result makes sense.

Solution

Solving for ℓ' in Equation 36-6 gives us $\ell' = n_2 R \div [(n_2 - n_1)\ell - n_1 R]$. If the object is at the center of a glass sphere ($\ell = |R|$ and R is negative for a concave surface toward the object, i.e., $R = -|R|$) and viewed from the air outside ($n_2 = 1$), one obtains $\ell' = -|R|^2/[(1 - n_1)|R| + n_1|R|] = -|R|$. This represents a virtual image, on the same side of the surface as the object, a distance $|R|$ from the surface, or at the center, coincident with the object. This is to be expected, because all the rays from an object at the center of curvature, strike the surface normally and are not refracted. The diagram is hardly necessary to understand this.

Problem 33 Solution.

Problem

35. Rework Example 36-6 for a fish 15 cm from the *far* wall of the tank.

Solution

As in Example 36-6, use Equation 36-6 with $R = -35$ cm, $n_1 = 1.333$, $n_2 = 1$, but with $\ell = 70$ cm $-$ 15 cm $= 55$ cm (as distance from the near wall). Then $\ell' = [(1 - 1.333)/(-35 \text{ cm}) - 1.333/55 \text{ cm}]^{-1} = -67.9$ cm. In this case, the object is closer to the refracting surface than its image (see sketch and compare with Fig. 36-25b).

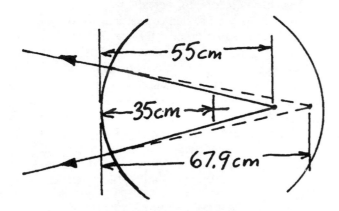

Problem 35 Solution.

Problem

37. Two specks of dirt are trapped in a crystal ball, one at the center and the other halfway to the surface. If you peer into the ball on a line joining the two specks, the outer one appears to be only one-third of the way to the other. What is the refractive index of the ball?

Solution

The outer speck appears $\frac{1}{3}$ the distance to the center of the ball, $\ell' = -|R|/3$, since the speck at the center appears at the center (see Problem 33.) The actual distance of the outer speck is given as $\ell = |R|/2$, so Equation 36-6 (with n_1 for the ball's material, $n_2 = 1$ for air, and $R = -|R|$ for a concave surface toward the object) gives $(n_1/\frac{1}{2}|R|) + (1/(-\frac{1}{3}|R|)) = (1 - n_1) \div (-|R|)$. This simplifies to $2n_1 - 3 = n_1 - 1$, or $n_1 = 2$.

Problem

39. A contact lens is in the shape of a convex meniscus (see Fig. 36-28); the inner surface is curved to fit the eye, with a curvature radius of 7.80 mm. The lens is made from plastic with refractive index $n = 1.56$. If it's to have a focal length of 44.4 cm, what should be the curvature radius of its outer surface?

Solution

The lensmaker's formula (Equation 36-8) relates the four quantities mentioned in this problem. The sign conventions used here for a convex meniscus lens require $R_1 < R_2$ and $R_1 R_2 > 0$; i.e., R_1 and R_2 are either both positive or both negative, depending on whether one takes the light coming from the left, as we choose here, or right side of the lens, in Fig. 36-28. Then $R_2 = 7.80$ mm, and $R_1^{-1} = (n-1)^{-1}f^{-1} + R_2^{-1} = (0.56 \times 44.4 \text{ cm})^{-1} + (7.80 \text{ mm})^{-1} = (7.56 \text{ mm})^{-1}$. (To test your understanding of the sign conventions, try taking the light coming from the right, using $R_1 = -7.80$ mm. Also note that for a concave meniscus lens, $R_1 > R_2$ and $R_1 R_2 > 0$.)

Problem

41. An object is 28 cm from a double convex lens with $n = 1.5$ and curvature radii 35 cm and 55 cm. Where and what type is the image?

Solution

The focal length of the lens, $f^{-1} = (n-1)(R_1^{-1} - R_2^{-1}) = (1.5 - 1)[(35 \text{ cm})^{-1} + (55 \text{ cm})^{-1}] = (42.8 \text{ cm})^{-1}$ (from Equation 36-8) is greater than the object distance, so the lens equation gives a virtual, erect image located at $\ell' = \ell f/(\ell - f) = (28 \times 42.8 \text{ cm}) \div (28 - 42.8) = -81.1$ cm (negative ℓ' being on the same side of the lens as the object).

Problem

43. A plano-convex lens has curvature radius 20 cm and is made from glass with $n = 1.5$. Use the generalized lensmaker's formula given in Problem 73 to find the focal length when the lens is (a) in air, (b) submerged in water ($n = 1.333$) and (c) embedded in glass with $n = 1.7$. Comment on the sign of your answer to (c).

Solution

In the generalized lensmaker's formula, $n_r = n_{\text{lens}} \div n_{\text{ext}}$ is the index of refraction of the lens relative to the external medium. For a plano-convex lens, $R_1 = 20$ cm and $R_2 = \infty$ (or $R_1 = \infty$ and $R_2 = -20$ cm), so $f = 20 \text{ cm}/(n_r - 1)$. (a) In air, the relative index of refraction of the lens is 1.5/1, so $f = 40$ cm. (b) In water, $n_r = 1.5/1.333$ and $f = 160$ cm. (c) In a medium of higher index of refraction, the relative index is less than one, so the lens acts as a diverging lens with $f = 20 \text{ cm}/[(1.5/1.7) - 1] = -170$ cm.

Problem

45. Two plano-convex lenses are geometrically identical, but one is made from crown glass

(n = 152), the other from flint glass. An object at 45 cm from the lens focuses to a real image at 85 cm with the crown-glass lens and at 53 cm with the flint-glass lens. Find (a) the curvature radius (common to both lenses) and (b) the refractive index of the flint glass.

Solution

From the lens equation, we calculate the focal lengths of the lenses to be $f_c^{-1} = (45 \text{ cm})^{-1} + (85 \text{ cm})^{-1} = (29.4 \text{ cm})^{-1}$ for the crown glass and $f_f^{-1} = (45 \text{ cm})^{-1} + (53 \text{ cm})^{-1} = (24.3 \text{ cm})^{-1}$ for the flint glass. The lens maker's formula, with the crown glass data, gives the radius of curvature, $R = (n_c - 1)f = (1.52 - 1)(29.4 \text{ cm}) = 15.3 \text{ cm}$ (see solution to Problem 43). The same formula gives the index of refraction of the flint glass, with data for this lens, as $n_f = 1 + R/f = 1 + 15.3/24.3 = 1.63$.

Problem

47. An object placed 15 cm from a plano-convex lens made of crown glass focuses to a virtual image twice the size of the object. If the lens is replaced with an identically shaped one made from diamond, what type of image will appear and what will be its magnification? See Table 35-1.

Solution

Equations 36-5 and 4 for the magnification (which is positive for a virtual image) give $M_g = 2 = -\ell'/\ell = f_g/(f_g - \ell)$ for the crown glass lens, so $f_g = 2\ell = 30$ cm. The focal length of a diamond lens with the same radii of curvature is $f_d = (n_g - 1)f_g/(n_d - 1) = 30 \text{ cm } (1.520 - 1)/(2.419 - 1) = 11.0 \text{ cm}$ (use Equation 36-8 and Table 35-1). An object 15 cm from the diamond lens produces a real, inverted image (negative M) magnified by $M_d = 11.0/(11.0 - 15) = -2.74$.

Section 36-5:   Optical Instruments

Problem

49. Grandma's new reading glasses have 3.8-diopter lenses to provide full correction of her farsightedness. Her old glasses were 2.5 diopters. (a) Where is the near point for her unaided eyes? (b) Where will be the near point if she wears her old glasses?

Solution

(a) The new correction is designed to make an object placed at the standard near point, $\ell = 25$ cm, appear (as a virtual image) to be at the near point for unaided vision, $\ell' = -$near point. Therefore, $(f_{\text{cor}})^{-1} =$

$(25 \text{ cm})^{-1} + (-\text{near point})^{-1}$, or $(\text{near point})^{-1} = (0.25 \text{ m})^{-1} - 3.8 \text{ diopters} = 0.2 \text{ diopters} = (5 \text{ m})^{-1}$. (b) The old correction, $(f_{\text{cor}})^{-1} = 2.5 \text{ diopters}$, would not bring an object, placed at the standard 25 cm, to a virtual image at the near point of 5 m, but only from $\ell^{-1} = (f_{\text{cor}})^{-1} - \ell'^{-1} = 2.5 \text{ diopters} - (-5 \text{ m})^{-1} = (37.0 \text{ cm})^{-1}$, which is the near point when wearing the old correction.

Problem

51. A camera's zoom lens covers the focal length range from 38 mm to 110 mm. You point the camera at a distant object and photograph it first at 38 mm and then with the camera zoomed out to 110 mm. Compare the sizes of its images on the two photos.

Solution

The image size can be determined from the lens equation, $h'/h = -\ell'/\ell = -f/(\ell - f)$. The ratio of the image sizes, for two different focal lengths, is $h_1'/h_2' = [-f_1/(\ell - f_1)]/[-f_2/(\ell - f_2)] \approx f_1/f_2$, if $\ell \gg f_1$ or f_2 (as for a distant object). Then the image size at 110 mm is approximately $110/38 = 2.89$ times larger than at 38 mm.

Problem

53. The maximum magnification of a simple magnifier occurs with the image at the 25-cm near point. Show that the angular magnification is then given by $m = 1 + \frac{25 \text{ cm}}{f}$, where f is the focal length.

Solution

Another way of using a magnifier is to arrange for the virtual image to be at the near point, as shown (instead of at ∞, as in Fig. 36-38b). From the diagram and the lens equation, $\beta \approx \tan \beta = h'/(-\ell') = h/\ell = h[(1/f) - (1/\ell')] = h[(1/f) + (1/25 \text{ cm})]$. Thus, with the same definition of angular magnification as before (Equation 36-9), $m = \beta/\alpha = h[(1/f) + (1/25 \text{ cm})] \div (h/25 \text{ cm}) = 1 + (25 \text{ cm}/f)$. This is the maximum magnification obtainable with a simple magnifier, because the image can't be seen clearly if it's closer

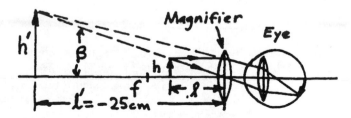

Problem 53 Solution.

than the near point (i.e., $\beta \approx h'/|\ell'|$ and $|\ell'| \geq 25$ cm). Incidentally, the most effective way to use a magnifier is to hold it close to your eye, moving the object up to it until a sharp image is seen.

Problem

55. A 300-power compound microscope has a 4.5-mm-focal-length objective lens. If the distance from objective to eyepiece is 10 cm, what should be the focal length of the eyepiece?

Solution

For a 300×microscope, with $f_0 = 4.5$ mm and $L = 100$ mm, we can solve Equation 36-10 for $f_e = (100/4.5)(25 \text{ cm}/300) = 1.85$ cm.

Problem

57. A Cassegrain telescope like that shown in Fig. 36-42b has 1.0-m focal length, and the convex secondary mirror is located 0.85 m from the primary. What should be the focal length of the secondary in order to put the final image 0.12 m behind the front surface of the primary mirror?

Solution

Reference to Fig. 36-42b shows that parallel rays reflected by the objective mirror converge toward a point, $1 \text{ m} - 0.85 \text{ m} = 15$ cm behind the secondary mirror, and behave as if they came from a virtual object, with $\ell = -15$ cm in the mirror equation. The final image is located a distance $\ell' = 85$ cm + 12 cm = 97 cm from the secondary mirror, whose required focal length is therefore $f^{-1} = \ell^{-1} + \ell'^{-1} = (-15 \text{ cm})^{-1} + (97 \text{ cm})^{-1} = (-17.7 \text{ cm})^{-1}$. (Recall that a convex mirror has negative focal length.)

Paired Problems

Problem

59. (a) How far from a 1.2-m-focal-length concave mirror should you place an object in order to get an inverted image 1.5 times the size of the object? (b) Where will the image be?

Solution

(a) Equation 36-1 for the mirror's magnification can be combined with Equation 36-2 to yield $-1/M = (\ell/f) - 1$, so $\ell = f(1 - 1/M)$. For a concave mirror with $f = 1.2$ m > 0, and a real image with $M = -1.5$, one finds $\ell = (1.2 \text{ m})(1 + 1/1.5) = 2.0$ m. (b) From Equation 36-1, $\ell' = -M\ell = 1.5(2.0 \text{ m}) = 3.0$ m. (A real image is, of course, in front of the mirror.)

Problem

61. Find the focal length of a concave mirror if an object 15 cm from the mirror has a virtual image 2.5 times the object's actual size.

Solution

The equation in part (a) of the solution to Problem 59, with M positive for an upright, virtual image, gives $f = \ell/(1 - 1/M) = (15 \text{ cm})/(1 - 1/2.5) = 25$ cm.

Problem

63. How far from a 1.6-m-focal-length concave mirror should you place an object to get an upright image magnified by a factor of 2.5?

Solution

The analysis in the solutions to Problems 59 and 61 shows that $\ell = f(1 - 1/M) = (1.6 \text{ m})(1 - 1/2.5) = 96$ cm.

Problem

65. An object and its lens-produced real image are 2.4 m apart. If the lens has 55-cm focal length, what are the possible values for the object distance and magnification?

Solution

If the distance between the object and the real image is $L = \ell + \ell'$ (all positive), the lens equation, $\ell^{-1} + (L - \ell)^{-1} = f^{-1}$, can be rewritten as $\ell(L - \ell) = Lf$. This quadratic in ℓ has solutions $\ell = (L/2) \pm \sqrt{(L/2)^2 - Lf} = 1.2 \text{ m} \pm \sqrt{(1.2 \text{ m})^2 - (2.4 \text{ m})(0.55 \text{ m})} = 85.4$ cm and 154.6 cm. The corresponding magnifications are $M = -f \div (\ell - f) = -1.81$ and -0.552, respectively (see the similar equation in the solution to Problem 5(a)). (Since the image and object distances satisfy the same quadratic, $\ell(L - \ell) = (L - \ell')\ell' = Lf$, their numerical values are conjugates. Compare with the solution to problem 29.)

Problem

67. An object is 68 cm from a plano-convex lens whose curved side has curvature radius 26 cm. The refractive index of the lens is 1.62. Where and of what type is the image?

Solution

The focal length of the lens is given by Equation 36-8, with $R_1 = 26$ cm and $R_2 = \infty$ (or $R_1 = \infty$ and $R_2 = -26$ cm), so $f^{-1} = (n - 1)/R_1 = (1.62 - 1)/26$ cm $= (41.9 \text{ cm})^{-1}$. An object at $\ell = 68$ cm is imaged at

$\ell'^{-1} = f^{-1} - \ell^{-1} = (41.9 \text{ cm})^{-1} - (68 \text{ cm})^{-1} = (109 \text{ cm})^{-1}$. This is a real, inverted image, on the opposite side of the lens from the object.

Supplementary Problems

Problem

69. My contact lens prescription calls for +2.25-diopter lenses with an inner curvature radius of 8.6 mm to fit my cornea. (a) If the lenses are made from plastic with $n = 1.56$, what should be the outer curvature radius? (b) Wearing these lenses, I hold a newspaper 30 cm from my eyes. Where is its image as viewed through the lenses?

Solution

(a) This prescription calls for a converging lens (positive dioptric power) of focal length $1/(2.25 \text{ m}^{-1}) = 44.4$ cm, with a convex meniscus shape (Fig. 36-28). The analysis in the solution to Problem 39 shows that the outer curvature radius (of the first surface to intercept light coming to the eye) is $R_1^{-1} = (n-1)^{-1} f^{-1} + R_2^{-1} = (0.56)^{-1}(2.25 \text{ m}^{-1}) + (8.60 \text{ mm})^{-1} = (8.31 \text{ mm})^{-1}$. (b) The lens equation gives the distance of the image of the newspaper formed by just the contact lens as $(\ell')^{-1} = f^{-1} - \ell^{-1} = 2.25 \text{ m}^{-1} - (30 \text{ cm})^{-1} = (-92.3 \text{ cm})^{-1}$. The negative sign indicates a virtual, erect image in front of the lens. (Of course, the contact lens and the eye together form a real image on the retina.)

Problem

71. Show that identical objects placed equal distances on either side of the focal point of a concave mirror or converging lens produce images of equal size. Are the images of the same type?

Solution

A concave mirror or a converging lens are both represented by positive focal lengths in the lens or mirror equations. For either, the object and image sizes are related by $h'/h = -\ell'/\ell = -f/(\ell - f)$. One can easily see that for $\ell - f = \pm x$ (where $0 < x < f$ is implicit in the statement of this problem), the image size is the same. However, for $\ell = f + x$, the image is real ($h' < 0$), and for $\ell = f - x$, it is virtual ($h' > 0$). (A lens has two symmetrically placed focal points; this problem makes sense for the one on the same side of the lens as the object.)

Problem

73. Generalize the derivation of the lensmaker's formula (Equation 36-8) to show that a lens of

refractive index n_{lens} in an external medium with index n_{ext} has focal length given by

$$\frac{1}{f} = \left(\frac{n_{\text{lens}}}{n_{\text{ext}}} - 1\right)\left(\frac{1}{R_1} - \frac{1}{R_2}\right).$$

Solution

Refraction at the two lens surfaces in Fig. 36-27, when the surrounding medium has index of refraction n_{ext} (instead of $n_1 = 1$), is described by equations analogous to the two preceding Equation 36-7:

$$\frac{n_{\text{ext}}}{\ell_1} + \frac{n_{\text{lens}}}{\ell_1'} = \frac{n_{\text{lens}} - n_{\text{ext}}}{R_1},$$

and

$$\frac{n_{\text{lens}}}{t - \ell_1'} + \frac{n_{\text{ext}}}{\ell_2'} = \frac{n_{\text{ext}} - n_{\text{lens}}}{R_2}.$$

(These are just Equation 36-6 applied to the left- and right-hand surfaces.) For $t \to 0$, there is no distinction between distances measured from either surface, so adding the equations and dropping the subscripts 1 and 2, we find $(n_{\text{ext}}/\ell) + (n_{\text{ext}}/\ell') = (n_{\text{lens}} - n_{\text{ext}}) \times (R_1^{-1} - R_2^{-1})$. Division by n_{ext} gives the sought-for generalization of Equation 36-8. Note that $n_{\text{lens}}/n_{\text{ext}}$ is the relative index of refraction.

Problem

75. A Newtonian telescope like that of Fig. 36-42c has a primary mirror with 20-cm diameter and 1.2-m focal length. (a) Where should the flat diagonal mirror be placed to put the focus at the edge of the telescope tube? (b) What shape should the flat mirror have to minimize blockage of light to the primary?

Solution

(a) The focal point of the primary mirror is 1.2 m from the mirror apex, and the radius of the telescope tube (presumably the same diameter as the mirror) is 0.1 m. Since the total distance of the light path from the primary mirror to the focus is not changed by the insertion of the secondary mirror, the secondary mirror should be (at least) 0.1 m closer than the focal point, or 1.1 m from the primary mirror apex. (b) The cone of rays reflected by the circular primary mirror is sliced by the plane of the secondary mirror into an ellipse (recall the definition of conic sections), which is the shape which blocks the least amount of incoming light.

Problem

77. Just before Equation 36-7 are two equations describing refraction at the two surfaces of a lens with thickness t. Combine these equations to show

that the object distance ℓ and image distance ℓ' for such a lens are related by

$$\frac{1}{\ell} + \frac{1}{\ell'} - \frac{[(n-1)\ell - R_1]^2 t}{\ell R_1[t(\ell + R_1) + n\ell(R_1 - t)]}$$
$$= (n-1)\left(\frac{1}{R_1} - \frac{1}{R_2}\right).$$

Solution

Combine the two equations by adding them to obtain

$$\frac{1}{\ell_1} + \frac{1}{\ell_2'} + n\left(\frac{1}{\ell_1'} + \frac{1}{t - \ell_1'}\right) = \frac{1}{\ell_1} + \frac{1}{\ell_2'} + \frac{nt}{\ell_1'(t - \ell_1')}$$
$$= (n-1)\left(\frac{1}{R_1} - \frac{1}{R_2}\right).$$

Eliminate ℓ_1' from the third term in the middle member of the above equation, by using either of the original two equations, with the following result:

$$\frac{nt}{\ell_1'(t - \ell_1')} = -\frac{[(n-1)\ell_1 - R_1]^2 t}{\ell_1 R_1[t(\ell_1 + R_1) + n\ell_1(R_1 - t)]}$$
$$= -\frac{[(n-1)\ell_2' + R_2]^2 t}{\ell_2' R_2[t(R_2 - \ell_2') + n\ell_2'(R_2 + t)]}.$$

Then the desired relation between ℓ_1 and ℓ_2' (with t, n, R, and R_2 as parameters) is achieved. (Note that the object distance $\ell_1 = \ell$, and the image distance $\ell_2' = \ell'$, are measured from different lens surfaces. The subscripts 1 and 2 are retained as a reminder of this.) For example, from the first of the original equations,

$$\frac{n}{\ell_1'} = \frac{n-1}{R_1} - \frac{1}{\ell_1} = \frac{(n-1)\ell_1 - R_1}{\ell_1 R_1}.$$

Then

$$\ell_1' = \frac{n\ell_1 R_1}{[(n-1)\ell_1 - R_1]}$$

and

$$t - \ell_1' = -\frac{n\ell_1(R_1 - t) + t(R_1 + \ell_1)}{[(n-1)\ell_1 - R_1]}.$$

Multiplication and division by nt produces the first expression for the third term above. The other expression comes from a similar treatment of the second original equation,

$$\frac{n}{t - \ell_1'} = -\frac{(n-1)}{R_2} - \frac{1}{\ell_2'}.$$

Then

$$t - \ell_1' = -\frac{n\ell_2' R_2}{[(n-1)\ell_2' + R_2]},$$

and

$$\ell_1' = \frac{n\ell_2'(R_2 + t) + t(R_2 - \ell_2')}{[(n-1)\ell_2' + R_2]},$$

and the second expression for $nt/\ell_1'(t - \ell_1')$ follows.

CHAPTER 37   INTERFERENCE AND DIFFRACTION

ActivPhysics can help with these problems:
Activities in Section 16, Physical Optics

Section 37-2:   Double-Slit Interference

Problem

1. A double-slit system is used to measure the wavelength of light. The system has slit spacing $d = 15$ μm and slit-to-screen distance $L = 2.2$ m. If the $m = 1$ maximum in the interference pattern occurs 7.1 cm from screen center, what is the wavelength?

Solution

The experimental arrangement and geometrical approximations valid for Equation 37-2a are satisfied for the situation and data given, so $\lambda = y_{\text{bright}}\, d/mL = $ (7.1 cm/2.2 m)(15 μm/1) = 484 nm. (In particular, $\lambda \ll d$ and $\theta_1 = 3.23 \times 10^{-2} = 1.85°$ is small.)

Problem

3. A double-slit experiment has slit spacing 0.12 mm. (a) What should be the slit-to-screen distance L if the bright fringes are to be 5.0 mm apart when the slits are illuminated with 633-nm laser light? (b) What will be the fringe spacing with 480-nm light?

Solution

The particular geometry of this type of double-slit experiment is described in the paragraphs preceding Equations 37-2a and b. (a) The spacing of bright fringes on the screen is $\Delta y = \lambda L/d$, so $L = $ (0.12 mm)(5 mm)/(633 nm) = 94.8 cm. (b) For two different wavelengths, the ratio of the spacings is $\Delta y'/\Delta y = \lambda'/\lambda$; therefore $\Delta y' = (5$ mm)(480/633) = 3.79 mm.

Problem

5. The green line of gaseous mercury at 546 nm falls on a double-slit apparatus. If the fifth dark fringe is at 0.113° from the centerline, what is the slit separation?

Solution

The interference minima fall at angles given by Equation 37-1b; therefore $d = (4 + \frac{1}{2})\lambda/\sin\theta = $

4.5(546 nm)/$\sin 0.113° = 1.25$ mm. (Note that $m = 0$ gives the first dark fringe.)

Problem

7. Light shines on a pair of slits whose spacing is three times the wavelength. Find the locations of the first- and second-order bright fringes on a screen 50 cm from the slits. *Hint:* Do Equations 37-2 apply?

Solution

Since $d = 3\lambda$, the angles are not small, and Equations 37-2 do not apply. The interference maxima occur at angles given by Equation 37-1a, $\theta = \sin^{-1}(m\lambda/d) = \sin^{-1}(m/3)$, so only two orders are present, for values of $m = 1$ and 2 ($\theta < 90°$). If we assume that the slit/screen geometry is as shown in Fig. 37-6, then $y = L\tan\theta = L\tan(\sin^{-1}(m/3)) = Lm/\sqrt{9 - m^2}$. (Consider a right triangle with hypotenuse of 3 and opposite side m, or use $\tan\theta = \sin\theta/\sqrt{1 - \sin^2\theta}$.) For $m = 1$ and 2, and $L = 50$ cm, this gives $y_1 = (50$ cm)$(1/\sqrt{8}) = 17.7$ cm, and $y_2 = (50$ cm)$(2/\sqrt{5}) = 44.7$ cm.

Problem

9. For a double-slit experiment with slit spacing 0.25 mm and wavelength 600 nm, at what angular position is the path difference equal to one-fourth of the wavelength?

Solution

If we set the path difference equal to a quarter wavelength, we obtain $d\sin\theta = \lambda/4$, or $\theta \approx \sin\theta = $ 600 nm/4(0.25 mm) $= 6 \times 10^{-4}$ rad $\simeq 0.0344°$.

Problem

11. Laser light at 633 nm falls on a double-slit apparatus with slit separation 6.5 μm. Find the separation between (a) the first and second and (b) the third and fourth bright fringes, as seen on a screen 1.7 m from the slits.

Solution

Since $d \sim 10\lambda$ for this interference process, the small-angle approximation is not particularly accurate, especially for higher orders. The angular position and position on the screen (for the usual slit/screen configuration) for bright fringes are $\theta_m = \sin^{-1}(m\lambda/d)$ and $y_m = L\tan\theta_m$, so the separation of two bright fringes on the screen is $\Delta y_{m_1 m_2} = y_{m_2} - y_{m_1} = L[\tan(\sin^{-1}(m_2\lambda/d)) - \tan(\sin^{-1}(m_1\lambda/d))]$. (a) For $L = 1.7$ m, $d = 6.5$ μm, $\lambda = 633$ nm, $m_1 = 1$ and $m_2 = 2$, one finds $\Delta y_{12} = 17.1$ cm. (b) For $m_1 = 3$ and $m_2 = 4$, and the same other data, one finds $\Delta y_{34} = 20.0$ cm. (The approximate separation implied by Equation 37-2a is $\Delta y = \lambda L/d = 16.6$ cm.)

Section 37-3: Multiple-Slit Interference and Diffraction Gratings

Problem

13. In a 5-slit system, how many minima lie between the zeroth-order and first-order maxima?

Solution

In an N-slit system with slit separation d (illuminated by normally incident plane waves), the main maxima occur for angles $\sin\theta = m\lambda/d$, and minima for $\sin\theta = m'\lambda/Nd$ (excluding m' equal to zero or multiples of N). Between two adjacent maxima, say $m' = mN$ and $(m+1)N$, there are $N-1$ minima. (The number of integers between mN and $(m+1)N$ is $(m+1)N - mN - 1 = N-1$, because the limits are not included.) For $N = 5$, the number of minima is 4.

Problem

15. A 5-slit system with 7.5-μm slit spacing is illuminated with 633-nm light. Find the angular positions of (a) the first two maxima and (b) the third and sixth minima.

Solution

(a) Primary maxima occur at angles $\theta = \sin^{-1}(m\lambda/d)$. The first two (after the central peak, $m = 0$) are for $m = 1$ and 2 at $\theta_1 = \sin^{-1}(633 \text{ nm}/7.5 \text{ μm}) = 4.84°$ and $\theta_2 = \sin^{-1}(2\times633 \text{ nm}/7.5 \text{ μm}) = 9.72°$.
(b) Minima occur at angles $\theta' = \sin^{-1}(m'\lambda/Nd)$, where $m' = \pm1, \pm2, \ldots$, but excluding multiples of $N = 5$, in this case. The third minimum is for $m' = 3$, and the sixth for $m' = 7$ (because $m' = 5$ doesn't count). Then $\theta'_3 = \sin^{-1}(3\lambda/5d) = 2.90°$ and $\theta'_7 = \sin^{-1}(7\lambda/5d) = 6.79°$. (These minima would be difficult to observe because the secondary maxima between them are faint.)

Problem

17. Green light at 520 nm is diffracted by a grating with 3000 lines per cm. Through what angle is the light diffracted in (a) first and (b) fifth order?

Solution

For light normally incident on a diffraction grating, maxima occur at angles $\theta = \sin^{-1}(m\lambda/d)$, where d is the grating spacing (equal to the reciprocal of the number of lines per meter), and m is the order. (a) In first order, $\theta_1 = \sin^{-1}(520 \text{ nm} \times 3000/\text{cm}) = 8.97°$, and (b) in fifth order, $\theta_5 = \sin^{-1}(5 \sin 8.97°) = 51.3°$.

Problem

19. Light is incident normally on a grating with 10,000 lines per cm. What is the maximum order in which (a) 450-nm and (b) 650-nm light will be visible?

Solution

The grating condition is $\sin\theta = m\lambda/d$, and, of course, for the diffracted light to be visible, $\theta < 90°$, or $m\lambda/d < 1$. Therefore, the highest order visible is the greatest integer m less than d/λ. For this grating, $d = 1 \text{ cm}/10^4 = 10^3$ nm, so for $\lambda = 450$ nm and 650 nm, the highest visible orders are less than $10^3/450 = 2.22$ and $10^3/650 = 1.54$, or second and first, respectively.

Problem

21. A solar astronomer is studying the Sun's 589-nm sodium spectral line with a 2500-line/cm grating spectrometer whose fourth-order dispersion puts the wavelength range from 575 nm to 625 nm on a detector. The astonomer is interested in observing simultaneously the so-called calcium-K line, at 393 nm. What order dispersion will put this line also on the detector?

Solution

In fourth order on a grating with spacing $d = 1 \text{ cm}/2500 = 4000$ nm, the wavelength range 575 nm to 625 nm lies in the angular range $\theta = \sin^{-1}(4\lambda/d) = \sin^{-1}(0.575) = 35.1°$ to $\sin^{-1}(0.625) = 38.7°$. The Ca-K line of order m also lies in the range if $0.575 < m\,\lambda_K/d = m(0.393/4) < 0.625$, or $5.85 < m < 6.36$, which is satisfied for $m = 6$.

Problem

23. Estimate the number of lines per cm in the grating used to produce Fig. 37-15.

Solution

The number of lines per cm ($1/d$ in cm^{-1}) is easily estimated from the angular position of the central 550-nm line in a particular order, as shown in the figure; that is, $1/d = \sin\theta/m\lambda$. For example, in fifth order, this line is at $\theta = 61°$ (average of right and left values), so $1/d = \sin 61°/5(550 \text{ nm}) = 3.18 \times 10^3/\text{cm}$ or about 3200 lines/cm.

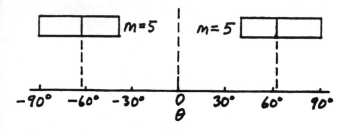

FIGURE 37-15 Problem 23 Solution.

Problem

25. When viewed in sixth-order, the 486.1-nm hydrogen-β spectral line is flanked by another line that appears at the position of 484.3 nm in the sixth-order spectrum. Actually the line is from a different order of the spectrum. What are the possible visible wavelengths of this line?

Solution

To appear at the same angular position as wavelength 484.3 nm in sixth order, the possible other wavelengths and orders must satisfy $m\lambda = 6(484.3 \text{ nm})$. (Then $d \sin \theta$ is the same.) If λ is in the visible range, then $400 \text{ nm} < \lambda = 6(484.3 \text{ nm})/m < 700 \text{ nm}$, or $6(484.3 \div 700) = 4.15 < m < 7.26 = 6(484.3/400)$. Thus, $m = 5$ and 7 are possible orders, corresponding to wavelengths $6(484.3 \text{ nm})/m = 415 \text{ nm}$, and 581 nm respectively.

Problem

27. Echelle spectroscopy uses relatively course gratings in high order. Compare the resolving power of an 80-line/mm echelle grating used in twelfth order with a 600-line/mm grating used in first order, assuming the two have the same width.

Solution

If the echelle and grating have the same width, then the number of lines in each is proportional to the given spacings, $N/N' = 80/600$. The ratio of the resolving powers (Equation 37-6) is then $mN/m'N' = 12 \times 80/1 \times 600 = 1.6$, so the echelle has about 60% greater resolving power than the grating.

Problem

29. You wish to resolve the calcium-H line at 396.85 nm from the hydrogen-ε line at 397.05 nm

in a first-order spectrum. To the nearest hundred, how many lines should your grating have?

Solution

From Equation 37-6, $N = \lambda/m \, \Delta\lambda \approx 397/(1 \times 0.2) = 1985 \approx 2000$ lines. (This is the number of lines that must be illuminated.)

Section 37-4: Thin Films and Interferometers

Problem

31. Find the minimum thickness of a soap film ($n = 1.33$) in which 550-nm light will undergo constructive interference.

Solution

The condition for constructive interference from a soap film is Equation 37-8a, in which the minimum thickness corresponds to the integer $m = 0$. Thus, $2nd_{min} = \frac{1}{2}\lambda$, or $d_{min} = \lambda/4n = 550 \text{ nm}/4(1.33) = 103 \text{ nm}$. (Recall that Equation 37-8a applies to normal incidence on a thin film in air.)

Problem

33. Monochromatic light shines on a glass wedge with refractive index 1.65, and enhanced reflection occurs where the wedge is 450 mm thick. Find all possible values for the wavelength in the visible range.

Solution

Equation 37-8a gives the condition for constructive interference (enhanced reflection) from a given thickness of glass surrounded by air, so $\lambda = 4nd/(2m + 1) = 4(1.65)(450 \text{ nm})/(2m + 1) = 2970 \text{ nm}/(2m + 1)$. Integers giving wavelengths in the visible range (400 to 700 nm) are $m = 2$ and 3, corresponding to $\lambda = 594 \text{ nm}$ and 424 nm, respectively.

Problem

35. As a soap bubble ($n = 1.33$) evaporates and thins, the reflected colors gradually disappear. (a) What is its thickness just as the last vestige of color vanishes? (b) What is the last color seen?

Solution

The minimum thickness of the bubble, which produces interference colors, is $d_{min} = \lambda_{min}/4n$, where λ_{min} is the shortest visible wavelength, normally 400 nm violet light. (See the solution to Problem 31.) Thus, $d_{min} = 400 \text{ nm}/4(1.33) = 75.2 \text{ nm}$.

Problem

37. Light reflected from a thin film of acetone ($n = 1.36$) on a glass plate ($n = 1.5$) shows maximum reflection at 500 nm and minimum at 400 nm. Find the minimum possible film thickness.

Solution

The index of refraction of acetone is less than that of glass, so there are 180° phase changes at both boundaries of the film. Then, for normal incidence, Equation 37-8a describes destructive and Equation 37-8b constructive interference. Therefore $2nd = (m + \frac{1}{2})400$ nm and $2nd = m'(500$ nm$)$. These are consistent for $(m + \frac{1}{2})400 = m'(500)$, or $2m + 1 = 5m'/2$. The smallest integer values satisfying this are $m' = 2$ and $m = 2$. When either is substituted in the interference conditions, along with $n = 1.36$, one finds $d = 1.25(400$ nm$)/1.36 = 500$ nm$/1.36 = 368$ nm.

Problem

39. An oil film with refractive index 1.25 floats on water. The film thickness varies from 0.80 μm to 2.1 μm. If 630-nm light is incident normally on the film, at how many locations will it undergo enhanced reflection?

Solution

In a thin film of oil between air and water ($n_{\text{air}} < n_{\text{oil}} < n_{\text{water}}$), there are 180° phase changes for reflection at both boundaries. Therefore, for normally incident light, Equation 37-8b gives the condition for constructive interference, $d = m\lambda/2n = m(630$ nm$) \div 2(1.25) = (0.252\ \mu\text{m})m$. Varying thickness of 0.80 μm $\leq d \leq 2.1\ \mu$m implies that $3.17 \leq m \leq 8.33$. Since m is an integer, $4 \leq m \leq 8$, or 5 bright maxima occur at locations corresponding to the allowed integers from 4 to 8.

Problem

41. Two perfectly flat glass plates are separated at one end by a piece of paper 0.065 mm thick. A source of 550-nm light illuminates the plates from above, as shown in Fig. 37-43. How many bright bands appear to an observer looking down on the plates?

Solution

Equation 37-8a applies to constructive interference for normally incident light on the thin, wedge-shaped film of air between glass surfaces (although in this case, the 180° phase change affects rays reflected from the bottom surface of the film). Thus, $d = (m + \frac{1}{2}) \times \lambda/2n_{\text{air}} = (m + \frac{1}{2})(550$ nm$)/2 = (m + \frac{1}{2})(275$ nm$)$. The

thickness of the film varies between 0 and 0.065 mm, so m varies between 0 and $[(0.065$ mm$/275$ nm$) - \frac{1}{2}] = 235$. Thus, there are 236 bright bands visible from above (since $m = 0$ counts as the first bright fringe).

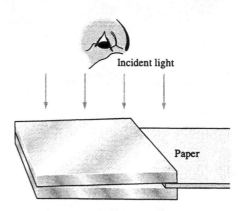

FIGURE 37-43 Problems 41, 42, 43, 72.

Problem

43. You apply a slight pressure with your finger to the upper of a pair of glass plates forming an air wedge as in Fig. 37-43. The wedge is illuminated from above with 500-nm light, and you place your finger where, initially, there is a dark band. If you push gently so the band becomes light, then dark, then light again, by how much have you deflected the plate?

Solution

The difference in the thickness of air in the wedge between a bright and an adjacent dark band is $\frac{1}{4}\lambda$ (one-quarter wavelength in air), so the upper plate was depressed by $\frac{3}{4}\lambda = 375$ nm.

Problem

45. What is the wavelength of light used in a Michelson interferometer if 550 bright fringes go by a fixed point when the mirror moves 0.150 mm?

Solution

Each bright fringe shift corresponds to a path difference of one wavelength. The path changes by twice the distance moved by the mirror. Thus, $550\lambda = 2 \times 0.15$ mm, or $\lambda = 545$ nm.

Problem

47. The evacuated box of the previous problem is filled with chlorine gas, whose refractive index is 1.000772. How many bright fringes pass a fixed point as the tube fills?

Solution

Since the wavelength of the light is different in a gas (e.g., chlorine or air) and in vacuum ($\lambda_{gas} = \lambda_{vac}/n_{gas}$), there is a difference in the number of wave cycles in the enclosed interferometer arm when the box is evacuated or filled with gas. The light travels the length of the arm twice, out and back, and each cycle of difference results in one fringe shift. Thus, the number of fringes in the shift is

$$\frac{2\times42.5 \text{ cm}}{\lambda_{gas}} - \frac{2\times42.5 \text{ cm}}{\lambda_{vac}} = \frac{(n_{gas} - 1)(85 \text{ cm})}{(641.6 \text{ nm})} \approx 1022,$$

where $n_{gas} - 1 = 7.72\times10^{-4}$ for chlorine gas, and we dropped approximately three quarters of a fringe.

Sections 37-6 and 37-7:   Single-Slit Diffraction and the Diffraction Limit

Problem

49. For what ratio of slit width to wavelength will the first minima of a single-slit diffraction pattern occur at $\pm90°$

Solution

When $\theta = 90°$, in Equation 37-9, and $m = 1$ for the first minimum, then $a/\lambda = 1$.

Problem

51. A beam of parallel rays from a 29-MHz citizen's band radio transmitter passes between two electrically conducting (hence opaque to radio waves) buildings located 45 m apart. What is the angular width of the beam when it emerges from between the buildings?

Solution

Take the width of the diffracted beam to be the angular separation between the first minima. These occur at $\theta = \pm\sin^{-1}(\lambda/a)$ (see Exercise following Example 37-6). Thus, $\Delta\theta = 2|\theta| = 2\sin^{-1}[(3\times10^{-8} \text{ m/s})/(29 \text{ MHz})(45 \text{ m})] = 26.6°$.

Problem

53. Find the intensity as a fraction of the central peak intensity for the second secondary maximum in single-slit diffraction, assuming the peak lies midway between the second and third minima.

Solution

The second and third minima lie at angles $\sin\theta_2 = 2\lambda/a$ and $\sin\theta_3 = 3\lambda/a$. If we take the mid-value (as in Example 37-7) to be at $\sin\theta = 5\lambda/2a$, then the intensity at this angle, relative to the central intensity, is $\bar{S}/\bar{S}_0 = [\sin(5\pi/2)/(5\pi/2)]^2 = 4/25\pi^2 = 1.62\times10^{-2}$.

Problem

55. The movie *Patriot Games* has a scene in which CIA agents use spy satellites to identify individuals in a terrorist camp. Suppose that a minimum resolution for distinguishing human features is about 5 cm. If the spy satellite's optical system is diffraction-limited, what diameter mirror or lens is needed to achieve this resolution from an altitude of 100 km? Assume a wavelength of 550 nm.

Solution

The angle subtended by a human feature 5 cm across at 100 km is $\theta_{min} = 5 \text{ cm}/100 \text{ km} = 5\times10^{-7}$ (radians). The Rayleigh criterion for a diffraction-limited telescope, using light of wavelength $\lambda = 550$ nm, requires an aperture of $D = 1.22\lambda/\theta_{min} = 1.22\times(550 \text{ nm})/(5\times10^{-7}) = 1.34$ m (see Equation 37-13b). (Atmospheric turbulence would limit the resolution to no better than $\frac{1}{2}'' = 2.4\times10^{-6}$ radians.)

Problem

57. A camera has an $f/1.4$ lens, meaning that the ratio of focal length to lens diameter is 1.4. Find the smallest spot diameter (defined as the diameter of the first diffraction minimum) to which this lens can focus parallel light with 580-nm wavelength.

Solution

The diffraction limit for a lens opening of diameter D, focusing light of wavelength λ is $\theta_{min} = 1.22 \lambda/D$. The radius of a spot, at the focal length of the lens, with this angular spread, is $r = f\theta_{min}$ (the spot radius equals the distance between the central maximum and first minimum). The minimum spot diameter is, therefore, $2f\theta_{min} = 2(1.22)\lambda f/D = 2(1.22)(550 \text{ nm})\times 1.4 \simeq 2.0$ μm (since f/D is the f-ratio).

Problem

59. While driving at night, your eyes' irises have dilated to 3.1-mm diameter. If your vision were diffraction-limited, what would be the greatest distance at which you could see as distinct the two headlights of an oncoming car, which are spaced 1.5 m apart? Take $\lambda = 550$ nm.

Solution

If we use the Rayleigh criterion (Equation 37-13b for small angles) to estimate the diffraction-limited angular resolution of the eye, at a pupil diameter of 3.1 mm, in light of wavelength 550 nm, we obtain $\theta_{min} = 1.22(550 \text{ nm})/(3.1 \text{ mm}) = 2.16\times10^{-4} \approx 45''$.

(Actually, the wavelength inside the eye is different, $\lambda' = \lambda/n$, because of the average index of refraction of the eye.) This angle corresponds to a linear separation of $y = 1.5$ m at a distance of $r = y/\theta_{\min} = 1.5$ m$/2.16 \times 10^{-4} = 6.93$ km $\simeq 4$ mi. Although other factors determine visual acuity, this is a reasonable ballpark estimate.

Problem

61. Under the best conditions, atmospheric turbulence limits the resolution of ground-based telescopes to about 1 arc second (1/3600 of a degree) as shown in text Fig. 37-45. For what aperture sizes is this limitation more severe than that of diffraction at 550 nm? Your answer shows why large, ground-based telescopes do not produce better images than small ones, although they do gather more light.

Solution

The aperture satisfying the Rayleigh criterion (Equation 37-13b) for $\theta_{\min} = 1'' = \pi/(180 \times 3600) = 4.85 \times 10^{-6}$ (radians), at the given wavelength, is $D = 1.22(550$ nm$)/4.85 \times 10^{-6} = 13.8$ cm, or about $5\frac{1}{2}$ in. The resolution of all larger-diameter ground-based telescopes is limited by atmospheric conditions at this wavelength.

Paired Problems

Problem

63. Find the total number of lines in a 2.5-cm-wide diffraction grating whose third-order spectrum has the 656-nm hydrogen-α spectral line at an angular position of $37°$.

Solution

The grating condition for normally incident light (same as Equation 37-1a) and the given data imply a grating spacing of $d = m\lambda/\sin\theta = 3(656$ nm$)\div \sin 37° = 3.27$ μm, or a grating constant of $d^{-1} = 3.06 \times 10^3$ lines/cm. On a grating 2.5 cm wide, the total number of lines is $(3.06 \times 10^3$/cm$)(2.5$ cm$) = 7.65 \times 10^3$ (=7645, to within a hundredth of a line).

Problem

65. A 400-line/mm diffraction grating is 3.5 cm wide. Two spectral lines whose wavelengths average to 560 nm are just barely resolved in the fourth-order spectrum of this grating. What is the difference between their wavelengths?

Solution

The resolving power of a grating is $\lambda/\Delta\lambda = mN$ (see Equation 37-6). If the entire width of the grating is illuminated, $N = (400$/mm$)(3.5$ cm$)$ and $\Delta\lambda = 560$ nm$/4(14,000) = 0.01$ nm.

Problem

67. A thin film of toluene ($n = 1.49$) floats on water. What is the minimum film thickness if the most strongly reflected light has wavelength 460 nm?

Solution

Since $n_{\text{toluene}} > n_{\text{water}} > n_{\text{air}}$, there is a $180°$ phase change for reflection only at the air/toluene interface, and not at the toluene/water interface, of the film. Then Equation 37-8a applies for constructive interference (of normally incident rays) and $d = (m + \frac{1}{2})\lambda/2n = (m + \frac{1}{2})(460$ nm$)/(2 \times 1.49) = (2m+1)(77.2$ nm$)$. The minimum thickness is 77.2 nm, although odd multiples of this are also possible.

Problem

69. What diameter optical telescope would be needed to resolve a Sun-sized star 10 light-years from Earth? Take $\lambda = 550$ nm. Your answer shows why stars appear as point sources in optical astronomy.

Solution

The angular size of the sun, at a distance of 10 ly, is only $\theta = 2(6.96 \times 10^5$ km$)/10(9.46 \times 10^{12}$ km$) = 1.47 \times 10^{-8}$ (see Appendix E), so even a diffraction-limited space telescope would need an aperture of $D = 1.22\lambda/\theta = 1.22(550$ nm$)\div (1.47 \times 10^{-8}) = 45.6$ m to resolve it in visible light (see Equation 37-13b). (However, ground-based optical interferometers are currently being developed.)

Supplementary Problems

Problem

71. White light shines on a 250-nm-thick layer of diamond ($n = 2.42$). What wavelength *of visible light* is most strongly reflected.

Solution

If the diamond layer is surrounded by air (or material of lesser index of refraction) then Equation 37-8a is the condition for constructive interference for normally incident light. Thus, maximum intensity occurs for wavelengths $\lambda = 4nd/(2m+1) = 4(2.42)(250$ nm$)\div (2m+1) = 2420$ nm$/(2m+1)$. The only integer which gives a visible wavelength is $m = 2$ (i.e., $2420/700 < 2m+1 < 2420/400$), for which $\lambda = 2420$ nm$/5 = 484$ nm.

Problem

73. In Fig. 37-23 the mth Newton's ring appears a distance r from the center of the lens. Show that the curvature radius of the lens is given approximately by $R = r^2/(m + \frac{1}{2})\lambda$, where the approximation holds when the thickness of the air space is much less than the curvature radius.

Solution

As explained in the solution to Problem 41, there is constructive interference between rays reflected (normally) from the upper and lower surfaces of the film of air which separates the lens and the glass plate, producing a bright Newton's ring when $2d = (m + \frac{1}{2})\lambda$ (Equation 37-8a with $n_{air} = 1$). The thickness of the air film at the position of the ring (a distance r from the central axis of the lens) is $d = R - R\cos\theta$, where $\sin\theta = r/R$. If $d \ll R$, then $\theta \ll 1$, and the small-angle approximation gives $\theta \approx \sin\theta = r/R$, and $\cos\theta = \sqrt{1 - \sin^2\theta} \approx 1 - \frac{1}{2}\theta^2$. Then $d \approx R(1 - \cos\theta) \approx R\cdot\frac{1}{2}(r/R)^2 = r^2/2R$. Substituting this into the interference condition, one gets $R = r^2/(m + \frac{1}{2})\lambda$.

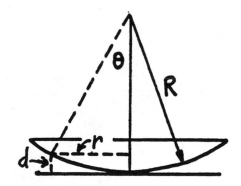

Problem 73 Solution.

Problem

75. How many rings would be seen if the system of the preceding problem were immersed in water $(n = 1.33)$?

Solution

When the interference pattern of Newton's rings, described in Problems 73 and 74, is generated by a thin wedge of water between two glass surfaces $(n = 1.33 < n_{glass})$, one need only replace λ by λ/n in the result in Problem 73. The maximum distance of a bright fringe from the central axis of the lens in Problem 74 is still $r_{max} = 1.25$ cm; therefore $m + \frac{1}{2} \leq nr^2_{max}/\lambda R = (1.33)(1.25 \text{ cm})^2/(500 \text{ nm})\times (7.5 \text{ cm}) = 5.54\times10^3$. There is a bright fringe for each integer m from 0 to 5.54×10^3, provided the incident

light is sufficiently coherent. The small-angle approximation used in Problem 73 is barely justified here, since $\theta_{max} = \sin^{-1}(r_{max}/R) = 9.6°$. The more exact result is $m + \frac{1}{2} \leq 2nR(1 - \cos\theta_{max})/\lambda = 5.58\times10^3$.

Problem

77. The signal from a 103.9-MHz FM radio station reflects off a building 400 m away, effectively producing two sources of the same signal. You're driving at 60 km/h along a road parallel to a line between the station's antenna and the building and located a perpendicular distance of 6.5 km from them. How often does the signal appear to fade when you're driving roughly opposite the transmitter and building?

Solution

If we assume a constant phase difference between the direct and reflected waves, essentially a two-slit interference pattern is produced along the road traveled by the car, with minima (or maxima) spaced approximately $\Delta y = \lambda L/d$ apart, along the road and near the perpendicular bisector of the line between the sources. (A constant phase difference in Equations 37-2a and b cancels out in calculating the spacing.) The time between dead spots on this section of road, for a passing car with speed v, is $\Delta t = \Delta y/v = \lambda L/vd$. Using $\lambda = c/f$ and the numbers given, we find $\Delta t = (3\times10^8 \text{ m/s})(6.5 \text{ km})(3.6 \text{ s/60 m})\div (400 \text{ m})(103.9 \text{ MHz}) = 2.82$ s. (The actual positions of the maxima and minima would, of course, depend on the phase difference between the direct and reflected waves.)

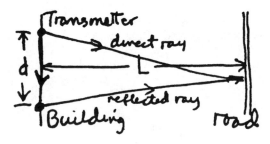

Problem 77 Solution.

Problem

79. The component of a star's velocity in the radial direction relative to Earth is to be measured using the doppler shift in the hydrogen-β spectral line, which appears at 486.1 nm when the source is stationary relative to the observer. What is the minimum speed that can be detected by observing

in first order with a 10,000-line/cm grating 5.0 cm across? *Hint:* See Equation 17-9.

Solution

The resolving power of this grating, in first order, if fully illuminated, is $mN = 1(10^4/\text{cm})(5 \text{ cm}) = 5 \times 10^4$ (see Equation 37-6). This means that two closely spaced wavelengths are resolvable if $|\lambda/\Delta\lambda|$ is less than the resolving power, or equivalently, $|\Delta\lambda/\lambda| > (mN)^{-1} = 2 \times 10^{-5}$. The Doppler shift from a light source moving along the line of sight (the radial direction) with speed v is $|\Delta\lambda/\lambda| \approx v/c$. (This approximate formula is valid if $v \ll c$.) Therefore, the Doppler shift is resolvable with this grating if $v/c > (mN)^{-1}$, or $v > (2 \times 10^{-5})(3 \times 10^5 \text{ km/s}) = 6 \text{ km/s}$. (The resolving power and the Doppler formula both depend only on the relative shift, so this result is independent of the observed wavelength.)

Problem

81. In a double-slit experiment, a thin glass plate with refractive index 1.56 is placed over one of the slits.

The fifth bright fringe now appears where the second dark fringe previously appeared. How thick is the plate if the incident light has wavelength 480 nm?

Solution

Without the glass plate, the second dark fringe appears at an angular position given by $d \sin\theta = 3\lambda/2$ (Equation 37-1b with $m = 1$). The glass plate, with thickness Δ, introduces an additional optical path difference of $(\Delta/\lambda_{\text{glass}}) - (\Delta/\lambda) = (n-1)(\Delta/\lambda)$, where $n = \lambda/\lambda_{\text{glass}}$ is the refractive index of the plate. The fifth bright fringe occurs at an angular position for which the total path difference is five wavelengths, or $(d \sin\theta/\lambda) + (n-1)(\Delta/\lambda) = 5$ (this is a modified Equation 37-1a with $m = 5$). Since the angle is the same in both cases, we can substitute $d \sin\theta/\lambda = \frac{3}{2}$ to obtain $5 = (\frac{3}{2}) + (n-1)(\Delta/\lambda)$, or $\Delta = 7\lambda/2(n-1) = 7(480 \text{ nm})/2(0.56) = 3.00 \ \mu\text{m}$.

PART 5   CUMULATIVE PROBLEMS

Problem

1. A *grism* is a grating ruled onto a prism, as shown in Fig. 1. The grism is designed to transmit undeviated one wavelength of the spectrum in a given order, as refraction in the prism compensates for the deviation at the grating. Find an equation relating the separation d of the grooves that constitute the grating, the wedge angle α of the prism, the refractive index n, the undeviated wavelength λ_0, and order m_0.

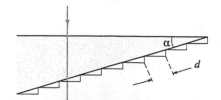

FIGURE 1 A grism (Cumulative Problem 1).

Solution

Consider the interference of rays of wavelength λ, from the same plane wavefront incident normally on the

upper surface of the grism in Fig. 1, which emerge from adjacent slits of the grating (grooves). The two rays will be in phase (producing an intensity maximum) when their path lengths, in wavelengths, differ by an integer, m. From the sketch, $|PQ/\lambda_n -$

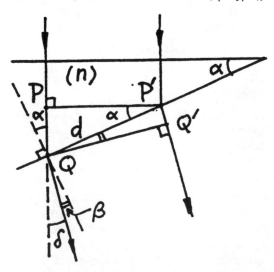

Cumulative Problem 1 Solution.

$P'Q'/\lambda| = m$, where $\lambda_n = \lambda/n$ is the wavelength in the grism, and n is its refractive index. Since $PQ = d\sin\alpha$, $P'Q' = d\sin\beta$, where $d = P'Q$ is the groove separation, and $\delta = \alpha - \beta$ is the deviation, this condition can be written as $|n\sin\alpha - \sin(\alpha - \delta)| = m\lambda/d$ (the grism equation). That a given wavelength (λ_0) be transmitted undeviated ($\delta = 0$) requires $(n-1)\sin\alpha = m_0\lambda_0/d$, which is the sought-for equation.

Problem

3. A closed cylindrical tube whose glass walls have negligible thickness measures 5.0 cm long by 5.0 mm in diameter. It is filled with water, initially at 15°C, and placed with its long dimension in one arm of a Michelson interferometer. The water is not perfectly transparent, and it absorbs 3.2% of the light energy incident on it. The laser power incident on the water is 50 mW, and the wavelength is 633 nm. The refractive index of water in the vicinity of 15°C is given approximately by $n = 1.335 - 8.4 \times 10^{-5}\,T$, where T is the temperature in °C. As the water absorbs light energy, how long does it take the interference pattern to shift by one whole fringe?

Solution

The interference pattern shifts by one fringe when the optical path length through the water (the distance out and back across the tube, divided by the wavelength in water) changes by unity. (The phase difference is the optical path length times 2π radians.) The wavelength in water depends on the temperature, through the refractive index, i.e., $\lambda(T) = \lambda/n(T)$, and $n(T) = a - bT$, with a and b given. Therefore, a shift of one fringe implies:

$$\frac{2\ell}{\lambda(15°C)} - \frac{2\ell}{\lambda(T)} = 1 = \frac{2\ell}{\lambda}(n(15°C) - n(T)) = \frac{2\ell}{\lambda}b\,\Delta T,$$

or $\Delta T = \lambda/2b\ell$, where ΔT is the change in temperature from 15°C, $\ell = 5$ cm is the length of the tube, $\lambda = 633$ nm, and $b = 8.4 \times 10^{-5}/°C$. From Equation 19-5, $\Delta T = \Delta Q/mc_W$, where $\Delta Q = \mathcal{P}_{abs}\Delta t = $ (power absorbed from laser)(time), $m = \frac{1}{4}\pi D^2\ell\rho = $ (volume of tube) × (density of water), and c_W is the specific heat of water. Thus, the time required for a shift of one fringe is $\Delta t = mc_W\Delta T/\mathcal{P}_{abs} = \frac{1}{4}\pi D^2\rho c_W\lambda/2b\mathcal{P}_{abs}$. Numerically,

$$\Delta t = \frac{\frac{1}{4}\pi(5\text{ mm})^2(10^3\text{ kg/m}^3)(4186\text{ J/kg·K})(633\text{ nm})}{2(0.032\times50\text{ mW})(8.4\times10^{-5}/\text{ K})}$$

$$= 193\text{ s} = 3.22\,\text{min}.$$

Problem

5. In one type of optical fiber, called a *graded-index fiber*, the refractive index varies in a way that results in light rays being guided along the fiber on curved trajectories, rather than undergoing abrupt reflections. Figure 3 shows a simple model that demonstrates this effect; it also describes the basic optical effect in mirages. A slab of transparent material has refractive index $n(y)$ that varies with position y perpendicular to the slab face. A light ray enters the slab at $x = 0$, $y = 0$, making an angle θ_0 with the normal just inside the slab. The refractive index at this point is $n(y = 0) = n_0$. (a) By writing $\sin\theta$ in Snell's law in terms of the components dx and dy of the ray path, show that that path (written in the form of x as a function of y) is given by

$$x = \int_0^y \frac{n_0\sin\theta_0}{\sqrt{[n(y)]^2 - n_0^2\sin^2\theta_0}}\,dy.$$

(b) Suppose $n(y) = n_0(1 - ay)$, where $n_0 = 1.5$ and $a = 1.0 = \text{mm}^{-1}$. If $\theta_0 = 60°$, find an explicit expression for x as a function of y, and plot your result to give the actual ray path. Explain the shape of your curve in terms of what happens when the ray reaches a point where $n(y) = n_0\sin\theta_0$. What happens beyond this point?

Solution

(a) Snell's law (Equation 35-3) relates the refractive index and the direction of the ray path, at any y, to the values at $y = 0$; that is, $n\sin\theta = n_0\sin\theta_0$ (n and θ are both functions of y). The slope of the ray path (with the y-axis) is $\tan\theta = dx/dy = \sin\theta/\sqrt{1 - \sin^2\theta}$, so substitution and separation of variables yields $dx = n_0\sin\theta_0[n^2 - n_0^2\sin^2\theta_0]^{-1/2}\,dy$. The given expression for x as a function of y follows immediately

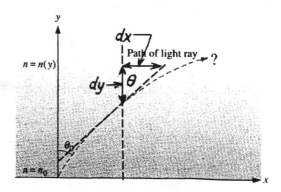

FIGURE 3 Cumulative Problem 5 Solution.

by integration from the origin ($x = 0 = y$). (b) If $n(y) = n_0(1 - ay)$, the integral in part (a) can be evaluated by using the last entry in the first column of the integral table in Appendix A. An intermediate substitution of $z = (1 - ay)/\sin\theta_0$ facilitates this, with the following result:

$$x = \int_0^y \frac{\sin\theta_0 \, dy}{\sqrt{(1 - ay)^2 - \sin^2\theta_0}} = \left(\frac{\sin\theta_0}{a}\right) \int_z^{1/\sin\theta_0} \frac{dz}{\sqrt{z^2 - 1}}$$

$$= \left(\frac{\sin\theta_0}{a}\right) \left[\ln\left(\frac{1 + \cos\theta_0}{\sin\theta_0}\right) - \ln\left(z + \sqrt{z^2 - 1}\right)\right].$$

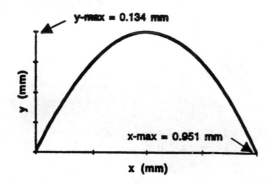

Cumulative Problem Part 5-5 Chart 1.

The given value $\theta_0 = 60°$ (or $\sin\theta_0 = \frac{1}{2}\sqrt{3}$) reduces this further to $x = (\sqrt{3}/2a)[\ln\sqrt{3} - \ln(z + \sqrt{z^2 - 1})]$, where $z = 2(1 - ay)/\sqrt{3}$. (Note that if the ln's are combined and the identity $1/(z + \sqrt{z^2 - 1}) = z - \sqrt{z^2 - 1}$ is used, the expression in the "Answers to Odd-Numbered Problems" is obtained, since

$$\frac{\sqrt{3}}{z + \sqrt{z^2 - 1}} = \sqrt{3}\left[\frac{2(1 - ay)}{\sqrt{3}} - \sqrt{\frac{4(1 - ay^2)}{3} - 1}\right]$$

$$= 2\left[1 - ay - \sqrt{\frac{1}{4} - 2ay + a^2 y^2}\right].$$

However, this expression is not well suited for plotting.) At this point, it is convenient to exploit the relation between the natural logarithm and inverse hyperbolic functions, namely, $\ln(z + \sqrt{z^2 - 1}) = \cosh^{-1} z$, which allows one to express y(or z) as a function of x. The result is $z = \cosh(\ln\sqrt{3} - 2ax/\sqrt{3})$, or $ay = 1 - (\sqrt{3}/2)\cosh(\ln\sqrt{3} - 2ax/\sqrt{3})$, which is plotted below, for $a = 1$ mm^{-1}. Since $\sin\theta = (n/n_0)\sin\theta_0 \le 1$, the above expressions are valid for $0 \le y \le (1 - \sin\theta_0)/a = 0.134$ mm (or $1 \le z \le 1/\sin\theta_0 = 2/\sqrt{3}$). The ray path is bent back towards the x-axis after reaching its maximum penetration, where $n = n_0 \sin\theta_0$.

PART 6   MODERN PHYSICS

CHAPTER 38   RELATIVITY

Section 38-2:   Matter, Motion, and the Ether

Problem

1. Consider an airplane flying at 800 km/h airspeed between two points 1800 km apart. What is the round-trip travel time for the plane (a) if there is no wind? (b) if there is a wind blowing at 130 km/h perpendicular to a line joining the two points? (c) if there is a wind blowing at 130 km/h along a line joining the two points? Ignore relativistic effects. (Why are you justified in doing so?)

Solution

Since the velocities are small compared to c, we can use the non-relativistic Galilean transformation of velocities in Equation 3-10, $\mathbf{u} = \mathbf{u'} + \mathbf{v}$, where $\mathbf{u}$ is the velocity relative to the ground (S), $\mathbf{u'}$ is that relative to the air (S'), and $\mathbf{v}$ is that of S' relative to S (in this case, the wind velocity). We used a notation consistent with that in Equations 38-11 and 12. (a) If $\mathbf{v} = 0$ (no wind), then $\mathbf{u} = \mathbf{u'}$ (ground speed equals air speed), and the round-trip travel time is $t_a = 2d/u = 2(1800 \text{ km})/(800 \text{ km/h}) = 4.5$ h. (b) If $\mathbf{v}$ is perpendicular to $\mathbf{u}$, then $u'^2 = u^2 + v^2$, or $u = \sqrt{800^2 - 130^2}$ km/h $= 789$ km/h, and the round-trip travel time is $t_b = 2d/u = 4.56$ h. (c) If $\mathbf{v}$ is parallel or antiparallel to $\mathbf{u}$ on alternate legs of the round-trip, then $u = u' \pm v$ and the travel time is

$$t_c = \frac{d}{u' + v} + \frac{d}{u' - v} = \frac{1800 \text{ h}}{800 + 130} + \frac{1800 \text{ h}}{800 - 130}$$
$$= 4.62 \text{ h}.$$

Note that $t_a < t_b < t_c$, as mentioned following Equation 38-2.

Problem

3. Figure 38-30 shows a plot of James Bradley's data on the aberration of light from the star γ Draconis, taken in 1727–1728. (a) From the data, determine the magnitude of Earth's orbital velocity. (b) The data very nearly fit a perfect sine curve. What does this say about the shape of Earth's orbit?

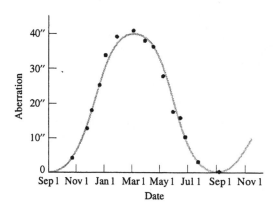

FIGURE 38-30 Problem 3.

Solution

(a) The angle of aberration, α, is defined in the reference frame of the Sun, S, whose motion we need not consider. Since $v/c \ll 1$ for the orbital speed of the Earth, we can use non-relativistic expressions, such as Equation 3-10, which says that the apparent velocity of starlight in the Earth's frame, S', is the vector difference of its velocity in S and the orbital velocity, or $\mathbf{c'} = \mathbf{c} - \mathbf{v}$. The shift in the position of

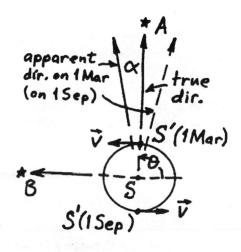

Problem 3 Solution. (1)

distant stars (with no measurable parallax) observed perpendicular to **v** (say on 1 Mar), relative to stars seen parallel to **v**, is α, which reverses after six months. The maximum annual difference, $\alpha - (-\alpha) = 2\alpha$, is about $40''$ in Fig. 38-30. From the vector diagram, $v/c = \tan\alpha \approx \alpha = 20''$. Therefore, $v = (3\times10^5 \text{ km/s})(20'')(1 \text{ rad}/206,265'') = 29 \text{ km/s}$. (b) If the Earth's orbit were circular, the component of its velocity perpendicular to the direction of stars at A would vary as $v\sin\theta$ over a year ($\theta = 90°$ on 1 Mar, as shown). Therefore, the aberration would vary as $\alpha = (v/c)\sin\theta$.

Problem 3 Solution. (2)

Problem

5. Suppose the speed of light differed by 100 m/s in two perpendicular directions. How long should the arms be in a Michelson interferometer if this difference is to cause the interference pattern to shift one-half cycle (i.e., a light fringe shifts to where a dark one was) relative to the pattern if there were no speed difference? Assume 550-nm light.

Solution

For equal length arms, different velocities of light, say $c_\| \neq c_\perp$, introduce a phase difference of half a cycle when $\pm\pi = \Delta\phi = k_\|2L - k_\perp 2L = 2\omega L(c_\|^{-1} - c_\perp^{-1})$ (recall that $k = \omega/c$). Since $\Delta c = |c_\| - c_\perp| \ll c$, this can be written as $2\omega L\,\Delta c/c^2 = \pi$. If we use $\omega = 2\pi f = 2\pi c/\lambda$ (think of λ and c as average values for the two directions), then $2(2\pi c/\lambda)L\,\Delta c/c^2 = \pi$, or $L = \lambda c \div (4\,\Delta c) = (550 \text{ nm}\times3\times10^8 \text{ m/s})/(4\times100 \text{ m/s}) = 41.3$ cm.

Section 38-4:   Space and Time in Relativity
Problem

7. How long would it take a spacecraft traveling at 65% of the speed of light to make the 5.8×10^9 km journey from Earth to Pluto according to clocks (a) on Earth and (b) on the spacecraft?

Solution

(a) On Earth (system S, where the velocity of Pluto is much less than $0.65c$) $\Delta t = \Delta x/v = (5.8\times10^9 \text{ km}) \div (0.65\times3\times10^8 \text{ m/s}) = 2.94\times10^4 \text{ s} = 8.26$ h. (b) Time dilation (Equation 38-4) means that the time interval measured in the spacecraft (system S') is $\Delta t' = \Delta t\times\sqrt{1 - v^2/c^2} = (2.94\times10^4 \text{ s})\sqrt{1 - (0.65)^2} = 2.26\times10^4 \text{ s} = 6.28$ h. (Note that $\Delta t' = (\Delta x/v)\times\sqrt{1 - v^2/c^2} = \Delta x'/v$, so Lorentz contraction could have been used to reach the same result.)

Problem

9. Earth and Sun are 8.3 light-minutes apart, as measured in their rest frame. (a) What is the speed of a spacecraft that makes the trip in 5.0 min according to its on-board clocks? (b) What is the trip time as measured by clocks in the Earth-Sun frame?

Solution

Note that the distance is given in the system S, where the Earth and the Sun are practically at rest (the orbital speed of the earth is very small compared to the speed of light or the speed of the spacecraft), but the time interval is given in system S', where the spacecraft is at rest; that is, $\Delta x = 8.3$ c·min, and $\Delta t' = 5$ min. (One light-minute, the distance light travels in one minute, equals c multiplied by a minute, so c· min is a convenient notation for this light-unit.) (a) Equations 38-4 and 5, for time dilation and Lorentz contraction, relate the given quantities to $\Delta x'$ and Δt, so that the spacecraft's speed, $v = \Delta x/\Delta t = \Delta x'/\Delta t'$, can be determined. For example, if we use Equation 38-8 for shorthand notation, Equations 38-4 and 5 become $\Delta x = \gamma\Delta x'$ and $\Delta t = \gamma\Delta t'$. Then $v = \Delta x \div (\gamma\Delta t') = (\Delta x/\gamma)/\Delta t' = (8.3 \text{ c·min})/(\gamma 5 \text{ min})$, or $\gamma v/c = 8.3/5 = 1.66$, which is easily solved for $v/c = 0.857$. (Square the equation and substitute for γ: $\gamma^2 \times (v/c)^2 = (v/c)^2/(1 - v^2/c^2) = (1.66)^2$. Then $(v/c)^2 = (1.66)^2/[1 + (1.66)^2]$.) (b) The equation $\gamma v/c = 1.66$ is also easily solved for $\gamma = \sqrt{(1.66)^2 - 1} = 1.32$. (Use the fact that $(v/c)^2 = 1 - 1/\gamma^2$.) Then $\Delta t = (1.32)\times (5 \text{ min}) = 9.69$ min.

Problem

11. How fast would you have to move relative to a meter stick for its length to measure 99 cm in your frame of reference?

Solution

The distance you measure in a system S', moving (in a direction parallel to the length of the meter stick) with speed v, is $\Delta x' = 99$ cm, while the proper length of the meter stick is $\Delta x = 100$ cm (in system S). These are related by Equation 38-5, which gives $\sqrt{1 - v^2/c^2} = 0.99$. Therefore, $v/c = \sqrt{1 - (0.99)^2} = 0.141$.

Problem

13. You wish to travel to a star N light-years from Earth. How fast must you go if the one-way journey is to occupy N years of your life?

Solution

This problem is similar to Problem 9(a), with $\Delta x = N$ ly and $\Delta t' = N$ y. Thus, $\gamma v/c = \Delta x/c \,\Delta t' = N$ ly/cN y $= 1$, and $v/c = 1/\sqrt{2}$.

Problem

15. Twins A and B live on Earth. On their 20th birthday, Twin B climbs into a spaceship and makes a round-trip journey at $0.95c$ to a star 30 light-years distant, as measured in the Earth-star frame of reference. What are their ages when the twins are reunited?

Solution

Twin A must wait $\Delta t = 2(30\text{ ly})/(0.95c) = 63.2$ y for twin B to return, so A is 83.2 y old. Because of time dilation, twin B ages for $\Delta t' = \Delta t \sqrt{1 - (0.95)^2} = 19.7$ y during the trip, so he/she returns at age 39.7 y.

Problem

17. Two distant galaxies are receding from Earth at $0.75c$, in opposite directions. How fast does an observer in one galaxy measure the other to be moving?

Solution

Our galaxy (S') is moving with speed $v = 0.75c$ relative to one of the galaxies mentioned in the question (S), and the other galaxy is moving with speed $u' = 0.75c$ relative to us. (All velocities are assumed to be along the common x-x' axes.) Then the relativistic velocity addition formula (Equation 38-11) gives $u = (u' + v)/(1 + u'v/c^2) = (0.75c + 0.75c)/[1 + (0.75)^2] = 0.960c$.

Problem

19. Muons traveling vertically downward at $0.994c$ relative to Earth are observed from a rocket traveling upward at $0.25c$. What speed does the rocket's crew measure for the muons?

Solution

Take $u = 0.994c$ for the velocity of muons relative to Earth (system S, positive downward), and $v = -0.25c$

for the velocity of the rocket relative to Earth (system S' relative to S). Then Equation 33-12 gives $u' = (u - v)/(1 - uv/c^2) = (0.994c + 0.25c)/(1 + 0.994 \times 0.25) = 0.996c$.

Problem

21. Earth and Sun are 8.33 light-minutes apart. Event A occurs on Earth at time $t = 0$, and event B on the Sun at time $t = 2.45$ min, as measured in the Earth-Sun frame. Find the time order and time difference between A and B for observers
    (a) moving on a line from Earth to Sun at $0.750c$,
    (b) moving on a line from Sun to Earth at $0.750c$, and (c) moving on a line from Earth to Sun at $0.294c$.

Solution

The relative speed of the Earth and the Sun is small compared to c ($v/c \approx 10^{-4}$), so we may consider the Sun to be approximately at rest, a distance 8.33 c·min, from Earth (system S). Events A and B have coordinates $x_A = 0$, $t_A = 0$, $x_B = 8.33$ c·min, and $t_B = 2.45$ min in S, where we chose the Earth as the origin and the x-axis in the direction of the Sun. An observer in system S', moving along the x-axis with speed v, sees these events separated by a time interval $\Delta t' = t'_B - t'_A = \gamma[t_B - x_B(v/c^2) - t_A + x_A(v/c^2)] = \gamma[2.45\text{ min} - 8.33\text{ min }(v/c)]$. (Coordinates in S and S' are related by the Lorentz transformation; see Table 38-1.) (a) If $v = 0.75c$ and $\gamma = 1.51$, then $\Delta t' = -5.74$ min, or event B occurs before A in S'. (b) If $v = -0.75c$, then $\Delta t' = 13.1$ min, and B occurs after A in S'. (c) Since $2.45/8.33 \approx 0.294$, $\Delta t' = 0$ if $v = 0.294c$, and A and B are essentially simultaneous in S'.

Problem

23. Repeat the preceding problem, now assuming that civilization B lags A by 1 million years in the galaxy frame of reference.

Solution

Denote the frame of the galaxy by S. The spacecraft are launched at $t_A = 0$ and $t_B = 10^6$ y, from $x_A = 0$ and $x_B = 10^5$ ly (we choose the origin of S at civilization A for simplicity). Let S' be the frame of the traveler from C. The Lorentz transformation between S and S' is summarized in Table 38-1, where $v = 0.99c$ (positive x and x' axes from A to B), and $\gamma = [1 - (1 - 0.01)^2]^{-1/2} \approx 10/\sqrt{2}$. In S', $t'_A = \gamma(t_A - x_A v/c^2) = 0$, and $t'_B = (10/\sqrt{2})[10^6\text{ y} - (10^5 c\cdot\text{y}) \times (0.99c)/c^2] = 6.39 \times 10^6$ y. Evidently, the observer from C assigns priority to civilization A by this amount.

[Note: precisely how the being from C observes events in A and B is a subject for speculation by science fiction buffs.]

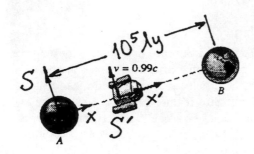

FIGURE 38-31 Problems 22 and 23 Solution.

Problem

25. Derive the Lorentz transformations for time, Equations 38-9 and 38-10, from the transformations for space.

Solution

If we solve Equation 38-7 for t', and substitute for x' from Equation 38-6, we obtain Equation 38-9:

$$t' = \frac{x}{\gamma v} - \frac{x'}{v} = \frac{x}{\gamma v} - \frac{\gamma(x - vt)}{v}$$

$$= \gamma\left[\left(1 - \frac{v^2}{c^2}\right)\frac{x}{v} - \frac{x}{v} + t\right] = \gamma\left(t - \frac{vx}{c^2}\right).$$

By a similar procedure, one may solve for t and obtain Equation 38-10.

Problem

27. Two spaceships are each 25 m long, as measured in their rest frames (Fig. 38-32). Ship A is approaching Earth at $0.65c$. Ship B is approaching Earth from the opposite direction at $0.50c$. Find the length of ship B as measured (a) in Earth's frame of reference and (b) in ship A's frame of reference.

Solution

The length of ship B, measured in a system in which its velocity (parallel to its length along common x-axes) is v_B, is given by Equation 38-5 (Lorentz contraction) as $(25\text{ m})\sqrt{1 - v_B^2/c^2}$, where 25 m is its proper length. We therefore need to find v_B in the systems specified. (a) In the Earth's frame (system S, with x-axis in the direction of motion of ship B) the velocity of ship B is $u_B = 0.5c$, so substituting this for v_B in Equation 38-5 gives $\Delta x_B = (25\text{ m}) \times \sqrt{1 - (0.5)^2} = 21.7\text{ m}$. (b) The velocity of ship A (system S') relative to Earth is $v = -0.65c$ (since ship

A is approaching along the x-axis from the opposite direction as ship B), so the velocity of ship B relative to ship A can be found from Equation 38-12: $u'_B = (u_B - v)/(1 - u_B v/c^2) = [0.5c - (-0.65c)]/(1 + 0.5 \times 0.65) = 0.868c$. When this is substituted for v_B in Equation 38-5, we find $\Delta x'_B = (25\text{ m})\sqrt{1 - (0.868)^2} = 12.4\text{ m}$ for the length of ship B measured in S'.

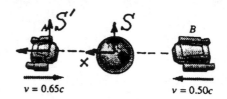

FIGURE 38-32 Problem 27 Solution.

Section 38-5:   Energy and Momentum in Relativity

Problem

29. At what speed will the momentum of a proton (mass 1 u) equal that of an alpha particle (mass 4 u) moving at $0.5c$?

Solution

The momenta of the proton and alpha particle are equal when $m_p u_p/\sqrt{1 - u_p^2/c^2} = m_\alpha u_\alpha/\sqrt{1 - u_\alpha^2/c^2}$; see Equation 38-12. Square and solve for u_p to obtain:

$$\frac{u_p}{c} = \left[1 + \left(\frac{m_p}{m_\alpha}\right)^2\left(\frac{c^2}{u_\alpha^2} - 1\right)\right]^{-1/2}$$

$$= \left[1 + \left(\frac{1}{4}\right)^2\left(\frac{1}{0.5^2} - 1\right)\right]^{-1/2}$$

$$= \frac{1}{\sqrt{1 - 3/16}} = 0.918.$$

Problem

31. A particle is moving at $0.90c$. If its speed increases by 10%, by what factor does its momentum increase?

Solution

A general formula for $\Delta p/p$ in terms of $\Delta u/u$ is not particularly useful, so it is easiest to evaluate the momenta in this problem directly from Equation 38-13, $p = \gamma mu$. For $u_1 = 0.9c$, $\gamma_1 = (1 - 0.81)^{-1/2} = 2.29$, and $p_1 = 2.06mc$. For $u_2 = (1 + 10\%)u_1 = 0.99c$, $\gamma_2 = 7.09$, and $p_1 = 7.02mc$. Therefore, $p_2/p_1 = 3.40$.

Problem

33. Find (a) the total energy and (b) the kinetic energy of an electron moving at $0.97c$.

Solution

For this electron, $v/c = 0.97$, $\gamma = 4.11$, and $m_e c^2 = 0.511$ MeV. (a) From Equation 38-15, $E = \gamma mc^2 = (4.11)(0.511 \text{ MeV}) = 2.10$ MeV. (b) From Equation 38-14, $K = (\gamma - 1)mc^2 = (3.11) \times (0.511 \text{ MeV}) = 1.59$ MeV.

Problem

35. At what speed will the relativistic and Newtonian expressions for kinetic energy differ by 10%?

Solution

The fractional difference between Equation 38-14 and the Newtonian expression is $[(\gamma - 1)mc^2 - \frac{1}{2}mv^2] \div (\gamma - 1)mc^2 = 1 - v^2/2c^2(\gamma - 1) = 0.1$, or $v^2/c^2 = 1.8 \times (\gamma - 1) = 1 - 1/\gamma^2 = (\gamma - 1)(\gamma + 1)/\gamma^2$, where we used the definition of γ (Equation 38-8) and factorization. This is equivalent to a quadratic equation in γ, $1.8\gamma^2 - \gamma - 1 = 0$, with positive solution $\gamma = (1 + \sqrt{1 + 7.2}) \div 3.6 = 1.07$. The corresponding speed is $v/c = \sqrt{1 - 1/\gamma^2} = 0.363$.

Problem

37. Among the most energetic cosmic rays ever detected are protons with energies around 10^{20} eV. Find the momentum of such a proton, and compare with that of a 25-mg insect crawling at 2 mm/s (Fig. 38-33).

Solution

This energy is so much greater than the proton's rest energy ($mc^2 = 938$ MeV) that $p \approx E/c = 10^{20}$ eV/c. (See Equation 38-16.) The atomic unit of momentum in conventional SI units is 1 eV/c $= (1.602 \times 10^{-19} \text{ J}) \div (2.998 \times 10^8 \text{ m/s}) = 5.34 \times 10^{-28}$ kg·m/s, so $p \approx 5.34 \times 10^{-8}$ kg·m/s. This is about equal to the momentum of the insect, $mv = (25 \times 10^{-6} \text{ kg}) \times (2 \times 10^{-3} \text{ m/s}) = 5 \times 10^{-8}$ kg·m/s.

Problem

39. In a nuclear fusion reaction, two deuterium nuclei (^{2}H) combine to give a helium nucleus (^{3}He) plus a neutron. The energy released in the process is 3.3 MeV. By how much do the combined masses of the helium nucleus and neutron differ from the combined masses of the original deuterium nuclei?

Solution

The mass-equivalent of the released energy is $\Delta m = \Delta E/c^2 = 3.3$ MeV/c^2. In conventional SI units, 1 MeV/$c^2 = 1.602 \times 10^{-13}$ J/$(2.998 \times 10^8 \text{ m/s})^2 = 1.78 \times 10^{-30}$ kg $= 1.0731 \times 10^{-3}$ u, so $\Delta m = 0.00354$ u $= 5.88 \times 10^{-30}$ kg (see Appendix C).

Problem

41. Find the speed of an electron with kinetic energy (a) 100 eV. (b) 100 keV, (c) 1 MeV, (d) 1 GeV. Use suitable approximations where possible.

Solution

For the electron, $m_e c^2 = 511$ keV, so γ corresponding to kinetic energy $K = (\gamma - 1)m_e c^2$ is $\gamma = 1 + K \div 511$ keV, from which $v/c = \sqrt{1 - 1/\gamma^2}$ can be found (see Equations 38-14 and 8). (a) If $K \ll 511$ keV, the non-relativistic expression, $v/c = \sqrt{2K/mc^2}$, can be used (see solution to Problem 43), so $v/c = \sqrt{2(100 \text{ eV})/511 \text{ keV}} = 1.98 \times 10^{-2}$. (b) If $K = 100$ keV, then $\gamma = 1.20$ and $v/c = 0.548$, while (c) if $K = 1$ MeV, $\gamma = 2.96$ and $v/c = 0.941$. (d) If $K \gg 511$ keV, γ is large and $v/c \approx 1$. If one expands the square root in $v/c = \sqrt{1 + (1 + K/m_e c^2)^{-2}}$, in powers of $m_e c^2/K \ll 1$, one obtains $v/c \approx 1 - \frac{1}{2}(m_e c^2/K)^2 = 1 - \frac{1}{2}(0.511 \times 10^{-3})^2 = 1 - 1.31 \times 10^{-7}$.

Problem

43. Use the binomial approximation (Appendix A) to show that Equation 38-14 reduces to the Newtonian expression for kinetic energy in the limit $u \ll c$.

Solution

Equation 38-14, for $u/c \ll 1$, can be expanded to yield: $K = mc^2[(1 - u^2/c^2)^{-1/2} - 1] = mc^2[1 + (u^2/2c^2) + (3u^4/8c^4) + \cdots - 1] = \frac{1}{2}mu^2(1 + 3u^2/4c^2 + \cdots)$. Sometimes, it is useful to have the next term in the expansion.

Section 38-6:   What Is Not Relative

Problem

45. Show from the Lorentz transformations that the spacetime interval of Equation 38-18 has the same value in all frames of reference.

Solution

Consider two frames, S and S', related by the Lorentz transformations in Table 38-1. (Since the equations are linear, they apply to differences of coordinates also.) We have $c^2\Delta t'^2 - \Delta x'^2 - \Delta y'^2 - \Delta z'^2 = c^2\gamma^2 \times (\Delta t - v\,\Delta x/c^2)^2 - \gamma^2(\Delta x - v\,\Delta t)^2 - \Delta y^2 - \Delta z^2 = $

$\gamma^2(c^2\Delta t^2 - 2v\,\Delta x\,\Delta t + v^2\Delta x^2/c^2 - \Delta x^2 + 2v\,\Delta x\,\Delta t - v^2\Delta t^2) - \Delta y^2 - \Delta z^2 = \gamma^2(1 - v^2/c^2)(c^2\,\Delta t^2 - \Delta x^2) - \Delta y^2 - \Delta z^2 = c^2\,\Delta t^2 - \Delta x^2 - \Delta y^2 - \Delta z^2$. Therefore, $\Delta s'^2 = \Delta s^2$, and the spacetime interval is invariant.

Problem

47. Use Equation 38-18 to calculate the square of the spacetime interval between the events (a) of Problem 22 and (b) of Problem 23. Comment on the signs of your answers in relation to the possibility of a causal relation between the events.

Solution

(a) The square of the spacetime interval between the first launchings by civilizations A and B in Problem 22 is most easily calculated in the frame of the galaxy, S. Then $\Delta s^2 = c^2(t_B - t_A)^2 - (x_B - x_A)^2 = (1\text{ ly/y})^2\times (5\times10^4\text{ y})^2 - (10^5\text{ ly})^2 = -0.75\times10^{10}\text{ ly}^2$. (Since the spacetime interval is Lorentz-invariant, the same result would be found in any frame moving with constant velocity relative to S.) (b) For the events A and B in Problem 23, $\Delta s^2 = (1\text{ ly/y})^2(10^6\text{ y})^2 - (10^5\text{ ly})^2 = 99\times10^{10}\text{ ly}^2$. If the square of the spacetime interval between two events is positive (called a timelike separation), then the events can be causally connected. If Δs^2 is negative (a spacelike separation), the events cannot be causally connected, and there is a Lorentz frame in which they occur simultaneously.

Paired Problems

Problem

49. An extraterrestrial spacecraft passes Earth and 4.5 s later, according to its clocks, it passes the Moon. Find its speed.

Solution

(Note: This and the next problem are essentially the same as Problem 9a, whose result, $v/c = (\Delta x/c\,\Delta t') \div \sqrt{1 + (\Delta x/c\,\Delta t')^2}$, can be used directly. However, a repetition of the reasoning may be useful.) In the spacecraft's frame (S') the distance from the Earth to the Moon appears Lorentz-contracted from the proper distance (3.85×10^5 km in the Earth-Moon frame, S), $\Delta x' = \Delta x\sqrt{1 - v^2/c^2} = \Delta x/\gamma$, so its speed is $v = \Delta x'/\Delta t' = (\Delta x/\gamma)/(4.5\text{ s})$. Thus $\gamma v/c = (3.85\times 10^5\text{ km})/(3\times10^5\text{ km/s})(4.5\text{ s}) = 0.285$ and $v/c = 0.285/\sqrt{1 + (0.285)^2} = 0.274$ (square and solve for v/c). (Alternatively, time dilation gives $\Delta t = \gamma\Delta t'$ in the Earth-Moon frame, and $v = \Delta x/\Delta t$ leads to the same result.)

Problem

51. An electron moves down a 1.2-km-long particle accelerator at $0.999992c$. In the electron's frame, (a) how much time does the trip take and (b) how long is the accelerator?

Solution

The accelerator's frame is S and the electron's frame is S'. The same reasoning as in Problem 9 and the previous pair of problems applies to this and the next problem; that is, $\Delta t' = \Delta x'/v$ and $\Delta x' = \Delta x/\gamma$, where now Δx and v (hence also γ) are given. With a calculator of sufficient accuracy, one finds $\gamma = 1 \div \sqrt{1 - (0.999992)^2} = 250$, so (b) $\Delta x' = 1.2\text{ km}/250 = 4.80$ m, and (a) $\Delta t' = (4.80\text{ m})(0.999992)^{-1}\times (3\times10^8\text{ m/s})^{-1} = 16.0$ ns. Of course, since $v/c = 1 - 8\times10^{-6} \equiv 1 - \varepsilon$, one can expand γ in powers of $\varepsilon \ll 1$ to obtain $\gamma = (2\varepsilon)^{-1/2}(1 + \varepsilon/4 + \cdots) \approx 1/\sqrt{2\varepsilon} = 1 \div \sqrt{2\times8\times10^{-6}} = 250$, as before.

Problem

53. Event A occurs at $x = 0$ and $t = 0$ in a frame of reference S. Event B occurs at $x = 3.8$ light-years, $t = 1.6$ years in S. Find (a) the distance and (b) the time between A and B in a frame S' moving at $0.80c$ along the x-axis of S.

Solution

The coordinates of the events in S and S', are related by the Lorentz transformation in Table 38-1, with $v/c = 0.8$ and $\gamma = 5/3$. (a) $x'_B - x'_A = \gamma[x_B - x_A - v(t_B - t_A)] = (5/3)[3.8\text{ ly} - (0.8c)(1.6\text{ y})] = (5/3)(3.8 - 1.28)\text{ ly} = 4.20$ ly. (b) $t'_B - t'_A = \gamma[t_B - t_A - (v/c^2)\times (x_B - x_A)] = (5/3)[1.6\text{ y} - (0.8/c)(3.8\text{ ly})] = -2.40$ y; that is, B occurs before A in S'. (Since the light-travel time from the position of A to that of B is greater than the magnitude of the time difference 3.8 y versus 1.6 y in S, or 4.2 y versus 2.4 y is S', the events are not causally connected.)

Problem

55. When a particle's speed doubles, its momentum increases by a factor of 3. What was the original speed?

Solution

The given conditions are $v_2 = 2v_1$ and $p_2 = 3p_1$. Use Equation 38-13 and divide these equations to obtain: $v_2/p_2 = v_2/\gamma_2 m v_2 = 2v_1/3\gamma_1 m v_1$, or $3/\gamma_2 = 2/\gamma_1$. Square this and use the first condition again to solve for the original speed: $9/\gamma_2^2 = 9(1 - v_2^2/c^2) = 4(1 - v_1^2/c^2) = 9(1 - 4v_1^2/c^2)$, or $v_1/c = \sqrt{5/32} = 0.395$.

Supplementary Problems

Problem

57. How fast would you have to travel to reach the Crab Nebula, 6500 light-years from Earth, in 20 years? Give your answer to 7 significant figures.

Solution

As in Problems 9 and 49, the distance to the Crab Nebula, $\Delta x = 6500$ ly, is specified in the Earth's system (S), while the time interval, $\Delta t' = 20$ y, is in the traveler's system (S'), so either Lorentz contraction ($\Delta x = \gamma \Delta x'$) or time dilation ($\Delta t = \gamma \Delta t'$) give $v = \Delta x / \Delta t = \Delta x' / \Delta t' = \Delta x / \gamma \Delta t'$, or $\gamma v / c = \Delta x / c \, \Delta t' = 6500$ ly/$20c\cdot$y $= 325$. Then $v/c = 325 \div \sqrt{1 + (325)^2}$. A good calculator gives $v/c = 0.99999527$, but one can also use the binomial expansion to find $v/c = [1 + (1/325)^2]^{-1/2} \approx 1 - \frac{1}{2} \times (1/325)^2 = 1 - 4.73 \times 10^{-6}$.

Problem

59. A cosmic ray proton with energy 20 TeV is heading toward Earth. What is Earth's diameter measured in the proton's frame of reference?

Solution

The phenomenon of Lorentz contraction makes the Earth's diameter, along the direction of motion of the proton, appear to be $2R' = 2R_E / \gamma$ in the proton's frame. (The Earth's diameter measured perpendicular to this would still be R_E.) For a proton of energy 20 TeV, $\gamma = E/m_p c^2$, so $2R' = 2(6.37 \times 10^6 \text{ m}) \times (938 \text{ MeV})/(20 \text{ TeV}) = 598$ m. (Note: At this high energy, it makes little difference whether 20 TeV is the kinetic or total energy. See the appendices for R_E and $m_p c^2$.)

Problem

61. When the speed of an object increases by 5%, its momentum goes up by a factor of 5. What was the original speed?

Solution

As in Problem 55, we write $v_2 = 1.05 v_1$ and $p_2 = 5p_1$. Following the same procedure, we find $1.05/\gamma_1 = 5/\gamma_2$, or $(1.05)^2(1 - v_1^2/c^2) = 5^2(1 - v_2^2/c^2) = 5^2(1 - (1.05)^2 \times v_1^2/c^2)$, or $v/c = \sqrt{[25 - (1.05)^2]/24(1.05)^2} = 0.950c$.

Problem

63. A source emitting light with frequency f moves toward you at speed u. By considering both time dilation and the effect of wavefronts "piling up" as shown in Fig. 17-20, show that you measure a

Doppler-shifted frequency given by

$$f' = f\sqrt{\frac{c+u}{c-u}}.$$

Use the binomial approximation to show that this result can be written in the form of Equation 17-10 for $u \ll c$.

Solution

In order to correspond to the notation in Section 17-7 and the wording of this problem, let S be the rest system of a source of light waves (with frequency and wavelength $\lambda f = c$) which is moving with speed u towards an observer in S' (who measures $\lambda' f' = c$). Suppose that N waves are emitted in S in a time interval Δt. The first wavefront has expanded to a distance $c\,\Delta t$ in S, so the wavelength (i.e., the distance between surfaces of constant phase) is $\lambda = c\,\Delta t/N$. In S', however, the wavefronts are "piled-up" into a smaller distance, due to the motion of S, so the wavelength is $\lambda' = (c\,\Delta t' - u\Delta t')/N = (c - u)\Delta t'/N$ (see sketch). Therefore $\lambda'/\lambda = (1 - u/c)\Delta t'/\Delta t$. Except for time dilation, this would be Equation 17-9a. But Δt (the proper time interval in the source's rest system) is related to $\Delta t'$ (the time interval measured in a system where the source is moving) by time dilation, $\Delta t' = \gamma \Delta t$ (Equation 38-4 with altered notation), so $\lambda'/\lambda = \gamma(1 - u/c)$, or in terms of frequency, $f/f' = \gamma(1 - u/c)$. Since $\gamma = 1 \div \sqrt{1 - u^2/c^2}$, this can be written as $f'/f = \sqrt{(1 - u/c)(1 + u/c)}/(1 - u/c) = \sqrt{(1 + u/c)/(1 - u/c)}$ which is the radial Doppler shift (i.e., along the line of sight) in special relativity, with u positive for approach and negative for recession (note the difference in signs with Equation 17-10). For $u/c \ll 1$, $\sqrt{(1 + u/c)/(1 - u/c)} \to (1 + \frac{1}{2}u/c)(1 - \frac{1}{2}(-u/c)) \to 1 + u/c$, which, allowing for the difference in signs, is the same as the limit of Equation 17-10 for $u/c \ll 1$. (Note: it is more customary to write this limit as $\Delta f/f = (f' - f)/f = u/c$.)

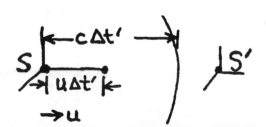

Problem 63 Solution.

Problem

65. Consider a line of positive charge with line charge density λ, as measured in a frame S at rest with respect to the charges. (a) Show that the electric field a distance r from this charged line has magnitude $E = \lambda/2\pi\varepsilon_0 r$, and that there is no magnetic field (no relativity needed here). Now consider the situation in a frame S' moving at speed v parallel to the line of charge. (b) Show that the line charge density as measured in S' is given by $\lambda' = \gamma\lambda$, with $\gamma = 1/\sqrt{1 - v^2/c^2}$. (c) Use the result of (b) to find the electric field in S'. Since the line of charge is moving with respect to S', there is a current in S'. (d) Find an expression for this current and (e) for the magnetic field it produces. Show that the quantities (f) $\mathbf{E}\cdot\mathbf{B}$ and (g) $E^2 - c^2 B^2$ are the same in both frames of reference. (In fact, these quantities are always invariant.)

Solution

(a) In frame S, there are just static charges, and no current. The electric field points radially away from the x-axis (taken along the line of charge) with magnitude $E = \lambda/2\pi\varepsilon_0 r$ (Example 24-4), and the magnetic field is zero. (b) The amount of charge, $dq = dq'$, on a length $d\ell$ in S, appears to lie on a Lorentz-contracted length $d\ell' = d\ell\sqrt{1 - v^2/c^2} = d\ell/\gamma$

in S'. Thus, $dq = \lambda d\ell = \lambda\gamma d\ell' = dq' = \lambda' d\ell'$ implies $\lambda' = \gamma\lambda$. (c) The electric field in S' is $E' = \lambda'/2\pi\varepsilon_0 r' = \gamma\lambda/2\pi\varepsilon_0 r$. ($r = \sqrt{y^2 + z^2}$, measured perpendicular to the line of charge, is not changed by the Lorentz transformation, so $r' = r$.) (d) In S', the charge moving with speed $v = dx'/dt'$, along the negative x'-axis, constitutes a current, $I' = dq'/dt' = \lambda' dx'/dt' = \lambda' v = \gamma\lambda v$ in that direction. (e) The magnetic field from such a current encircles the x'-axis according to the right-hand rule, with magnitude $B' = \mu_0 I'/2\pi r' = \mu_0\gamma\lambda v/2\pi r$ (Equation 30-8). (f) In S, $\mathbf{B} = 0$, so $\mathbf{E}\cdot\mathbf{B} = 0$. In S', $\mathbf{E}'$ is radial and $\mathbf{B}'$ is tangential, so $\mathbf{E}'\cdot\mathbf{B}' = 0$ also. (g) In S, $E^2 - c^2 B^2 = (\lambda/2\pi\varepsilon_0 r)^2$. In S', $E'^2 - c^2 B'^2 = (\lambda'/2\pi\varepsilon_0 r')^2 - (1/\mu_0\varepsilon_0)(\mu_0 I'/2\pi r')^2 = (1/2\pi\varepsilon_0 r)^2 \times (\gamma^2\lambda^2 - \mu_0\varepsilon_0\gamma^2\lambda^2 v^2) = (\lambda/2\pi\varepsilon_0 r)^2\gamma^2(1 - v^2/c^2) = (\lambda/2\pi\varepsilon_0 r)^2$ also, where we used $\lambda' = \gamma\lambda$, $I' = \gamma\lambda v$, and $c^2 = 1/\mu_0\varepsilon_0$.

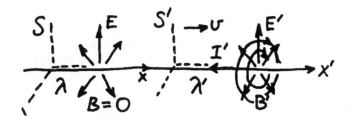

Problem 65 Solution.

CHAPTER 39   INSIDE ATOMS AND NUCLEI

ActivPhysics can help with these problems:
Activities 17.3–17.6

Section 39-1:   Toward the Quantum Theory

Useful numerical values of Planck's constant, in SI and atomic units, are:
$h = 6.626 \times 10^{-34}$ J·s $= 4.136 \times 10^{-15}$ eV·s $= 1240$ eV·nm/c, and $\hbar = h/2\pi = 1.055 \times 10^{-34}$ J·s $= 6.582 \times 10^{-16}$ eV·s $= 197.3$ MeV·fm/c.

Problem

1. Find the energy in electron-volts of (a) a 1.0-MHz radio photon, (b) a 5.0×10^{14}-Hz optical photon, and (c) a 3.0×10^{18}-Hz X-ray photon.

Solution

The energy of a photon (Equation 39-1) is $E_\gamma = hf$, where h is Planck's constant. For the frequencies given, the photon energies in atomic units are (a) 4.14 neV (b) 2.07 eV, and (c) 12.4 keV.

Problem

3. A microwave oven uses electromagnetic radiation at 2.4 GHz. (a) What is the energy of each microwave photon? (b) At what rate does a 625-W oven produce photons?

Solution

(a) $E_\gamma = hf = (4.136 \times 10^{-15}$ eV·s$)(2.4$ GHz$) = 9.93$ μeV (or 1.59×10^{-24} J), as in Problem 1.
(b) $\mathcal{P}/E_\gamma = (625$ J/s$)/(1.59 \times 10^{-24}$ J/photon$) = 3.94 \times 10^{26}$ photons/s is the power output in photons.

Problem

5. Find the rate of photon production by (a) a radio antenna broadcasting 1.0 kW at 89.5 MHz, (b) a laser producing 1.0 mW of 633-nm light, and (c) and X-ray machine producing 0.10-nm X rays with a total power of 2.5 kW.

Solution

The rate of photon emission is the power output (into photons) divided by the photon energy, $\mathcal{P}/E_\gamma = \mathcal{P}/hf = \mathcal{P}\lambda/hc$. For the devices specified, rates are
(a) 1 kW/$(6.626 \times 10^{-34}$ J·s $\times 89.5$ MHz$) = 1.69 \times 10^{28}$ s^{-1}, (b) (1 mW $\times$ 633 nm)/$(6.626 \times$

10^{-34} J·s $\times 3 \times 10^8$ m/s$) = 3.18 \times 10^{15}$ s^{-1}, and (c) 1.26×10^{18} s^{-1}.

Problem

7. Which spectral line of the hydrogen Paschen series $(n_2 = 3)$ has wavelength 1282 nm?

Solution

The wavelengths in the Paschen series for hydrogen are given by Equation 39-3, with $n_2 = 3$ and $n_1 = 4, 5, 6, \ldots$. Thus, $\lambda = (1/R_H)9n_1^2/(n_1^2 - 9)$. If one substitutes $\lambda = 1282$ nm and $R_H = 0.01097$ (nm)$^{-1}$, one finds $9n_1^2 = 0.01097 \times 1282(n_1^2 - 9)$, or $n_1 = \sqrt{(9 \times 14.1)/(14.1 - 9)} = 5$, so this is the second line in the series.

Problem

9. A Rydberg hydrogen atom makes a downward transition to the $n = 225$ state. If the photon emitted has energy 9.32 μeV, what was the original state?

Solution

Combining Equations 39-1 and 3, one can write the energy of a photon emitted in a hydrogen atom transition between states $n_1 \to n_2$ as $E_\gamma = hc/\lambda = hcR_H(n_2^{-2} - n_1^{-2})$. (In a Rydberg atom, n_1 and n_2 are both very large; see Example 39-1.) Solving for n_1, we find $n_1 = [n_2^{-2} - (E_\gamma/hcR_H)]^{-1/2} = [(225)^{-2} - (9.32 \; \mu\text{eV}/13.6 \text{ eV})]^{-1/2} = 229$. (The constant $hcR_H = (1240$ eV·nm$)(0.01097(\text{nm})^{-1}) = 13.6$ eV, called the ionization energy, is the energy difference between an ionized hydrogen atom, $n_1 = \infty$, and the ground state, $n_2 = 1$.)

Problem

11. How slowly must an electron be moving for its de Broglie wavelength to equal 1 mm?

Solution

For a non-relativistic electron, the de Broglie wavelength is $\lambda = h/mv$ (see Example 39-2), so $v = h/m\lambda = (6.63 \times 10^{-34}$ J·s$)(9.11 \times 10^{-31}$ kg$)^{-1} \times (10^{-3}$ m$)^{-1} = 72.8$ cm/s.

Problem

13. Electron microscopes can usually resolve smaller objects than optical microscopes because they illuminate their subjects with electrons whose de Broglie wavelengths are much smaller than those of light (Fig. 39-37). What is the minimum electron speed that would make an electron microscope superior to an optical microscope using 450-nm light?

Solution

$\lambda = h/p < 450$ nm implies $p = \gamma mv > h/450$ nm, or $\gamma v/c > hc/(mc^2)(450 \text{ nm}) = (1240 \text{ eV·nm}) \times (511 \text{ keV})^{-1}(450 \text{ nm})^{-1} = 5.39 \times 10^{-6}$. Then $v/c > (5.39 \times 10^{-6})/\sqrt{1 + (5.39 \times 10^{-6})^2} \approx 5.39 \times 10^{-6}$, or $v > (5.39 \times 10^{-6})(3 \times 10^5 \text{ km/s}) = 1.62$ km/s (see Chapter 38). (Since $\lambda \gg h/mc = 0.00243$ nm, for an electron, if you used the non-relativistic expression for momentum, $p = mv$, then you would get the same result, $v > h/m\lambda = (6.63 \times 10^{-34} \text{ J·s})(9.11 \times 10^{-31} \text{ kg})^{-1}(450 \text{ nm})^{-1} = 1.62$ km/s, see Example 39-2.)

Section 39-2: Quantum Mechanics
Problem

15. Is it possible to measure an electron's velocity to an accuracy of ± 1.0 m/s while simultaneously finding its position to an accuracy of ± 1.0 μm? What about a proton?

Solution

If we take $\Delta v = 2$ m/s and $\Delta x = 2$ μm (the full range of uncertainty in both measurements), and use the non-relativistic expression for momentum (so that $\Delta p = m \Delta v$), the uncertainty principle requires that $\Delta x \Delta v \gtrsim h/2\pi m$, or $m \gtrsim \hbar/\Delta x \Delta v = (1.055 \times 10^{-34} \text{ J·s})(4 \times 10^{-6} \text{ m}^2/\text{s})^{-1} = 2.64 \times 10^{-29}$ kg. (Here, $\hbar = h/2\pi$ is pronounced "h-bar.") (a) For the electron $(m_e = 9.11 \times 10^{-31}$ kg) this is not possible, but (b) for the proton $(m_p = 1.67 \times 10^{-27}$ kg) it is.

Problem

17. An electron is moving in the $+x$ direction with speed measured at 5.0×10^7 m/s, to an accuracy of $\pm 10\%$. What is the minimum uncertainty in its position?

Solution

The non-relativistic expression, $p = mv$, gives a rough estimate of Δx in Equation 39-5 (since $v/c = 1/6$ and $\gamma = 1.014$). It is $\Delta x \gtrsim \hbar/m \Delta v = (1.055 \times 10^{-34} \text{ J·s}) \times (9.11 \times 10^{-31} \text{ kg})^{-1}(10^7 \text{ m/s})^{-1} = 11.6$ pm, where

$\Delta v = 20\%$ of 5×10^7 m/s is the full range of uncertainty in v. (If the relativistic momentum $p = \gamma mv$ is used, then $\Delta p = m \Delta(\gamma v) = 9.50 \times 10^{-24}$ kg·m/s, and $\Delta x \gtrsim \hbar/\Delta p = 11.1$ pm.)

Problem

19. An electron beam is accelerated at the back of a TV tube and then heads toward the center of the screen with a horizontal velocity of 2.2×10^7 m/s. As the electrons leave the acceleration region, their vertical position is known to within ± 45 nm. Find the minimum angular spread in the beam, as set by the uncertainty principle.

Solution

An uncertainty in the vertical position of the electrons of $\Delta y = +45$ nm $- (-45$ nm$) = 90$ nm, leads to a minimum uncertainty in their vertical momentum of $\Delta p_y \gtrsim \hbar/\Delta y$ (see Equation 39-5), as they head towards the center of the tube (with momentum p_x). Thus, their angular spread is $\Delta\theta = \Delta p_y/p_x = \hbar/mv_x \Delta y = (1.055 \times 10^{-34} \text{ J·s})(9.11 \times 10^{-31} \text{ kg})^{-1} \times (2.2 \times 10^7 \text{ m/s})^{-1}(90 \text{ nm})^{-1} = (5.85 \times 10^{-5})(180°/\pi) = 3.35 \times 10^{-3}$ degrees.

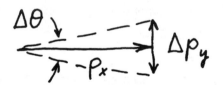

Problem 19 Solution.

Section 39-3: Nuclear Physics
Problem

21. What is the half-life of a radioactive material if 75% of it decays in 5.0 hours?

Solution

If 75% decays in 5 h, then 25% remains after 5 h. Since 25% = 50% × 50%, 5 h is two half-lives, or $t_{1/2} = 2.5$ h. [In general, the amount remaining after time t is $N/N_0 = (1/2)^{t/t_{1/2}}$, where N_0 is the original amount at $t = 0$. In this problem, $t_{1/2} = t \ln(1/2) \div \ln(1/4) = \frac{1}{2}(5 \text{ h})$.]

Problem

23. Oxygen-15 is a radioactive isotope with a 2-minute half-life, widely used in medical studies because its short half-life ensures relative safety. Approximately how long will it take for 99.9% of the nuclei in a sample of O-15 to decay?

Solution

The reasoning in the solution to Problem 21 shows that $1 - 0.999 = 10^{-3} = (1/2)^{t/t_{1/2}}$, so $t = t_{1/2} \ln(10^{-3})/\ln(1/2) = (2 \text{ min})(\ln 10^3 / \ln 2) = 19.9 \text{ min}$.

Problem

25. The energy released in the fission of a single uranium-235 nucleus is about 200 MeV. Estimate the mass of uranium that fissioned in the Hiroshima bomb, shown in Fig. 39-39, whose explosive yield was 12.5 kilotons (see Appendix C for a useful conversion factor).

Solution

The explosive yield of 1 megaton is about 4.18×10^{15} J (from Appendix C), so the Hiroshima bomb released the energy in about $(12.5)(4.18 \times 10^{12} \text{ J})/(200 \text{ MeV}) \times (1.6 \times 10^{-13} \text{ J/MeV}) = 1.63 \times 10^{24}$ fissions of U^{235}. This number of atoms has a mass of about $(1.63 \times 10^{24}) \times (235 \text{ u})(1.66 \times 10^{-27} \text{ kg/u}) = 0.637$ kg. (Of course, the uranium in the bomb wasn't enriched to 100% U^{235}, so the actual mass would have been somewhat greater.)

Section 39-4:   Elementary Particles

Problem

27. The lambda particle (Λ) consists of an up quark, a down quark, and a strange quark. What is its charge?

Solution

The charges of the up, down, and strange quarks are $\frac{2}{3}, -\frac{1}{3},$ and $-\frac{1}{3}$, respectively (in units of e), so the net charge of the $\Lambda = uds$ is zero.

Problem

29. What is the quark composition of the antiproton?

Solution

Since a proton is formed from the combination uud of quarks, an antiproton is $\overline{uud}$.

Problem

31. One of two charged mesons called pions is formed from an up quark and a down antiquark. What is this meson's charge?

Solution

The u-quark has charge $\frac{2}{3}e$, and the $\bar{d}$-quark has charge (opposite to the d-quark) of $\frac{1}{3}e$, so the positive pion $\pi^+ = u\bar{d}$ has charge $\frac{2}{3}e + \frac{1}{3}e = e$. (The negative pion, $\pi^- = \bar{u}d$, is the antiparticle of the π^+ and has charge $-e$.)

Paired Problems

Problem

33. (a) Find the energy of the highest-energy photon that can be emitted as the electron jumps between two adjacent energy levels in the Bohr hydrogen atom. (b) Which energy levels are involved?

Solution

The energy of the emitted photon in a hydrogen atom transition between adjacent states $(n_1 \to n_2 = n_1 - 1)$ is $E_\gamma = hc/\lambda = hcR_H[(n_1 - 1)^{-2} - n_1^{-2}] = hcR_H \times (2n_1 - 1)n_1^{-2}(n_1 - 1)^{-2}$ (see Equations 39-1 and 3). This is a maximum for the smallest allowed $n_1 > n_2 \geq 1$, which is $n_1 = 2$. Then (a) $E_{\gamma,\max}(\frac{3}{4})hcR_H = (\frac{3}{4})(13.6 \text{ eV}) = 10.2$ eV (see solution to Problem 9 for the constant), and (b) $n_2 = n_1 - 1 = 1$. (This is the Lyman alpha photon at 122 nm.)

Problem

35. In which of the following will wave properties be more evident: an electron moving at 10 Mm/s or a proton moving at 1 km/s?

Solution

The de Broglie wavelengths of the two particles, calculated from Equation 39-4 and the non-relativistic momentum $p = mv$, are $\lambda = h/mv = 7.28 \times 10^{-11}$ m for the electron, and 3.97×10^{-10} m for the proton. With a larger wavelength, the wave aspects of this proton would be more apparent.

Problem

37. An electron is known to be within ± 0.05 nm of the nucleus of an atom. What is the uncertainty in its velocity?

Solution

The minimum uncertainty in momentum (Equation 39-5) is $\Delta p = \hbar/\Delta x = 1.055 \times 10^{-34}$ J·s $\div 10^{-10}$ m $= 1.055 \times 10^{-24}$ kg·m/s $= (197.3 \text{ eV·nm/c}) \div (0.1 \text{ nm}) = 1.973$ keV/c [if $\Delta x = 2(0.05 \text{ nm})$]. This is small compared to $mc = 511$ keV/c for an electron, so non-relativistic expressions can be used. Then $\Delta v = \Delta p/m = (1.055 \times 10^{-24} \text{ kg·m/s})/(9.11 \times 10^{-31} \text{ kg}) = 1.16 \times 10^6$ m/s $= (1.973 \text{ keV/c})/(511 \text{ keV/c}^2) = 3.86 \times 10^{-3}$ c. (We used both SI and atomic units for pedagogical reasons.)

Supplementary Problems

Problem

39. An electron is accelerated from rest through a 4.5-kV potential difference. What is its de Broglie wavelength?

Solution

The electron's kinetic energy, $K = e\Delta V = 4.5$ keV, is small compared to $mc^2 = 511$ keV, so the non-relativistic expression $K = p^2/2m$ can be used to write the de Broglie wavelength as $\lambda = h/p = h/\sqrt{2mK} = hc/\sqrt{2mc^2K} = 1240$ eV·nm$/\sqrt{2(511 \text{ keV})(4.5 \text{ keV})} = 18.3$ pm (or 0.183 Å). (We used atomic units to simplify the numerical calculation.)

Problem

41. Use the result of the preceding problem to estimate the minimum possible kinetic energy for (a) an electron confined to a region of atomic dimensions, about 0.1 nm and (b) a proton confined to a region of nuclear dimensions, about 1 fm. Your answers show the order of magnitude of the energies to be expected in atomic and nuclear reactions, respectively.

Solution

For a particle confined to a one-dimensional region of size Δx, the minimum momentum (the uncertainty of which is given in the previous problem) is $\Delta p = 2p \geq \hbar/\Delta x$ (see Equation 39-5). If $pc \ll mc^2$ (the particle's rest energy), then the non-relativistic kinetic energy must satisfy $K = p^2/2m \geq (\hbar/2\ \Delta x)^2/2m = \hbar^2/8m\ \Delta x^2 = (\hbar c)^2/8(mc^2)\Delta x^2$. (a) For an electron ($mc^2 = 511$ keV) confined to atomic dimensions ($\Delta x = 0.1$ nm), $K \geq (197.3 \text{ eV·nm})^2/8(511 \text{ keV})(0.1 \text{ nm})^2 = 0.952$ eV $\simeq 1$ eV (typical of atomic transitions). (b) For a proton ($mc^2 = 938$ MeV) confined to nuclear

dimensions ($\Delta x \simeq 1$ fm), $K \geq (197.3 \text{ MeV·fm})^2 \div 8(938 \text{ MeV})(1 \text{ fm})^2 = 5.19$ MeV $\simeq 5$ MeV (typical of nuclear transitions).

Problem

43. The expansion of the universe is described by the so-called Hubble relation, $v = H_0 d$, where v is the recession speed of two galaxies a distance d apart, and where H_0 is called the Hubble constant. Estimates for the value of H_0 range from 50 km/s/megaparsec to 100 km/s/megaparsec. (See Appendix C for the distance unit parsec). If the expansion rate had been the same from the beginning, how long ago would two galaxies have been together? Give two answers, derived from the two extreme values for H_0. Your answers are estimates for the age of the universe.

Solution

If the recession of distant galaxies is due to the expansion of the universe at a roughly constant rate, then $d = vt$, where $t = d/v$ is time the universe has been expanding, or its approximate age. From Hubble's law, $v = H_0 d$, therefore $t = 1/H_0$. The smaller value of H_0 gives

$$t = \frac{3.09 \times 10^{19} \text{ km/Mpc}}{50 \text{ km/s/Mpc}} \cdot \frac{1y}{3.156 \times 10^7 \text{ s}} = 19.6 \text{ Gy},$$

and the larger gives half this, so t is in the range of 10 to 20 billion years.

PHYSICS WITH MODERN PHYSICS

CHAPTER 39X LIGHT AND MATTER: WAVES OR PARTICLES?

ActivPhysics can help with these problems: Activities 17.3–17.6

Useful constants and combinations, in SI and atomic units, for the problems in Chapters 39X–45 are:

$h = 6.626 \times 10^{-34}$ J·s $= 4.136 \times 10^{-15}$ eV·s
$\quad = 1240$ eV·nm/c
$\hbar = h/2\pi = 1.055 \times 10^{-34}$ J·s $= 6.582 \times 10^{-16}$ eV·s
$\quad = 197.3$ MeV·fm/c
$c = 2.998 \times 10^8$ m/s, $\quad ke^2 = 1.440$ eV·nm,

$1u = 1.661 \times 10^{-27}$ kg $= 931.5$ MeV/c^2
$k_B = 1.381 \times 10^{-23}$ J/K $= 8.617 \times 10^{-5}$ eV/K

Section 39X-2: Blackbody Radiation

Note: Most answers in this section are very sensitive to the exact values used for the constants h and c.

Problem

1. If you increase the temperature of a blackbody by a factor of 2, by what factor does its radiated power increase?

Solution

The Stefan-Boltzmann law (Equation 39X-1) says that the total radiated power, or luminosity, of a blackbody is proportional to T^4, so doubling the absolute temperature increases the luminosity by a factor of $2^4 = 16$.

Problem

3. At what wavelength does Earth, approximated as a 286-K blackbody, radiate the most energy?

Solution

The wavelength at which a blackbody at a given temperature radiates the maximum power is given by Wien's displacement law (Equation 39X-2): $\lambda_{\max} = (2.898 \text{ mm·K})/(286 \text{ K}) = 10.1 \ \mu\text{m}$, in the infrared.

Problem

5. According to Planck's theory, what is the minimum non-zero energy of a molecule with vibration frequency 3.4×10^{14} Hz?

Solution

In Planck's treatment, the energy of the lowest nonzero oscillator state is $hf = (4.136 \times 10^{-15} \text{ eV·s}) \times (3.4 \times 10^{14} \text{ Hz}) = 1.41 \text{ eV}$.

Problem

7. The Sun approximates a blackbody at 5800 K. Find the wavelength at which the Sun emits the most energy.

Solution

As in Problem 3, $\lambda_{\max} = 2.898 \text{ mm·K}/5800 \text{ K} = 500 \text{ nm}$.

Problem

9. Find the temperature range over which a blackbody will radiate the most energy at a wavelength in the visible range (400 nm–700 nm).

Solution

The Wien displacement law, for 400 nm $< \lambda_{\max} <$ 700 nm, gives $2.898 \text{ mm·K}/700 \text{ nm} = 4.14 \times 10^3 \text{ K} < T < 7.25 \times 10^3 \text{ K} = 2.898 \text{ mm·K}/400 \text{ nm}$. (See Equation 39X-2.)

Problem

11. Treating the Sun as a 5.8-kK blackbody, how does its ultraviolet radiance at 200 nm compare with its visible radiance at its peak wavelength of 500 nm?

Solution

As in Example 39X-1, the ratio of the blackbody radiances is

$$\frac{R(\lambda_2, T)}{R(\lambda_1, T)} = \left(\frac{\lambda_1}{\lambda_2}\right)^5 \left(\frac{e^{hc/\lambda_1 kT} - 1}{e^{hc/\lambda_2 kT} - 1}\right)$$
$$= \left(\frac{5}{2}\right)^5 \left(\frac{146.9}{2.66 \times 10^5}\right) = 5.39 \times 10^{-2},$$

where $\lambda_1 = 500$ nm, $\lambda_2 = 200$ nm, $T = 5800$ K, and $hc/k = 1.449 \times 10^{-2}$ m·K.

Problem

13. Use the series expansion for e^x (Appendix A) to show that Planck's law (Equation 39X-3) reduces to the Rayleigh-Jeans law (Equation 39X-5) when $\lambda \gg hc/kT$.

Solution

For $\lambda \gg hc/kT$, the exponent in Planck's law (Equation 39X-3) is small, and $e^x - 1 \approx x = hc/\lambda kT$. Then the radiance becomes $R(\lambda, T) \approx 2\pi hc^2/\lambda^5 \times (hc/\lambda kT) = 2\pi ckT/\lambda^4$, which is the Rayleigh-Jeans law.

Section 39X-3:   Photons
Problem

15. What is the wavelength of a 6.5-eV photon? In what spectral region is this?

Solution

Equation 39X-6 can be solved for the wavelength, $\lambda = c/f = hc/E_\gamma = (1240 \text{ eV·nm})/(6.5 \text{ eV}) = 191 \text{ nm}$. This photon is in the ultraviolet region of the electromagnetic spectrum.

Problem

17. A red laser at 650 nm and a blue laser at 450 nm emit photons at the same rate. How do their total power outputs compare?

Solution

The ratio of the photon energies (Equation 39X-6) is $E_{\text{blue}}/E_{\text{red}} = f_{\text{blue}}/f_{\text{red}} = \lambda_{\text{red}}/\lambda_{\text{blue}} = 650/450 = 1.44$. Since the lasers emit photons at the same rate, this is also the ratio of their power outputs.

Problem

19. Find the rate of photon production by (a) a radio antenna broadcasting 1.0 kW at 89.5 MHz, (b) a laser producing 1.0 mW of 633-nm light, and (c) an X-ray machine producing 0.10-nm X rays with a total power of 2.5 kW.

Solution

The rate of photon emission is the power output (into photons) divided by the photon energy, $\mathcal{P}/E_\gamma = \mathcal{P}/hf = \mathcal{P}\lambda/hc$. For the devices specified, this rate is (a) $1 \text{ kW}/(6.626\times10^{-34} \text{ J·s} \times 89.5 \text{ MHz}) = 1.69\times 10^{28} \text{ s}^{-1}$, (b) $(1 \text{ mW} \times 633 \text{ nm})/(6.626\times10^{-34} \text{ J·s} \times 3\times10^8 \text{ m/s}) = 3.18\times10^{15} \text{ s}^{-1}$, and (c) $1.26\times10^{18} \text{ s}^{-1}$.

Problem

21. Electrons in a photoelectric experiment emerge from an aluminum surface with maximum kinetic energy of 1.3 eV. What is the wavelength of the illuminating radiation?

Solution

Einstein's equation for the photoelectric effect (Equation 39X-7) gives $K_{\max} = (hc/\lambda) - \phi$, or $\lambda = hc/(K_{\max} + \phi) = 1240 \text{ eV·nm}/(1.3 + 4.28) \text{ eV} = 222 \text{ nm}$. (We used the work function listed in Table 39X-1.)

Problem

23. (a) Find the cutoff frequency for the photoelectric effect in copper. (b) Find the maximum energy of the ejected electrons if the copper is illuminated with light of frequency 1.8×10^{15} Hz.

Solution

(a) At the cutoff frequency, the photon energy equals the work function (i.e., $K_{\max} = 0$), or $\phi = hf_{\text{cutoff}} = E_\gamma$. Then $f_{\text{cutoff}} = \phi_{\text{Cu}}/h = (4.65 \text{ eV})/(4.136\times 10^{-15} \text{ eV·s}) = 1.12\times10^{15} \text{ Hz}$ (see Table 39X-1). (b) Einstein's photoelectric effect equation (Equation 39X-7) gives $K_{\max} = hf - \phi = h(f - f_{\text{cutoff}}) = (4.136\times10^{-15} \text{ eV·s})(1.8 - 1.12)\times10^{15} \text{ Hz} = 2.79 \text{ eV}$.

Problem

25. Which materials in Table 39X-1 exhibit the photoelectric effect *only* for wavelengths shorter than 275 nm?

Solution

The energy of a photon of wavelength 275 nm is $hc/\lambda = 4.51 \text{ eV}$. Such photons cannot eject photoelectrons from materials whose work functions are greater than this energy, which includes copper, silicon, and nickel in Table 39X-1. (Higher energy, or shorter wavelength, is necessary.)

Problem

27. A photon with wavelength 15 pm Compton-scatters off an electron, its direction of

motion changing by 110°. What fraction of the photon's initial energy is lost to the electron?

Solution

The fraction of the initial photon energy (hc/λ) lost is $(E - E')/E = 1 - E'/E = 1 - \lambda/\lambda' = \Delta\lambda/(\lambda + \Delta\lambda)$. From Equation 39X-8, $\Delta\lambda = \lambda_c(1 - \cos\theta) = 2.43 \text{ pm}(1 - \cos 110°) = 3.26 \text{ pm}$, and $\lambda = 15 \text{ pm}$; therefore the fraction lost is $3.26/18.26 = 17.9\%$.

Problem

29. The maximum electron energy in a photoelectric experiment is 2.8 eV. When the wavelength of the illuminating radiation is increased by 50%, the maximum electron energy drops to 1.1 eV. Find (a) the work function of the emitting surface and (b) the original wavelength.

Solution

The photoelectric effect equations (Equation 39X-7) for the two experimental runs are $K_{\max} = (hc/\lambda) - \phi = 2.8 \text{ eV}$, and $K'_{\max} = (hc/\lambda') - \phi = (hc/1.5 \lambda) - \phi = 1.1 \text{ eV}$. (b) Subtracting these, one gets $hc/3\lambda = 1.7 \text{ eV}$, or $\lambda = 1240 \text{ eV·nm}/5.1 \text{ eV} = 243 \text{ nm}$. (a) Substituting this wavelength into either photoelectric effect equation, one finds $\phi = (5.1 - 2.8) \text{ eV} = (3.4 - 1.1) \text{ eV} = 2.3 \text{ eV}$.

Problem

31. An electron is initially at rest. What will be its kinetic energy after a 0.10-nm X-ray photon scatters from it at 90° to its original direction of motion?

Solution

The kinetic energy of the recoil electron, in Compton scattering, equals the energy lost by the photon, $K = E - E' = (hc/\lambda) - (hc/\lambda') = hc(\lambda' - \lambda)/\lambda\lambda' = E\lambda_c(1 - \cos\theta)/[\lambda + \lambda_c(1 - \cos\theta)]$, where we used Equation 39X-8 for $\Delta\lambda = \lambda' - \lambda$. For the given data ($\theta = 90°$, $\lambda = 100 \text{ pm}$, and $E = hc/\lambda = 12.4 \text{ keV}$), we find $K = E\lambda_c/(\lambda + \lambda_c) = (12.4 \text{ keV})(2.43/102.43) = 294 \text{ eV}$.

Problem

33. What is the minimum photon energy for which it is possible for the photon to lose half its energy undergoing Compton scattering with an electron?

Solution

The solution to Problem 31 shows that if the photon loses half its energy, then $E - E' = \frac{1}{2}E = E\Delta\lambda\div$

$(\lambda + \Delta\lambda)$, or $\Delta\lambda = \lambda = hc/E$. Thus, $E = hc/\Delta\lambda = hc/\lambda_c(1 - \cos\theta)$. Since $\cos\theta \geq -1$, and $\lambda_c = h/m_ec$, the minimum energy is $hc/(h/m_ec)(1 - (-1)) = \frac{1}{2}m_ec^2 = \frac{1}{2}(511 \text{ keV}) = 255 \text{ keV}$.

Section 39X-4:   Atomic Spectra and the Bohr Atom

Problem

35. Which spectral line of the hydrogen Paschen series ($n_2 = 3$) has wavelength 1282 nm?

Solution

The wavelengths in the Paschen series for hydrogen are given by Equation 39X-9, with $n_2 = 3$ and $n_1 = 4, 5, 6, \ldots$. Thus, $\lambda = R_H^{-1}9n_1^2/(n_1^2 - 9)$. If one substitutes $\lambda = 1282$ nm and $R_H = 0.01097$ (nm)$^{-1}$, one finds $9n_1^2 = (1282\times 0.01097)(n_1^2 - 9)$, or $n_1 = \sqrt{9\times 1.41/(14.1 - 9)} = 5$, corresponding to the second line in this series.

Problem

37. What is the maximum wavelength of light that can ionize hydrogen in its ground state? In what spectral region is this?

Solution

The energy of the ground state of hydrogen is -13.6 eV (Equation 39X-12b), which has the same magnitude as the ionization energy. A photon with at least this energy has wavelength $\lambda \leq hc/|E| = 1240$ eV·nm/13.6 eV $= 91.2$ nm. This is the same as the Lyman series limit (Equation 39X-9 with $n_2 = 1$ and $n_1 = \infty$)$R_H^{-1} = hc/13.6$ eV, and lies in the ultraviolet.

Problem

39. It takes 4.9 eV of energy to excite electrons to a higher energy level in mercury atoms. What will be the wavelength of the photon emitted when an excited electron drops back to the lower level?

Solution

This transition in mercury atoms releases a photon of energy $\Delta E = 4.9$ eV $= hc/\lambda$, or $\lambda = 1240$ eV·nm$\div$ 4.9 eV $= 253$ nm. (Note: In the famous Franck-Hertz experiment in 1914, this transition dramatically demonstrated quantized atomic energy states, when the energy in a beam of electrons was absorbed in a tube filled with mercury vapor.)

Problem

41. At what energy level does the Bohr hydrogen atom have diameter 5.18 nm?

Solution

The diameter of a hydrogen atom in the Bohr model is (Equation 39X-13) $2r_n = 2n^2a_0$, so $n^2 = 5.18$ nm$\div$ 2(0.0529 nm) $= 49.0$, or $n = 7$. This is the sixth excited state.

Problem

43. Ultraviolet light with wavelength 75 nm shines on a gas of hydrogen atoms in their ground states, ionizing some of the atoms. What is the energy of the electrons freed in this process?

Solution

The final kinetic energy of the electron (actually the ionized atom with the nucleus assumed at rest) is its initial ground state energy (-13.6 eV) plus the energy absorbed from the photon (hc/λ), or -13.6 eV + 1240 eV·nm/75 nm $= 2.93$ eV.

Problem

45. Helium with one of its two electrons removed acts very much like hydrogen, and the Bohr model successfully describes it. Find (a) the radius of the ground-state electron orbit and (b) the photon energy emitted in a transition from the $n = 2$ to the $n = 1$ state in this singly ionized helium. *Hint:* The nuclear charge is $2e$.

Solution

Modifying the treatment of the Bohr atom in the text for singly ionized helium (He II), by replacing the nuclear charge with $2e$, one gets $r_n = -2ke^2/2E_n$ and $E_n = -(2ke^2)^2m/2\hbar^2n^2$. Thus, $E_n = -4(ke^2\div 2n^2a_0) = -2^2(13.6 \text{ eV})/n^2$, and $r_n = \frac{1}{2}n^2a_0$. (In general, replacing the nuclear charge with Ze gives results for any one-electron Bohr atom.) (a) The radius of the ground state of He II is $\frac{1}{2}a_0 = 0.0265$ nm. (b) The energy released in the transition $n = 2$ to $n = 1$ is $\Delta E = 4(13.6 \text{ eV})(1 - \frac{1}{4}) = 40.8$ eV. (There is also a small change in m for helium, different from the correction in hydrogen, for the motion of the nucleus.)

Section 39X-5:   Matter Waves

Problem

47. Find the de Broglie wavelength of (a) Earth, in its 30-km/s orbital motion, and (b) an electron moving at 10 km/s.

Solution

For non-relativistic momentum, Equation 39X-14 becomes $\lambda = h/mv$. (a) Using the values given for the

Earth, one finds $\lambda = (6.626 \times 10^{-34}$ J·s$)(5.97 \times 10^{24}$ kg$)^{-1}(30$ km/s$)^{-1} = 3.70 \times 10^{-63}$ m (much smaller than the smallest physically meaningful distance). (b) For the given electron, $\lambda = (6.626 \times 10^{-34}$ J·s$)(9.11 \times 10^{-31}$ kg$)^{-1}(10$ km/s$)^{-1} = 72.7$ nm.

Problem

49. A proton and an electron have the same de Broglie wavelength. How do their speeds compare, assuming both are much less than that of light?

Solution

The same de Broglie wavelength means the same momentum, so at non-relativistic speeds, $m_p v_p = m_e v_e$, or $v_e/v_p = m_p/m_e = 1836$ (use the "best known" values of physical constants given on the inside front cover).

Problem

51. Through what potential difference should you accelerate an electron from rest so its de Broglie wavelength will be the size of a hydrogen atom, about 0.1 nm?

Solution

Since 0.1 nm $= \lambda \gg \lambda_c = 2.43$ pm, non-relativistic expressions can be used to write the kinetic energy in terms of the de Broglie wavelength, $K = p^2/2m = (h/\lambda)^2/2m = h^2/2m\lambda^2 = (hc)^2/2mc^2\lambda = (1.24$ keV·nm$)^2/2(511$ keV$)(0.1$ nm$)^2 = 150$ eV. Since the kinetic energy gained in an acceleration from rest equals the potential energy difference, then $K = e\,\Delta V$, and $\Delta V = 150$ V.

Problem

53. A Davisson-Germer type experiment using nickel gives peak intensity in the reflected beam when the incident and reflected beams have an angular separation of 100°. What is the electron energy?

Solution

The Bragg angle in the Davisson-Germer experiment, for a deflection of 100°, is $\theta = \frac{1}{2}(180° - \phi) = \frac{1}{2}(180° - 100°) = 40°$ (see Fig. 39X-18). The Bragg condition (for the first-order intensity peak) gives $\lambda = 2d\sin\theta$, while the kinetic energy (for a non-relativistic electron) is $K = p^2/2m = (h/\lambda)^2/2m = (hc/\lambda)^2/(2mc^2)$. For scattering from a nickel crystal, $d = 91$ pm, so $K = (hc/2d\sin\theta)^2/(2mc^2) = (1.24$ keV·nm$/91$ pm $\sin 40°)^2/8(511$ keV$) = 110$ eV.

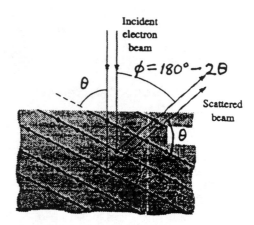

FIGURE 39X-18 Problem 53 Solution.

Section 39X-6: The Uncertainty Principle

Problem

55. A proton is confined to a space 1 fm wide (about the size of the atomic nucleus). What is the minimum uncertainty in its velocity?

Solution

From the uncertainty principle (Equation 39X-15) with $\Delta x = 1$ fm, $\Delta p \gtrsim \hbar/\Delta x = (197.3$ MeV·fm/$c) \div 1$ fm $= 197.3$ MeV/c. Although this is barely small enough compared to $mc = 938$ MeV/c to justify using the non-relativistic relation $p = mv$, this is good enough for approximate purposes, so $\Delta v = \Delta p/m \geq 197.3c/938 = 0.21c = 6.3 \times 10^7$ m/s.

Problem

57. A proton has velocity $\mathbf{v} = (1500 \pm 0.25)\hat{\imath}$ m/s. What is the uncertainty in its position?

Solution

Take the uncertainty in velocity to be the full range of variation given; that is, $\Delta v = 0.25$ m/s $- (-0.25$ m/s$) = 0.5$ m/s, and use $\Delta p = m\,\Delta v$ in Equation 39X-15 (since $v/c \ll 1$). Then

$$\Delta x \gtrsim \frac{\hbar}{m\,\Delta v} = \frac{(1.055 \times 10^{-34}\text{ J·s})}{(1.67 \times 10^{-27}\text{ kg})(0.5\text{ m/s})} = 126\text{ nm}.$$

Problem

59. Find the minimum energy for a neutron in a uranium nucleus whose diameter is 15 fm.

Solution

If one retraces the reasoning in Example 39X-8 for a neutron ($mc^2 \approx 940$ MeV) confined to a uranium nucleus ($\Delta x \approx 15$ fm) one finds a kinetic energy $K \geq (\hbar c/2\,\Delta x)^2/2mc^2 = (197.3$ MeV·fm/30 fm$)^2 \div$

$(2\times940$ MeV$) \simeq 23.0$ keV. (This is smaller than the 5 MeV estimated for the nucleon in Example 39X-8 by a factor of 15^2, since Δx is 15 times larger. Most estimates of nuclear energies for single-particle states, based on the uncertainty principle, give values of order 1 MeV, consistent with experimental measurements.)

Problem

61. A proton is moving along the x-axis with speed $v = (1500 \pm 0.25)$ m/s, but its direction (+ or $-x$) is unknown. Find the uncertainty in its position. *Hint:* What is the difference between the two extreme possibilities for the *velocity?*

Solution

The uncertainty in the proton's momentum is $\Delta p_x = 2mv = 2(1.67\times10^{-27}$ kg$)(1500$ m/s$) = 5.01\times10^{-24}$ kg·m/s, since p_x could vary between $+mv$ and $-mv$ if the direction is unknown. The Equation 39X-15 gives $\Delta x \gtrsim \hbar/\Delta p_x = (1.055\times10^{-34}$J·s$)\div(5.01\times10^{-24}$ kg·m/s$) = 21.1$ pm.

Problem

63. An electron is moving at 10^6 m/s, and you wish to measure its energy to an accuracy of $\pm0.01\%$. What is the minimum time necessary for the measurement?

Solution

The electron is non-relativistic ($v/c = 1/300 \ll 1$), and the uncertainty in its (kinetic) energy is $\Delta E = 2(0.01\%)\frac{1}{2}mv^2 = 10^{-4}mc^2(v/c)^2 = 10^{-4}(511$ keV$)\times(1/300)^2 = 5.68\times10^{-4}$ eV. An energy measurement of this precision (from Equation 39X-16) requires a time $\Delta t \simeq \hbar/\Delta E = 6.582\times10^{-16}$ eV·s$/5.68\times10^{-4}$ eV $= 1.16$ ps.

Problem

65. The lifetimes of unstable particles set energy-time uncertainty limits on the accuracy with which the rest energies—and hence the masses—of those particles can be known. The particle known as the neutral pion has rest energy 135 MeV and lifetime 8×10^{-17} s. Find the uncertainty in its rest energy (a) in eV and (b) as a fraction of the rest energy.

Solution

(a) The maximum time for a measurement of the pion's rest mass is about one lifetime, $\Delta t = 8\times10^{-17}$ s, so Equation 39X-16 gives $\Delta E \gtrsim \hbar/\Delta t = 6.582\times10^{-16}$ eV·s$/8\times10^{-17}$ s $= 8.23$ eV (this is the width of the neutral pion). (b) As a fraction of the rest energy, this is $\Delta E/E = 8.23$ eV$/135$ MeV $= 6.10\times10^{-8}$.

Paired Problems

Problem

67. (a) Find the wavelength at which a 2200-K blackbody has its peak radiance. (b) How does the radiance for 690-nm red light compare with the peak radiance?

Solution

(a) Wien's displacement law (Equation 39X-2) gives $\lambda_{max} = 2.898$ mm·K$/2200$ K $= 1.32$ μm (in the infrared). (b) From Planck's blackbody distribution law (Equation 39X-3), the ratio of the radiance at these two wavelengths is

$$\frac{R(0.69\ \mu m)}{R(1.32\ \mu m)} = \left(\frac{1.32}{0.69}\right)^5 \left(\frac{e^{6.59/1.32} - 1}{e^{6.59/0.69} - 1}\right) = 0.267,$$

where we used $hc/k = 1.449$ cm·K and $T = 2200$ K.

Problem

69. A photocathode ejects electrons with maximum energy 0.85 eV when illuminated with 430-nm blue light. Will it eject electrons when illuminated with 633-nm red light, and if so what will be the maximum electron energy?

Solution

Einstein's photoelectric effect equation, and the data for the blue light, give the work function of the photocathode material, $\phi = E_\gamma - K_{max} = hc/\lambda - K_{max} = (1240$ eV·nm$/430$ nm$) - 0.85$ eV $= 2.03$ eV. The energy of a photon of the red light is only $hc/\lambda = (1240$ eV·nm$)/(633$ nm$) = 1.96$ eV, and is insufficient to eject photoelectrons. (633 nm is greater than the cutoff wavelength of $\lambda_{cutoff} = hc/\phi = 610$ nm for the photocathode material.)

Problem

71. Find the initial wavelength of a photon that loses 20% of its energy when it Compton-scatters from an electron through an angle of 90°.

Solution

When a photon Compton scatters through 90°, the wavelength shift is $\Delta\lambda = \lambda_c$ (see Equation 39X-8), and the scattered photon has wavelength $\lambda' = \lambda + \lambda_c$. The energy of the scattered photon is 80% of the incident photon's, so $E'_\gamma = 0.8\ E_\gamma$, or $(\lambda + \lambda_c)/hc = \lambda/0.8hc$. Thus, $\lambda = 0.8\ \lambda_c/(1 - 0.8) = 0.8\times2.43$ pm$/0.2 = 9.72$ pm, where $\lambda_c = 2.43$ pm is the electron's Compton wavelength.

Problem

73. (a) Find the energy of the highest energy photon that can be emitted as the electron jumps between two adjacent allowed energy levels in the Bohr hydrogen atom. (b) Which energy levels are involved?

Solution

The energy of the photon emitted in a hydrogen atom transition between adjacent states ($n_1 \to n_2 = n_1 - 1$) is $E_\gamma = hc/\lambda = hcR_{\mathrm{H}}[(n_1 - 1)^{-2} - n_1^{-2}] = hcR_{\mathrm{H}} \times (2n_1 - 1)n_1^{-2}(n_1 - 1)^{-2}$ (see Equations 39X-6 and 9 and the discussion of the Bohr atom in the text). This is a maximum for the smallest allowed $n_1 > n_2 \geq 1$, which is $n_1 = 2$. Then (a) $E_{\gamma,\mathrm{max}} = \left(\frac{3}{4}\right) hcR_{\mathrm{H}} = \left(\frac{3}{4}\right)(13.6 \text{ eV}) = 10.2 \text{ eV}$, and (b) $n_2 = n_1 - 1 = 1$. (This is the Lyman alpha photon at 122 nm.)

Problem

75. In which of the following will wave properties be more evident: an electron moving at 10 Mm/s or a proton moving at 1 km/s?

Solution

The de Broglie wavelengths of the two particles, calculated from Equation 39X-14 and the non-relativistic momentum, $p = mv$, is $\lambda = h/mv = 7.28 \times 10^{-11}$ m for the electron, and 3.97×10^{-10} m for the proton. With a larger wavelength, the wave aspects of this proton would be more apparent, other things being equal.

Problem

77. An electron is known to be within about 0.1 nm of the nucleus of an atom. What is the uncertainty in its velocity?

Solution

We suppose that "within 0.1 nm" means ±0.1 nm, so we may take the uncertainty in the electron's position to be $\Delta x = 0.2$ nm (as in Example 39X-7). Then the minimum uncertainty in momentum is $\Delta p \gtrsim \hbar/\Delta x = 1.055 \times 10^{-34} \text{ J·s}/2 \times 10^{-10}$ m $= (197.3 \text{ eV·nm}/c) \div 0.2$ nm $\simeq 5 \times 10^{-25}$ kg·m/s $\simeq 1$ keV/c. This is small compared to $mc = 511$ keV/c, so non-relativistic expressions can be used. Then $\Delta v = \Delta p/m \approx (5 \times 10^{-25} \text{ kg·m/s})/(9.11 \times 10^{-31} \text{ kg}) \simeq 6 \times 10^5$ m/s $\simeq (1 \text{ keV}/c)/(511 \text{ keV}/c^2) \simeq 2 \times 10^{-3} c$. (We included both SI and atomic units for pedagogical reasons.)

Supplementary Problems

Problem

79. An electron is accelerated from rest through a 4.5-kV potential difference. What is its de Broglie wavelength?

Solution

The electron acquires a kinetic energy of $K = e \Delta V = 4.5$ keV, which is small compared to its rest energy ($mc^2 = 511$ keV) so its momentum is $p = \sqrt{2mK}$ and its de Broglie wavelength is $\lambda = h/p = hc/\sqrt{2mc^2 K} = 1.24 \text{ keV·nm}/\sqrt{2(511 \text{ keV})(4.5 \text{ keV})} = 18.3$ pm.

Problem

81. A photon's wavelength is equal to the Compton wavelength of a particle with mass m. Show that the photon's energy is equal to the particle's rest energy.

Solution

The Compton wavelength of a particle is $\lambda_c = h/mc$. The energy of a photon with this wavelength is $E_\gamma = hc/\lambda_c = hc/(h/mc) = mc^2$, the same as the particle's rest energy.

Problem

83. Solve the energy and momentum conservation equations given in the discussion of the Compton effect to verify Equation 39X-8. Show also that the electron's recoil angle ϕ can be written $\tan \phi = \frac{\cot(\theta/2)}{1 + \lambda_c/\lambda_0}$.

Solution

The conservation of momentum (for components parallel and perpendicular to the incident photon direction) gives $(E_0/c) = (E/c)\cos\theta + p_e \cos\phi$ and $0 = (E/c)\sin\theta - p_e \sin\phi$; therefore $\tan\phi = p_e \sin\phi \div p_e \cos\phi = E\sin\theta/(E_0 - E\cos\theta) = \sin\theta/(\lambda/\lambda_0 - \cos\theta)$. (Here, $E_0 = hc/\lambda_0$ and $E = hc/\lambda$ are the energies of the initial and scattered photon, E_0/c and E/c their momenta, p_e is the momentum of the recoil electron, and θ and ϕ are the scattering and recoil angles shown in Fig. 39X-10b.) The Compton scattering formula (Equation 39X-8), verified below, gives $\lambda/\lambda_0 = 1 + (\lambda_c/\lambda_0)(1 - \cos\theta) = 1 + (E_0/mc^2)(1 - \cos\theta)$, where $\lambda_c = h/mc$ is the electron's Compton wavelength. When this is substituted in the above equation for $\tan\phi$, one finds $\tan\phi = \sin\theta/(1 - \cos\theta) \times (1 + E_0/mc^2)$. But, $\sin\theta/(1 - \cos\theta) = \cot\frac{1}{2}\theta$ (use trigonometric identities in Appendix A; $\sin\theta = 2\sin\frac{1}{2}\theta\cos\frac{1}{2}\theta$ and $1 - \cos\theta = 2\sin^2\frac{1}{2}\theta$), and $E_0/mc^2 = hc/\lambda_0 mc^2 = \lambda_c/\lambda_0$, so the given expression for $\tan\phi$ follows. Now, Equation 39X-8 follows from the conservation of energy and momentum. The difference in the initial and scattered photon energies is the recoil kinetic energy of the electron, $E_0 - E = K_e = \sqrt{p_e^2 c^2 + m^2 c^4} - mc^2$. Eliminating $p_e c$ and ϕ from this and the momentum equations leads one to $p_e^2 c^2 (\sin^2\phi + \cos^2\phi) = E^2 + E_0^2 - 2E_0 E\cos\theta = p_e^2 c^2 = (E_0 - E + mc^2)^2 - m^2 c^4$. After cancelling terms and rearranging, one gets $E_0 E(1 - \cos\theta) = mc^2(E_0 - E)$, which is equivalent to Equation 39X-8 when $E_0 - E = hc\,\Delta\lambda/\lambda\lambda_0$ and $E_0 E = h^2 c^2/\lambda\lambda_0$ are inserted.

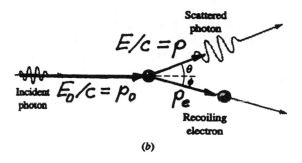

FIGURE 39X-10(b) Problem 83 Solution.

Problem

85. A photon undergoes a 90° Compton scattering off a stationary electron, and the electron emerges with *total* energy $\gamma m_e c^2$, where γ is the relativistic factor introduced in Chapter 38. Find an expression for the initial photon energy.

Solution

For Compton scattering at 90°, Equation 39X-8 gives $\lambda = \lambda_0 + \lambda_c$. In terms of the photon energies ($\lambda_0 = hc/E_0, \lambda = hc/E$) and the electron's Compton wavelength ($\lambda_c = hc/mc^2$), this can be written as $1/E = 1/E_0 + 1/mc^2$, or $E = E_0 mc^2/(E_0 + mc^2)$. The recoil electron's kinetic energy is $K_e = (\gamma - 1)mc^2 = E_0 - E = E_0 - E_0 mc^2/(E_0 + mc^2) = E_0^2/(E_0 + mc^2)$. This is a quadratic equation in E_0, namely $E_0^2 - (\gamma - 1)mc^2(E_0 + mc^2) = 0$. The positive solution for E_0

is $\frac{1}{2}[(\gamma - 1)mc^2 + \sqrt{(\gamma - 1)^2 m^2 c^4 + 4(\gamma - 1)m^2 c^4}] = \frac{1}{2}mc^2[(\gamma - 1) + \sqrt{(\gamma - 1)(\gamma + 3)}]$.

Problem

87. The total power per unit area emitted by a blackbody is given by integrating the power per area per unit wavelength (Equation 39X-3) over all wavelengths. Carry out this integration to show that the radiated power per unit area is proportional to the fourth power of the temperature. Show that your result also implies that the Stefan-Boltzmann constant is given by $\sigma = 2\pi^5 k^4/15c^2 h^3$. *Hint:* Use the quantity $u = hc/\lambda kT$ as your integration variable, and consult a table of definite integrals.

Solution

The power emitted per unit area is $\mathcal{P}/A = \int_0^\infty R(\lambda, T)\,d\lambda$. The integration is facilitated by introducing the dimensionless variable $x = hc/\lambda kT$. As λ goes from 0 to ∞, x goes from ∞ to 0, but $dx/x = -d\lambda/\lambda$ introduces a minus sign which switches the limits again, so

$$\int_0^\infty R(\lambda, T)\,d\lambda = 2\pi hc^2 \int_0^\infty \frac{1}{\lambda^4}\left(\frac{1}{e^{hc/\lambda kT} - 1}\right)\frac{d\lambda}{\lambda}$$
$$= 2\pi hc^2 \int_\infty^0 \left(\frac{kTx}{hc}\right)^4 \left(\frac{1}{e^x - 1}\right)\left(\frac{-dx}{x}\right)$$
$$= \frac{2\pi k^4 T^4}{c^2 h^3}\int_0^\infty \frac{x^3\,dx}{(e^x - 1)}.$$

The integral in this equation has the value $\pi^4/15$ (some tables of definite integrals may include it, but see below), so $\mathcal{P}/A = (2\pi^5 k^4/15c^2 h^3)T^4 = \sigma T^4$, which is Equation 39X-1 with the constant evaluated.

To establish the value of the above integral, use the binomial expansion to write $(e^x - 1)^{-1} = e^{-x}(1 - e^{-x})^{-1} = \sum_{n=1}^\infty e^{-nx}$. Then $\int_0^\infty x^3(e^x - 1)^{-1}\,dx = \sum_{n=1}^\infty(\int_0^\infty x^3 e^{-nx}\,dx)$. These integrals are in every table (and in fact can easily be found by integration by parts), $\int_0^\infty x^3 e^{-nx}\,dx = 3!/n^4$, and their sum (six times a zeta function, $\zeta(4)$) is also known to be $6\sum_{n=1}^\infty(1/n^4) = 6(\pi^4/90) = \pi^4/15$.

CHAPTER 40   QUANTUM MECHANICS

ActivPhysics can help with these problems:
Activities in Section 20, Quantum Mechanics

Section 40-2:   The Schrödinger Equation

Problem

1. What are the units of the wave function $\psi(x)$ in a one-dimensional situation?

Solution

The one-dimensional wave function is related to the probability by Equation 40-2, $dP = \psi^2(x)\,dx$. Since probability (a pure number) is dimensionless, the units of ψ must be the square root of the inverse of the units of x (a length), or $(\text{meters})^{-1/2}$.

Problem

3. The solution to the Schrödinger equation for a particular potential is $\psi = 0$ for $|x| > a$, and $\psi = A\sin(\pi x/a)$ for $-a \le x \le a$, where A and a are constants. In terms of a, what value of A is required to normalize ψ?

Solution

The wave function is like the first excited state of an infinite square well found in Section 40-3, except that $-a \le x \le a$ rather than $0 \le x' \le L$ as in Fig. 40-8. The correspondence is explicit if one takes $a = L/2$ and $x = x' - L/2$. Then $\sin(\pi x/a) = \sin[(2\pi x'/L) - \pi] = -\sin(2\pi x'/L)$, which is the wave function in Equation 40-5 for $n = 2$, except for an overall phase. The normalization constant for this wave function is $A = \sqrt{2/L} = \sqrt{1/a}$ (see Equation 40-7). (Of course,

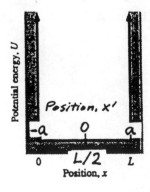

FIGURE 40-8 Problem 3 Solution.

A can be determined by repeating this integration,

$$1 = A^2 \int_{-a}^{a} \sin^2(\pi x/a)\,dx = aA^2.)$$

Problem

5. Use a table of definite integrals or symbolic math software to help evaluate the normalization constant A in the wave function of Problem 2.

Solution

The normalization condition for $\psi = Ae^{-x^2/a^2}$ is $1 = \int_{-\infty}^{\infty} A^2 e^{-2x^2/a^2}\,dx = A^2\sqrt{\pi/(2/a^2)} = A^2 a\sqrt{\pi/2}$. (See, for example, the CRC table of definite integrals.) Then $A = (2/\pi)^{1/4} a^{-1/2}$.

Problem

7. Describe the potential, in terms of x and the particle energy E, that would result in the wave function of Problem 4.

Solution

The Schrödinger equation, with the wave function in Problem 4, gives

$$-\frac{\hbar^2}{2m}\frac{\partial^2}{\partial x^2}[A(b^2 - x^2)] = -\frac{\hbar^2}{2m}(-2A)$$
$$= (E - U)A(b^2 - x^2),$$

so $U(x) = E - \hbar^2/m(b^2 - x^2)$. This holds, of course, for $|x| \le b$. For $|x| > b, U = \infty$, since ψ vanishes in this region.

Section 40-3:   The Infinite Square Well

Problem

9. What is the width of an infinite square well in which a proton cannot have an energy less than 100 eV?

Solution

The lowest possible energy of the proton (in a one-dimensional infinite square well) is its ground-state energy, which we take to be 100 eV. We can solve for the well-width from Equation 40-6, with $n = 1: L = h/\sqrt{8\,mE_1} = (1240 \text{ eV·nm/c}) \div \sqrt{8(938 \text{ MeV/c}^2)(100 \text{ eV})} = 1.43 \text{ pm}$.

Problem

11. A particle is confined to an infinite square well of width 1.0 nm. If the energy difference between the ground state and the first excited state is 1.13 eV, is the particle an electron or a proton?

Solution

From Equation 40-6, $\Delta E = (4 - 1)h^2/8mL^2 = 3(hc/L)^2/8(mc^2)$ is the energy difference between the first excited state ($n = 2$) and the ground-state ($n = 1$) of a one-dimensional infinite square well. Using given values, we find that $mc^2 = (3/8)(hc/L)^2 \div \Delta E = 3(1240 \text{ eV·nm}/1 \text{ nm})^2/8(1.13 \text{ eV}) = 510 \text{ keV}$, which is very close to the electron's rest energy.

Problem

13. An electron drops from the $n = 7$ to the $n = 6$ level of an infinite square well 1.5 nm wide. Find (a) the energy and (b) the wavelength of the photon emitted.

Solution

(a) $\Delta E = (n_i^2 - n_f^2)h^2/8mL^2 = (7^2 - 6^2) \times (1240 \text{ eV·nm}/c)^2/8(511 \text{ keV}/c^2)(1.5 \text{ nm})^2 = 2.17 \text{ eV}$ (from Equation 40-6). (b) $\lambda = hc/\Delta E = 8mc^2L^2 \div (n_i^2 - n_f^2)hc = 571 \text{ nm}$.

Problem

15. An electron is in a narrow molecule 4.4 nm long, a situation that approximates a one-dimensional infinite square well. If the electron is in its ground state, what is the maximum wavelength of electromagnetic radiation that can cause a transition to an excited state?

Solution

The smallest transition energy is to the first excited state ($n = 1$ to $n = 2$), so the maximum wavelength that can be absorbed is $\lambda = hc/\Delta E = hc \div [(4 - 1)h^2/8mL^2] = 8mc^2L^2/3hc = 8(511 \text{ keV}) \times (4.4 \text{ nm})^2/3(1240 \text{ eV·nm}) = 21.3 \ \mu\text{m}$.

Problem

17. Repeat Example 40-1 for a proton trapped in a nuclear-size square well of width 1 fm.
Comparison with the result of Example 40-1 gives a rough estimate of the energy difference between nuclear and chemical reactions.

Solution

The ground-state energy of a proton in an infinite square well of width 1 fm can be estimated from

Equation 40-6:

$$E_1 = \frac{h^2}{8mL^2} = \frac{(6.63 \times 10^{-34} \text{ J·s})^2}{8(1.67 \times 10^{-27} \text{ kg})(10^{-15} \text{ m})^2}$$
$$= 3.29 \times 10^{-11} \text{ J}.$$

In atomic units:

$$E_1 = \frac{(hc)^2}{8mc^2L^2} = \frac{(1240 \text{ MeV-fm})^2}{8(938 \text{ MeV})(1 \text{ fm})^2}$$
$$= 205 \text{ MeV}.$$

Comparison with Example 40-1 shows that the energy scale for nuclear reactions is millions of times greater than for chemical reactions.

Problem

19. An electron drops from the $n = 5$ level to the $n = 2$ level of an infinite square well, emitting a 1.4-eV photon in the process. Find the width of the square well.

Solution

The expression for the energy difference between two states in a one-dimensional infinite square well in the solution to Problem 13 gives $L^2 = (n_i^2 - n_f^2) \times (hc)^2/8mc^2\Delta E$, or $L = (1240 \text{ eV·nm}) \times \sqrt{(5^2 - 2^2)/8(511 \text{ keV})(1.4 \text{ eV})} = 2.38 \text{ nm}$.

Problem

21. Sketch the probability density for the $n = 2$ state of an infinite square well extending from $x = 0$ to

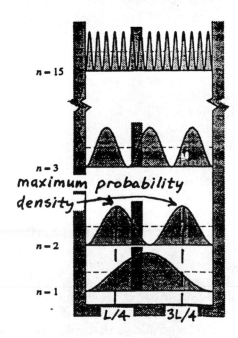

FIGURE 40-11 Problem 21 Solution.

$x = L$, and determine where the particle is most likely to be found.

Solution

The probability density for the $n = 2$ state of a one-dimensional infinite square well is $\psi_2(x)^2 = (2/L)\sin^2(2\pi x/L)$ (see Equation 40-7), a graph of which is shown in Fig. 40-11. The probability density has maxima when $2\pi x/L = \pi/2$ and $3\pi/2$, or at $x = L/4$ and $3L/4$, where the particle is most likely to be found (in a measurement with given Δx).

Problem

23. A particle is in the ground state of an infinite square well. What is the probability of finding the particle in the left-hand third of the well?

Solution

The ground-state wave function for an infinite square well is $\psi_1(x) = \sqrt{2/L}\sin(\pi x/L)$, and the left-hand third of the well extends from $x = 0$ to $x = L/3$. The probability of finding the particle in this interval is (see Example 40-2):

$$P = \int_0^{L/3} \frac{2}{L}\sin^2\left(\frac{\pi x}{L}\right) dx = \frac{2}{L}\left|\frac{x}{2} - \frac{\sin(2\pi x/L)}{(4\pi/L)}\right|_0^{L/3}$$

$$= \frac{1}{3} - \frac{\sin(2\pi/3)}{2\pi} = \left(\frac{1}{3}\right) - \left(\frac{\sqrt{3}}{4\pi}\right) = 0.196.$$

Sections 40-4, 40-5, and 40-6: The Harmonic Oscillator, Quantum Tunneling, and Finite Potential Wells

Problem

25. What is the ground-state energy for a particle in a harmonic oscillator potential whose classical angular frequency is $\omega = 1.0 \times 10^{17}$ s^{-1}?

Solution

The ground-state energy of a one-dimensional harmonic oscillator is $E_0 = \frac{1}{2}\hbar\omega = \frac{1}{2}(6.582 \times 10^{-16}$ eV·s$)(1.0 \times 10^{17}$ s$^{-1}) = 32.9$ eV (see Equation 40-11 with $n = 0$).

Problem

27. A hydrogen chloride molecule may be modeled as a hydrogen atom (mass 1.67×10^{-27} kg) on a spring; the other end of the spring is attached to a rigid wall (the massive chlorine atom). If the minimum photon energy that will promote this molecule to its first excited state is 0.358 eV, find the "spring constant."

Solution

The minimum energy difference between two states of a one-dimensional harmonic oscillator is $\Delta E = \hbar\omega$. The "spring constant" is therefore $k = m\omega^2 = m(\Delta E/\hbar)^2 = (1.67 \times 10^{-27}$ kg$)(0.358$ eV$/6.582 \times 10^{-16}$ eV·s$)^2 = 494$ N/m.

Problem

29. For what quantum numbers is the spacing between adjacent energy levels in a harmonic oscillator less than 1% of the actual energy?

Solution

The energy levels are $E_n = (n + \frac{1}{2})\hbar\omega$ (Equation 40-11), so the spacing of adjacent levels is $\Delta E = \hbar\omega$. Thus, $\Delta E/E_n = 1/(n + \frac{1}{2}) < 0.01$ for $n + \frac{1}{2} > 100$, or $n > 99$.

Problem

31. When the square well width L of Fig. 40-32, the well depth U_0, and the particle mass m are related by $\sqrt{mU_0L^2/2\hbar^2} = 4$, a detailed analysis shows that the ground-state energy is given by $E_1 = 0.098\,U_0$, and the corresponding wave function is approximately

$$\psi_1 = 17.9\frac{1}{\sqrt{L}}e^{7.60x/L} \qquad (x \leq -\tfrac{1}{2}L)$$

$$\psi_1 = 1.26\frac{1}{\sqrt{L}}\cos(2.50x/L) \qquad (-\tfrac{1}{2}L \leq x \leq \tfrac{1}{2}L)$$

$$\psi_1 = 17.9\frac{1}{\sqrt{L}}e^{-7.60x/L} \qquad (x \geq \tfrac{1}{2}L).$$

Show that both ψ_1 and its first derivative are continuous at the well edge. (That is, show that the second and third expressions above have the same value and the same derivative at $x = \frac{1}{2}L$.) Because the functions given are approximate, your results won't be exactly the same, but should agree to two significant figures. *Hint:* Be sure to set your calculator to radians before computing the trig functions! (b) Plot the function $\sqrt{L}\psi_1$ as a function of x/L, indicating the well edges on your plot.

Solution

(a) This wave function is symmetric about $x = 0$ (i.e., $\psi_1(x) = \psi_1(-x)$), so it is sufficient to consider only the edge at $x = L/2$. Then from inside the edge, $\psi_1(x = L/2) = 1.26\cos(2.50/2) = 0.397$, while from outside, $\psi_1(x = L/2) = 17.9e^{-7.60/2} = 0.400$ (the same to two significant figures), so ψ_1 is continuous. Inside the well, $\psi_1' = -(1.26)(2.50)L^{-3/2}\sin(2.50x/L)$, and outside $\psi_1' = \pm(7.60)(17.9)L^{-3/2}e^{\pm 7.60x/L}$ (where the positive signs are for $x < -L/2$ and the negative ones for $x > L/2$). At $x = L/2$, $(1.26)(2.50)\sin(1.25) = 2.99 \approx (7.60)(17.9)e^{-3.80} = 3.04$ (again approximately equal) so ψ_1' is also continuous. (b) This wave function is shown in Fig. 40-26.

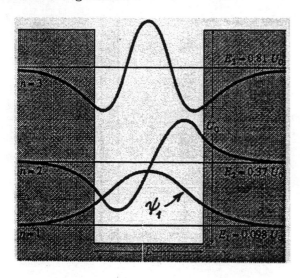

FIGURE 40-26 Problem 31 Solution.

Problem

33. The probability that a particle of mass m and energy $E < U$ will tunnel through the potential barrier of Fig. 40-16 is approximately

$$P = e^{-2\sqrt{2m(U-E)}L/h},$$

where L is the barrier width. Evaluate this probability (a) for a 2.8-eV electron incident on a 1.0-nm wide barrier 4.0 eV high and (b) for a 1200-kg car moving at 15 m/s striking a 1.0-m-thick stone wall requiring, classically, 150 kJ to breach it.

Solution

(a) For the electron, we use atomic units to calculate the probability. $P = \exp\{-2\sqrt{2mc^2(U-E)}L/hc\} = \exp\{-2\sqrt{2(511 \text{ keV})(4.0-2.8)\text{eV}}(1.0 \text{ nm})\div 1240 \text{ eV·nm}\} = e^{-1.79} = 0.168$. (b) For the car, SI units are more appropriate, and $E = \frac{1}{2}mv^2 = 135$ kJ. The probability of penetrating a 150-kJ-high barrier, 1 m thick, is $P = \exp\{-2\sqrt{2(1200 \text{ kg})(150-135)\text{kJ}}(1 \text{ m})/(6.626\times10^{-34} \text{ J·s})\} = \exp\{-1.81\times10^{37}\} = 10^{-7.87\times10^{36}}$, which is vanishingly small. (We used $e^x = 10^{x/\ln 10}$ to express the exponential as a power of ten.)

Section 40-7: Quantum Mechanics in Three Dimensions

Problem

35. A very crude model for an atomic nucleus is a cubical box about 1 fm on a side. What would be the energy of a gamma ray emitted if a proton in such a nucleus made a transition from its first excited state to the ground state?

Solution

The difference in energy between the first excited state (one quantum number equal to 2, the other two equal to 1, in Equation 40-12) and the ground-state (all three quantum numbers equal 1), for a proton (mass 938 MeV/c^2) in a cubical box (side length 1 fm) is $\Delta E = (2^2 + 1^2 + 1^2 - 1^2 - 1^2 - 1^2)h^2/8mL^2 = 3(hc/L)^2/8mc^2 = 3(1240 \text{ MeV·fm}/1 \text{ fm})^2 \div 8(938 \text{ MeV}) = 615$ MeV.

Solution

The quantum numbers , energy, and degeneracy of the first six levels in the three-dimensional infinite square well are most conveniently summarized in tabular form, or diagrammatically; see the accompanying table and diagram.

Problem

37. The generalization of the Schrödinger equation to three dimensions is

$$-\frac{\hbar^2}{2m}\left(\frac{\partial^2\psi}{\partial x^2} + \frac{\partial^2\psi}{\partial y^2} + \frac{\partial^2\psi}{\partial z^2}\right) + U(x,y,z)\,\psi = E\psi.$$

(a) For a particle confined to the cubical region $0 \le x \le L$, $0 \le y \le L$, $0 \le z \le L$, show by direct substitution that the equation is satisfied by wave functions of the form $\psi(x,y,z) = A\sin(n_x\pi x/L)\sin(n_y\pi y/L)\sin(n_z\pi z/L)$, where the n's are integers and A is a constant. (b) In the process of working part (a), verify that the energies E are given by Equation 40-12.

Solution

For the given wave function $\psi(x,y,z)$, the second partial derivatives are $\partial^2\psi/\partial x^2 = -(n_x\pi/L)^2\psi$, $\partial^2\psi/\partial y^2 = -(n_y\pi/L)^2\psi$, and $\partial^2\psi/\partial z^2 = -(n_z\pi/L)^2\psi$. Substituting into the Schrödinger equation, with a potential for a cubical box ($U = 0$ inside and $U = \infty$ outside), we find

$$-\frac{\hbar^2}{2m}\left(\frac{\partial^2\psi}{\partial x^2} + \frac{\partial^2\psi}{\partial y^2} + \frac{\partial^2\psi}{\partial z^2}\right)$$

$$= -\frac{1}{2m}\left(\frac{h}{2\pi}\right)^2\left[-\left(\frac{n_x\pi}{L}\right)^2 - \left(\frac{n_y\pi}{L}\right)^2\right.$$

$$-\left(\frac{n_z \pi}{L}\right)^2\bigg]\psi$$

$$= (h^2/8mL^2)(n_x^2 + n_y^2 + n_z^2)\psi = E\psi,$$

thus, (a) demonstrating that $\psi(x, y, z)$ is a solution, and (b) displaying the energy. This wave function also satisfies the boundary conditions appropriate for confinement ($\psi = 0$ at $x = 0$ or L, $y = 0$ or L, and $z = 0$ or L, points where $U = \infty$), since n_x, n_y and n_z are integers.

Section 40-8:   Relativistic Quantum Mechanics
Problem

39. Early in the history of the universe, the temperature was so high that thermal energy resulted in frequent pair creation. Find the approximate temperature such that the thermal energy kT would be enough to create (a) an electron-positron pair and (b) a proton-antiproton pair.

Solution

(a) The approximate temperature for electron-positron pair production is $kT = 2(mc^2)$, or $T = 2(511 \text{ keV}) \div (8.617 \times 10^{-5} \text{ eV} \cdot \text{K}^{-1}) = 1.19 \times 10^{10}$ K (use Boltzmann's constant in atomic units). (b) For proton-antiproton pair production, this estimate is $T = 2(938 \text{ MeV})/(8.617 \times 10^{-5} \text{ eV} \cdot \text{K}^{-1}) = 2.18 \times 10^{13}$ K.

Paired Problems
Problem

41. An alpha particle (mass 4 u) is trapped in a uranium nucleus of diameter 15 fm. Treating the system as a one-dimensional square well, what would be the minimum energy for the alpha particle?

Solution

Equation 40-6, with $n = 1$, gives the lowest alpha-particle energy, in a one-dimensional infinite square well of width 15 fm, as $E = h^2/8mL^2 = (hc/L)^2/8(mc^2) = (1240 \text{ MeV} \cdot \text{fm}/15 \text{ fm})^2 \div 8(4 \times 931.5 \text{ MeV}) = 0.229$ MeV (use 1 u = 931.5 MeV/c^2).

Problem

43. What is the probability of finding a particle in the central 80% of an infinite square well, assuming it's in the ground state?

Solution

The ground-state wave function is $\sqrt{2/L}\sin(\pi x/L)$, so the probability of finding the particle in the central

80% of the well ($0.1L \le x \le 0.9L$) is found, as in Example 40-2, to be:

$$P = \int_{0.1L}^{0.9L} \frac{2}{L}\sin^2\left(\frac{\pi x}{L}\right) \, dx$$

$$= \frac{2}{L}\left|\frac{x}{2} - \frac{\sin(2\pi x/L)}{(4\pi/L)}\right|_{0.1L}^{0.9L}$$

$$= 0.8 - (2\pi)^{-1}(\sin(2\pi - 36°) - \sin 36°)$$

$$= 0.8 + (\sin 36°)/\pi = 0.987.$$

(See also Problem 23.)

Problem

45. A harmonic oscillator emits a 1.1-eV photon as it undergoes a transition between adjacent states. What is its classical oscillation frequency?

Solution

For a one-dimensional harmonic oscillator, $\Delta E = \hbar\omega = hf$ for transitions between adjacent states (see Equation 40-11), so $f = \Delta E/h = 1.1 \text{ eV}/4.136 \times 10^{-15}$ eV·s $= 2.66 \times 10^{14}$ Hz.

Problem

47. An electron is confined to a cubical box. For what box width will a transition from the first excited state to the ground state result in emission of a 950-nm infrared photon?

Solution

The energy difference is $\Delta E = 3h^2/8mL^2 = hc/\lambda$ (see solution to Problem 35), so $L = \sqrt{3hc\lambda/8mc^2} = \sqrt{3(1240 \text{ eV} \cdot \text{nm})(950 \text{ nm})/8(511 \text{ keV})} = 0.930$ nm.

Supplementary Problems
Problem

49. For what quantum state is the probability of finding a particle in the left-hand quarter of an infinite square well equal to 0.303?

Solution

A straightforward generalization of Example 40-2 shows that the probability of finding a particle, in the quantum state n, in the left-hand quarter of a one-dimensional infinite square well is just

$$P = \int_0^{L/4} \psi_n^2(x) \, dx = \frac{2}{L}\left|\frac{x}{2} - \frac{\sin(2n\pi x/L)}{(4n\pi/L)}\right|_0^{L/4}$$

$$= \frac{1}{4} - \frac{\sin(n\pi/2)}{2\pi n}.$$

The probability is equal to $\frac{1}{4}$ for any even n, and is greater than $\frac{1}{4}$ for $n = 3 + 4n'$ (or less than $\frac{1}{4}$ for

$n = 1 + 4n'$) where $n' = 0, 1, 2, \ldots$ (This follows from the fact that $\sin(n\pi/2)$ alternates between ± 1 for integer n.) In this problem, the probability is greater than $\frac{1}{4}$, so $0.303 - 0.25 = \frac{1}{2}\pi n$, or $n = \frac{1}{2}\pi(0.053) = 3.00$, for the third quantum state.

Problem

51. A large number of electrons are all confined to infinite square wells 1.2 nm wide. They are undergoing transitions among all possible states. How many (a) visible (400 nm to 700 nm) and (b) infrared lines will there be in the spectrum emitted by this ensemble of square-well systems?

Solution

The transition energies and corresponding photon wavelengths for an electron confined to a one-dimensional infinite square well were given in the solutions to Problems 12 and 13. For this particular well, $\lambda = 8mc^2L^2/(n_i^2 - n_f^2)hc = 8(511\text{ keV})(1.2\text{ nm})^2 \div (n_i^2 - n_f^2)(1240\text{ eV·nm}) = 4.75\ \mu\text{m}/(n_i^2 - n_f^2)$, where $n_i > n_f$ are the quantum numbers for the initial and final states, (b) Infrared photons, with $\lambda > 0.7\ \mu\text{m}$, occur for transitions with $n_i^2 - n_f^2 < 4.75/0.7 = 6.78$. The only positive integers $n_i > n_f \geq 1$ satisfying this condition are for $2 \to 1$ and $3 \to 2$ transitions, so there are only two infrared photons in the spectrum. (Note: the spectrum includes only infrared and shorter wavelengths, so we did not need to consider the upper limit of the infrared region.) (a) For visible photons, $0.4\ \mu\text{m} < \lambda < 0.7\ \mu\text{m}$, or $6.78 < n_i^2 - n_f^2 < 11.9 = 4.75/0.4$. There are four transitions satisfying this condition: $3 \to 1$, $4 \to 3$, $5 \to 4$, and $6 \to 5$.

Problem

53. Consider an infinite square well with a steplike potential in the bottom, as shown in Fig. 40-33. Without solving any equations, sketch what you

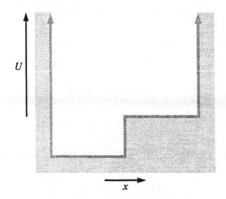

FIGURE 40-33 Problem 53.

think the wave function should look like for a particle whose energy is (a) less than the step height and (b) greater than the step height.

Solution

The solutions of the Schrödinger equation ($\psi'' = -[2\,m(E-U)/\hbar^2]\,\psi$) are oscillatory in regions where the kinetic energy of the particle is positive (ψ'' has the opposite sign to ψ if $E - U > 0$, so ψ always curves towards the x-axis), while the solutions are exponential in classically forbidden regions (ψ'' has the same sign as ψ if $E - U < 0$, so ψ always curves away from the x-axis). In regions where $E > U$, the wave function oscillates more rapidly when $E - U$ is greater (i.e., the deBroglie wavelength, $\lambda = h/\sqrt{2m(E-U)}$, is smaller). Except where the potential is infinite (in which case $\psi \equiv 0$), ψ and ψ' are continuous. In general, where oscillatory wave functions join, as at the step in part (b), the amplitude of the oscillations is greater where the wavelength is greater (classically, the probability density is greater where the velocity is smaller). (If the functions join at a point where $\psi' = 0$, the amplitudes are the same, indicative of a resonance condition.) Typical wave functions are sketched below for the potential in Fig. 40-33.

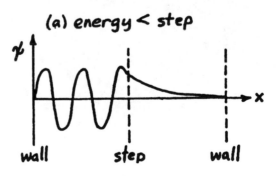

Problem 53 (a) Solution.

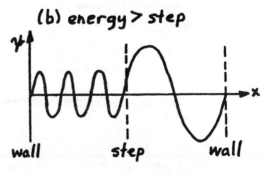

Problem 53 (b) Solution.

Problem

55. The next three problems illustrate the matching procedure necessary in solving the Schrödinger equation for potential wells of finite depth. Consider a particle of mass m in the semi-infinite well shown in Fig. 40-34. It will be convenient to work with dimensionless forms of the particle energy E and well depth U_0, in units of $\hbar^2/2mL^2 : \varepsilon \equiv 2mL^2E/\hbar^2$ and $\mu \equiv 2mL^2U_0/\hbar^2$. Assume that $E < U_0$ or, equivalently, $\varepsilon < \mu$. Show by substitution that the following wave functions satisfy the Schrödinger equation in the regions indicated:

$$\psi_1 = A\sin(\sqrt{\varepsilon}x/L)(0 \leq x \leq L)$$

and

$$\psi_2 = Be^{-\sqrt{\mu-\varepsilon}x/L} \ (x \geq L),$$

where A and B are arbitrary constants.

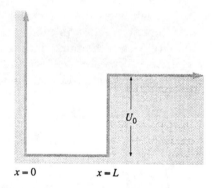

FIGURE 40-34 Problem 55–57.

Solution

See Problem 57.

Problem

57. The equation derived in the preceding problem cannot be solved algebraically since the unknown ε appears in a trig function and under the square root. It can be solved graphically, by plotting both sides on the same graph and determining where they intersect. It can also be solved by trial and error on a calculator, by using a calculator with a root-finding routine, or by computer. Use one of these methods to find all possible values of ε for (a) $\mu = 2$, (b) $\mu = 20$, and (c) $\mu = 50$. *Note:* The number of solutions varies with μ; there may be no solutions, meaning no bound states are possible, or there may be one or more bound states.

Solution

(The solutions of Problems 55–57 have been consolidated below.) In the interior of the well, $U(x) = 0$, while in the exterior, $U(x) = U_0$. After multiplication by $2m/\hbar$ and a rearrangement of terms, Equation 40-1 becomes:

$$\frac{d^2\psi}{dx_2} = \begin{cases} -(2mE/\hbar^2)\psi, & \text{for} \quad 0 \leq x \leq L \\ (2m(U_0 - E)/\hbar^2)\,\psi, & \text{for} \quad L \leq x \leq \infty. \end{cases}$$

Since $E > 0$, ψ is oscillatory inside the well. The solution which satisfies the condition $\psi = 0$ at $x = 0$ (where $U(x) = \infty$) is $\psi_{\text{in}} = A\sin(\sqrt{2mE}\,x/\hbar)$. Outside the well, the type of solution (exponential or oscillatory) depends on whether $U_0 - E$ is positive or negative. The former corresponds to a bound state and the latter to an unbound or scattering state. Here, we have assumed that $E < U_0$ (exponential ψ), so the normalizable solution outside is $\psi_{\text{out}} = Be^{-\sqrt{2m(U_0-E)}x/\hbar}$. Direct substitution of $\varepsilon/L^2 = 2mE/\hbar^2$ and $\mu/L^2 = 2mU_0/\hbar^2$ yields $\psi_{\text{in}} = A\sin \times (\sqrt{\varepsilon}x/L)$ and $\psi_{\text{out}} = Be^{-\sqrt{\mu-\varepsilon}x/L}$.

At $x = L$, $\psi_{\text{in}} = \psi_{\text{out}}$; therefore $A\sin\sqrt{\varepsilon} = Be^{-\sqrt{\mu-\varepsilon}}$, which is the first equation. $|d\psi_{\text{in}}/dx|_{x=L} = (\sqrt{\varepsilon}A/L)\cos\sqrt{\varepsilon} = |d\psi_{\text{out}}/dx|_{x=L} = -(\sqrt{\mu-\varepsilon}B/L)\times e^{-\sqrt{\mu-\varepsilon}}$, which, after multiplication by L, yields the second equation: $\sqrt{\varepsilon}A\cos\sqrt{\varepsilon} = -\sqrt{\mu-\varepsilon}Be^{-\sqrt{\mu-\varepsilon}}$. These equations can be combined, as indicated in the problem, to yield:

$$A\sin\sqrt{\varepsilon} = Be^{-\sqrt{\mu-\varepsilon}} = -\frac{\sqrt{\varepsilon}A\cos\sqrt{\varepsilon}}{\sqrt{\mu-\varepsilon}},$$

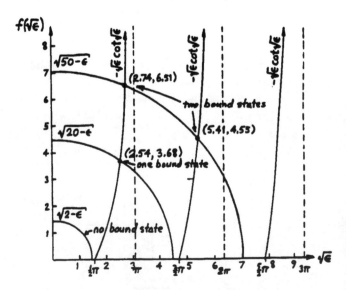

Problem 57 Solution.

or

$$\tan \sqrt{\varepsilon} = -\sqrt{\frac{\varepsilon}{\mu - \varepsilon}}.$$

Before a graphical solution is attempted, it is convenient to rewrite the above condition for the bound-state energies ($0 < \varepsilon < \mu$) as $-\sqrt{\varepsilon} \cot \sqrt{\varepsilon} = \sqrt{\mu - \varepsilon}$. The right-hand side, as a function of $\sqrt{\varepsilon}$, represents a quarter of a circle of radius $\sqrt{\mu}$. The left-hand side is a function which ranges from 0 to ∞ in the intervals $\frac{1}{2}\pi \le \sqrt{\varepsilon} \le \pi$, $\frac{3}{2}\pi \le \sqrt{\varepsilon} \le 2\pi$, $\frac{5}{2}\pi \le$

$\sqrt{\varepsilon} \le 3\pi$, etc., for which physical solutions are possible. From the sketch, it can be seen that there are no bound states for $\sqrt{\mu} < \frac{1}{2}\pi$ (or $\mu < 2.47$), there is one bound state for $\frac{1}{2}\pi < \sqrt{\mu} < \frac{3}{2}\pi$ (or $2.47 < \mu < 22.2$), there are two bound states for $\frac{3}{2}\pi < \sqrt{\mu} < \frac{5}{2}\pi$ (or $22.2 < \mu < 61.7$), etc. Values of ε can be found from the graph or by numerical computation. For the specified cases, they are: (a) none for $\mu = 2$, (b) $\varepsilon = 6.44$ for $\mu = 20$, and (c) $\epsilon = 7.52$ and 29.3 for $\mu = 50$.

CHAPTER 41   ATOMIC PHYSICS

ActivPhysics can help with these problems:
Activities 18.2, 18.4

Section 41-1:   The Hydrogen Atom

Problem

1. Using physical constants accurate to four significant figures (see the inside front cover), verify the numerical values of the Bohr radius a_0 and the hydrogen ground-state energy E_1.

Solution

$$a_0 = \frac{4\pi\varepsilon_0\,\hbar^2}{me^2}$$
$$= \frac{4\pi(8.854\text{ pF/m})(6.626\times10^{-34}\text{ J·S}/2\pi)^2}{(9.109\times10^{-31}\text{ kg})(1.602\times10^{-19}\text{ C})^2}$$
$$= 5.293\times10^{-11}\text{ m,}$$

while

$$hE_1 = -\frac{\hbar^2}{2ma_0^2}$$
$$= \frac{-(6.626\times10^{-34}\text{ J·s}/2\pi)^2}{2(9.109\times10^{-31}\text{ kg})(5.292\times10^{-11}\text{ m})^2}$$
$$= -2.180\times10^{-18}\text{ J} = -13.61\text{ eV.}$$

Problem

3. Find (a) the probability density and (b) the radial probability density at $r = \frac{1}{2}a_0$ for the hydrogen ground state. Comment on the relative sizes of these quantities.

Solution

(a) At $r = a_0/2$, the probability density for the hydrogen ground state (wave function given by Equation 41-3 and Example 41-1) is $\psi^2(a_0/2) = A^2 e^{-2(a_0/2)/a_0} = e^{-1}/\pi a_0^3 = e^{-1}/\pi(5.29\times10^{-11}\text{ m})^3 = 7.91\times10^{29}\text{ m}^{-3}$, while (b) the radial probability density is $4\pi(a_0/2)^2 = \pi a_0^2$ times this (see Equation 41-6) or $e^{-1}/a_0 = 6.95\times10^9\text{ m}^{-1}$.

Problem

5. What is the maximum possible magnitude for the orbital angular momentum of an electron in the $n = 7$ state of hydrogen?

Solution

The quantum number ℓ can take integer values from 0 to $n - 1$, so its maximum value is 6. From Equation 41-10, $L = \sqrt{6\times7}\,\hbar = \sqrt{42}\,\hbar$.

Problem

7. The orbital angular momentum of the electron in a hydrogen atom has magnitude 2.585×10^{-34} J·s. What is the minimum possible value for its energy?

Solution

From Equation 41-10, $\ell(\ell + 1) = (L/\hbar)^2 = (2.585/1.054)^2 = 6.01 \approx 6$. Therefore $\ell = 2$. Since $\ell \leq n - 1$, the smallest value of n is 3; thus the minimum energy is $-13.6\text{ eV}/(3)^2 = -1.51\text{ eV}$ (from Equation 41-7).

Problem

9. Determine the principal and orbital quantum numbers for a hydrogen atom whose electron has energy -0.850 eV and orbital angular momentum of magnitude $\sqrt{12}\,\hbar$.

Solution

Equations 41-9 and 10 can be written as $n^2 = -13.6\text{ eV}/E$, and $\ell(\ell + 1) = (L/\hbar)^2$. For the given state, $n^2 = 13.6/0.850 = 16 = (4)^2$, and $\ell(\ell + 1) = 12 = 3(4)$, so this is a $4f$-state ($n = 4$, $\ell = 3$).

Problem

11. A 1200-kg car rounds a turn of radius 150 m at a speed of 10 m/s. Assuming its angular momentum is quantized according to Equation 41-10, find the approximate value for ℓ.

Solution

As in Example 41-3, $L = mvr = (1200\text{ kg})(10\text{ m/s})\times(150\text{ m}) = 1.8\times10^6$ J·s $= \sqrt{\ell(\ell+1)}\,\hbar \approx \ell\hbar$, so $\ell \approx 1.8\times10^6$ J·s$/1.055\times10^{-34}$ J·s $\approx 1.7\times10^{40}$ (again overwhelmingly classical).

Problem

13. Give a symbolic description for the state of the electron in a hydrogen atom when the total energy

is -1.51 eV and the orbital angular momentum is $\sqrt{6}\,\hbar$.

Solution

From Equation 41-9, $n = \sqrt{13.6/1.51} = 3.00$, and from Equation 41-10, $\ell(\ell+1) = 6 = 2(3)$. Thus, $n = 3$ and $\ell = 2$, and this is a $3d$ state.

Problem

15. Which of the following pairs of energy and magnitude of orbital angular momentum are possible for a hydrogen atom, and to what n and ℓ values do they correspond? (a) -0.544 eV, 3.655×10^{-34} J·s; (b) -1.51 eV, 3.655×10^{-34} J·s; (c) -1.51 eV, 5.842×10^{-34} J·s; (d) -3.4 eV, 1.492×10^{-34} J·s.

Solution

Equations 41-9 and 10 require that -13.6 eV$/E = n^2$ and $(L/\hbar)^2 = \ell(\ell+1)$, where n and ℓ are non-negative integers such that $\ell \le n-1$. (a) The first pair of values gives $13.6/0.544 = 25.0$ and $(3.655/1.054)^2 = 12.0$, so this is a possible state with $n = 5$ and $\ell = 3$. (b) The energy -1.51 eV corresponds to $n = 2$, which is not compatible with $\ell = 3$. This is not a possible state for a hydrogen atom. (c) Similarly, $L = 5.842\times10^{-34}$ J·s corresponds to a greater ℓ-value than n ($\ell = 5 > n = 2$), so this is also impossible. (d) The final pair of values is possible, since $13.6/3.4 = 4.00$ and $(1.492/1.054)^2 = 2.00$ correspond to $n = 2$ and $\ell = 1$, respectively.

Problem

17. A hydrogen atom has energy $E = -0.850$ eV. What are the maximum possible values for (a) the magnitude of its orbital angular momentum and (b) the component of that angular momentum on a chosen axis?

Solution

Equation 41-9 shows that $n = \sqrt{-13.6\text{ eV}/E} = \sqrt{13.6/0.850} = 4$ for this hydrogen atom state. (a) The fact that $\ell_{max} = n - 1$ and Equation 41-10 imply $L_{max} = \sqrt{12}\,\hbar = 3.65\times10^{-34}$ J·s. (b) From Equation 41-11, with $m_{\ell,\,max} = \ell_{max}$, one finds $L_{z,\,max} = 3\hbar = 3.16\times10^{-34}$ J·s.

Problem

19. Substitute Equation 41-3 for ψ in Equation 41-4, and carry out the indicated differentiation to show that the result is Equation 41-5.

Solution

For the wave function $\psi = Ae^{-r/a_0}$, $d\psi/dr = -(A/a_0)e^{-r/a_0} = -\psi/a_0$, and $d^2\psi/dr^2 = \psi/a_0^2$. Substituting, after using the product rule to expand Equation 41-4, we obtain Equation 41-5:

$$-\frac{\hbar^2}{2mr^2}\left(r^2\frac{d^2\psi}{dr^2} + 2r\frac{d\psi}{dr}\right) - \frac{ke^2}{r}\psi$$

$$= -\frac{\hbar^2}{2mr^2}\left(\frac{r^2}{a_0^2} - \frac{2r}{a_0}\right)\psi - \frac{ke^2}{r}\psi$$

$$= \left(-\frac{\hbar^2}{2ma_0^2} + \frac{\hbar^2}{mra_0} - \frac{ke^2}{r}\right)\psi = E_1\psi.$$

Section 41-2:   Electron Spin

Problem

21. Verify the value of the Bohr magneton in Equation 41-15.

Solution

Values from the inside front cover substituted into Equation 41-15 yield:

$$\mu_B = \frac{e\hbar}{2m} = \frac{(1.602\times10^{-19}\text{ C})(6.626\times10^{-34}\text{ J·s})}{4\pi(9.109\times10^{-31}\text{ kg})}$$

$$= 9.273\times10^{-24}\text{ A·m}^2$$

Problem

23. Some very short-lived particles known as delta resonances have spin $\frac{3}{2}$. Find (a) the magnitude of their spin angular momentum and (b) the number of possible spin states.

Solution

(a) For $s = \frac{3}{2}$, Equation 41-12 gives $S = \sqrt{\frac{3}{2}\cdot\frac{5}{2}}\,\hbar = \frac{1}{2}\sqrt{15}\,\hbar$. (b) There are $2s + 1 = 2(\frac{3}{2}) + 1 = 4$ values of m_s, corresponding to states with $m_s = \pm\frac{1}{2}$ and $\pm\frac{3}{2}$.

Problem

25. What are the possible j-values for a hydrogen atom in the $3D$ state?

Solution

For the $3D$ state, $\ell = 2$ and $s = \frac{1}{2}$ (hydrogen has one electron), so the possible j-values are $j = \ell - \frac{1}{2} = \frac{3}{2}$ and $j = \ell + \frac{1}{2} = \frac{5}{2}$.

Problem

27. Draw vector diagrams similar to Fig. 41-12 for spin-orbit coupling of an electron in the $\ell = 3$ state.

Solution

The quantum numbers $\ell = 3$ and $s = \frac{1}{2}$ combine to give $j = |\ell - s| = \frac{5}{2}$ and $j = |\ell + s| = \frac{7}{2}$. For the lower state, $j = \frac{5}{2}$ and $J = \sqrt{j(j+1)}\,\hbar = \sqrt{35}\,\hbar/2$; for the upper, $j = \frac{7}{2}$ and $J = \sqrt{63}\,\hbar/2$. (Vector model diagrams add little to the understanding of angular momentum.)

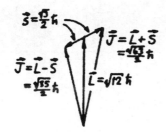

Problem 27 Solution.

Section 41-3: The Pauli Exclusion Principle

Problem

29. Suppose you put five electrons into an infinite square well. (a) How do the electrons arrange themselves to achieve the lowest total energy? (b) Give an expression for this energy in terms of the electron mass m, the well width L, and Planck's constant.

Solution

(a) The energy levels for a one-dimensional infinite square well are spacially non-degenerate (see Fig. 40-10), so the Pauli principle allows, at most, two electrons per level (one with spin up, one with spin down). Thus, two electrons may occupy the ground state $(n = 1)$, two the first excited state $(n = 2)$, and one the second excited state $(n = 3)$. (b) The energy of this configuration is $2E_1 + 2E_2 + E_3 = 2E_1 + 2(2)^2 E_1 + (3)^2 E_1 = 19E_1 = 19h^2/8mL^2$, where we used Equation 40-6 for the energies, with E_1 denoting the ground-state energy. (For a higher dimensional potential, degeneracy alters these conclusions accordingly.)

Problem

31. A harmonic oscillator potential of natural frequency ω contains a number of electrons and is in the state of lowest energy. If that energy is $6.5\,\hbar\omega$, (a) how many electrons are in the potential well and (b) what is the energy of the highest-energy electron?

Solution

The energy levels of single electrons in a one-dimensional harmonic oscillator potential are $E_n = (n + \frac{1}{2})\hbar\omega$ (Equation 40-11), and the Pauli principle limits each state to no more than 2 electrons. Since the total energy is $(\frac{13}{2})\hbar\omega = 2(\frac{1}{2}\hbar\omega) + 2(\frac{3}{2}\hbar\omega) + 1(\frac{5}{2}\hbar\omega)$, there must be 5 electrons, with the highest energy level, $\frac{5}{2}\hbar\omega$, singly occupied.

Section 41-4: Multielectron Atoms and the Periodic Table

Problem

33. Write the full electronic structure of scandium.

Solution

Scandium $(Z = 21)$, the first of the transition metals, has one $3d$-electron in addition to the configuration of the preceding element, calcium $(Z = 20)$. (See explanation in the text just before Example 41-7.) Thus $1s^2 2s^2 2p^6 3s^2 3p^6 4s^2 3d^1$ is the full electronic configuration for scandium.

Problem

35. Determine the electronic configuration of copper.

Solution

As mentioned in the text just before Example 41-6, the electronic configuration of copper is $1s^2 2s^2 2p^6 3s^2 3p^6 4s^1 3d^{10}$, instead of $\ldots 4s^2 3d^9$. The closed $3d$-subshell (with total angular momentum zero) has enough of a lower energy (in spin-orbit, orbit-orbit and spin-spin interactions) to compensate for the $4s - 3d$ difference.

Section 41-5: Transitions and Atomic Spectra

Problem

37. An electron in a highly excited state of hydrogen $(n_1 \gg 1)$ drops into the state $n = n_2$. What is the lowest value of n_2 for which the emitted photon will be in the infrared $(\lambda > 700 \text{ nm})$?

Solution

The wavelength emitted in a hydrogen atom transition from states n_1 to n_2 is given by $\Delta E = hc/\lambda = 13.6\,\text{eV} \times (n_2^{-2} - n_1^{-2})$ (subtract energy levels in Equation 41-9 or see Chapter 39). When $n_1 \gg 1$, $n_1^{-2} \approx 0$ and $hc/\lambda \approx 13.6 \text{ eV}/n_2^2$. Therefore, for photons in the infrared $(\lambda > 700 \text{ nm})$, $n_2 = \sqrt{\lambda(13.6 \text{ eV})/hc} > \sqrt{(700 \text{ nm})(13.6 \text{ eV})/(1240 \text{ eV·nm})} = 2.77$, or $n_2 \geq 3$. (Under similar conditions, i.e., $n_1 \approx \infty$, $n_2 \geq 105$ for microwave photons, $\lambda > 1$ mm.)

Problem

39. The $4f \rightarrow 3p$ transition in sodium produces a spectral line at 567.0 nm. What is the energy difference between these two levels?

Solution

This wavelength corresponds to a transition energy of $\Delta E = hc/\lambda = 1240$ eV·nm/567.0 nm $= 2.19$ eV.

Problem

41. The $4s \rightarrow 3p$ transition in sodium produces a doublet spectral line at 1138.1 nm and 1140.4 nm. Combine this fact with the discussion in Example 41-7 to find an accurate value for the energy difference between the $3s$ and $4s$ states in sodium.

Solution

Because the s-levels have no fine structure splitting, the transitions and photon energies given in this problem and in Example 41-7 can be combined to give the $4s - 3s$ energy difference. The transitions are $4s \xrightarrow{1140.4 \text{ nm}} 3P_{3/2} \xrightarrow{589.0 \text{ nm}} 3s$, and $4s \xrightarrow{1138.1 \text{ nm}} 3P_{1/2} \xrightarrow{589.6 \text{ nm}} 3s$, and the energy difference is $\Delta E = hc(\lambda_{4s-3p}^{-1} + \lambda_{3p-3s}^{-1})$. From either set of transitions:

$$1240 \text{ eV·nm} \left(\frac{1}{1140.4 \text{ nm}} + \frac{1}{589.0 \text{ nm}} \right)$$

$$= 1240 \text{ eV·nm} \left(\frac{1}{1138.1 \text{ nm}} + \frac{1}{589.6 \text{ nm}} \right) = 3.193 \text{ eV}.$$

Problem

43. Estimate the wavelength of the $K\alpha$ x-ray line in calcium.

Solution

Equation 41-20 can be used. The transition involves an electron jumping from the L-shell ($n_1 = 2$) to a vacancy in the K-shell ($n_2 = 1$). The nuclear charge is screened by one K-electron, so $Z_{\text{eff}} \approx Z - 1 = 20 - 1 = 19$ for calcium. Therefore,

$$\lambda = \frac{hc}{\Delta E} = \left(\frac{1240 \text{ eV·nm}}{13.6 \text{ eV}} \right) (19)^{-2} \left(\frac{1}{1^2} - \frac{1}{2^2} \right)^{-1}$$

$$= 0.337 \text{ nm}.$$

Problem

45. What is the approximate minimum accelerating voltage for an x-ray tube with an iron target to produce the $L\alpha$ line? *Hint:* See Problem 44.

Solution

In the operation of an x-ray tube, the bombarding electrons (with kinetic energy equal to the electronic charge times the accelerating voltage, $K_e = e\,\Delta V$) effectively eject an inner shell electron from the target atoms (as hinted at in the previous problem). The ionization energy of an L-shell electron in iron can be estimated (as in Example 41-8) from Equation 41-9, with $Z_{\text{eff}} = 26 - 9 = 17$, or $(13.6 \text{ eV})(17)^2/2^2 = 983$ eV. Thus, about 980 V accelerating voltage is necessary.

Problem

47. Use information from Fig. 41-35(a) to find the energy of the level E_3 and (b) to determine the maximum fraction of the energy delivered to excite neon atoms in a He-Ne laser that actually ends up as laser light.

Solution

(a) The energy of the metastable excited state in neon is given in the text (see Application: The Laser) as $E_2 = 20.66$ eV, above the ground state. The lasing transition energy follows from the wavelength given in Fig. 41-35, $E_2 - E_3 = hc/\lambda = 1240$ eV·nm/632.8 nm $= 1.96$ eV. Therefore, $E_3 = E_2 - hc/\lambda = 20.66$ eV $- 1.96$ eV $= 18.70$ eV, above the ground state. (b) The energy required to excite one neon atom, by pumping a helium atom to the desired state, is 20.61 eV, so only about $1.96/20.61 \approx 9.5\%$ of this appears in the laser radiation. (This ignores the energy required to maintain the gases in thermal equilibrium, so the actual quantum efficiency of the laser is much less than this.)

Problem

49. An ensemble of square-well systems of width 1.17 nm all contain electrons in highly excited states. They undergo all possible transitions in dropping toward the ground state, obeying the selection rule that Δn must be odd. (a) What wavelengths of visible light are emitted? (b) Is there any infrared emission? If so, how many spectral lines fall in the infrared?

Solution

The energy levels for an electron in an infinite square well are given by Equation 40-6. In atomic units, $E_n = n^2(hc)^2/8\,mc^2L^2 = n^2(1240 \text{ eV·nm})^2/8(0.511 \text{ MeV}) \times (1.17 \text{ nm})^2 = n^2(0.275 \text{ eV})$. The transition energies are proportional to the differences of squares of integers, $\Delta E_{fi} = (n_i^2 - n_f^2)(0.275 \text{ eV})$, where $\Delta n = n_i - n_f$ must be odd. (a) Photons in the visible region have energies between $hc/700$ nm $= 1.77$ eV and $hc/400$ nm $= 3.10$ eV, so $n_i^2 - n_f^2 = (n_i + n_f) \times (n_i - n_f) = (2n_i - \Delta n) \Delta n$ should lie between $1.77/0.275 = 6.45$ and $3.10/0.275 = 11.3$. For $\Delta n = 1$ transitions, this implies $6.45 < 2n_i - 1 < 11.3$, or $3.72 < n_i < 6.14$. Thus, $n_i = 4$, 5, or 6, and the transitions $4 \rightarrow 3$, $5 \rightarrow 4$, or $6 \rightarrow 5$ lead to visible photons with wavelengths $\lambda = hc/\Delta E_{fi} = 1240$ eV·nm$/(2n_i - 1)(0.275$ eV$) = 645$ nm, 501 nm, and 410 nm, respectively. The lowest energy $\Delta n = 3$ transition ($n_i^2 - n_f^2 = 4^2 - 1^2 = 15 > 11.3$) lies outside the visible region, in the ultraviolet. (b) The two $\Delta n = 1$ transitions $3 \rightarrow 2$ and $2 \rightarrow 1$ ($n_i^2 - n_f^2 = 5$ or $3 < 6.45$) both lie in the infrared at wavelengths 0.903 μm and 1.50 μm, respectively.

Paired Problems

Problem

51. Find the probability that the electron in the hydrogen ground state will be found in the radial distance range $r = a_0 \pm 0.1 a_0$.

Solution

The probability is given by the integral in Example 41-1, but with limits of $0.9 a_0$ to $1.1 a_0$:

$$P = \int_{0.9a_0}^{1.1a_0} 4\pi r^2 \psi^2 \, dr$$

$$= 4\pi \left(\frac{1}{\pi a_0^3} \right)$$

$$\times \left| \frac{r^2 e^{-2r/a_0}}{(-2/a_0)} - \frac{2}{(-2/a_0)} \left[\frac{e^{-2r/a_0}}{(-2/a_0)^2} \left(-\frac{2r}{a_0} - 1 \right) \right] \right|_{0.9a_0}^{1.1a_0}$$

$$= 2\{(0.9)^2 e^{-1.8} - (1.1)^2 e^{-2.2}$$

$$+ \frac{1}{2}[-(3.2)e^{-2.2} + (2.8)e^{-1.8}]\} = 0.108.$$

Problem

53. Find the spacing between adjacent orbital angular momentum values for hydrogen in the $n = 2$ state.

Solution

For the $n = 2$ states in hydrogen, the ℓ-values can be 0, or 1. The orbital angular momentum values are $L = \sqrt{\ell(\ell + 1)}\,\hbar = 0$, or $\sqrt{2}\,\hbar$, and their spacing is $\sqrt{2}\,\hbar$.

Problem

55. (a) Write the alpha-numeric description of a hydrogen atom whose energy is one-ninth of the ground-state energy (recall that the latter is

negative, so this is an excited state), whose orbital angular momentum has magnitude $\sqrt{6}\,\hbar$, and whose spin and orbital angular momentum are as nearly aligned as possible. (b) How many m_j values are there for this state?

Solution

(a) Since $E_n = E_1/n^2$ (where $E_1 = -13.6$ eV is the hydrogen ground-state energy), an energy of $E_1/9$ corresponds to the $n = 3$ shell. $L = \sqrt{\ell(\ell + 1)}\,\hbar = \sqrt{6}\,\hbar$ implies an orbital quantum number of $\ell = 2$ (a D-state), and "parallel" spin and orbital angular momenta means $j = \ell + \frac{1}{2} = \frac{5}{2}$. The spectroscopic notation for this state is $3D_{5/2}$. (b) There are $2j + 1 = 2(\frac{5}{2}) + 1 = 6$ values of m_j (ranging from $-\frac{5}{2}$ to $+\frac{5}{2}$ in integer steps).

Problem

57. Estimate the energy of the $L\alpha$ X-ray transition in arsenic.

Solution

For the $L\alpha$ line in As ($Z = 33$), we may use $Z_{\text{eff}} = Z - 9$ (as in Example 41-8) to estimate the transition energy $\Delta E(L\alpha) = (13.6$ eV$)(33 - 9)^2 \times (2^{-2} - 3^{-2}) = 1.09$ keV.

Supplementary Problems

Problem

59. Verify that the normalization constant $1/4\sqrt{2\pi a_0^3}$ in Equation 41-8 is correct.

Solution

Introduce the dimensionless variable $x = r/a_0$ to express the radial probability density for the hydrogen $2s$-wave function in Equation 41-8 as

$$P(r) \, dr = 4\pi r^2 \psi_{2s}^2 \, dr$$

$$= \frac{4\pi r^2 (2 - r/a_0)^2}{16(2\pi a_0^3)} e^{-r/a_0} a_0 d\left(\frac{r}{a_0} \right)$$

$$= \frac{1}{8} x^2 (2 - x)^2 e^{-x} \, dx = P(x) \, dx.$$

The normalization constant is verified if $\int_0^\infty P(x) \, dx = 1$. But $\int_0^\infty x^m e^{-x} \, dx = m!$ for any finite non-negative integer m (see any table of definite integrals, the Gamma Function, or use integration by parts), so $\frac{1}{8} \int_0^\infty x^2 (4 - 4x + x^2) e^{-x} \, dx = \frac{1}{8}(4 \times 2! - 4 \times 3! + 4!) = 1$, as expected. [Note: Integration by parts (see Appendix A) can be used to generalize the above definite integrals to

$$\int_{x_0}^{\infty} x^m e^{-x}\, dx = e^{-x_0} \sum_{r=0}^{m} \frac{m!}{(m-r)!} x_0^{m-r},$$

which is useful in the solution to the next problem.]

Problem

61. Substitute the ψ_{2s} wave function (Equation 41-8) into Equation 41-4 to verify that the equation is satisfied and that the energy is given by Equation 41-7 within $n = 2$.

Solution

First, we find the derivatives of $\psi_{2s} = Ae^{-r/2a_0} \times (2 - r/a_0)$, where $A = 1/4\sqrt{2\pi a_0^3}$, using the product rule and collecting terms:

$$d\psi_{2s}/dr = -Ae^{-r/2a_0}(2 - r/2a_0)/a_0,$$
$$d^2\psi_{2s}/dr^2 = Ae^{-r/2a_0}(3 - r/2a_0)/2a_0^2.$$

Next, we expand the derivatives on the left-hand side of Equation 41-4, substitute and factorize:

$$-\frac{\hbar^2}{2m}\left(\frac{d^2\psi_{2s}}{dr^2} + \frac{2}{r}\frac{d\psi_{2s}}{dr}\right)$$
$$= -\frac{\hbar^2}{2m}Ae^{-r/2a_0}\left[\frac{1}{2a_0^2}\left(3 - \frac{r}{2a_0}\right) - \frac{2}{ra_0}\left(2 - \frac{r}{2a_0}\right)\right]$$
$$= \frac{\hbar^2}{2m}Ae^{-r/2a_0}\frac{1}{ra_0}\left(4 - \frac{r}{2a_0}\right)\left(1 - \frac{r}{2a_0}\right)$$
$$= \frac{\hbar^2}{4mra_0}\left(4 - \frac{r}{2a_0}\right)\psi_{2s}.$$

When this is substituted into the full Equation 41-4, we can cancel a common factor of ψ_{2s} and use $\hbar^2/ma_0 = ke^2$:

$$\frac{\hbar^2}{4mra_0}\left(4 - \frac{r}{2a_0}\right) - \frac{ke^2}{r} = \frac{\hbar^2}{mra_0} - \frac{\hbar^2}{8ma_0^2} - \left(\frac{\hbar^2}{ma_0}\right)\frac{1}{r}$$
$$= \frac{-\hbar^2}{8ma_0^2} = E.$$

This is just the left-hand part of Equation 41-7 with $n = 2$.

Problem

63. (a) Verify Equation 41-9 by considering a single-electron atom with nuclear charge Ze instead of e. (b) Calculate the ionization energies for single-electron versions of helium, oxygen, iron, lead, and uranium.

Solution

(a) The substitution of Ze for the nuclear charge replaces ke^2 with Zke^2 in Equations 41-1, 2, 4 and 5, so a_0 is replaced by a_0/Z for the radius of the ground state. When this radius is inserted into Equation 41-7, the result is Equation 41-9. (b) The ionization energy of a one-electron atom is the magnitude of Equation 41-9 with $n = 1$ (see paragraph following Example 39X-4). Numerical values for the named nuclei are:

Z	2	8	26	82	92
$Z^2(13.6 \text{ eV})$	54.4 eV	870 eV	9.19 keV	91.4 keV	115 keV

CHAPTER 42   MOLECULAR AND SOLID-STATE PHYSICS

Section 42-2:   Molecular Energy Levels

Problem

1. Find the energies of the first four rotational states of the HCl molecules described in Example 42-1.

Solution

The energies of rotational states (above the $j = 0$ state) are given by Equation 42-2, where for the HCl molecule, $\hbar^2/I = 2.63$ meV (from Example 42-1). Thus, $E_{rot} = j(j+1)\hbar^2/2I = \frac{1}{2}j(j+1)2.63$ meV. For $j = 0$, 1, 2, and 3, $E_{rot} = 0$, 2.63 meV, 7.89 meV, and 15.78 meV.

Problem

3. A molecule drops from the $j = 2$ to the $j = 1$ rotational level, emitting a 2.50-meV photon. If the molecule then drops to the rotational ground state, what energy photon will be emitted?

Solution

The energy of a photon emitted in a transition between rotational levels with $\Delta j = -1$ is shown in Example 42-1 to be $\Delta E_{j \to (j-1)} = j\hbar^2/I$. Using the given data for the first transition, $j = 2$ to $j = 1$, we find $\hbar^2/I = 2.50 \times 10^{-3}$ eV$/2 = 1.25 \times 10^{-3}$ eV. This is also equal to the energy of a photon in the $j = 1$ to $j = 0$ transition.

Problem

5. Photons of wavelength 1.68 cm excite transitions from the rotational ground state to the first rotational excited state in a gas. What is the rotational inertia of the gas molecules?

Solution

The energy of the absorbed photon equals the difference in energy between the $j = 1$ and $j = 0$ rotational levels, which is (see Example 42-1) $\hbar^2/I = \Delta E_{1 \to 0} = hc/\lambda = 1240$ eV·nm/1.68 cm $= 7.38 \times 10^{-5}$ eV $= 1.18 \times 10^{-23}$ J. Therefore, $I = (1.055 \times 10^{-34}$ J·s$)^2/(1.18 \times 10^{-23}$ J$) = 9.41 \times 10^{-46}$ kg·m^2.

Problem

7. Find an expression for the energy of a photon required for a transition from the $(j-1)$th level to the jth level in a molecule with rotational inertia I.

Solution

This energy difference is calculated in Example 42-1. The energy difference between two adjacent rotational levels is proportional to the upper j-value.

Problem

9. The rotational spectrum of diatomic oxygen (O_2) shows spectral lines spaced 0.356 meV apart in energy. Find the atomic separation in this molecule. *Hint:* See Example 42-1, but remember that the oxygen atoms have equal mass.

Solution

The separation of the rotational spectral lines in energy is $\Delta(\Delta E) = \hbar^2/I$ (see Example 42-1), or $\hbar^2/I = 0.356$ meV for O_2. In a diatomic molecule, with equal-mass atoms and atomic separation R, each atom rotates about the center of mass at a distance of $R/2$, so $I = 2(m_O)(R/2)^2 = (8$ u$)R^2$, where the mass of an oxygen atom is about $m_O = 16$ u. Then $R^2 = I/8$ u $= \hbar^2/(8$ u$)(0.356$ meV$)$, or $R = \hbar c \div \sqrt{(8 \text{ uc}^2)(0.356 \text{ meV})} = (197.3 \text{ eV·nm}) \div \sqrt{(8 \times 931.5 \text{ MeV})(0.356 \text{ meV})} = 0.121$ nm.

Problem

11. For the HCl molecule of Example 42-2, determine (a) the energy of the vibrational ground state and (b) the energy of a photon emitted in a transition between adjacent vibrational levels. Assume the rotational quantum number does not change.

Solution

(a) The vibrational ground-state energy is $\frac{1}{2}\hbar\omega = \frac{1}{2}hf = \frac{1}{2}(4.136 \times 10^{-15}$ eV·s$)(8.66 \times 10^{13}$ Hz$) = 0.179$ eV (see Equation 42-3 with $n = 0$). (b) The photon energy for allowed transitions ($\Delta n = 1$) is $\hbar\omega$, which is twice the zero point energy, or 0.358 eV.

Problem

13. The energy between adjacent vibrational levels in diatomic nitrogen is 0.293 eV. What is the classical vibration frequency of this molecule?

Solution

From $\Delta E_{\text{vib}} = \hbar\omega = hf$, for adjacent levels, we find $f = 0.293 \text{ eV}/4.136 \times 10^{-15} \text{ eV·s} = 7.08 \times 10^{13}$ Hz for the N_2-molecule.

Problem

15. An oxygen molecule is in its vibrational and rotational ground states. It absorbs a photon of energy 0.19653 eV and jumps to the $n = 1$, $j = 1$ state. It then drops to the $n = 0$, $j = 2$ level, emitting a 0.19546-eV photon. Find (a) the classical vibration frequency and (b) the rotational inertia of the molecule.

Solution

For the $n = 0$, $j = 0$ to $n = 1$, $j = 1$ transition, the energy difference is 0.19653 eV $= \Delta E_{\text{vib}} + \Delta E_{\text{rot}} = hf + (\hbar^2/I)$, and for the $n = 1$, $j = 1$ to $n = 0$, $j = 2$ transition, 0.19546 eV $= hf - 2(\hbar^2/I)$ (see Equations 42-2 and 3). We can solve these equations simultaneously for hf and $\hbar^2/I$ to find f and I.
(a) $2(0.19653 \text{ eV}) + (0.19546 \text{ eV}) = 3hf$, or $f = 4.74 \times 10^{13}$ Hz. (b) $(0.19653 - 0.19546) \text{ eV} = 3(\hbar^2/I)$, or $I = 1.95 \times 10^{-46}$ kg·m².

Section 42-3:   Solids

Problem

17. Express the 7.84-eV ionic cohesive energy of NaCl in kilocalories per mole of ions.

Solution

Using conversion factors from Appendix C, we find: $(7.84 \text{ eV})(1.602 \times 10^{-19} \text{ J/eV})(1 \text{ kcal}/4184 \text{ J})(6.022 \times 10^{23}/\text{mol}) = 181$ kcal/mol.

Problem

19. Determine the constant n in Equation 42-4 for potassium chloride (KCl), for which $r_0 = 0.315$ nm and $U_0 = -7.21$ eV. The crystal structure is the same as for NaCl.

Solution

As shown in Example 42-3, $n = (1 + U_0 r_0/\alpha ke^2)^{-1}$. Since the crystal structures of KCl and NaCl are the same, $\alpha = 1.748$. Therefore:

$$n = \left[1 + \frac{(-7.21 \text{ eV})(0.315 \text{ nm})}{(1.748)(1.44 \text{ eV·nm})}\right]^{-1} = 10.2,$$

where we used a convenient value of ke^2 in atomic units:

$$ke^2 = (9 \times 10^9 \text{ N·m}^2/\text{C}^2)(1.6 \times 10^{-19} \text{ C})^2$$
$$\times (1 \text{ eV}/1.6 \times 10^{-19} \text{ J}) = 1.44 \text{ eV·nm}.$$

Problem

21. (a) Differentiate Equation 42-4 to obtain an expression for the force on an ion in an ionic crystal. (b) Use your result to find the force on an ion in NaCl if the crystal could be compressed to half its equilibrium spacing (see Example 42-3 for relevant parameters). Compare with the electrostatic attraction between the ions at a separation of $\frac{1}{2}r_0$. Your result shows how very "stiff" this ionic crystal is.

Solution

(a) The force in the r direction (positive is repulsive) is:

$$F_r = -\frac{dU}{dr} = \alpha\frac{ke^2}{r_0}\frac{d}{dr}\left[\frac{r_0}{r} - \frac{1}{n}\left(\frac{r_0}{r}\right)^n\right]$$
$$= \alpha\frac{ke^2}{r_0}\left[-\frac{r_0}{r^2} - \frac{n}{n}\left(\frac{r_0}{r}\right)^{n-1}\left(-\frac{r_0}{r^2}\right)\right]$$
$$= \alpha\frac{ke^2}{r_0^2}\left[\left(\frac{r_0}{r}\right)^{n+1} - \left(\frac{r_0}{r}\right)^2\right].$$

(b) If $r = \frac{1}{2}r_0$, $F_r = (\alpha ke^2/r_0^2)(2^{n+1} - 2^2) = \alpha(2^{n-1} - 1)F_{\text{el}}$, where $F_{\text{el}} = 4ke^2/r_0^2$ is the magnitude of the electrostatic attraction between a Na^+ and Cl^- ion at a distance of $\frac{1}{2}r_0$. If we use values from Example 42-3 ($\alpha = 1.748$, $r_0 = 0.282$ nm, and $n = 8.22$), the repulsive force is $(1.748)(2^{7.22} - 1) = 259$ times the electrostatic attractive force. Since $F_{\text{el}} = 4(9 \times 10^9 \text{ N·m}^2/\text{C}^2)(1.6 \times 10^{-19} \text{ C})^2/(0.282 \text{ nm})^2 = 1.16 \times 10^{-8}$ N, $F_r = 259(1.16 \times 10^{-8} \text{ N}) = 3.00 \times 10^{-6}$ N.

Problem

23. Determine the Fermi energy for calcium, which has 4.60×10^{28} conduction electrons per cubic meter.

Solution

The equation for n in Example 42-4 can be solved for the Fermi energy. With the aid of constants expressed in atomic units, the result for Ca is:

$$E_F = \frac{(hc)^2}{8mc^2}\left(\frac{3n}{\pi}\right)^{2/3}$$
$$= \frac{(1240 \text{ eV·nm})^2}{8(0.511 \text{ MeV})}\left(\frac{3 \times 4.60 \times 10^{28} \text{ m}^{-3}}{\pi}\right)^{2/3}$$
$$= 4.68 \text{ eV}.$$

Problem

25. Suppose the charge carriers in a material were protons, with density 10^{28} m^{-3}—comparable to that of electrons in a metal. What would be the order of magnitude of the Fermi energy?

Solution

Since the Fermi energy is proportional to the reciprocal of the charge carrier's mass (see solution to Problem 23), we expect the magnitude for protons to be 1/1836 times the Fermi energy for electrons (with comparable density), or about three orders of magnitude smaller (i.e., ~meV). In fact, if one uses $n = 10^{28}$ m^{-3} and $mc^2 = 938$ MeV (for protons) in the expression for E_F in Example 42-4, one finds $E_F = (3n/\pi)^{2/3}(hc)^2/8(mc^2) = (3\times10^{28}$ m$^{-3}/\pi)^{2/3}\times(1.24\times10^{-6}$ eV·m$)^2/8(938\times10^6$ eV$) = 9.22\times10^{-4}$ eV $\simeq$ 1 meV.

Problem

27. The *Fermi temperature* is defined by equating the thermal energy kT to the Fermi energy, where k is Boltzmann's constant. Calculate the Fermi temperature for silver ($E_F = 5.48$ eV), and compare with room temperature.

Solution

$T_F = E_F/k_B = 5.48$ eV$/(8.617\times10^{-5}$ eV·K$^{-1}) = 6.36\times10^4$ K, about 212 times room temperature (300 K).

Problem

29. What is the shortest wavelength of light that could be produced by electrons jumping the band gap in a material from Table 42-1? What is the material?

Solution

The wavelength emitted depends on the energy gap, since $\lambda = hc/E_{\text{gap}}$. The maximum wavelength for the materials in Table 42-1 (corresponding to the smallest gap) is for InAs, $\lambda = 1240$ eV·nm/0.35 eV = 3.54 μm (in the infrared). The minimum is for ZnS, $\lambda = 1240$ eV·nm/3.6 eV = 344 nm (in the ultraviolet).

Problem

31. A common light-emitting diode is made from a combination of gallium, arsenic, and phosphorous (GaAsP) and emits red light at 650 nm. What is its band gap?

Solution

The band gap corresponding to this wavelength is $E_{\text{gap}} = hc/\lambda = 1240$ eV·nm/650 nm = 1.91 eV.

Problem

33. The Sun radiates most strongly at about 500 nm, the peak of its Planck curve. The semiconductor zinc sulfide has a band gap of 3.6 eV. (a) What is the maximum wavelength absorbed by ZnS? (b) Would ZnS make a good photovoltaic cell? Why or why not?

Solution

(a) The photon wavelength corresponding to the band gap in ZnS is 344 nm (see solution to Problem 29), so photons with greater wavelengths (less energy) would not be readily absorbed (see statement of Problem 32). (b) ZnS would be a poor photovoltaic material since most of the energy in the solar spectrum is near the peak wavelength at 500 nm, and would not be absorbed.

Problem

35. A blue-green semiconductor laser being developed for long-playing compact discs emits at 447 nm (see Application: From CD to DVD, in Chapter 37). What is the band gap in this laser?

Solution

The band gap corresponding to photons of the given wavelength is $E_{\text{gap}} = hc/\lambda = 1240$ eV·nm/447 nm = 2.77 eV.

Section 42-4: Superconductivity

Problem

37. The critical magnetic field in niobium-titanium superconductor is 15 T. What current is required in a 5000-turn solenoid 75 cm long to produce a field of this strength?

Solution

The magnetic field inside a long thin solenoid is $B = \mu_0 nI$ (see Equation 30-11), so $I = 15$ T$/(5000\div0.75$ m$)(4\pi\times10^{-7}$ T·m/A$) = 1.79$ kA.

Paired Problems

Problem

39. The atomic spacing in diatomic hydrogen (H$_2$) is 74 pm. Find the energy of a photon emitted in a transition from the first rotational excited state to the ground state.

Solution

The rotational inertia of an H$_2$-molecule is $I = 2m_H \times (R/2)^2 = \frac{1}{2}m_H R^2$, since both atoms rotate about the center-of-mass, a distance $R/2$ from each atom, where R is the interatomic spacing (see Problems 9 and 52). The energy of the first rotational state $j = 1$ (above the ground state $j = 0$ and $E_0 = 0$) is $\Delta E_{1,0} = j(j+1)\hbar^2/2I = 2\hbar^2 m_H R^2 = 2(\hbar c/R)^2/m_H c^2 = 2(197.3 \text{ eV·nm}/74 \text{ pm})^2/(938 \text{ MeV}) = 15.2 \text{ meV}$, which equals the photon energy released in this transition. (We took the mass of a hydrogen atom approximately equal to a proton's mass.)

Problem

41. What wavelength of infrared radiation is needed to excite a transition between the $n = 0$, $j = 3$ state and the $n = 1$, $j = 2$ state in KCl, for which the rotational inertia is 2.43×10^{-45} kg·m^2 and the classical vibration frequency is 8.40×10^{12} Hz?

Solution

The difference in energy between these vibrational-rotational levels is $\Delta E = \Delta E_{\text{vib}} + \Delta E_{\text{rot}} = hf - 3\hbar^2/I = (4.136 \times 10^{-15} \text{ eV·s})(8.40 \times 10^{12} \text{ Hz}) - 3(6.582 \times 10^{-16} \text{ eV·s})^2(1.602 \times 10^{-19} \text{ J/eV}) \div (2.43 \times 10^{-45} \text{ J·s}^2) = 34.7 \text{ meV}$, corresponding to a photon wavelength of $\lambda = hc/\Delta E = 35.8$ μm. (See Example 42-2 for ΔE_{rot} between the $j = 2$ and $j = 3$ levels, and Equation 42-3 for ΔE_{vib} between the $n = 1$ and $n = 0$ levels. Note that $\Delta n = -\Delta j$ in this transition.)

Problem

43. Lithium chloride, LiCl, has the same structure and therefore the same Madelung constant as NaCl. The equilibrium separation in LiCl is 0.257 nm, and $n = 7$ in Equation 42-4. Find the ionic cohesive energy of the LiCl crystal.

Solution

As calculated in Example 42-3, the ionic cohesive energy for LiCl is $U_0 = -(\alpha ke^2/r_0)(1 - 1/n) = -(1.748)(1.44 \text{ eV·nm}/0.257 \text{ nm})(1 - 1/7) = -8.40 \text{ eV}$.

Supplementary Problems
Problem

45. What would be the Fermi energy in a one-dimensional infinite square well 10 nm wide and holding 100 electrons? Assume two electrons (with opposite spins) per energy level.

Solution

With two electrons of opposite spin in each level (which is non-degenerate for the one-dimensional infinite square well), the highest filled level for

100 electrons is $n = 50$. Thus, the Fermi energy is $E_F = n^2(hc/L)^2/8mc^2 = (50 \times 1240 \text{ eV·nm}/10 \text{ nm})^2 \div 8(511 \text{ keV}) = 9.40 \text{ eV}$ (see Equation 40-6).

Problem

47. Figure 21-23 shows that diatomic hydrogen acts like a monotomic gas at temperatures below about 100 K. Use the fact that the rotational inertia of H$_2$ is 4.6×10^{-48} kg·m^2 to show that this low-temperature behavior makes sense. *Hint:* Compare the thermal energy kT with the minimum energy needed to excite molecular rotation.

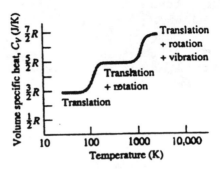

FIGURE 21-23 For reference.

Solution

The minimum energy needed to excite the first non-zero rotational level (ΔE for $j = 0$ to $j = 1$) in the H$_2$- molecule is $\hbar^2/I = (6.582 \times 10^{-16} \text{ eV·s}) \div (1.055 \times 10^{-34} \text{ J·s})/(4.6 \times 10^{-48} \text{ kg·m}^2) = 15.1 \text{ meV}$. A typical thermal energy at a temperature of 100 K is $k_B T = (8.617 \times 10^{-5} \text{ eV/K})(100 \text{ K}) = 8.62 \text{ meV}$, which is enough smaller than 15.1 meV so that one would not expect any rotational levels to be occupied. Thus, the specific heat of H$_2$ (and all gases which don't first liquify) should approach the monatomic value at sufficiently low temperatures, as shown in Fig. 21-23.

Problem

49. The transition from the ground state to the first rotational excited state in diatomic oxygen (O$_2$) requires about 356 μeV. At what temperature would the thermal energy kT be sufficient to set diatomic oxygen into rotation? Would you ever find diatomic oxygen exhibiting the specific heat of a monatomic gas at normal pressure?

Solution

With reference to the solution of Problem 47, O$_2$ would behave like a monatomic gas if there were insufficient thermal energy to excite rotational states.

This implies that $k_B T \lesssim 356$ μeV, or $T \lesssim 356$ μeV$\div$ (86.17 μeV/K) $\simeq 4.1$ K. Such behavior is not observed, however, since the normal boiling point of oxygen (below which O_2 liquifies) is 90.2 K (see Table 20-1).

Problem

51. The HCl bond-length calculation of Example 42-1 assumed the molecule was not vibrating. But because the ground-state vibrational energy is not zero, the bond is actually stretched slightly by its vibration. To estimate this stretching (a) use the classical vibration frequency of 8.66×10^{13} Hz to find the ground-state vibrational energy. (b) Use the result of part (a) and the fact that the effective "spring constant" for HCl is about 480 N/m to estimate the stretching of the molecular bond due to ground-state vibration. Assume that half the energy is potential energy of the stretched "spring." (c) Compare your result with the bond length found in Example 42-1.

Solution

(a) The ground-state vibrational energy corresponding to the given frequency is $E_0 = \frac{1}{2} hf = \frac{1}{2}(6.626 \times 10^{-34}$ J·s$)(8.66 \times 10^{13}$ Hz$) = 2.87 \times 10^{-20}$ J $= 0.179$ eV. (b) If half of this energy is associated with the potential energy of stretching the bond, then $\frac{1}{2} E_0 = \frac{1}{2} k \, \Delta x^2$, or $\Delta x = \sqrt{2.87 \times 10^{-20} \text{ J}/480 \text{ N/m}} = 7.73 \times 10^{-12}$ m ≈ 8 pm. (c) In this case, Δx is about 6% of the bond length found in Example 42-1, which is a small correction.

Problem

53. The Madelung constant (Section 42-3) is notoriously difficult to calculate because it is the sum of an alternating series of nearly equal terms. But it can be calculated for a hypothetical one-dimensional crystal consisting of a line of alternating positive and negative ions, evenly spaced (Fig. 42-39). Show that the potential energy of an ion in this "crystal" can be written

$$U = -\alpha \frac{ke^2}{r_0},$$

where the Madelung constant α has the value $2 \ln 2$. *Hint:* Study the series expansions listed in Appendix A.

FIGURE 42-39 Problem 53.

Solution

For any ion in Fig. 42-39, there are two oppositely charged ions at distances of r_0, two similarly charged ions at distances of $2r_0$, two opposite ions at $3r_0$, etc. Thus, the electrostatic potential energy of any ion is

$$U = -\frac{2ke^2}{r_0} + \frac{2ke^2}{2r_0} - \frac{2ke^2}{3r_0} + \cdots$$
$$= -\frac{2ke^2}{r_0}\left(1 - \frac{1}{2} + \frac{1}{3} - \cdots\right) = -\alpha \frac{ke^2}{r_0}.$$

Comparison of this with the series expansion of $\ln(1 + x)$ in Appendix A shows that $\alpha = 2 \ln 2$ for this "crystal". (Note: The convergence of this series, which needs special consideration for $x = 1$, is discussed in many first-year calculus text-books.)

CHAPTER 43 NUCLEAR PHYSICS

Section 43-1: Discovery of the Nucleus

Problem

1. In a head-on collision of a 9.0-MeV α particle and a nucleus in a gold foil, what is the minimum distance before electrical repulsion reverses the α particle's direction? Assume the gold nucleus remains at rest. Your answer shows how Rutherford set upper limits on the size of nuclei.

Solution

In Rutherford scattering of α-particles (charge $2e$) from approximately infinitely massive gold nuclei (charge Ze, with $Z = 79$ and no recoil assumed), we may equate the Coulomb potential energy at the distance of closest approach, $k(2e)(Ze)/r_{min}$, to the initial kinetic energy, 9 MeV in this case (see Equation 26-1 and use the conservation of energy). Therefore, 9 MeV $= 2\,Zke^2/r_{min}$, or $r_{min} = 2(79)\times$ (1.44 MeV·fm)/(9 MeV) = 25.3 fm. This is about 3.6 times the radius of a gold nucleus (Equation 43-1 gives $R = 1.2$ fm $(197)^{1/3} \simeq 7.0$ fm). However, for lighter nuclei (where r_{min} is comparable to R), deviations from the Rutherford scattering law, due to nuclear forces, allow an upper limit to be set on the nuclear size.

Section 43-2: Building Nuclei: Elements, Isotopes, and Stability

Problem

3. Three of the isotopes of radon ($Z = 86$) have 125, 134, and 136 neutrons, respectively. Write symbols for each.

Solution

With the number of protons ($Z = 86$ for all radon isotopes) and neutrons ($N = A - Z$) given, the mass numbers of the three isotopes are $A = Z + N = 86 + 125 = 211$, 220, and 222, respectively. The nuclear symbols are $^{211}_{86}\mathrm{Ra}$, etc.

Problem

5. How do (a) the number of nucleons and (b) the nuclear charge compare in the two nuclei $^{35}_{17}\mathrm{Cl}$ and $^{35}_{19}\mathrm{K}$?

Solution

(a) The mass number (number of nucleons) is $A = 35$ for both, but (b) the charge, Ze, of a potassium nucleus, $Z = 19$, is two electronic charge units greater than that for a chlorine nucleus, $Z = 17$.

Problem

7. Use Fig. 43-4 to find atomic numbers $Z < 83$ for which there are no stable isotopes. What elements are these?

Solution

Technetium, $Z = 43$, and promethium, $Z = 61$, have no stable isotopes.

Section 43-3: Properties of the Nucleus

Problem

9. Write the symbol for a boron nucleus with twice the radius of ordinary hydrogen ($^1_1\mathrm{H}$).

Solution

Since $R \sim A^{1/3}$ in Equation 43-1, a nucleus with twice the radius has eight times the mass number of $^1_1\mathrm{H}$, or $A = 8$. Reference to the periodic table shows $Z = 5$ for boron, so the symbol for this nucleus is $^8_5\mathrm{B}$.

Problem

11. A uranium-235 nucleus splits into two roughly equal-size pieces. What is their common radius?

Solution

Two fission products as equal as possible would have $A = 117$ or 118, and radii of about $R = (1.2 \text{ fm})A^{1/3} \simeq$ 5.9 fm.

Problem

13. Find the energy needed to flip the spin state of a proton in Earth's magnetic field, whose magnitude is about 30 μT.

Solution

As explained in Example 43-2, $\Delta U = 2\mu_p B =$ 2(30 μT)(1.41×10^{-26} J/T) = 8.46×10^{-31} J = 5.28× 10^{-12} eV. (The frequency of a photon with this energy, 1.28 kHz, is in the audible range!)

Problem

15. The permanent magnet from an old NMR spectrometer has a field strength of 1.4 T. What frequency of electromagnetic radiation was used in the instrument?

Solution

The energy of photons from the transmitter coil equals the spin-flip transition energy in the unit's magnetic field, or $hf = \Delta U = 2\mu_p B$. Thus, $f = 2\mu_p B/h = 2(1.41\times10^{-26}\ \text{J/T})(1.4\ \text{T})/(6.626\times10^{-34}\ \text{J}\cdot\text{s}) = 59.6$ MHz.

Section 43-4: Binding Energy

Problem

17. Determine the atomic mass of nickel-60, given that its binding energy is very nearly 8.8 MeV/nucleon.

Solution

The total nuclear binding energy of $^{60}_{28}\text{Ni}$ is the number of nucleons ($A = 60$) times the given binding energy per nucleon, or $E_b = 60(8.8\ \text{MeV})(1\ \text{u}\cdot c^2 \div 931.5\ \text{MeV}) = 0.567\ \text{u}\cdot c^2$. If we express Equation 43-3 in terms of atomic masses, by adding $Z = 28$ electron rest energies ($m_e c^2$) to both sides, and neglect atomic binding energies (as mentioned in the text following Example 43-3), we obtain:

$$M(^{60}_{28}\text{Ni}) = 28M(^1_1\text{H}) + (60 - 28)m_n - E_b/c^2$$
$$= 28(1.00783\ \text{u}) + 32(1.00867\ \text{u}) - 0.567\ \text{u}$$
$$= 59.930\ \text{u}.$$

(The actual binding energy of ^{60}Ni is so close to 8.8 MeV/nucleon that the accuracy of the atomic mass just calculated is better than one might expect from data given to two figures.)

Problem

19. Find the total binding energy of sodium-23, given its atomic mass of 22.989767 u.

Solution

From Equation 43-3, converted to atomic masses as in the solution to Problem 17, one finds $E_b = [ZM(^1_1\text{H}) + (A - Z)m_n - M(^A_Z X)]c^2 = [11(1.00783\ \text{u}) + 12\times (1.00867\ \text{u}) - 22.98977\ \text{u}]c^2 = (0.20040\ \text{u})\times (931.5\ \text{MeV/u}) = 187$ MeV, for $^{23}_{11}\text{Na}$. (Note: We rounded off the masses given in Table 43-1 and the problem, since we only desired three-figure accuracy in the final result.)

Problem

21. The mass of a lithium-7 nucleus is 7.01435 u. Find the binding energy per nucleon.

Solution

The binding energy per nucleon can be found from Equation 43-3 as written, since the nuclear mass of ^7_3Li is given. Then $E_b/A = [3(1.00728\ \text{u}) + (7 - 3)\times (1.00867\ \text{u}) - 7.01435\ \text{u}]\times(931.5\ \text{MeV/u})/7 = 5.61$ MeV/nucleon. (See note in solution of Problem 19 on accuracy.) The binding energy per nucleon, for very light nuclides, is low because the nuclear force is not yet saturated for so few nucleons.

Problem

23. By what percentage is the binding energy of $^{64}_{30}\text{Zn}$ different from that of $^{64}_{29}\text{Cu}$, another isotope with the same mass number? Use atomic masses of 63.92915 for ^{64}Zn and 63.92977 for ^{64}Cu.

Solution

The binding energies of $^{64}_{30}\text{Zn}$ and $^{64}_{29}\text{Cu}$ (calculated with atomic masses, as in the solution to Problem 19) are: $[30(1.00783\ \text{u}) + 34(1.00867\ \text{u}) - 63.92915\ \text{u}]c^2 = 0.60053\ \text{u}c^2$, and $[29(1.00783\ \text{u}) + 35(1.00867\ \text{u}) - 63.92977\ \text{u}]c^2 = 0.60075\ \text{u}c^2$. These differ by only 0.037%. Although ^{64}Zn and ^{64}Cu have almost the same binding energy, ^{64}Zn has a lower mass. Thus, ^{64}Cu is unstable, decaying to ^{64}Zn or ^{64}Ni with half-life at 12.7 h. The lower mass for ^{64}Zn is due to the nuclear pairing energy. In terms of the numbers of protons and neutrons, ^{64}Zn is an even-even nucleus, while ^{64}Cu is odd-odd.

Section 43-5: Radioactivity

Problem

25. The decay constant for argon-46 is 0.0835 s^{-1} What is its half-life?

Solution

Equation 43-6 gives $t_{1/2} = \ln 2/0.0835\ \text{s}^{-1} = 8.30$ s.

Problem

27. Referring to Fig. 43-21, write equations describing the decays of (a) radon-222 and (b) lead-214.

Solution

For the α-decay of radon-222 ($Z = 86$), an α particle is emitted as A decreases by four and Z by two: $^{222}_{86}\text{Rn} \rightarrow ^{218}_{84}\text{Po} + \alpha$ (see Equation 43-7). (b) In the β^--decay of lead-214 ($Z = 82$), A stays the same, Z increases by one, and a lepton-antilepton pair is created:

$^{214}_{82}\text{Pb} \rightarrow \, ^{214}_{83}\text{Bi} + e^- + \bar{\nu}_e$ (see Equation 43-8a). (The electron neutrino is different from two other types of neutrinos; see Chapter 45.)

Problem

29. A milk sample shows an iodine-131 activity level of 450 pCi/L. What is its activity in Bq/L?

Solution

Since 1 Ci $= 3.7 \times 10^{10}$ Bq, the activity of the milk sample in SI units is (450 pCi/L)(3.7$\times 10^{10}$ Bq/Ci) $=$ 16.7 Bq/L.

Problem

31. In the 4.5-billion-year lifetime of the Earth, how many half-lives have passed of (a) carbon-14, (b) uranium-238, and (c) potassium-40? For each isotope, give the number of atoms remaining today from a sample of 10^6 atoms at Earth's formation.

Solution

The age of the earth in half-lives of the isotopes specified (given in Table 43-2) is $n = t/t_{1/2} = $ (a) $4.5 \times 10^9/5730 = 7.85 \times 10^5$ for ^{14}C, (b) $4.5 \times 10^9/4.46 \times 10^9 = 1.01$ for ^{238}U, and (c) $4.5 \times 10^9/1.25 \times 10^9 = 3.60$ for ^{40}K. The number of atoms remaining from an original sample of 10^6 is $N = N_0/2^n$ (see Example 43-5). This is (a) $10^6/2^{7.85 \times 10^5} = 10^{-2.36 \times 10^5}$ (practically zero) for ^{14}C, (b) $10^6/2^{1.01} = 4.97 \times 10^5$ (one half) for ^{238}U, and (c) $10^6/2^{3.60} = 8.25 \times 10^4$ for ^{40}K.

Problem

33. How many atoms in a 1-g sample of U-238 decay in 1 minute? *Hint:* This time is so short compared with the half-life that you can consider the activity essentially constant.

Solution

Since N is essentially constant over time intervals much shorter than the half-life, the number decaying in the sample is just the activity, λN. (Consider Equation 43-4 for $dt = 1$ min and $\lambda = \ln 2/(4.46 \times 10^9$ y); dN/N is so small that N is essentially constant). In 1 g of ^{238}U, the number of atoms is approximately Avogadro's number divided by the atomic weight, $N \approx 6.02 \times 10^{23}/238 = 2.53 \times 10^{21}$, and λ from Table 43-2 and Equation 43-6 is $\lambda = \ln 2 \div (4.46 \times 10^9$ y$) = 2.95 \times 10^{-16}$ min^{-1}. Thus, the number decaying is $\lambda N = 7.47 \times 10^5$ decays/min.

Problem

35. The Swedish standard for I-131 in milk is 2000 Bq/liter. How long would Swedes have waited following the Chernobyl accident for their milk to be considered safe, assuming the same initial contamination level as in Example 43-6?

Solution

If the limit of safe activity in Example 43-6 is changed to 2000 Bq/L, the waiting time becomes $t = nt_{1/2} = $ (8.04 d) $\times \ln(2900/2000)/\ln 2 = 4.31$ d.

Problem

37. Nitrogen-13 has a 10-min half life. A sample of ^{13}N contains initially 10^5 atoms. Plot the number of atoms as a function of time from $t = 0$ to $t = 1$ hour. Make your horizontal axis (time) linear, but your vertical axis logarithmic. Why is the curve a straight line? What is the significance of its slope?

Solution

The number of nuclei remaining after time t is, from Equation 43-5b, $N(t) = N_0 2^{-t/t_{1/2}}$, so $\log N = \log N_0 - t(\log 2/t_{1/2}) = 5 - t(0.0301$ min^{-1}), where $N_0 = 10^5$ and $t_{1/2} = 10$ min as given. (Note that logarithmic graphs are normally base ten.) This is a straight line with slope $-\log 2/t_{1/2}$ and intercept $\log N_0$, i.e., the smaller the half-life, the steeper the slope. The time range specified in this case covers six half-lives.

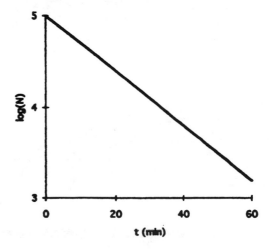

Problem 37 Solution.

Problem

39. How much cobalt-60 ($t_{1/2} = 5.24$ years) must be used to make a laboratory source whose activity will exceed 1 GBq for a period of 2 years?

Solution

Since the activity is $|dN/dt| = \lambda N$ (from Equation 43-4), Equations 43-5a and b show that activity decays with the same decay constant and half-life as the number of nuclei, i.e., $\lambda N = \lambda N_0 e^{-\lambda t} = \lambda N_0 2^{-t/t_{1/2}}$. If the activity of ^{60}Co must meet the specified condition after $t = 2$ y, $1 \text{ GBq} \le \lambda N = N_0 \times [\ln 2/(5.24 \times 3.156 \times 10^7 \text{ s})]2^{-2/5.24}$, or $N_0 \ge 3.11 \times 10^{17}$ nuclei, where we used Equation 43-6 for the decay rate in terms of the given $t_{1/2}$. The atomic weight of ^{60}Co is about 60 g and contains Avogadro's number of atoms, so the sample should have a mass of at least $(60 \text{ g})(3.11 \times 10^{17}/6.02 \times 10^{23}) = 31.0 \ \mu\text{g}$.

Problem

41. Marie Curie and Pierre Curie won the 1903 Nobel Prize for isolating 0.1 g of radium-226 chloride (^{226}RaCl$_2$). What was the activity of their sample, in Bq and in curies?

Solution

The activity, or the number of decays per unit time, of a sample of radioactive material is the magnitude of Equation 43-4, or λN. The decay constant for ^{226}Ra is given by Equation 43-6 and its half-life in Table 43-2: $\lambda = \ln 2/t_{1/2} = 0.693/1600 \text{ y} = 1.37 \times 10^{-11} \text{ s}^{-1}$. The number of ^{226}Ra nuclei in 10^{-1} g of RaCl$_2$ is one-tenth of Avogadro's number divided by the molecular weight, or about $N = 10^{-1} \times 6.022 \times 10^{23} \div (226 + 2 \times 35.45) \approx 2.03 \times 10^{20}$. Thus, the activity of this sample was $\lambda N = (1.37 \times 10^{-11} \text{ s}^{-1})(2.03 \times 10^{20}) = 2.78 \text{ GBq} = (2.78 \times 10^9 \text{ Bq})(1 \text{ Ci}/3.7 \times 10^{10} \text{ Bq}) = 75.2 \text{ mCi}$. (Note that the activity of 1 g of ^{226}Ra is approximately $1.37 \times 10^{-11} \text{ Bq} \times 6.02 \times 10^{23}/226 = 3.65 \times 10^{10} \text{ Bq} \approx 1 \text{ Ci}$.)

Problem

43. A mixture initially contains twice as many atoms of sodium-24 ($t_{1/2} = 15$ h) as it does of potassium-43 ($t_{1/2} = 22.3$ h). Plot the decay curves for the two isotopes, and use your graph to determine when the numbers of atoms are equal.

Solution

From Equation 43-5b and the given data, the numbers of ^{24}Na and ^{43}K-atoms are $N(^{24}\text{Na}) = 2N_0 2^{-t/15 \text{ h}}$ and $N(^{43}\text{K}) = N_0 2^{-t/22.3 \text{ h}}$. These are equal when the corresponding powers of 2 are equal, $1 - t/15 \text{ h} = -t/22.3 \text{ h}$, or $t = (15 \times 22.3 \text{ h})/(22.3 - 15) = 45.8 \text{ h}$. The decay curves ($N/N_0$ versus t) can be plotted and the time of their intersection estimated.

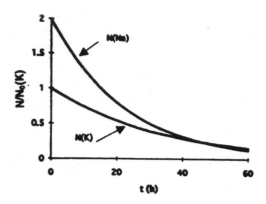

Problem 43 Solution.

Problem

45. Analysis of a moon rock shows that 82% of its initial K-40 has decayed to Ar-40, a process with a half-life of 1.2×10^9 years. How old is the rock?

Solution

If 82% of the original ^{40}K decayed, then 18% remains in a rock of age t. From Equation 43-5b and the given half-life, $t = t_{1/2} \ln(N_0/N)/\ln 2 = (1.2 \text{ Gy}) \ln \times (1/0.18)/\ln 2 = 2.97 \text{ Gy}$. (A type of lunar highlands rock rich in potassium (K), rare earth elements (REE), and phosphorus (P), is called KREEP norite.)

Problem

47. An excited state of the isotope technetium-99, designated Tc-99*, has a 6.01-hour half-life and is widely used in nuclear medicine. Tc-99* decays to Tc-99, an unstable isotope with a 2.13×10^5-year half-life. A 0.10-μg sample initially contains pure Tc-99*. Find its activity level at (a) $t = 0$, (b) $t = 1$ day, (c) $t = 1$ week, (d) $t = 1$ year, and (e) $t = 10$ years. The atomic weight of Tc-99 is 98.9 u.

Solution

(a) A sample of 0.1 μg of $^{99}_{43}$Tc* contains $N_0^* = 0.1 \ \mu\text{g}/98.9 \text{ u} = (10^{-7}/98.9)6.02 \times 10^{23} = 6.09 \times 10^{14}$ nuclei, and the decay constant of ^{99}Tc* is $\lambda^* = \ln 2 \div t_{1/2}^* = \ln 2/6.01 \text{ h} = 3.20 \times 10^{-5} \text{ s}^{-1}$, so the initial activity is $\lambda^* N_0^* = 19.5 \text{ GBq}$. The activity of the parent species decays with half-life $t_{1/2}^*$, so at time t, it is $\lambda^* N^* = \lambda^* N_0^* e^{-\lambda^* t} = (19.5 \text{ GBq})2^{-t/t_{1/2}^*}$ (see solution to Problem 39). However, the daughter nuclei, ^{99}Tc, are also radioactive (with half-life $t_{1/2} = 2.13 \times 10^5$ y and decay constant $\lambda = 1.03 \times 10^{-13} \text{ s}^{-1} \ll \lambda^*$) so they also contribute to the activity. The number of daughter nuclei that decay in a time interval dt is $-\lambda N \, dt$ (as in Equation 43-4) but a daughter nucleus is produced for each parent nucleus which decays, so the net change is $dN = -\lambda N\,dt + \lambda^* N^*\,dt = (-\lambda N + \lambda^* N_0^* e^{-\lambda^* t})dt$. The solution of this differential equation [i.e., the number of ^{99}Tc-nuclei as a function of time, $N(t)$], which satisfies the initial condition $N(0) = 0$ (pure parent means no daughters at $t = 0$), is $N(t) = \lambda^* N_0^* (e^{-\lambda^* t} - e^{-\lambda t})/(\lambda - \lambda^*)$ (see below). Therefore, the total activity of this sample is $(\lambda N)_{\text{tot}} = \lambda^* N^* + \lambda N = \lambda^* N_0^* [e^{-\lambda^* t} + (e^{-\lambda t} - e^{-\lambda^* t})\lambda/(\lambda^* - \lambda)] \approx \lambda^* N_0^* [e^{-\lambda^* t} + (\lambda/\lambda^*)e^{-\lambda t}] = (19.5\text{ GBq})[2^{-t/t^*_{1/2}} + (3.22 \times 10^{-9})2^{-t/t_{1/2}}]$, where we used the approximation that $\lambda/\lambda^* = t^*_{1/2} = 6.01$ h $\div$ 2.13×10^5 y is very small just in the coefficients of the exponentials. In this case, the first term dominates for times less than about $20 t^*_{1/2} \simeq 5$ d (see tip following Example 43-6) and the second term dominates for times greater than about $40 t^*_{1/2} \simeq 10$ d; in between these times, both terms contribute. (Note also that for times on a human scale, $t \ll t_{1/2} = 2.13 \times 10^5$ y, and $2^{-t/t_{1/2}} \approx 1$.) Therefore, (b) for $t = 1$ d, $(\lambda N)_{\text{tot}} \simeq (19.5\text{ GBq})(6.28 \times 10^{-2}) = 1.22$ GBq, (c) for $t = 7$ d, $(\lambda N)_{\text{tot}} \simeq (19.5\text{ GBq})(3.85 \times 10^{-9} + 3.22 \times 10^{-9}) = 138$ Bq, (d) and (e) for $t = 1$ y and 10 y, $(\lambda N)_{\text{tot}} \approx (19.5\text{ GBq})(3.22 \times 10^{-9}) = 62.8$ Bq.

The general solution of a linear first-order differential equation, like $dN/dt = -\lambda N + \lambda^* N_0^* e^{-\lambda^* t}$, can be found by adding a particular solution, and a general solution to the corresponding homogeneous equation (just the terms with the function N and its derivative, i.e., $dN/dt = -\lambda N$). Since the derivative of $e^{-\lambda^* t}$ is proportional to $e^{-\lambda^* t}$, a particular solution is easily found by trying $Ce^{-\lambda^* t}$, where C is a constant. Substituting this, we find $d(Ce^{-\lambda^* t})/dt = -\lambda^* Ce^{-\lambda^* t} = -\lambda(Ce^{-\lambda^* t}) + \lambda^* N_0^* e^{-\lambda^* t}$, from which C is determined to be $C = \lambda^* N_0^*/(\lambda - \lambda^*)$. The solution of the homogeneous equation is just constant $\times\, e^{-\lambda t}$, so the general solution to the original equation is $N(t) = Ce^{-\lambda^* t} + \text{constant} \times e^{-\lambda t}$. If we take the constant $= -C$, the initial condition $N(0) = 0$ is satisfied, and $N(t) = \lambda^* N_0^* (e^{-\lambda^* t} - e^{-\lambda t})/(\lambda - \lambda^*)$ as above.

Problem

49. Today, uranium-235 comprises only 0.72% of natural uranium; essentially all the rest is U-238. Use the half-lives given in Table 43-2 to determine the percentage of uranium-235 in natural uranium when Earth formed, about 4.5 billion years ago.

Solution

Suppose that when the Earth formed ($t = 0$), natural uranium consisted of just the two longest-lived isotopes in Table 43-2 (^{234}U has an abundance of

0.0057%). Then the percentage of ^{235}U today ($t = 4.5$ Gy) is $0.0072 = N^{235}/(N^{235} + N^{238})$, or $N^{238} = 138\ N^{235}$. The original amounts of the two isotopes are given by Equation 43-5b, with half-lives from Table 43-2, as $N_0^{235} = N^{235}2^{4.5/0.704}$ and $N_0^{238} = N^{238}2^{4.5/4.46}$, so that the original percentage must have been

$$\frac{N_0^{235}}{N_0^{235} + N_0^{238}}$$
$$= \frac{1}{1 + (N_0^{238}/N_0^{235})}$$
$$= \frac{1}{1 + (N^{238}/N^{235})2^{(4.5/4.46) - (4.5/0.704)}}$$
$$= \frac{1}{1 + (138)2^{-5.38}} = \frac{1}{1 + 3.30} = 23.2\%$$

(Actually, from current data on nuclear reactions and models of nucleosynthesis in supernova explosions, one can predict the isotopic abundances of U^{235} and U^{238} when they were produced. By reversing the above argument, one can then estimate the age of the elements in the nebula from which the solar system formed.)

Section 43-6: Models of Nuclear Structure

Problem

51. For each of the magic numbers, give an isotope that has a magic number of protons and one with a magic number of neutrons.

Solution

The magic numbers of protons or neutrons in the shell model (for which there are stable nuclei) are 2, 8, 20, 28, 50, 82, and 126. Examination of the chart of nuclei (Fig. 43-4), or an equivalent table, shows the following nuclides with magic numbers:

Magic No.	Protons (Z)	Neutrons (N = A − Z)
2	$^{3,4}_{2}$He	$^{4}_{2}$He
8	$^{16,17,18}_{8}$O	$^{15}_{7}$N, $^{16}_{8}$O
20	$^{40,42,43,44,46,48}_{20}$Ca	$^{38}_{18}$A, $^{39}_{19}$K, $^{40}_{20}$Ca
28	$^{58,60,61,62,64}_{28}$Ni	$^{48}_{20}$Ca, $^{50}_{22}$Ti, $^{51}_{23}$V, $^{52}_{24}$Cr, $^{54}_{26}$Fe
50	$^{112,114 \text{ to } 120,122,}_{}$ $^{124}_{50}$Sn	$^{86}_{36}$Kr, $^{87}_{37}$Rb,* $^{88}_{38}$Sr, $^{89}_{39}$Y, $^{90}_{40}$Zr, $^{92}_{42}$Mo
82	$^{204,206,207,208}_{82}$Pb	$^{136}_{54}$Xe, $^{138}_{56}$Ba, $^{139}_{57}$La, $^{140}_{58}$Ce, $^{141}_{59}$Pr, $^{142}_{60}$Nd, $^{144}_{62}$Sm
126	none	$^{208}_{82}$Pb, $^{209}_{83}$Bi**

*($t_{1/2} = 5 \times 10^{10}$ y)

**(may be $t_{1/2} \simeq 2 \times 10^{17}$ y)

(Note that $^4_2$He, $^{16}_8$O, $^{40}_{20}$Ca and $^{208}_{82}$Pb are doubly magic.)

Paired Problems

Problem

53. The nuclear mass of $^{48}_{22}$Ti is 47.9359 u. Find the binding energy per nucleon.

Solution

As in the solution to Problem 21, $E_b/A = [22 \times (1.00728) + 26(1.00867) - 47.9359](931.5 \text{ MeV})/48 = 8.72$ MeV/nucleon.

Problem

55. Geologists are looking for underground sites that could store nuclear wastes securely for 250,000 years. What fraction of plutonium-239 initially in such waste would remain after that time? See Table 43-2.

Solution

The fraction of ^{239}Pu (half-life 24,110 y) remaining is $N/N_0 = 2^{-t/t_{1/2}} = 2^{-2.5 \times 10^5/24,110} = 7.56 \times 10^{-4} = 0.0756\%$ (see Equation 43-5b).

Problem

57. A sample of oxygen-15 ($t_{1/2} = 2.0$ min) is produced in a hospital's cyclotron. What should be the initial activity concentration if it takes 3.5 min to get the O-15 to a patient undergoing a PET scan for which an activity of 0.5 mCi/L is necessary?

Solution

Multiplying Equation 43-5b by the decay rate, λ, one finds that the activity produced must be $(\lambda N_0) = (\lambda N)2^{t/t_{1/2}} = (0.5 \text{ mCi/L})2^{3.5/2.0} = 1.68$ mCi/L (see solution to Problem 39.)

Supplementary Problems

Problem

59. If a solid sphere with the 1.9-cm radius of a ping-pong ball were made entirely of nuclear matter, what would be the acceleration, due to gravity at its surface? Compare with that at Earth's surface.

Solution

The surface gravity (see Example 9-1) on a uniform sphere of radius $R = 1.9$ cm and density $\rho = M/\frac{4}{3}\pi R^3$ of nuclear matter (see Example 43-1) is $g = GM/R^2 =$

$\frac{4}{3}\pi\rho GR = \frac{4}{3}\pi(2.3 \times 10^{17} \text{ kg/m}^3)(6.67 \times 10^{-11} \text{ N·m}^2 \div \text{kg}^2)(0.019 \text{ m}) = 1.22 \times 10^6 \text{ m/s}^2$, or about 1.25×10^5 times the Earth's surface gravity, $g_E = 9.8 \text{ m/s}^2$.

Problem

61. How cool would you have to get a material before the thermal energy kT was insufficient to excite protons in a 35-T magnetic field from their lower to upper spin state?

Solution

The thermal energy is less than the spin-flip energy when $k_B T < \Delta U = 2\mu_p B$ (see Example 43-2), or $T < 2(1.41 \times 10^{-26} \text{ J/T})(35 \text{ T})/(1.38 \times 10^{-23} \text{ J/K}) = 71.5$ mK.

Problem

63. The atomic masses of uranium-238 and thorium-234 are 238.050784 u and 234.043593 u, respectively. Find the energy released in the alpha decay of ^{238}U.

Solution

The energy released in the decay is the difference in the mass-energy of the initial nucleus and that of the decay products, or $E_\alpha/c^2 = M(^A_Z X) - M(^{A-4}_{Z-2}Y) - M(^4_2\text{He})$. [This is analogous to the binding energy for a stable nucleus, which has the opposite sign. In nuclear reactions, the difference in mass-energy between the initial and final reactants is called the Q-value, which is positive for exothermic (energy releasing) and negative for endothermic (energy absorbing) reactions.] Note that atomic or nuclear masses can be used to calculate E_α, since in α-decay, the mass of the Z-electrons cancels. The neglect of atomic binding energies is also less serious here than in Equation 43-3 because only differences enter. (The atomic masses of $^{238}_{92}$U and $^{234}_{90}$Th are given; for $^4_2$He, from Table 43-1, we calculate $4.001506 \text{ u} + 2(0.000548579 \text{ u}) = 4.002603$ u.) Thus $E_\alpha/c^2 = M \times (^{238}_{92}\text{U}) - M(^{234}_{90}\text{Th}) - M(^4_2\text{He}) = 238.050784 \text{ u} - 234.043593 \text{ u} - 4.002603 \text{ u} = 0.004588 \text{ u} = 4.27 \text{ MeV}/c^2$.

Problem

65. Some human lung cancers in smokers may be caused by polonium-210, which arises in the decay series of uranium-238 that occurs naturally in fertilizers used on tobacco plants. Write equations for (a) the production and (b) the decay of ^{210}Po. (c) How might the health effects of Po-210 differ if its half-life were 1 day or 10 years instead of the actual 138 days?

Solution

From Fig. 43-21, one sees that polonium-210 is (a) produced by the β-decay of bismuth-210, $^{210}_{83}\text{Bi} \rightarrow {}^{210}_{84}\text{Po} + e^- + \bar{\nu}$, and (b) decays by α-decay into lead-206, $^{210}_{84}\text{Po} \rightarrow {}^{206}_{82}\text{Pb} + \alpha$. (c) Biological effects depend on the activity and energy release of the decay, as well as on the type of radiation and the properties of the absorbing tissue. In a radioactive decay series, where the half-life of the progenitor is much greater than that of any decay product, the activity of each product in the chain is approximately the same (secular equilibrium). Since a smoker replenishes the supply of ^{210}Po to his or her lungs on a daily basis, it is unlikely that a change in half-life, within the given range, will make much difference.

Problem

67. In the preparation of a transuranic nucleus, $^{209}_{83}\text{Bi}$ and $^{54}_{24}\text{Cr}$ interact to form a heavy nucleus plus a neutron. Identify the heavy nucleus.

Solution

The reaction forming the nucleus is $^{209}_{83}\text{Bi} + {}^{54}_{24}\text{Cr} \rightarrow {}^{A}_{Z}\text{X} + {}^{1}_{0}n$. Since the atomic and mass numbers must balance on both sides, $209 + 54 = A + 1$ and $83 + 24 = Z$, or $Z = 107$ and $A = 262$. In August 1997, the International Union of Pure and Applied Chemistry finally resolved the dispute over the names of elements 102–109. These are listed in Appendix D. The heavy nucleus found in this problem is bohrium-262, $^{262}_{107}\text{Bh}$. Names for elements 110–112 have not yet been officially determined.

Problem

69. Nickel-65 beta decays by electron emission with decay constant $\lambda = 0.275$ h^{-1}. (a) What is the daughter nucleus? (b) In a sample that is initially pure Ni-65, how long will it be before there are twice as many daughter nuclei as parent nuclei?

Solution

(a) In a β^--decay, the atomic number of the parent nucleus increases by one while the mass number stays the same ($Z \rightarrow Z + 1$ and $A \rightarrow A$). Thus, the daughter nucleus has $Z = 28 + 1 = 29$ and $A = 65$, which is copper-65 (or $^{65}_{29}\text{Cu}$). (b) The number of parent nuclei at time t (in an initially pure sample at $t = 0$) is $N = N_0 e^{-\lambda t}$ (Equation 43-5a), or $t = \lambda^{-1} \ln(N_0/N)$. The number of daughter nuclei is $N_d = N_0 - N$, so when $N_d = 2N$, $N_0/N = 3$ and $t = \ln 3/(0.275 \text{ h}^{-1}) = 3.99$ h.

Problem

71. The **mean lifetime**, τ, of a radioactive nucleus is the average of the times from $t = 0$ to $t = \infty$, weighted by the number dN of nuclei that decay in each time interval dt, that is,

$$\tau = \frac{\int_{t=0}^{\infty} t \, dN}{\int_{t=0}^{\infty} dN}.$$

Evaluate this quantity to show that $\tau = 1/\lambda$. *Hint:* Write dN as $|dN/dt|dt$.

Solution

In exponential decay, dN is the number of nuclei that decay after having "lived" for a time t, and is thus proportional to the probability a nucleus has a lifetime of t. The mean life-time $\tau = \int_0^\infty t \, dN / \int_0^\infty dN$ follows from the usual definition for the average. From Equation 43-5a, $dN = -\lambda N_0 e^{-\lambda t} dt$, so $\tau = \int_0^\infty \times t e^{-\lambda t} dt / \int_0^\infty e^{-\lambda t} dt$. The integral in the numerator, after integrating by parts, is $\left|t(-e^{-\lambda t}/\lambda)\right|_0^\infty - \int_0^\infty \times (-e^{-\lambda t}/\lambda) \, dt$, which equals $1/\lambda$ times the integral in the denominator ($te^{-\lambda t} = 0$ at $t = 0$ and ∞ since $\lambda > 0$), so $\tau = 1/\lambda$ as stated above.

CHAPTER 44   NUCLEAR ENERGY: FISSION AND FUSION

Section 44-1:   Energy from the Nucleus

Problem

1. The masses of the neutron, the deuterium nucleus, and the ^{3}He nucleus are 1.008665 u, 2.013553 u, and 3.014932 u, respectively. Use the Einstein mass-energy relation to verify the 3.27-MeV energy release in the D-D fusion reaction of Equation 44-4a.

Solution

The energy release (called the Q-value for the reaction) is the difference in the mass-energy of the initial and final reactants, $Q = [2M(^2_1\text{H}) - M(^3_2\text{He}) - M(^1_0n)]c^2 = [2 \times 2.013553 - 3.014932 - 1.008665]uc^2 = (0.003509) \times (931.5 \text{ MeV}) = 3.27 \text{ MeV}$.

Problem

3. Using the values 200 MeV for ^{235}U fission and 17.6 MeV for D–T fusion, calculate and compare the energy release per kilogram for each reaction.

Solution

To obtain a rough estimate, approximate values of the nuclear masses can be used, namely, 235 u, 2 u, and 3 u, for ^{235}U, ^{2}H, and ^{3}H, respectively. Recalling that one ^{235}U-nucleus is consumed per fission, whereas both a D and T-nucleus is consumed per fusion, we find that

$$\left(\begin{array}{c}\text{energy-release} \\ \text{per kilogram}\end{array}\right) = \frac{\text{energy-release per reaction}}{\text{kilograms consumed per reaction}}$$

$$= \frac{(200 \text{ MeV})(1.6 \times 10^{-13} \text{ J/MeV})}{(235 \text{ u})(1.66 \times 10^{-27} \text{ kg/u})} = 8.2 \times 10^{13} \text{ J/kg},$$

$$= \frac{(17.6 \text{ MeV})(1.6 \times 10^{-13} \text{ J/MeV})}{(2 \text{ u} + 3 \text{ u})(1.66 \times 10^{-27} \text{ kg/u})} = 3.4 \times 10^{14} \text{ J/kg},$$

for these two reactions. D–T fusion releases about four times as much energy per kilogram as ^{235}U-fission.

Sections 44-2 and 44-3:   Nuclear Fission and its Applications

Problem

5. Neutron-induced fission of ^{235}U results in the fission fragments iodine-139 and yttrium-95. How many neutrons are released?

Solution

In this reaction, the conservation of charge (atomic number Z) is the same as the conservation of the number of protons, so the conservation of the number of nucleons (mass number A) is equivalent to the conservation of the number of neutrons. The numbers of neutrons in the nuclei ^{235}U, ^{139}I, and ^{95}Y are $A{-}Z = 235 - 92 = 143$, $139 - 53 = 86$, and $95 - 39 = 56$, respectively, so the number of neutrons released in the reaction is $1 + 143 - 86 - 56 = 2$.

Problem

7. Assuming 200 MeV per fission, determine the number of fission events occurring each second in a reactor whose thermal power output is 3200 MW.

Solution

If all of the energy released in fissions goes into thermal power output, the fission rate is:

$$\frac{3200 \text{ MW}}{(200 \text{ MeV/fission})(1.6 \times 10^{-13} \text{ J/MeV})}$$
$$= 1.00 \times 10^{20} \text{ fissions/s.}$$

Problem

9. Find the explosive yield in equivalent tonnage of TNT for the chain reaction analyzed in Example 44-3.

Solution

The supercritical chain reaction (or explosion) of Example 44-3 released 1.1×10^{15} J of energy, equivalent to $(1.1 \times 10^{15} \text{ J})(1 \text{ kt}/4.18 \times 10^{12} \text{ J}) = 263 \text{ kt}$ of TNT. (We used the energy equivalent for kilotons of TNT given in Section 44-3 and Appendix C.)

Problem

11. How much uranium-235 would be consumed in a fission bomb with 20-kt explosive yield?

Solution

Using the data in Appendix C, we find that a 1 kt explosive yield is equivalent to the energy released in the fission of about (1 kt) $(4.18 \times 10^{12} \text{ J/kt}) \div$

$(8.21 \times 10^{13}$ J/kg of ^{235}U$) = 50.9$ g of ^{235}U. Then a 20 kt yield is equivalent to the fissioning of 20×50.9 g $= 1.02$ kg of ^{235}U.

Problem

13. The 1974 Threshold Test Ban Treaty limited underground nuclear tests to a maximum yield of 150 kt. (a) How much ^{235}U would be needed for a device with this yield, assuming 30% efficiency? (b) What would be the diameter of this mass, assembled into a sphere? The density of uranium is 18.7 g/cm^3.

Solution

(a) Use of the conversion factor in the solution to Problem 11 gives the amount of ^{235}U as $(150$ kt$) \times (50.9$ g/kt$)/30\% = 25.5$ kg. (b) A uniform sphere of mass M and density ρ has diameter $D = (6M/\pi\rho)^{1/3} = (6 \times 25.5$ kg$/\pi \times 18.7 \times 10^3$ kg/m$^3)^{1/3} = 13.7$ cm. (The volume of a sphere is $\frac{1}{6}\pi D^3$.)

Problem

15. The effective multiplication factor in a typical nuclear weapon is about 1.5, and the generation time is about 10 ns. Under these conditions, (a) how many generations would it take to fission a 10-kg mass of ^{235}U? (b) How long would the process take?

Solution

(a) A 10 kg mass of ^{235}U contains approximately $N = 10$ kg$/(235 \times 1.66 \times 10^{-27}$ kg$) = 2.56 \times 10^{25}$ nuclei. This is related to the multiplication factor and the number of generations by Equation 44-2. Solving for n, and neglecting 1 compared to $N(k-1)$ (as in Example 44-3), we find $n = -1 + \ln N(k-1)/\ln k = -1 + \ln[(2.56 \times 10^{25})(1.5-1)]/\ln 1.5 = 142$. (b) Since each generation takes $\tau = 10$ ns, the entire process takes $n\tau = 1.42$ μs.

Problem

17. The temperature in a typical reactor core is 600 K. What is the thermal speed of a neutron at this temperature? *Hint:* See Section 20-1.

Solution

The thermal speed can be calculated from Equation 20-3:

$$v_{rms} = \sqrt{\frac{3k_B T}{m}} = \sqrt{\frac{3(1.38 \times 10^{-23} \text{ J/K})(600 \text{ K})}{1.67 \times 10^{-27} \text{ kg}}}$$

$$= 3.86 \text{ km/s}.$$

Problem

19. A neutron collides elastically and head-on with a stationary deuteron in a reactor moderated by heavy water. How much of its kinetic energy is transferred to the deuteron? *Hint:* Consult Chapter 11.

Solution

The particles involved in the fission reactions discussed all have had non-relativistic energies, so Equation 11-9a or b, for a head-on elastic collision between a neutron (mass m_n) and a deuteron initially at rest $(v_{di} = 0)$, gives $v_{df} = 2m_n v_{ni}/(m_n + m_d)$. The fraction of the neutron's initial kinetic energy transferred to the deuteron is therefore $K_{df}/K_{ni} = m_d v_{df}^2/m_n v_{ni}^2 = 4m_n m_d/(m_n + m_d)^2$. Since $m_n \simeq 1$ u and $m_d \simeq 2$ u, the fraction is $\frac{8}{9} = 88.9\%$.

Problem

21. If a reactor is shut down abruptly, neutron absorption by ^{135}Xe prevents rapid start-up. The xenon has a 9.2-h half-life. How long must reactor operators wait until the ^{135}Xe level drops to one-tenth of its peak value? (Your answer neglects formation of additional ^{135}Xe from the decay of ^{135}I.)

Solution

From Equation 43-5b, the time for decay to one tenth the initial amount of ^{135}Xe (with no production assumed) is $t = t_{1/2} \ln(N_0/N)/\ln 2 = (9.2$ h$) \ln 10 \div \ln 2 = 30.6$ h.

Problem

23. One reason breeder reactors are considered potentially dangerous is that the reaction in a breeder is sustained by fast neutrons alone; there is no moderator. The generation time for prompt neutrons is then 100 ns, down from 100 ms in a thermal reactor. For the conditions of Example 44-5, compute the power-doubling time in a breeder with $\tau = 100$ ns.

Solution

As explained in Example 44-5, a supercritical reactor with $k = 1.001$ doubles its power output in 693 generations. If the generation time for prompt neutron induced fission in a breeder reactor is 100 ns, the power doubling time is 693×100 ns $= 69.3$ μs. Since this is much shorter than a human operator's reaction time, such a situation is extremely dangerous.

Problem

25. In the dangerous situation of prompt criticality in a fission reactor, the generation time drops to 100 μs as prompt neutrons alone sustain the chain reaction. If a reactor goes prompt critical with $k = 1.001$, how long does it take for a 100-fold increase in reactor power?

Solution

Following the same reasoning as in Example 44-5 (or in the solution to the previous problem), we have $k^n = 10^2$, or $n \log k = 2$. The time required for this increase in power is the number of generations times the generation time, or $n\tau = (2/\log 1.001)100 \ \mu$s = 0.461 s.

Problem

27. Operators seek to double the power output of a nuclear reactor over a period of 1 hour. If the generation time is 120 ms, by how much should the multiplication factor k be increased from 1 during this period?

Solution

The 1 h period corresponds to $n = t/\tau = $ 1 h/120 ms = 3×10^4 generations. The multiplication factor which doubles the power (which is proportional to the number of fissions per generation) is $k^n = 2$, so $\ln k = \ln 2/n = 2.31\times10^{-5}$. Since $e^x \approx 1 + x$ for small x, $k = e^{\ln k} \approx 1 + 2.31\times10^{-5}$, or k should be increased by $\Delta k = 2.31\times10^{-5}$.

Section 44-4: Nuclear Fusion

Problem

29. Fusion researchers often express temperature in energy units, giving the value of kT rather than T. What is the temperature in kelvin of a 2-keV plasma?

Solution

A thermal energy of $kT = 2$ keV corresponds to a temperature of $T = 2$ keV$(1.6\times10^{-16}$ J/keV$)\div$ 1.38×10^{-23} J/K) = 2.3×10^7 K.

Problem

31. In a magnetic-confinement fusion device with confinement time 0.5 s, what density would be required to meet the Lawson criterion for D–T fusion?

Solution

The Lawson criterion for D–T fusion is $n\tau >$ 10^{20} s/m^3 (Equation 44-5). For $\tau = 0.5$ s, a particle

density of $n > 10^{20}$ s/m^2/0.5 s = 2×10^{20} m^{-3} would be required.

Problem

33. How much heavy water (deuterium oxide, ^{2}H$_2$O or D$_2$O) would be needed to power a 1000-MW D–D fusion power plant for 1 year?

Solution

In one year, a 1 GW power plant produces 1 GW·y = $(10^9$ J/s$)(3.156\times10^7$ s$)(1$ MeV$/1.602\times10^{-13}$ J$) =$ 1.97×10^{29} MeV of energy. If we use 7.2 MeV/deuteron as the average energy release in a D–D reactor (see Example 44-6), then 2.74×10^{28} deuterons are required. The molecular weight of D$_2$O is about 20 u, so about $\frac{1}{2}(20$ u$)(1.66\times10^{-27}$ kg/u$)(2.74\times10^{28}) = 454$ kg of heavy water (each molecule of which contains two deuterons) would be needed.

Problem

35. Inertial-confinement schemes generally involve confinement times on the order of 0.1 ns. What is the corresponding density needed to meet the Lawson criterion for D–T fusion?

Solution

For a confinement time as short as 0.1 ns, the Lawson criterion (Equation 44-5) demands a particle density of $n > 10^{20}$ s/m^2/0.1 ns = 10^{30} m^{-3}. This is 30 times the particle density of water molecules under ordinary conditions.

Problem

37. The proton-proton cycle consumes four protons while producing about 27 MeV of energy. (a) At what rate must the Sun consume protons to produce its power output of about 4×10^{26} W? (b) The present phase of the Sun's life will end when it has consumed about 10% of its original protons. Estimate how long this phase will last, assuming the Sun's 2×10^{30} kg mass was initially 71% hydrogen.

Solution

(a) The number of protons consumed per second is the power output divided by the energy release per proton. This is about $\left(\frac{4\times10^{26} \text{ J/s}}{27 \text{ MeV/4 protons}}\right)\left(\frac{1 \text{ MeV}}{1.6\times10^{-13} \text{ J}}\right) =$ 3.7×10^{38} protons/s. (Note: four protons are consumed in releasing 27 MeV.) (b) 10% of the Sun's original protons is about $(0.1)(0.71)(2\times10^{30}$ kg$/1.67\times$ 10^{-27} kg/proton$) = 8.5\times10^{55}$ protons. The consumption of this many protons at the rate found in part (a) would take about $(8.5 \times 10^{55}$ protons$) \div$ $(3.7 \times 10^{38}$ protons/s$) = 2.3 \times 10^{17}$ s $= 7.3$ billion years. The present age of the Sun is about 4.5 billion years.

Problem

39. About 0.015% of hydrogen nuclei are actually deuterium. (a) How much energy would be released if all the deuterium in a gallon of water underwent fusion? (Use an average of 7.2 MeV per deuteron; see preceding problem.) (b) In terms of energy content, to how much gasoline does a gallon of water correspond? Gasoline's energy content is 36 kWh/gal, as listed in Appendix C.

Solution

(a) The number of D-atoms in a gallon of water (mostly H_2O) is (1 gal $\times$ 3.786×10^{-3} m^3/gal)$\times$ $(10^3$ kg/m$^3)$ (2 atoms per molecule) (0.00015 D/H)$\div$ $(18 \times 1.66 \times 10^{-27}$ kg/molecule) $= 3.80 \times 10^{22}$ D-atoms. If all of these release energy via D-D fusion reactions, an average of $(3.80 \times 10^{22})(7.2$ MeV$)\times$ $(1.602 \times 10^{-13}$ J/MeV$) = 4.38 \times 10^{10}$ J would be produced. (See Example 44-6.) (b) This is equivalent to the energy content of $(4.38 \times 10^{10}$ J$)\div$ $(36 \times 10^3 \times 3600$ J/gal$) = 338$ gallons of gasoline.

Paired Problems

Problem

41. The total power generated in a nuclear power reactor is 1500 MW. How much ^{235}U does it consume in a year?

Solution

The number of ^{235}U-atoms which undergo fission to produce the energy generated in one year is $(1.5 \times 10^9$ W$)(3.156 \times 10^7$ s$)/(200$ MeV/fission$)\times$ $(1.602 \times 10^{-13}$ J/MeV$) = 1.48 \times 10^{27}$. These atoms have a total mass of about $(235$ u$)(1.48 \times 10^{27})(1.66 \times 10^{-27}$ kg/u$) = 576$ kg. (See Example 44-2).

Problem

43. Two fission chain reactions have $\tau = 20$ ns, but reaction A has $k = 1.3$ and reaction B has $k = 1.4$. Compare the total number of nuclei that fission in the first μs of each reaction.

Solution

Since the generation time is the same for both reactors, so is the number of generations in 1 μs, i.e., $n = 1 \; \mu$s/20 ns $= 50$. From Equation 44-2, with the

approximation k_A^n and $k_B^n \gg 1$,

$$\frac{N_B}{N_A} = \left(\frac{k_B^{n+1} - 1}{k_B - 1} \right) \left(\frac{k_A - 1}{k_A^{n+1} - 1} \right) \approx \left(\frac{k_B}{k_A} \right)^{n+1} \left(\frac{k_A - 1}{k_B - 1} \right)$$

$$= \left(\frac{1.4}{1.3} \right)^{51} \left(\frac{0.3}{0.4} \right) = 32.8.$$

Problem

45. This problem and the next one explore the differences between magnetic and inertial-confinement fusion. (a) What is the mean thermal speed of a deuteron in the PLT device of Problem 40? (b) At this speed, how long would it take a deuteron to cross the plasma column? Using this value as a "confinement time" in the absence of magnetic confinement and the density given in Problem 40, calculate the Lawson parameter and show that it falls far short of the value needed for D–T fusion.

Solution

(a) We can use Equation 20-3 to find the thermal speed of the deuterons $(m \approx 2$ u): $v_{rms} = \sqrt{3k_B T/m} = $ $\sqrt{3(1.38 \times 10^{-23} \text{ J/K})(6 \times 10^7 \text{ K})/(2 \times 1.66 \times 10^{-27} \text{ kg})} =$ 865 km/s. (b) It would take only (90 cm)$\div$ (865 km/s) $= 1.04 \; \mu$s to cross the plasma column at this speed, if there were no collisions or other confinement mechanism. Without confinement, the Lawson parameter would be $n\tau = (4 \times 10^{19}$ m$^{-3})\times$ $(1.04 \; \mu$s$) = 4 \times 10^{13}$ s/m^2, about 4×10^{-7} times smaller than the desired criterion for D–T fusion (Equation 44-5).

Supplementary Problems

Problem

47. A laser-fusion fuel pellet has mass 5.0 mg and is composed of equal parts (by mass) of deuterium and tritium. (a) If half the deuterium, and an equal number of tritium nuclei, participate in D–T fusion, how much energy is released? (b) At what rate must pellets be fused in a power plant with 3000 MW thermal power output? (c) What mass of fuel would be needed to run the plant for 1 year? Compare with the 3.6×10^6 tons of coal needed to fuel a comparable coal-burning power plant.

Solution

(a) Half of the number of deuterons in one fuel pellet is $\frac{1}{2}(2.5$ mg$)/(2$ u $\times 1.66 \times 10^{-27}$ kg/u$) = 3.77 \times 10^{20}$ (since there is 2.5 mg of deuterium in a pellet and the mass of a deuteron is approximately 2 u). This number of D–T fusion reactions would release $(3.77 \times 10^{20})(17.6 \text{ MeV/fusion})(1.602 \times 10^{-13} \text{ J/MeV}) = 1.06$ GJ of energy. (b) The thermal power output, 3 GW, is equal to the energy release per pellet times the rate that pellets are consumed, hence this rate is 3 GW/1.06 GJ = 2.83 s^{-1}. (c) 5 mg pellets, consumed at this rate for a year, have a total mass of $(5 \text{ mg}) \times (2.83 \text{ s}^{-1})(3.156 \times 10^7 \text{ s}) = 446$ kg. A comparable coal-burning power plant uses more than 7 million times this mass for its fuel.

Problem

49. Use a graphical or numerical method to find the multiplication factor necessary to fission 10^{22} uranium-235 nuclei in 2.0 μs, if the generation time is 10 ns.

Solution

The number of generations is $n = 2 \text{ μs}/10 \text{ ns} = 200$, so Equation 44-2 gives $10^{22}(k-1) = k^{201} - 1 \approx k^{201}$. A numerical solution can be obtained from Newton's method, applied to the function $f(k) = k^{201} - 10^{22}(k-1)$, with a first guess of $k_0 = 10^{22/201} = 1.2866$. Then $f(k_0) = 10^{22}(2 - k_0)$ and $f'(k_0) = 201 k_0^{200} - 10^{22} = 10^{22}(201 - k_0)/k_0$, so the next approximation is $k_1 = k_0 - f(k_0)/f'(k_0) = 199 k_0 \div (201 - k_0) = 1.2820$. Another iteration gives $k_2 = k_1 - f(k_1)/f'(k_1) = 1.2793$, so the root seems to be converging toward 1.279.

A more accurate value of k can be obtained by graphical methods, if a suitable PC and software are handy. Plotting $10^{22}(k-1)$ and $k^{201} - 1$ versus k on the same axes, as shown above, one obtains $k = 1.27847$.

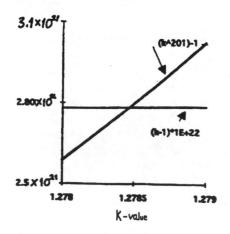

Problem 49 Solution.

Problem

51. Roughly half the yield of a thermonuclear weapon comes from D–T fusion and the rest from ^{238}U fission induced by fast neutrons from the D–T fusion. Each D–T reaction releases about 18 MeV and each fission about 200 MeV. Estimate the masses of fusion fuel and ^{238}U in a 1-Mt thermonuclear bomb, assuming both types fuel react completely.

Solution

Half the yield of a 1 Mt bomb is equivalent to about 2.09×10^{15} J (see Section 44-3 or Appendix C). The number of each type of reaction is this energy divided by the energy release per reaction, or $(2.09 \times 10^{15} \text{ J}) \times (1 \text{ MeV}/1.602 \times 10^{-13} \text{ J})/(18 \text{ MeV}) = 7.25 \times 10^{26}$ D–T fusions and 6.52×10^{25} ^{238}U-fissions, respectively. The mass of the reactants is approximately $2 \text{ u} + 3 \text{ u} = 5 \text{ u}$ per D–T fusion, and 238 u per ^{238}U-fission, therefore the amount of each type of fuel consumed is: $(5 \times 1.66 \times 10^{-27} \text{ kg})(7.25 \times 10^{26}) = 6.02$ kg for D–T fusion, and $(238 \times 1.66 \times 10^{-27} \text{ kg}) \times (6.52 \times 10^{25}) = 25.8$ kg for ^{238}U fission.

Problem

53. The volume of the fireball produced in a nuclear explosion is roughly proportional to the weapon's explosive energy yield. For weapons exploded at ground level, how would the land area subject to a given level of damage scale with the weapon's yield? Your result shows one reason military strategists favor multiple smaller warheads.

Solution

If the volume of the fireball is proportional to the yield, then the radius of the fireball goes like the cube root of the yield, i.e., $R \sim (\text{yield})^{1/3}$. The land area affected by the fireball goes like the square of its radius, i.e., $R^2 \sim (\text{yield})^{2/3}$. Thus, doubling the yield increases the effected land area by $2^{2/3} = 1.59$, which is less area than would be affected by exploding two bombs on non-overlapping targets.

Problem

55. Of the neutrons emitted in each fission event in a light-water reactor, an average of 0.6 neutrons are absorbed by ^{238}U, leading to the formation of ^{239}Pu. (a) Assuming 200 MeV per fission, how much ^{239}Pu forms each year in a 30% efficient nuclear power plant whose electric power output is 1000 MW? (b) With careful design, a fission explosive can be made from 5 kg of ^{239}Pu. How many potential bombs are produced each year in

the power plant of part (a)? (Extracting the ^{239}Pu is not an easy job.)

Solution

(a) The number of fission events in a power plant, whose thermal output is 1 GW/30% = 3.33 GW, in a year of operation, is (3.33 GW)(3.156×10^7 s)÷ (200 MeV/event)(1.602×10^{-13} J/Mev) = 3.29×10^{27}.

(This is the total energy released in a year divided by the energy release in one fission event, as in Example 44-2.) Then the amount of ^{239}Pu formed (atomic weight approximately 239 u) is (3.29×10^{27})× (0.6)(239 u)(1.66×10^{-27} kg/u) = 783 kg. (b) This is enough to make 783/5 = 157 bombs, if all of the ^{239}Pu could be extracted from the spent reactor fuel.

CHAPTER 45   FROM QUARKS TO THE COSMOS

Section 45-1:   Particles and Forces

Problem

1. How long could a virtual photon of 633-nm red laser light exist without violating conservation of energy?

Solution

In order to test the conservation of energy in a process involving one virtual photon, a measurement of energy with uncertainty less than the photon's energy ($\Delta E < hc/\lambda$), must be performed in a time interval less than the virtual photon's lifetime ($\Delta t < \tau$). Thus, $\Delta E \, \Delta t < hc\tau/\lambda$. But Heisenberg's principle limits the product of these uncertainties to $\Delta E \, \Delta t \gtrsim \hbar$, so $hc\tau/\lambda > \hbar$ or $\tau > \lambda/2\pi c = (633 \times 10^{-9} \text{ m})/(2\pi \times 3 \times 10^8 \text{ m/s}) \simeq 3 \times 10^{-16}$ s, for the given wavelength. In other words, if the lifetime of a virtual photon of wavelength 633 nm were less than 3×10^{-16} s, no measurement showing a violation of conservation of energy would be possible.

Problem

3. The mass of the photon is assumed to be zero, but experiments put only an upper limit of 5×10^{-63} kg for the photon mass. What would be the range of the electromagnetic force if the photon mass were actually at this upper limit?

Solution

According to the version of Yukawa's argument in the text, the relation between the mass of a field particle and the range of the force it mediates is $\Delta x \simeq c \, \Delta t \simeq \hbar c/\Delta E \simeq \hbar c/mc^2 = \hbar/mc$ (i.e., the range of the force is approximately the Compton wavelength of the mediating field particle). For $m = 5 \times 10^{-63}$ kg, $\Delta x \simeq (1.055 \times 10^{-34} \text{ J·s})/(5 \times 10^{-63} \text{ kg} \times 3 \times 10^8 \text{ m/s}) \simeq 7 \times 10^{19}$ m, less than a tenth the diameter of our galaxy (see Table 1-2).

Section 45-2:   Particles and More Particles

Problem

5. Use Table 45-1 to find the total strangeness before and after the decay $\Lambda^0 \rightarrow \pi^- + p$, and use your answer to determine the force involved in this reaction.

Solution

The Λ^0 has strangeness -1, while the p and π^- both have zero strangeness. Thus $\Delta S = 1$ for this decay (final minus initial strangeness). Since strangeness is conserved in strong and electromagnetic interactions, the decay must be a weak interaction.

Problem

7. Are either or both of the following decay schemes possible for the tau particle? (a) $\tau^- \rightarrow e^- + \bar{\nu}_e + \nu_\tau$, (b) $\tau^- \rightarrow \pi^- + \pi^0 + \nu_\tau$.

Solution

Reference to Table 45-1 shows that both decays satisfy conservation of charge, electron-lepton number, and tau-lepton number (the negatively charged leptons are particles). None of the particles are baryons, weak decays don't have to conserve strangeness, and the tau's rest energy is greater than that of either final state. Therefore, both decays are possible.

Problem

9. Which of the following reactions (a) $\Lambda^0 \rightarrow \pi^+ + \pi^-$ and (b) $K^0 \rightarrow \pi^+ + \pi^-$ is not possible, and why?

Solution

Decay (a) violates the conservation of baryon number (and angular momentum, since the spin of Λ^0 is $\frac{1}{2}$ and that of the pions is 0). Decay (b) is an observed weak interaction.

Problem

11. Grand unification theories suggest that the decay $p \rightarrow \pi^0 + e^+$ may be possible, in which case all matter may eventually become radiation. Are (a) baryon number and (b) electric charge conserved in this hypothetical proton decay?

Solution

The hypothetical decay $p \rightarrow \pi^0 + e^+$ (a) does not conserve baryon number (which is 1 for the proton and 0 for mesons and leptons), nor does it conserve lepton number, although (b) it does conserve charge.

Problem

13. What happens to the sign of the wave functions proportional to the terms (a) xy^2t and (b) xy^2t^2 under the operation PT?

Solution

Under space-inversion, both terms (a) and (b) change sign (they are parity reversing): $P[xy^2t] \to (-x) \times (-y)^2t = -xy^2t$, and $P[xy^2t^2] \to -xy^2t^2$. Under time-inversion, the first (a) changes sign while the second (b) does not (the first is T-reversing, the second is T-conserving): $T[xy^2t] \to xy^2(-t) = -xy^2t$, and $T[xy^2t^2] \to xy^2t^2$. Under the combined transformation, the first term (a) is PT-conserving while the second (b) is PT-reversing: $PT[xy^2t] \to P[-xy^2t] \to xy^2t$, and $PT[xy^2t^2] \to P[xy^2t^2] \to -xy^2t^2$. (Note: $PT = TP$, i.e., P and T-transformations commute.)

Section 45-3: Quarks and the Standard Model

Problem

15. The Eightfold Way led Gell-Mann to predict a baryon with strangeness -3. What must be its quark composition?

Solution

A baryon composed of three strange quarks has strangeness $S = -3$ and charge $-e$. It is the $\Omega^- = sss$.

Problem

17. The J/ψ particle is an uncharmed meson that nevertheless includes charmed quarks. What is its quark composition?

Solution

The J/ψ must have quark content $c\bar{c}$ in order to have zero net charm.

Section 45-4: Unification

Problem

19. Estimate the volume of the 50,000 tons of water used in the Super Kamio Kande experiment of Fig. 45-15.

Solution

The mass of 50,000 tons of water is $(10^8 \text{ lb}) \times (0.454 \text{ kg/lb}) = 4.54 \times 10^7$ kg. At the ordinary density of 10^3 kg/m^3, this amount of water occupies a volume of $V = M/\rho = 4.54 \times 10^4$ m$^3 = 1.20 \times 10^6$ gal. (This is the volume of a cube of side length $(4.54 \times 10^4)^{1/3}$ m $= 35.7$ m $= 117$ ft. See Appendix C for the conversion factors.)

Problem

21. Repeat the preceding problem for the 10^{15} GeV energy of grand unification.

Solution

The temperature corresponding to the energy 10^{15} GeV $= 10^{24}$ eV $= k_BT$, where the strong and electro-weak forces unify, is about $T = 10^{24}$ eV $\div (8.617 \times 10^{-5}$ eV·K$^{-1}) \approx 10^{28}$ K.

Problem

23. (a) What would be the relativistic factor γ for a 7-TeV proton in the Large Hadron Collider? (b) Find an accurate value for the proton's speed.

Solution

(a) A proton with 7 TeV kinetic energy is extremely relativistic $(E = K + mc^2 \gg mc^2)$, therefore $\gamma = E/mc^2 = 1 + K/mc^2 \simeq 7$ TeV/938 MeV $= 7.46 \times 10^3$. (b) For γ so large, v/c is very close to 1. More precisely, $v/c = \sqrt{1 - 1/\gamma^2} \approx 1 - 1/2\gamma^2 = 1 - 0.5 \times (7.46 \times 10^3)^{-2} = 1 - 8.98 \times 10^{-9} = 0.999\,999\,9910$. (See Section 38-5 for relativistic kinematic expressions.)

Section 45-5: The Evolving Universe

Note: In working these problems, take $H_0 = 15$ km/s/Mly unless otherwise indicated.

Problem

25. Express the Hubble constant in SI units.

Solution

From Example 45-3, one sees that H_0 in SI units is the reciprocal of the Hubble time in seconds, i.e., $H_0 = (20$ Gy $\times 3.156 \times 10^7$ s/y$)^{-1} = 1.58 \times 10^{-18}$ s^{-1}, where the value of H_0 used is the lower end of the currently accepted range. [Direct conversion of units, using 1 Mly $= (10^6)(3 \times 10^5$ km/s$)(3.156 \times 10^7$ s) gives the same result: $H_0 = (15$ km/s$)/(9.47 \times 10^{18}$ km).] Another common distance unit used by astronomers is the parsec (1 pc $= 3.26$ ly, see Appendix D). $H_0 = 15$ km/s/Mly ≈ 50 km/s/Mpc.

Problem

27. What is the recession speed of a galaxy 300 Mly from Earth?

Solution

From the Hubble Law (Equation 45-1) and the suggested value of the Hubble constant, one finds $v = H_0d = (15$ km/s/Mly$) \times (300$ Mly$) = 4.5 \times 10^3$ km/s $= 0.015c$. (Thus, the cosmological red-shift is about 1.5% for this galaxy.)

Problem

29. A widely used value for H_0 is 17 km/s/Mly. What age does this value imply for the universe under the simple assumptions of Example 45-3?

Solution

Alteration of the value of H_0 in Example 45-3 to 17 km/s/Mly changes the Hubble time to $(17 \text{ km/s/Mly})^{-1}[3 \times 10^5 \text{ km/s/(ly/y)}] = 18 \text{ Gy}$.

Problem

31. Find the critical density if the Hubble constant is 34 km/s/Mly.

Solution

The expression for the critical density obtained in Example 45-4 gives $\rho_c = 3H_0^2/8\pi G = (3/8\pi) \times [(34 \text{ km/s/Mly})/(9.47 \times 10^{18} \text{ km/Mly})]^2/(6.67 \times 10^{-11} \text{ m}^3/\text{kg·s}^2) = 2.31 \times 10^{-26} \text{ kg/m}^3$, for the specified value of H_0. (See solution to Problem 25 for the conversion of units. Note that since this value of H_0 is about twice that used in Example 45-3, ρ_c is about $2^2 = 4$ times the value in the Exercise following Example 45-4.)

Supplementary Problems

Problem

33. A so-called muonic atom is a hydrogen atom with the electron replaced by a muon; the muon's mass is 207 times the electron's. Find (a) the size, and (b) the ground-state energy of a muonic atom.

Solution

We can use the results for the Bohr atom, with $m_\mu = 207 m_e$ replacing m_e (see Equations 39X-13 and 12a). For the ground state ($n = 1$), (a) $r_1 = a_0/207 = 0.0529 \text{ nm}/207 = 256 \text{ fm}$, and (b) $E_1 = 207(-ke^2 \div 2a_0) = 207(-13.6 \text{ eV}) = -2.81 \text{ keV}$.

Problem

35. How long would it take the proton in Problem 24 to circle the LHC, as measured in the proton's

frame? Assume you can apply special relativity even though the path is curved.

Solution

The equations for time dilation (or Lorentz contraction, as used in the solution to Problem 38-9), applied to the proton in Problems 23 and 24 (whose speed $v \simeq c$) yields $\Delta t' = \Delta t/\gamma = \Delta x/\gamma v = (27 \text{ km})/(7.46 \times 10^3)(3 \times 10^5 \text{ km/s}) = 12.1 \text{ ns}$.

Problem

37. Use Wien's law (Equation 39X-2) to find the wavelength at which the 2.7-K cosmic microwave background has its peak intensity.

Solution

If the more accurate value for the temperature of the cosmic background radiation from Fig. 45-23 is used in Wien's displacement law (Equation 39X-2), the wavelength at peak intensity found is $\lambda_{\text{max}} = 2.898 \text{ mm·K}/2.726 \text{ K} = 1.063 \text{ mm}$.

Problem

39. Many particles are far too short-lived for their lifetimes to be measured directly. Instead, tables of particle properties often list "width," measured in energy units, and indicating the width of distribution of measured rest energies for the particle. For example, the Z^0 has a mass of 91.18 GeV and a width of 2.5 GeV. Use the energy-time uncertainty relation to estimate the corresponding lifetime.

Solution

In the energy-time uncertainty relation (Equation 39X-16), $\Delta t = \tau$ can be taken to be the lifetime of the particle (i.e., the time available for the measurement) and $\Delta E = \Gamma$ to be its width (i.e., the spread in measured rest-energies), so $\tau = \hbar/\Gamma$. (This is, infact, the definition of the width.) For the Z^0, a width $\Gamma = 2.5 \text{ GeV}$ implies a lifetime $\tau = 6.582 \times 10^{-16} \text{ eV·s}/2.5 \text{ GeV} = 2.6 \times 10^{-25} \text{ s}$.